Brain Aromatase, Estrogens, and Behavior

Oxford Series in Behavioral Neuroendocrinology

Series Editors

Gregory F. Ball, Jacques Balthazart, and Randy J. Nelson

Hormones and Brain Plasticity
Luis Miguel Garcia-Segura

Biology of homosexuality
Jacques Balthazart

Brain Aromatase, Estrogens, and Behavior
Edited by Jacques Balthazart and Gregory F. Ball

Brain Aromatase, Estrogens, and Behavior

Edited by
JACQUES BALTHAZART AND
GREGORY F. BALL

OXFORD
UNIVERSITY PRESS

OXFORD
UNIVERSITY PRESS

Oxford University Press is a department of the University of Oxford.
It furthers the University's objective of excellence in research, scholarship,
and education by publishing worldwide.

Oxford New York
Auckland Cape Town Dar es Salaam Hong Kong Karachi
Kuala Lumpur Madrid Melbourne Mexico City Nairobi
New Delhi Shanghai Taipei Toronto

With offices in
Argentina Austria Brazil Chile Czech Republic France Greece
Guatemala Hungary Italy Japan Poland Portugal Singapore
South Korea Switzerland Thailand Turkey Ukraine Vietnam

Published in the United States of America by
Oxford University Press
198 Madison Avenue, New York, NY 10016

Library of Congress Cataloging-in-Publication Data
Brain aromatase, estrogens, and behavior / edited by Jacques Balthazart and Gregory F. Ball.
p. cm. — (Oxford series in behavioural neuroendocrinology)
Includes bibliographical references and index.
ISBN 978-0-19-984119-6 (alk. paper)
1. Estrogen. 2. Estrogen—Physiological effect. 3. Aromatase. 4. Neuroendocrinology.
I. Balthazart, J. (Jacques), 1949- II. Ball, Gregory
F. (Gregory Francis)
QP572.E85B73 2013
616.8'0461—dc23 2012022944

[9 8 7 6 5 4 3 2 1]
Printed in the United States of America
on acid-free paper

Contents

Contributors

Elizabeth Adkins-Regan
Departments of Psychology/
Neurobiology & Behavior
Cornell University
Ithaca, New York

Maria Angeles Arevalo
Instituto Cajal
CSIC
Madrid, Spain

Iñigo Azcoitia
Cell Biology
Faculty of Biology
Complutense University of Madrid
Madrid, Spain

Julie Bakker
GIGA Neurosciences
University of Liège
Liege, Belgium

Gregory F. Ball
Johns Hopkins University
Department of Psychological and
Brain Sciences
Baltimore, Maryland

Jacques Balthazart
University of Liège
GIGA Neuroscience
Research Group in Behavioral
Neuroendocrinology
Liège, Belgium

Pascal Benquet
Laboratoire de Traitement du Signal
et de l'Image
Université de Rennes 1
Campus de Beaulieu
Rennes, France

Trish Berger
Department Psychology and Animal
Science
University of California
Davis, California

Anat Biegon
Medical Department
Brookhaven National Lab.
Upton, New York

François Brion
Unitéd' Écotoxicologie
INERIS
Verneuil-en-Halatte, France

Gloria V. Callard
Department of Biology
Boston University
Boston, Massachussetts

Thierry D. Charlier,
University of Liege
GIGA Neurosciences
Liège, Belgium

Lynwood G. Clemens
Professor of Zoology and
Neuroscience
Michigan State University
East Lansing, Michigan

Rachel E. Cohen
Department of Psychology
University of Washington
Seattle, Washington

Alan Conley
Department of Population Health & Reproduction
School of Veterinary Medicine
University of California
Davis, California

Charlotte A. Cornil
University of Liège
GIGA-Neurosciences
Research Group in Behavioral Neuroendocrinology
Liège, Belgium

Kellie A. Cotter
Department of Biology
Boston University
Boston, Massachussets

Molly J. Dickens
University of Liège
GIGA-Neurosciences
Research Group in Behavioral Neuroendocrinology
Liège, Belgium

Nicolas Diotel
Institut de Recherche en Santé Environnement et Travail
Université de Rennes 1
Rennes, France

Kelli A. Duncan
Department of Biology
Vassar College
Poughkeepsie, New York

H. Bobby Fokidis
Department of Psychology
University of British Columbia
Vancouver, Canada

Joanna Fowler
Medical Department
Brookhaven National Laboratory
Upton, New York

Luis Miguel Garcia-Segura
Instituto Cajal, CSIC
Madrid, Spain

Glenda E. Gillies
Division of Brain Sciences
Department of Medicine
Imperial College London
Hammersmith Hospita
London, United Kingdom

Sarah R. Greytak
Department of Biology
Boston University
Boston, Massachusetts

Nobuhiro Harada
Department of Biochemistry
Fujita Health University School of Medicine
Toyoake, Japan

Sarah A. Heimovics
Department of Psychology
University of British Columbia
Vancouver, Canada

Shin-ichiro Honda
Department of Biochemistry
Faculty of Pharmaceutical Sciences
Fukuoka University
Fukuoka, Japan

Austin Hughes
Department of Biological Sciences
University of South Carolina
Columbia, South Carolina

Olivier Kah
Institut de Recherche en Santé Environnement et Travail
Université de Rennes 1
Rennes, France

Sunny Kim
National Institute on Alcohol and Alcohol Abuse
Bethesda, Maryland

Randolph W. Krohmer
Department of Biological Sciences
Saint Xavier University
Chicago, Illinois

Anne T. M. Konkle
Interdisciplinary School of Health Sciences
University of Ottawa
Ottawa, Canada

Sarah A. Laredo
Department of Psychology
University of California, Davis
Davis, California

Jean Logan
Medical Department
Brookhaven National Laboratory
Upton, New York

Lisa A. Mangiamele
Department of Psychology
Program in Neuroscience
Bowdoin College
Brunswick, Maine

Margaret M. McCarthy
Department of Pharmacology
Department of Physiology and Program in Neuroscience
University of Maryland School of Medicine
Baltimore, Maryland

Rebecca Meyer
Department of Biology
Boston University
Boston, Massachusetts

Frederick Naftolin
New York University Langone Medical Center
Departments of Environmental Medicine and
Obstetrics and Gynecology
New York, New York

Apolonia Novillo
Department of Ecology and Evolutionary Biology
Boston University
Boston, Massachusetts

Veronica N. Orr
Department of Psychology
University of California
Davis, California

Deborah Pareto
Magnetic Resonance Unit (IDI)
Vall d'Hebron University Hospital
Barcelona, Spain

Elisabeth Pellegrini
Institut de Recherche en Santé Environnement et Travail
Université de Rennes 1
Rennes, France

Raphael Pinaud
Department of Neurobiology
Northwestern University
Evanston, Illinois

Luke Remage-Healey
Department of Psychology
University of Massachusetts,
Amherst, Massachusetts

Charles E. Roselli
Department of Physiology and Pharmacology L334
Oregon Health and Sciences University
Portland, Oregon

Colin J. Saldanha
Department of Psychology
Department of Biology
American University
Washington, District of Columbia

Barney A. Schlinger
Department of Integrative Biology and Physiology
Department of Ecology and Evolutionary Biology
The Laboratory for Neuroendocrinology
University of California—Los Angeles
Los Angeles, California

Per Södersten
Karolinska Institutet
Section of Applied Neuroendocrinology
Sweden

Kiran K. Soma
Department of Psychology
Department of Zoology
Graduate Program in Neuroscience
University of British Columbia
Vancouver, Canada

Richmond R. Thompson
Kanbar Hall
Bowdoin College
Brunswick, Maine

Brian C. Trainor
Department of Psychology
University of California
Animal Behavior Graduate Group
Center for Neuroscience
University of California
Davis, California

Liisa A. Tremere
Department of Neurobiology
Northwestern University
Evanston, Illinois

Colette Vaillant
Institut de Recherche en Santé Environnement et Travail
Université de Rennes 1
Rennes, France

Juli Wade
Department of Psychology
Department of Zoology
Neuroscience Program
Michigan State University
East Lansing, Michigan

Gene-Jack Wang
Medical Department
Brookhaven National Laboratory
Upton, New York

Christopher L. Wright
Department of Physiology and Program in Neuroscience
University of Maryland School of Medicine
Baltimore, Maryland

Introduction

Gregory F. Ball and Jacques Balthazart

WHY THIS TOPIC AND WHY IS IT IMPORTANT?

The title of this book: Brain Estrogens, Aromatase and Behavior may suggest to some that a substantial number of pages have been written about a quite selective subject matter. The book does indeed focus on a single enzyme, aromatase (or estrogen synthase), and one class of hormones it produces, estrogens, when provided the appropriate substrate, androgens. We delineate our focus even further by considering the distribution and action of aromatase in one organ, the brain and its consequences for one brain function, behavior. Given the widespread distribution of the aromatase enzyme in the body the diverse and complex actions of estrogens in many tissues, why this particular focus?

We argue that one of the most important discoveries of late twentieth century neuroendocrinology is that androgens such as testosterone are essentially prohormones and their actions in the brain are often potentiated, amplified, or dissipated by the fact that they are metabolized at their local site of action via enzymatic transformation to an estrogenic, an androgenic, or an inactive metabolite. This local metabolism represents an entire level of endocrine regulation in addition to the well-known level of glandular secretion into the blood. Local metabolism means that the availability of particular steroids can be fine-tuned at target sites as needed and that the release of a general endocrine signal does not mean that all possible targets will experience the same concentration of the hormone (Balthazart, 1989; Schmidt et al., 2008). This is important for several reasons. Steroid hormones such as androgen and estrogens are potent transcription factors (Tsai & O'Malley, 1994) with beneficial as well as detrimental effects depending on the dose and time course that a tissue experiences with the hormone (Wingfield & Silverin, 2002; Wingfield, 2005). Thus the principle of local hormone metabolism that is perhaps best understood in the case of androgens being converted to estrogens is a fundamental feature of endocrine regulation. As it became apparent in the 1970s that the phenomena of the metabolism of androgens to estrogens in the brain is widespread among vertebrates, it also became clear that one needed to understand how and when the aromatase enzyme was regulated if one were going to elucidate the causal mechanisms underlying how androgens such as testosterone exerted their effects on behavior (Balthazart, 1989; Baum et al., 1990; Balthazart et al., 2004). These findings revealed that many effects of androgens on behavior and its associated brain plasticity were mediated via estrogens such as 17β-estradiol acting on estrogen receptors. Such observations generated a substantial amount of interest in estrogen action in the brain of *both* males and females. These studies in turn provided, part of the set of findings that formed the basis for another revolution in late twentieth century neuroendocrinology, namely that steroids, especially estradiol, can have rapid actions mediated at the membrane in addition to their well known genomic actions (e.g., (Kelly et al., 1977; Schumacher, 1990)). In this brief introduction, we will first discuss the history and current understanding of the concept androgens being metabolized to estrogens in the brain. We will then introduce the idea of aromatase in relation to the genomic and non-genomic actions of estrogens. Finally we will define the scope and rationale of the book.

HISTORY AND CURRENT STATUS OF THE IDEA OF BRAIN METABOLISM OF ANDROGENS INTO ESTROGENS

The discovery of androgens being metabolized into estrogens in the brain first required that there was an understanding of steroidogenesis in general. As reviewed by Callard in this volume, although estrone was isolated in the 1920s it was not until the 1950s, when high specific activity radiolabeled steroids became available, that the biosynthetic pathway from the C19 androgens to the C18 estrogens were elucidated by Engel, Ryan, and colleagues (Engel, 1974).

Because androgens and estrogens were synthesized before the biosynthetic pathways for steroids were understood (see figure I–1), many seemingly unexplainable results were obtained. After "sex" steroid hormones were identified, it was thought by some that there were "female" and "male" hormones that might mediate sex-typical effects (e.g., Steinach, 1940; see Feder, 1981 for a review). Early experiments implanted ovaries into males and testes into females and tried to assess whether various sex-typical attributes could be reversed (e.g., Steinach, 1916; Goodale, 1918)). With the development of purified estrogens and androgens, experiments were designed that involve administering estrogens to males and androgens to females and many exceptions to the rule that female sex hormones promote female-typical traits and male sex hormones promote male-typical traits were discovered. For example, in the field of behavioral endocrinology Josephine Ball observed that sexual behavior in castrated male rats could be activated by androgens and estrogens (Ball, 1937). Frederick Naftolin describes in his introductory chapter to this volume how his studies with Geoffrey Harris resulted in a similar conundrum. Male and female rats exhibit marked differences in feedback properties on luteinizing hormone in the hypothalamus. Females can exhibit positive and negative effects in response to steroids, whereas males only exhibit negative effects. Surprisingly, Naftolin and others found that a neonatal female could be masculinized by treatment with either an androgen or an estrogen. He further observed that ring A-reduced androgen metabolites were ineffective suggesting that an estrogenic metabolite of testosterone might be the critical factor. This observation resulted in the postulation of the aromatase hypothesis in

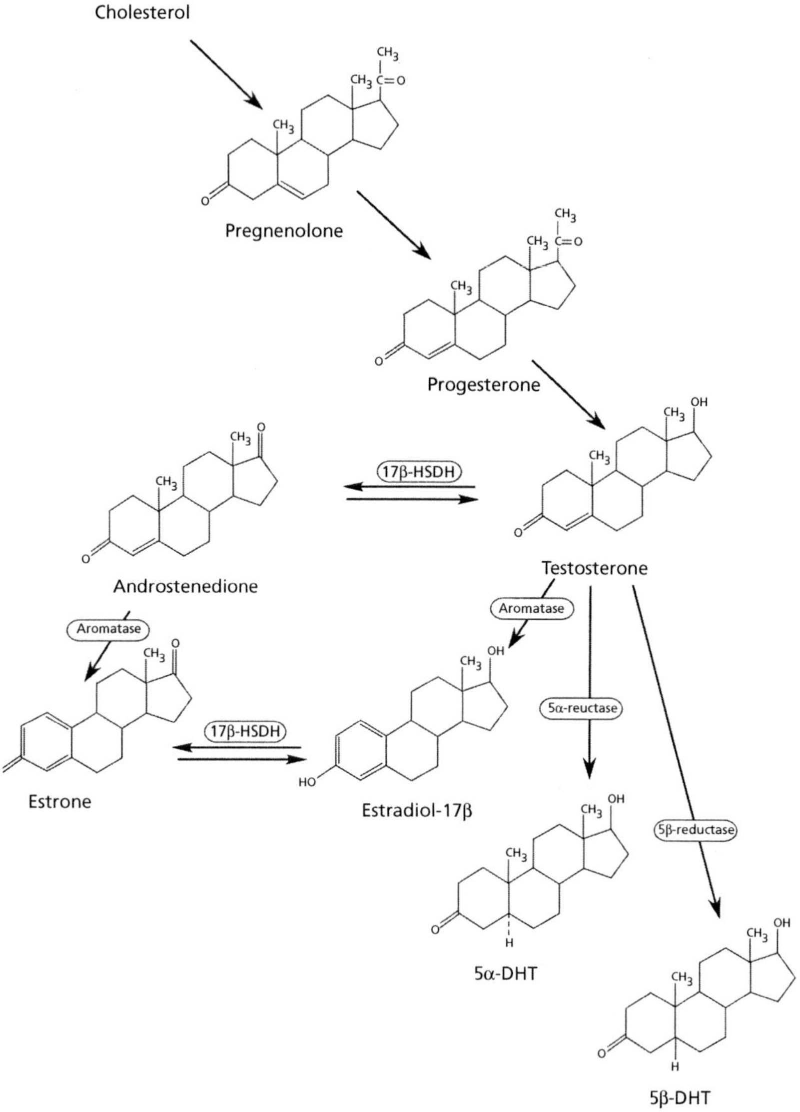

Figure I–1. Simplified view of the metabolic pathways leading to the synthesis of testosterone and to its metabolism into estrogens (aromatization) or into 5α- or 5β-reduced metabolites. Note that only selected steroids are presented and only their trivial names are used. Irreversible and reversible transformations are represented by single and double arrows respectively. Several of the *arrows* involve multiple transformations catalyzed by several enzymes (e.g. the conversion of cholesterol into pregnenolone). The name of key enzymes are associated with the *arrows*. 17β-HSDH = 17β-hydroxysteroid dehydrogenase.

the 1970s that provided the framework for many of the studies that provide the foundation for the chapters assembled in this book.

Although the aromatase hypothesis is at least 40 years old, it still seems to constitute "news" in many circles, such as the popular press, but even within scientific circles at times. For example, a news report in the Economist (2010, April 29th) on an important finding by Nirao Shah that described transgenic mice (Wu et al., 2009) that highlighted cells expressing the aromatase gene was entitled "Oestrogen, not testosterone is what makes a male act like a male." The article then went on to say that it would be "expected" that the effects of testosterone on aggression would be via androgen receptors and the article focused primarily on the fact that androgens seemed to be converted to estrogens in the male brain! Even in a sophisticated book by Rebecca Jordan-Young entitled *Brainstorms* for the educated layperson as well as scientists that critiqued scientific studies on sex differences in the human brain, the author spends nearly an entire page (Jordan-Young 2010, p. 16) discussing the novelty of the fact that androgens as well as estrogens can act in males and females! There is no mention that this has been an organizing paradigm of research since the 1970s, but rather these observations are presented as if scientists are still generally uninformed of this fact.

An even more recent revolution in endocrine physiology is the recognition that steroid hormones, including estrogens, can act in another faster mode than the well-documented genomic action that requires hours to days and even weeks for the full effect of steroid action to be apparent. Steroid hormones were not supposed to be able to act on the scale of seconds to minutes. E.O. Wilson (1975, p. 6) famously referred to hormones as being "concerned with the crude tuning devices of nervous activity." Because of this fact, endocrinology would not in the future be as relevant to the study of the mechanistic control of behavior as compared with the field of integrative neurophysiology (Wilson, 1975). Over the course of the past 40 years, this view of endocrine action in the brain has changed with very exciting implications for the study of endocrine function. Resistance to the idea that steroid hormones can indeed act more quickly in a manner akin to neurotransmitters is understandable because right from the start of the academic field of endocrinology, Bayliss and Starling advocated that chemical secretions were regulated in a manner different from and independent of the nervous system (Bayliss & Starling, 1902). It is useful to remember that some electrophysiologists questioned even whether neurotransmitters existed and asserted that such chemical messengers could not be involved in neurophysiological action because chemical communication would be too slow to explain fast muscle contractions and other similar responses to environmental and behavioral stimuli (Valenstein, 2005)! Estradiol now exhibits so many functions similar to certain classes of neurotransmitter/neuromodulators that we have argued that it would have been classified as such if its actions in the brain had been discovered before its actions in the periphery (Balthazart & Ball, 2006).

SUMMARY OF THE GENOMIC AND NON-GENOMIC ACTIONS OF ESTROGENS

Our current understanding of the diverse actions of brain estradiol is summarized in figure I–2 (see also chapter 10, this volume). Figure I–2A illustrates the "classic" view of testosterone metabolism and hormone action. Testosterone either coming from the periphery or synthesized in the brain passes through the cell membrane and is then transformed via aromatase in the cytoplasm to 17β-estradiol (E2).

E2 can then bind to intracellular estrogen receptors as is illustrated in Figure I–2A or E2 can diffuse from the cell it is formed in to an adjacent cell and act in a paracrine fashion on ER available in this cell (Fig. I–2B). Aromatase is also known to be available in the synaptic bouton and estrogens formed in this part of a neuron can be released into the synapse and then interact with postsynaptic receptors on this postsynaptic neuron or an adjacent neuron (Fig. I–2C). These different actions of E2 require different degrees of modulation of aromatase. Long-acting genomic actions of estrogens are generally associated with differences in reproductive state that involve significant differences in the availability of the aromatase protein that occur over relatively long time periods. Shorter actions of estrogens at the synapse require changes in estrogen availability over a short time course that is implemented via changes in the phosphorylation state of the aromatase protein that can occur rather quickly.

SPECIES DIVERSITY AND THE QUESTION OF BRAIN AROMATASE IN PRIMATES INCLUDING HUMANS

The field of brain aromatase and estrogens has benefited since the early days of the formulation of the aromatase hypothesis from scientists investigating these questions in a variety of species. Prominent species differences in the distribution and activity of brain aromatase meant that the enzyme could be studied more easily in for example fish and birds than in mammals. Much early work therefore focused on fish and birds (e.g., Callard et al., 1978; Steimer & Hutchison, 1981) but as methods improved in sensitivity, a series of significant mammalian studies were completed notably in rats, ferrets, and monkeys (Roselli et al., 1985; Tobet et al., 1985; Roselli & Resko, 1989; Roselli et al., 2009). There is clear evidence for aromatase activity in the brain of primates (see Roselli this volume) including humans (Azcoitia et al., 2011; Biegon, this volume). Less is known about the functions of this aromatase in primates because experimental manipulations are more difficult than they are in rodents and ferrets. So, all the principles of aromatase regulation and function gleaned from studies of other species may not apply specifically to humans. General theories about brain aromatase will doubtlessly be useful as we learn more about the distribution and function of this enzyme in the primate brain. As noted in the chapter by Glenda Gillies in this volume, there are

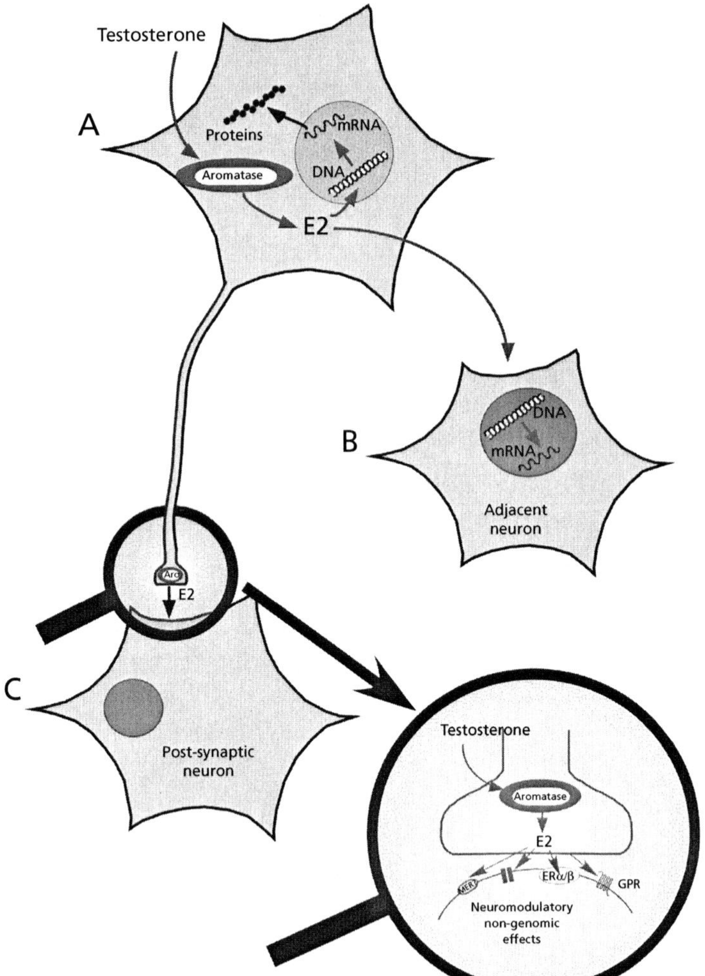

Figure I–2. Schematic view of the different potential modes of action of estradiol (E2) produced in the brain by aromatization of testosterone coming either from the periphery or synthesized in the brain. E2 can bind to intracellular estrogen receptors located in the cell where it has been produced by aromatization (*A*) or E2 can diffuse from the cell where it has been formed to an adjacent cell and act in a paracrine fashion on nuclear estrogen receptors available in this cell (*B*). Alternatively, E2 can also be synthesized by aromatase present in presynaptic boutons and be released into the synapse so that it can then interact with post-synaptic receptors on a postsynaptic neuron (*C*). These membrane receptors potentially include intracellular receptors translocated to the membrane (ER α/β), g-protein coupled receptors (GPR) or other form of membrane estrogen receptors (mER) as indicated under the magnifying glass showing detail of an estrogen-synthesizing synapse.

prominent sex differences in the action of estrogens in the diseased and normal human brain. Understanding basic processes of estrogen synthesis and action in the brain can help design rational therapies related to the manipulation of estrogens.

RATIONALE AND SCOPE OF THE BOOK

We were very pleased that the range of authors we approached about contributing to the book all accepted our invitations. The one exception was Nirao Shah who has recently made some important contributions to the field based on the use of transgenic mice to precisely localize aromatase in the mouse brain and to study aspects of its function in relation to aggression (Wu *et al.*, 2009) and who could not provide a chapter because of other personal commitments. Some of his work is reviewed in the chapter by Roselli in this volume.

We have organized the book into five major sections. Our goal is to capture studies of aromatase distribution and regulation in the brain as well as the major functions of brain estradiol that results from the activity of aromatase. The book starts with a perspective piece by Professor Federick Naftolin who first identified the transformation of androgens into estrogens in the brain in the early 1970s. The second section starts with a consideration of the brain distribution and regulation of aromatase. The number of investigators working on diverse species is apparent from the selection of species that ranges from fish to mammals. The range of techniques utilized to study localization and function is also featured with enzymatic assays, chemical neuroanatomical techniques of various sorts of molecular methods all being featured by the different authors. The next section considers behavioral effects of brain estradiol that occur in adulthood and involve the genomic action of estradiol. In this section we start with a historical and conceptual foundation of this topic written by Professor Lynwood Clemens, who was one of the pioneers in this field. There are then a series of chapters again taking into account phylogenetic diversity and ending with a consideration of these effects in humans. The fourth section takes a developmental perspective and reviews the role of brain aromatization in the sexual differentiation of the brain. This section also starts with an introductory chapter by Per Södersten who provides a provocative perspective on the complex field of the sexual differentiation of the brain. He reviews an older literature that concerns whether or not there are robust sex differences in the response of the brain to steroids such as estradiol. The fifth section concerns an emerging concept in brain estradiol namely that it has neuroprotective effects. Work in mammals and in birds has illustrated how aromatase will be upregulated by insults to the brain and increases in estradiol can largely diminish brain damage. Finally, the last section of the book considers the growing field of the fast actions of estradiol on behavior. This includes the effects of estradiol on a wide range of behaviors including aggression and sex as well as sensory processes such as audition.

Our hope is that this volume will provide a view of the state of this field that has relevance for basic research in endocrinology, neuroscience, reproduction as well as a variety of behaviors. It is also relevant to those interested in any aspect of the actions of estrogens.

ACKNOWLEDGMENTS

We would like to thank the many mentors, colleagues and friends who we have interacted with over the years in our attempts to understand the hormonal control of behavior from the perspective of brain mechanisms. Many appear as authors in this volume. In early stages of our career, we both spent time at the Institute of Animal Behavior at Rutgers in the 1970–80s where discussions with Mei Cheng, Harvey Feder and Barry Komisaruk were very useful. Rae Silver was also a very positive influence at this time. An early paper by Francesco Dessi-Fulgheri demonstrating for the first time a correlation between brain aromatase and behavior (Aggressive Behavior 1976, 2: 223–231) played a key role in developing our initial interest in the possible role played by brain aromatase specifically in the control of behavior and should be acknowledged here. Discussions with John Wingfield and Bruce McEwen at Rockefeller were especially helpful in honing ideas about the neuroendocrine control of behavior. The intellectual and practical support of Nobuhiro Harada and Fred Naftolin for developing the first detailed anatomical studies of brain aromatase was also greatly appreciated. We have been very lucky to have many very gifted students and post-doctoral fellows who have worked with us specifically on the function of aromatase. It is unfortunately impossible to list them all due to space limitations but they can be assured of our gratitude. Two collaborators, Michael Schumacher and Michelle Baillien, must however be singled out because they respectively established in the laboratory of Liege the product-formation and the tritiated water assays for aromatase activity and thus played a key role in many discoveries reported in this book. Our work on aromatase and behavior has been supported by a grant from the NIMH (MH 50388) and we are very grateful for this support.

REFERENCES

Azcoitia, I., Yague, J. G., & Garcia-Segura, L. M. (2011). Estradiol synthesis within the human brain. *Neuroscience, 191,* 139–147.

Ball, J. (1937) Sex activity of castrated male rats increased by estrin administration. *J Comp Psychol*, 24, 135–144.

Balthazart, J. (1989). Steroid metabolism and the activation of social behavior. In: Balthazart, J. (ed..), *Advances in Comparative and Environmental Physiology,* vol 3. Berlin: Springer Verlag,, pp. 105–159.

Balthazart, J., Baillien, M., Cornil, C. A., & Ball, G. F. (2004). Estrogen synthase (aromatase) in the preoptic area modulates the expression of male sexual behavior: slow and fast mechanisms of action. *Physiol Behav, 83,* 247–270.

Balthazart, J., & Ball, G. F. (2006). Is brain estradiol a hormone or a neurotransmitter? *Trends Neurosci, 29,* 241–249.
Baum, M. J., Carroll, R. S., Cherrv, J. A., & Tobet, S. A. (1990). Steroidal control of behavioural, neuroendocrine and brain sexual differentiation: studies in a carnivore, the ferret. *J Neuroendocrinology, 2,* 401–418.
Bayliss, W. M., & Starling, E. H. (1902). The mechanism of pancreatic secretion. *J Physiol Lond, 28,* 325–353.
Callard, G. V., Petro, Z., & Ryan, K. J. (1978). Phylogenetic distribution of aromatase and other androgen-converting enzymes in the central nervous system. *Endocrinology, 103,* 2283–2290.
Engel, L. (1974) The biosynthesis of estrogen. In: Greep, R. O. (ed.), *Handbook of Physiology* Washington, DC: American Physiological Society, pp. 467–483.
Feder, H. H. (1981). Perinatal hormones and their role in the development of sexually dimorphic behaviors. Neuroendocrinology of reproduction. In: Adler, N. (ed.), New York: Plenum Press, pp. 127–158.
Goodale, H. R. (1918). Feminized male birds. *Genetics, 3,* 276–298.
Jordan-Young, R. M. (2010). Brain storm: The flaws in the Science of Sex Differences. Cambridge, MA: Harvard University Press, pp. 408.
Kelly, M. J., Moss, R. L., & Dudley, C. A. (1977). The effects of microelectrophoretically applied estrogen, cortisol and acetylcholine on medial preoptic-septal unit activity throughout the estrous cycle of the female rat. *Exp Brain Res, 30,* 53–64.
Roselli, C. E., Horton, L. E., & Resko, J. A. (1985). Distribution and regulation of aromatase activity in the rat hypothalamus and limbic system. *Endocrinology, 117,* 2471–2477.
Roselli, C. E., Liu, M., & Hurn, P. D. (2009). Brain aromatization: classic roles and new perspectives. *Sem Reprod Med, 27,* 207–217.
Roselli, C. E., & Resko, J. A. (1989). Testosterone regulates aromatase activity in discrete brain areas of male rhesus macaques. *Biol Reprod, 40,* 929–934.
Schmidt, K. L., Pradhan, D. S., Shah, A. H., Charlier, T. D., Chin, E. H., & Soma, K. K. (2008). Neurosteroids, immunosteroids, and the Balkanization of endocrinology. *Gen Comp Endocrinol, 157,* 266–274.
Schumacher, M. (1990). Rapid membrane effects of steroid hormones: an emerging concept in neuroendocrinology. *Trends Neurosci, 13,* 359–362.
Steimer, T., & Hutchison, J. B. (1981). Androgen increases formation of behaviourally effective oestrogen in dove brain. *Nature, 292,* 345–347.
Steinach, E. (1916). Pubertäsdrüsen und Zwitterbildung. *Arch. f. Entwicklungsmech. d. organs, 42,* 1–26.
Steinach, E. (1940). *Sex and Life.* New York: Viking Press.
Tobet, S. A., Baum, M. J., Tang, H. B., Shim, J. H., & Canick, J. A. (1985). Aromatase activity in the perinatal forebrain: effects of age, sex and intrauterine position. *Dev Brain Res, 23,* 171–178.
Tsai, M. J., & O'Malley, B. W. (1994). Molecular mechanisms of action of steroid/thyroid receptor superfamily members. *Annu Rev Biochem, 63,* 451–486.
Valenstein, E. S. (2005). *The War of the Soups and the Sparks. The Discovery of Neurotransmitters and the Dispute over How Nerves Communicate.* New York: Columbia University Press.
Wilson, E. O. (1975). *Sociobiology: The New Synthesis.* Cambridge, MA: Belknap Press of Harvard University Press,

Wingfield, J. C. (2005). A continuing saga: the role of testosterone in aggression. *Horm Behav, 48,* 253–255; discussion 256–258.

Wingfield, J. C., & Silverin, B. (2002). Ecophysiological studies of hormone-behavior relations in birds. In: Pfaff, D. W., Arnold, A. P., Etgen, A. M., Fahrbach, S. E., Rubin, R. T. (eds.), *Hormones, Brain and Behavior.* San Diego: Academic Press, pp. 587–647.

Wu, M. V., Manoli, D. S., Fraser, E. J., Coats, J. K., Tollkuhn, J., Honda, S., et al. (2009) Estrogen masculinizes neural pathways and sex-specific behaviors. *Cell,* 139, 61–72.

Part I

Historical Origins and Conceptual Basis of the Field

Chapter 1

The Aromatization Hypothesis and the Discovery of Estrogen Synthesis by Brain Cells

Frederick Naftolin

INTRODUCTION

The discovery of estrogen synthatase (aromatase) in the brain was driven by the need to explain the molecular basis of sexual differentiation gonadotrophin regulation by the brain. Because the aromatization and brain sexual differentiation are intimately entwined, this chapter presents the developments in a chronological manner.

PREPARING TO STUDY BRAIN SEXUAL DIFFERENTIATION

The aromatization hypothesis resulted from my time as a graduate student and NIH Fogarty Fellow in the Department of Human Anatomy at Oxford University. My Oxford mentors were Geoffrey Harris and Keith Brown-Grant. They were extremely kind and welcoming to a displaced American reproductive endocrinologist. My thesis work examined brain sexual differentiation in the rat.

Decades earlier, Geoffrey Harris and collaborators had refined the work of Pfeiffer, identifying the role of the hypothalamus, rather than the pituitary, in sex-specific gonadotrophin regulation; females have both negative and positive feedback control of gonadotrophins, whereas males have only negative feedback control (Harris & Naftolin, 1970). During my graduate studies I worked to further clarify how this came about. Since objective, sensitive, precise, and practical hormonal measurements were becoming available and I had helped develop a radioimmunoassay (RIA) for human luteinizing hormone (LH) as an endocrinology fellow with C. Alvin Paulsen, I was assigned to develop and apply an immunoassay for rat LH to studies on brain sexual differentiation.

This was the subject of my thesis (Naftolin, 1970). The laboratory worked mainly on rats, though we occasionally studied LH in other species, including humans (Corker et al., 1969; Brown-Grant et al., 1971; Katongole et al., 1971; Naftolin et al., 1971a, 1972b). Complementary blood estradiol levels were measured using a binding assay that was concurrently developed by a fellow graduate student, Charles Corker (Corker et al., 1970).

BRAIN SEXUAL DIFFERENTIATION

Rats are born during the period of sexual differentiation of the regulation of gonadotrophins. This allows study of developmental programming of the brain. In rats this is the time ("the critical period"), when the hypothalamic circuitry for hormonal suppression of gonadotrophin releasing hormone (GnRH; negative feedback) and for the GnRH release during the preovulatory surge of circulating estradiol (positive feedback), or for blocking positive feedback are laid down. In adult genetic females, both negative and positive feedback are possible; this is the default circuitry. In males, a further elaboration of the developmental program blocks the hypothalamus'ability to respond to a surge in circulating estradiol with an increase in GnRH secretion (Harris & Naftolin, 1970; MacLusky & Naftolin, 1981).

The pioneering work of Pfeiffer showed that the organization of the central axis that silences positive feedback in rats arises from the testis and that this testis-dependent effect occurs during the perinatal period in genetic males but not females; the presence of the ovary is of no moment in this aspect of development (Pfeiffer, 1936). Parenthetically, Pfeiffer thought that this effect was at the level of the adenohypophysis. This bias was the result of the enormous influence at the time of Harvey Cushing and his preoccupation with the pituitary as the main controlling influence over the regulation of endocrine function (Carroll Pfeiffer, personal communication). Harris and Jacobson showed that the organization of positive feedback and its expurgation by the testis is independent of the pituitary; the location of sexual differentiation is in the hypothalamus. Because the GnRH neurons that secrete into the portal vessels are located in the hypothalamus and are targeted by hypothalamic neurons, brain sexual differentiation must involve developmental regulation of this circuitry (Harris & Naftolin, 1970; MacLusky & Naftolin, 1981).

The timing of the "critical period" for brain sexual differentiation is species dependent and usually between E15 and 35. Since rats are born at E21–22, so there is a window of opportunity to study experimentally the mechanism of the testes'effect on brain sexual differentiation that extends from E~17 to P~14 in rats (MacLusky & Naftolin, 1981).

WHAT IS THE ACTIVE AGENT IN BRAIN SEXUAL DIFFERENTIATION?

Studies administering sex steroids rather than minced testis showed that testosterone or its metabolites will masculinize a rat's regulation of gonadotrophins as an adult, irrespective of its genetic sex (Jaggard & Bradbury, 1961). However, the molecular mechanism the hypothalamic action of testosterone was not known. This was especially relevant because evidence had also been reported indicating that natural and synthetic estrogens and partially aromatized androgens also could masculinize the genetic female rat's control of gonadotrophins (MacLusky & Naftolin, 1981). So, at the end of the 1960s, the active agent regulating brain cell development was not known.

RING A-REDUCED ANDROGENS DO NOT REGULATE BRAIN SEXUAL DIFFERENTIATION

In the late 1960s it was determined that the active agents of androgen action are the ring A-reduced androgen metabolites of testosterone and androstenedione (Hiipakka & Liao, 1998). These end-metabolites of androgens cannot be converted to their parent compounds. This new understanding of androgen action raised the question of whether ring A-reduced metabolites of androgens could be the active principles in brain sexual differentiation. Therefore it was not surprising that when Alan Munck of Dartmouth University visited the Department of Human Anatomy for his sabbatical period, he brought along ring A-reduced compounds to test for their ability to masculinize the genetic female rat's brain. Munck and Keith Brown-Grant developed a group of studies designed to test the hypothesis that ring A-reduced androgens are the active principles of brain sexual differentiation. I came along because of my ability to measure gonadotrophins. The experiments used neonatal treatment with dihydrotestosterone propionate or androstanediol propionate to attempt the blockade of ovarian cycling (positive feedback) when the rats reached puberty. There was, however, no statistically significant difference in the presence of positive feedback between treated females and vehicle treated females (Brown-Grant et al., 1971); androgen-induced masculinization of the brain does not require ring A-reduction of testosterone. These findings were later confirmed by another group (McDonald & Doughty, 1972).

SO, WHAT IS THE ACTIVE PRINCIPLE FROM THE TESTIS THAT ORGANIZES SEXUALLY DETERMINED CONTROL OF THE GONADOTROPHINS?

The lack of a response to ring A-reduced androgens was not a dead end. Rather, it highlighted the facts that androgens are aromatized to estrogens, and estrogens had been shown to cause brain sexual differentiation (Ciric, 1973). As well, an intermediate of estrogen formation, specifically 19-hydroxytestosterone, had been reported to have the same effect (McDonald & Doughty, 1974). As well, we had performed experiments that indicated stilbestrol, a weak estrogen compared with E_2 that is not bound by rat alpha fetoprotein (MacLusky & Naftolin, 1981), was as powerful at causing brain sexual differentiation (Slaughter et al., 1977). For me, all this indicated that estrogen was the likely active principle causing hypothalamic organization that excluded positive feedback. I argued this point with only lukewarm success during the writing of the paper describing the failure of ring A-reduced androgens to masculinize the brain; however, this conclusion seemed inescapable. I continued to ruminate on it as my graduate studies ended and I returned to the United States to begin my academic career.

Before leaving Oxford I discussed with Professor Harris the possibility that testosterone was aromatized in the cells of the brain that were involved in brain sexual differentiation and he assured me that this was not possible—he had discussed it with chemists and "...there is not enough energy (for this metabolism)."

The metabolism of testosterone by brain had indeed been studied and found to result in the formation of ring A-reduced metabolites; however, the formation of estrogens had not been shown. To be sure, these efforts were not aimed at finding phenolic (aromatized) products. As well, they were based on the analysis of C^{14}-labeled metabolites and 1% or more of the C^{14} remained unaccounted for after the experiments. Thus, a conversion of approximately 1% of testosterone into phenolic metabolites could have escaped attention (Sholiton et al., 1966; Jaffe, 1969). In fact, in one report there had been some counts that might have represented estrogens but that was not proved (Knapstein et al., 1968).

GERMINATION OF THE AROMATIZATION HYPOTHESIS

At this point, my studies in Oxford ended and I joined the Department of Obstetrics, Gynecology, and Reproductive Sciences at the University of California, San Diego. Fatefully, my chairman was Kenneth John Ryan, who by fortunate coincidence was one of the leading experts in placental aromatization of androgens. Furthermore, my clinical duties included development of the mid-trimester pregnancy termination service, which gave me access to brain material from newly deceased fetuses. I pleaded the aromatization hypothesis to Dr. Ryan and the possibility that estrogen formation could be shown in the human brain! Averred that adequate amounts of androgens were present for such metabolism to occur during human gestation. In fact, as a resident physician I had worked with Hans Simmer on the 16-hydroxylation of fetal dehydroepiandrosterone, whose aromatized product is estriol (Easterling et al., 1966).

Although he was extremely skeptical of the idea of brain aromatization of androgens, Dr. Ryan agreed to allow me to test for brain aromatization. He asked his technician Zoltan Petro to assist me; I had two weeks to prove the aromatization hypothesis. The wager of a Coca Cola sweetened the bargain.

AROMATIZATION BY BRAIN TISSUE IS A FACT

Zoltan Petro and I began our work. We studied dissected hypothalamic tissue from mid-trimester human female and male fetuses. Promptly after delivery of fetuses whose gestation that had been terminated by saline infusion into the amniotic sac it was determined that there were no signs of life and the brain was removed and bloc-dissected to furnish the anterior and posterior hypothalamus. I included the temporal lobe as a putative negative control. The method of proving the presence of aromatization in the tissue was the same as the Ryan lab had been using to study placental aromatization of androgens. This method is based on incubation of the homogenized tissue with radiolabeled androgen (^{14}C-androstenedione) and performing a phenolic extraction to isolate the estrogens. The extract is then treated to form phenolic conjugates, acetates in this case, which are then extracted and recrystallized to constant specific activity. The products are compared with authentic estrogen acetates to prove that newly formed estrogen has been made by the microsomes in the (brain) homogenates (Naftolin et al., 1971b). While I wrote the play, Zolie acted all of the important roles; the biochemical work was all his.

Our first run ended at about 2 in the morning, when Zoltan and I were alone in the lab. We whooped and danced about when the chromatograph brought forth generous peaks of constant specific activity estrone acetate from all hypothalamic pieces. Less, but define quantities were present in the temporal lobe tissues. The next morning I rushed in to tell Dr. Ryan of our results and to collect my Coke. He asked to witness the next experiment. When it showed smaller peaks of the same pattern I was crestfallen. He chided me, saying that others waited their whole professional lives to see any peaks, at all. Because the "control" temporal lobe had also shown aromatase activity, we specifically tested to confirm that aromatase was present and showed the presence of aromatase in this area that contains the hippocampus (Naftolin et al., 1972a). We went on to confirm the presence of aromatase in rat, rabbit, and monkey brain (Naftolin et al., 1975).

The calculation of the formation of estrogen from androgen in the human, rat, and monkey brains hovered below 1% or less. This supported incomplete accounting for 100% of the radioactivity added to the incubates obscures the presence of aromatase in previous studies. The low yields of estrogens is not of concern since, any production of estrogen within brain cells is of great moment since mole for mole estrogens are on average 100- to 1,000-fold more potent in causing biological activity than androgens.

The aromatization hypothesis of brain sexual differentiation had been proved (Naftolin et al., 1975). Further, others showed that a selective estrogen receptor modulator can block brain sexual differentiation by androgen (McDonald & Doughty, 1972).

HOW DOES ESTROGEN AFFECT BRAIN SEXUAL DIFFERENTIATION?

In the 1970s, Matsumoto and Arai performed pioneering work showing synaptogenic effects of estrogen on the newborn rat's hypothalamic arcuate nucleus (AN), the "junction box" of the GnRH delivery system (Matsumoto & Arai, 1976). Almost a decade later we began to study the AN of rats and monkeys to better understand the control of GnRH by estrogen. Estradiol was confirmed to regulate the number, type (excitatory or inhibitory), neurotransmitter content, and membrane composition of GnRH-targeting connections in the rat AN (Naftolin et al., 2007). Our group discovered a leash of endorphin-expressing processes that connect the AN with the GnRH neurons in the POA, explaining the mechanism of the Rat's internuclear control of GnRH (Leranth et al., 1986). We showed in monkeys that estradiol is synaptogenic during its negative feedback and synaptolytic during positive feedback control of gonadotrophins (Zsarnovszky et al., 2001).

As a result of these and other studies it was possible to develop a picture of the changes induced by estrogen during the critical period, the synaptology during the ovarian cycle and in adult males, and the mechanism of hypothalamic failure in the aging rat as it becomes acyclic. It is now clear that during the critical period the circuitry of the female rat's GnRH delivery system is organized to allow the AN neurons to undergo both estrogen-induced synaptogenesis and synaptolysis, whereas because of androgen sectered during the critical period by the testers males are limited to synaptogenesis (Naftolin et al., 1975). In other studies we showed that estrogen-induced astroglial spreading is necessary to the synaptolytic effect of high estrogen levels (Naftolin et al., 2007). Others have shown that the astroglia of males are sexually differentiated during the critical period (McCarthy et al., 2003). That might explain the failure of estrogen-induced astroglial spreading and synaptolysis that characterizes the male rat AN.

OTHER INTERESTING ASPECTS OF OUR EARLY WORK ON AROMATIZATION BY BRAIN CELLS

The original work on human fetal brain was followed by similar testing of rat fetal and adult brains. In both humans and rats, weight for weight the males'conversion reliably exceeded that of the females (Naftolin et al., 1972a, 1975). This has not yet been explained though studies have shown some of the factors that regulate aromatase.

1. With the collaboration of Robert White of Case Reserve School of Medicine we showed that the completely isolated monkey brain forms estrogen from androgen. This proved that there is no need for other tissues or blood from other organs to furnish estrogen in the brain (Naftolin et al., 1975).
2. Aromatase mapping was begun with the collaboration of Nobuhiro Harada and Jacques Balthazart (Naftolin et al., 2001). This resulted in the first atlas of brain aromatization, in the Japanese quail brain. This work was enabled by the outstanding skill of Dr. Harada, who developed some of the first antiaromatase antibodies.
3. The proof of the presence of aromatase in neurons was substantiated by the discovery of aromatase in synaptosomes for rat brain. While not yet applied scientifically, this finding is very challenging because it implies additional effects of aromatase in the brain other than as would result from the product estrogen. We have commented on the interesting possibilities and hope someday to discover the true meaning of this interesting finding (Naftolin et al., 1996).

CONCLUSIONS

The discovery of aromatization by brain cells was a byproduct of studies on brain sexual differentiation that led to the Aromatization Hypothesis. The demonstration of this novel biology led to important changes in the understanding of brain metabolism and action. There remain many unexplained ramifications of this work (Naftolin et al., 2007; Panzica et al., 2012). Regardless, the discovery that set them off would not have been possible without the support of many seasoned scientists for a naïve young academic. This reprise is dedicated to their mentorship and generosity.

REFERENCES

Brown-Grant, K., Munck, A., Naftolin, F., & Sherwood, M. (1971). The effects of the administration of testosterone propionate alone or with phenobarbitone and of testosterone metabolites to neonatal female rats. *Horm Behav, 2,* 173–182.

Ciric, O. (1973). The production of permanent anoestrous sterile rats by injections of estradiol dipropionate early after birth. *Iugoslavica physiologica et pharmacologica acta, 9,* 93–96.

Corker, C.S., Exley, D., & Naftolin, F. (1970). Assay of 17beta-oestradiol by competitive protein binding methods. *Acta endocrinologica. Supplementum, 147,* 305–319.

Corker, C. S., Naftolin, F., & Exley, D. (1969). Interrelationship between plasma luteinizing hormone and oestradiol in the human menstrual cycle. *Nature, 222,* 1063.

Easterling, W. E., Jr., Simmer, H. H., Dignam, W. J., Frankland, M. V., & Naftolin, F. (1966). Neutral C19-steroids and steroid sulfates in human pregnancy. II. Dehydroepiandrosterone sulfate, 16-alpha-hydroxydehydroepiandrosterone, and 16-alpha-hydroxydehydroepiandrosterone sulfate in maternal and fetal blood of pregnancies with anencephalic and normal fetuses. *Steroids, 8,* 157–178.

Harris, G. W., & Naftolin, F. (1970). The hypothalamus and control of ovulation. *Br Med Bull, 26,* 3–9.

Hiipakka, R. A., & Liao, S. (1998). Molecular mechanism of androgen action. *Trends Endocrin Met TEM, 9,* 317–324.

Jaffe, R. B. (1969). Testosterone metabolism in target tissues. Hypothalamic and pituitary tissues of the adult rat and human fetus, and the immature rat epiphysis. *Steroids, 14,* 483–498.

Jaggard, J., & Bradbury, J. T. (1961). Local and systemic effects of intratesticular testosterone and estradiol pellets in immature rats. *Endocrinology, 68,* 759–766.

Katongole, C. B., Naftolin, F., & Short, R. V. (1971). Relationship between blood levels of luteinizing hormone and testosterone in bulls, and the effects of sexual stimulation. *J Endocrinol, 50,* 457–466.

Knapstein, P., David, A., Wu, C. H., Archer, D. F., Flickinger, G. L., & Tochstone, J. C. (1968). Metabolism of free and sulfoconjugated DHEA in brain tissue in vivo and in vitro. *Steroids, 11,* 885–896.

Leranth, C., MacLusky, N. J., & Naftolin, F. (1986). Interconnections between neurotransmitter- and neuropeptide-containing neurons involved in gonadotropin release in the rat. In: Moody, T. (ed.), *Neural and Endocrine Peptides and Receptors.* New York: Plenum Press, pp. 177–193.

MacLusky, N. J., & Naftolin, F. (1981). Sexual differentiation of the central nervous system. *Science, 211,* 1294–1303.

Matsumoto, A., & Arai, Y. (1976). Effect of estrogen of early postnatal development of synaptic formation in the hypothalamic arcuate nucleus of female rats. *Neurosci Letts, 2,* 79–82.

McCarthy, M. M., Todd, B. J., & Amateau, S. K. (2003). Estradiol modulation of astrocytes and the establishment of sex differences in the brain. *Ann N Y Acad Sci, 1007,* 283–297.
McDonald, P. G., & Doughty, C. (1972). Inhibition of androgen-sterilization in the female rat by administration of an antioestrogen. *J Endocrinol, 55,* 455–456.
McDonald, P. G., & Doughty, C. (1974). Effect of neonatal administration of different androgens in the female rat: correlation between aromatization and the induction of sterilization. *J Endocrinology, 61,* 95–103.
Naftolin, F. (1970). *Application of Radioimmunoassay and Competitive Protein Binding Methodology to the Study of Neuroendocrine Function.* Oxford: Oxford University Press.
Naftolin, F., Garcia-Segura, L. M., Horvath, T. L., Zsarnovszky, A., Demir, N., Fadiel, A., et al. (2007). Estrogen-induced hypothalamic synaptic plasticity and pituitary sensitization in the control of the estrogen-induced gonadotrophin surge. *Reprod Sci, 14,* 101–116.
Naftolin, F., Harris, G. W., & Bobrow, M. (1971a). Effect of purified luteinizing hormone releasing factor on normal and hypogonadotrophic anosmic men. *Nature, 232,* 496–497.
Naftolin, F., Horvath, T. L., & Balthazart, J. (2001). Estrogen synthetase (aromatase) immunohistochemistry reveals concordance between avian and rodent limbic systems and hypothalami. *Exp Biol Med (Maywood),* 226, 717–725.
Naftolin, F., Horvath, T. L., Jakab, R. L., Leranth, C., Harada, N., & Balthazart, J. (1996). Aromatase immunoreactivity in axon terminals of the vertebrate brain. An immunocytochemical study on quail, rat, monkey and human tissues. *Neuroendocrinology, 63,* 149–155.
Naftolin, F., Ryan, K. J., Davies, I. J., Reddy, V. V., Flores, F., Petro, Z., et al. (1975). The formation of estrogens by central neuroendocrine tissues. *Recent Prog Horm Res, 31,* 295–319.
Naftolin, F., Ryan, K. J., & Petro, Z. (1971b). Aromatization of androstenedione by the diencephalon. *J Clin Endocrinol Metab, 33,* 368–370.
Naftolin, F., Ryan, K. J., & Petro, Z. (1972a). Aromatization of androstenedione by the anterior hypothalamus of adult male and female rats. *Endocrinology, 90,* 295–298.
Naftolin, F., Yen, S. S., & Tsai, C. C. (1972b). Rapid cycling of plasma gonadotrophins in normal men as demonstrated by frequent sampling. *Nat New Biol, 236,* 92–93.
Panzica, G. C., Balthazart, J., Frye, C. A., Garcia-Segura, L. M., Herbison, A. E., Mensah-Nyagan, A. G., et al. (2012). Milestones on steroids and the nervous system: 10 years of basic and translational research. *J Neuroendocrinol, 24,* 1–15.
Pfeiffer, C. (1936). Sexual differences of hypophyses and their determination by gonads. *Am J Anat, 58,* 195.
Sholiton, L. J., Marnell, R. T., & Werk, E. E. (1966). Metabolism of testosterone-4-C14 by rat brain homogenates and subcellular fractions. *Steroids, 8,* 265–275.
Slaughter, M., Wilen, R., Ryan, K. J., & Naftolin, F. (1977). The effects of low dose diethylstilbestrol administration in neonatal female rats. *J Steroid Biochem, 8,* 621–623.
Zsarnovszky, A., Horvath, T. L., Garcia-Segura, L. M., Horvath, B., & Naftolin, F. (2001). Oestrogen-induced changes in the synaptology of the monkey (Cercopithecus aethiops) arcuate nucleus during gonadotropin feedback. *J Neuroendocrinol, 13,* 22–28.a

Part II

Aromatase, its Brain Distribution and Regulation

Chapter 2

Brain Aromatase in Fishes: Historical Perspective and Comparative Approaches

Gloria V. Callard, Sarah R. Greytak, Apolonia Novillo, Kellie A. Cotter and Rebecca K. Meyer

BRAIN AROMATASE: AN HISTORICAL PERSPECTIVE

Classically, estrogens are regarded as circulating hormones derived from the gonads and placenta, but certain actions of estrogens are dependent on the aromatization of C_{19} androgens to C_{18} estrogens in close proximity to target cells in the brain and many other glandular (ovary, testis) and nonglandular (fat, bone) tissues. In this context, estrogens have an additional autocrine or paracrine signaling role, defined by the proximity and topographic relationship of the cytochrome P450 aromatase enzyme complex and the cellular targets of estrogen action (Fig. 2–1). Research into the nature and biochemistry of "sex steroids" began in the 1920s and '30s with the isolation of estrone, but the pathway to estrogen biosynthesis was largely unknown until the development of high specific activity radiolabeled steroids in the 1950s. This technical advance led to a series of studies that revolutionized our understanding of steroid biosynthetic pathways from small acyclic precursors, and demonstrated the universality of steroidogenesis from fish to humans (Engel, 1974; Tata, 2005; Santen et al., 2009). Nonetheless, the debate continued as to the ability of eukaryotic cells to carry out aromatization of the A-ring of androgens, the final step in estrogen biosynthesis. Using human placenta and ovary, major sources of secreted estrogens, Engel et al. isolated the aromatase enzyme complex in microsomal subfractions; demonstrated that the androgens (testosterone and androstenedione) are the immediate and obligatory precursors of the estrogens (estradiol and estrone); defined oxygen and cofactor requirements for aromatization; and identified several reaction intermediates (Ryan, 1959; Engel, 1974). This assay, measuring the production of estrogens from radiolabeled androgens, remains the gold standard for previously unstudied or low aromatase tissue types, but has been streamlined to measure radiolabeled water (a byproduct of aromatization) in which aromatase activity is high or previously validated. Ultimately, the enzyme complex catalyzing estrogen biosynthesis (estrogen synthetase) was identified as a nicotinamide adenine dinucleotide phosphate (NADPH)-dependent cyctochrome P450 reductase and

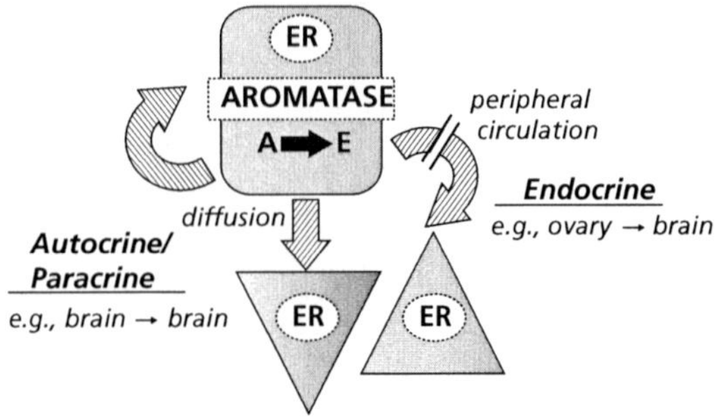

Figure 2–1. Estrogens have multiple signaling roles in the brain. Two core components of the estrogen signaling pathway are cytochrome P450 aromatase, the enzyme catalyzing conversion of androgens (A) to estrogens (E), and estrogen receptors (ER), ligand activated transcription factors that transduce the estrogen signal into a biological response. The proximity and topographical relationship of cells expressing aromatase and ER determine whether the estrogen signaling properties are endocrine, paracrine, or autocrine in nature. All three signaling modes are operational in brain. This scheme does not rule out additional estrogen actions mediated by extranuclear ER or other receptor classes (see text).

a cytochrome P450 aromatase, a member of the large cytochrome P450 superfamily of monooxygenases.

Within the next decade, Naftolin and Ryan obtained definitive evidence that the human fetal hypothalamus is capable of synthesizing estrogens from aromatizable substrate (Naftolin et al., 1971). Low estrogen yields in vitro initially raised questions about the biological relevance of brain-formed (neuro-) estrogens, but subsequent in vivo studies demonstrated estrogen biosynthesis and secretion using an isolated perfused rhesus brain preparation (Flores et al., 1973) and a spatial and temporal correspondence between aromatization and estrogen receptor occupancy (Lieberburg & McEwen, 1977). Most significant, Naftolin recognized that the ability of estrogens to mimic many of the organizational and activational effects of androgens on sex behavior and neuroendocrine functions could be explained by androgen to estrogen conversion in situ (reviewed in MacLusky & Naftolin, 1981; Naftolin et al., 1975). Over the next few decades, the "aromatization hypothesis" provided the framework for hundreds of behavioral and physiological studies that strengthened the concept and eventually extended its applicability to higher neural functions such as cognition, aging, and neurodegeneration (see other chapters in this book).

Among the second generation studies in the Ryan lab was a phylogenetic survey reporting that the ability to convert androgens to estrogens is an ancient and highly conserved characteristic of the vertebrate brain (Callard et al., 1978, 1980). Although gonadal aromatase is already well developed in amphioxus, an invertebrate that is a member of the phylum Chordata but not the subphylum Vertebrata and is therefore generally regarded as an archetype of vertebrates, no enzyme activity is detectable in the head (brain) segments of this species or in the gonads or brain of any other invertebrates tested (Callard,

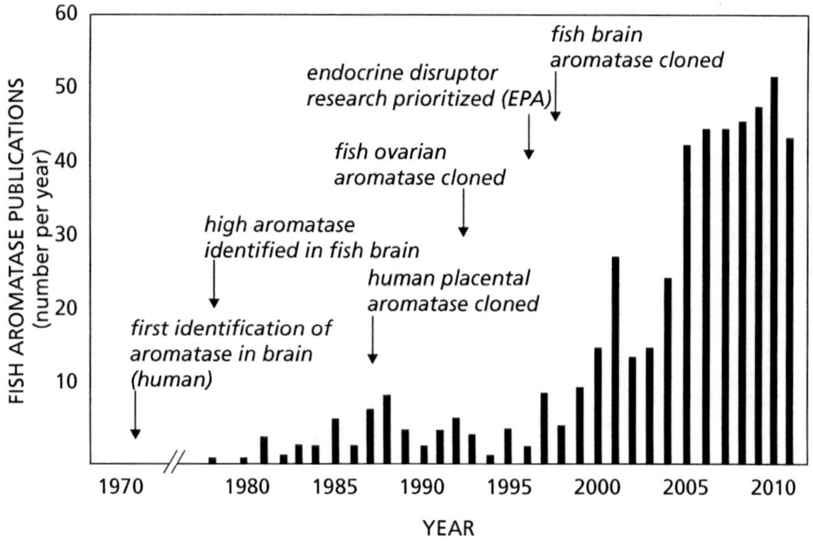

Figure 2–2. Number of fish aromatase publications per year (1978–present). Data are taken from PubMed and shown in relation to the timing of key advances in the field.

et al., 1984). This phylogenetic distribution suggests that a functional form of the enzyme first evolved in the brain of a basal jawed vertebrate (~400 million years ago, mya), not long after its origin in the urogenital ridge of a common vertebrate ancestor (~500 mya) (Callard & Tchoudakova, 1997 Callard et al., 2011), and provides a sound scientific rationale for a wide choice of animals for brain aromatase research. Here we review the history of of brain aromatase research in fishes, specifically, their utility as laboratory models for advancing our understanding of the basic biology of neural aromatase; and the applicability of the predominant brain form of aromatase in fishes as a biomarker of their general reproductive health and as an indicator of pollutants in the natural environment. Figure 2–2 illustrates a timeline showing the progressive increase in the use of fish in aromatase studies as related to key advances in the field.

NEURAL AROMATASE ENZYME ACTIVITY IN FISHES: LOCALIZATION AND ESTROGEN REGULATION

As compared with all other vertebrate groups, including the cartilaginous fishes (sharks and skates), the bony fish have exceptionally high levels of brain aromatase activity, suggesting that it is a specialization of the teleostean lineage (Callard et al., 1978, 1981; Callard, 1983). This point can be illustrated by considering that estrogen yields from radiolabeled androgens are 100- to 1,000-fold higher in goldfish preoptic area/hypothalamus than in corresponding regions of rat, rabbit, and hamster brain under the same assay conditions and 10-fold higher than in the ovary of the same fish species (Pasmanik & Callard, 1985, 1988). Pituitary aromatase is correspondingly high, especially in regions with somatotrophs (Callard, 1981; Callard et al., 1981, 1988). As well, authentic estrogen biosynthesis is easily measured at somewhat lower levels outside the reproductive brain, in areas such as the midbrain, hindbrain, spinal cord, and retina (Gelinas & Callard, 1993, 1997).

Because of their high estrogen biosynthetic potential, teleost fish have been technically advantageous to demonstrate a region-by-region correspondence between aromatase activity in vitro and the retention of formed estrogens after perfusion of the isolated fish brain with radiolabeled androgens under simulated in vivo conditions (Callard et al., 1981). These results indicate that high estrogen yields are not just a peculiarity of the in vitro assay carried out under optimized substrate and cofactor conditions, and further suggest that aromatization per se (or receptor binding of neuroestrogen) is an effective mechanism for determining estrogen concentrations in specific neuroanatomic areas. As is the case in rhesus monkeys (Flores et al., 1973), the isolated perfused teleostean brain secretes radiolabeled free and conjugated estrone and estradiol into the perfusate (Callard et al., 1981), further evidence that the brain of taxonomically diverse species can contribute quantitatively important amounts of estrogens to the general circulation. Interestingly, free and conjugated steroids have been implicated as pheromones in fish (Stacey et al., 2003). Many findings in fish were subsequently verified in mammals or other vertebrates. For example, an early report in fish showing that aromatase activity is enriched in synaptosomal subfractions of the brain (pinched off nerve endings) (Mak et al., 1985) was confirmed in quail (Schlinger & Callard, 1989) and the aromatase peptide subsequently was immunolocalized in axon terminals of the quail, rhesus, rodent, and human brain via electron microscopy methods (Naftolin et al., 1996), leading to the current "synaptocrine" hypothesis of neuroestrogen action (Schlinger & Remage-Healey, 2011); also see Chapters 6 and 10 in this volume.

Significantly, the human placental polyclonal antibody that was used to localize aromatase in the axonic terminals of quail, rhesus, rodent and human (Naftolin et al., 1996) also cross-reacts with the aromatase protein in goldfish brain, wherein it immunolabels cells that are neuronal in appearance. Evidence that this antibody is specific for goldfish aromatase protein was obtained by immunoprecipitating the in vitro transcription/translation product of the isolated goldfish brain aromatase cDNA (Gelinas et al, 1998). Neurons with immunoreactive aromatase are most numerous in hypophysiotrophic regions (mediate positive feedback) and in medial and ventral telencephalic areas (critical for male courtship and spawning) (Gelinas & Callard, 1993, 1997). Additional cells are labeled in the olfactory bulbs, fiber tracts, and fibrous layers in auditory pathways, perikarya, and processes of premotor neurons known to integrate sensory input (reticulospinal neurons, Mauthner cells), and neurons of the cerebellum resembling Purkinje cells. Particularly striking is the extent of labeling in the goldfish visual system. In the retina, a subset of horizontal, bipolar, amacrine, and retinal ganglion cells (RGCs), some fibers of the outer and inner synaptic layers, and some projections of the optic nerve and tract (but not photoreceptors) are labeled. Immunolabeling is also found in central visual processing areas, specifically, small neurons of the stratum periventriculare and fibers of two other strata of the optic tectum. Teleost fish are known to be highly visual and rely on photic input and visual cues for reproduction. Visual approach behavior toward a female is associated with an androgen surge in male goldfish and the behavior is rapidly stimulated by treatment with estrogens and aromatizable androgens (Lord et al, 2009). Also, a recent study found that treatment with aromatase inhibitors or estrogen receptor (ER) blockers significantly diminishes maturation and growth rates in developing zebrafish retina (Hamad et al, 2007).

In general, the neuronal localization and neuroanatomic distribution of aromatase expressing cells is consistent with measured enzyme levels in goldfish and confirms earlier studies or anticipates findings in rodents. For example, aromatase activity and immunoactive protein have now been identified in Purkinje cells in the neonatal rat, wherein neuroestrogen is found to promote dendritic growth and synaptogenesis (Sakamoto et al., 2003). Also, consistent with identification of aromatase in rodent retina (Salyeret al., 2001), estrogens are reported to have a neuroprotective effect on axotomized RGCs through the ERK signal transduction pathway (Nakazawa et al., 2006).

Paradoxically, a localization study in the plainfin midshipman reports localization of reaction product exclusively in radial glial cells, but an absence of immunolabeling in neurons in the same fish or in the goldfish (Forlano et al., 2001). The synthetic peptide used to generate the antibody used in this study, however, was designed to target a common sequence in the AroA and AroB proteins of the plainfin midshipman, and it is not clear how reactivity with native protein was accomplished. Similar results are reported using homologous antipeptide antibodies in rainbow trout and zebrafish (Pellegrini et al., 2007). The reason for discrepant results with polyclonal and antipeptide aromatase antibodies in different fish species, as compared with localization of aromatase in birds and mammals, is difficult to explain. What remains to be resolved is whether different antibodies react differently with the fish AroA and AroB isoforms (see later), or perhaps recognize one of the other members of the large P450 enzyme superfamily (Goldstone et al., 2010) that are also expressed in fish brain.

An association between neuroestrogen synthesis and seasonal reproduction is supported by studies in goldfish showing that 2- to 6-fold variations are superimposed on high constitutive aromatase activity and correspond to cyclical gonadal growth and circulating sex steroid levels with a peak at spawning in both sexes (Pasmanik & Callard, 1988). The seasonal peak can be induced in seasonally inactive fish by peripheral administration of estrogens or aromatizable androgens, indicating an autoregulatory positive feedback loop driven by neuroestrogen (Pasmanik et al., 1988) (Fig. 2–3). It is relevant here that the Km of the aromatization reaction in goldfish brain is 2 to 5 nM, a point midway between the seasonal extremes of circulating androgens in vivo (0.5–25 nM) (Callard & Pasmanik, 1987). This predicts that neuroestrogen production would be substrate limited during reproductive inactivity when the gonads are regressed and androgen secretion is low. At the onset of the breeding season, neuroestrogen production would progressively increase, first, as a consequence of androgen secretion by growing gonads and, second, because of autoregulatory upregulation of the aromatase enzyme. This positive feedback loop would culminate in a gonadotropin surge prior to ovulation/spermiation (Trudeau, 1997). This neural estrogen surge in fish is reminiscent of the ovarian estrogen surge involved in the positive feedback regulation of LH secretion and ovulation in mammals (see Fig. 2–3).

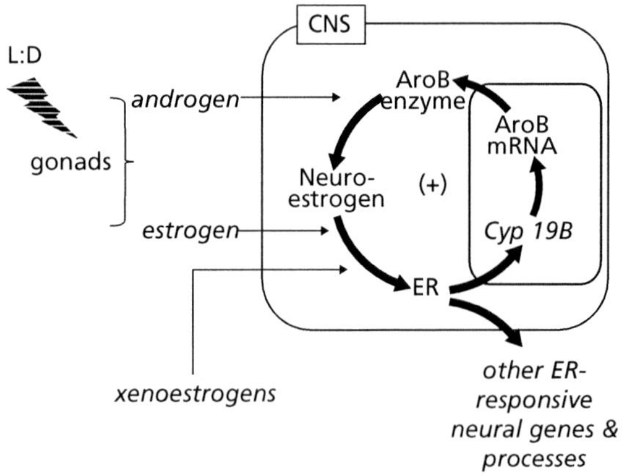

Figure 2–3. Expression of the predominant CYP19B gene in the teleostean brain is part of an autoregulatory positive feedback loop. The rate of accumulation of aromatase (AroB) mRNA and enzyme would be driven by (1) availability of aromatizable androgens; (2) neuroestrogens derived from in situ aromatization of androgens (autocrine/paracrine signaling mode); (3) estrogen synthesized and secreted by peripheral glandular tissues (endocrine signaling mode); and (4) exogenous hormonal estrogens and estrogen-like environmental chemicals. Estrogenic effects are mediated by estrogen receptors (ER). The onset and offset of seasonal reproductive activity is driven by photoperiod (L:D) operating through the hypothalamic-pituitary-gonadal axis (see text). In seasonally inactive fish, photoperiod operating through the hypothalamic pituitary gonadal axis would be a check on gonadal growth and the availability of aromatizable androgen. The onset of the breeding season is triggered by photic-induced gonadal growth and increased gonadal androgen and estrogen secretion which, in turn, would drive ever increasing levels of AroB mRNA and enzyme activity. Adapted from Callard, G. V., Tchoudakova, A. V., Kishida, M., & Wood, E. (2001). Differential tissue distribution, developmental programming, estrogen regulation and promoter characteristics of cyp19 genes in teleost fish. *J Steroid Biochem Mol Biol, 79* (1–5), 305–314.

CYP19 GENES IN FISHES

Gene Duplicates with Subdivided Expression Domains, Developmental Programming and Estrogen Regulation

A powerful tool in the study of aromatase has been the isolation of the aromatase protein (Mendelson et al., 1985) and the cloning of a cDNA from human placenta (Mendelsonet al., 1985; Chen et al., 1988; Corbin et al., 1988; Harada, 1988;). These advances led to the cloning of the first fish aromatase cDNA from the rainbow trout ovary (Tanaka et al., 1992), and then the first full-length aromatase cDNA from the goldfish brain (Gelinas et al., 1998) (see Fig. 2–2). Northern analysis using the brain-derived goldfish probe confirmed and extended results of the earlier enzyme assays by showing exceptionally high transcript levels in neural tissues, distinct neuroanatomical differences (forebrain > mid/hindbrain = retina), and seasonal variations with a peak at spawning. In the same study, aromatase mRNA was dramatically upregulated in seasonally inactive goldfish after administration of estrogens or aromatizable androgens and correspondingly downregulated in seasonally active goldfish by the aromatase inhibitor ATD (see Fig. 2–3).

Surprisingly, the brain-derived probe failed to hybridize with ovarian RNA under any condition, which ultimately led to identification of a second, ovary-derived cDNA in goldfish (Tchoudakova & Callard, 1998). The two cDNAs (AroB and AroA, in brain and ovary, respectively) have only 62% amino acid identity and are encoded by separate and unique *CYP19* loci, termed *CYP19B (CYP19A1b)* and *CYP19A (CYP19A1a)*, which differ in their estrogen responsiveness (B only). Northern blot and reverse transcription polymerase chain reaction analysis show that AroB is restricted to the brain and retina, which also have high levels of mRNA and enzyme activity; whereas AroA is the only form expressed in the ovary, which has relatively low mRNA and aromatase activity. Interestingly, the two genes have overlapping expression in neural tissues, but the B:A ratios differ in the brain (14:1) and retina (1:25), as measured by Northern blot in goldfish (Tchoudakova & Callard, 1998).

The authors and others (Kishida & Callard, 2001; Trant et al., 2001) soon identified cDNAs encoding orthologs of *CYP19A* and *CYP19B* in zebrafish, a premier model for studying development. Consistent with expansion and morphogenesis of the CNS and gonads during embryogenesis and the early larval period, AroB and AroA mRNAs accumulate progressively in embryos and larvae between 24 and 120 hours postfertilization (hpf), but the two genes are differentially programmed, as measured by the time of onset of transcription (B, 24–48 hpf; A, 48–72 hpf), the temporal and quantitative patterns of mRNA accumulation (B >A) and their response estrogen regulation (B only) (Kishida & Callard, 2001; Sawyer et al., 2006). Moreover, the mechanism regulating estrogen-mediated upregulation of AroB mRNA in the brain of adult goldfish is already operative in 24- to 48-hpf zebrafish embryos and possibly even earlier, as indicated by suppression of a component of AroB mRNA expression by the

ER antagonist ICI164.384 between 2 and 48 hpf in untreated embryos (Kishida & Callard, 2001). Also, ICI blocks effects of added estrogens, indicating that the autoregulatory loop shown in Figure 2–3 is ER-dependent.

Seasonal patterns and estrogen dependence of aromatase expression have been described in other fish species: the red-spotted grouper (Li et al., 2007); the European sea bass (Gonzalez & Piferrer, 2003); the channel catfish (Kazeto & Trant, 2005); and the killifish (Greytak et al., 2005). Seasonal and estrogen regulated changes are superimposed on already high constitutive neural aromatase expression and, therefore, do not per se explain high neural aromatase in fish.

Not surprisingly, xenoestrogens such as diethylstilbestrol and bisphenol-A that are known to interact with ER mimic the dose-related upregulation of AroB mRNA by estradiol in zebrafish embryos with the expected relative potency (DES > E2 > BPA) (Kishida et al., 2001),. as do ethinylestradiol, nonylphenol, and benzo(a)pyrene (Kazeto et al., 2004). As shown in Figure 2–3, these and many other studies in adult and developing fish indicate that AroB mRNA is a sensitive, robust, and valid marker of estrogen effect in fishes with utility for screening and testing chemicals and environmental samples in a laboratory setting and for monitoring exposure of fish to estrogenic contaminants in the natural environment.

There is good evidence that the autoregulatory loop controlling *CYP19B* in fish brain is conserved to some extent in the *CYP19* genes of birds and mammals. Although androgens acting via androgen receptors upregulate brain aromatase mRNA and enzyme in the adult rodent (Roselli & Resko, 1984), estrogens induce a permanent increase in brain levels of aromatase in perinatal rodents (Roselli & Klosterman, 1998) and upregulate brain aromatase in birds (Harada et al., 1992). Estrogens also upregulate transcription through estrogen response elements (EREs) in mouse hypothalamic cell lines (Yilmaz et al., 2009), and the placenta promoter of the human gene is estrogen/ER inducible through an ERE-like element (Kumar et al, 2009). Elements known to function in indirect ERα binding (SP1, AP-1) are present in multiple human *CYP19* promoters (Chen et al., 2009; Santen et al., 2009), suggesting that all these promoters could be direct targets of hormonal and environmental estrogens.

Phylogeny and Evolution

Figure 2–4 shows a phylogenetic tree produced by alignment and analysis of representative vertebrate aromatase proteins and includes the cephalochordate amphioxus *CYP19* for an evolutionary perspective (Callard et al., 2011). Consistent with the results of enzyme assays (Callard et al., 1978), *CYP19* orthologs have not been identified within the sequenced genomes of any invertebrate (urochordates, echinoderms) or bilaterian animals (Markov et al., 2009; Reitzel & Tarrant, 2010) other than amphioxus, nor have phylogenetic analyses of the *CYP* gene superfamily revealed close relationships of *CYP19* with any other family members (Nelson, 1998; Castro et al., 2005); thus, it is not currently possible to trace the origin of *CYP19* aromatase from ancestral CYPs that may have served other metabolic functions.

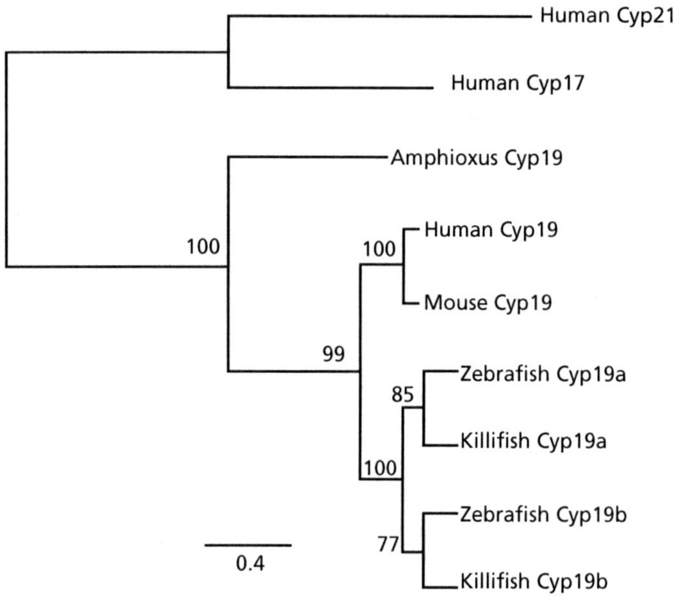

Figure 2–4. Phylogeny of aromatase proteins

Deduced amino acid sequences of teleost duplicates, zebrafish CYP19A (CYP19A1a, AA65788.1), zebrafish CYP19B (CYP19A1b, AAK00642.1), killifish CYP19A (CYP19A1a, AAR97268.1), and killifish CYP19B (CYP19A1b, AAR97269.1), were aligned with the single aromatases of amphioxus (ABA47317.1), human (NP_112503.1), and mouse (P28649.1). The maximum likelihood tree was rooted with human CYP17 (AAA36405) and CYP21 (NP_000491) sequences. Numbers indicate percentage of 100 bootstrap replicates supporting each node. Adapted from Callard, G. V., Tarrant, A. M., Novillo, A., Yacci, P., Ciaccia, L., Vajda, S., et al. (2011). Evolutionary origins of the estrogen signaling system: Insights from amphioxus. *J Steroid Biochem Mol Biol 127*(3–5), 76–88.

Consistent with earlier trees (Tchoudakova & Callard, 1998; Kishida & Callard, 2001; Greytak et al., 2005) and known taxonomic relationships, Figure 2–4 shows two *CYP19* clades in teleost fish (zebrafish, killifish) but only a single clade in amphioxus and tetrapods (mouse, human). Whole genome duplication events are thought to have occurred at several points during early vertebrate evolution (~500 mya), and then later (~100 mya) in an ancestor of the ray-finned teleosts, shortly after their divergence from the lobe-finned fish lineage leading to tetrapods (Amores et al., 1998; Meyer & Schartl, 1999). Over the course of evolution, some duplicates are lost (diploidization), others attain new functions (neofunctionalization) or subfunctions (expression domains) or both. Zebrafish as well as other fish species are believed to have retained duplicates for as many as 20% of their genes (Postlethwait et al., 2000). Among these are the two *CYP19* genes (A and B) and the genes encoding the estrogen receptor beta subtypes (ERβA and βB). Interestingly, the ERα homologue occurs as a single copy in teleosts, signifying the loss of a duplicate over evolutionary time. Although pigs have three *CYP19* genes, they share 87% to 93% sequence identity, indicative of much more recent tandem duplication events (Choi et al., 1997; Conley et al., 1997) as compared with the duplication event that led to the *CYP19A* and *-B* paralogues of fish, which are only about 60% identical (Tchoudakova & Callard, 1998).

To date, full cDNA sequences of both A and B forms have been characterized in 30 different teleostean species (Table 2–1) and full or partial sequences are available in GenBank for an additional 59 fishes. Only one species, the European eel (but not the Japanese eel), has been reported to have a single *CYP19* gene, perhaps because of the secondary loss of a gene duplicate (Tzchori et al., 2004). Interestingly, this fish is described as having relatively low brain aromatase activity compared with the exaggerated brain enzyme levels and/or mRNA expression in all other fish tested (Jeng et al., 2005). Sequence and phylogenetic analysis show that the single eel gene is not more closely related to the A or B forms of other fish, and although it is expressed in both brain and ovary, it is estrogen upregulated in brain only (Jeng et al., 2011). Two copies of the *CYP19B*-type (I and II) have been reported in the rainbow trout, but one of the two lacks the untranslated first exon, the first coding exon (II) and all but one intron of the paralogous gene. Perhaps this transcript is a remnant of a relatively recent extra duplication event (15–20 mya) in the salmonid lineage (Brunet et al., 2006) and it could be a pseudogene (Dalla Valle et al., 2005). Based on the estimated time of the whole genome duplication event in ancient fishes, it can be predicted that both A and B forms of *CYP19* would be present in the basal ray-finned teleosts (paddlefish, garpike & sturgeon) but not in the lobe finned fishes (coelocanths, lungfishes), which are monophyletic with tetrapods (Brunet et al., 2006). Unfortunately, no sequences are yet available from these fish groups.

Catalytic Properties

Formal proof that each of the two *CYP19* genes in goldfish (Tchoudakova & Callard, 1998) and zebrafish (Tong et al., 2001) encodes an authentic aromatase isozyme is based on the ability of the recombinant proteins to catalyze the transformation of androgens to estrogens in nonsteroidogenic COS cells. Consistent with the relatively low degree of overall identity of the two forms (~62%), and the number of differences in the amino acid sequence in putative functional domains, it is not surprising that the A and B isozymes have different reaction properties and inhibitor characteristics (Zhao et al., 2001). Specifically, AroB compared to other vertebrates, including human, is distinguished by its higher affinity for androgens (lower *K*m values). Additionally, in agreement with measured brain and ovarian enzyme activity, the brain isozyme when expressed at similar levels in the same cell line, has a significantly higher *V*max than the ovary isozyme. This could be a partial explanation for the ~10-fold higher

Table 2–1. Cyp19 aromatase genes in representative fishes*

			Database Identifier			
Species	Common Name	Gene	cDNA	Protein	Protein (aa)	Reference
Elasmobranchs						
Dasyatis sabina	Atlantic stingray	—	AF097513	AAF04617	511	Ijiri et al., 2000
Squalus acanthias	spiny dogfish shark	—	DQ223101	AAG35637	527	(unpubl.)
Teleosts						
Carassius auratus	goldfish	A	AF020704	AAC14013	518	Tchoudakova & Callard, 1998
		B	U18974	AAB39408	510	Gelinas et al., 1998
Clarias gariepinus	North African catfish	A	GU220075	ADH29765	516	Rasheeda et al., 2010
		B	GU220076	ADH29766	497	Rasheeda et al., 2010
Cromoleptes altivelis	humpback grouper	A	AY684255	AAV91178	515	Gardner et al., 2005
		B	AY684258	AAV91181	507	Gardner et al., 2005
Cynoglossus semilaevis	tongue sole	A	EF134716	ABL74474	526	Deng et al., 2009
		B	EF198239	ABM90641	498	Deng et al., 2009
Cyprinus carpio	common carp	A	EU375455	ACB13197	517	Barney et al., 2008
		B	EU375456	ACB13198	510	Barney et al., 2008
Danio rerio	zebrafish	A	AF004521	AAB65788	517	Kishida & Callard, 2001
		B	AF226619	AAK00642	511	Kishida & Callard, 2001
Dicentrarchus labrax	European sea bass	A	AJ318516	CAC86561	517	Dalla Valle et al., 2002
		B	AY138522	AAM95455	505	Blazquez & Piferrer, 2004
Epinephelus akaara	red-spotted grouper	A	AY547354	AAS58448	518	(unpubl.)

Continued

Table 2–1. Continued

Species	Common Name	Gene	Database Identifier cDNA	Protein	Protein (aa)	Reference
		B	AY547353	AAS58447	509	(unpubl.)
Epinephelus coioides	orange-spotted grouper	A	AY510711	AAR97601	518	Zhang et al., 2004
		B	AY510712	AAR97602	509	Zhang et al., 2004
Fugu rubripes	pufferfish	A	AB330138	BAI63583	519	Rashid et al., 2007
		B	AB330137	BAF93507	505	Rashid et al., 2007
Fundulus heteroclitus	killifish	A	AY428665	AAR97268	517	Greytak et al., 2005
		B	AY428666	AAR97269	499	Greytak et al., 2005
Gobiocypris rarus	rare gudgeon	A	GU220394	ADB29065	520	Wang et al., 2010
		B	GU220393	ADB44882	507	Wang et al., 2010
Gobiodon histrio	broad-barred goby	A	AY684254	AAV91177	537	Gardner et al., 2005
		B	AY684257	AAV91180	495	Gardner et al., 2005
Halichoeres tenuispinis	protogynous wrasse	A	AY489061	AAR37048	518	Choi et al., 2005
		B	AY489060	AAR37047	495	Choi et al., 2005
Hippoglossus hippoglossus	Atlantic halibut	A	AJ410171	CAC36394	523	van Nes et al., 2005
		B	AY999686	AAY26901	508	van Nes et al., 2005
Ictalurus punctatus	channel catfish	A	S75715	AAB32613	524	Trant et al., 1994
		B	AF417239	AAL14612	500	(unpubl.)
Kryptolebias marmoratus	mangrove killifish	A	DQ339107	ABC68614	516	Lee et al., 2006
		B	DQ339106	ABC68613	505	Lee et al., 2006

Continued

Table 2–1. Continued

Species	Common Name	Gene	cDNA	Protein	Protein (aa)	Reference
			Database Identifier			
Larimichthys crocea	large yellow croaker	A	FJ800566	ACO35041	518	(unpubl.)
		B	FJ800567	ACO35042	500	(unpubl.)
Micropogonias undulatus	Atlantic croaker	A	DQ184486	ABA26927	518	Nunez et al., 2006
		B	JF300170	AEL31294	502	(unpubl.)
Monopterus albus	swamp eel	A	EF640940	ABR37296	517	Yu et al., 2008
		B	EU252488	ABX45102	501	(unpubl.)
Mugil cephalus	striped mullet	A	AY859425	AAW72732	523	Nocillado et al., 2007
		B	AY859423	AAW72730	505	Nocillado et al., 2007
Odontesthes bonariensis	pejerrey fish	A	EF030342	ABK30807	518	Karube et al., 2007
		B	AY380061	AAQ88434	501	Strobl-Mazzulla, et al. 2005
Oncorhynchus mykiss	rainbow trout	A	—	1806325A	522	Tanaka et al., 1992
		B	AJ311937	CAC84574	488	DallaValle et al., 2002
Oreochromis mossambicus	Mozambique tilapia	A	AF135851	AAD31031	521	(unpubl.)
		B	AF135850	AAD31030	493	(unpubl.)
Oreochromis niloticus	Nile tilapia	A	AF472620	AAO62625	521	Chang et al., 2005
		B	AF306786	AAG18458	480	Kwon et al., 2001
Oryzias latipes	medaka	A	D82968	BAA11656	518	Tanaka et al., 1995
		B	AY319970	AAP83449	499	Kuhl et al., 2005
Pseudolabrus japonicus	bambooleaf wrasse	A	DQ298135	ABB96485	522	(unpubl.)
		B	DQ298136	ABB96486	494	(unpubl.)

Continued

Table 2–1. Continued

		Database Identifier				
Species	Common Name	Gene	cDNA	Protein	Protein (aa)	Reference
Rhabdosargus sarba	tarwhine	A	DQ345786	ABC70869	519	(unpubl.)
		B	DQ345785	ABC70868	500	(unpubl.)
Rutilus rutilus	roach	A	AB190291	BAD91037	517	Lange et al., 2008
		B	AB190292	BAD91038	507	Lange et al., 2008
Silurus meridionalis	Southern catfish	A	AY325908	AAP83133	540	(unpubl.)
		B	AY325907	AAP83132	507	(unpubl.)

Teleostean species with complete coding sequences for both the A and B genes were selected from a total of 88 GenBank entries. For additional confirmatory or variant sequences, see GenBank and text. All available sequences are representative of the actinopterygian (ray-finned) fishes that emerged after the whole genome duplication event. There are no entries for sarcopterygian (lobe-finned) fish, the lineage leading to tetrapods, or for the earlier evolving jawless (cyclostome) fishes. The complete coding sequence for the single Cyp19 gene in elasmobranch (cartilaginous) fishes has been cloned from only two species.

aromatase activity in fish brain than in fish ovary, but still does not negate the significance of large differences in mRNA expression across tissue type (AroB > AroA mRNA) (Tchoudakova & Callard, 1998; Kishida & Callard, 2001; Sawyer et al., 2006). In addition, the testing of seven different aromatase inhibitors showed that the AroA isozyme is much more sensitive than the AroB isoform. Assuming that inhibitor uptake by the ovary and brain are similar, it should be possible to selectively suppress the isozyme in ovary by selecting the appropriate inhibitor concentrations. Also, because the ovary isozyme has been shown to be more sensitive than the brain isozyme to environmental estrogens such as chrysin and 8-dihydroxyflavone, it may be possible to screen for and categorize environmental pollutants based on their selective impact on estrogen formation in fish ovary rather than brain.

Sequence Differences

Comparison of deduced A and B proteins in a given fish species generally shows that the B isoform is overall larger than A (see Table 2–1) but that A has an elongated membrane spanning domain when compared with B or with nonteleostean aromatases (Tchoudakova & Callard, 1998; Kishida & Callard, 2001; Greytak et al., 2005). Like all vertebrate cytochromes P450, the aromatases contain an absolutely conserved cysteine in the heme-binding domain (HBD), which is known to be essential for catalytic function (Graham-Lorence et al., 1995). When

Amphioxus	LGVRSCVGRTIAP
Stingray	FGPRSCVGKYVAM
Dogfish	CGPRSCVGKYIAM
Medaka B	CGPRSCVGKHIAM
Killifish B	CGPRSCVGKHIAM
Pufferfish B	CGPRSCVGKHIAM
Tilapia B	CGPRSCVGKHIAM
Zebrafish B	CGPRACVGKHIAM
Eel	SGPRSCVGKHISM
Medaka A	SGPRACVGRHIAM
Killifish A	SGPRACVGKHIAM
Pufferfish A	SGPRACVGKHIAM
Sea Bream A	SGPRACVGKHIAM
Tilapia A	SGPRACIGKHMAM
Trout A	SGPRSCVGKHIAM
Zebrafish A	SGPRSCVGKHIAM
Turtle	FGPRGCVGKFIAM
Alligator	FGPRACVGKFIAM
Chicken	FGPRGCVGKFIAM
Zebra Finch	FGPRSCVGKFIAM
Mouse	FGPRGCAGKYIAM
Rat	FGPRSCAGKYIAM
Pig	FGPRACAGKYIAM
Cow	FGPRGCAGKYIAM
Human	FGPRGCAGKYIAM
	* : * . * * : . : * :

Figure 2–5. Sequence alignment of Heme-Binding domains (HBD) of fish aromatase A and B proteins in comparison with amphioxus and representative reptilian, avian and mammalian aromatases. Insert full stop boundaries of the HBD are based on human aromatase (residues 430–442). Identical and similar amino acid residues are marked by *asterisks* and *dots*, respectively. Shaded residues highlight phylogenetic patterns discussed in text. For GenBank entries, see Table 2–1, legend to Fig. 2–4 and sea bream A(AAL27699); eel (AAS47028.1); frog (BAA90529); turtle (AAG09376); alligator (AAK31803); chicken (AAA48738); zebra finch (AAB32404); mouse (BAA00551); rat (AAA41044); cow (AAA62244); pig (AAB51387); and human (AAA52132). Adapted from Greytak, S. R., champlin, D., & callard, G. V. (2005). Isolation and characterization of two cytochrome P450 aromatase forms in killifish (Fundulus heteroclitus): differential expression in fish from polluted and unpolluted environments. *Aquat toxicol, 71*(4), 371–389.

the core HBDs of aromatases from amphioxus and selected vertebrate species are aligned, 5 of 13 positions are identical, but a clear phylogenetic pattern emerges by inspection of three of the non-conserved positions (Figure 2–5). First, five residues upstream of the conserved cysteine, all teleost fish B-forms have a second cysteine residue, whereas all A-forms have a serine in this position and tetrapod aromatases have a phenylalanine. Note that the single European eel aromatase is more like the ovarian isoforms at this position. Second, all but one (sea bass B) of the A- and B-aromatases of teleost fish have a histidine four residues downstream of the conserved cysteine, whereas birds and reptiles have a phenylalanine and mammals have a tyrosine in this position. Interestingly, dogfish also have a tyrosine, but stingrays have a phenylalanine. Finally, with one exception (tilapia A), all nonmammalian aromatases have a valine after the conserved cysteine, whereas mammals have an alanine in this position. This pattern suggests that the substitution of specific residues in the critical HBD of the vertebrate aromatases occurred at distinct points in evolution, perhaps optimizing enzyme function by tissue type (brain and ovary) and in conjunction with the acquisition of a specific trait (e.g., homeothermy) or a general increase in functional complexity.

The crystal structure of the human aromatase protein has recently been determined (Ghosh et al., 2009). Although A and B isozymes of fish have not yet been compared by modeling, Figure 2–6 shows a model of the amphioxus aromatase catalytic cleft and the simulated docking of androstenedione in comparison to the human aromatase (Callard et al., 2011). All residues that are critical for the binding of small molecules are highly conserved, despite a distance of greater than 500 million years of evolutionary time between cephalochordates and humans.

Exon-Intron Organization

Like the human *CYP19* gene (Santen et al., 2009)), amphioxus and all fish *CYP19* genes have nine coding exons, and these are well conserved in size and sequence, especially in key functional domains (Callard et al., 2011) (Figure 2–7).

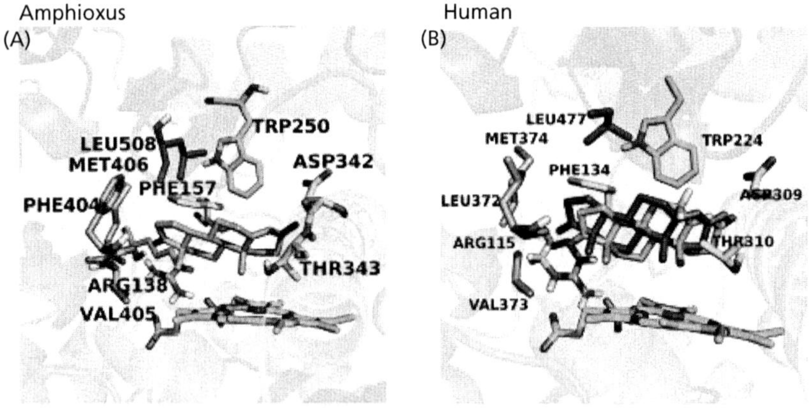

Figure 2–6. Homology model of androstenedione docked within the active site of aromatases of (*A*) amphioxus and (*B*) human. The model supports other lines of evidence indicating that a functional form of the enzyme originated <500 mya and has been conserved through the evolution of vertebrates. See text. Adapted from Callard, G. V., Tarrant, A. M., Novillo, A., Yacci, P., Ciaccia, L., Vajda, S., et al. (2011). Evolutionary origins of the estrogen signaling system: Insights from amphioxus. *J Steroid Biochem Mol Biol 127*(3–5), 76–88.

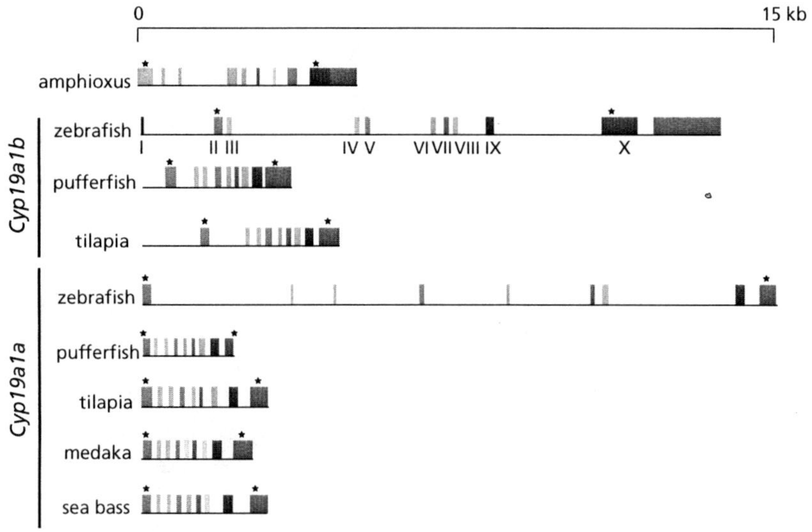

Figure 2–7. Comparison of Genomic Organization of the CYP19A and CYP19B Genes in Teleost Fishes. Through interrogation of publicly available databases and genomic and cDNA sequence information previously reported in the literature, the genomic organization of the coding region the two CYP19 genes was compared in representative teleosts: zebrafish A and B (Yukinori Kazeto et al., 2001); pufferfish A and B (Rashid et al., 2007); tilapia A and B (Chang et al., 2005); medaka A (Tanaka et al., 1995); sea bass A (Dalla Valle et al., 2002). The single amphioxus CYP19 is depicted for evolutionary perspective (Callard et al., 2011). Comparison with the single human CYP19 gene is described in the text. Coding exons are labeled II–X in zebrafish and correspondingly shaded in other species. The translation initiation and stop codons are indicated (*). Note that the untranslated first exon in amphioxus and zebrafish CYP19A is contiguous with coding sequence in exon II, like that of the promoter (PII)/untranslated first exon of the human CYP19 used in ovary. By contrast, the untranslated first exon of zebrafish CYP19B is further upstream of the start codon in exon II and in that respect is like that of the human I.f promoter/first exon used in brain. Also, the very long exon X containing the 3′-untranslated region of zebrafish CYP19B has a ~250-bp region deleted from the mRNA. Adapted from Callard, G. V., Tarrant, A. M., Novillo, A., Yacci, P., Ciaccia, L., Vajda, S., et al. (2011). Evolutionary origins of the estrogen signaling system: insights from amphioxus. *J Steroid Biochem Mol Biol* 127(3–5), 76–88.

As initially reported for zebrafish (Kazeto et al., 2001), the single amphioxus *CYP19* gene and the predominant ovarian aromatases in fish (*CYP19A)* most closely resemble the organization of the human gene in which the PII (ovarian) promoter and the untranslated first exon are contiguous with and immediately upstream of the ATG site in exon II. In contrast, *CYP19B,* the predominant brain aromatase of teleostean fish, has an untranslated first exon farther upstream and, in this respect, resembles the human ortholog, in which multiple promoters and untranslated first exons, located as far as -93 kb from the translation initiation site, are alternatively spliced in a tissue-specific manner to a common site in exon II, such that the aromatase protein is identical in all tissues (Simpson et al., 1994; Bulun et al., 2004).Entirely consistent with the idea that aromatase evolved first in the urogenital ridge (gonadal primordium) and secondarily in the brain, gene structure suggests that *CYP19A* is closest to the ancestral form. A corollary of this hypothesis suggests that a whole genome duplication event resulted in the acquisition of a second *CYP19B* locus in fishes, whereas the tissue-specific promoters of the single copy human *CYP19* were acquired sequentially during the course of organ system evolution (ovary > brain > placenta).

From the ATG in exon II, the amphioxus *CYP19* is approximately 7 kb, much smaller than either zebrafish *CYP19B* (15 kb) or *CYP19A* (12 kb), owing primarily to shorter introns (see Fig. 2–7). Interestingly, exon X of the zebrafish *CYP19B* gene, which encodes the very long 3′-untranslated region (UTR), has a ~250-bp sequence that is deleted from the transcript (designated XA and XB). Retained introns are common in vertebrate 3′-UTRs, but the occurrence of an intron in the very long exon X of the zebrafish *CYP19B* is not seen in the other fish *CYP19* genes.

Promoter Regions

Conservation of *CYP19* gene duplicates with subdivided expression domains implies that neural and gonadal regulatory mechanisms have evolved over ~100 million years of evolutionary time. To determine whether high neural aromatase expression is an intrinsic property of the *CYP19B* gene promoter, and to identify cis-elements and cellular factors involved in the correct tissue-specific and regulated expression, we and others sought to compare the 5 -flanking DNA from the *CYP19* duplicates in goldfish (Tchoudakova et al., 2001), zebrafish (Kazeto et al., 2001); Nile tilapia (Chang et al., 2005); medaka (Tanaka et al., 1995; Kuhl et al., 2005); humpback grouper, broad barred goby, barramundi perch (Gardner et al., 2005); flathead mullet (Nocillado et al., 2007); rainbow trout (Toffolo et al., 2007); killifish (Dong & Willett, 2008), and rare minnow (Wang et al., 2010). Putative promoter sequences for one or the other *CYP19* gene in additional species are listed in GenBank and are available for analysis in the whole genome databases of two pufferfish species (*Tetraodon, Takifugu*).

Although the overall size and sequence identity in the 5 -flanking regions is low, statistically overrepresented motifs corresponding to known cis-elements are identifiable. The *CYP19B* gene includes a TATA box, one or more EREs or ERE half-sites. The *CYP19A* gene has TATA and CAAT boxes, binding sites for steroidogenic factor-1 (SF-1), a cAMP response element, and one or more recognition sites for the aryl hydrocarbon receptor (AhR)/ARNT heterodimer. Additional potential cis-elements of interest in the B-gene promoter include a consensus binding site (NBRE) for nerve growth factor inducible-B protein (NGFI-B/Nur77; goldfish), and in the A-gene promoter a peroxisome proliferator-activated receptor α/retinoid X receptor alpha heterodimer-responsive element (PPARα/

RXRa; zebrafish). Interestingly, the amphioxus *CYP19* promoter has a TATA box and at least five potential transcription factor binding sites, each of which has been identified in the aromatase promoters of other taxa (Callard et al., 2011). Although we did not identify EREs upstream of the amphioxus *CYP19*, three largely conserved putative ERE half-sites were found (designated by MATCH as ERR and SF-1 binding sites).

The putative cis-elements identified in fish *CYP19* genes are consistent with the regulation of *CYP19B* by estrogens and the regulation of *CYP19A* by gonadotropin/cAMP, but few of the isolated promoter regions have actually been tested for their transcriptional activity. One exception is the goldfish, in which *CYP19A*- and *CYP19B*/luciferase constructs were found to have basal promoter activity in aromatase-expressing rodent pituitary (GH3) cells (Tchoudakova et al., 2001). Nonetheless, differences in transactivational activities of the two isoforms (A > B) did not reflect relative expression levels in the goldfish pituitary in vivo (B > A). One interpretation is that high transcriptional activity is not intrinsic to the *CYP19* gene promoter itself. Alternatively, rodent GH3 cells may be lacking in the tissue-specific factors essential for the high levels of AroB expression in the goldfish pituitary. Consistent with the immunolocalization of aromatase in neurons (see previous), the microinjection of a green fluorescent protein (GFP) construct driven by the goldfish *CYP19B* (but not the *CYP19A*) promoter labels neuron-like cells and processes in the eye, forebrain, and optic tectum of zebrafish embryos (Tchoudakova et al., 2001), and similar results have been reported using a homologous zebrafish *CYP19B*/GFP construct (Lassiter & Linney, 2007). Although ligand-activated recombinant ERα is reported to upregulate transcription from a zebrafish AroB/luciferase construct (Menuet et al., 2005), a recent study in which we used ER subtype-specific morpholinos showed that ERβb is necessary and sufficient for estrogen induction of AroB mRNA in living zebrafish embryos (Burnam & Callard, unpublished data).

Although the molecular mechanism of high neural aromatase expression in teleost fish is still unresolved, a recent study in grey mullet could provide new insight. Through the use of serially deleted promoter/luciferase constructs in COS-7 and other cell lines, the enhancing and silencing roles of the different putative transcription factor binding sites were demonstrated and, most significant, enhancer activity was identified in a region downstream of the TATA box within the first intron of *CYP19B* (Nocillado et al., 2007). Sequence analysis of this region, which has no counterpart in the *CYP19A* gene, reveals two AP1-like sites, a consensus initiator sequence, and a downstream promoter element, which have been shown to function together as a downstream core promoter unit. The authors of this study suggest that alternative promoters are utilized in tissue-specific and temporal expression of *CYP19B* and note that multiple transcriptional start sites have been reported for *CYP19B*, but not *CYP19A*, in medaka (Tanaka et al., 1995), Nile tilapia (Chang et al., 2005), goldfish (Tchoudakova et al., 2001), and zebrafish (Kazeto et al., 2001; Tong & Chung, 2003). It is relevant here that at least five *CYP19B* transcripts with different 5′-UTRs have been identified in the ovary and testis of rainbow trout (von Schalburg et al., 2010), and sequence analysis of intron I suggests that binding elements for FOXL2 and SF-1 may be involved in the tissue specific expression of these variants. These new studies indicate a heretofore unrecognized level of complexity in the regulation of *CYP19B* expression in fishes. Thus, focus on this region, and the functional properties of different 5′-end sequences, perhaps will be productive in explaining the exaggerated levels of aromatase activity in teleostean brain.

Finally, it is important to note that the transcriptional regulation of the *CYP19* genes of fishes does not rule out the role of rapid, nongenomic mechanisms of aromatase enzyme regulation, for example, by posttranslational mechanisms such as phosphorylation (Charlier et al., 2011) (see subsequent chapters in this book). As reported for the avian and mammalian *CYP19* genes, a number of protein kinase C phosphorylation (PKC) sites at residues 118, 168, and 469, and casein kinase II phosphorylation (CKII) sites at 291 and 325 (based on the zebrafish B-isoform), are distinct to B-isoforms, whereas a PKC site at position 392 and a CKII site at 190 (based on the zebrafish A-isoform) are distinct to most fish A-isoforms (Kishida & Callard, 2001; Tchoudakova et al., 2001). Although no information is yet available for fishes, a recent study using microdialysis for sampling and administration of pharmacologic agents shows that neuroestrogen accumulation is controlled by specific, depolarization-sensitive, calcium-dependent events at presynaptic terminal within the zebra finch forebrain (Remage-Healey et al., 2011).

CONCLUSIONS

It has been more than three decades since exaggerated aromatase enzyme levels were first identified in fish brain and hypothetically linked to their remarkable potential for lifelong brain growth and neuroregeneration (Callard, 1983), but the functional significance and adaptive value of this trait in fishes remains to be proven. The pattern of aromatase-expressing cells in fishes and diverse other taxa is consistent with the idea that neuroestrogen has important functions outside classical reproductive centers, perhaps as a trophic or a neuroprotective factor in support of postmitotic neurons that transmit, process and relay sensory and motor information. Alternatively, localization of aromatase protein or mRNA radial glia cell progenitors suggests involvement in neurogenesis, but clear cause-and-effect relationships remain to be demonstrated. It is significant to note that neurogenesis and neuroregeneration occurs in veretebrate groups other than fishes (e.g., amphibia), despite their relatively low neural aromatase levels. Most important, the discrepancies between cell-type localization studies need to be resolved. The accessibility and relative organizational simplicity of the teleostean visual system when compared with the brain suggests that it could be an advantageous model for recognition of neuroestrogen controlled genes and physiological processes. Knowledge of the basic biology of the retina (Hitchcock & Raymond, 2004) and the availability of a large dataset of genes and processes regulated during the regrowth of RGC axons after optic nerve section (McCurley & Callard, 2010) could provide an entry point to explain the role of neuroestrogens in neural tissues.

From the preceding sections and published reviews (Blazquez et al, 1998; Piferrer et al., 2005; Mouriec et al., 2008; Blazquez & Somoza, 2010; Le Page et al., 2011), it is evident that aromatase somehow participates in the control of reproduction and reproductive behaviors, brain and gonadal sex differentiation, neurodevelopment, and other basic neural processes in fishes, but the unique functions of the *CYP19* gene duplicates are not always clearly separable. Future studies using gene-specific morpholinos, zinc finger nucleases, overexpression strategies, or isoform selective aromatase inhibitors can be expected to clarify this issue.

Still unexplored is the possibility that high neural aromatase expression in fishes is somehow associated with a preponderance of rapid (sec–min) membrane-mediated nongenomic effects, as opposed to the slower nuclear-mediated genomic effects. In situ estrogen synthesis, especially at synapses, would seem to be optimal for generating high local levels and rapid fluctuations in ligand availability for receptor binding. Although the nature of the membrane-associated receptors involved in rapid estrogen actions is debated (Vasudevan & Pfaff, 2008; Christensen et al., 2011), an estrogen-binding G-protein coupled receptor was recently characterized in fish and humans and was found to mediate paracrine signaling by estrogens in the fish ovary (Pang & Thomas, 2010; Thomas et al., 2010). This protein is a plausible candidate for further investigating neuroestrogen actions in the CNS.

A long history of research in fish shows that they are excellent models for obtaining new information of general relevance (Cossins & Crawford, 2005; Burnett et al., 2007). They have basic vertebrate endocrine, reproductive and nervous systems and a similar (but ex utero) developmental program that allows manipulation of the embryo directly, without the complications of maternal effects. The sequencing and annotation of the genomes of zebrafish, two pufferfish species, medaka, killifish, and several other teleosts are works in progress. Expansion of our knowledge base, and the increased availability of research tools, can be expected to facilitate extrapolations across taxa and advance our mechanistic understanding of the molecular biology and physiology of the predominant brain and gonadal aromatases.

In addition to the value of fish species as laboratory models, free ranging fish populations provide an important research opportunity. Reproductive abnormalities first observed in wild caught fish gave significant impetus to the endocrine disruptor hypothesis and to the study of brain and ovarian aromatases in fishes (see Fig. 2–2). AroB and AroA mRNAs have been widely used as reliable indicators of chemical exposure and reproductive condition in fishes in the natural environment, but the presence or absence of changes in biomarkers is not *per se* an indication of environmental impact. Fish are highly polymorphic and readily adapt to changing environmental conditions. For example, a population of Atlantic killifish at a highly polluted site has evolved resistance to dioxin-like chemicals as measured by toxicity and biomarker (*CYP1A*) inducibility (Nacci et al., 2002; Hahn, 2011). The same population continues to reproduce, despite evidence of exposure to environmental estrogens, as measured by the induction of *CYP19B* expression and an impact on the hypothalamic-pituitary-gonadal axis (Greytak et al., 2005).). Interestingly, fish at the polluted site are hyporesponsive to estradiol, as measured by inducibility of AroB and ERα mRNAs (Greytak et al., 2010). To understand the ecological, physiological, genetic and epigenetic processes that evolve in all living organisms in response to environmental change, it is important to continue and expand studies in natural fish populations.

ACKNOWLEDGMENTS

The authors acknowledge support by grants from the National Institutes of Health (NIEHS P42 ES07381 to the Boston University Superfund Research Program), the U.S. Environmental Protection Agency (STAR-RD831301), the National Sea Grant College Program (R/P-68 22227323), an NICHD predoctoral traineeship (NICHD 2T32 HD 073897; SG), and a NATO Fellowship (AN).

REFERENCES

Amores, A., Force, A., Yan, Y. L., Joly, L., Amemiya, C., Fritz, A., et al. (1998). Zebrafish hox clusters and vertebrate genome evolution. *Science, 282*(5394), 1711–1714.

Barney, M. L., Patil, J. G., Gunasekera, R. M., & Carter, C. G. (2008). Distinct cytochrome P450 aromatase isoforms in the common carp (Cyprinus carpio): Sexual dimorphism and onset of ontogenic expression. *Gen Comp Endocrinol, 156*(3), 499–508.

Blazquez, M., Bosma, P. T., Fraser, E. J., Van Look, K. J., & Trudeau, V. L. (1998). Fish as models for the neuroendocrine regulation of reproduction and growth. *Comp Biochem Physiol C Pharmacol Toxicol Endocrinol, 119*(3), 345–364.

Blázquez, M., & Piferrer, F. (2004). Cloning, sequence analysis, tissue distribution, and sex-specific expression of the neural form of P450 aromatase in juvenile sea bass (Dicentrarchus labrax). *Mol Cell Endocrinol, 219*(1–2), 83–94.

Blázquez, M. & Somoza, G. M. (2010). Fish with thermolabile sex determination (TSD) as models to study brain sex differentiation. *Gen Comp Endocrinol, 166*(3), 470–477.

Brunet, F. G., Roest Crollius, H., Paris, M., Aury, J. M., Gibert, P., Jaillon, O., et al. (2006). Gene loss and evolutionary rates following whole-genome duplication in teleost fishes. *Mol Biol Evol, 23*(9), 1808–1816.

Bulun, S. E., Takayama, K., Suzuki, T., Sasano, H., Yilmaz, B., & Sebastian, S. (2004). Organization of the human aromatase p450 (CYP19) gene. *Semin Reprod Med, 22*(1), 5–9.

Burnett, K. G., Bain, L. J., Baldwin, W. S., Callard, G. V., Cohen, S., Di Giulio, R. T., et al. (2007). Fundulus as the premier teleost model in environmental biology: opportunities for new insights using genomics. *Comp Biochem Physiol Part D Genomics Proteomics, 2*(4), 257–286.

Callard, G. V. (1981). Aromatization is cyclic AMP-dependent in cultured brain cells. *Brain Research, 204*(2), 461–464.

Callard, G. V. (1983). Androgen and estrogen actions in the vertebrate brain. *Amer Zool, 23,* 607–620.
Callard, G. V., Pasmanik, M. (1987). The role of estrogen as a parahormone in brain and pituitary. *Steroids,* 50(4-6), 475-493.
Callard, G. V., Petro, Z. & Ryan, K. J. (1978). Phylogenetic distribution of aromatase and other androgen-converting enzymes in the central nervous system. *Endocrinology, 103*(6), 2283–2290.
Callard, G. V., Petro, Z., & Ryan, K. J. (1980). Aromatization and 5α-reduction in brain and nonneural tissues of a cyclostome, Petromyzon marinus. *Gen Comp Endocrinol, 42*(2), 155–159.
Callard, G. V., Petro, Z., & Ryan, K. J. (1981). Biochemical evidence for aromatization of androgen to estrogen in the pituitary. *Gen Comp Endocrinol, 44*(3), 359–364.
Callard, G. V., Pudney, J. A., Kendall, S. L., & Reinboth, R. (1984). In vitro conversion of androgen to estrogen in amphioxus gonadal tissues. *Gen Comp Endocrinol, 56*(1), 53–58.
Callard, G. V., Specker, J. L., Knapp, J., Nishioka, R. S., & Bern, H. A. (1988). Aromatase is concentrated in the proximal pars distalis of tilapia pituitary. *Gen Comp Endocrinol, 71*(1), 70–79.
Callard, G. V., Tarrant, A. M., Novillo, A., Yacci, P., Ciaccia, L., Vajda, S., et al. (2011). Evolutionary origins of the estrogen signaling system: insights from amphioxus. *J Steroid Biochem Mol Biol 127*(3–5), 76–88.
Callard, G. V., & Tchoudakova, A. (1997). Evolutionary and functional significance of two CYP19 genes differentially expressed in brain and ovary of goldfish. *J Steroid Biochem Mol Biol, 61*(3–6), 387–392.
Callard, G. V., Tchoudakova, A. V., Kishida, M., & Wood, E. (2001). Differential tissue distribution, developmental programming, estrogen regulation and promoter characteristics of cyp19 genes in teleost fish. *J Steroid Biochem Mol Biol, 79*(1–5), 305–314.
Castro, L. F., Santos, M., & Reis-Henriques, M. (2005). The genomic environment around the Aromatase gene: evolutionary insights. *BMC Evol Biol 5*(1), 43.
Chang, X., Kobayashi, T., Senthilkumaran, B., Kobayashi-Kajura, H., Sudhakumari, C. C., & Nagahama, Y. (2005). Two types of aromatase with different encoding genes, tissue distribution and developmental expression in Nile tilapia (Oreochromis niloticus). *Gen Comp Endocrinol, 141*(2), 101–115.
Charlier, T. D., Harada, N., Balthazart, J., & Cornil, C. A. (2011). Human and quail aromatase activity is rapidly and reversibly inhibited by phosphorylating conditions. *Endocrinology, 152*(11), 4199–4210.
Chen, D., Reierstad, S., Lu, M., Lin, Z., Ishikawa, H., & Bulun, S. E. (2009). Regulation of breast cancer-associated aromatase promoters. *Cancer Lett, 273*(1), 15–27.
Chen, S. A., Besman, M. J., Sparkes, R. S., Zollman, S., Klisak, I., Mohandas, T., et al. (1988). Human aromatase: cDNA cloning, Southern blot analysis, and assignment of the gene to chromosome 15. *DNA, 7*(1), 27–38.
Choi, I., Troyer, D. L., Cornwell, D. L., Kirby-Dobbels, K. R., Collante, W. R., & Simmen, F. A. (1997). Closely related genes encode developmental and tissue isoforms of porcine cytochrome P450 aromatase. *DNA Cell Biol, 16*(6), 769–777.
Choi, J. Y., Park, J. G., Jeong, H. B., Lee, Y. D., Takemura, A., & Kim, S. J. (2005). Molecular cloning of cytochrome P450 aromatases in the protogynous wrasse, Halichoeres tenuispinis. *Comp Biochem Physiol B Biochem Mol Biol, 141*(1), 49–59.

Christensen, A., Dewing, P., & Micevych, P. (2011). Membrane-initiated estradiol signaling induces spinogenesis required for female sexual receptivity. *J Neurosci, 31*(48), 17583–17589.

Conley, A., Corbin, J., Smith, T., Hinshelwood, M., Liu, Z. & Simpson, E. (1997). Porcine aromatases: studies on tissue-specific, functionally distinct isozymes from a single gene? *J Steroid Biochem Mol Biol, 61*(3–6), 407–413.

Corbin, C. J., Graham-Lorence, S., McPhaul, M., Mason, J. I., Mendelson, C. R., & Simpson, E. R. (1988). Isolation of a full-length cDNA insert encoding human aromatase system cytochrome P-450 and its expression in nonsteroidogenic cells. *Proc Natl Acad Sci U S A, 85*(23), 8948–8952.

Cossins, A. R. & Crawford, D. L. (2005). Fish as models for environmental genomics. *Nat Rev Genet, 6*(4), 324–333.

Dalla Valle, L., Lunardi, L., Colombo, L., & Belvedere, P. (2002) European sea bass (Dicentrarchus labrax L.) cytochrome P450arom: cDNA cloning, expression and genomic organization. *J Steroid Biochem Mol Biol, 80*(1), 25–34.

Dalla Valle, L., Toffolo, V., Vianello, S., Ikuo, H., Takashi, A., Belvedere, P., et al. (2005). Genomic organization of the CYP19b genes in the rainbow trout (Oncorhynchus mykiss Walbaum). *J Steroid Biochem Mol Biol, 94*(1–3), 49–55.

Deng, S. P., Chen, S. L., Xu, J. Y., & Liu, B. W. (2009). Molecular cloning, characterization and expression analysis of gonadal P450 aromatase in the half-smooth tongue-sole, Cynoglossus semilaevis. *Aquaculture, 287*(1–2), 211–218.

Dong, W., & Willett, K. L. (2008) Local expression of CYP19A1 and CYP19A2 in developing and adult killifish (Fundulus heteroclitus). *Gen Comp Endocrinol, 155*(2), 307–317.

Engel, L. (1974). The biosynthesis of estrogen. In: Greep, R. O. (ed.), *Handbook of Physiology,* vol. 2. Washington, DC: American Physiological Society, pp. 467–483.

Flores, F., Naftolin, F., Ryan, K. J., & White, R. J. (1973). Estrogen Formation by the Isolated Perfused Rhesus Monkey Brain. *Science, 180*(4090), 1074–1075.

Forlano, P. M., Deitcher, D. L., Myers, D. A., & Bass, A. H. (2001). Anatomical distribution and cellular basis for high levels of aromatase activity in the brain of teleost fish: aromatase enzyme and mRNA expression identify glia as source. *J Neurosci, 21*(22), 8943–8955.

Gardner, L., Anderson, T., Place, A. R., Dixon, B., & Elizur, A. (2005). Sex change strategy and the aromatase genes. *J Steroid Biochem Mol Biol, 94*(5), 395–404.

Gelinas, D., & Callard, G. V. (1993). Immunocytochemical and biochemical evidence for aromatase in neurons of the retina, optic tectum and retinotectal pathways in goldfish. *J Neuroendocrinol, 5*(6), 635–641.

Gelinas, D., & Callard, G. V. (1997). Immunolocalization of aromatase- and androgen receptor-positive neurons in the goldfish brain. *Gen Comp Endocrinol, 106*(2), 155–168.

Gelinas, D., Pitoc, G. A., & Callard, G. V. (1998). Isolation of a goldfish brain cytochrome P450 aromatase cDNA: mRNA expression during the seasonal cycle and after steroid treatment. *Mol Cell Endocrinol, 138*(1–2), 81–93.

Ghosh, D., Griswold, J., Erman, M., & Pangborn, W. (2009). X-ray structure of human aromatase reveals an androgen-specific active site. *J Steroid Biochem Mol Biol, 118*(4–5), 197–202.

Goldstone, J., McArthur, A., Kubota, A., Zanette, J., Parente, T., Jonsson, M., et al. (2010) Identification and developmental expression of the full complement of Cytochrome P450 genes in Zebrafish. *BMC Genomics, 11*(1), 643.

Gonzalez, A., & Piferrer, F. (2003) Aromatase activity in the European sea bass (Dicentrarchus labrax L.) brain. Distribution and changes in relation to age, sex, and the annual reproductive cycle. *Gen Comp Endocrinol, 132*(2), 223–230.

Graham-Lorence, S., Amarneh, B., White, R. E., Peterson, J. A., & Simpson, E. R. (1995). A three-dimensional model of aromatase cytochrome P450. *Protein Sci, 4*(6), 1065–1080.

Greytak, S. R., Champlin, D., & Callard, G. V. (2005). Isolation and characterization of two cytochrome P450 aromatase forms in killifish (Fundulus heteroclitus): Differential expression in fish from polluted and unpolluted environments. *Aquat Toxicol, 71*(4), 371–389.

Greytak, S. R., Tarrant, A. M., Nacci, D., Hahn, M. E., & Callard, G. V. (2010). Estrogen responses in killifish (Fundulus heteroclitus) from polluted and unpolluted environments are site- and gene-specific. *Aquat Toxicol, 99*(2), 291–299.

Hahn, M. E. (2011). Mechanistic research in aquatic toxicology: perspectives and future directions. *Aquat Toxicol 105*(3–4 Suppl), 67–71.

Hamad, A., Kluk, M., Fox, J., Park, M., & Turner, J. E. (2007). The effects of aromatase inhibitors and selective estrogen receptor modulators on eye development in the zebrafish (Danio rerio). *Curr Eye Res, 32*(10), 819–827.

Harada, N. (1988). Cloning of a complete cDNA encoding human aromatase: immunochemical identification and sequence analysis. *Biochem Biophys Res Commun, 156*(2), 725–732.

Harada, N., Yamada, K., Foidart, A., & Balthazart, J. (1992). Regulation of aromatase cytochrome P-450 (estrogen synthetase) transcripts in the quail brain by testosterone. *Brain Res Mol Brain Res, 15*(1–2), 19–26.

Hitchcock, P. F., & Raymond, P. A. (2004). The teleost retina as a model for developmental and regeneration biology. *Zebrafish, 1*(3), 257–271.

Ijiri, S., Berard, C., & Trant, J. M. (2000). Characterization of gonadal and extra-gonadal forms of the cDNA encoding the Atlantic stingray (Dasyatis sabina) cytochrome P450 aromatase (CYP19). *Mol Cell Endocrinol, 164*(1–2), 169–181.

Jeng, S.-R., Dufour, S., & Chang, C.-F. (2005). Differential expression of neural and gonadal aromatase enzymatic activities in relation to gonadal development in Japanese eel, Anguilla japonica. *J Exp Zool Part A: Comp Exp Biol, 303A*(9), 802–812.

Jeng, S. R., Pasquier, J., Yueh, W. S., Chen, G. R., Lee, Y. H., Dufour, S., et al. (2012). Differential regulation of the expression of cytochrome P450 aromatase, estrogen and androgen receptor subtypes in the brain-pituitary-ovarian axis of the Japanese eel (Anguilla japonica) reveals steroid dependent and independent mechanisms. *Gen Comp Endocrinol, 175*(1), 163–172.

Karube, M., Fernandino, J. I., Strobl-Mazzulla, P., Strüssmann, C. A., Yoshizaki, G., Somoza, G. M., et al. (2007). Characterization and expression profile of the ovarian cytochrome P-450 aromatase (cyp19A1) gene during thermolabile sex determination in pejerrey, Odontesthes bonariensis. *J Exp Zool Part A: Ecol Genet Physiol, 307A*(11), 625–636.

Kazeto, Y., Ijiri, S., Place, A. R., Zohar, Y., & Trant, J. M. (2001). The 5′-Flanking Regions of CYP19A1 and CYP19A2 in Zebrafish. *Biochem Biophys Res Commun, 288*(3), 503–508.

Kazeto, Y., Place, A. R., & Trant, J. M. (2004). Effects of endocrine disrupting chemicals on the expression of CYP19 genes in zebrafish (Danio rerio) juveniles. *Aquat Toxicol, 69*(1), 25–34.

Kazeto, Y., & Trant, J. M. (2005). Molecular biology of channel catfish brain cytochrome P450 aromatase (CYP19A2): cloning, preovulatory induction of gene expression, hormonal gene regulation and analysis of promoter region. *J Mol Endocrinol, 35*(3), 571–583.

Kishida, M., & Callard, G. V. (2001). Distinct cytochrome P450 aromatase isoforms in zebrafish (Danio rerio) brain and ovary are differentially programmed and estrogen regulated during early development. *Endocrinology, 142*(2), 740–750.

Kishida, M., McLellan, M., Miranda, J. A., & Callard, G. V. (2001. Estrogen and xenoestrogens upregulate the brain aromatase isoform (P450aromB) and perturb markers of early development in zebrafish (Danio rerio). *Comp Biochem Physiol Part B: Biochem Mol Biol, 129*(2–3), 261–268.

Kuhl, A. J., Manning, S., & Brouwer, M. (2005) Brain aromatase in Japanese medaka (Oryzias latipes): molecular characterization and role in xenoestrogen-induced sex reversal. *J Steroid Biochem Mol Biol, 96*(1), 67–77.

Kumar, P., Kamat, A., & Mendelson, C. R.. (2009). Estrogen receptor (ER) mediates stimulatory effects of estrogen on aromatase (CYP19) gene expression in human placenta. *Mol Endocrinol, 23*(6), 784–793.

Kwon, J. Y., Haghpanah, V., Kogson-Hurtado, L. M., McAndrew, B. J., & Penman, D. J. (2000). Masculinization of genetic female Nile tilapia (Oreochromis niloticus) by dietary administration of an aromatase inhibitor during sexual differentiation. *J Exp Zool, 287*(1), 46–53.

Lange, A., Katsu, Y., Ichikawa, R., Paull, G. C., Chidgey, L. L., Coe, T. S., et al. (2008). Altered sexual development in roach (Rutilus rutilus) exposed to environmental concentrations of the pharmaceutical 17α-ethinylestradiol and associated expression dynamics of aromatases and estrogen receptors. *Toxicol Sci, 106*(1), 113–123.

Lassiter, C. S., & Linney, E. (2007). Embryonic expression and steroid regulation of brain aromatase cyp19a1b in zebrafish (Danio rerio). *Zebrafish, 4*(1), 49–57.

Le Page, Y., Diotel, N., Vaillant, C., Pellegrini, E., Anglade, I., Merot, Y., et al. (2011). Aromatase, brain sexualization and plasticity: the fish paradigm. *Eur J Neurosci, 32*(12), 2105–2115.

Lee, Y.-M., Seo, J. S., Kim, I.-C., Yoon, Y.-D., & Lee, J.-S. (2006). Endocrine disrupting chemicals (bisphenol A, 4-nonylphenol, 4-tert-octylphenol) modulate expression of two distinct cytochrome P450 aromatase genes differently in gender types of the hermaphroditic fish Rivulus marmoratus. *Biochem Biophys Res Commun, 345*(2), 894–903.

Li, G.-L., Liu, X.-C. & Lin, H.-R. (2007). Seasonal changes of serum sex steroids concentration and aromatase activity of gonad and brain in red-spotted grouper (Epinephelus akaara). *Anim Reprod Sci, 99*(1–2), 156–166.

Lieberburg, I. & McEwen, B. S. (1977). Brain cell nuclear retention of testosterone metabolites, 5alpha-dihydrotestosterone and estradiol-17beta, in adult rats. *Endocrinology, 100*(2), 588–597.

Lord, L.-D., Bond, J., & Thompson, R. R. (2009). Rapid steroid influences on visually guided sexual behavior in male goldfish. *Horm Behav, 56*(5), 519–526.

MacLusky, N. J., & Naftolin, F. (1981). Sexual differentiation of the central nervous system. *Science, 211*(4488), 1294–1302.

Mak, P., Zenn, R., & Callard, G. V. (1985). *Subcellular distribution and reaction kinetics of aromatase in goldfish brain.* Paper presented at the Proceedings of the 67th Annual Meeting, The Endocrine Society, Baltimore.

Markov, G. V., Tavares, R., Dauphin-Villemant, C., Demeneix, B. A., Baker, M. E., & Laudet, V. (2009). Independent elaboration of steroid hormone signaling pathways in metazoans. *Proc Natl Acad Sci U S A, 106*(29), 11913–11918.

McCurley, A. T., & Callard, G. V. (2010). Time course analysis of gene expression patterns in zebrafish eye during optic nerve regeneration. *J Exp Neurosci, 2010*(4), 17–33.

Mendelson, C. R., Wright, E. E., Evans, C. T., Porter, J. C., & Simpson, E. R. (1985). Preparation and characterization of polyclonal and monoclonal antibodies against human aromatase cytochrome P-450 (P-450AROM), and their use in its purification. *Arch Biochem Biophys, 243*(2), 480–491.

Menuet, A., Pellegrini, E., Brion, F., Gueguen, M.-M., Anglade, I., Pakdel, F., et al. (2005) Expression and estrogen-dependent regulation of the zebrafish brain aromatase gene. *J Comp Neurol,* 485(4), 304–320.

Meyer, A. & Schartl, M. (1999) Gene and genome duplications in vertebrates: the one-to-four (-to-eight in fish) rule and the evolution of novel gene functions. *Curr Opin Cell Biol,* 11(6), 699–704.

Mouriec, K., Pellegrini, E., Anglade, I., Menuet, A., Adrio, F., Thieulant, M. L., et al. (2008) Synthesis of estrogens in progenitor cells of adult fish brain: evolutive novelty or exaggeration of a more general mechanism implicating estrogens in neurogenesis? *Brain Res Bull,* 75(2–4), 274–280.

Nacci, D. E., Champlin, D., Coiro, L., McKinney, R. & Jayaraman, S. (2002) Predicting the occurrence of genetic adaptation to dioxinlike compounds in populations of the estuarine fish Fundulus heteroclitus. *Environ Toxicol Chem,* 21(7), 1525–1532.

Naftolin, F., Horvath, T. L., Jakab, R. L., Leranth, C., Harada, N., Balthazar, J. (1996) Aromatase immunoreactivity in axon terminals of the vertebrate brain. An immoncytochemical study on quail, rat, monkey and human tissues. *Neuroendocrinology,* 63(2), 149–155.

Naftolin, F., Ryan, K. J., Davies, I. J., Reddy, V. V., Flores, F., Petro, Z., et al. (1975) The formation of estrogens by central neuroendocrine tissues. *Recent Prog Horm Res,* 31, 295–319.

Naftolin, F., Ryan, K. J. & Petro, Z. (1971) Aromatization of androstenedione by the diencephalon. *J Clin Endocrinol Metab,* 33(2), 368–370.

Nakazawa, T., Takahashi, H. & Shimura, M. (2006) Estrogen has a neuroprotective effect on axotomized RGCs through ERK signal transduction pathway. *Brain Res,* 1093(1), 141–149.

Nelson, D. R. (1998) Metazoan cytochrome P450 evolution. *Comp Biochem Physiol C Pharmacol Toxicol Endocrinol,* 121(1–3), 15–22.

Nocillado, J. N., Elizur, A., Avitan, A., Carrick, F. & Levavi-Sivan, B. (2007) Cytochrome P450 aromatase in grey mullet: cDNA and promoter isolation; brain, pituitary and ovarian expression during puberty. *Mol Cell Endocrinol,* 263(1–2), 65–78.

Nunez, B. S. & Applebaum, S. L. (2006) Tissue- and sex-specific regulation of CYP19A1 expression in the Atlantic croaker (Micropogonias undulatus) *Gen Comp Endocrinol,* 149(2), 205–216.

Pang, Y. & Thomas, P. (2010) Role of G protein-coupled estrogen receptor 1, GPER, in inhibition of oocyte maturation by endogenous estrogens in zebrafish. *Dev Biol,* 342(2), 194–206.

Pasmanik, M. & Callard, G. V. (1985) Aromatase and 5α-reductase in the teleost brain, spinal cord, and pituitary gland. *Gen Comp Endocrinol,* 60(2), 244–251.

Pasmanik, M. & Callard, G. V. (1988) Changes in brain aromatase and 5α-reductase activities correlate significantly with seasonal reproductive cycles in goldfish (Carassius auratus) *Endocrinology,* 122(4), 1349–1356.

Pasmanik, M., Schlinger, B. A. & Callard, G. V. (1988) In vivo steroid regulation of aromatase and 5α-reductase in goldfish brain and pituitary. *Gen Comp Endocrinol,* 71(1), 175–182.

Pellegrini, E., Mouriec, K., Anglade, I., Menuet, A., Le Page, Y., Gueguen, M.-M., et al. (2007) Identification of aromatase-positive radial glial cells as progenitor cells in the ventricular layer of the forebrain in zebrafish. *J Comp Neurol,* 501(1), 150–167.

Piferrer, F., Blazquez, M., Navarro, L. & Gonzalez, A. (2005) Genetic, endocrine, and environmental components of sex determination and differentiation in the European sea bass (Dicentrarchus labrax L.). *Gen Comp Endocrinol,* 142(1–2), 102–110.

Postlethwait, J. H., Woods, I. G., Ngo-Hazelett, P., Yan, Y. L., Kelly, P. D., Chu, F., et al. (2000) Zebrafish comparative genomics and the origins of vertebrate chromosomes. *Genome Res,* 10(12), 1890–1902.

Rasheeda, M. K., Sridevi, P. & Senthilkumaran, B. (2010) Cytochrome P450 aromatases: Impact on gonadal development, recrudescence and effect of hCG in the catfish, Clarias gariepinus. *Gen Comp Endocrinol,* 167(2), 234–245.

Rashid, H., Kitano, H., Hoon Lee, K., Nii, S., Shigematsu, T., Kadomura, K., et al. (2007) Fugu (Takifugu rubripes) sexual differentiation: CYP19 regulation and aromatase inhibitor induced testicular development. *Sex Dev,* 1(5), 311–322.

Reitzel, A. M. & Tarrant, A. M. (2010) Correlated evolution of androgen receptor and aromatase revisited. *Mol Biol Evol,* 27(10), 2211–2215.

Remage-Healey, L., Dong, S., Maidment, N. T. & Schlinger, B. A. (2011) Presynaptic control of rapid estrogen fluctuations in the songbird auditory forebrain. *J Neurosci,* 31(27), 10034–10038.

Roselli, C. E. & Klosterman, S. A. (1998) Sexual differentiation of aromatase activity in the rat brain: effects of perinatal steroid exposure. *Endocrinology,* 139(7), 3193–3201.

Roselli, C. E. & Resko, J. A. (1984) Androgens regulate brain aromatase activity in adult male rats through a receptor mechanism. *Endocrinology,* 114(6), 2183–2189.

Ryan, K. J. (1959) Biological aromatization of steroids. *J Biol Chem,* 234(2), 268–272.

Sakamoto, H., Mezaki, Y., Shikimi, H., Ukena, K. & Tsutsui, K. (2003) Dendritic growth and spine formation in response to estrogen in the developing Purkinje cell. *Endocrinology,* 144(10), 4466–4477.

Salyer, D. L., Lund, T. D., Fleming, D. E., Lephart, E. D., & Horvath, T. L. (2001). Sexual dimorphism and aromatase in the rat retina. *Brain Res Dev Brain Res, 126*(1), 131–136.

Santen, R. J., Brodie, H., Simpson, E. R., Siiteri, P. K., & Brodie, A. (2009). History of aromatase: saga of an important biological mediator and therapeutic target. *Endocr Rev, 30*(4), 343–375.

Sawyer, S. J., Gerstner, K. A., & Callard, G. V. (2006). Real-time PCR analysis of cytochrome P450 aromatase expression in zebrafish: gene specific tissue distribution, sex differences, developmental programming, and estrogen regulation. *Gen Comp Endocrinol, 147*(2), 108–117.

Schlinger, B. A., & Callard, G. V. (1989). Localization of aromatase in synaptosomal and microsomal subfractions of quail (Coturnix coturnix japonica) brain. *Neuroendocrinology, 49*(4), 434–441.

Schlinger, B. A., & Remage-Healey, L. (2012). Neurosteroidogenesis: insights from studies of songbirds. *J Neuroendocrinol, 24*(1), 16–21.

Simpson, E. R., Mahendroo, M. S., Means, G. D., Kilgore, M. W., Hinshelwood, M. M., Graham-Lorence, S., et al. (1994). Aromatase cytochrome P450, the enzyme responsible for estrogen biosynthesis. *Endocr Rev, 15*(3), 342–355.

Stacey, N., Chojnacki, A., Narayanan, A., Cole, T., & Murphy, C. (2003) Hormonally derived sex pheromones in fish: exogenous cues and signals from gonad to brain. *Can J Physiol Pharmacol, 81*(4), 329–341.

Strobl-Mazzulla, P. H., Moncaut, N. P., Lopez, G. C., Miranda, L. A., Canario, A. V. M., & Somoza, G. M. (2005). Brain aromatase from pejerrey fish (Odontesthes bonariensis): cDNA cloning, tissue expression, and immunohistochemical localization. *Gen Comp Endocrinol,* 143(1), 21–32.

Tanaka, M., Fukada, S., Matsuyama, M., & Nagahama, Y. (1995). Structure and promoter analysis of the cytochrome P-450 aromatase gene of the teleost fish, medaka (Oryzias latipes). *J Biochem, 117*(4), 719–725.

Tanaka, M., Telecky, T. M., Fukada, S., Adachi, S., Chen, S., & Nagahama, Y. (1992). Cloning and sequence analysis of the cDNA encoding P-450 aromatase (P450arom) from a rainbow trout (Oncorhynchus mykiss) ovary; relationship between the amount of P450arom mRNA and the production of oestradiol-17β in the ovary. *J Mol Endocrinol, 8*(1), 53–61.

Tata, J. R. (2005). One hundred years of hormones. *EMBO Rep, 6*(6), 490–496.

Tchoudakova, A., & Callard, G. V. (1998). Identification of multiple CYP19 genes encoding different cytochrome P450 aromatase isozymes in brain and ovary. *Endocrinology, 139*(4), 2179–2189.

Tchoudakova, A., Kishida, M., Wood, E., & Callard, G. V. (2001). Promoter characteristics of two cyp19 genes differentially expressed in the brain and ovary of teleost fish. *J Steroid Biochem Mol Biol, 78*(5), 427–439.

Thomas, P., Alyea, R., Pang, Y., Peyton, C., Dong, J., & Berg, A. H. (2010). Conserved estrogen binding and signaling functions of the G protein-coupled estrogen receptor 1 (GPER) in mammals and fish. *Steroids, 75*(8–9), 595–602.

Toffolo, V., Belvedere, P., Colombo, L., & Valle, L. D. (2007). Tissue-specific transcriptional initiation of the CYP19 genes in rainbow trout, with analysis of splicing patterns and promoter sequences. *Gen Comp Endocrinol, 153*(1–3), 311–319.

Tong, S. K., Chiang, E. F., Hsaio, P. H., & Chung, B. (2001). Phylogeny, expression and enzyme activity of zebrafish cyp19 (P450 aromatase) genes. *J Steroid Biochem Mol Biol, 290*(7), 709–714.

Tong, S.-K., & Chung, B.-c. (2003). Analysis of zebrafish cyp19 promoters. *J Steroid Biochem Mol Biol, 86*(3–5), 381–386.

Trant, J. M. (1994). Isolation and characterization of the cDNA encoding the channel catfish (Ictalurus punctatus) form of cytochrome P450arom. *Gen Comp Endocrinol, 95*(2), 155–168.

Trant, J. M., Gavasso, S., Ackers, J., Chung, B.-C., & Place, A. R. (2001). Developmental expression of cytochrome P450 aromatase genes (CYP19a and CYP19b) in zebrafish fry (Danio rerio). *J Exp Zool, 290*(5), 475–483.

Trudeau, V. L. (1997). Neuroendocrine regulation of gonadotrophin II release and gonadal growth in the goldfish, Carassius auratus. *Rev Reprod, 2*(1), 55–68.

Tzchori, I., Degani, G., Hurvitz, A., & Moav, B. (2004). Cloning and developmental expression of the cytochrome P450 aromatase gene (CYP19) in the European eel (Anguilla anguilla). *Gen Comp Endocrinol, 138*(3), 271–280.

van Nes, S., Moe, M., & Andersen, Ø. (2005). Molecular characterization and expression of two cyp19 (P450 aromatase) genes in embryos, larvae, and adults of Atlantic halibut (Hippoglossus hippoglossus). *Mole Reprod Dev, 72*(4), 437–449.

Vasudevan, N., & Pfaff, D. W. (2008). Non-genomic actions of estrogens and their interaction with genomic actions in the brain. *Front Neuroendocrinol, 29*(2), 238–257.

von Schalburg, K. R., Yasuike, M., Davidson, W. S., & Koop, B. F. (2010). Regulation, expression and characterization of aromatase (cyp19b1) transcripts in ovary and testis of rainbow trout (Oncorhynchus mykiss). *Comp Biochem Physiol B Biochem Mol Biol, 155*(2), 118–125.

Wang, J., Liu, X., Wang, H., Wu, T., Hu, X., Qin, F., et al. (2010) .Expression of two cytochrome P450 aromatase genes is regulated by endocrine disrupting chemicals in rare minnow Gobiocypris rarus juveniles. *Comp Biochem Physiol C Toxicol Pharmacol, 152*(3), 313–320.

Yilmaz, M. B., Wolfe, A., Cheng, Y.-H., Glidewell-Kenney, C., Jameson, J. L., & Bulun, S. E. (2009). Aromatase promoter I.f is regulated by estrogen receptor alpha (ESR1) in mouse hypothalamic neuronal cell lines. *Biol Reprod, 81*(5), 956–965.

Yu, J.-H., Tang, Y.-K., & Li, J.-L. (2008). Cloning, structure, and expression pattern of the P-450 aromatase gene in rice field eel (Monopterus albus). *Biochem Genet, 46*(5), 267–280.

Zhang, Y., Zhang, W., Zhang, L., Zhu, T., Tian, J., Li, X., et al. (2004) Two distinct cytochrome P450 aromatases in the orange-spotted grouper (Epinephelus coioides): cDNA cloning and differential mRNA expression. *J Steroid Biochem Mol Biol, 92*(1–2), 39–50.

Zhao, J., Mak, P., Tchoudakova, A., Callard, G., & Chen, S. (2001). Different catalytic properties and inhibitor responses of the goldfish brain and ovary aromatase isozymes. *Gen Comp Endocrinol, 123*(2), 180–191.

Chapter 3

The Distribution and Regulation of Aromatase in the Mammalian Brain: from Mice to Monkeys

Charles E. Roselli

INTRODUCTION

The aromatization hypothesis states that the brain is capable of synthesizing estrogens from androgens in the circulation via the enzyme aromatase and that these locally produced estrogens mediate many of the effects of androgens on brain development and function (Naftolin et al., 1975). This idea that estrogens could be produced in the brain was born from early physiologic studies that found many of the central neuroendocrine effects of androgens could be mimicked by estrogen and blocked by antiestrogen pretreatments (Gorski, 1963; McCann & Ramirez, 1964; Feder et al., 1974). Naftolin and his colleagues obtained the definitive evidence for the aromatase hypothesis when they found that neuroendocrine structures in the brain contained aromatase enzymatic activity (Naftolin et al., 1975). Originally the aromatization hypothesis was advanced to explain the mechanism whereby precursor androgens are aromatized to estrogens before working at the cellular level to regulate processes in key neuroendocrine tissues. However, over the intervening 30+ years since it was first proposed, the aromatization hypothesis has been examined from many experimental perspectives and it is clear that locally synthesized estrogens play a larger role in brain function than was conceived initially. Thus, it is important to elucidate the localization of aromatase in brain tissue in order to establish the possible sites of estrogen aromatization. This chapter will present an overview of the distribution, regulation, and functional significance of aromatase in the mammalian brain.

DISTRIBUTION OF BRAIN AROMATASE

Aromatase Activity

Aromatase was initially localized in the adult and fetal central nervous system (CNS) by radiometric assay designed to measure enzyme activity by identifying estrogen synthesis in vitro (Weisz & Gibbs, 1974; Naftolin et al., 1975; Reddy et al., 1974; George & Ojeda, 1982). These assays involved the incubation of tissue homogenates derived from regional brain dissections with either ^{3}H or ^{14}C-labeled testosterone or androstenedione. This step was then followed by the measurement of the production of labeled estradiol and estrone, respectively. The technique involved a laborious set of organic extractions, paper and thin layer chromatographies, and finally chemical identification through recrystallization to a constant specific activity. In an in vivo variation on the technique, ^{3}H-testosterone was infused into animals and ^{3}H-estradiol was recovered in cell nuclear fractions purified from regional brain dissections (Lieberburg & McEwen, 1977; Bonsall et al., 1989). End-product isolation assays were eventually replaced with a simple and highly sensitive radiometric enzyme assay based on the quantification of ^3H_2O formed from [1β-^{3}H] androstenedione (Weisz, 1982; Roselli et al., 1984). Using a variety of mammalian species, including rats (Weisz & Gibbs, 1974; Naftolin et al., 1975; George & Ojeda, 1982; Roselli & Resko, 1987), mice (Sinchak et al., 1996), hamsters (Callard et al., 1986; Negri-Cesi et al., 1989; Hutchison et al., 1991), guinea pigs (Negri-Cesi et al., 1989), rabbits (Reddy et al., 1973; George et al., 1978;), ferrets (Weaver & Baum, 1991), and primates (Ellinwood et al., 1984), studies using this method found that aromatase activity was greatest in areas of neuroendocrine activity, particularly the hypothalamus and amygdala. However, these levels are relatively modest compared with tissues classically known as estrogen-producing sources such as the ovary and placenta. Whole hypothalami were found to convert less than 0.05% of androgen substrates to estrogens in vitro; whereas the conversion is at least 10 times greater for placenta and 100 times for ovary. Despite this low conversion rate, it is abundantly clear that aromatization in the CNS plays an essential role in brain sexual differentiation and neuroendocrine function in the adult (Roselli, 2007).

It soon became clear that one of the reasons such low amounts of estrogens are formed in the brain is that aromatase is not uniformly distributed within cell populations of anatomically defined brain regions such as the hypothalamus, nor is it confined only to the cell body of neurons. The heterogeneous anatomical distribution of aromatase activity was first demonstrated by studies using the Palkovitz punch technique to dissect more precisely discrete subregions within the hypothalamus, amygdala, and other areas of the rat brain (Selmanoff et al., 1977; Roselli et al., 1984). Table 3–1 shows a comparison of aromatase activity in microdissected brain regions from rats, guinea pigs, sheep and rhesus monkey (Roselli et al., 1985, 1987, 1998; Connolly et al., 1991). These studies found that the highest levels of aromatase activity in the adult brain are found across species in the medial amygdala, cortical amygdala, and bed nucleus of the stria terminalis. Significant levels of aromatase activity are also found in the periventricular and medial zones of the hypothalamus, with the greatest levels in the periventricular preoptic area and the medial preoptic nucleus. These brain regions in large part encompass the medial subdivision of the extended amygdala (Heimer & Alheid, 1991), which contains an abundance of androgen- and estrogen-receptor–containing neurons (Simerly et al., 1990). Interestingly, it has been found that these areas are involved in the activation of several social and sexual behaviors, in particular male reproductive behavior (Newman, 1999).

Table 3–1. Species Comparison of Aromatase Activity in Brain Nuclei

Structure	Rat	Guinea Pig	Sheep	Rhesus Monkey
Diencephalon				
Septum	-	+	+	+
Bed n. of the stria terminalis	$++^{R/S}$	++	++	+++
Periventricular preoptic area	$++^{R/S}$	$++^{R}$	$+^{R}$	$+^{R}$
Medial preoptic n.	$++^{R/S}$	++	$+^{R}$	$++^{R}$
Lateral preoptic n.	+	+	+	++
Anterior hypothalamus	$+^{R/S}$	++	n.d.	+
Suprachiasmatic n.	+	n.d.	n.d.	+
Superoptic n.	+	n.d.	n.d.	+
Periventricular anterior hypothalamus	$+^{R/S}$	++	n.d.	n.d.
Ventromedial hypothalamic n.	$+^{R/S}$	$+^{R}$	+	++
Arcuate n./medium eminence	$+^{R/S}$	$+^{R}$	$+^{R}$	+
Dorsomedial hypothalamic n.	+		+	+
Lateral hypothalamus	$+^{R}$	+	+	$++^{R}$
Amygdala				
Medial amygdala	$++^{R}$	++	++	$+++^{R}$
Cortical amygdala	++	$+++^{R}$	++	+++

The pluses provide an indication of aromatase activity (pmol/h/mg protein) ranging from < 0.015 (-) nondectable, 0.015 to 0.25 (+) low, 0.25 to 2.0 (++) medium, and >2.0 (+++) high. Superscripts denote whether there is a sex difference (S) in activity and/or whether activity is regulated by testosterone (R). Abbreviations: n., nucleus; n.d., not determined. There were several additional regions that were assayed in rat where aromatase activity was not detectable: caudate-putamen, dentate gyrus, and CA1 region of the hippocampus. References: (Selmanoff et al., 1977; Roselli et al., 1985; 1987; 1998; Connolly et al., 1991).

Various other brain regions such as the hippocampus, cerebral cortex, septum, midbrain, sensory afferent neurons, and spinal cord (MacLusky et al., 1994; Horvath & Wikler, 1999; Hojo et al., 2004; Yague et al., 2008; Liu et al., 2011) also contain aromatase activity, suggesting that locally synthesized estrogens may be involved in other neural functions such as memory, cognition, and sensory perception. Of these extra-hypothalamic areas, aromatase activity has been most extensively studied in the hippocampus (Roselli et al., 1984; Abdelgadir et al., 1994; MacLusky et al., 1986, 1994). More recently, it was demonstrated that estradiol can be synthesized from cholesterol precursor in hippocampal slices and that this de novo synthesized estradiol stimulates the formation and maintenance of dendritic spines in hippocampus (Hojo et al., 2008). Aromatase also plays a role in neuroprotection after injury or ischemia (Garcia-Segura, 2008; Hojo et al., 2009; Fester et al., 2011). These findings are the first in mammalian brain to suggest that in addition to the production of neuroactive estrogens from serum androgen precursors, estrogens are neurosteroids synthesized directly from cholesterol in at least some regions of the nervous system.

Aromatase mRNA and Protein

The coding region of the aromatase gene spans 9 exons starting with exon II. Upstream of exon II there are a number of first exon Is that are spliced into the untranslated region of the transcript in a tissue specific fashion (Simpson et al., 2002). A brain specific promoter (1f) is the major 5′-untranslated transcript in amygdala and hypothalamus although some degree of differential promoter usage has been detected in brain (Honda et al., 1994, 1996; Yamada-Mouri et al., 1995; Kato et al., 1997; Vanselow et al., 2001; Simpson et al., 2002; Golovine et al., 2003; Chow et al., 2009). Northern blot analysis of aromatase mRNA prepared from perinatal rat hypothalamus identified a single 2.7 kb mRNA transcript (Lephart et al., 1995). A highly sensitive and specific ribonuclease protection assay was used first to quantify aromatase mRNA in the rat and monkey brain (Abdelgadir et al., 1994, 1997). These studies found high levels of aromatase mRNA expression within the hypothalamus, preoptic area, and limbic system and demonstrated that levels of aromatase mRNA significantly and positively correlated with aromatase activity, providing evidence that brain aromatase is regulated transcriptionally. Furthermore, these studies identified a novel mRNA variant that diverges 5′ to exon IV (Roselli et al., 1998). This variant mRNA was subsequently characterized as *cortex type* variant and shown to encode a potential protein that has different kinetic properties from the full-length transcript (Yamada-Mouri et al., 1997; Kao et al., 1999).

Neuroanatomical studies using in situ hybridization histochemistry (Lauber & Lichtensteiger, 1994; Wagner & Morrell, 1996; Roselli et al., 1998, 2000, 2001) generally agree with the regional pattern of enzymatic activity in the mammalian CNS (Table 3–2 vs. Table 3–1). Brain regions with high levels of aromatase activity, such as the bed nucleus of the stria terminalis and the medial amygdala, also contain the highest expression of aromatase mRNA. Regions with relatively lower levels of activity (e.g., medial preoptic nucleus and ventromedial hypothalamic nucleus) contain fewer aromatase mRNA-expressing cells. These observations agree with biochemical studies that also demonstrate strong transcriptional regulation of aromatase in these brain regions (Abdelgadir et al., 1994, 1997). In rats, musk shrews, and sheep aromatase mRNA is highly expressed within a sexually dimorphic preoptic nucleus

Table 3–2. Species Comparison of Aromatase mRNA in Diencephalic and Limbic Nuclei

Brain Areas	Rat	Sheep	Monkey
Diencephalon			
Septum	+	n.d.	+
Bed n. of the strial terminalis, anterior	++	-	+
Bed n. of the strial terminalis, medial	++	-	+
Bed n. of the strial terminalis,encapsulated	$+++^{R/S}$	+++	+++
Periventricular preoptic area	+	+	+
Medial preoptic n.	$++^{R/S}$	++	++
Medial preoptic n., central part	++	+++	+++
Anterior hypothalamic area	++	-	+
Ventrolateral ventromedial hypothalamus	$++^{R}$	+	++
Arcuate n.	+	-	+
Amygdala			
Medial amygdala	$+++^{R}$	++	+++
Cortical amygdala	++	++	+++
Basal amygdala	+	n.d.	+
Central amygdala	+	n.d.	

Pluses are used to provide a scale of relative abundance of mRNA and/or immunoreactivity. Scale = +++, high density; ++, moderate density, +, low density, -, not detected. Superscripts denote whether there is a sex difference (S) in activity and/or whether activity is regulated by testosterone (R). Abbreviations: n., nucleus; n.d., not determined. References: (Lauber & Lichtensteiger, 1994; Wagner & Morrell, 1996; Roselli et al., 1998, 2000, 2001).

(Lauber & Lichtensteiger, 1994; Roselli et al., 2000; Veney, et al., 2000;), and in sheep the volume of the sexually dimorphic nucleus as defined by the pattern of aromatase expression has been correlated with sexual partner preferences (Roselli et al., 2004).

The detection of aromatase-expressing cells and their projections in the mouse brain has recently been accomplished by genetically tagging the aromatase locus with the reporter genes nuclear LacZ and placental alkaline phosphatase (Wu et al., 2009). This experimental approach confirmed that the greatest numbers of cells expressing aromatase are confined to the bed nucleus of the stria terminalis and medial amygdala where there are more aromatase positive neurons in males than in females. Although the anterior hypothalamus and ventromedial hypothalamus had few aromatase-expressing cells, they contained significantly more projections from aromatase-expressing neurons and this also was sexually dimorphic.

There is some indication from in situ hybridization studies that aromatase mRNA may be expressed in areas where activity has not been examined or has not been found via other methods, such as the mediodorsal thalamus and cingulate cortex (Wagner & Morrell, 1996). These disparities may be caused by alternate splicing of the primary aromatase mRNA transcript as mentioned (Yamada-Mouri et al., 1997). Other discrepancies exist for the lateral preoptic area, anterior hypothalamic area, suprachiasmatic nucleus, and dorsomedial hypothalamus, which all contain low to moderate aromatase activity but exhibit little or no aromatase mRNA expression. Although these discrepancies may be caused by differences in sensitivity or anatomical specificity between techniques being used to compare aromatase distribution, it is also possible that within these areas aromatase activity exists in subcellular compartments other than the cell body wherein aromatase mRNA is expressed. To support this, there is evidence that in addition to the neuronal cell body, aromatase protein and activity is distributed within processes and axon terminals (Naftolin et al., 1996; Wu et al., 2009), although to date most of this evidence exists in nonmammalian species (Saldanha, 2011).

Reports of the distribution of aromatase-immunoreactivity visualized in adult mammals using immunohistochemistry have varied depending on the antiaromatase antiserum employed (Shinoda et al., 1989; Balthazart et al., 1991; Sanghera et al., 1991; Jakab et al., 1993; Dellovade et al., 1994). A moderate density of aromatase-immunoreactive cells is detected in the paraventricular and arcuate nuclei, the CA1–CA3 regions and dentate gyrus of the hippocampus, and the dorsomedial and lateral hypothalamic nuclei. All of these early studies failed to detect immunoreactivity in the bed nucleus of the stria terminalis, medial preoptic area, and ventromedial hypothalamic nucleus, regions that contain moderate to high aromatase activity and the presence of many cells expressing high levels of aromatase mRNA as discussed previously. Better correspondence was observed between brain aromatase immunoreactivity and aromatase activity, when mice or musk shrews were pretreated with the aromatase inhibitor vorozole prior to performing immunohistochemistry (Foidart et al., 1995; Rissman et al., 1996). Vorozole inhibits aromatase activity by irreversibly binding with the enzyme. In particular, treatment with vorozole revealed aromatase immunoreactive cells and fibers in several important areas, including the bed nucleus of the stria terminalis, medial preoptic, and ventromedial hypothalamic nuclei. Vorozole appears to enhance aromatase protein content either through altering turnover or a regulatory feedback mechanism (Choate & Resko, 1996; Foidart et al., 1994), which likely explains the enhanced immunostaining observed.

The aromatase inhibitor vorozole has been modified to emit positrons upon decay (i.e., [^{11}C] vorozole) for use as a positron emission tomography (PET) tracer to localize aromatase in vivo with high accuracy and sensitivity. Initial studies using autoradiography demonstrated specific [^{11}C] vorozole binding in the bed nucleus of the stria terminalis, medial preoptic area, and medial amygdala that was enhanced by pretreatment with the anabolic androgenic steroid nandrolone and reduced after treatment with the antiandrogen flutamide (Takahashi et al., 2007; Takahashi et al., 2008). In vivo PET imaging in monkeys found high and specific binding of [^{11}C] vorozole in the hypothalamus and amygdala (Takahashi et al., 2006, 2011). Binding was enhanced in the hypothalamus after pretreatment with nandrolone but not in the amygdala. The brain distribution identified by in vitro and in vivo binding studies are in good agreement with regions found to be enriched in enzyme activity and mRNA expression as discussed above.

Aromatase is predominantly expressed in neurons under normal conditions (Negri Cesi, et al., 1992; Beyer et al., 1994; Jakab et al., 1994). However, recent studies using double-label immunohistochemical techniques of specific markers have demonstrated that aromatase is expressed by radial glia and intermediate progenitor cells of the developing cortex in mice (Martinez-Cerdeno et al., 2006). This is particularly interesting because actively dividing radial glia that differentiate into neurons have been shown to express aromatase in fish brain implicating aromatase in neurogenesis (Mouriec et al., 2008). Although aromatase has been detected within astrocytes of the human temporal cortex (Yague et al., 2006), it appears to be greatly upregulated after brain injury and ischemia and appears to play a role in neuroprotection (Garcia-Segura et al., 1999; Garcia-et al., 2003; McCullough et al., 2003; Carswell et al., 2005). In vitro, aromatase is constitutively expressed in astrocytes (Zwain et al., 1997; Azcoitia et al., 2003) and plays a role in cell viability after oxygen glucose deprivation (OGD); a commonly used model of ischemic injury. Astrocytes from female rats are more resistant to OGD than are cells from males in part because they express higher levels of aromatase (Liu et al., 2007) and are able to generate greater local levels of cytoprotective estradiol.

REGULATION OF BRAIN AROMATASE

Ontogeny of Brain Aromatase

The conversion of androgens to estrogens by aromatase in target areas of the brain is a key step in the mechanism by which androgens regulate many physiological functions throughout an animal's life. The regulation of aromatase over the lifespan is not completely understood. It has been shown to depend on age, to be specific for the brain region examined, and to involve both transcriptional and posttranscriptional mechanisms.

During development, aromatase plays an essential role in the sexual differentiation and maturation of the brain in many mammalian species (Wallen & Baum, 2002). In short-gestation species (i.e., rodents, rabbits, and ferrets) both enzyme activity and mRNA expression in the preoptic area/hypothalamus peak late in gestation and early neonatal life corresponding to the critical period for sexual differentiation of the brain (George et al., 1978; George & Ojeda, 1982;

MacLusky et al., 1985; Tobet et al., 1985; Harada & Yamada, 1992; Lephart et al., 1992; Yamada et al., 1994; Colciago et al., 2005). Transiently higher levels of aromatase activity (Tobet et al., 1985; Lephart, 1996) and greater serum concentrations of testosterone (Weisz & Ward, 1980; Baum et al., 1990) are present in males than in females during this perinatal interval, suggesting that the male brain is exposed to higher levels of estrogens than the female brain. Neuroanatomical studies demonstrated that aromatase mRNA and protein expression is restricted to a small number of specific hypothalamic and limbic brain nuclei comprising much, but not all, of the same circuitry found in adults (Lauber & Lichtensteiger, 1994; Shinoda et al., 1994; Zhao et al., 2007).

Very little is known about what factors regulate aromatase in the fetal and neonatal brain. In rodents, the activation of cAMP appears to inhibit aromatase expression (Canick et al., 1987; Lephart et al., 1992;) and there are contradictory reports on the effects of androgens (Paden & Roselli, 1987; Lephart et al., 1992; Lauber, et al., 1997). Other studies in rats and mice show that regional aromatase expression (activity, protein, and mRNA) peaks earlier in the diencephalon than in the limbic system or cortex, suggesting an even more complex system of regulation that has not yet been completely explained (Tobet et al., 1985; Yamada et al., 1994; Zhao et al., 2007).

In long-gestation species such as nonhuman primates, sheep, cattle, and guinea pigs, the highest levels of aromatase are observed during the first half of gestation, a time that also overlaps with the period for brain sexual differentiation in these species and when circulating testosterone is elevated in males (Roselli & Resko, 1986; Sholl & Kim, 1990; Connolly et al., 1994; Roselli et al., 2003; Peruffo, et al., 2008; Mura et al., 2010). However, an obvious role for aromatase has not been identified for brain sexual differentiation of long-gestation as opposed to short-gestation species. In nonhuman primates and guinea pigs, testosterone itself and other nonaromatizable androgens are more effective than estradiol for brain differentiation (Resko & Roselli, 1997; Wallen & Baum, 2002). Likewise, aromatase activity has been measured in the hypothalamus and limbic system of the human fetus (Naftolin et al., 1971a, 1971b), but does not seem to be required for male-typical brain differentiation (Rochira et al., 2005). Both androgens and estrogens may play roles in brain sexual differentiation in sheep depending on the neuroendocrine or behavioral endpoint examined (Foster et al,, 2007). Similar to other mammalian species, aromatase activity is most abundant within subregions of the hypothalamus and amygdala in fetal lambs (Roselli et al., 2003) and the expression of mRNA occurs earlier for the hypothalamus-preoptic area than for the limbic system (Roselli & Stormshak, 2011). No sex differences exist in aromatase activity or mRNA expression in sheep fetuses (Roselli et al., 2003; Roselli & Stormshak, 2011). However, a sex difference in the volume of the aromatase-containing ovine sexually dimorphic nucleus of the preoptic area is apparent late in gestation (Roselli et al., 2007), persists into adulthood and is correlated with sex-typical mate preferences (Roselli et al., 2004). In the amygdala and frontal cortex of nonhuman primates, male fetuses exhibit higher levels of enzyme activity than female fetuses (Roselli & Resko, 1986). Testosterone administration during gestation exerts species-specific effects on the regulation of fetal aromatase in long-gestation animals. In guinea pigs, prenatal testosterone induces aromatase activity in the hypothalamus and cortex (Connolly et al., 1994), whereas prenatal castration and testosterone treatment did not alter brain aromatase activity in non-human primate fetuses (Roselli & Resko, 1986). Table 3–3 summarizes what is currently known about the ontogeny of hypothalamic and limbic aromatase expression and its regulation by androgens in mammals.

Aromatase enzyme activity and mRNA expression is also detectable during perinatal development in several brain areas outside of the hypothalamus and amygdala, including the hippocampus, cerebral cortex, midbrain, pons, medulla, olfactory bulb, and striatum (Roselli & Resko, 1986; Lauber & Lichtensteiger, 1994; MacLusky et al., 1994; Raab et al., 1995; Küppers & Beyer, 1998; Ivanova & Beyer, 2000; Roselli et al., 2003; Roselli & Stormshak, 2011). The levels of aromatase activity are considerably lower in these areas than those observed in the fetal hypothalamus and amygdala (MacLusky et al., 1994; Roselli et al., 2003;). In contrast to the developing hypothalamus (see Table 3–3), aromatase expression in the midbrain, hippocampus, and cortex does not peak until the first or second week after birth in rodents and with the possible exception of the hippocampus does not exhibit sex differences (MacLusky et al., 1994; Raab et al., 1995; Ivanova & Beyer, 2000;). Studies in rodents and sheep found that aromatase is expressed in the developing striatum, and in mice expression has been shown to persist into adulthood (Küppers & Beyer, 1998; Roselli & Stormshak, 2011). In the hippocampus and cerebral cortex there is some evidence that estrogen is synthesized de novo from cholesterol instead of being derived from peripheral androgen sources (Amateau et al., 2004). The physiological importance of local estrogen biosynthesis in the development of extrahypothalamic brain areas remains to be established. Estrogen appears to have trophic effects on hippocampal and striatal development (Solum & Handa, 2002; Stroppolo et al., 2004), and recent evidence in aromatase knockout mice shows deficits on hippocampal-dependent tasks (Martin et al., 2003). Thus, brain aromatase could be an important source of estrogen that acts throughout the developing brain to regulate processes important for neuronal survival, differentiation and maturation that may not be directly related to sexual function.

Regulation of Aromatase in the Adult Brain

The earliest information about regulation of aromatase in the adult mammalian brain was gleaned from studies of the hypothalamus and limbic system. The first comprehensive studies were performed using enzymatic assays on rabbits by Naftolin and colleagues who clearly showed that "central aromatization is responsive to sex, gonadal state, steroid treatment, and steroidal enzyme inhibitors" (Naftolin et al., 1975, p. 312). Subsequent studies of enzyme regulation have been performed in several mammalian species, including rats (Roselli & Resko, 1987), mice (Sinchak et al., 1996), hamsters

Table 3–3. Ontogeny of Aromatase in the Hypothalamus and Amygdala of Mammals

Species	Duration of Gestations (days)	Peak Aromatase Activity	Peak Aromatase mRNA Expression	Sex Difference	Regulated	Reference
Mouse	18–21	E17–20	PN0, PN5, PN15	Yes higher in % HT	Yes by androgen	(Harada & Yamada, 1992; Karolczak et al., 1998; Hutchison et al., 1999)
Rat	21–23	E19	E19, PN2	Yes higher in % HT & Amyg	Yes by androgen	(Paden & Roselli, 1987 Yamada et al., 1994; Lauber et al., 1997; Colciago et al., 2005)
Rabbit	30	E19–25	n.d.a.	No	n.d.a.	(George et al., 1978)
Ferret	42	E37	n.d.a.	Yes higher in % HT	No	(Tobet et al., 1985)
Guinea Pig	72	E35–40	n.d.a.	No	Yes by androgen	(Connolly et al., 1994)
Sheep	145	n.d.a.	E65	No	n.d.a.	(Roselli et al., 2003; Roselli & Stormshak, 2011)
Cow	280	n.d.a.	E120	n.d.a.	n.d.a.	(Peruffo et al., 2008)
Rhesus monkey	150–175	E80	n.d.a.	Yes higher in % Amyg	No	(Roselli & Resko, 1986; Sholl & Kim, 1990)

Abbreviations: Amyg, amygdala; E, embryonic; HT, hypothalamus; PN, postnatal; n.d.a., no data available.

Modified from Peruffo, A., Cozzi, B., & Ballarin, C. (2008). Ontogenesis of brain aromatase P450 expression in the bovine hypothalamus. *Brain Res Bull, 75,* 60–65.

(Hutchison et al., 1991), guinea pigs (Connolly et a., 1990), ferrets (Connolly et al., 1990), sheep (Roselli et al., 1998) and monkeys (Resko, et al., 2000). It is now clear that in mammals, with the possible exception of the rabbit, aromatase activity in the adult hypothalamus is induced primarily by androgens acting through the androgen receptor. Treatment of rats with androgen antagonists or studies using a genetically androgen receptor-deficient rat strain (i.e., testicular-feminized rats) confirm this conclusion (Roselli & Resko, 1984; Roselli et al., 1987;). Moreover, in large part aromatase is regulated by androgens at the transcriptional level (Abdelgadir et al., 1994). Extensive colocalization of androgen receptor and aromatase is observed in the mammalian hypothalamus (Veney & Rissman, 2000; Roselli et al., 2001) and a potential androgen-response element is present in the brain specific aromatase promoter (Honda et al., 1994). Estrogens have also been found to regulate hypothalamic aromatase in rats and rabbits (Naftolin et al., 1975; Honda et al., 1994; Zhao et al., 2007) and have been shown to synergize with dihydrotestosterone to induce preoptic aromatase activity (Roselli, 1991b). Interestingly, both estrogen- and androgen-response elements have been identified upstream of the brain aromatase gene promoter in lower vertebrates, such as fish (Gardner et al., 2005).

Male rats have a higher concentration of aromatase in the preoptic area and hypothalamus due to greater stimulation from higher serum levels of testosterone. When rats are gonadectomized as adults, aromatase activity and mRNA concentrations are reduced and the sex difference is lost (Roselli et al., 1985; Roselli et al., 1996;). However, exogenous testosterone induces preoptic aromatase to a greater extent in males than in females (Steimer & Hutchison, 1990; Roselli, 1991a; Beyer et al., 1994; Roselli et al., 1996;) and it has been demonstrated that not only is the induction of aromatase sexually dimorphic within the hypothalamus-preoptic area, it is also sexually differentiated by testosterone and estrogen exposure during perinatal and peripubertal development (Roselli & Klosterman, 1998). In mice, neonatal estrogen exposure is sufficient to masculinize the numbers of aromatase-expressing neurons and expression of territorial behaviors in females (Wu et al., 2009)

Regulation of aromatase expression in the adult brain is region specific. In contrast with aromatase expression in the nuclei of the preoptic area and hypothalamus, gonadal steroids do not regulate aromatase in most other brain areas including parts of the amygdala, hippocampus and cerebral cortex (see Table 3–1). This observation suggests that there are at least two populations of aromatase-positive neurons in the adult brain: a steroid-dependent and steroid-independent population. Region-specific regulation could depend on differential promoter usage as suggested by studies of brain aromatase mRNA structure (Yamada-Mouri et al., 1995; Kato et al., 1997; Honda et al., 1999; Vanselow et al., 2001) supporting evidence that additional factors, including neurotransmitters or neuropeptides (Abe-Dohmae et al., 1996; Canick et al., 1987; Lephart et al., 1992), are involved in regulating brain aromatase depending on age, sex, specific cell type, or population involved. Finally, recent studies demonstrate that human and quail aromatase is rapidly and reversibly inhibited by direct phosphorylation of specific amino acid residues (Cornil, 2009; Charlier et al., 2011). Fast changes in the local concentration of estrogens within synaptic terminals could in turn rapidly modulate neuronal physiology and behavior (Balthazart & Ball, 2006).

CONCLUSIONS

Aromatase plays an important role in the brain to mediate testosterone's effects on sexual differentiation, neuronal maturation, gonadotropin control, and sexual behavior. As such, the enzyme is enriched in areas of the brain such as the hypothalamus and amygdala that regulate these important reproductive functions. Changes in enzyme activity often reflect changes in concentrations of enzyme protein that is regulated in large part transcriptionally. In adults, androgens and/or estrogens appear to be the predominant factors regulating transcript, but in the fetus neurotransmitters and neuropeptides may be involved. More rapid control of activity has been demonstrated most convincingly in quail brain, although rapid inhibition of aromatase activity caused by phosphorylation of enzyme protein has been found to affect human aromatase as well (Charlier et al., 2011). Thus the regulation of aromatase activity, like the functions served by locally synthesized estrogens in the brain, must be evaluated temporally and spatially (i.e., in relation to signaling occurring through transcription or mediated at the level of the membrane or synapse). Until relatively recently, aromatase's role was thought to be exclusively that of translating peripheral hormone signals to modulate neural function. It now appears clear that specific brain areas under certain conditions are capable of synthesizing estradiol de novo from cholesterol and that the estrogens formed play important roles in synaptic plasticity and neuroprotection. Under physiological conditions, aromatase is predominantly expressed in mammalian neurons and has also been identified in radial glia, but under pathological conditions enzyme can be expressed by astroglia to play a role in neuroprotection. Together, these observations suggest that brain aromatase has a broader role to play in brain physiology and pathophysiology than has been recognized before.

ACKNOWLEDGMENTS

This work was supported by NIH R01 OD011047 (CER).

REFERENCES

Abdelgadir, S. E., Resko, J. A., Ojeda, S. R., Lephart, E. D., McPhaul, M. J., & Roselli, C. E. (1994). Androgens regulate aromatase cytochrome P450 messenger ribonucleic acid in rat brain. *Endocrinology, 135*, 395–401.

Abdelgadir, S. E., Roselli, C. E., Choate, J. V. A., & Resko, J. A. (1997). Distribution of aromatase cytochrome P450 messenger ribonucleic acid in adult rhesus monkey brains. *Biol Reprod, 57*, 772–777.

Abe-Dohmae, S., Takagi, Y., & Harada, N. (1996). Neurotransmitter-mediated regulation of brain aromatase: protein kinase C- and G-dependent induction. *J Neurochem, 67,* 2087–2095.

Amateau, S. K., Alt, J. J., Stamps, S. L., & McCarthy, M. M. (2004). Brain estradiol content in newborn rats: sex differences, regional heterogeneity and possible *de novo* synthesis by the female telencephalon. *Endocrinology, 145,* 2906–2917.

Azcoitia, I., Sierra, A., & E Garcia-Segura, L. M. (2003). Aromatase expression by reactive astroglia is neuroprotective. *Ann N Y Acad Sci, 1007,* 298–305.

Balthazart, J. & Ball, G. F. (2006). Is brain estradiol a hormone or a neurotransmitter? *Trends Neurosci, 29,* 241–249.

Balthazart, J., Foidart, A., Surlemont, C., & Harada, N. (1991). Distribution of aromatase-immunoreactive cells in the mouse forebrain. *Cell Tissue Res, 263,* 71–79.

Baum, M. J., Carroll, R. S., & Tobet, S. A. (1990). Steroidal control of behavioural, neuroendocrine and brain sexual differentiation: studies in a carnivore, the ferret. *J Neuroendocrinol, 2,* 401–418.

Beyer, C., Green, S. J., Barker, P. J., Huskisson, N. S., & Hutchison, J. B. (1994). Aromatase-immunoreactivity is localised specifically in neurones in the developing mouse hypothalamus and cortex. *Brain Res, 638,* 203–210.

Beyer, C., Green, S. J., & Hutchison, J. B. (1994). Androgens influence sexual differentiation of embryonic mouse hypothalamic aromatase neurons in vitro. *Endocrinology, 135,* 1220–1226.

Bonsall, R. W., Rees, H. D., & Michael, R. P. (1989). Identification of radioactivity in cell nuclei from brain, pituitary gland and genital tract of male rhesus monkeys after the administration of [^{3}H]testosterone. *J Steroid Biochem, 32,* 599–608.

Callard, G. V., Mak, P., & Solomon, D. J. (1986). Effects of short days on aromatization and accumulation of nuclear estrogen receptors in the hamster brain. *Biol Reprod, 35,* 282–291.

Canick, J. A., Tobet, S. A., Baum, M. J., Vaccaro, D. E., Ryan, K. J., Leeman, S. E., et al. (1987). Studies on the role of catecholamines in the regulation of the developmental pattern of hypothalamic aromatase. *Steroids, 50,* 509–521.

Carswell, H. V. O., Dominiczak, A. F., Garcia-Segura, L. M., Harada, N., Hutchison, J. B., & Macrae, I. M. (2005). Brain aromatase expression after experimental stroke: topography and time course. *J Steroid Biochem Mol Biol, 96,* 89–96.

Charlier, T. D., Harada, N., Balthazart, J., & Cornil, C. A. (2011). Human and quail aromatase activity Is rapidly and reversibly inhibited by phosphorylating conditions. *Endocrinology, 152,* 4199–4210.

Choate, J. V. A., & Resko, J. A. (1996). Paradoxical effect of an aromatase inhibitor, CGS 20267, on aromatase activity in guinea pig brain. *J Steroid Biochem Mol Biol, 58,* 411–415.

Chow, J. D. Y., Simpson, E. R., & Boon, W. C. (2009). Alternative 5'-untranslated first exons of the mouse Cyp19A1 (aromatase) gene. *J Steroid Biochem Mol Biol, 115,* 115–125.

Colciago, A., Celotti, F., Pravettoni, A., Mornati, O., Martini, L., & Negri-Cesi, P. (2005). Dimorphic expression of testosterone metabolizing enzymes in the hypothalamic area of developing rats. *Dev Brain Res, 155,* 107–116.

Connolly, P. B., Roselli, C. E., & Resko, J. A. (1990). Aromatase activity in adult guinea pig brain is androgen dependent. *Biol Reprod, 43,* 698–703.

Connolly, P. B., Roselli, C. E., & Resko, J. A. (1991). Androgen-dependent and -independent aromatase activity coexists with androgen receptors in male guinea-pig brain. *J Neuroendocrinol, 3,* 679–684.

Connolly, P. B., Roselli, C. E., & Resko, J. A. (1994). Aromatase activity in developing guinea pig brain: ontogeny and effects of exogenous androgens. *Biol Reprod, 50,* 436–441.

Cornil, C. A. (2009). Rapid regulation of brain oestrogen synthesis: the behavioural roles of oestrogens and their fates. *J Neuroendocrinol, 21,* 217–226.

Dellovade, T. L., Harada, N., & Rissman, E. F. (1994). Distribution and steroid dependence of aromatase enzyme immunoreactivity in limbic nuclei of the female musk shrew brain. *Brain Res, 634,* 141–149.

Ellinwood, W. E., Hess, D. L., Roselli, C. E., Spies, H. G., & Resko, J. A. (1984). Inhibition of aromatization stimulates luteinizing hormone and testosterone secretion in adult male rhesus monkeys. *J Clin Endocrinol Metab, 59,* 1088–1096.

Feder, H. H., Naftolin, F., & Ryan, K. J. (1974). Male and female sexual responses in male rats given estradiol benzoate and 5α-androstan-17b-ol-3-one propionate. *Endocrinology, 94,* 136–141.

Fester, L., Prange-Kiel, J., Jarry, H., & Rune, G. (2011). Estrogen synthesis in the hippocampus. *Cell Tissue Res, 345,* 285–294.

Foidart, A., Harada, N., & Balthazart, J. (1994). Effects of steroidal and non steroidal aromatase inhibitors on sexual behavior and aromatase-immunoreactive cells and fibers in the quail brain. *Brain Res, 657,* 105–123.

Foidart, A., Harada, N., & Balthazart, J. (1995). Aromatase-immunoreactive cells are present in mouse brain areas that are known to express high levels of aromatase activity. *Cell Tissue Res, 280,* 561–574.

Foster, D. L., Jackson, L. M., & Padmanabhan, V. (2007). Novel concepts about normal sexual differentiation of reproductive neuroendocrine function and the developmental origins of female reproductive dysfunction: the sheep model. *Soc Reprod Fertil Suppl, 64,* 83–107.

Garcia-Segura, L. M. (2008). Aromatase in the brain: not just for reproduction anymore. *J Neuroendocrinol, 20,* 705–712.

Garcia-Segura, L. M., Veiga, S., Sierra, A., Melcangi, R. C., & Azcoitia, I. (2003). Aromatase: a neuroprotective enzyme. *Prog Neurobiol, 71,* 31–41.

Garcia-Segura, L. M., Wozniak, A., Azcoitia, I., Rodriguez, J. R., Hutchison, R. E., & Hutchison, J. B. (1999). Aromatase expression by astrocytes after brain injury: Implications for local estrogen formation in brain repair. *Neurosci, 89,* 567–578.

Gardner, L., Anderson, T., Place, A. R., Dixon, B., & Elizur, A. (2005). Sex change strategy and the aromatase genes. *J Steroid Biochem Mol Biol, 94,* 395–404.

George, F. W., & Ojeda, S. R. (1982). Changes in aromatase activity in the rat brain during embryonic, neonatal, and infantile development. *Endocrinology,* i 522–529.

George, F. W., Tobleman, W. T., Milewich, L., & Wilson, J. D. (1978). Aromatase activity in the developing rabbit brain. *Endocrinology, 102*(1), 86–91.

Golovine, K., Schwerin, M., & Vanselow, J. (2003). Three different promoters control expression of the aromatase cytochrome p450 gene (cyp19) in mouse gonads and brain. *Biol Reprod, 68,* 978–984.

Gorski, R. A. (1963). Modification of ovulatory mechanisms by postnatal administration of estrogen to the rat. *Am J Physiol, 205,* 842–844.

Harada, N., & Yamada, K. (1992). Ontogeny of aromatase messenger ribonucleic acid in mouse brain: fluorometrical quantitation by polymerase chain reaction. *Endocrinology, 131,* 2306–2312.

Heimer, L., & Alheid, G. (1991). Piecing together the puzzle of basal forebrain anatomy. In: Napier, T. C., Kalivas, P. W., & Hanin, I. (eds.), *The Basal Foebrain: Anatomy to Function*. New York: Plenum Press, pp. 1–42.

Hojo, Y., Hattori, T. A., Enami, T., Furukawa, A., Suzuki, K., Ishii, H. T., et al. (2004). Adult male rat hippocampus synthesizes estradiol from pregnenolone by cytochromes P45017α and P450 aromatase localized in neurons. *Proc Natl Acad. Sci U S A, 101,* 865–870.

Hojo, Y., Higo, S., Ishii, H., Ooishi, Y., Mukai, H., Murakami, G., et al. (2009). Comparison between hippocampus-synthesized and circulation-derived sex steroids in the hippocampus. *Endocrinology, 150,* 5106–5112.

Hojo, Y., Murakami, G., Mukai, H., Higo, S., Hatanaka, Y., Ogiue-Ikeda, M., et al. (2008). Estrogen synthesis in the brain—role in synaptic plasticity and memory. *Mol Cell Endocrinol, 290,* 31–43.

Honda, S., Harada, N., Abe-Dohmae, S., & Takagi, Y. (1999). Identification of *cis*-acting elements in the proximal promoter region for brain-specific exon 1 of the mouse aromatase gene. *Mol Brain Res, 66,* 122–132.

Honda, S., Harada, N., & Takagi, Y. (1994). Novel exon 1 of the aromatase gene specific for aromatase transcripts in human brain. *Biochem Biophys Res Commun, 198,* 1153–1160.

Honda, S., Harada, N., & Takagi, Y. (1996). The alternative exons 1 of the mouse aromatase cytochrome P-450 gene. *Biochim Biophys Acta, 1305,* 145–150.

Horvath, T. L., & Wikler, K. C. (1999). Aromatase in developing sensory systems of the rat brain. *J Neuroendocrinol, 11,* 77–84.

Hutchison, J. B., Wozniak, A., Beyer, C., Karolczak, M., & Hutchison, R. E. (1999). Steroid metabolising enzymes in the determination of brain gender. *J Steroid Biochem Mol Biol, 69,* 85–96.

Hutchison, R. E., Hutchison, J. B., Steimer, T., Steel, E., Powers, J. B., Walker, A. P., et al. (1991). Brain aromatization of testosterone in the male Syrian hamster: effects of androgen and photoperiod. *Neuroendocrinology, 53,* 194–203.

Ivanova, T., & Beyer, C. (2000). Ontogenetic expression and sex differences of aromatase and estrogen receptor-α/β mRNA in the mouse hippocampus. *Cell Tissue Res, 300,* 231–237.

Jakab, R. L., Harada, N., & Naftolin, F. (1994). Aromatase- (estrogen synthetase) immunoreactive neurons in the rat septal area. A light and electron microscopic study. *Brain Res, 664,* 85–93.

Jakab, R. L., Horvath, T. L., Leranth, C., Harada, N., & Naftolin, F. (1993). Aromatase immunoreactivity in the rat brain: gonadectomy-sensitive hypothalamic neurons and an unresponsive "limbic ring" of the lateral septum-bed nucleus-amygdala complex. *J Steroid Biochem Mol Biol, 44,* 481–498.

Kao, Y.-C., Higashiyama, T., Yarborough, C., Osawa, Y., & Chen, S. (1999). Functional characterization of 102-amino acid-deleted form of human aromatase (D102-aromatase). *Steroids, 64,* 422–429.

Karolczak, M., Küppers, E., & Beyer, C. (1998) Developmental expression and reguation of aromatase- and 5alpha-reductase type I mRNA in the male and female mouse hypothalamus. *J Neuroendocrinol, 10,* 267–274.

Kato, J., Yamada-Mouri, N., & Hirata, S. (1997). Structure of aromatase mRNA in the rat brain. *J Steroid Biochem Mol Biol, 61,* 381–385.

Küppers, E., & Beyer, C. (1998). Expression of aromatase in the embryonic and postnatal mouse striatum. *Mol Brain Res, 63,* 184–188.

Lauber, M. E., & Lichtensteiger, W. (1994). Pre- and postnatal ontogeny of aromatase cytochrome P450 messenger ribonucleic acid expression in the male rat brain studied by *in situ* hybridization. *Endocrinology, 135,* 1661–1668.

Lauber, M. E., Sarasin, A., & Lichtensteiger, W. (1997). Sex differences and androgen-dependent regulation of aromatase (CYP19) mRNA expression in the developing and adult rat brain. *J Steroid Biochem Mol Biol, 61,* 359–364.

Lephart, E. D. (1996). A review of brain aromatase cytochrome P450. *Brain Res Rev, 22,* 1–26.

Lephart, E. D., Herbst, M. A., & McPhaul, M. J. (1995). Characterization of aromatase cytochrome P-450 mRNA in rat perinatal brain, ovary, and a Leydig tumor cell line: evidence for the existence of brain specific aromatase transcripts. *Endocrine, 3,* 25–31.

Lephart, E. D., Simpson, E. R., McPhaul, M. J., Kilgore, M. W., Wilson, J. D., & Ojeda, S. R. (1992). Brain aromatase cytochrome P-450 messenger RNA levels and enzyme activity during prenatal and perinatal development in the rat. *Mol Brain Res, 16,* 187–192.

Lephart, E. D., Simpson, E. R., & Ojeda, S. R. (1992). Effects of cyclic AMP and androgens on in vitro brain aromatase enzyme activity during prenatal development in the rat. *J Neuroendocrinol, 4,* 29–35.

Lieberburg, I., & McEwen, B. S. (1977). Brain cell nuclear retention of testosterone metabolites 5α-dihydrotestosterone and estradiol-17β, in adult rats. *Endocrinology, 100,* 588–597.

Liu, M., Hurn, P. D., Roselli, C. E, & Alkayed, N. J. (2007). Role of P450 aromatase in sex-specific astrocytic cell death. *J Cereb Blood Flow Metab, 27,* 135–141.

Liu, N. J., Chakrabarti, S., Schnell, S., Wessendorf, M., & Gintzler, A. R. (2011). Spinal synthesis of estrogen and concomitant signaling by membrane estrogen receptors regulate spinal κ- and μ-opioid receptor heterodimerization and female-specific spinal morphine antinociception. *J Neurosci, 31,* 11836–11845.

MacLusky, N. J., Naftolin, F., & Goldman-Rakic, P. S. (1986). Estrogen formation and binding in the cerebral cortex of the developing rhesus monkey. *Proc Natl Acad Sci U S A, 83,* 513–516.

MacLusky, N. J., Philip, A., Hurlbut, C., & Naftolin, F. (1985). Estrogen formation in the developing rat brain: sex differences in aromatase activiy during early post-natal life. *Psychoneuroendocrinology, 10,* 355–361.

MacLusky, N. J., Walters, M. J., Clark, A. S., & Toran-Allerand, C. D. (1994). Aromatase in the cerebral cortex, hippocampus, and mid-brain: ontogeny and developmental implications. *Mol Cell Neurosci, 5,* 691–698.

Martin, S., Jones, M., Simpson, E., & van den Buuse, M. (2003). Impaired spatial reference memory in aromatase-deficient (ArKO) mice. *Neuroreport, 14,* 1979–1982.

Martinez-Cerdeno, V., Noctor, S. C., & Kriegstein, A. R. (2006). Estradiol stimulates progenitor cell division in the ventricular and subventricular zones of the embryonic neocortex. *Eur J Neurosci, 24,* 3475–3488.

McCann, S. M., & Ramirez, V. D. (1964). The neuroendocrine regulation of hypophyseal luteinizing hormone secretion. *Rec Prog Horm Res, 30,* 131–181.

McCullough, L. D., Blizzard, K., Simpson, E. R., Oz, O. K., & Hurn, P. D. (2003). Aromatase cytochrome p450 and extragonadal estrogen play a role in ischemic neuroprotection. *J Neurosci, 23,* 8701–8705.

Mouriec, K., Pellegrini, E., Anglade, I., Menuet, A., Adrio, F., Thieulant, M. L., et al. (2008). Synthesis of estrogens in progenitor cells of adult fish brain: evolutive novelty or exaggeration of a more general mechanism implicating estrogens in neurogenesis? *Brain Res Bull, 75,* 274–280.

Mura, A., Gadau, S., Lepore, G., Balzano, F., Zedda, M., Mura, E., et al. (2010). Expression and distribution of P450-aromatase in the ovine hypothalamus at different stages of fetal development. *Neuroendocrinol Lett, 31,* 690–699.

Naftolin, F., Horvath, T. L., Jakab, R. L., Leranth, C., Harada, N., & Balthazart, J. (1996). Aromatase immunoreactivity in axon terminals of the vertebrate brain—an immunocytochemical study on quail, rat, monkey and human tissues. *Neuroendocrinology, 63,* 149–155.

Naftolin, F., Ryan, K. J., Davies, I. J., Reddy, V. V., Flores, F., Petro, Z., et al. (1975). The formation of estrogens by central neuroendocrine tissues. *Rec Prog Horm Res, 31,* 295–319.

Naftolin, F., Ryan, K. J., & Petro, Z. (1971a). Aromatization of androstenedione by limbic system tissue from human foetuses. *J Endocrinol, 51,* 795–796.

Naftolin, F., Ryan, K. J., & Petro, Z. (1971b). Aromatization of androstenedione by the diencephalon. *J Clin Endrocinol Metab, 31,* 368–370.

Negri-Cesi, P., Celotti, F., & Martini, L. (1989). Androgen metabolism in the male hamster-2. Aromatization of androstenedione in the hypothalamus and in the cerebral cortex; kinetic parameters and effect of exposure to different photoperiods. *J Steroid Biochem, 32,* 65–70.

Negri Cesi, P., Cosimo Melcangi, R., Celotti, F., & Martini, L. (1992). Aromatase activity in cultured brain cells: difference between neurons and glia. *Brain Res, 589,* 327–332.

Newman, S. W. (1999). The medial extended amygdala in male reproductive behavior. A node in the mammalian social behavior network. *Ann N Y Acad Sci, 877,* 242–257.

Paden, C. M., & Roselli, C. E. (1987). Modulation of aromatase activity by testosterone in transplants of fetal rat hypothalamus-preoptic area. *Dev Brain Res, 33,* 127–133.

Peruffo, A., Cozzi, B., & Ballarin, C. (2008). Ontogenesis of brain aromatase P450 expression in the bovine hypothalamus. *Brain Res Bull, 75,* 60–65.

Raab, H., Beyer, C., Wozniak, A., Hutchison, J. B., Pilgrim, C., & Reisert, I. (1995). Ontogeny of aromatase messenger ribonucleic acid and aromatase activity in the rat midbrain. *Mol Brain Res, 34,* 333–336.

Reddy, V. V. R., Naftolin, F., & Ryan, K. J. (1973). Aromatization in the central nervous system of rabbits: effects of castration and hormone treatment. *Endocrinology, 92*(2), 589–594.

Reddy, V. V. R., Naftolin, F., & Ryan, K. J. (1974). Conversion of androstenedione to estrogen by neural tissues from fetal and neonatal rats. *Endocrinology, 94,* 117–121.

Resko, J. A., Pereyra-Martinez, A. C., Stadelman, H. L., & Roselli, C. E. (2000). Region-specific regulation of cytochrome P450 aromatase messenger ribonucleic acid by androgen in brains of male rhesus monkeys. *Biol Reprod, 62,* 1818–1822.

Resko, J. A., & Roselli, C. E. (1997). Prenatal hormones organize sex differences of the neuroendocrine reproductive system: observations on guinea pigs and nonhuman primates. *Cell Mol Neurobiol, 17,* 627–648.

Rissman, E. F., Harada, N., & Roselli, C. E. (1996). Effect of vorozole, an aromatase enzyme inhibitor, on sexual behavior, aromatase activity and neural immunoreactivity. *J Neuroendocrinol, 8,* 199–210.

Rochira, V., Granata, A. R. M., Madeo, B., Zirilli, L., Rossi, G., & Carani, C. (2005). Estrogens in males: what have we learned in the last 10 years? *Asian J Androl, 7,* 3–20.

Roselli, C. E. (1991a). Sex differences in androgen receptors and aromatase activity in microdissected regions of the rat brain. *Endocrinology, 128,* 1310–1316.

Roselli, C. E. (1991b). Synergistic induction of aromatase activity in the rat brain by estradiol and 5α-dihydrotestosterone. *Neuroendocrinology, 53,* 79–84.

Roselli, C. E. (2007). Brain aromatase: roles in reproduction and neuroprotection. *J Steroid Biochem Mol Biol, 106,* 143–150.

Roselli, C. E., Abdelgadir, S. E., Jorgensen, E., & Resko, J. A. (1996). Sex differences in androgen regulated cytochrome P450 aromatase mRNA in the rat brain. *Endocrine, 5,* 59–65.

Roselli, C. E., Abdelgadir, S. E., Ronnekleiv, O. K., & Klosterman, S. A. (1998). Anatomic distribution and regulation of aromatase gene expression in the rat brain. *Biol Reprod, 58,* 79–87.

Roselli, C. E., Ellinwood, W. E., & Resko, J. A. (1984). Regulation of brain aromatase activity in rats. *Endocrinology, 114,* 192–200.

Roselli, C. E., Horton, L. E. & Resko, J. A. (1985) Distribution and regulation of aromatase activity in the rat hypothalamus and limbic system. *Endocrinology, 117,* 2471–2477.

Roselli, C. E., & Klosterman, S. A. (1998). Sexual differentiation of aromatase activity in the rat brain: effects of perinatal steroid exposure. *Endocrinology, 139,* 3193–3201.

Roselli, C. E., Klosterman, S. A., & Fasasi, T. A. (1996). Sex differences in androgen responsiveness in the rat brain: regional differences in the induction of aromatase activity. *Neuroendocrinology, 64,* 139–145.

Roselli, C. E., Klosterman, S. A., & Resko, J. A. (2001). Anatomic relationships between aromatase and androgen receptor mRNA expression in the hypothalamus and amygdala of adult male cynomolgus monkeys. *J Comp Neurol, 439,* 208–223.

Roselli, C. E., Larkin, K., Resko, J. A., Stellflug, J. N., & Stormshak, F. (2004). The volume of a sexually dimorphic nucleus in the ovine medial preoptic area/ anterior hypothalamus varies with sexual partner preference. *Endocrinology, 145,* 478–483.

Roselli, C. E., & Resko, J. A. (1984). Androgens regulate brain aromatase activity in adult male rats through a receptor mechanism. *Endocrinology, 114,* 2183–2189.

Roselli, C. E., & Resko, J. A. (1986). Effects of gonadectomy and androgen treatment on aromatase activity in the fetal monkey brain. *Biol Reprod, 35,* 106–112.

Roselli, C. E., & Resko, J. A. (1987). The distribution and regulation of aromatase acitivity in the central nervous system. *Steroids, 50,* 496–508.

Roselli, C. E., Resko, J. A., & Stormshak, F. (2003). Estrogen synthesis in fetal sheep brain: effect of maternal treatment with an aromatase inhibitor. *Biol Reprod, 68,* 370–374.

Roselli, C. E., Salisbury, R. L., & Resko, J. A. (1987). Genetic evidence for androgen-dependent and independent control of aromatase activity in the rat brain. *Endocrinology, 121,* 2205–2210.

Roselli, C. E., Stadelman, H., Horton, L. E., & Resko, J. A. (1987). Regulation of androgen metabolism and luteinizing hormone-releasing hormone content in discrete hypothalamic and limbic areas of male rhesus macaques. *Endocrinology, 120,* 97–106.

Roselli, C. E., Stadelman, H., Reeve, R., Bishop, C. V., & Stormshak, F. (2007). The ovine sexually dimorphic nucleus of the medial preoptic area is organized prenatally by testosterone. *Endocrinology, 148,* 4450–4457.

Roselli, C. E., & Stormshak, F. (2012). Ontogeny of cytochrome P450 aromatase mRNA expression in the developing sheep brain. *J Neuroendocrinol, 24,* 443–452.

Roselli, C. E., Stormshak, F., & Resko, J. A. (1998). Distribution and regulation of aromatase activity in the ram hypothalamus and amygdala. *Brain Res, 811,* 105–110.

Roselli, C. E., Stormshak, F., & Resko, J. A. (2000). Distribution of aromatase mRNA in the ram hypothalamus: an *In Situ* hybridization study. *J Neuroendocrinol, 12,* 656–664.

Saldanha, C. J., Remage-Healey, L., & Schlinger, B. A. (2011). Synaptocrine signaling: steroid synthesis and action at the synapse. *Endocrine Rev, 32,* 532–549.

Sanghera, M. K., Simpson, E. R., McPhaul, M. J., Kozlowski, G., Conley, A. J., & Lephart, E. D. (1991). Immunocytochemical distribution of aromatase cytochrome P450 in the rat brain using peptide-generated polyclonal antibodies. *Endocrinology, 129,* 2834–2844.

Selmanoff, M. K., Brodkin, L. D., Weiner, R. I., & Siiteri, P. K. (1977). Aromatization and 5α-reduction of androgens in discrete hypothalamic and limbic regions of the male and female rat. *Endocrinology, 101,* 841–848.

Shinoda, K., Nagano, M., & Osawa, Y. (1994). Neuronal aromatase expression in preoptic, strial, and amygdaloid regions during late prenatal and early postnatal development in the rat. *J Comp Neurol, 343,* 113–129.

Shinoda, K., Yagi, H., Fijita, H., Osawa, Y., & Shiotani, Y. (1989). Screening of aromatase-containing neurons in rat forebrain: an immunohistochemical study with antibody against human placental antigen X-P2 (hPAX-P2). *J Comp Neurol, 290,* 502–515.

Sholl, S. A., & Kim, K. L. (1990) Aromatase, 5-alpha-reductase, and androgen receptor levels in the fetal monkey brain during early development. *Neuroendocrinology, 52,* 94–98.

Simerly, R. B., Chang, C., Muramatsu, M., & Swanson, L. W. (1990). The distribution of androgen and estrogen receptor mRNA-containing cells in the rat brain: an in situ hybridization study. *J Comp Neurol, 294,* 76–95.

Simpson, E. R., Clyne, C., Rubin, G., Boon, W. C., Robertson, K., Britt, K., et al. (2002). Aromatase—a brief overview. *Annu Rev Physiol, 64,* 93–127.

Sinchak, K., Roselli, C. E., & Clemens, L. G. (1996). Levels of serum steroids, aromatase activity, and estrogen receptors in preoptic area, hypothalamus, and amygdala of B6D2F1 male house mice that differ in the display of copulatory behavior after castration. *Behav Neurosci, 110,* 593–602.

Solum, D. T., & Handa, R. J. (2002). Estrogen regulates the development of brain-derived neurotrophic factor mRNA and protein in the rat hippocampus. *J Neurosci, 22,* 2650–2659.

Steimer, T., & Hutchison, J. B. (1990). Is androgen-dependent aromatase activity sexually differentiated in the rat and dove preoptic area? *J Neurobiol, 21,* 787–795.

Stroppolo, A., Tian, C., Guinea, B., Olm, V., Sheffield, R., Sommer, J., et al. (2004). 17β-Estradiol promotes striatal medium size spiny neuronal maturation in vitro. *Neuroendocrinology, 79,* 259–267.

Takahashi, K., Bergstrom, M., Frandberg, P., Vesstrom, E. L., Watanabe, Y., & Langstrom, B. (2006). Imaging of aromatase distribution in rat and rhesus monkey brains with [^{11}C]vorozole. *Nuc Med Biol, 33,* 599–605.

Takahashi, K., Hallberg, M., Magnusson, K., Nyberg, F., Watanabe, Y., Langstrom, B., et al. (2007). Increase in [^{11}C]vorozole binding to aromatase in the hypothalamus in rats treated with anabolic androgenic steroids. *Neuroreport, 18,* 171–174.

Takahashi, K., Kayo, O., Hisashi, D., Hiroko, N., Gen, Y., Takamitsu, H., et al. (2011). Increase in hypothalamic aromatase in macaque monkeys treated with anabolic-androgenic steroids: PET study with [^{11}C]vorozole. *Neuroreport, 22,* 326–330.

Takahashi, K., Yasuhisa, T., Yasuyoshi, W., Bengt, L., & Mats, B. (2008). Alteration in [^{11}C]vorozole binding to aromatase in neuronal cells of rat brain induced by anabolic androgenic steroids and flutamide. *Neuroreport, 19,* 431–435.

Tobet, S. A., Baum, M. J., Tang, H. B., Shim, J. H., & Canick, J. A. (1985). Aromatase activity in the perinatal rat forebrain: effects of sex, age and intrauterine position. *Dev Brain Res, 23,* 171–178.

Tobet, S. A., Shim, J. H., Osiecki, S. T., Baum, M. J., & Canick, J. A. (1985). Androgen aromatization and 5alpha-reduction in ferret brain during perinatal development: effects of sex and testosterone manipulation. *Endocrinology, 116,* 1869–1877.

Vanselow, J., Furbass, R., Zsolnai, A., Kalbe, C., Said, H. M., & Schwerin, M. (2001). Expression of the aromatase cytochrome P450 encoding gene in cattle and sheep. *J Steroid Biochem Mol Biol, 79,* 279–288.

Veney, S. L., & Rissman, E. F. (2000). Immunolocalization of androgen receptors and aromatase enzyme in the adult musk shrew brain. *Neuroendocrinology, 72,* 29–36.

Veney, S. L., Rissman, E. F., & Freeman, L. M. (2000). Perinatal organization of a sexually dimorphic aromatase enzyme- containing immunoreactive nucleus. *Neuroreport, 11,* 3409–3412.

Wagner, C. K., & Morrell, J. I. (1996). Distribution and steroid hormone regulation of aromatase mRNA expression in the forebrain of adult male and female rats: a cellular-level analysis using in situ hybridization. *J Comp Neurol, 370,* 71–84.

Wallen, K., & Baum, M. J. (2002). Masculinization and defeminization in altricial and precocial mammals: comparative aspects of steroid hormone action. In: Pfaff, D. W., Arnold, A. P., Etgen, A. M., Fahrbach, S. E., & Rubin, R. T. (eds.), *Hormones, Brain and Behavior.* San Diego: Elsevier Science, pp. 385–423.

Weaver, C. E., & Baum, M. J. (1991). Differential regulation of brain aromatase by androgen in adult and fetal ferrets. *Endocrinology, 128,* 1247–1254.

Weisz, J. (1982). In vitro assays of aromatase and their role in studies of estrogen formation in target tissues. *Cancer Res, 42,* 3295s–3298s.

Weisz, J., & Gibbs, C. (1974). Conversion of testosterone and androstenedione to estrogens in vitro by the brain of female rats. *Endocrinology, 94,* 616–620.

Weisz, J., & Ward, I. L. (1980). Plasma testosterone and progesterone titers of pregnant rats, their male and female fetuses, and neonatal offspring. *Endocrinology, 106,* 306–316.

Wu, M. V., Manoli, D. S., Fraser, E. J., Coats, J. K., Tollkuhn, J., Honda, S. I., et al. (2009). Estrogen masculinizes neural pathways and sex-specific behaviors. *Cell, 139,* 61–72.

Yague, J. G., Munoz, A., Monasterio-Schrader, P., DeFelipe, J., Garcia-Segura, L. M., & Azcoitia, I. (2006). Aromatase expression in the human temporal cortex. *Neuroscience, 138,* 389–401.

Yague, J. G., Wang, A. C. J., Janssen, W. G. M., Hof, P. R., Garcia-Segura, L. M., Azcoitia, I., et al. (2008). Aromatase distribution in the monkey temporal neocortex and hippocampus. *Brain Res, 1209,* 115–127.

Yamada, N. M., Hirata, S., & Kato, J. (1994). Distribution and postnatal changes of aromatase mRNA in the female rat brain. *J Steroid Biochem Mol Biol, 48,* 529–533.

Yamada-Mouri, N., Hirata, S., Hayashi, M., & Kato, J. (1995). Analysis of the expression and the first exon of aromatase mRNA in monkey brain. *J Steroid Biochem Mol Biol, 55,* 17–23.

Yamada-Mouri, N., Hirata, S., Kato, J., & Hoshi, K. (1997). Expression and distribution of cortical type aromatase mRNA variant in the adult rat brain. *J Steroid Biochem Mol Biol, 60,* 325–329.

Zhao, C., Fujinaga, R., Tanaka, M., Yanai, A., Nakahama, K., & Shinoda, K. (2007). Region-specific expression and sex-steroidal regulation on aromatase and its mRNA in the male rat brain: immunohistochemical and in situ hybridization analyses. *J Comp Neurol, 500,* 557–573.

Zwain, I. H., Yen, S. S., & Cheng, C. Y. (1997). Astrocytes cultured in vitro produce estradiol-17beta and express aromatase cytochrome P-450 (P-450 AROM) mRNA. *Biochim Biophys Acta, 1334,* 338–348.

Chapter 4

Multiple Forms of Aromatase in Suiformes: Tissue Distribution, Regulation, and Functional Significance

Alan Conley, Veronica N. Orr, Brian C. Trainor, Trish Berger and Austin Hughes

INTRODUCTION

Estrogen synthesis is required for normal vertebrate reproductive development, maturation, function, and fertility, and its synthesis is totally and most immediately dependent on the enzyme, aromatase cytochrome P450 (P450arom). The P450arom enzyme is encoded by CYP19, which is a single copy gene in most mammals with one known taxonomic exception, the pig and related species in the Suiforme suborder (Gaucher et al., 2004). This unusual duplication and reduplication, perhaps unique among mammals, has provided an opportunity for gene evolution and adaptation of the function of P450arom, relating to both how it is expressed and how catalytic function itself might have changed. The involvement of P450arom in the control of reproductive function at every level of the physiological axis to an unparalleled degree among the steroidogenic enzymes suggests that any such change cannot depress fertility significantly if it is to survive selection, and may have even promoted fecundity. This chapter will attempt to summarize available knowledge of the porcine aromatase gene cluster, and how tissue-specifically expressed isozymes may have shaped the life history of this species.

In suiformes, duplication of CYP19 in an ancestor, and reduplication in domestic pigs at least, has given rise to two or three functional paralogues of P450arom (Conley et al., 2008). In fact, though some teleost fishes have two, the domestic pig, *Sus scrofa domesticus*, is currently the only vertebrate known to carry three CYP19 genes (Fig. 4–1) and all encode functional enzymes (protein paralogues). A highly conserved but ancient gene, CYP19 is among the super-family of cytochrome P450 enzymes (Conley et al., 2008). The substrate binding or active site is the smallest known among the P450s that have been crystallized to date (Ghosh et al., 2009), consistent with a conserved and relatively restricted number of potential substrates (Thompson & Siiteri, 1973; Simpson et al., 1994). The metabolism of androgens to estrogens involves three sequential oxygenation steps, the last culminating in aromatization and estrogen formation. The 19-hydroxy and 19-oxo-intermediates that arise from the first two oxygenations are therefore also substrates of P450arom. These intermediates can be released in variable amounts depending on the species

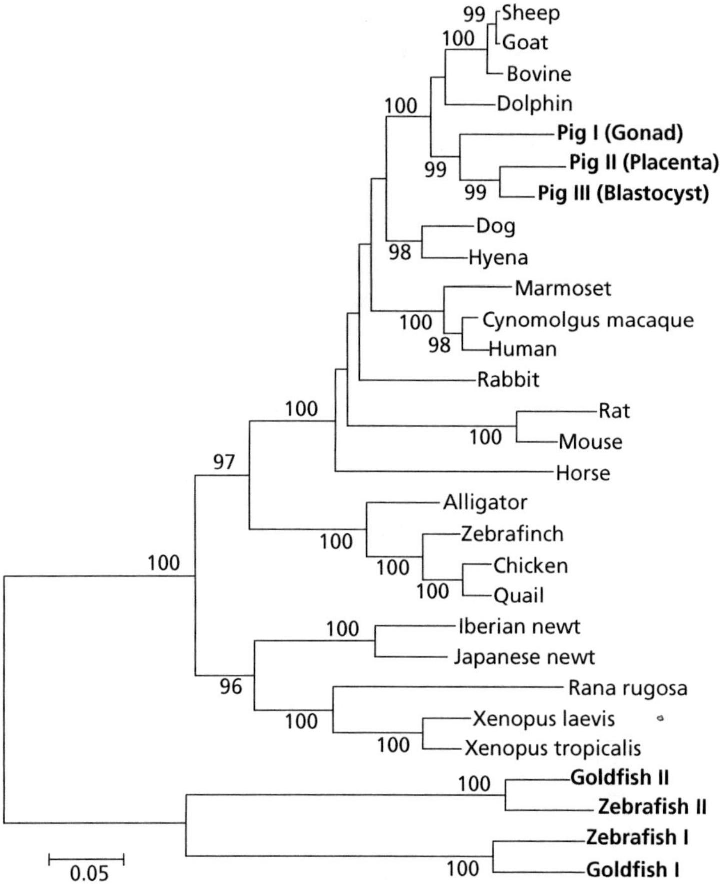

Figure 4–1. Vertebrate aromatase (CYP19) phylogeny constructed by the neighbor-joining method based on the JTT distance showing multiple CYP19 genes (*bolded*) in the domestic pig and teleosts (goldfish and zebrafish). Teleost sequences were used as an outgroup to root the tree. Phylogenies using maximum parsimony, ML, and Bayesian methods yielded similar topologies.

and, in the case of the pig, the particular paralogue of porcine P450arom (Corbin et al., 2001). Attention has focused on the proximate substrate androgens and the final estrogen products, even though some of the intermediates themselves such as 19OH-androstenedione have hypertensive properties (Sekihara, 1983), because the physiological or clinical significance of intermediate synthesis and release remains unclear and largely unstudied.

THE PORCINE CYP19 GENE CLUSTER ENCODES FUNCTIONALLY DIVERSE PARALOGUES OF P450AROM

The CYP19 gene cluster (Graddy et al., 2000) that gives rise to the three functional paralogues of P450arom in pigs is unique among vertebrates studied to date, and it is unique among the genes encoding steroidogenic P450 enzymes (Conley et al., 2008). Examples of gene duplication such as this are recognized to provide opportunities for the evolution of protein function (Hughes, 1994; Hughes, 2005; Hughes & Friedman, 2005), and we have shown duplication of the ancestral porcine CYP19 resulted in catalytic diversity in the aromatase system of the domestic pig (Conley et al., 1997; Conley & Hinshelwood, 2001). The porcine paralogues of P450arom are expressed in a highly tissue-specific fashion in the gonads [testis (Conley et al., 1996) and ovary (Corbin et al., 1995)], placenta (Corbin et al., 1995; Hinshelwood et al., 1995) and preimplantation blastocyst (Choi et al., 1996). Estrogen synthesis in each of these sites serves functions that are common among many species (Lange et al., 2002), but evidence suggests that the porcine gonadal and placental P450arom paralogues have adapted functionally. The porcine gonadal P450arom synthesizes a novel testosterone metabolite, 1OH-testosterone (1OH-T) in addition to estradiol (Fig. 4–2).

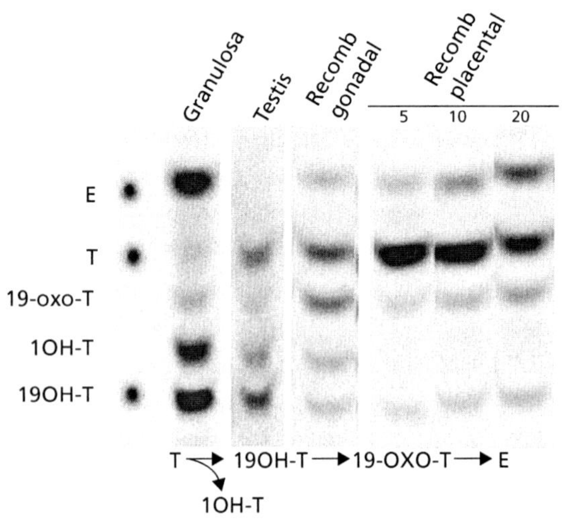

Figure 4–2. Testosterone (T) metabolism by porcine follicular granulosa cells and testis tissue microsomes, and purified recombinant porcine gonadal and placental P450arom enzymes, showing metabolites resolved by thin layer chromatography (100% ethyl acetate). 1OH-T synthesis is evident as a product in all reactions except that involving the recombinant porcine placental P450arom paralogue, even when analyzed at different time points in the reaction (5, 10, and 20 minutes). The known intermediates in estradiol (E) synthesis, 19OH- and 19-oxo-T, are evident in all reactions. Ovarian follicular granulosa and testicular microsomes (100 μg each), and fully reconstituted recombinant gonadal and placental P450arom (2 pmol), were incubated with 14C-testosterone (1 μM). Incubations extended over 5, 10, and 20 minutes with recombinant placental P450arom as indicated above, for 15 minutes with recombinant gonadal P450arom and 2 hours for tissue microsomes. Standards shown on the *left*; reaction sequences shown *below*. Reprinted with permission from Corbin et al. (2004). Paralogues of porcine aromatase cytochrome P450: a novel hydroxylase activity is associated with the survival of a duplicated gene. *Endocrinology, 145*, 2157–2164. Copyright 2004, The Endocrine Society.

Synthesis of 1OH-T by gonadal P450arom has been confirmed in incubations of testosterone with gonadal tissue microsomes as well as with purified recombinant porcine gonadal P450arom in reconstituted reactions (Fig 4-2). Hydroxylation at the C-1 carbon is not evident when androstenedione is the substrate, nor does the porcine placental P450arom catalyze 1-hydroxylation with either testosterone or androstenedione, so the reaction is paralogue- and substrate-specific. In fact, no other vertebrate P450arom studied to date exhibits this catalytic property. The porcine gonadal P450arom synthesizes 1OH-T in amounts almost equal to those of estradiol when reactions go to completion. This is because 1OH-T is nonaromatizable and therefore it accumulates as a major product (Corbin et al., 2004). More importantly, 1OH-T is a bioactive androgen with comparable bioactivity in vitro to the better-known and potent, nonaromatizable androgen, dihydrotestosterone (DHT; Fig. 4–3) (Corbin et al., 2004). Administration of DHT markedly increases ovulation rate in pigs, an increase that is blocked by flutamide and is therefore mediated through androgen receptor activation (Cardenas et al., 2002) This response to androgen is mediated at least in part by elevated FSH secretion (Jimenez et al., 2008). However, porcine preovulatory follicular granulosa synthesizes 1OH-T but not DHT (Corbin et al., 2004). Therefore, we hypothesize that 1OH-T synthesis by gonadal P450arom acted physiologically in pigs to increase ovulation rate by activation of androgen receptors, increasing FSH secretion as the function of this isozyme diverged after CYP19 duplication, but this hypothesis remains to be tested experimentally. This suggests in turn that the diversification of the catalytic ability of the porcine gonadal P450arom, making 1OH-T in addition to estradiol, promoted an increase in ovulation rate and litter size over their ancestors (Conley et al., 2008).

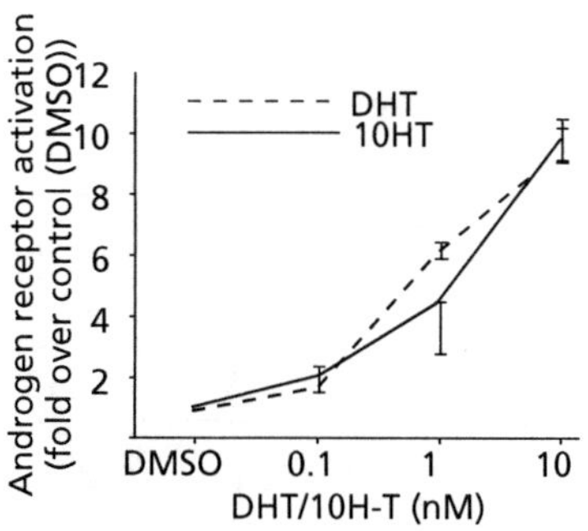

Figure 4–3. Androgenic potency of the novel metabolite 1OH-testosterone (1OH-T) and dihydrotestosterone (DHT) in an androgen receptor-dependent luciferase reporter assay conducted in LnCAP (prostate) cells. Note equipotent bioactive androgenic stimulation by 1OH-T and DHT. Redrawn with permission from Corbin et al. (2004). Paralogues of porcine aromatase cytochrome P450: a novel hydroxylase activity is associated with the survival of a duplicated gene. *Endocrinology, 145,* 2157–2164. Copyright 2004, The Endocrine Society.

Peccaries (species of the Tayassuid family) are the most distant relatives of the domestic pig in the Suiforme suborder (Fig. 4–4), and have an average litter size of just over two (Sowls, 1966). Collared (*Pecari tayacu*) and white-lipped (*Tayassu pecari*) peccaries have separate CYP19 genes encoding gonadal and placental forms of P450arom, consistent with CYP19 duplication in the suiforme ancestor (Corbin et al., 2007). However, unlike the gonadal P450arom of the domestic pig, we have been unable to detect 1OH-T synthesis by gonadal tissues. This is consistent with our overall hypothesis, which predicts that 1OH-T would not be synthesized in a suiforme that typically has a small litter size, ovulating one or two follicles per cycle (Corbin et al., 2007). Increasing reproductive output by gestating litters also incurs potential costs (Oksanen et al., 2002), especially for female fetuses that may be androgenized by male siblings in utero (Ryan & Vandenbergh, 2002). The placental paralogue of porcine P450arom may also have adapted to minimize the possible effect of male fetuses on female litter mates in utero, having a much higher catalytic efficiency for testosterone metabolism than the gonadal paralogue (Corbin et al., 1999). Thus, the high catalytic efficiency of androgen metabolism by the porcine placental P450arom (twice that of the gonadal/hypothalamic enzyme [Corbin et al., 2003]) may help to minimize virilization of female fetal pigs by androgens from male litter mates (Conley et al., 2008). If true, the functional diversification of the porcine gonadal and placental P450arom enzymes represents an example of coevolution that promoted successful litter bearing, first by increasing ovulation as explained, and second by providing a measure of protection of female fetuses from testosterone secreted by male litter mates.

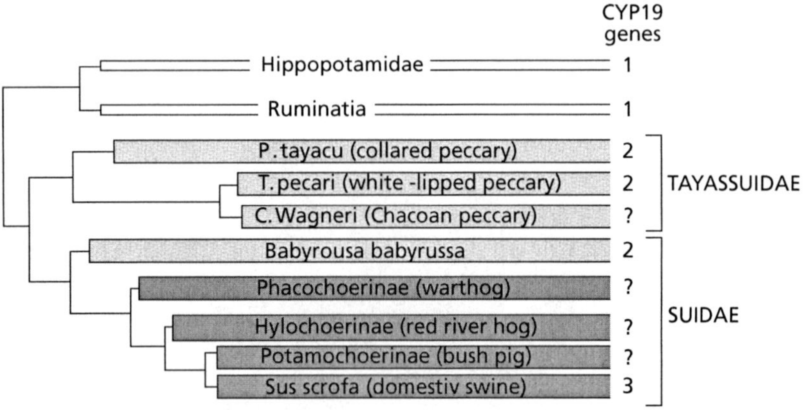

Figure 4–4. Proposed phylogeny among artiodactyls, showing relationships within the Suiforme suborder (Suidae and Tayassuidae), along with ruminants and hippopotamus. CYP19 gene numbers are as indicated at right where known, and ovulation rates are indicated by the shading of the bars; white, 1 per cycle, light grey, 2 per cycle and dark grey, 3 or more per cycle.

THE BRAIN AROMATASE HYPOTHESIS

Life-long physiological consequences result from events or conditions experienced during development in utero, a phenomenon now referred to as *fetal programming*. Sexual differentiation is one of the earliest recognized examples (Lillie, 1917), and decades of study confirm the essential role of fetal testicular androgens in the morphological differentiation of the male internal and external genitalia (Wilson et al., 1995), as well as the brain. Among the most recognizable differences between the male and female brain is the histologically recognizable sexually dimorphic nucleus (SDN) in the medial preoptic area (MPOA) of the hypothalamus (Simerly, 2002; Forger, 2006) that is also induced by sex steroids' action during embryonic or early postnatal life (Wu & Shah, 2011). The MPOA is significantly larger in male than female rats, is decreased by neonatal castration and is increased in females by either testosterone or estradiol given on the fourth postnatal day or earlier (Gorski et al., 1978). A sexually dimorphic nucleus has also been identified in sheep (referred to as the ovine SDN (oSDN), which is larger in males than females (Roselli et al., 2004). Similar to rats, prenatal testosterone injections increase the size of the oSDN (Roselli et al., 2007), whereas testosterone at physiological concentrations had no effect on the size of the oSDN in adult rams or ewes (Roselli et al., 2009).

There are two basic levels of differentiation of male brains (Phoenix et al., 1959): (1) In utero androgen exposure *organizes* neural development that enables (2) male reproductive behaviors to be *activated* by testosterone at reproductive maturation (Goy & McEwen, 1980; Ford & D'Occhio, 1989). Many of the organizational and activational effects of testosterone are mediated by the conversion of testosterone to estradiol (Beach, 1942), with variation existing among species (Wallen & Baum, 2002). The aromatase hypothesis suggests that the male hypothalamus is defeminized by the aromatization of androgens to estrogens in the fetal or neonatal brain (Naftolin et al., 1971; Fournet-Dulguerov et al., 1987; MacLusky et al., 1987; Lephart, 1996; Roselli, 2007). The genitalia and brains of female mammals develop in the relative absence of fetal gonadal hormones by a pathway therefore seen as "default," but female brains are also less sensitive to the organization of copulatory behavior induced by prenatal testosterone (Gerall & Ward, 1966). This may reflect differences in levels and sites of expression of sex steroid receptors among species (McAbee & DonCarlos, 1998; Scott et al., 2000). For example, in rodents there are sex differences in the relative expression of steroid receptors in several regions of the hypothalamus (Shah et al., 2004; Cao & Patisaul, 2011). However, reduced sensitivity among females might also reflect differences in hypothalamic P450arom expression, which is higher in males than in females quite consistently across species. In rats, this is already evident by gestation day 20 (George & Ojeda, 1982), notably in functionally relevant nuclei (Roselli & Resko, 1993; Lauber et al., 1997), peaking just before birth and declining progressively thereafter. Although some aspects of sexual differentiation of the brain are independent of gonadal steroids

(Abel & Rissman, 2011), the aromatase hypothesis has become one of the cornerstones of our understanding of brain sexual differentiation (Wallen & Baum, 2002). Because testosterone is the proximate substrate for estradiol synthesis by P450arom, this enzyme has unique potential to influence the balance of both estrogens and androgens in situ and thus likely plays a particularly influential role in brain sexual differentiation.

HYPOTHALAMIC AROMATASE EXPRESSION IN PIGS

Given the unique synthesis of 1OH-T by the gonadal paralogue, we were especially interested in discerning which of the paralogues might be expressed in the porcine hypothalamus and whether expression levels differed between the sexes as in other species (Naftolin et al., 1975). Several independent approaches were taken, all demonstrating that the porcine hypothalamus expresses what has been called previously the gonadal paralogue of porcine P450arom (Corbin et al., 2009). The porcine "gonadal" P450arom must therefore be renamed and referred to as the "gonadal/hypothalamic" paralogue of porcine P450arom. This was demonstrated by sequencing amplicons from hypothalamic mRNA, and confirmed by additional functional analyses (Fig. 4–5). Specifically, microsomes isolated from hypothalamic tissue were shown to synthesize 1OH-T and this activity was inhibited by letrozole, a well-known potent P450arom inhibitor. In addition, hypothalamic P450arom activity was inhibited by etomidate (Corbin et al., 2009), a characteristic shown previously to be an unusual property of the gonadal paralogue (Corbin et al., 1995; Conley et al., 2002). Hypothalamic expression of P450arom was four-fold higher in mature males than in females and

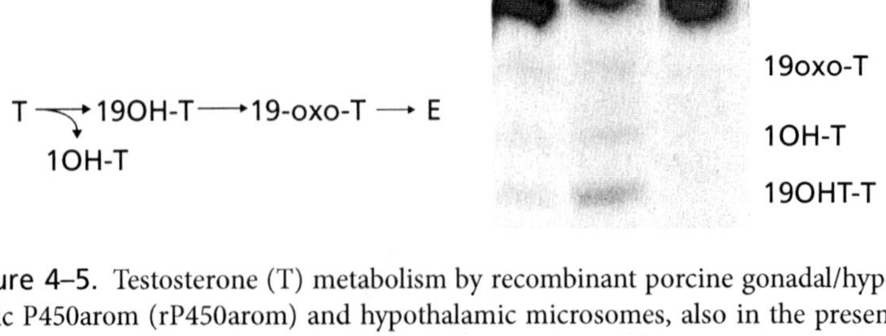

Figure 4–5. Testosterone (T) metabolism by recombinant porcine gonadal/hypothalamic P450arom (rP450arom) and hypothalamic microsomes, also in the presence of the P450arom inhibitor, letrozole (LET), demonstrating inhibitor-sensitive synthesis of intermediates and 1OH-T in the branching reaction shown.

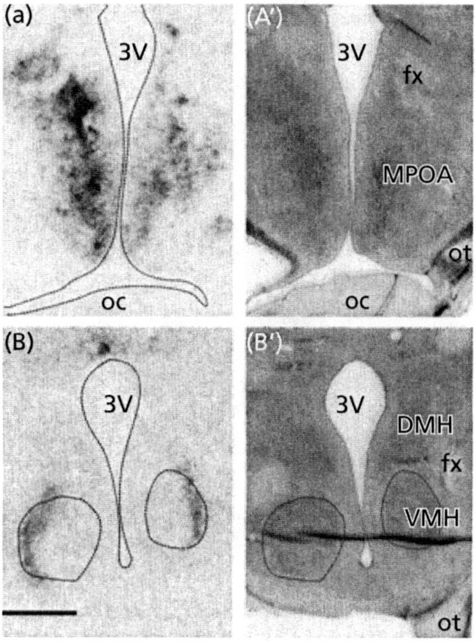

Figure 4–6. Hypothalamic expression of P450arom a detected by in situ hybridization. Shown are digitized autoradiographs from representative sections of the medial preoptic area (*A*) and ventromedial hypothalamus (*B*) from a boar using 32P-antisense P450arom cRNA, control autoradiographs from brain sections preincubated with RNase (not shown), and thionin histology (A' and B' respectively). The digital images were all scanned at the same magnification; bar length in lower left = 2 mm. Borders have been drawn to delineate the ventricles in the autoradiographs. The VMH has been outlined to illustrate that P450arom mRNA expression is confined to the lateral region of this nucleus. Abbreviations: 3V, third ventricle; DMH, dorsomedial hypothalamic n.; fx, fornix; MPOA, medial preoptic n.; oc, optic chiasm; ot, optic tract; VMH,ventromedial hypothalamic n. From Corbin, C. J., Berger, T., Ford, J. J., Roselli, C. E., Sienkiewicz, W., Trainor, B. C., et al. (2009). Porcine hypothalamic aromatase cytochrome P450: isoform characterization, sex-dependent activity, regional expression and regulation by enzyme inhibition in neonatal boars. *Biol Reprod 81,* 388–395.

expression within the hypothalamus included both the MPOA and the lateral borders of the ventromedial nucleus (Fig. 4–6) (Corbin et al., 2009). Thus, the higher expression of P450arom in the male than female hypothalamus has not been greatly altered by the evolution of 1OH-T synthesis by the gonadal/hypothalamic form of porcine P450arom.

There is a general paucity of data on the sexual differentiation of the porcine hypothalamus compared with that in other species. However, the authors have conducted preliminary studies on level of hypothalamic aromatase activity in fetal and neonatal pigs. Aromatase activity is detectable in the fetal hypothalamus as early as gestation day 50 and is considerably higher than it is at gestation day 70, but there is no difference in the levels between males and females. However, males appear to have developed higher levels of aromatase activity at one day of age and levels continue to rise in 3-week-old males (Fig. 4–7). How the sexual divergence of hypthalamic aromatase activity relates to the development of sexual differentiation in other areas of the brain is unknown at this time and requires that careful localization studies be conducted. The supraoptic nucleus becomes sexually dimorphic during development in pigs (van Eerdenburg et al., 1992; van Eerdenburg & Swaab, 1994), and is a site of ERα expression (van Leeuwen et al., 1995). Gonadotropin releasing hormone (GnRH) expression is highest in the MPOA of the pig hypothalamus (Kineman et al., 1988). In contrast, kisspeptin expression is highest in the periventricular and arcuate nuclei (Tomikawa et al., 2010), which is not radically different from the sites of expression in other species, the anteroventral and preoptic periventricular regions in rats (Kauffman et al., 2007), mice (Clarkson & Herbison, 2006) and sheep (Pompolo, et al., 2006). The kisspeptin receptor, GPR54, has yet to be localized in pigs, but it is possible that kisspeptin and P450arom expression colocalize in the porcine paraventricular nucleus based on the above-mentioned observations and our data on P450arom expression in boars. Clearly, more extensive studies are needed to map the expression patterns of P450arom, GnRH, kisspeptin, GPR54, and steroid receptors at a minimum, and these must be conducted in male and female pigs, prenatally and postnatally to begin to understand their functional significance. Regardless, the evolutionary selection of the gonadal/hypothalamic form of porcine P450arom must necessarily have included neuroendocrine as well as gonadal function, and we believe both supported increased ovulation rate and litter bearing.

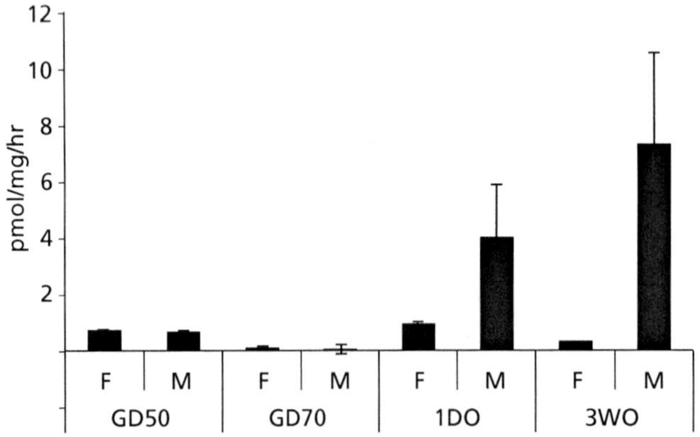

Figure 4–7. Hypothalamic aromatase activity (pmol/mg/hr) in fetal and neonatal pigs. Shown are means and standard errors of the means for female (F) and male (M) piglets at 50 (GD50) and 70 (GD70) days of gestation, and at one day (1DO) and 3 weeks (3WO) of age.

ANDROGENIC AND ESTROGENIC INFLUENCES ON DIFFERENTIATION OF THE BRAIN

Notwithstanding the brain aromatase hypothesis, evidence indicates that sexual differentiation of the brain is influenced also by the activation of the androgen receptor (Zuloaga, et al., 2008). In rodents lacking functional androgen receptor (AR), male typical behavior is lost (Sato et al., 2004) and the formation of sexually dimorphic nuclei and brain morphology is altered (Zuloaga et al., 2008). Activation of AR also increases brain P450arom expression in adult male rats (Roselli, et al., 1987). Testosterone exposure of guinea pigs in utero from 30 to 50 days of pregnancy defeminized the luteinizing hormone (LH) surge mechanism (Connolly & Resko, 1994); estradiol did not (Hines & Goy, 1985). Dihydrotestosterone (nonaromatizable) restored copulation in male castrate guinea pigs (Alsum & Goy, 1974), and prenatal AR blockade increased female-typical behavior in females (Thornton, et al., 1991), suggesting they are defeminized to some degree by androgens in utero. Among primates, prenatal DHT treatment significantly defeminized receptivity and masculinized copulatory behavior of rhesus macaques (Thornton & Goy, 1986). Similarly, prenatal testosterone and antiandrogen treatments have stage-specific effects on postnatal reproductive development in female (Zehr et al., 2005) and male (Herman et al., 2006) rhesus. Men that lack functional ARs have normal or higher levels of testosterone but typically exhibit a strong bias toward female gender identity (Hines et al., 2003). It has been suggested that girls experiencing androgen excess from congenital adrenal hyperplasia may exhibit a tendency toward male-typical behavior (Hines & Kaufman, 1994; Berenbaum, et al., 2000), though aromatization within the brain could still be important in these cases. Conversely, the lack of a functional P450arom enzyme or estrogen receptors (ER) does not seem to affect gender identity in men, or their sexual behavior (Grumbach & Auchus, 1999), suggesting that androgen may play a preeminent role in human sexual development. In sum, it is likely that androgens and estrogens act together in directing some key components of brain sexual differentiation, and the relative importance of each type of steroid varies with species.

Sexual differentiation of the brain during development can have lasting consequences on behavior. In rats, sex differences in play behavior before puberty are known to be driven in part by gonadal steroids (Auger, et al., 2011). Piglets engage in several forms of play behavior including locomotor play, social play, and agonistic interactions. Locomotor play generally includes fast angular movements of the head and body, such as scampers, pivots, and head tosses. Social play behavior is thought to peak after 4 weeks of age, and includes actions such as noninjurious biting, pushing, and mounting (Donaldson et al., 2002). Immediately after weaning, we observed the behaviors of male piglets treated with letrozole or vehicle. Weekly treatment with letrozole induced a significant decrease in mounting behavior (Fig. 4–8; Wilcoxon Z = 2.03, $p < 0.05$) but had no effect on general activity or aggressive behaviors (all

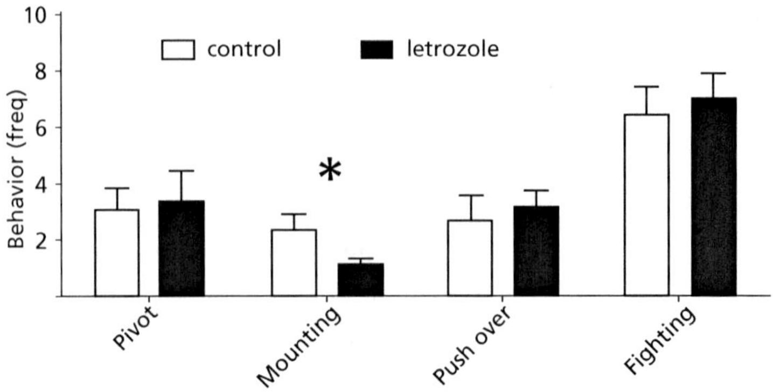

Figure 4–8. Effects of letrozole on various behaviors of male piglets immediately after weaning.

$p > 0.48$). Because males show significantly more mounting behavior than females at this age (Kranendonk et al., 2006), we observed a demasculinzing effect of letrozole treatment in these piglets. No significant differences in locomotor activity were observed between the groups, which is consistent with previous observations showing no sex differences in activity of this kind among piglets. Although previous studies have demonstrated that male piglets are more aggressive than females, this is usually observed after piglets are mixed with other litters. In our experiments, the piglets were not mixed with unfamiliar piglets. Since the effects of estrogens on aggression are often context dependent (Laredo & Trainor this volume; Trainor et al., 2006), it is possible that estrogens may only regulate aggressive behaviors when new dominance hierarchies are forming (Chaloupkov et al., 2007).

SEXUAL DIFFERENTIATION OF HYPOTHALAMIC–PITUITARY FUNCTION

Successful reproduction is entirely dependent on pituitary gonadotropin secretion stimulated by hypothalamic GnRH. Like genitalia and reproductive behavior, neuroendocrine function is "sexually dimorphic" in most species (Morris et al., 2004). The female brain differentiates in the absence of androgens, retaining a considerably more complex feedback system regulating GnRH release than male brains (Foster et al., 2006). Foster et al. (2007) summarized the sex-related differences in hypothalamic function, describing four separate feedback mechanisms in females, but just one in males. Pituitary gonadotropin release is regulated in females by both negative and positive feedback effects of estradiol on GnRH secretion, and the negative feedback of progesterone on both pulsatile and surge GnRH release. The male hypothalamus is described as exhibiting ostensibly only a negative feedback response to sex steroids (Foster et al., 2002), that is less sensitive than the female hypothalamus pre- and postpubertally (Wood & Foster, 1992; Robinson et al., 1999). However, the normal process of sexual differentiation of the brain has been examined in only a limited number of taxa, sheep being the best studied among artiodactyls (Roselli et al., 2003, 2007). In reality, the role of brain P450arom in the sexual differentiation of neuroendocrine feedback and function has yet to be clarified in any taxa, because most studies rely essentially on observational data that may suggest hypotheses but do not test them.

BRAIN SEXUAL DIFFERENTIATION IN PIGS

Much needs to be learned about sexual differentiation of the brain in pigs (Parvizi, 2000, 2006), but available data indicate that prenatal androgen excess defeminizes female neuroendocrine and reproductive function, consistent with data in sheep. Genital differentiation occurs between 30 and 50 days of gestation (Inomata et al., 1993), but sexual differentiation of the brain occurs later in pigs than in any other species investigated. It may not be complete in males until puberty, based on the organization of behavior (Ford, 1990) and the emergence of sexually dimorphic nuclei that include a vasopressin-oxytocin containing nucleus and the supraoptic nucleus (van Eerdenburg & Swaab, 1994). To date, castration and pharmacological steroid treatments have been the principal experimental paradigms used to study the differentiation of sexual behavior (Ford, 1990) and neuroendocrine function in pigs (Petric, et al., 2004). For instance, estradiol can induce receptive and proceptive behaviors in neonatally castrated boars, as it does in females (Ford, 1982, 1983), so males are not yet defeminized at birth. Receptivity is blocked in castrates given testosterone from 3 to 6 months of age (Ford & Christenson, 1987) and cannot be elicited in mature boars. In contrast, females exposed to testosterone prenatally already exhibit muted estradiol-induced LH release (Elsaesser & Parvizi, 1979), suggesting that positive estrogen feedback is sensitive to sex steroids and can be defeminized in utero. Because LH surges are induced by estradiol in normal neonatal female but not male piglets, this response is a sexually dimorphic trait already programmed by birth (Elsaesser et al., 1978). Testosterone exposure from 40 to 46 days of gestation decreased female pituitary LH-β mRNA levels (Petric et al., 2004), indicating negative feedback begins to develop after genital differentiation is complete (Inomata et al., 1993). Release of LH is pulsatile in fetal pigs as early as 81 days, but gonadectomy has no effect on LH levels, suggesting that the fetal pituitary may be inhibited by nongonadal steroids at this stage of gestation (Ponzilius et al., 1986). Therefore, hypothalamic feedback is a sexually dimorphic character in pigs, is established in females before birth, and can be disrupted (virilized) in females by prenatal androgen exposure resulting in infertility.

REGULATION OF HYPOTHALAMIC P450AROM EXPRESSION IN PIGS

Whether or not the effects of androgen exposure on hypothalamic-pituitary function are associated with, let alone mediated by, changes in P450arom expression in the porcine hypothalamus is unknown. However, recent studies by the authors have begun to elucidate how expression of P450arom in the hypothalamus can be regulated in the pig. Specifically, neonatal boars were treated weekly with the P450arom inhibitor letrozole to suppress estrogen synthesis, and hypothalamic and testicular P450arom expression was investigated. Letrozole suppressed testicular estrogen synthesis but more than doubled hypothalamic P450arom activity in neonatal boars (Corbin et al., 2009). Testicular P450arom activity was markedly inhibited to less than 10% of control levels, consistent with results of previous studies (At-Taras et al., 2006). Two accompanying observations were significant in the interpretation of these data. First, consistent with the results of several prior studies, treatment with letrozole was not accompanied by an increase in circulating testosterone concentrations. Thus, if hypothalamic P450arom is sensitive to androgen exposure as other mammals appear to be, it seems not to be dependent on an increase in systemic concentrations. If an increase in androgen concentration drives increased P450arom expression in the pig hypothalamus, it must be synthesized locally in neural tissue from locally derived androgen substrate (Compagnone, et al., 1995; Compagnone & Mellon, 1998). An alternative possibility is that circulating estradiol itself inhibits hypothalamic P450arom expression, and letrozole decreases that inhibition. The second observation of particular interest was that letrozole treatment had opposite effects on hypothalamic and testicular expression of P450arom, even though both tissues express the same CYP19 (gonadal/hypothalamic) gene. This suggests that the hypothalamus and the testis must utilize different promoters (Fig. 4–9), which is consistent with data from a variety of species (Furbasss et al., 1997; Golovine et al., 2003). An inhibition of testicular P450arom concurrent with an increase in hypothalamic P450arom expression in the neonatal boar also argues against any direct effect of the inhibitor at the protein level, perhaps stabilizing the enzyme as proposed by Choate and Resko (Choate & Resko, 1996). These investigators also observed a paradoxical increase in hypothalamic P450arom expression in intact and castrated male guinea pigs treated with letrozole without detectable changes in androgen concentrations. Further studies are required to elucidate the local synthesis of androgen in the hypothalamus, and the roles of androgen and estrogen in the control of hypothalamic P450arom expression in pigs.

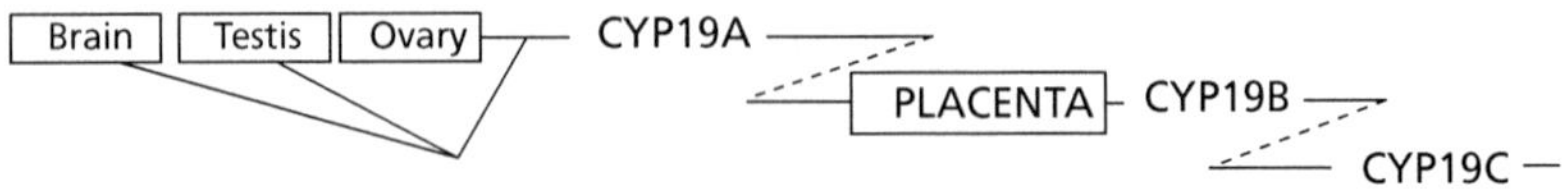

Figure 4–9. Schematic representation of the three porcine CYP19 genes (A, gonad/brain; B, placenta; C, blastocyst), and known or putative tissue-specific promoters (*boxed*) organization regulating expression of each gene in porcine tissues examined to date.

FETAL ANDROGEN EXPOSURE AND CONSEQUENCES OF LITTER BEARING

Androgen exposure has special physiological relevance in litter-bearing species, where female fetuses develop alongside male siblings in utero (Ryan & Vandenbergh, 2002). Unfortunately, few such species have been investigated and the common paradigm used to study in utero androgen excess has been administration of high doses of testosterone. Pigs may require exceptionally high levels of testosterone to induce the virilization of the genitalia of female fetuses. For instance, testosterone administration at levels required to induce masculinization were so high in one experimental study that abortion resulted in one of the treated sows (Petric et al., 2004). Considerable evidence exists suggesting that sexual behavior of female rats and mice is influenced by proximity of male siblings (vom Saal, 1981; vom Saal, et al., 1981; Sugawara, Pearson, Blanchard, & Ward, 2011), as is hypothalamic expression of ER-α (DonCarlos, 1996), but P450arom expression has not been assessed. Alpha-fetoprotein protects female fetal mice from circulating estrogens (Bakker & Baum, 2008) but does not bind estradiol avidly in pigs (Stone, 1981) or other species (Wallen & Baum, 2002), and its role in this regard in nonrodent species remains unclear. Rodents may simply be more sensitive to positional effects because they lack placental P450arom (Harada & Yamada, 1992) that might otherwise minimize testosterone exposure among fetuses (Corbin et al., 1999). Relatively high P450arom in sheep and pig placenta that would decrease the concentration of available androgens may be partly responsible for the large doses of testosterone required to masculinize the external genitalia of female fetuses of these species. Evidence of equivalent anatomical (anogenital distance, AGD) or behavioral masculinization of female pigs by their male siblings has not been seen among litters of piglets, but pubertal age decreased in females as the proportion of males increased, especially in small litters (Lamberson et al., 1988). Others reported that the number of males in a litter correlated positively with female AGD distance, and negatively with their fertility (Drickamer et al., 1997) and number of teats (Drickamer et al., 1999). A more detailed analysis on a smaller number of litters failed to find a positional effect of males on AGD of female fetuses at GD 110 (Rohde Parfet et al., 1990), suggesting that any possible effects are subtle. Differences in hypothalamic function arising from more subtle virilization of female by male pig fetuses has not been examined.

Studies were initiated examining the effects of inhibiting P450arom on female development, to test our hypothesis that P450arom expression shields female fetuses from virilization by male fetal androgens during pregnancy. We have shown that a single oral dose of letrozole (0.1 mg/kg BW) given to pregnant sows at GD30 (or at GD50) suppresses estradiol concentrations in maternal plasma measured 20 days later, from 24 to less than 5 pg/ml in control and treated animals, respectively. This is entirely consistent with our observations in boars where a single dose of letrozole inhibits circulating estrogen for 4 weeks or more (Berger et al., 2007). Fetal tissues were collected at slaughter,

20 days after sows received letrozole. Pools of blood from fetuses also indicate that estradiol levels were halved in males (371 vs. 139 pg/ml, control vs. treated) and females (173 vs. 76 pg/ml, control vs. treated). Testosterone in control male fetal plasma (202 pg/ml) was twice levels in control females (93 pg/ml) and was higher in letrozole-treated male (291 pg/ml) and female (134 pg/ml) fetuses. Collectively, these data confirm that a single dose of letrozole suppresses maternal and fetal estrogen levels for 3 weeks.

Fetuses and placental tissues were recovered from control and treated pregnancies at slaughter on GD50, 20 days after vehicle (control) or letrozole treatments. Fetal hypothalamic aromatase activity was not different between males and females (0.62±0.06 and 0.52±0.14 pmol/mg/hr, respectively). Letrozole induced a marked increase in aromatase activity in both male (3.2±0.44 pmol/mg/hr) and female (3.67±0.44 pmol/mg/hr) hypothalami. These levels exceed those in 6-week-old (2.64±0.43 pmol/mg/hr) and mature (1.97±0.04 pmol/mg/hr) boars. We have repeated this experiment in one control sow and one sow treated with letrozole at GD50. Both were killed at GD70 and tissues were harvested as before with essentially identical results. Although fetal hypothalamic aromatase activity (pmol/mg/hr) was much lower in male (0.10±0.05) and female (0.08±0.04) controls compared with GD50, it reached the same high levels in letrozole-treated male (3.17±0.24) and female (3.94±0.19) fetuses. These data indicate that a single oral dose of letrozole lowers estradiol and increases hypothalamic P450arom 5- to 50-fold, reaching supraphysiological levels detectable at least 3 weeks posttreatment in both male and female fetal pigs. Despite profound changes in hypothalamic enzyme activity, there was no change in aromatase activity of fetal testis at (0.51±0.20 vs. 0.50±0.22 pmol/mg/hr, control vs. letrozole, respectively). Fetal ovaries from control and letrozole-treated litters lacked measurable activity. The decline in hypothalamic P450arom from GD50 through GD70 occurs as estrogen concentrations in maternal plasma and fetal fluids rise (Robertson et al., 1985), suggesting that estrogen could inhibit P450arom expression in the developing hypothalamus. It is possible that letrozole has such potent effects in pigs because in tissues expressing the gonadal/hypothalamic P450arom, it inhibits production of a potent androgen (1OH-T) as well as estradiol, and therefore both androgenic (1OH-T) and estradiol-induced responses (Fig. 4–10).

Fetal testosterone concentrations are generally low before birth, but still higher in male than female fetuses (Ford et al., 1980) and therefore placental estradiol concentrations (from aromatization of testosterone) are correspondingly higher in males also at GD100 (Choong & Raeside, 1974). Hypothalamic ER expression is higher in sows than boars (van Leeuwen et al., 1995), but when this dimorphism arises is also unknown, perhaps with the rise in testosterone levels in male pigs soon after birth (Colenbrander et al., 1978; McCoard et al., 2003). Lower ER expression was seen in the hypothalamus of female rat fetuses adjacent to male fetuses in utero (DonCarlos, 1996), suggesting that sex steroids (androgens and/

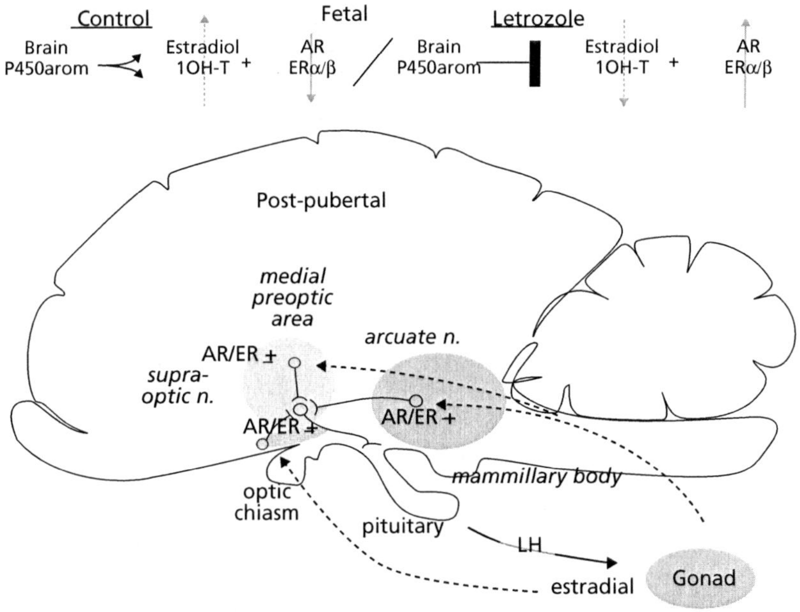

Figure 4–10. Proposed model for in utero (fetal) programming of the (postpubertal) reproductive axis by letrozole by changing steroid feedback on GnRH and gonadotropin release. Levels of estrogen receptors (ERα/β) are programmed by sex steroid exposure in utero when the female LH surge mechanism is organized. Letrozole treatment inhibits P450arom, decreasing both estradiol and 1OH-testosterone (1OH-T) in fetal brain. This increases (reprograms) ERs (α and/or β) and/or androgen receptors (AR) in critical hypothalamic nuclei which, at reproductive maturity, alters negative and positive (ER±) feedback sensitivity, decreases LH, delays puberty, and limits ovulation rate postpubertally. Reprogramming neuroendocrine feedback sensitivity may thereby alter reproductive function and success.

or estrogens) may suppress hypothalamic ER expression. Our data argue against the involvement of testicular androgens as the sole determinant of hypothalamic P450arom expression in utero because there is no sexual dimorphism at GD50 when male fetal androgen secretion is high or at GD70 when it is low (see Fig. 4–7). Furthermore, female fetuses respond equally to letrozole as the males regardless of stage of gestation or sex of adjacent fetuses (not shown). Thus, if the letrozole effect is through changes in androgen levels, it may well involve local production of androgens in the brain itself, because porcine fetal ovaries do not make androgens (Kaminski et al., 1999). This must remain a possibility because androgen synthesis does occur in mouse brain (Compagnone et al., 1995; Compagnone & Mellon, 1998; Mellon et al., 2001) and is further evidence of a possibly new and transforming paradigm for the development of sexual dimorphisms in the brain as proposed by others (Arnold et al., 2003).

CONCLUSIONS

The unusual occurrence of CYP19 gene duplication and reduplication in pigs may have allowed or even promoted a change in the reproductive strategy as the species evolved. This is proposed to have involved the emergence of a new catalytic activity and/or elevated levels of the resulting isozyme expressed in the gonads and the hypothalamus, leading to changes in ovarian and neuroendocrine function. Reproduction determines the balance and ultimate success of life history tradeoffs. Hormones such as testosterone (Hau, 2007) are known to influence tradeoffs, but mechanisms for selection through steroid hormones are unclear. Although hormones by their nature regulate physiology at multiple levels, feedback mechanisms like those controlling the reproductive axis have been seen to promote stasis, resist evolutionary change (Adkins-Regan, 2008), and thus constrain the phenotypic variation (Prud'homme et al., 2007) that might be facilitated by gene duplication. In general, litter bearing is costly for dams (Oksanen et al., 2002; Speakman, 2008) and problematic for fetuses, particularly females whose reproductive development may be jeopardized by proximity to males in utero (Ryan & Vandenbergh, 2002). Still, pigs evolved litters as they diverged from suiforme ancestors that were not litter bearers (Conley et al., 2008). The biochemical properties and physiological roles of the porcine gonadal/brain and placental paralogues of P450arom at least are likely to have coevolved functionally following duplication of the CYP19 gene, promoting the success of this reproductive strategy. Functional divergence and expression of P450arom paralogues in the placenta, the gonads and particularly the hypothalamus is likely to profoundly affect reproductive potential because estrogen synthesis (and testosterone metabolism) plays such a central role in all aspects of vertebrate reproductive function (Lange et al., 2002). Our recent data indicate that the porcine aromatases evolved much higher levels of expression in gonads and placentas at least than did their distant suiform relatives, the peccaries (Corbin et al., 2007), that have about two offspring on average (Sowls, 1966). Pigs also have much higher levels of P450arom expression in the hypothalamus than many other species. Therefore, we hypothesize that elevating P450arom expression, together with functional diversification of the gonadal/brain and placental paralogues, promoted an increased ovulation rate and litter bearing, in part by reprogramming hypothalamic feedback and function directed through altered sex steroids and their receptors. Further studies are required to test this hypothesis.

REFERENCES

Abel, J. L., & Rissman, E. F. (2011). Location, location, location: genetic regulation of neural sex differences. *Rev Endocr Metab Disord,* May 24. DOI 10.1007/s11154-011-9186-0 [Epub ahead of print].

Adkins-Regan, E. (2008). Review. Do hormonal control systems produce evolutionary inertia? *Philos Trans R Soc Lond B Biol Sci, 363,* 1599–1609.

Alsum, P., & Goy, R. W. (1974). Actions of esters of testosterone, dihydrotestosterone, or estradiol on sexual behavior in castrated male guinea pigs. *Horm Behav, 5,* 207–217.

Arnold, A. P., Rissman, E. F., & de Vries, G. J. (2003). Two perspectives on the origin of sex differences in the brain. *Ann N Y Acad Sci, 1007,* 176–188.
At-Taras, E. E., Conley, A. J., Berger, T., & Roser, J. F. (2006). Reducing estrogen synthesis does not affect gonadotropin secretion in the developing boar. *Biol Reprod, 74,* 58–66.
Auger, A. P., Jessen, H. M., & Edelmann, M. N. (2011). Epigenetic organization of brain sex differences and juvenile social play behavior. *Horm Behav, 59,* 358–363.
Bakker, J., & Baum, M. J. (2008). Role for estradiol in female-typical brain and behavioral sexual differentiation. *Front Neuroendocrinol, 29,* 1–16.
Beach, F. A. (1942). Copulatory behavior in prepuberally castrated male rats and its modification by estrogen administration. *Endocrinology, 31,* 679–683.
Berenbaum, S. A., Duck, S. C., & Bryk, K. (2000). Behavioral effects of prenatal versus postnatal androgen excess in children with 21-hydroxylase-deficient congenital adrenal hyperplasia. *J Clin Endocrinol Metab, 85,* 727–733.
Berger, T., McCarthy, M. A., Pearl, C. A., At-Taras, E., Roser, J. F., & Conley, A. J. (2008). Reducing endogenous estrogen during the early and late neonatal period affects reproductive tract development and sperm production in postpuberal boars. *Anim Reprod Sci, 109,* 218–235.
Cao, J., & Patisaul, H. B. (2011). Sexually dimorphic expression of hypothalamic estrogen receptors alpha and beta and Kiss1 in neonatal male and female rats. *J Comp Neurol, 519,* 2954–2977.
Cardenas, H., Herrick, J. R., & Pope, W. F. (2002). Increased ovulation rate in gilts treated with dihydrotestosterone. *Reproduction, 123,* 527–533.
Chaloupkova, H., Illmann, G., Neuhauserova, K., Tomanek, M., & Valis, L. (2007). Preweaning housing effects on behavior and physiological measures in pigs during the suckling and fattening periods. *J Anim Sci, 85,* 1741–1749.
Choate, J. V., & Resko, J. A. (1996). Paradoxical effect of an aromatase inhibitor, CGS 20267, on aromatase activity in guinea pig brain. *J Steroid Biochem Mol Biol, 58,* 411–415.
Choi, I., Simmen, R. C., & Simmen, F. A. (1996). Molecular cloning of cytochrome P450 aromatase complementary deoxyribonucleic acid from periimplantation porcine and equine blastocysts identifies multiple novel 5′-untranslated exons expressed in embryos, endometrium, and placenta. *Endocrinology, 137,* 1457–1467.
Choong, C. H., & Raeside, J. I. (1974). Chemical determination of oestrogen distribution in the foetus and placenta of the domestic pig. *Acta Endocrinol (Copenh), 77,* 171–185.
Clarkson, J., & Herbison, A. E. (2006). Postnatal development of kisspeptin neurons in mouse hypothalamus; sexual dimorphism and projections to gonadotropin-releasing hormone neurons. *Endocrinology, 147,* 5817–5825.
Colenbrander, B., de Jong, F. H., & Wensing, C. J. (1978). Changes in serum testosterone concentrations in the male pig during development. *J Reprod Fertil, 53,* 377–380.
Compagnone, N. A., Bulfone, A., Rubenstein, J. L., & Mellon, S. H. (1995). Steroidogenic enzyme P450c17 is expressed in the embryonic central nervous system. *Endocrinology, 136,* 5212–5223.
Compagnone, N. A., & Mellon, S. H. (1998). Dehydroepiandrosterone: a potential signalling molecule for neocortical organization during development. *Proc Natl Acad Sci U S A, 95,* 4678–4683.

Conley, A., Corbin, J., Smith, T., Hinshelwood, M., Liu, Z., & Simpson, E. (1997). Porcine aromatases: studies on tissue-specific, functionally distinct isozymes from a single gene? *J Steroid Biochem Mol Biol, 61,* 407–413.

Conley, A., & Hinshelwood, M. (2001). Mammalian aromatases. *Reproduction., 121,* 685–695.

Conley, A., Mapes, S., Corbin, C. J., Greger, D., & Graham, S. (2002). Structural determinants of aromatase cytochrome P450 inhibition in substrate recognition site-1. *Mol.Endocrinol, 16,* 1456–1468.

Conley, A. J., Corbin, C. J., Hinshelwood, M. M., Liu, Z., Simpson, E. R., Ford, J. J., et al. (1996). Functional aromatase expression in porcine adrenal gland and testis. *Biol Reprod, 54,* 497–505.

Conley, A. J., Corbin, C. J., & Hughes, A. L. (2008). Adaptive evolution of mammalian aromatases: lessons from Suiformes. *J Exp Zool Part A Ecol Gene Physiol, 309A,* 1–12.

Connolly, P. B., & Resko, J. A. (1994). Prenatal testosterone differentiates brain regions controlling gonadotropin release in guinea pigs. *Biol Reprod, 51,* 125–130.

Corbin C.J, Mapes, S. M., Lee, Y. M., & Conley, A. J. (2003). Structural and functional differences among purified recombinant mammalian aromatases: glycosylation, N-terminal sequence and kinetic analysis of human, bovine and the porcine placental and gonadal isozymes. *Mol Cell Endocrinol, 206,* 147–157.

Corbin, C. J., Berger, T., Ford, J. J., Roselli, C. E., Sienkiewicz, W., Trainor, B. C., et al. (2009). Porcine hypothalamic aromatase cytochrome P450: isoform characterization, sex-dependent activity, regional expression and regulation by enzyme inhibition in neonatal boars. *Biol Reprod 81,* 388–395.

Corbin, C. J., Hughes, A. L., Heffelfinger, J. R., Berger, T., Waltzek, T. B., Roser, J. F., et al. (2007). Evolution of suiform aromatases: ancestral duplication with conservation of tissue-specific expression in the collared peccary (Pecari tayassu). *J Mol Evol, 65,* 403–412.

Corbin, C. J., Khalil, M. W., & Conley, A. J. (1995). Functional ovarian and placental isoforms of porcine aromatase. *Mol Cell Endocrinol, 113,* 29–37.

Corbin, C. J., Mapes, S. M., Marcos, J., Shackleton, C. H., Morrow, D., Safe, S. et al. (2004). Paralogues of porcine aromatase cytochrome P450: a novel hydroxylase activity is associated with the survival of a duplicated gene. *Endocrinology, 145,* 2157–2164.

Corbin, C. J., Trant, J. M., & Conley, A. J. (2001). Porcine gonadal and placental isozymes of aromatase cytochrome P450: sub-cellular distribution and support by NADPH-cytochrome P450 reductase. *Mol Cell Endocrinol, 172,* 115–124.

Corbin, C. J., Trant, J. M., Walters, K. W., & Conley, A. J. (1999). Changes in testosterone metabolism associated with the evolution of placental and gonadal isozymes of porcine aromatase cytochrome P450. *Endocrinology, 140,* 5202–5210.

Donaldson, T. M., Newberry, R. C., Spinka, M., & Cloutier, S. (2002). Effects of early play experience on play behaviour of piglets after weaning. *Appl Anim Behav Sci, 79,* 221–231.

DonCarlos, L. L. (1996). Developmental profile and regulation of estrogen receptor (ER) mRNA expression in the preoptic area of prenatal rats. *Brain Res Dev.Brain Res, 94,* 224–233.

Drickamer, L. C., Arthur, R. D., & Rosenthal, T. L. (1997). Conception failure in swine: importance of the sex ratio of a female's birth litter and tests of other factors. *J Anim Sci, 75,* 2192–2196.

Drickamer, L. C., Rosenthal, T. L., & Arthur, R. D. (1999). Factors affecting the number of teats in pigs. *J Reprod Fertil, 115,* 97–100.

Elsaesser, F., & Parvizi, N. (1979). Estrogen feedback in the pig: sexual differentiation and the effect of prenatal testosterone treatment. *Biol Reprod, 20,* 1187–1193.

Elsaesser, F., Parvizi, N., & Ellendorff, F. (1978). Steroid feedback on luteinizing hormone secretion during sexual maturation in the pig. *J Endocrinol, 78,* 329–342.

Ford, J. J. (1982). Testicular control of defeminization in male pigs. *Biol Reprod, 27,* 425–430.

Ford, J. J. (1983). Postnatal differentiation of sexual preference in male pigs. *Horm Behav, 17,* 152–162.

Ford, J. J. (1990). Differentiation of sexual behaviour in pigs. *J Reprod Fertil Suppl, 40,* 311–321.

Ford, J. J., & Christenson, R. K. (1987). Influences of pre- and postnatal testosterone treatment on defeminization of sexual receptivity in pigs. *Biol Reprod, 36,* 581–587.

Ford, J. J., Christenson, R. K., & Maurer, R. R. (1980). Serum testosterone concentrations in embryonic and fetal pigs during sexual differentiation. *Biol Reprod, 23,* 583–587.

Ford, J. J., & D'Occhio, M. J. (1989). Differentiation of sexual behavior in cattle, sheep and swine. *J Anim Sci, 67,* 1816–1823.

Forger, N. G. (2006). Cell death and sexual differentiation of the nervous system. *Neuroscience, 138,* 929–938.

Foster, D. L., Jackson, L. M., & Padmanabhan, V. (2006). Programming of GnRH feedback controls timing puberty and adult reproductive activity. *Mol Cell Endocrinol, 254–255,* 109–119.

Foster, D. L., Jackson, L. M., & Padmanabhan, V. (2007). Novel concepts about normal sexual differentiation of reproductive neuroendocrine function and the developmental origins of female reproductive dysfunction: the sheep model. *Soc Reprod Ferti Suppl, 64,* 83–107.

Foster, D. L., Padmanabhan, V., Wood, R. I., & Robinson, J. E. (2002). Sexual differentiation of the neuroendocrine control of gonadotrophin secretion: concepts derived from sheep models. *Reprod.Suppl, 59,* 83–99.

Fournet-Dulguerov, N., MacLusky, N. J., Leranth, C. Z., Todd, R., Mendelson, C. R., Simpson, E. R., et al. (1987). Immunohistochemical localization of aromatase cytochrome P-450 and estradiol dehydrogenase in the syncytiotrophoblast of the human placenta. *J Clin Endocrinol Metab, 65,* 757–764.

Furbass, R., Kalbe, C., & Vanselow, J. (1997). Tissue-specific expression of the bovine aromatase-encoding gene uses multiple transcriptional start sites and alternative first exons. *Endocrinology, 138,* 2813–2819.

Gaucher, E. A., Graddy, L. G., Li, T., Simmen, R. C., Simmen, F. A., Schreiber, D. R., et al. (2004). The planetary biology of cytochrome P450 aromatases. *BMC Biol, 2,* 19.

George, F. W., & Ojeda, S. R. (1982). Changes in aromatase activity in the rat brain during embryonic, neonatal, and infantile development. *Endocrinology, 111,* 522–529.

Gerall, A. A., & Ward, I. L. (1966). Effects of prenatal exogenous androgen on sexual behavior of female albino rat. *J Comp Physiol Psychol, 62,* 370.

Ghosh, D., Griswold, J., Erman, M., & Pangborn, W. (2009). Structural basis for androgen specificity and oestrogen synthesis in human aromatase. *Nature, 457,* 219–223.

Golovine, K., Schwerin, M., & Vanselow, J. (2003). Three different promoters control expression of the aromatase cytochrome p450 gene (cyp19) in mouse gonads and brain. *Biol Reprod, 68,* 978–984.

Gorski, R. A., Gordon, J. H., Shryne, J. E., & Southam, A. M. (1978). Evidence for a morphological sex difference within the medial preoptic area of the rat brain. *Brain Res, 148,* 333–346.

Goy, R. W., & McEwen, B. S. (1980). *Sexual Differentiation of the Brain.* Cambridge, MA: MIT.

Graddy, L. G., Kowalski, A. A., Simmen, F. A., Davis, S. L., Baumgartner, W. W., & Simmen, R. C. (2000). Multiple isoforms of porcine aromatase are encoded by three distinct genes. *J Steroid Biochem Mol Biol, 73,* 49–57.

Grumbach, M. M., & Auchus, R. J. (1999). Estrogen: consequences and implications of human mutations in synthesis and action. *J Clin Endocrinol Metab, 84,* 4677–4694.

Harada, N. & Yamada, K. (1992). Ontogeny of aromatase messenger ribonucleic acid in mouse brain: fluorometrical quantitation by polymerase chain reaction. *Endocrinology, 131,* 2306–2312.

Hau, M. (2007). Regulation of male traits by testosterone: implications for the evolution of vertebrate life histories. *Bioessays, 29,* 133–144.

Herman, R. A., Zehr, J. L., & Wallen, K. (2006). Prenatal androgen blockade accelerates pubertal development in male rhesus monkeys. *Psychoneuroendocrinology, 31,* 118–130.

Hines, M., Ahmed, S. F., & Hughes, I. A. (2003). Psychological outcomes and gender-related development in complete androgen insensitivity syndrome. *Arch.Sex Behav, 32,* 93–101.

Hines, M., & Goy, R. W. (1985). Estrogens before birth and development of sex-related reproductive traits in the female guinea pig. *Horm Behav, 19,* 331–347.

Hines, M., & Kaufman, F. R. (1994). Androgen and the development of human sex-typical behavior: rough-and-tumble play and sex of preferred playmates in children with congenital adrenal hyperplasia (CAH). *Child Dev, 65,* 1042–1053.

Hinshelwood, M. M., Liu, Z., Conley, A. J., & Simpson, E. R. (1995). Demonstration of tissue-specific promoters in nonprimate species that express aromatase P450 in placentae. *Biol Reprod, 53,* 1151–1159.

Hughes, A. L. (1994). The evolution of functionally novel proteins after gene duplication. *Proc R Soc Lond B Biol Sci, 256,* 119–124.

Hughes, A. L. (2005). Gene duplication and the origin of novel proteins. *Proc Natl Acad Sci U S A, 102,* 8791–8792.

Hughes, A. L., & Friedman, R. (2005). Gene duplication and the properties of biological networks. *J Mol Evol, 61,* 758–764.

Inomata, T., Inoue, S., Sugawara, H., Kajihara, H., Shinomiya, T., Wagai, I., et al. (1993). Developmental changes in paramesonephric and mesonephric ducts and the external genitalia in swine fetuses during sexual differentiation. *J Vet Med Sci, 55,* 371–378.

Jimenez, E., Cardenas, H., & Pope, W. F. (2008). Effects of androgens on serum concentrations of gonadotropins and ovarian steroids in gilts. *Biol Reprod, 79,* 1148–1152.

Kaminski, M. A., Corbin, C. J., & Conley, A. J. (1999). Development and differentiation of the interstitial and tubular compartments of fetal porcine testes. *Biol Reprod, 60,* 119–127.

Kauffman, A. S., Gottsch, M. L., Roa, J., Byquist, A. C., Crown, A., Clifton, D. K., et al. (2007). Sexual differentiation of Kiss1 gene expression in the brain of the rat. *Endocrinology, 148,* 1774–1783.
Kineman, R. D., Leshin, L. S., Crim, J. W., Rampacek, G. B., & Kraeling, R. R. (1988). Localization of luteinizing hormone-releasing hormone in the forebrain of the pig. *Biol Reprod, 39,* 665–672.
Kranendonk, G., Hopster, H., Fillerup, M., Ekkel, E. D., Mulder, E. J., & Taverne, M. A. (2006). Cortisol administration to pregnant sows affects novelty-induced locomotion, aggressive behaviour, and blunts gender differences in their offspring. *Horm Behav, 49,* 663–672.
Lamberson, W. R., Blair, R. M., Rohde Parfet, K. A., Day, B. N., & Johnson, R. K. (1988). Effect of sex ratio of the birth litter on subsequent reproductive performance of gilts. *J Anim Sci, 66,* 595–598.
Lange, I. G., Hartel, A., & Meyer, H. H. (2002). Evolution of oestrogen functions in vertebrates. *J Steroid Biochem Mol Biol, 83,* 219–226.
Lauber, M. E., Sarasin, A., & Lichtensteiger, W. (1997). Sex differences and androgen-dependent regulation of aromatase (CYP19) mRNA expression in the developing and adult rat brain. *J Steroid Biochem Mol Biol, 61,* 359–364.
Lephart, E. D. (1996). A review of brain aromatase cytochrome P450. *Brain Res Brain Res Rev, 22,* 1–26.
Lillie, F. R. (1917). The free-martin: a study of the action of sex hormones in the foetal life of cattle. *J Exp Zool, 23,* 371–452.
MacLusky, N. J., Clark, A. S., Naftolin, F., & Goldman-Rakic, P. S. (1987). Estrogen formation in the mammalian brain: possible role of aromatase in sexual differentiation of the hippocampus and neocortex. *Steroids, 50,* 459–474.
McAbee, M. D. & DonCarlos, L. L. (1998). Ontogeny of region-specific sex differences in androgen receptor messenger ribonucleic acid expression in the rat forebrain. *Endocrinology, 139,* 1738–1745.
McCoard, S. A., Wise, T. H., & Ford, J. J. (2003). Endocrine and molecular influences on testicular development in Meishan and White Composite boars. *J Endocrinol, 178,* 405–416.
Mellon, S. H., Griffin, L. D., & Compagnone, N. A. (2001). Biosynthesis and action of neurosteroids. *Brain Res Brain Res Rev, 37,* 3–12.
Morris, J. A., Jordan, C. L., & Breedlove, S. M. (2004). Sexual differentiation of the vertebrate nervous system. *Nat Neurosci, 7,* 1034–1039.
Naftolin, F., Ryan, K. J., Davies, I. J., Reddy, V. V., Flores, F., Petro, Z., et al. (1975). The formation of estrogens by central neuroendocrine tissues. *Recent Prog Horm Res, 31,* 295–319.
Naftolin, F., Ryan, K. J., & Petro, Z. (1971). Aromatization of androstenedione by the diencephalon. *J Clin Endocrinol Metab, 33,* 368–370.
Oksanen, T. A., Koskela, E., & Mappes, T. (2002). Hormonal manipulation of offspring number: maternal effort and reproductive costs. *Evolution Int J Org Evolution, 56,* 1530–1537.
Parvizi, N. (2000). Neuroendocrine regulation of gonadotropins in the male and the female. *Anim Reprod Sci, 60–61,* 31–47.
Parvizi, N. (2006). Foetal and neonatal development of luteinising hormone and its regulatory systems in the pig. *Soc Reprod Fertil Suppl, 62,* 19–32.
Petric, N., Kato, Y., & Elsaesser, F. (2004). Influence of prenatal testosterone treatment on foetal and prepubertal LHbeta-subunit mRNA and plasma LH concentrations in the female pig. *Domest Anim Endocrinol, 27,* 25–38.

Phoenix, C. H., Goy, R. W., Gerall, A. A., & Young, W. C. (1959). The organizing action of an androgen administered prenatally on the tissues mediating mating behavior in the guinea pig. *Anat Rec, 133,* 323.

Pompolo, S., Pereira, A., Estrada, K. M., & Clarke, I. J. (2006). Colocalization of kisspeptin and gonadotropin-releasing hormone in the ovine brain. *Endocrinology, 147,* 804–810.

Ponzilius, K. H., Parvizi, N., Elaesser, F., & Ellendorff, F. (1986). Ontogeny of secretory patterns of LH release and effects of gonadectomy in the chronically catheterized pig fetus and neonate. *Biol Reprod, 34,* 602–612.

Prud'homme, B., Gompel, N., & Carroll, S. B. (2007). Emerging principles of regulatory evolution. *Proc Natl Acad Sci U S A, 104,* 8605–8612.

Robertson, H. A., Dwyer, R. J., & King, G. J. (1985). Oestrogens in fetal and maternal fluids throughout pregnancy in the pig and comparisons with the ewe and cow. *J Endocrinol, 106,* 355–360.

Robinson, J. E., Forsdike, R. A., & Taylor, J. A. (1999). In utero exposure of female lambs to testosterone reduces the sensitivity of the gonadotropin-releasing hormone neuronal network to inhibition by progesterone. *Endocrinology, 140,* 5797–5805.

Rohde Parfet, K. A., Lamberson, W. R., Rieke, A. R., Cantley, T. C., Ganjam, V. K., vom Saal, F. S., et al. (1990). Intrauterine position effects in male and female swine: subsequent survivability, growth rate, morphology and semen characteristics. *J Anim Sci, 68,* 179–185.

Roselli, C. E., Estill, C. T., Stadelman, H. L., & Stormshak, F. (2009). The volume of the ovine sexually dimorphic nucleus of the preoptic area is independent of adult testosterone concentrations. *Brain Res, 1249,* 113–117.

Roselli, C. E., Horton, L. E., & Resko, J. A. (1987). Time-course and steroid specificity of aromatase induction in rat hypothalamus-preoptic area. *Biol Reprod, 37,* 628–633.

Roselli, C. E., Larkin, K., Resko, J. A., Stellflug, J. N., & Stormshak, F. (2004). The volume of a sexually dimorphic nucleus in the ovine medial preoptic area/anterior hypothalamus varies with sexual partner preference. *Endocrinology, 145,* 478–483.

Roselli, C. E., & Resko, J. A. (1993). Aromatase activity in the rat brain: hormonal regulation and sex differences. *J Steroid Biochem Mol Biol, 44,* 499–508.

Roselli, C. E., Resko, J. A., & Stormshak, F. (2003). Estrogen synthesis in fetal sheep brain: effect of maternal treatment with an aromatase inhibitor. *Biol Reprod, 68,* 370–374.

Roselli, C. E., Stadelman, H., Reeve, R., Bishop, C. V., & Stormshak, F. (2007). The ovine sexually dimorphic nucleus of the medial preoptic area is organized prenatally by testosterone. *Endocrinology, 148,* 4450–4457.

Roselli, C. F. (2007). Brain aromatase: roles in reproduction and neuroprotection. *J Steroid Biochem Mol Biol, 106,* 143–150.

Ryan, B. C., & Vandenbergh, J. G. (2002). Intrauterine position effects. *Neurosci Biobehav Rev, 26,* 665–678.

Sato, T., Matsumoto, T., Kawano, H., Watanabe, T., Uematsu, Y., Sekine, K., et al. (2004). Brain masculinization requires androgen receptor function. *Proc Natl Acad Sci U S A, 101,* 1673–1678.

Scott, C. J., Tilbrook, A. J., Simmons, D. M., Rawson, J. A., Chu, S., Fuller, P. J., et al. (2000). The distribution of cells containing estrogen receptor-alpha (ERalpha) and ERbeta messenger ribonucleic acid in the preoptic area and hypothalamus of the sheep: comparison of males and females. *Endocrinology, 141,* 2951–2962.

Sekihara, H. (1983). 19-Hydroxyandrostenedione: a potent hypertensinogenic steroid in man. *J Steroid Biochem, 19,* 353–358.
Shah, N. M., Pisapia, D. J., Maniatis, S., Mendelsohn, M. M., Nemes, A., & Axel, R. (2004). Visualizing sexual dimorphism in the brain. *Neuron, 43,* 313–319.
Simerly, R. B. (2002). Wired for reproduction: organization and development of sexually dimorphic circuits in the mammalian forebrain. *Annu Rev Neurosci, 25,* 507–536.
Simpson, E. R., Mahendroo, M. S., Means, G. D., Kilgore, M. W., Hinshelwood, M. M., Graham-Lorence, S. et al. (1994). Aromatase cytochrome P450, the enzyme responsible for estrogen biosynthesis. *Endocr Rev, 15,* 342–355.
Sowls, L. K. (1966). Reproduction in the collared peccary (Tayassu tayacu). In: Rowlands, I. W. (ed.), *Comparative Biology of Reproduction in Mammals.* London: Academic Press, pp. 155–172.
Speakman, J. R. (2008). The physiological costs of reproduction in small mammals. *Philos Trans R Soc Lond B Biol Sci, 363,* 375–398.
Stone, R. T. (1981). In vitro liver synthesis and serum levels of alpha fetoprotein and albumin in the fetal pig. *Biol Reprod, 24,* 573–580.
Sugawara, A., Pearson, B. L., Blanchard, D. C., & Ward, M. A. (2012). Mouse females devoid of exposure to males during fetal development exhibit increased maternal behavior. *Psychoneuroendocrinology, 37,* 383–395.
Thompson, E. A., & Siiteri, P. K. (1973). Studies on the aromatization of C-19 androgens. *Ann N Y Acad Sci, 212,* 378–391.
Thornton, J. & Goy, R. W. (1986). Female-typical sexual behavior of rhesus and defeminization by androgens given prenatally. *Horm Behav, 20,* 129–147.
Thornton, J. E., Irving, S., & Goy, R. W. (1991). Effects of prenatal antiandrogen treatment on masculinization and defeminization of guinea pigs. *Physiol Behav, 50,* 471–475.
Tomikawa, J., Homma, T., Tajima, S., Shibata, T., Inamoto, Y., Takase, K. et al. (2010). Molecular characterization and estrogen regulation of hypothalamic kiss1 gene in the pig. *Biol Reprod,* 82, 313–319.
Trainor, B. C., Kyomen, H. H., & Marler, C. A. (2006). Estrogenic encounters: how interactions between aromatase and the environment modulate aggression. *Front Neuroendocrinol, 27,* 170–179.
van Eerdenburg, F. J., Lugard-Kok, C. M., & Swaab, D. F. (1992). The supraoptic nucleus in the pig hypothalamus: postnatal development and the effect of gonadal steroids. *J Endocrinol, 134,* 19–25.
van Eerdenburg, F. J., & Swaab, D. F. (1994). Postnatal development and sexual differentiation of pig hypothalamic nuclei. *Psychoneuroendocrinology, 19,* 471–484.
van Leeuwen, F. W., Chouham, S., Axelson, J. F., Swaab, D. F., & van Eerdenburg, F. J. (1995). Sex differences in the distribution of estrogen receptors in the septal area and hypothalamus of the domestic pig (Sus scrofa). *Neuroscience, 64,* 261–275.
vom Saal, F. S. (1981). Variation in phenotype due to random intrauterine positioning of male and female fetuses in rodents. *J Reprod Fertil, 62,* 633–650.
vom Saal, F. S., Pryor, S., & Bronson, F. H. (1981). Effects of prior intrauterine position and housing on oestrous cycle length in adolescent mice. *J Reprod Fertil, 62,* 33–37.
Wallen, K., & Baum, M. J. (2002). Masculinization and defeminization in altricial and precocial mammals: comparative aspects of steroid hormone action. In: Pfaff, D. W., Arnold, A.P., Etgen, A. M., Fahrbach, S. E., & Rubin, R. T. (eds.), *Hormones, Brain and Behavior.* New York: Elsevier, pp. 385–423.

Wilson, J. D., George, F. W., & Renfree, M. B. (1995). The endocrine role in mammalian sexual differentiation. *Recent Prog Horm Res, 50,* 349–364.

Wood, R. I., & Foster, D. L. (1992). Prenatal androgens and the timing of seasonal reproductive transitions in sheep. *Biol Reprod, 47,* 389–396.

Wu, M. V., & Shah, N. M. (2011). Control of masculinization of the brain and behavior. *Curr Opin Neurobiol, 21,* 116–123.

Zehr, J. L., Van Meter, P. E., & Wallen, K. (2005). Factors regulating the timing of puberty onset in female rhesus monkeys (Macaca mulatta): role of prenatal androgens, social rank, and adolescent body weight. *Biol Reprod, 72,* 1087–1094.

Zuloaga, D. G., Puts, D. A., Jordan, C. L., & Breedlove, S. M. (2008). The role of androgen receptors in the masculinization of brain and behavior: what we've learned from the testicular feminization mutation. *Horm Behav, 53,* 613–626.

Chapter 5

Regional Distribution of Aromatase in the Human Brain

Anat Biegon, Joanna Fowler, Sunny Kim, Jean Logan,
Deborah Pareto and Gene-Jack Wang

INTRODUCTION

Aromatase, the obligatory enzyme catalyzing the last step of estrogen biosynthesis, is found in the brain of all vertebrates including humans (Naftolin et al., 1996). Brain atromatase involvement in sexual behavior, aggression, cognition, memory, and neuroprotection has been documented in several animal species (Roselli, 2007; Garcia-Segura, 2008; Saldahana et al., 2009), although these findings have not yet been translated to the human/clinical arena. However, the discovery of in situ estrogen production by aromatase in the breast, which is increased in a large percentage of estrogen receptor positive tumors and other peripheral disorders (Bulun & Simpson 2008; Fedele et al., 2008; Miceli et al., 2009), as well as in bone and fat tissue (Purohit et al., 1992; Cleland et al., 1985), has provided a rationale for clinical and investigational use of aromatase inhibitors in a variety of disorders in women, men, and children (Budzar & Howell 2001; Cohen et al., 2002; Shulman et al., 2008; de Ronde & de Jong, 2011).

Inhibitors of aromatase activity (AI) are a relatively new class of drugs, currently approved for the hormonal treatment of breast cancer in the adjuvant setting as well as first-line therapy (Budzar & Howell 2001; Cohen et al., 2002; Arora & Potter, 2004). Also AI are used recreationally by body builders who self-administer these drugs to avoid feminizing effects of excess testosterone, the precursor to estrogen (Hartgens and Kuipers 2004). Despite the fact that the current uses of aromatase inhibitors do not target the brain, these drugs do cross the blood-brain barrier and thus provide powerful tools for the investigation of brain aromatase in vivo using positron emission tomography (PET). Modern PET scanners support the detection and quantification of molecular targets in the brain with high sensitivity and good spatial resolution (~2.5mm), enabling target quantification in the majority of brain regions known to subserve brain functions, which are modulated by estrogen.

Noninvasive scrutiny of aromatase availability throughout the primate brain has thus become possible following the radiosynthesis, initial primate studies (Lidstrom et al., 1998; Kim et al., 2009; Biegon et al., 2010a), and validation in humans (Biegon et al., 2010b) of AI radiotracers suitable for PET.

The AI fall into two categories: (1) steroidal AI such as formestane and exemestane, which bind irreversibly to the active site in aromatase, and (2) non-steroidal AI such as aminoglutethimide, fadrozole, anastrozole, letrozole, and vorozole, the binding of which is competitive and reversible (Budzar and Howell 2001). Among the AI, vorozole ((S)-6-[(4-chlorophenyl)(1H-1,2,4-triazol-1-yl)methyl]-1-methyl-1H-benzotriazole), $K_i = 0.7$ nM, (Van den Bossche et al., 1990) and letrozole (Iveson et al., 1993; Cohen et al., 2002) have been labeled with carbon-11 using [^{11}C]methyl iodide and evaluated as radiotracers for in vivo imaging of brain aromatase in primates (Lidstrom et al., 1998; Takahashi et al., 2006; Kil et al., 2009; Kim et al., 2009). Although [^{11}C]letrozole failed to show specific displaceable binding in vivo (Kil et al., 2009), [^{11}C]vorozole brain scans revealed high specific binding in the rhesus amygdala, similar to results obtained with autoradiography of the rat brain (Takahashi et al., 2006). We have recently reinvestigated and modified the radiosynthesis and purification of [^{11}C]vorozole (Kim et al., 2009). In female baboons, the pure tracer showed heterogeneous brain distribution, with the highest binding occurring in the amygdala and hypothalamic/preoptic area (Kim et al., 2009).

AROMATASE DISTRIBUTION IN THE HUMAN BRAIN

PET Studies

The most comprehensive study of aromatase distribution in the human brain to date, performed using PET and [^{11}C]vorozole (Biegon et al., 2010a), revealed a highly specific and heterogenous pattern that appears to be unique to humans. The highest levels were seen in the thalamus, although distribution was not uniform either: within the thalamus; the highest levels were found in the dorsomedial and pulvinar nuclei, with lower density in lateral and ventral thalamic nuclei (Figs. 5–1, 5–2). Very high levels were also found in the paraventricular hypothalamic nucleus.

Moderately high levels of aromatase were noted in amygdala and preoptic area/anterior hypothalamus and in the medulla (inferior olive). Basal ganglia levels were relatively low, with visibly higher levels in the ventral striatum/nucleus accumbens (Biegon et al, 2010a). All cortical regions bound the tracer, with hippocampus indistinguishable from the temporal cortex. The distribution volume values derived from a two-compartment model (Gunn et. al., 2001; Logan, 2003; Logan et al., 2011) in both men and women (regardless of menstrual cycle) followed the rank order: thalamus > amygdala = preoptic area > medulla (inferior olive) > cortex = hippocampus, putamen, cerebellum, and white matter (Fig. 5–2).

The specificity of the PET signal was confirmed by "blocking" experiments, in which a pharmacological dose of the specific aromatase inhibitor letrozole, known to inhibit aromatase activity by more than 90%, was injected just before the administration of tracer doses (less than 1% of the pharmacological dose) of the labeled vorozole. Imaging under these conditions provides the regional

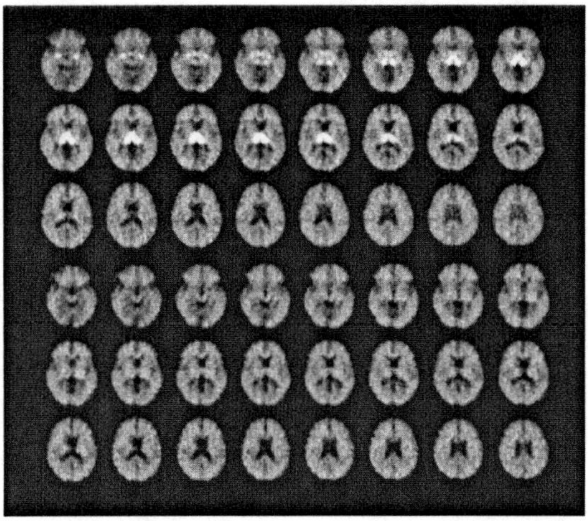

Baseline
("total binding")

Blocked
("non-specific binding")

Figure 5–1. Distribution of Aromatase in the Human Brain. The first three rows depict the mean distribution volume images from four subjects in the horizontal plane, from ventral (*top left*) to dorsal (*bottom right*). Highest distribution volume is in white and the lowest values are in black. The last three rows show the mean distribution volume of the same individuals obtained 2 hours after ingestion of a blocking dose of the aromatase inhibitor letrozole (Femara), 2.5 mg p.o. The bright spots in the first 3 slices correspond to amygdala. The bright area near the midline in slices 4 to 6 corresponds to the preoptic area. The pulvinar nucleus of the thalamus can be seen in slices 5 to 15, and mediodorsal thalamus is prominent in slices 8 to 15. Note the relatively uniform distribution in other regions, including cortex (all slices) and cerebellum (slices 1–5), and the uniformly low levels in the slices from blocking experiments, illustrating nonspecific tracer binding.

distribution of nonspecific tracer binding, because the specific and saturable binding sites on the enzyme are fully saturated by the large excess of unlabeled letrozole. Pretreatment with letrozole reduced tracer uptake in all of the regions examined, resulting in a homogenous distribution across regions (see Fig. 5–1). The size of the reduction was region dependent, ranging from ~70% blocking in thalamus and preoptic area to ~10% in cerebellum (Biegon et al., 2010). This indicates that the heterogeneous distribution of tracer binding in the brain is indeed indicative of region-dependent differences in the distribution of aromatase. Furthermore, the fact that non-specific binding was lower than total binding throughout the brain indicates that aromatase expression is quite ubiquitous, and most, if not all, human brain regions express aromatase.

2.2. Effects of Sex, Age, Plasma Hormone Levels and Body Mass Index son Human Brain Aromatase

Aromatase availability was calculated from PET scans of 28 healthy individuals (10 men, 18 women, age range 23–67, body mass index [BMI] range 19–34). Sex differences and age-related decline in aromatase availability did not reach

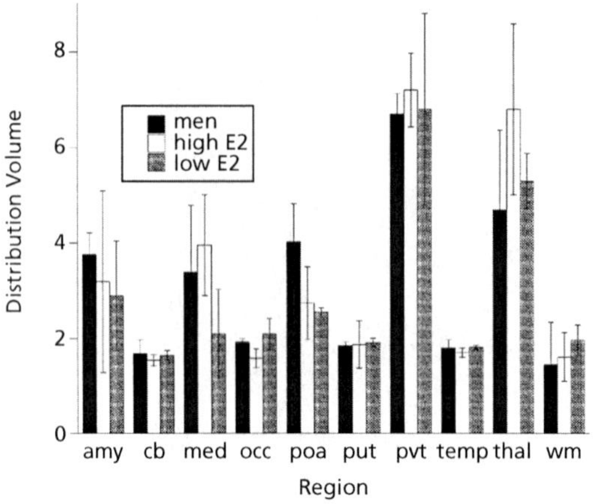

Figure 5–2. Effect of Sex and Menstrual Cycle on Aromatase Distribution. The figure depicts mean and standard deviation of distribution volume from PET scans of men and women less than 45 years old. The women, who were not using hormonal contraceptives, were scanned twice: One scan was scheduled around midcycle (11–15 days from the last menstrual flow), when estrogen levels are at their highest (highE2), and another study was scheduled 2 weeks later during the premenstrual phase of the cycle, when estrogen levels are low (lowE2). There were no statistically significant differences among the groups in any region. Abbreviations: amy, amygdala; cb, cerebellum; med, medulla (inferior olive); occ, occipital cortex; poa, preoptic area; put, putamen; pvt, periventrcular hypothalalmus; temp, temporal cortex; thal, thalamus (pulvinar and mediodorsal); wm, cortical white matter.

statistical significance with this sample size and age range. Similarly, there was no statistically significant correlation among total testosterone or estrogen in plasma, menstrual cycle phase, and distribution volume in any of the regions examined, similar to the findings in the initial cohort of three men and three women (see Fig. 5–2) (Biegon et al., 2010a).

Interestingly, a statistically significant negative correlation between aromatase levels and BMI was found in the human amygdala ($R = -0.55$, $p < .015$) (Wang et al., 2011). A comparison between the subjects with BMI in the normal range (18–25) and those in the overweight or obese range (BMI >25) also revealed lower aromatase availability in the high BMI groups. The mean distribution volume in the amygdala of the normal weight group was 2.9 (SD 0.5) compared with 2.0 (SD 0.2) in the high BMI group ($p < .05$).

Gene Expression, Immunoreactivity and Enzymatic Activity of Aromatase in the Human Brain

Aromatase gene expression was examined in postmortem samples from eight brain regions (Sasano et al., 1998). The amount of aromatase mRNA determined by RT-PCR assay in six cases (four men, two women) was highest in pons, thalamus, hypothalamus, and hippocampus. Analysis of multiple exons 1 revealed that exons 1f, considered specific for brain, as well as 1b (fibrolast type) and 1d (gonadal type), were expressed in the brain; 1d and 1f tended to be utilized in hypothalamus, thalamus, and amygdala. The amount of overall mRNA expression was also higher in hypothalamus, thalamus, and amygdala than in other regions of the brain. There were no differences of utilization of exons 1 and mRNA expression of aromatase between female and male brain. The authors conclude that their results demonstrate that aromatase is expressed widely in human brain tissues in both men and women. The presence of aromatase transcripts in human temporal cortex, frontal cortex, and hippocampus was also confirmed by Stoffel-Wagner et al. (1999).

Aromatase immunoreactivity was found in hypothalamus, amygdala, preoptic area, and (cholinergic) ventral forebrain nuclei by Ishunina et al. (2005). More recent studies confirmed aromatase immunoreactivity in temporal cortex, hippocampus, and prefrontal cortex (Yague et al., 2006, 2008). Immunohistochemistry was also used to examine the cellular and subcellular distribution of aromatase in the human brain, establishing the presence of aromatase immunoreactivity in neurons as well as in glia. Thus, cortical and hippocampal aromatase was detected in pyramidal cells, granule cells, and interneurons; in perikarya, dendrites, axons, and axon terminals (Naftolin et al., 1996; Yague et al., 2006, 2008). The presence of glial aromatase was confirmed in prefrontal cortex, temporal cortex and hippocampus, where it was associated with astocytes (Yague et al., 2006, 2008).

Aromatase enzymatic activity was first described in the fetal human limbic system by Naftolin et al. (1971), followed by reports on activity in the adult brain and temporal cortex (Naftolin et al., 1996; Steckelbroeck et al., 1999). In agreement with the results from other methods described above, there were no differences in aromatase enzymatic activity between men and women and no significant effect of aging (Steckelbroeck et al., 1999)

HUMAN BRAIN AROMATASE IN BRAIN DISORDERS

Region-specific disease-related changes in brain aromatase were detected in postmortem samples from subjects with Alzheimer's disease (AD) (Ishunina et al., 2005; Hiltunen et al., 2006), and autism (Sarachana et al., 2011). No change was detected in epilepsy (Yague et al., 2008).

Thus, enhanced aromatase immunoreactivity was observed in the basal forebrain nucleus basalis of Meynert in AD (Ishunina et al., 2005) with the opposite effect seen in the supraoptic, infundibular, and mammillary nuclei. In addition, aromatase immunoreactivity was decreased in ependymal and choroid plexus cells in AD. In prefrontal cortex, quantitative PCR studies showed increased mRNA levels in Alzheimer's disease patients, compared with matched controls (Luchetti et al., 2011).

A significant decrease in aromatase immunoreactivity was reported in a postmortem study of the frontal cortex samples from subjects with autism and sex and age matched controls (Sarachana et al., 2011).

BRAIN AROMATASE IN HUMANS: COMPARISON WITH OTHER SPECIES

Neuroanatomical Distribution

Some level of aromatase expression was found in all investigated brain regions in rodents, nonhuman primates, and humans. However the regional distribution pattern of aromatase in the human brain is strikingly different from the distribution reported in rodents and nonhuman primates using survey methods such as PET and autoradiography. In rodents, the aromatase "hot spots" were found in amygdala and the bed nucleus of the stria terminalis (Takahashi et al., 2006) with low though significant levels in other regions (Sierra et al., 2003). In both rhesus monkeys and baboons, the highest levels were found in the amygdala and preoptic area, whereas the thalamus and medulla contained low levels of aromatase (Lidstrom et al., 1998; Roselli & Resko, 2001; Takahashi et al., 2006; Kim et al., 2009). Although difference between rodents and primates in brain receptor and enzyme distribution are common, primate and human brain usually show similar regional distribution patterns (Osterlund et al., 2000, 2001) making this an unexpected finding and suggesting a unique role for thalamic and medullar aromatase in humans.

At present, we do not know which brain functions are served by the estrogens produced locally in the thalamus and inferior olive or the identity of the relevant estrogen receptor subtypes. Although the exact functions subserved by the human inferior olive are not clear, available information implicates this nucleus as well as specific thalamic nuclei in cognitive aspects of sensory and motor information processing, which may be modulated by estrogens (Ward & Arend, 2007; Liu et al., 2008).

As for the estrogen receptors, both ERα and ERβ receptors are expressed in a subtype-specific manner in the human brain, with ERα mostly restricted to hypothalamus and amygdala whereas ERβ expression is more widespread, and found also in hippocampus, cortex, and thalamus (Osterlund et al., 2000; also see review by Azcoitia et al., 2011). It is also possible that estrogens synthesized in thalamus and medulla interact with the more recently characterized membrane estrogen receptors (Toran-Allerand, 2004; Qiu et al., 2008), some of which are expressed in regions lacking classical ERα and ERβ such as striatum and medulla in rats (Brailoiu et al., 2007). However, the regional distribution of membrane estrogen receptors in the human brain has not been reported to date.

Effects of Sex, Age and Hormonal Environment

No significant sex differences have been observed in human brain aromatase in any of the regions in which the sexes were compared. This is supported by findings from different methodologies including PET, RT-PCR, and immunohistochemistry (Sasano et al., 1998; Steckelbroeck et al., 1999; Stoffel-Wagner et al., 1999; Ishunina et al., 2005; Biegon et al., 2010) reporting similar levels of brain aromatase activity and gene expression in men and women. Conversely, results from animal studies demonstrated higher levels of brain aromatase in males and suggested that testosterone was a positive modulator of aromatase in the hypothalamic-preoptic area (Roselli et al., 1984; Abdelgadir et al., 1994).

Physiological changes in ovarian hormone levels in plasma do not appear to have a significant effect on brain aromatase in humans or animals. Thus, brain aromatase did not appear to be significantly regulated by the estrous cycle in rodents (Roselli et al., 1984) and was not affected by menopause in monkeys (Higaki et al., 2012), matching the PET findings from women imaged at opposite phases of the menstrual cycle (Biegon et al., 2010a). The significant negative correlation between aromatase availability and BMI described above has not been examined before in either animals or humans; however, it resonates with a substantial body of evidence in humans and rodents, linking estrogen in limbic brain regions with suppression of feeding behavior (Nunez et al., 1980; Butera & Beikirch, 1989; Dagnault & Richard, 1997; Asarian 2006).

CONCLUSIONS

Taken together, findings from various methods, brain regions and species support the notion that brain aromatase expression is regulated in a species- and region-selective manner (Roselli et al., 1984). Such specific regulation may be the result of tissue-specific aromatase promoters, which were identified in animal and human tissues (Golovine et al., 2003; Honda et al., 1994; Jones et al., 2006). Because other promoters besides the brain specific exon 1.f (Sasano et al., 1998) are expressed in the human brain, this heterogeneity may provide the basis for brain region–specific regulation and expression of aromatase in humans.

Clearly, there is still a substantial amount of work that needs to be done to gain a more complete understanding of the regulation of human brain aromatase in health and disease and its response to common environmental and pharmacological agents (e.g., Biegon et al., 2010b). However, with [N-methyl-^{11}C]vorozole and future tracers for aromatase being developed for PET studies of aromatase in the human brain (Erlandsson et al., 2007), it is now possible to perform noninvasive assessment of aromatase availability and regulation in the healthy brain as well as in various brain disorders and harness the results towards development of new, sex-specific disease treatment or prevention strategies (Gillies & McArthur, 2010).

ACKNOWLEDGMENTS

Supported in part by NIH R01 NS050285 (AB) and K05DA020001 (JSF).

REFERENCES

Abdelgadir, S. E., Resko, J. A., Ojeda, S. R., Lephart, E. D., McPhaul, M. J., & Roselli C. E. (1994). Androgens regulate aromatase cytochrome P450 messenger ribonucleic acid in rat brain. *Endocrinology, 135,* 395–401.

Arora, A., & Potter, J. F. (2004). Aromatase inhibitors: current indications and future prospects for treatment of post-menopausal breast cancer. *J Am Geriatr Soc, 52*(4), 611–616.

Asarian L., & Geary N. (2006). Modulation of appetite by gonadal steroid hormones. *Philos Trans R Soc Lond B Biol Sci, 361,* 1251–1263.

Azcoitia, I., Yague, J. G., & Garcia-Segura L. M. (2011). Estradiol synthesis within the human brain. *Neuroscience, 191,* 139–147.

Biegon, A., Kim, S. W., Alexoff, D. L., Jayne, M., Carter, P., Hubbard, B., et al. (2010a). Unique distribution of aromatase in the human brain: in vivo studies with PET and [N-methyl-^{11}C]vorozole. *Synapse, 64,* 801–807.

Biegon A., Kim, S. W., Logan J., Hooker J. M., Muench L., & Fowler, J. S. (2010b). Nicotine blocks brain estrogen synthase (aromatase): in vivo PET studies in female baboons. *Biol Psychiat, 67*(8), 774–777.

Brailoiu, E., Dun, S. L., Brailoiu, G. C., Mizuo, K., Sklar, L. A., Oprea, T. I., et al. (2007). Distribution and characterization of estrogen receptor G protein-coupled receptor 30 in the rat central nervous system. *J Endocrinol, 193,* 311–321.

Budzar, A., & Howell, A. (2001). Advances in aromatase inhibition: Clinical efficacy and tolerability in the treatment of breast cancer. *Clin Cancer Res, 7,* 2620–2635.

Bulun, S. E., & Simpson, E. R. (2008). Aromatase expression in women's cancers. *Adv Exp Med Biol, 630,* 112–132.

Butera, P. C., & Beikirch, R. J. (1989). Central implants of diluted estradiol: independent effects on ingestive and reproductive behaviors of ovariectomized rats. *Brain Res,* 491, 266–273.

Cleland, W. H., Mendelson, C. R., & Simpson, E. R. (1985). Effects of aging and obesity on aromatase activity of human adipose cells. *J Clin Endocrinol Metab, 60*(1), 174–177.

Cohen, M. H., Johnson, J. R., Li, N., Chen, G., & Pazdur, R. (2002). Approval summary: Letrozole in the treatment of postmenopausal women with advanced breast cancer. *Clin Cancer Res, 8,* 665–669.

Dagnault, A., & Richard, D. (1997). Involvement of the medial preoptic area in the anorectic action of estrogens. *Am J Physiol, 272,* R311–R317.

de Ronde, W., & de Jong, F. H. (2011). Aromatase inhibitors in men: effects and therapeutic options. *Reprod Biol Endocrinol, 21,* 9:93.

Erlandsson, M. Karimi, F., Takahashi, K., & Langstrom B. (2007). 18F-Labelled vorozole analogues as PET tracer for aromatase. *J Labelled Compd Rad, 51,* 207–212.

Fedele, L., Somigliana, E., Frontino, G., Benaglia, L., & Vigano, P. (2008). New drugs in development for the treatment of endometriosis. *Expert Opin Investig Drugs, 17,* 1187–1202.

Garcia-Segura, L. M. (2008). Aromatase in the brain: not just for reproduction anymore. *J Neuroendocrinol, 20,* 705–712.

Gillies, G. E., & McArthur, S. (2010). Estrogen actions in the brain and the basis for differential action in men and women: a case for sex-specific medicines. *Pharmacol Rev, 62*(2), 155–198.

Golovine, K., Schwerin, M., & Vanselow, J. (2003). Three different promoters control expression of the aromatase cytochrome p450 gene (cyp19) in mouse gonads and brain. *Biol Reprod, 68,* 978–984.

Gunn, R. N., Gunn, S. R., & Cunningham V. J. (2001). Positron emission tomography compartmental models. *J Cereb Blood Flow Metab, 21,* 635–652.

Hartgens, F., & Kuipers, H. (2004). Effects of androgenic-anabolic steroids in athletes. *Sports Med,* 34, 513–554.

Higaki, S., Takumi, K., Itoh, M., Watanabe, G., Taya, K., Shimizu, K., et al. (2012). Response of ERβ and aromatase expression in the monkey hippocampal formation to ovariectomy and menopause. *Neurosci Res, 72,* 148–154

Hiltunen, M., Iivonen, S., & Soininen, H. (2006). Aromatase enzyme and Alzheimer's disease. *Minerva Endocrinol, 31,* 61–73.

Honda, S., Harada, N., & Takagi, Y. (1994). Novel exon 1 of the aromatase gene specific for aromatase transcripts in human brain. *Biochem Biophys Res Commun, 198*(3), 1153–1160.

Ishunina, T. A., van Beurden, D., van der Meulen, G., Unmehopa, U. A., Hol, E. M., Huitinga, I., et al. (2005). Diminished aromatase immunoreactivity in the hypothalamus, but not in the basal forebrain nuclei in Alzheimer's disease. *Neurobiol Aging, 26,* 173–194.

Iveson, T. J., Smith, I. E., Ahern, J., Smithers, D. A., Trunet, P. F., & Dowsett, M. (1993). Phase I study of the oral nonsteroidal aromatase inhibitor CGS 20267 in healthy postmenopausal women. *J Clin Endocrinol Metab, 77,* 324–331.

Jones, M. E., Boon, W. C., Proietto, J., & Simpson, E. R. (2006). Of mice and men: the evolving phenotype of aromatase deficiency. *Trends Endocrinol Metab, 17,* 55–64.

Kil, K. E., Biegon, A., Ding, Y. S., Fischer, A., Ferrieri R. A., Kim, S. W., et al. (2009). Synthesis and PET studies of [(11)C cyano]letrozole (Femara), an aromatase inhibitor drug. *Nucl Med Biol, 36,* 215–223.

Kim, S. W., Biegon, A., Katsamanis, Z., Erlich, C. W., Hooker, J. M., Shea, C., et al. (2009). Reinvestigation of the Synthesis and Evaluation of [N-methyl-11C] Vorozole, a Radiotracer Targeting Cytochrome P450 Aromatase. *Nucl Med Biol, 36,* 323–334.

Lidström, P., Bonasera, T. A., Kirilovas, D., Lindblom, B., Lu, L., Bergström, E., et al. (1998) Synthesis, in vivo rhesus monkey biodistribution and in vitro evaluation of a 11C-labelled potent aromatase inhibitor: [N-methyl-11C]vorozole. *Nucl Med Biol 25,* 497–501

Liu, T., Xu, D., Ashe, J., Bushara, K. (2008). Specificity of inferior olive response to stimulus timing. *J Neurophysiol, 100,* 1557–1561

Logan, J. (2003). A review of graphical methods for tracer studies and strategies to reduce bias. *Nucl Med Biol, 30,* 833–844.

Logan, J., Biegon A., Telang F., Kim, S. W., Jayne, M., Wang, G. J., et al. (2011). Comparison of methods for evaluating aromatase (AR) in the human brain with PET and [N-methyl-11C]vorozole ([11C]VOR). *J Nucl Med, 52,* 2029P

Luchetti S, Bossers K, Van de Bilt S, Agrapart V, Morales R. R., Frajese G. V., et al. (2011). Neurosteroid biosynthetic pathways changes in prefrontal cortex in Alzheimer's disease. *Neurobiol Aging, 32,* 1964–1976.

Miceli, V., Cervello, M., Azzolina, A., Montalto, G., Calabrò, M., & Carruba, G. (2009). Aromatase and amphiregulin are correspondingly expressed in human liver cancer cells. *Ann N Y Acad Sci, 1155,* 252–256.

Naftolin, F., Horvath, T. L., Jakab, R. L., Leranth, C., Harada, N., & Balthazart, J. (1996). Aromatase immunoreactivity in axon terminals of the vertebrate brain. An immunocytochemical study on quail, rat, monkey and human tissues. *Neuroendocrinology, 63,* 149–155.

Naftolin, F., Ryan, K. J., & Petro, Z. (1971). Aromatization of androstenedione by limbic system tissue from human fetuses. *J Endocrinol, 51,* 795–796.

Nunez, A. A., Gray, J. M., & Wade, G. N. (1980). Food intake and adipose tissue lipoprotein lipase activity after hypothalamic estradiol benzoate implants in rats. *Physiol Behav, 25,* 595–598.

Osterlund, M. K., Gustafsson, J. A., Keller, E., & Hurd, Y. L. (2000). Estrogen receptor beta (ERbeta) messenger ribonucleic acid (mRNA) expression within the human forebrain: distinct distribution pattern to ERalpha mRNA. *J Clin Endocrinol Metab, 85,* 3840–3846.

Österlund, M. K., & Hurd Y. L. (2001). Estrogen receptors in the human forebrain and the relation to neuropsychiatric disorders. *Prog Neurobiol, 64,* 251–267

Purohit, A., Flanagan, A. M., & Reed, M. J. (1992). Estrogen synthesis by osteoblast cell lines. *Endocrinology, 131*(4), 2027–2029.

Qiu, J., Ronnekleiv, O. K., & Kelly, M. J. (2008). Modulation of hypothalamic neuronal activity through a novel G-protein-coupled estrogen membrane receptor. *Steroids, 73,* 985–991.

Roselli, C. E. (2007). Brain aromatase: Roles in reproduction and neuroprotection. *J Steroid Biochem Mol Biol, 106,* 143–150.

Roselli, C. E., Ellinwood, W. E., & Resko, J. A. (1984). Regulation of brain aromatase activity in rats. *Endocrinology, 114,* 192–200.

Roselli, C. E., & Resko, J. A. (2001). Cytochrome P450 aromatase (CYP19) in the non-human primate brain: distribution, regulation, and functional significance. *J Steriod Biochem Mol Biol, 79,* 247–253.

Saldahana, C. J., Duncan, K. A., & Walters, B. J. (2009). Neuroprotective actions of brain aromatase. *Front Neurondocrinol, 30,* 106–118.

Sarachana, T., Xu, M., Wu, R. C., & Hu, V. W. (2011). Sex hormones in autism: androgens and estrogens differentially and reciprocally regulate RORA, a novel candidate gene for autism. *PLoS One, 6*(2), e17116.

Sasano, H., Takahashi, K., Satoh, F., Nagura, H., & Harada N. (1998). Aromatase in the human central nervous system. *Clin Endocrinol,* i 325–329.

Shulman, D. I., Francis, G. L., Palmert, M. R., Eugster, E. A., et al. (2008). Use of aromatase inhibitors in children and adolescents with disorders of growth and adolescent development. *Pediatrics, 121,* e975–e983.

Sierra, A. Azcoitia, I., & Garcia-Segura, L. (2003). Endogenous estrogen formation is neuroprotective in model of cerebellar ataxia. *Endocrine, 21,* 43–51.

Steckelbroeck, S., Heidrich, D. D., Stoffel-Wagner, B., Hans, V. H. J., Schramm, J., Bidlingmaier, F., et al. (1999). Characterization of aromatase cytochrome P450 activity in the human temporal lobe. *J Clin Endocrinol Metab, 84,* 2795–2801.

Stoffel-Wagner, B., Watzka, M., Schramm, J., Bidlingmaier, F., & Klingmüller, D. (1999). Expression of CYP19 (aromatase) mRNA in different areas of the human brain. *J Steroid Biochem Mol Biol, 70,* 237–241.

Takahashi, K., Bergström, M., Frändberg, P., Vesström, E-L., Watanabe, Y., & Långström, B. (2006). Imaging of aromatase distribution in rat and rhesus monkey brains with [11C]vorozole. *Nucl Med Biol, 33,* 599–605.

Toran-Allerand, C. D. (2004). Minireview: A plethora of estrogen receptors in the brain: where will it end? *Endocrinology, 145,* 1069–1074.

Van den Bossche, H., Willemsens, G., Roels, I., Bellens, D., Moereels, H., Coene, M-C., et al. (1990). R76713 and enantiomers: selective nonsteroidal inhibitors of the cytochrome P450-dependent oestrogen synthesis. *Biochem Pharmacol, 40,* 1707–1718.

Wang, G-J., Biegon, A., Telang, F., Logan, J., Kim, S-W., Jayne, M., et al. (2011). Reduced brain aromatase availability in overweight humans. *J Nuc Med, 52,* 101P.
Ward, R. & Arend I. (2007). An object-based frame of reference within the human pulvinar. *Brain, 130,* 2462–2469.
Yague, J. G., Munoz, A., de Monasterio-Schrader, P., Defelipe, J., Garcia-Segura, L. M., & Azcoitia I. (2006). Aromatase expression in the human temporal cortex. *Neuroscience, 138,* 389–401.
Yague, J. G., Wang, A. C., Janssen, W. G., Hof, P. R., Garcia-Segura, L. M., Azcoitia, I., et al. (2008). Aromatase distribution in the monkey temporal neocortex and hippocampus. *Brain Res, 1209,* 115–127.

Chapter 6

Neuroanatomical Distribution of Aromatase in Birds: Cellular and Subcellular Analyses

Colin J. Saldanha, Luke Remage-Healey and Barney A. Schlinger

INTRODUCTION

Our understanding about the estrogenic modulation of brain and behavior is well informed by the identification of peripheral and central sites of estrogen synthesis. Birds, in particular, have offered us many insights into the expression, localization, and regulation of estrogen synthesis and action. In birds, as in most other vertebrates, cytochrome P450 enzyme P aromatase is the product of single or multiple copies of the *CYP19a* gene, and is best known for its role in the conversion of C19 androgens into C18 estrogens (Simpson et al., 1994). Although aromatase expression differs across species, it is not limited to classical endocrine glands, but rather, is present in many different tissues; most notably the brain (Simpson et al., 2002). Studies of birds, in particular, were crucial for unraveling some of the fundamental questions about estrogenic activation of male copulatory behavior (Balthazart & Ball, 1998b; Balthazart et al., 2004) and have continued to serve as excellent models towards understanding the pluripotent and dramatic influence of estrogens on vertebrate brain structure and function (see chapter 10). Importantly, many of the behaviors that we now understand to be estrogen dependent appear to be modulated by the site-specific and discrete synthesis of estrogen at circumscribed neural loci via the expression of aromatase (Schlinger, 1997; Saldanha et al., 2000a).

This chapter will address the considerable diversity in the neural distribution of aromatase in several species of birds, and, where appropriate, highlight some of the behaviors that this distribution subserves. However, because many of these general issues are covered in considerable depth by other authors in this book, we will discuss the overall neural distribution of aromatase with some generalities. Our focus will be on a novel, exciting, and emerging subarea of estrogen provision in the avian brain, that is the subcellular distribution and regulation of aromatase that emphasizes the significance of estrogen synthesis within synaptic terminals.

DISTRIBUTION OF AROMATASE IN AVIAN BRAIN—DISCRETE, LOCALIZED SITES OF ESTROGEN SYNTHESIS

The distribution of estrogen synthesizing cells in the avian brain has largely been studied with three main methodologies: biochemical assays of aromatase activity, direct detection of the aromatase transcript via in situ hybridization or, more recently, qPCR, and direct observation of the protein via immunocytochemistry. Whereas each technique affords its own unique advantages and disadvantages, in general they confirm one another in the identification of neural sites that express the aromatase enzyme.

Across vertebrates, the expression, regulation, and function of aromatase are best understood in the hypothalamic preoptic area (HPOA). The local conversion of testosterone into estradiol owing to HPOA aromatization is necessary and sufficient to modulate copulatory behaviors in several species of birds, but is best described in galliformes (Balthazart & Ball, 1998a, 1998b). Indeed, every species examined expresses aromatase in the HPOA, underscoring its fundamental and conserved role in the control of avian reproductive behavior. Another conserved site of aromatase expression is nucleus taeniae, the avian homolog of the mammalian medial amygdala, but its function in taeniae is not fully characterized. Species examined include members of several orders including Columbiformes, Galliformes, Apodiformes, Strigiformes, Charadriiformes, Psittaciformes, Anatiformes, and Passeriformes (Table 6–1).

Of interest is the observation that, although aromatase expression is detectable in the hypothalamus and nucleus taeniae in every species studied, its distribution diverges from this restricted limbic distribution in the mallard drake (*Anas platyrhynchos*) and in all the passerine species, including the suboscine, the golden-collared manakin (*Manacus vitellinus*). In the drake, aromatase expression is abundant in neuronal perikarya and fibers not only in the HPOA and nucleus taeniae but is readily visible in neurons within the bed nucleus of the stria terminalis, globus pallidus, the red nucleus, and in varicosities of fibers in the hippocampus (Saldanha et al., 2010). Members of the order Passeriformes are unique in demonstrating a widespread distribution of aromatase including brain areas like the hippocampus (HP), caudomedial nidopallium (NCM), pallium, ventral tegmentum, and lateral nidopallium. Thus, again in very general terms, the most constrained neural distribution of aromatase is observed in the galliform and columbiform brain. Although this distribution is somewhat more widespread and includes more telencephalic loci in the anatiform brain, the passerine brain appears to contain the most widespread and abundant distribution of estrogen synthetic areas. Interestingly, the distribution of neural aromatase in the suboscine manakin (Saldanha et al., 2000a) is quite similar to that of the true oscine songbirds, albeit at a much lower magnitude, suggesting it represents an intermediate stage of elevated neural aromatase expression in the passeriform order (see Table 6–1).

Identifying the function of encephalic aromatization in the passerine brain has been thwarted somewhat by two confusing, yet provocative characteristics. These have to do with (a) a mismatch in some brain areas in the distributions of aromatase and estrogen-sensitive cells, that is, those expressing estrogen-receptor isoforms, and (b) a similar mismatch in some brain areas in the patterns of aromatase expression between transcript, immunoreactive protein, and enzyme activity. In the first case, there is a striking discordance in

Table 6–1. Neural Distribution of Aromatase in the Avian Brain.

			POA/Dien/WB	VMN
Non-songbirds	Japanese Quail	*Coturnix japonica*	x	x
	Ring dove	*Streptopelia risoria*	x	x
	Domestic Chicken	*Gallus gallus*	x	x
	Wilson's phalarope	*Phalaropus tricolor*	x	
	King quail	*Coturnix chinensis*	x	
	Budgerigar	*Melopsittacus undulatus*	x	x
	Grey partridge	*Perdix perdix*	x	
	Common swift	*Apus apus*	x	x
	Barn owl	*Tito alba*	x	x
	Mallard duck	*Anas platyrhynchos*	x	x
Sub-oscine songbirds	Golden-collared manakin	*Manacus vitellinus*	x	x
Oscine songbirds	Zebra finch	*Taeniopygia guttata*	x	x
	Brown-headed cowbird	*Molothrus ater*	x	x
	House sparrow	*Passer domesticus*	x	
	Canary	*Serenis canarius*	x	x
	Pygmy nuthatch	*Sitta pygmaea*	x	
	House wren	*Troglodytes aedon*	x	
	Mountain chickadee	*Poecile gambeli*	x	
	Lapland longspur	*Calcarius lapponicus*	x	x
	Pied flycatcher	*Ficedula hypoleuca*	x	
	Great tit	*Parus major*	x	
	Willow warbler	*Phylloscopus trochilus*	x	
	Chaffinch	*Fringilla coelebs*	x	
	Song sparrow	*Melospiza melodia*	x	x
	Cave swallow	*Petrochelidon fulva*	x	
	Cliff swallow	*Petrochelidon phyrrhonota*	x	
	Black-capped chickadee	*Poecile atricapillus*	x	x

The distribution of aromatase has been studied using enzyme assays on whole brain (WB), diencephalic (Dien) and telencephalic (Telen) homogenates, microdissection of brain areas, and direct observation of the aromatase transcript or immunoreactive protein. Although aromatase is consistently detected at hypothalamic areas and limbic areas across all studied species, it's distribution widens considerably in oscine songbirds to include telencephalic loci such as the nucleus taeniae of the amygdala (TnA), the bed nucleus of the stria terminalis (BnST), hippocampus (HP), globus pallidus (GP), medial portion of the magnocellular nucleus of the nidopallium (mMAN), caudomedial nidopallium (NCM), the dorsal lamina of the arcopallium (LAD), lateral nidopallium (LN), and the ventral tegmentum (AVT). Note: absence of detection does not necessarily mean that the enzyme is not expressed in a given structure, but rather may reflect that research technique utilized may not have had the spatial resolution or sensitivity to detect restricted or low levels present.

TnA/Telen	BnST	HP	GP	mMAN	NCM	LAD	LN	AVT	
x	x								Schumacher & Balthazart, 1984
x									Hutchison & Steimer, 1986
x									Callard et al., 1984
									Schlinger et al., 1992
									Saldanha et al., 1998
									Saldanha et al., 1998
									Metzdorf et al., 1999
									Metzdorf et al., 1999
									Metzdorf et al., 1999
x	x	x	x		x				Saldanha et al., 2010
x	x	x			x			x	Saldanha et al., 2000
x	x	x	x	x	x	x	x	x	Shen et al., 1995
x	x	x	x	x	x	x	x	x	Saldanha & Schlinger, 1997
x		x			x				Saldanha et al., 1998
x	x	x	x	x	x	x	x		Balthazart et al., 1998
x		x			x				Saldanha et al., 1999
x		x			x				Saldanha et al., 1999
x		x			x		x		Saldanha et al., 1999
x		x			x				Bindra et al., 1999
x									Silverin et al., 2000
x									Silverin et al., 2000
x	x		x	x		x	x	x	Silverin et al., 2000
x									Silverin et al., 2000
x		x			x	x			Soma et al., 2003
	x						x	x	Sitzlar et al.. 2009
			x	x					Sitzlar et al., 2009
x		x			x		x		Saldanha et al., 2011

aromatase and endoplasmic reticulum expression in zebra finches in nucleus HVC as well as in the mediocaudal telelencephalon (Jacobs et al., 1996). As for the second case, whereas aromatase activity was reported in the zebra finch HVC (Vockel et al., 1990a, 1990b) in situ hybridization showed little to no aromatase mRNA expression (Shen et al., 1995), a result confirmed by light-level immunocytochemical studies of the zebra finch HVC. HVC is a demonstrated site of estrogen action in control of singing behavior (Meitzen et al., 2007) and the supply of estrogen to HVC has been a puzzle. In a second case, aromatase activity and mRNA are both abundant in the songbird HP (Saldanha et al., 1998), a region with a demonstrated role for estrogen mediation of spatial memory (Oberlander et al., 2004), but immunocytochemical studies show few aromatase-positive somata that stain only weakly for aromatase. These examples for a lack of conformity between aromatase transcript and protein raise questions such as: Where was the mRNA or protein going? Or if aromatase protein was undetectable with antibodies, why was aromatase activity readily detectable? The concept of synaptic aromatase helps resolve these paradoxical observations.

The Synaptocrine Hypothesis—a Novel Form of Estrogen Delivery

Aromatase in the CNS appears exclusively neuronal in homeotherms; neurons are the only cells immunostained in the brain and spinal cords of unmanipulated, uninjured birds and mammals (Foidart et al., 1994; Naftolin et al., 1996; Saldanha et al., 2000b). Staining is cytoplasmic, and is visible over somata and in processes. The detection of immunoreactive aromatase in somal cytoplasm is consistent with the localization of aromatase to endoplasmic reticulum a subcellular organelle distributed widely in the somal and dendritic cytoplasm.

Under light-field microscopy, however, in addition to immunoreactive somata, dendrites, and putative axonal processes, some aromatase immunostaining appears punctate (Saldanha et al., 2000b; Peterson et al., 2005). Of importance is the observation that this punctate staining is observable both in areas rich in aromatase-expressing soma, and in areas apparently devoid of somal aromatase (Saldanha et al., 2000b). These puncta could be synaptic terminals, but the limits of light-microscopy preclude their unequivocal identification. Two techniques; differential centrifugation and immunoelectron microscopy were utilized to further investigate the ultrastructural compartmentalization of aromatase in neurons. Indeed, both techniques revealed a surprising, novel compartmentalization of aromatase, and presumably, estrogen provision, in the avian brain. Specifically, when brain homogenates are subject to differential centrifugation (Schlinger & Callard, 1989; Rohmann et al., 2007; Remage-Healey et al., 2009), nerve terminals pinch-off and reseal. These synaptosomes can be isolated and purified for neurochemical analyses. Aromatase activity was enriched in purified synaptosomal preparations of the quail hypothalamus (Schlinger & Callard, 1989) and passerine telencephalon (Schlinger & Arnold, 1992b;

Rohmann et al., 2007; Remage-Healey et al., 2009). Together, these studies established the localization of aromatase within terminals.

True synaptic localization of this enzyme was confirmed and extended by studies that used immunoelectron microscopy (EM) on tissue from galliform, rodent, and primate brains (Naftolin et al., 1996). Aromatase positive synaptic terminals were documented in the hypothalamus, bed nucleus of the stria terminalis and amygdala. Although the identification of synaptic aromatase was unequivocal, its function remained mysterious, because all these areas were rife with somal aromatase and prevented an independent assessment of somal and synaptic estrogen provision. This distinction has proved easier to study in the passerine brain.

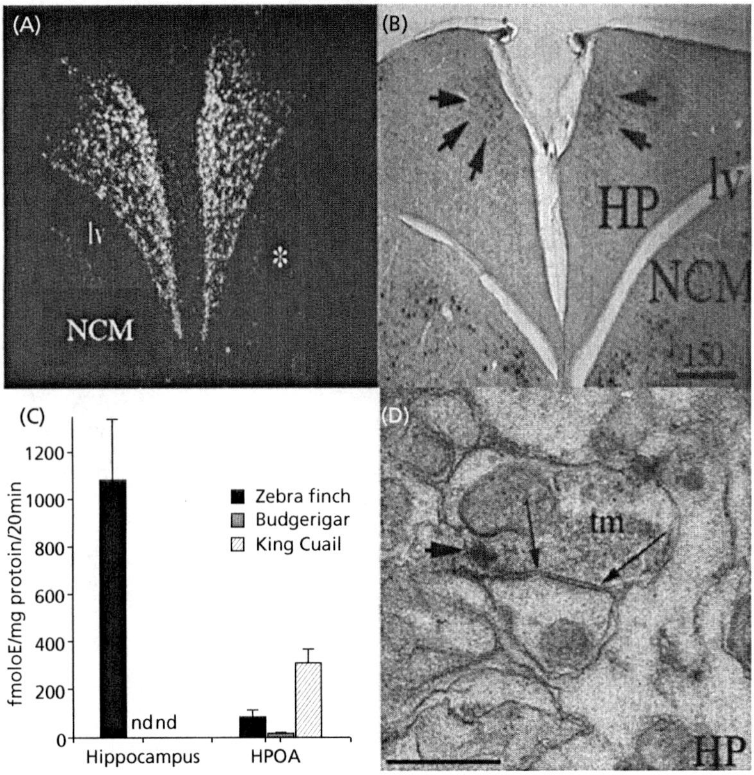

Figure 6–1. Aromatase Expression in the Hippocampus (HP) of the Zebra Finch. Whereas transcript (*A*) is abundant at this locus, visible immunoproduct (*B*) is surprisingly sparse. This is despite readily detectable aromatase activity (*C*). Resolution of this paradox may lie in the observation that much of the aromatase in the zebra finch HP is sequestered in presynaptic boutons (*D*) (Magnification in *A* and *B* are equivalent, magnification bar in *D* = 1 micron). Data presented herein are adapted from (Saldanha et al., 1998b, et al., 2000b; Peterson et al., 2005).

In the zebra finch, large numbers of aromatase-expressing punctuate structures are visible under light-microscopy, particularly in brain areas wherein somal immunoproduct is either lacking or extremely sparse. Since unequivocal identification of synaptic profiles requires EM, we developed an immunoEM protocol using fixation techniques compatible with an antibody specific to zebra finch aromatase to visualize the subcellular distribution of this enzyme (Peterson et al., 2005). These studies showed that aromatase is expressed in presynaptic boutons in every brain area examined including areas rich in somal aromatase such as the preoptic area, NCM, HP, and HVC.

These studies were strongly supported by experiments examining the subcellular localization of aromatase activity using cellular subfractionation techniques. Although aromatase is largely bound to endoplasmic reticulum (microsomal) in subcellular locations in the HPOA of quail (Schlinger & Callard, 1989) and in the brains of mammals (Roselli & Resko, 1984) its distribution is nearly equal in synaptic boutons (synaptosomes) and microsomes in the zebra finch telencephalon (Rohmann et al., 2007). Thus, a substantial portion of telencephalic aromatase (about 50% of total neural aromatase) is present in presynaptic boutons, underscoring the potential importance of synaptic estrogens-provision in songbirds and greatly increasing our chances of understanding its physiological relevance.

SOMAL AND SYNAPTIC AROMATASE IS DISTRIBUTED, REGULATED, AND MAY FUNCTION INDEPENDENTLY

In addition to the abundance of synaptosomal aromatase in the passerine brain, its distribution relative to somal aromatase makes synaptocrine aromatization easier to study in this animal model. Notably, although many somata immunoreactive for aromatase were detected in the HPOA and NCM, the ventral HP contained very low numbers of immunoreactive somata but abundant presynaptic boutons with aromatase immunoproduct (Saldanha et al., 2000b; Peterson et al., 2005). This pattern of distribution was even more striking in the HVC, in which immunoreactive somata were undetectable but immunoreactive presynaptic boutons are abundant. Thus, in the zebra finch, areas with low or even undetectable *somal* aromatase nonetheless contain high levels of *synaptic* aromatase. These data suggest the possibility that the major source of estradiol within these areas is located in presynaptic boutons.

The zebra finch HP bears similarities with HVC in having elevated aromatase activity (Saldanha et al., 1998; Saldanha et al., 2000b), many aromatase-positive synaptic profiles and a relatively low number of aromatase-positive somata (Fig, 6–1). These somata are spatially restricted to a small population of neurons at the dorsomedial aspect (Saldanha et al., 2000a). Neurons in the dorsomedial HP project almost exclusively within the HP itself (Székely & Krebs, 1996), supporting the hypothesis that aromatase-positive neurons in the dorsomedial HP are interneurons that transport aromatase protein down collaterals to terminals within the HP. A comparison of aromatase mRNA expression using in situ hybridization (Shen et al., 1995; Saldanha et al., 1998) with aromatase protein expression using ICC (Saldanha et al., 2000b) shows more cells with mRNA than with protein. This observation raises the possibility that some neurons may transport most or all of their aromatase from somata into terminals. Although this latter possibility remains to be tested, given the overall low somal but elevated presynaptic aromatase immunoreactivity seen in the HP (Peterson et al., 2005), we believe that estrogen provision within the songbird hippocampus is largely synaptic.

These studies confirm predictions of the synaptocrine hypothesis (Saldanha et al., 2011) that the steroidogenic enzyme aromatase is present in synapses and these synapses reside where other sources of aromatase are absent. Hopefully, these observations for synaptic estrogen production in songbirds will provoke similar studies in other vertebrate taxa.

SEASONAL REGULATION OF AROMATASE

In most animals, breeding is restricted to a season when the likelihood of reproductive success is maximal. As a consequence, the performance of sex-specific reproductive behavior is also restricted to the appropriate breeding season. In temperate zone breeding animals, photoperiod, the primary cue timing seasonal reproduction, activates the hypothalamic-pituitary-gonadal axis stimulating the release of gonadal steroids that, in turn, activate appetitive and consummatory sexual behaviors. In males of many species, the increased testicular secretion of testosterone acts within the hypothalamus to upregulate neuronal aromatase expression (Pasmanik et al., 1988; Balthazart et al., 1990; Roselli, 1991; Fusani et al., 2003;); this enhanced conversion of testosterone to estradiol at this time of year is required for the activation of masculine sex behavior (Schlinger & Callard, 1989; Fusani et al., 2003; Roselli et al., 2003; Balthazart et al., 2004; Portillo et al., 2007; Wu et al., 2009).

In the well-studied Japanese quail (Balthazart & Ball, 1998a), hypothalamic aromatase is present in both microsomal and synaptosomal preparations. Whereas exposure to photostimulatory long days stimulates an overall increase in hypothalamic aromatase, activity in cellular subfractions is elevated differentially; microsomal aromatase activity increases approximately 2.5-fold, whereas synaptosomal aromatase increases about 6-fold (Schlinger & Callard, 1989). It is unknown whether the increase in synaptosomal enzyme results from an increase in the number of aromatase-positive synapses, an increase in the amount of aromatase per synapse, or in the amount of biologically active aromatase per synapse. Nevertheless, in all likelihood, seasonal increases in synaptic aromatase may be essential to providing the requisite amounts of bioactive estrogen to activate masculine reproductive behavior in this species.

SEX DIFFERENCES IN AROMATASE

Sexually dimorphic expression of synaptic aromatase may also be the basis for some sex differences in steroidal signaling in the songbird brain. The zebra finch is a long-proven model in behavioral neuroendocrinology in part because of its dramatic sexual dimorphisms in singing behavior (Immelman, 1969) and in brain structure (Nottebohm & Arnold, 1976). Estrogens are implicated in at least some aspects of the masculine development of the song circuitry (Gurney & Konishi, 1980; Holloway & Clayton, 2001), in the plasticity underlying the masculine learning of song (Bottjer et al., 1997), in the seasonal growth of song circuitry in seasonal breeders (Tramontin et al., 2003), and in the functional activation of this circuitry for the production of male song (Meitzen et al., 2007). Nevertheless, despite numerous attempts to describe sex differences in estrogen provision in zebra finches that are the basis of these estrogen-dependent phenotypes, including studies of circulating estradiol (Hutchison et al., 1984; Adkins-Regan et al., 1990; Schlinger & Arnold, 1992a), gonadal estrogen synthesis (Schlinger & Arnold, 1991, 1992a), neural aromatase activity and expression (Vockel et al., 1990a, 1990b; Schlinger & Arnold, 1991; Shen et al., 1995; Wade et al., 1995), and aromatase immunoproduct (Schlinger, 2005), the *only* masculine biased feature so far consistently identified is the sexually dimorphic number of aromatase-positive fibers and terminals (Saldanha et al., 2000b; Peterson et al., 2005; Rohmann et al., 2007).

In keeping with measures of aromatase activity (Vockel et al., 1990a, 1990b), immunostaining with a zebra finch-specific antibody showed no sex differences in aromatase-positive somata in several brain regions, including the HPOA and NCM (Saldanha et al., 2000b). Nevertheless, males possessed significantly higher numbers of aromatase-positive fibers in these brain areas, suggesting that aromatase expressing neurons were more branched in males or that the aromatase protein was selectively compartmentalized into fibers in the male zebra finch brain (Saldanha et al., 2000b).

These studies were conducted using conventional light microscopy, preventing unequivocal inferences as to the dendritic and/or axonal nature of the fibers in question. Based on these observations, as part of our ultrastructural analysis of the subcellular localization of aromatase, we counted the number of synaptic terminals containing aromatase-immunoreactivity. We found that aromatase expressing presynaptic boutons were more numerous in the male brain relative to females (Peterson et al., 2005). Indeed, both the absolute and the relative (as a function of total synaptic profiles) incidence of aromatase-expressing presynaptic boutons were male biased. These studies were significant in revealing that the male zebra finch brain may be provided with higher levels of estradiol at various telencephalic areas relative to the female. Notably, somal and dendritic aromatase expression were not dimorphic (Peterson et al., 2005), supporting the concept that synaptocrine, and not paracrine, estradiol provision better explains neural estrogen provision in these brain regions of this species.

To confirm these immunohistochemical observations and to verify that the immunoreactivity we detected was translated into functional protein, we reasoned that subfractionation of zebra finch brain homogenates should parallel the results of EM studies, that is, synaptosomal preparations of brain homogenates should have greater aromatase activity in males than in females whereas microsomal preparations, presumably better reflecting somal aromatase, should show no sex differences. As predicted, synaptosomal preparations of the male telencephalon had significantly higher levels of aromatase activity than did similar preparations of the female telencephlon; microsomal preparations as well as measures of whole homogenates themselves demonstrated no sex differences (Rohmann et al., 2007). Given the significant overall enrichment of aromatase in synaptosomal preparations versus microsomal preparations, it is unclear why we do not detect the overall sex difference in whole homogenates. Perhaps proteases and other steroid-metabolic enzymes present in homogenates obscure detection of sex differences. Measures of the activity in purified subfractions, however, empirically matches our counts of aromatase-positive profiles at the EM level, providing independent confirmation of our conclusion that estrogenic synaptocrine signaling is elevated in the brain of male songbirds as compared with females and forms the basis for sex differences in neuroestrogenic signaling.

As described previously, aromatase in brain is found in two primary subcellular compartments cytoplasmic endoplasmic reticulum (microsomes subfraction) and specialized synapses (synaptosomal subfraction). In the zebra finch forebrain approximately half of the total aromatase is localized to the synaptosomal fraction (Rohmann et al., 2007), providing additional evidence for the anatomical localization of aromatase in presynaptic terminals (Peterson et al., 2005). Synaptosomal aromatase is particularly enriched in the forebrain of adult males relative to females (Rohmann et al., 2007), suggesting a link to male-typical behavior and/or neural processing in this species, such as singing, song learning and/or auditory feedback.

SYNAPTIC AROMATASE AND SINGING BEHAVIOR

The interconnected "song system" nuclei in the caudal forebrain of zebra finches receives projections that contain aromatase-positive synaptic boutons (Saldanha et al., 2000b; Peterson et al., 2005; London et al., 2009). Thus there has been a recent focus on singing and potential synaptocrine regulation of aromatase in the male zebra finch brain. In addition, brain aromatase activity is altered rapidly during changes in behavioral state in vertebrates (Black et al., 2005; Cornil et al., 2005; Cornil & Charlier, 2010).

To examine the regulation of synaptic aromatase, male zebra finches were briefly exposed to natural stimuli and brains were immediately collected. Males were housed singly in sound-attenuation chambers and exposed to females for 30 minutes, and some males sang to the females (singers) whereas others did not sing (nonsingers). Following rapid decapitation, brains were carefully dissected to separate the posterior telencephalon (containing several important song nuclei enriched in presynaptic aromatase) from the anterior telencephalon (fewer aromatase-positive cells and little to no terminal aromatase). Aromatase activity was elevated more than two-fold in the singers relative to the nonsingers, and this aromatase upregulation was specifically localized to the synaptosomal fraction within the posterior telencephalon (Remage-Healey et al., 2009). Playback of zebra finch song versus control white noise to males in chambers did not cause similar differences in synaptosomal aromatase activity, indicating that the association between singing and synaptic estrogen production likely did not depend on auditory self-stimulation. Courtship singing was thus associated with an upregulation of aromatase activity specifically in synaptic terminals, although it is unclear whether this is a short-term phenomenon occurring during the brief behavioral interactions, or a longer-term phenomenon in which males with elevated terminal aromatase are more likely to sing. These experiments provided evidence that the vertebrate CNS has the capacity to regulate aromatase in synaptic terminals independent of somal aromatase in a behaviorally relevant context.

CONCLUSIONS

Studies of avian brain aromatase have unveiled new perspectives on neurosteroidal signaling. Estrogen synthetic neurons are not only a feature of circuits controlling copulatory and aggressive behaviors, but they can be a crucial element of additional sensory-motor circuits. The subcellular distribution of aromatase in brain gives this enzyme the potential to provide neuroestrogen signals via paracrine mechanisms as well as via targeted synaptic connections. This latter form of synaptocrine signaling, first identified in the avian brain, may represent a conserved neuromodulatory mechanism in other vertebrates as well as in other brain regions. Studies of birds will likely continue to shed new light on neurosteroidal modulation of brain and behavior for years to come.

ACKNOWLEDGMENTS

We thank members of the Saldanha and Schlinger laboratories for their assistance on many of the experiments described herein. This work was supported by the National Institutes of Health; F32NS058009, R01MH061944, RO1NS042767 and K99/R00NS066179.

REFERENCES

Adkins-Regan, E., Abdelnabi, M., Mobarak, M., & Ottinger, M. A. (1990). Sex steroid levels in developing and adult male and female zebra finches (Poephila guttata). *Gen Comp Endocrinol, 78*(1), 93–109.

Balthazart, J., Baillien, M., Cornil, C. A., & Ball, G. F. (2004). Preoptic aromatase modulates male sexual behavior: slow and fast mechanisms of action. *Physiol Behav, 83*(2), 247–270.

Balthazart, J., & Ball, G. F. (1998a). The Japanese quail as a model system for the investigation of steroid-catecholamine interactions mediating appetitive and consummatory aspects of male sexual behavior. *Annu Rev Sex Res, 9*, 96–176.

Balthazart, J., & Ball, G. F. (1998b). New insights into the regulation and function of brain estrogen synthase (aromatase). *Trends Neurosci, 21*(6), 243–249.

Balthazart, J., Ballien, M., Cornil, C. A. & Ball, G. F. (2004). Preoptic aromatase modulates male sexual behavior: slow and fast mechanisms of action. *Physiol Behav, 83*(2): 247–270.

Balthazart, J., Foidart, A., Surlemont, C., Vockel, A., & Harada, N. (1990). Distribution of aromatase in the brain of the Japanese quail, ring dove, and zebra finch: an immunocytochemical study. *J Comp Neurol, 301*(2), 276–288.

Black, M. P., Balthazart, J., Baillien, M., & Grober, M. S. (2005). Socially induced and rapid increases in aggression are inversely related to brain aromatase activity in a sex-changing fish, Lythrypnus dalli. *Proc Biol Sci, 272*(1579), 2435–2440.

Bottjer, S. W., Roselinsky, H., & Tran, N. B. (1997). Sex differences in neuropeptide staining of song-control nuclei in zebra finch brains. *Brain Behav Evol, 50*(5), 284–303.

Cornil, C. A., & Charlier, T. D. (2010). Rapid behavioural effects of oestrogens and fast regulation of their local synthesis by brain aromatase. *J Neuroendocrinol, 22*(7), 664–673.

Cornil, C. A., Dalla, C., Papadopoulou-Daifoti, Z., Baillien, M., Dejace, C., Ball, G. F., & Balthazart, J. (2005). Rapid decreases in preoptic aromatase activity and brain monoamine concentrations after engaging in male sexual behavior. *Endocrinology, 146*(9), 3809–3820.

Foidart, A., Harada, N., & Balthazart, J. (1994). Effects of steroidal and nonsteroidal aromatase inhibitors on sexual behavior and aromatase-immunoreactive cells and fibers in the quail brain. *Brain Res, 657*, 105–123.

Fusani, L., Van't Hof, T., & Hutchison, J. B. (2003). Season-related changes in circulating androgen, brain aromatase, and perch-calling in male ring doves. *Gen Comp Endocrinol,* i(2), 142–147.

Gurney, M. E., & Konishi, M. (1980). Hormone-induced sexual differentiation of brain and behavior in zebra finches. *Science, 208*(4450), 1380–1383.

Holloway, C. C., & Clayton, D. F. (2001). Estrogen synthesis in the male brain triggers development of the avian song control pathway in vitro. *Nat Neurosci, 4*(2), 170–175.

Hutchison, J. B., Wingfield, J. C., & Hutchison, R. E. (1984). Sex differences in plasma concentrations of steroids during the sensitive period for brain differentiation in the zebra finch. *J Endocrinol, 103*(3), 363–369.

Immelman, K. (1969). Song development in the zebra finch and other estrildid finches. In: Hinde R. (ed.), *Bird Vocalizations*. Cambridge, UK: Cambridge University Press, pp. 61–74.

Jacobs, E. C., Arnold, A. P., & Campagnoni, A. T. (1996). Zebra finch estrogen receptor cDNA: cloning and mRNA expression. *J Steroid Biochem Mol Biol, 59*(2), 135–145.

London, S. E., Remage-Healey, L., & Schlinger, B. A. (2009). Neurosteroid production in the songbird brain: a re-evaluation of core principles. *Front Neuroendocrinol, 30*(3), 302–314.

Meitzen, J., Moore, I. T., Lent, K., Brenowitz, E. A., & Perkel, D. J. (2007). Steroid hormones act transsynaptically within the Forebrain to regulate neuronal phenotype and song stereotypy. *J Neurosci, 27*(44), 12045–12057.

Naftolin, F., Horvath, T. L., Jakab, R. L., Leranth, C., Harada, N., & Balthazart, J. (1996). Aromatase immunoreactivity in axon terminals of the vertebrate brain. An immunocytochemical study on quail, rat, monkey and human tissues. *Neuroendocrinology, 63*(2), 149–155.

Nottebohm, F., & Arnold, A. P. (1976). Sexual dimorphism in vocal control areas of the songbird brain. *Science, 194*(4261), 211–213.

Oberlander, J. G., Schlinger, B. A., Clayton, N. S., & Saldanha, C. J. (2004). Neural aromatization accelerates the acquisition of spatial memory via an influence on the songbird hippocampus. *Horm Behav, 45*(4), 250–258.

Pasmanik, M., Schlinger, B. A., & Callard, G. V. (1988). In vivo steroid regulation of aromatase and 5 alpha-reductase in goldfish brain and pituitary. *Gen Comp Endocrinol, 71*(1), 175–182.

Peterson, R. S., Yarram, L., Schlinger, B. A., & Saldanha, C. J. (2005). Aromatase is pre-synaptic and sexually dimorphic in the adult zebra finch brain. *ProcBiol Sci, 272*(1576), 2089–2096. doi: 10.1098/rspb.2005.3181.

Portillo, W., Castillo, C. G., Retana-Marquez, S., Roselli, C. E., & Paredes, R. G. (2007). Neuronal activity of aromatase enzyme in non-copulating male rats. *J Neuroendocrinol, 19*(2), 139–141.

Remage-Healey, L., Maidment, N. T., & Schlinger, B. A. (2008). Forebrain steroid levels fluctuate rapidly during social interactions. *Nat Neurosci 11*(11), 1327–1334.

Remage-Healey, L., Oyama, R. K., & Schlinger, B. A. (2009). Elevated aromatase activity in forebrain synaptic terminals during song. *J Neuroendocrinol, 21*(3), 191–199.

Rohmann, K. N., Schlinger, B. A., & Saldanha, C. J. (2007). Subcellular compartmentalization of aromatase is sexually dimorphic in the adult zebra finch brain. *Dev Neurobiols, 67*(1), 1–9.

Roselli, C. E. (1991). Sex differences in androgen receptors and aromatase activity in microdissected regions of the rat brain. *Endocrinology, 128*(3), 1310–1316.

Roselli, C. E., & Resko, J. A. (1984). Androgens regulate brain aromatase activity in adult male rats through a receptor mechanism. *Endocrinology, 114*(6), 2183–2189.

Roselli, C. E., Resko, J. A., & Stormshak, F. (2003). Estrogen synthesis in fetal sheep brain: effect of maternal treatment with an aromatase inhibitor. *Biol Reprod, 68*(2), 370–374.

Saldanha, C. J., Popper, P., Micevych, P. E., & Schlinger, B. A. (1998). The passerine hippocampus is a site of high aromatase: inter- and intraspecies comparisons. *Horm Behav, 34*(2), 85–97.

Saldanha, C. J., Remage-Healey, L., & Schlinger, B. A. (2011). Synaptocrine signaling: steroid synthesis and action at the synapse. *Endocr Rev, 32*(4), 532–549.

Saldanha, C. J., Schultz, J. D., London, S. E., & Schlinger, B. A. (2000a). Telencephalic aromatase but not a song circuit in a sub-oscine passerine, the golden collared manakin (Manacus vitellinus). *Brain Behav Evol, 56*(1), 29–37.

Saldanha, C. J., Tuerk, M. J., Kim, Y. H., Fernandes, A. O., Arnold, A. P., & Schlinger, B. A. (2000b). Distribution and regulation of telencephalic aromatase expression in the zebra finch revealed with a specific antibody. *J Comp Neurol, 423*(4), 619–630.

Saldanha, C. J., Walters, B. J., & Fraley, G. S. (2010). Neurons that co-localize aromatase- and kisspeptin-like immunoreactivity may regulate the HPG axis of the Mallard drake (Anas platyrhynchos). *Gen Comp Endocrinol, 166*(3), 606–613.

Schlinger, B. A. (1997). Sex steroids and their actions on the birdsong system. *J Neurobiol, 33*(5), 619–631.

Schlinger, B. A. (2005). Behavioral neuroendocrinology evolving: contributions of comparative and field studies. *Horm Behav, 48*(4), 349–351.

Schlinger, B. A., & Arnold, A. P. (1991). Brain is the major site of estrogen synthesis in a male songbird. *Proc Natl Acad Sci U S A, 88*(10), 4191–4194.

Schlinger, B. A., & Arnold, A. P. (1992a). Circulating estrogens in a male songbird originate in the brain. *Proc Nat Acad Sci U S A, 89,* 7650–7653.

Schlinger, B. A., & Arnold, A. P. (1992b). Plasma sex steroids and tissue aromatization in hatchling zebra finches: implications for the sexual differentiation of singing behavior. *Endocrinology, 130,* 289–299.

Schlinger, B. A., & Callard, G. V. (1989). Localization of aromatase in synaptosomal and microsomal subfractions of quail (Coturnix coturnix japonica) brain. *Neuroendocrinology, 49*(4), 434–441.

Schumacher, M., & Balthazart, J. (1984). The postnatal demasculinization of sexual behavior in the Japanese quail (Coturnix coturnix japonica). *Horm Behav, 18*(3), 298–312. doi: 0018-506X(84)90018-7 [pii].

Shen, P., Schlinger, B. A., Campagnoni, A. T., & Arnold, A. P. (1995). An atlas of aromatase mRNA expression in the zebra finch brain. *J Comp Neurol, 360*(1), 172–184. doi: 10.1002/cne.903600113.

Simpson, E. R., Mahendroo, M. S., Means, G. D., Kilgore, M. W., Hinshelwood, M. M., Graham-Lorence, S., et al. (1994). Aromatase cytochrome P450, the enzyme responsible for estrogen biosynthesis. *Endocr Rev, 15*(3), 342–355.

Simpson, E. R., Clyne, C., Rubin, G., Boon, W. C., Robertson, K., Britt, K., Speed, C., & Jones, M. 2002. *Annual Rev Physiol, 64,* 93-127.

Székely, A. D., & Krebs, J. R. (1996). Efferent connectivity of the hippocampal formation of the zebra finch (Taenopygia guttata): an anterograde pathway tracing study using Phaseolus vulgaris leucoagglutinin. *J Comp Neurol, 368,* 198–214.

Tramontin, A. D., Wingfield, J. C., & Brenowitz, E. A. (2003). Androgens and estrogens induce seasonal-like growth of song nuclei in the adult songbird brain. *J Neurobiol, 57*(2), 130–140. doi: 10.1002/neu.10263.

Vockel, A., Prove, E., & Balthazart, J. (1990a). Effects of castration and testosterone treatment on the activity of testosterone-metabolizing enzymes in the brain of male and female zebra finches. *J Neurobiol, 21*(5), 808–825.

Vockel, A., Prove, E., & Balthazart, J. (1990b). Sex- and age-related differences in the activity of testosterone-metabolizing enzymes in microdissected nuclei of the zebra finch brain. *Brain Res, 511*(2), 291–302.

Wade, J., Schlinger, B. A., & Arnold, A. P. (1995). Aromatase and 5 beta-reductase activity in cultures of developing zebra finch brain: an investigation of sex and regional differences. *J Neurobiol, 27*(2), 240–251.

Wu, M. V., Manoli, D. S., Fraser, E. J., Coats, J. K., Tollkuhn, J., Honda, S., et al. (2009). Estrogen masculinizes neural pathways and sex-specific behaviors. *Cell, 139*(1), 61–72.

Chapter 7

Expression, Regulation, and Potential Functions of Aromatase in Radial Glial Cells of the Fish Brain

Elisabeth Pellegrini, Colette Vaillant, Nicolas Diotel, Pascal Benquet, François Brion and Olivier Kah

INTRODUCTION

Studies of aromatase in the brain of fish has for decades produced surprising findings that challenge our assumptions about brain aromatase in comparison with what is known in other vertebrate models. This stems from the pioneering studies by Callard and colleagues back in the late '70s and early '80s, showing that the brain of adult fish exhibits an unusually elevated aromatase activity, evaluated as being one hundred to one thousand times higher than in any other vertebrates (Callard et al., 1978; Pasmanik & Callard, 1985, 1988b). This feature was first highlighted in the goldfish, the Atlantic salmon, the stickleback, and the African catfish, in which the forebrain consistently exhibited a high aromatase activity (Pasmanik & Callard, 1985; Timmers & Lambert, 1987; Borg et al., 1989).

Since then, aromatase in the fish brain has been an exciting research field leading to many surprising discoveries that did not comply with the accepted views regarding brain aromatase. First, in contrast to most vertebrates, teleost fishes generally have two aromatase genes, one of them being responsible for the high brain aromatase activity (see later). Another striking feature that will be discussed in detail in this chapter concerns the specific expression of aromatase in a unique brain cell type, the radial glial cells (RGCs). Such cells play a key role in embryonic neurogenesis in the brain of all vertebrates (Kriegstein & Alvarez-Buylla, 2009). In adult fish, RGCs persist and sustain the adult neurogenesis that occurs in many brain regions (Ekström et al., 2001; Pellegrini et al., 2007; Rothenaigner et al., 2011b). Third, the high aromatase activity found in the brain of sexually mature fish is because of the fact that the *cyp19a1b* gene is exquisitely sensitive to estrogens (Menuet et al., 2005).

This chapter intends to summarize the most relevant and recent information on the sites of aromatase expression in the fish brain, its cell-specific regulation, and potential functions. Data regarding the historical aspects of aromatase activity in fish are dealt with in another chapter (see chapter 2, this volume). It is important to stress here at the outset that testosterone is not a male hormone in the traditional sense in fish. The fish testis produces a variety of 11-oxygenated nonaromatizable androgens that mediate many, if not all, of the male-specific sexual behaviors and other functions (Kobayashi & Nakanishi, 1999; Munakata & Kobayashi, 2010).

TWO AROMATASE GENES IN TELEOST FISHES

Until now, most teleost fishes that have been investigated have two aromatase genes that are officially named *cyp19a1a* and *cyp19a1b*, but that were before given different names: *cyp19a1a* (cyp19a, aromatase A, AroA, CYP19A, P450aromA) and *cyp19a1b* (cyp19b, aromatase B, AroB, CYP19B, P450aromB). These genes generate two structurally and functionally different isoforms that, in this review, will be referred to aromatase A and aromatase B, mostly expressed in the ovary and brain, respectively (Diotel et al., 2010a). It is believed that these two genes were generated by the teleost specific genome duplication, named 3R (third round) that occurred at the emergence of the teleost lineage (Fig. 7–1) (Steinke et al., 2006). Interestingly, whereas the eel genome was also duplicated as indicated by the fact that eels have 8 *hox* genes (Guo et al., 2010), only one *cyp19a1* gene has been retrieved so far from the European and Japanese eels (Ijiri et al., 2003; Tzchori et al., 2004). This unique gene is expressed in both the gonads and the brain and its expression submitted to alternative usage of different promoters (Ijiri et al., 2003). It is thus possible that one copy of the *cyp19a1* gene in eel was lost soon after the 3R duplication. Nevertheless, recent data in Japanese eel show that, similar to other teleosts (see the following), expression of aromatase mRNA and protein is limited to RGCs (Shan-Ru Jeng et al., unpublished data).

Tongue sole (ABM90641; ABL74474), *Cyprinus carpio*/Common carp (ACB13198; ACB13197), *Danio rerio*/Zebrafish (AAV41033; AAK00643), *Dicentrarchus labrax*/Sea bass (AAM95455; CAC21712), *Epinephelus akaara*/Hong Kong grouper (AAS58447; AAS58448), *Epinephelus coioides*/Orange-spotted grouper (AAR97602; AAR97601), *Fundulus heteroclitus*/Mummichog (AAR97269; AAR97268), *Gobiodon histrio*/ Green clown goby (AAV91180; AAV91177), *Halichoeres tenuispinis*/Chinese Wrasse (AAR37047; AAR37048), *Hippoglossus hippoglossus*/Atlantic halibut (AAY26901; CAC36394), *Ictalurus punctatus*/Channel catfish (AAL14612; AAB32613), *Kryptolebias marmoratus*/Mangrove rivulus (ABC68613; ABC68614), *Mugil cephalus*/Black mullet (AAW72730; AAW72732), *Odontesthes bonariensis*/Argentinian silverside (AAQ88434; ABK30807), *Oreochromis mossambicus*/Java tilapia (AAD31030; AAD31031), *Oryzias latipes*/Japanese medaka (AAP83449; BAA11656), *Pseudolabrus japonicas*/Wrasse (ABB96486; ABB96485), *Rhabdosargus sarba*/Goldlined seabream (ABC70868; ABC70869), *Silurus meridionalis*/Catfish (AAP83132; AAP83133), and *Takifugu rubripes*/Fugu rubripes (BAF93507; BAF93506).

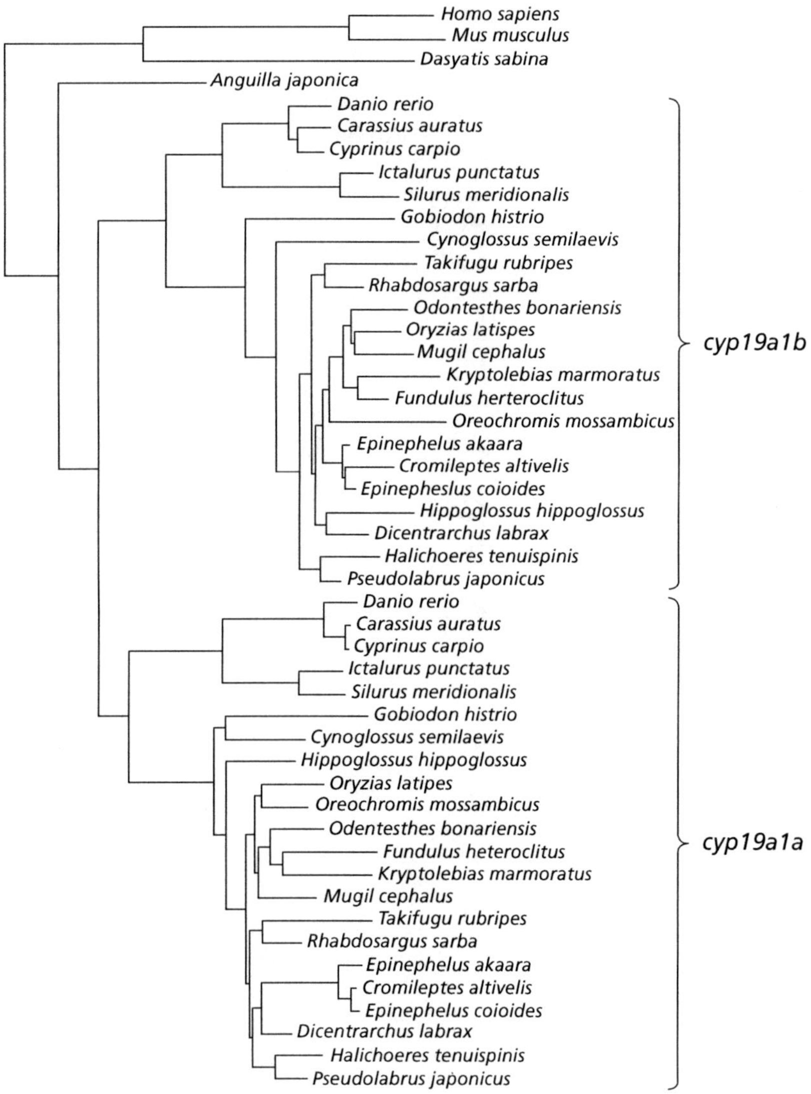

Figure 7–1. Phylogenic Analyses of Cyp19a1 Genes of Vertebrates. The tree shows that teleost fish have two genes with the exception of the Japanese eel (*4th line from top*). In contrast, Sarcopterygians have only one gene. The tree was constructed using online phylogeny program (http://www.phylogeny.fr). The mammalian Cyp19 proteins from *Homo sapiens* (AAA52141) and *Mus musculus* (BAA00551) served as outgroup to root the tree. Comparisons were made with the amino acid sequences (GenBank protein ID number respectively from brain and ovary when two genes were identified between brackets) of *Dasyatis sabina*/Atlantic stingray (AAF04617), *Anguilla japonica*/Japanese eel (AAS47028), *Carassius auratus*/Goldfish (BAA23757; BAA23758), *Cromileptes altivelis*/Humpback grouper (AAV91181; AAV91178), *Cynoglossus semilaevis*/

EXPRESSION OF THE AROMATASE MESSENGER AND PROTEIN IN THE BRAIN OF FISH

The first detailed information regarding aromatase distribution comes from studies in which aromatase activity was measured on brain sections (Timmers et al., 1987). The highest activity was detected in the preoptic area, followed by the mediobasal hypothalamus, the rostral part of the telencephalon, and the dorsal parts of the mesencephalon (tectum opticum and torus semicircularis). A moderate aromatase activity was observed in remaining parts of the brain (Timmers et al., 1987). The unique elevated levels of aromatase activity in fish brain are supported by the presence of a high abundance of the *cyp19a1b* transcripts in the brain homogenates of several teleosts species including the goldfish, the Nile tilapia, the zebrafish, the trout, and the common carp (Gelinas et al., 1998; Kishida et al., 2001; Valle et al., 2002; Chang et al., 2005; Tang et al., 2010). Those results imply that high steady-state mRNA levels of *cyp19a1b* in the fish brain are the determinant of the strong enzyme activity measured therein.

It is in the plainfin midshipman that the cellular localization of aromatase-expressing cells was achieved for the first time at both the mRNA and protein levels using homologous aromatase riboprobe and antibody, respectively (Forlano et al., 2001). Aromatase B immunoreactivity was consistently concentrated in ventricular position in the olfactory bulbs, the telencephalon, the preoptic area, and the hypothalamus. Aromatase B-positive cells were also observed more caudally within the sonic motor nucleus, around the fourth ventricle, and the cerebral aqueduct. The most remarkable outcome of this pioneer study was that cells expressing aromatase (mRNA and protein) did not express the neuronal marker Hu and thus did not correspond to neurons, unlike what has been mainly assumed in mammals and birds brain under normal condition (Balthazart & Ball, 1998; Yague et al., 2006; Zhao et al., 2007; Yague et al., 2008). In contrast, aromatase B-positive cells presented the typical morphology of RGCs with a small cell body bordering the ventricle and a long cytoplasmic extension entering the brain parenchyma (Rakic, 1978). In adult trout, similar to the plainfin midshipman, an intense expression of *cyp19a1b* mRNA was also found in cells lining the ventricles throughout the forebrain, particularly in the telencephalon, the preoptic region, the mediobasal hypothalamus, the torus semicircularis and the optic tectum (Menuet et al., 2003). Immunohistochemistry using the antibodies designed in midshipman clearly labeled cells with the typical morphology of RGCs, which also coexpressed protein S100ß (Pellegrini et al., 2005). Similar data based on the same antibodies were obtained in other species such as the pejerrey, *Odontesthes bonariensis*, and the bluehead wrasse, *Thalassoma bifasciatum*. Immunohistochemistry thus revealed a conserved pattern of aromatase B cells distribution within the brain with immunoreactive cells bordering the ventricles in the telencephalon, the preoptic area and the hypothalamus with the characteristic morphology of RGCs (Strobl-Mazzulla et al., 2005; Marsh et al., 2006; Strobl-Mazzulla et al., 2008).

The distribution of cells expressing aromatase has also been extensively studied in the brain of developing and adult zebrafish by in situ hybridization

(Goto-Kazeto et al., 2004; Menuet et al., 2005; Pellegrini et al., 2005; Chiang et al., 2001), immunohistochemistry (Menuet et al., 2005; Kallivretaki et al., 2007; Pellegrini et al., 2007; Mouriec et al., 2009; März et al., 2010; Diotel et al., 2011), and using the transgenic fish tg(*cyp19a1b)*-GFP (Tong et al., 2009). The results consistently demonstrated a very high expression in RGCs mainly in the anterior ventral part of the brain (i.e. ventral olfactory bulbs, pallium, preoptic area and hypothalamus), but also in the optic tectum and the midbrain tegmentum (Fig. 7–2). Intriguingly, the signal generated by in situ hybridization consistently appeared in the parenchyma away from the ventricles. This signal appears very similar to the signal yielded by immunohistochemistry or *cyp19a1b*-GFP expression (see Fig. 7–2). This indicates *cyp19a1b* mRNA export from the soma of RGCs to the radial processes with sometimes accumulation in the endfeet at the periphery of the brain (Goto-Kazeto et al., 2004; Pellegrini et al., 2007; Diotel et al., 2010b; Diotel et al., 2011). The significance

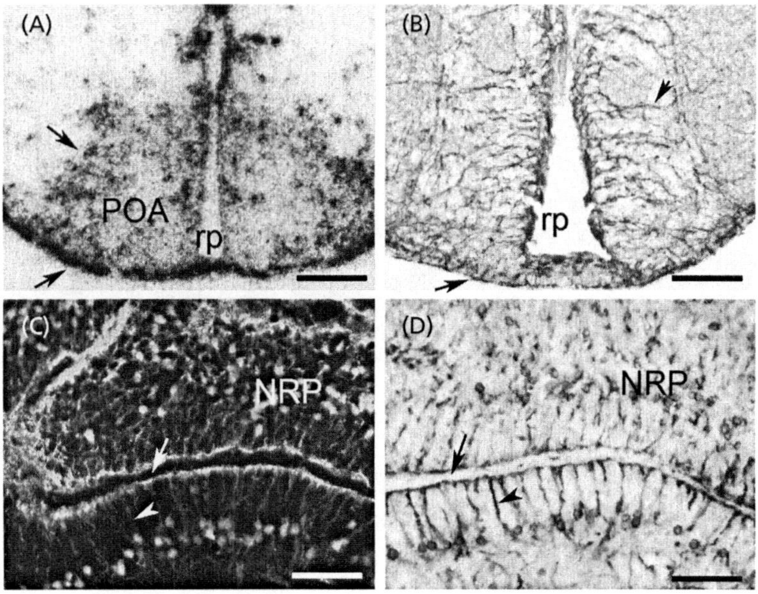

Figure 7–2. Expression of *cyp19a1b* mRNA (*A*), *cyp19a1b* protein (*B* and *D*), and *cyp19a1b*-GFP (*C*) in the preoptic area (POA) and around the posterior recess in the brain of the zebrafish. (*A*) and (*B*) show the same level around the preoptic recess (rp). Note the strong hybridization signal in the parenchyma and at the periphery of the brain (arrows). This corresponds to mRNA export in the radial processes and accumulations in the endfeet (arrows). (*C*) and (*D*) show the same level of the nucleus of the posterior recess (NRP) surrounding the posterior recess (*arrow*) of the hypothalamus. One can see that the *cyp19a1b*-GFP signal in (*C*) perfectly matches the distribution of the aromatase B protein in (*D*). At this level the radial glial cells have distal processes (*arrowheads*) longer than in other parts of the brain. (*A*) and (*B*) Bar = 100μm; (*C*) and (*D*) Bar = 50μm.

of this mRNA transport is not known, but is not restricted to *cyp19a1b* mRNAs because it also concerns *cxcr4, 3bhsd, cyp17,* or *p450scc* transcripts (Diotel et al, 2010b; Diotel et al., 2011).

Strikingly, *cyp19a1b*-GFP fluorescence consistently appeared in the tail of transgenic zebrafish during larval development suggesting potential expression of aromatase B at this level. Tail sections of embryos, larvae, and adults confirmed that aromatase B and *cyp19a1b*-GFP expression overlap in RGCs in the caudal portion of the spinal cord, including adults. Previous studies showed the presence of glial cells in the caudal portion of the caudal neurosecretory system in the blueback herring, *Pomolobus aestivalis* (Kriebel et al., 1979). This neurohemal organ only present in fish could constitute a neurosecretory system anatomically and functionally similar to the hypothalamic-pituitary complex of mammals. At this level, urotensin I and II neurosecretory cells project their axons in the vicinity of blood capillaries of the urophysis and release their neuropeptides (Parmentier et al., 2006). Further studies are necessary to clarify the anatomical and functional relationship between aromatase-positive RGCs and the urophysis.

The demonstration of the radial glia nature of the cells expressing aromatase B in teleost fish brain does not only rely on morphological features. Indeed, aromatase B or *cyp19a1b*-GFP–positive cells also express astroglial markers such as GFAP (glial fibrillary acid protein), Protein S100ß, nestin, BLBP (brain lipid binding protcin) (Forlano et al., 2001; Pellegrini et al., 2005; Lam et al., 2009; Tong et al., 2009; März et al., 2010; Diotel et al., 2010a) and never overlapped with neuronal markers such as Hu or acetylated tubuline. Interestingly, Cxcl12 and Cxcr4, a chemokine/chemokine receptor pair involved in neurogenesis (Tran et al., 2007), are coexpressed with aromatase B in RGCs (Diotel et al., 2010b).

REGULATION OF THE CELL SPECIFIC EXPRESSION OF AROMATASE B IN RADIAL GLIAL CELLS

Pioneer studies already pointed out the tight relationships between the sexual status of fish and the brain aromatase activity (Borg et al., 1987a, 1987b, 1989; Pasmanik & Callard, 1988a), suggesting that sex steroids could upregulate brain aromatase activity. More recent data have shown that estrogens regulate the *cyp19a1b* gene, but this regulation is cell context dependent and occurs only in RGCs.

Estrogens Upregulate the Cyp19a1b Gene via Estrogen Receptors

Thorough analyses in male European sea bass nicely demonstrated that brain aromatase activity increased in parallel with the gonadosomatic index and correlated well with estradiol, but above all, with testosterone concentrations in the blood (Gonzalez & Piferrer, 2003). This study also showed that aromatase activity did not strikingly differ in males and females. Following the characterization of the two duplicated aromatase genes, *cyp19a1a* and *cypa19a1b* with

		ERE	
Zebrafish	– 332	CTGGGTCAGTCTGACCTGC	– 314
Catfish	– 374	CTCGGTCAAGCTGACCTGC	– 355
Medaka	– 408	ACAGGTCAGCCTGACCCAG	– 390
Goldfish	– 358	CTGGGTCAGTCTGACCTGT	– 338
Grey mullet	– 355	ACGGGTCAGTCTGACCCAG	– 337
Tilapia	– 408	ACAGGTCAGTCTGACCCAG	– 390
Grouper	– 336	ACGGGTCAATCTGACCCAG	– 318
Barramundi	– 411	ACAGGTCAGTCTGACCCAG	– 393
Ricefield eel	– 302	ACAGATCAGTCTGACCAAG	– 284

Figure 7–3. Alignment of the ERE found in *cyp19a1b* promoters in different teleost species. One can see that the ERE is highly conserved over a wide range of species and that the consensus sequence resembles the consensus ERE sequence. Note that in the ricefield eel, there is a mutation in 5′ half ERE, which makes it nonfunctional. Origin of data: orange-spotted grouper (FJ914569), tilapia (AF472621), barramundi (AY686693), medaka (AY705086), grey mullet (AY859424), goldfish (AF324893), zebrafish (AF406756), and catfish (AY780360).

expression in gonad and brain, respectively (Tchoudakova & Callard, 1998), it rapidly appeared that the *cyp19a1b* gene promoter in most fish species carries an estrogen-response element (ERE) (Fig. 7–3), suggesting that estrogen receptors (ERs) could directly modulate aromatase B expression at the transcriptional level (Chiang et al., 2001; Kishida & Callard, 2001; Tong et al., 2001; Kazeto & Trant, 2005). Confirming the roles of estrogens and aromatizable androgens, E2 and testosterone caused a dramatic increase in *cyp19a1b* expression in the brain of zebrafish embryos and larvae (Kishida et al., 2001; Menuet et al., 2005; Mouriec et al., 2009; Tong et al., 2009). An excess of the selective E2 antagonist ICI 182,780 can block these effects, indicating that functional ERs are involved in this process and also demonstrating that testosterone effects are likely due to aromatization (Fig. 7–4).

Detailed promoter studies in vitro showed that, in the presence of one of the three zebrafish ER, E2 is able to transactivate a *cyp19a1b*-luciferase reporter, but only in astroglial cell contexts (Menuet et al., 2005; Le Page et al., 2006; Mouriec et al., 2009; Diotel et al., 2010a). Mutation or deletion of the ERE causes the total disappearance of the E2-induction (Menuet et al., 2005). Very interestingly, in the ricefield eel, which is not an eel but belongs to the evolved order Synbranchiforme, androgens appear to directly regulate the *cyp19a1b* gene. Indeed the ERE located on the proximal promoter is mutated and non-functional (see Fig. 7–3) (Zhang et al., 2011).

Importance of the Cell Context for the E2 Up-Regulation of Cyp19a1b in Radial Glial Cells

The fact that aromatase B expression is limited to RGCs in fish was surprising because, in mammals and birds, aromatase is mainly found in neurons (Schlinger

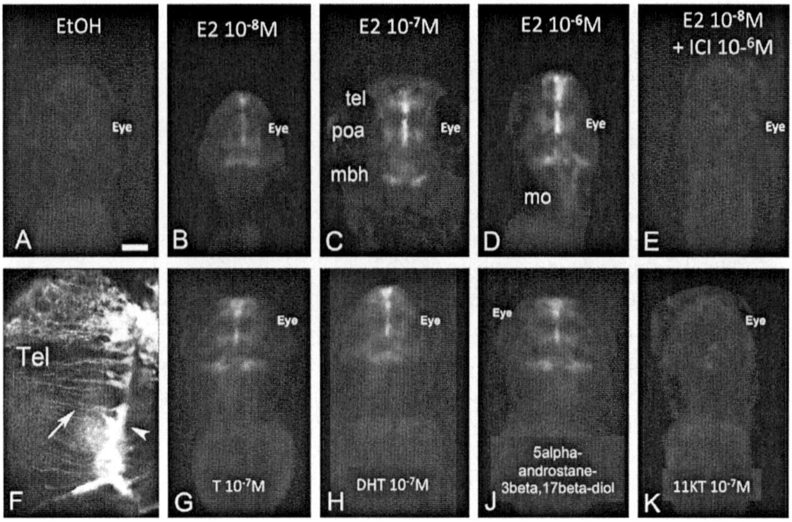

Figure 7–4. A.–E. *Cyp19a1b*-GFP in the forebrain of tg(*cyp19a1b*-GFP) transgenic zebrafish embryos treated for 48 hours with EtOH (*A*); estradiol at 10^{-8}M, 10^{-7}M, or 10^{-6}M (*B-D*); and estradiol (10^{-8}M) together with ICI (10^{-6}M) (*E*). Estradiol induces a dose-dependent increase of GFP in radial glial cells, an effect that is blocked by ICI 182,780. F. A confocal image of a tg(*cyp19a1b*–GFP) embryo treated with E2 (10^{-9}M) in which one can see that the signal corresponds to radial glial cells. Testosterone (10^{-7}M) (*G*), dihydrotestosterone (10^{-7}M) (*H*), and ß-diol (10^{-7}M) (*J*) also induce GFP expression, whereas 11 keto-testosterone (10^{-7}M) (*K*) has no effect. Tel: telencephalon; poa: preoptic area; mbh: mediobasal hypothalamus ; mo : medulla oblongata. Bar = 250 μm except in (*F*) where Bar corresponds to 25 μm.

et al., 1994; Balthazart & Ball, 1998) and to a lesser degree, after brain lesions, in astrocytes (Schlinger et al., 1994; Zwain & Yen, 1999; Saldanha et al., 2009; Yague et al., 2010). In plainfin midshipman, zebrafish, or pejerrey, aromatase B expression was never observed in neurons as demonstrated using double staining with neuronal markers (Forlano et al., 2001; Pellegrini et al., 2007; Strobl-Mazzulla et al., 2010). Also surprising was the fact that many cells in the brain parenchyma strongly express ER without expressing aromatase B (Anglade et al., 1994; Pakdel et al., 1994; Navas et al., 1995; Linard et al., 1996; Teitsma et al., 1998; Menuet et al., 2002; Andreassen et al., 2003; Servili et al., 2011).

In vitro experiments highlighted the importance of the cell context since a zebrafish *cyp19a1b*-reporter gene, cotransfected with any of the zebrafish ER, does not respond to E2 stimulation in many cell lines such as CHO, Hela, or PC12 cells (Menuet et al., 2005). In contrast, in the human astrocyte cell line U251-MG, the *cyp19a1b*-luciferase reporter is strongly upregulated by E2 (Menuet et al., 2005). Of particular interest are the mouse P19 embryonic carcinoma cells that, when undifferentiated, did not respond to E2 by increasing *cyp19a1b*-luciferase expression. However, when undifferentiated P19 cells were pretreated with retinoic acid, which leads to differentiation into both neurons and glial cells (Jones-Villeneuve et al., 1982), the E2-induced response was 10-fold higher than in undifferentiated P19 cells (Menuet et al., 2005). An artificial ERE-TK-luciferase promoter is inducible by E2 in all cell lines (Menuet et al., 2005). This demonstrates that the cell context is crucial for E2 transactivation of the *cyp19a1b* promoter. Further deletion/mutation experiments in U251-MG cells demonstrated that if the ERE is necessary, it is not sufficient for full estrogenic activation of the *cyp19a1b* promoter in a glial cell context. Indeed, the integrity of a so-called glia specific responsive element sequence upstream the ERE is absolutely mandatory for full induction. This was confirmed by gel shift experiments, showing that this sequence indeed binds factors present in astroglial cells such as the human astrocytes U251-MG or differentiated P19, but not from CHO, Hela, or PC12 cells 2 (Menuet et al., 2005; Le Page et al., 2008; Diotel et al., 2010a). Despite many attempts, a candidate gene approach failed to identify the astroglial specific factor interacting with ER to promote *cyp19a1b* expression in radial glia. In the medaka, *Oryzias latipes*, it was shown that liver receptor homologue-1, a member of the NR5A family, is involved in the regulation of *cyp19a1b* expression in the brain (Ohmuro-Matsuyama et al., 2007); however, given the distribution of LRH1, it is unlikely that it corresponds to the astroglial specific factor. It is worth noting here that, in mammals, radial glia gives birth to all brain cell types including the star-shaped astrocytes (Pinto & Gotz, 2007). Not surprising is the fact that many astrocyte markers also stain RGCs in fish whose forebrain apparently lacks astrocytes, in a strict sense (Menuet et al., 2005; Lam et al., 2009; März et al., 2010).

Because of its high sensitivity to estrogens, the *cyp19a1b* gene is also highly susceptible to endocrine disruptors acting as estrogenic compounds (Rotchell & Ostrander, 2003; Cheshenko et al., 2008; Hinfray et al., 2006; Le Page et al., 2006; Leanos-Castaneda et al., 2007; Hinfray et al., 2010; Orrego et al., 2010; Vosges et al., 2010). Undesirable exposure of fish to such compounds leads to increase of *cyp19a1b* expression in the radial glial cells (Vosges et al., 2010), which may not have profound effects in mature fish because expression is already very high, but certainly may exert influences in embryos, larvae, or immature fish. However, the consequences of such exposures are not known.

Androgens Upregulate Cyp19a1b Gene Expression via Estrogen Receptors

As mentioned above, when exposed to testosterone, zebrafish embryos react by strongly expressing aromatase B in their brains (see Fig. 7–4). This is not so surprising because testosterone is likely to be locally aromatized in estradiol as indicated by the fact that ICI 182,780 blocked this effect (see Fig. 7–4). However, more surprisingly, when exposed to the nonaromatizable androgen DHT (5alpha-dihydrotestosterone) zebrafish embryos also exhibited strong aromatase expression, suggesting a potential androgenic regulation of *cyp19a1b* through androgen receptors (ARs). To assess a putative direct regulation of the *cyp19a1b* gene by ARs, U251MG cells were transfected with zebrafish AR together with a luciferase reporter gene driven by 3,000 bp of the proximal *cyp19a1b* promoter containing the ERE and potential AREs (Mouriec et al., 2009). Interestingly, although zebrafish AR activated luciferase reporter genes controlled by AREs, they failed to induce the *cyp19a1b*-luciferase construct. These data suggest that the androgenic regulation of *cyp19a1b* does not involve AR. We further showed that regulation of the *cyp19a1b* gene by testosterone is, in fact, due to aromatization, whereas the effect of DHT involves conversion into 5alpha-androstane-3beta, 17beta-diol (betadiol), a metabolite of DHT with known estrogenic activity (see Fig. 7–4). The blockage of the androgen regulation of *cyp19a1b* expression using antiestrogens further confirmed the involvement of estrogen receptors in mediating these effects (Mouriec et al., 2009).

Relationships Between ERs and Aromatase B Expression

The previously mentioned data explain the high expression of aromatase B in RGCs that is due to a positive auto-regulatory loop through which *cyp19a1b* expression is stimulated by estrogens and aromatizable androgens. Although this explains the high aromatase activity in the brain of sexual mature fish, it implies that ERs must be expressed in RGCs, which has never been reported using conventional in situ hybridization and immunohistochemistry (Anglade et al., 1994; Menuet et al., 2002, 2003; Andreassen et al., 2003; Hawkins et al., 2005; Pellegrini et al., 2005). ERs were only found by RT-PCR in enriched glia primary culture from trout (Menuet et al., 2005). We suggest that glial cells may express low amounts of ERs, undetectable by conventional histological techniques but sufficient to induce the *cyp19b* gene. This is supported by the fact that new radioactive and fluorescence in situ hybridization experiments localized ERß1 and ERß2 mRNA along the brain ventricles (A. Servili, F. Adrio, and N. Diotel, unpublished data).

POTENTIAL FUNCTIONS OF BRAIN AROMATASE IN TELEOST FISHES

Aromatase B, Radial Glial Cells, and Neurogenesis

Recent data in zebrafish have highlighted the key role of RGCs in embryonic and adult neurogenesis (Pellegrini et al., 2007; März et al., 2010; Rothenaigner et al., 2011a). Identified markers of RGCs in zebrafish are canonical markers of the astroglial lineage. This is the case with nestin, protein S100ß, GFAP, or BLBP (März et al., 2010). It was rather surprising to discover that aromatase B is a marker of RGCs in fish (Forlano et al., 2001; Menuet et al., 2005; Diotel et al., 2010a), because most studies in mammals and birds reported aromatase expression mainly in neurons (Balthazart & Ball, 1998). More recently, there is increasing evidence for aromatase expression in astrocytes (Azcoitia et al., 2003; Garcia-Ovejero et al., 2005; Azcoitia et al., 2011; Gatson et al., 2011; Lepore et al., 2011), but there is only one report indicating that aromatase is expressed in radial progenitors in the cortex of the mouse (Martinez-Cerdeno et al., 2006). In the same report, it was found that RGCs in the mouse cortex express ERα and that blocking aromatase activity in a pregnant mouse causes reduced cell proliferation in the proliferation zone of the cortex. The situation in fish is that aromatase-expressing radial glial progenitors, if exposed to aromatizable androgens, produce estrogens at possibly elevated local concentrations. Interestingly, when examining the potential roles of estradiol on neurogenesis in zebrafish, it was found that estrogen has an inhibitory effect on the proliferation activity of RGCs and the migration speed of newborn neurons (Diotel et al., 2012). Blocking aromatase activity or treatment with anti-estrogens caused increased proliferation in many brain regions, whereas exposure to E2 caused a significant decrease in the number of PCNA or BrdU-positive cells. Similarly, estrogen treatment reduces the migration of newborn cells along the radial processes (Diotel et al., 2012). Given the complexity of estrogen signaling, we can only conclude that, under these conditions estrogens cause inhibition of cell proliferation in the brain of adult fish.

However, the hypothesis that estrogens or aromatizable androgens cause inhibition of cell proliferation in the brain of fish is not totally devoid of potential biological meaning. This could indeed lead to slowing down brain growth in sexually maturing or mature fish in which energy is preferentially driven towards gonadal growth. Indeed, most fish exhibit reduced growth at the time of reproduction (Lester et al., 2004).

Neuroendocrine Regulations

As is the case in other vertebrates, estradiol and testosterone via negative and positive feedback loops modulate the activity of the neuroendocrine circuits and the pituitary gonadotrophs, an effect that, in the case of testosterone, often involves aromatization. For instance, both estrogens and aromatizable androgens increased salmon GnRH contents in the brain of the triploid rainbow trout characterized by very low levels of circulating steroids. Testosterone is more active than estradiol, probably because of high aromatase activity in the GnRH neurons containing regions. Both steroids were also shown to increase salmon GnRH content in the pituitary, which in fish comes to the pituitary via a direct GnRH innervation (Breton & Sambroni, 1996; Breton et al., 1997). In the African catfish, testosterone accelerates the development of the catfish GnRH system, whereas hydroxy-androstenedione, a nonaromatizable androgen, has only minimal effects (Dubois et al., 2001). In rainbow trout, estrogens and aromatizable androgens are known to regulate the dopaminergic neurons that control inhibition of LH release (Linard et al., 1995; Vacher et al., 2000; Vetillard et al., 2003).

Aggressive and Sexual Behavior

Many teleost fishes are characterized by their unique abilities to change sex during their lifetime and even to exhibit simultaneous hermaphrodism, indicating a high plasticity of the neuronal substrates mediating male and female sexual behavior. In mammals or birds, the effects of estrogens on sexual behavior have been categorized into two main classes: organizational (irreversible genomic action during ontogeny) and activational (reversible genomic action and rapid nongenomic effects in adults). In these models, locally produced estrogens are important for brain maculinization in males (in mammals) and brain defeminization of females (in quail) (Balthazart et al., 2003; Balthazart et al., 2009). In adults, testosterone produced by the testes is converted into estrogens by brain aromatase, resulting in gene transcription through genomic effects implicating ER. In addition, brain aromatase activity can be modulated within minutes by calcium-dependent phosphorylations, triggered by variations in glutamatergic neurotransmission. Such rapid changes in aromatase activity affect male sexual behavior in quail and in mice (Balthazart et al., 2009; Charlier et al., 2010).

The role of aromatization in sexual behavior in fish is also well recognized. In many fish, androgen or estrogen treatment causes male-typical sexual behavior but the underlying cellular and molecular mechanisms are poorly understood (Munakata & Kobayashi, 2010). The plainfin midshipman is a fish model that has been extensively examined. This species uses vocal-acoustic communication and alternative male reproductive tactics (Forlano & Bass, 2005). The sonic motor nucleus in this species is sexually dimorphic. Measurements of brain aromatase mRNA expression were performed in the inter- and intrasexually dimorphic sonic motor nucleus and in the preoptic area in males and females in parallel with steroid hormone levels and sexual behaviors. The data shows clear seasonal and sex differences in brain aromatase mRNA expression suggesting a role for aromatization in the expression of vocal-acoustic and alternative male reproductive phenotypes (Forlano & Bass, 2005).

In goldfish, it was shown that testosterone influences approach responses towards females through nongenomic mechanisms. This effect is blocked by the aromatase inhibitor fadrozole, evidence that rapid stimulation of visual approach by testosterone relies on aromatization (Lord et al., 2009). This suggests that testosterone and aromatase may play a role in the dynamic regulation of social behavior, particularly in species in which social stimuli can rapidly influence circulating steroid levels.

Aromatase is also known for its involvement in aggressive behavior in fish. Fish treated with estrogens (Clotfelter & Rodriguez, 2006) or exposed to xenoestrogens exhibit reduced levels of agression. This finding is further supported by studies in the bluebanded goby, *Lythrypnus dalli*, a species that shows rapid sex changes. Removing the dominant male from a social group generates an immediate behavioral response, causing one female to become dominant and change sex to male. Increased testosterone concentrations and a reduction in brain aromatase activity accompany the increased aggressive behavior exhibited by the dominant female, whereas gonadal aromatase shows no change (Black et al., 2005). Interestingly, the decrease in brain aromatase activity follows, rather than precedes, the rapid increase in aggressive behavior (Black et al., 2011). This study provides a potential mechanism explaining the increase in androgens that follows aggressive interactions in many vertebrate species (Remage-Healey & Bass, 2005). This increase in androgens could then facilitate sex change notably in species changing from female to male (Black et al., 2011).

In addition, in birds and mammals, aromatase activity can be rapidly modulated by post-translation mechanisms, such as phosphorylation. Such fast changes can also possibly explain the rapid switch between male and female sex behavior in simultaneous hermaphrodites, and this should be further studied in teleost fishes, which provide a wealth of incredibly diverse models to study these mechanisms.

OTHER STEROIDOGENIC ENZYMES

As described above, aromatase B is strongly expressed in the brain and is regulated by estrogens and androgens. This raises the question of the origin of the substrates available for aromatization and aromatase up regulation. In addition to the steroids produced by peripheral organs, recent data indicate that the brain of fishes can produce neurosteroids.

Indeed, apart from aromatase, the activity of key enzymes such as P450scc (cytochrome P450 side-chain cleavage), 3β-HSD (3 beta-hydroxysteroid dehydrogenase/Delta5-Delta4-isomerase) and P450c17 (P450 17alpha-hydroxylase/c17–20 lyase) was reported in zebrafish. Real time RT-PCR experiments resulted in the detection of P450scc and P450c17 expression in the brain of zebrafish (Hsu et al., 2002; Wang & Ge, 2004), P450c17 in the fathead minnow (Halm et al., 2003) and in the brain of tongue sole (Chen et al., 2010). The presence of 3β-HSD transcripts and protein was also reported in the central nervous system of zebrafish (Sakamoto et al., 2001; Diotel et al., 2011). In addition 17ß-hydroxysteroid dehydrogenase (17ß-HSD) (Mindnich & Adamski, 2009) and 5α-reductase (Pasmanik & Callard, 1985) are also present and active.

At the present stage, the potential functions of neurosteroids in the brain of zebrafish remain to be discovered. It is interesting, however, to mention that expression of all these enzymes in radial glial cells is a possibility that cannot be excluded in view of the messenger distribution in the brain parenchyma (Diotel et al., 2011). However, until antibodies to those steroidogenic enzymes are available for fish, the question will remain open.

EVOLUTIONARY CONSIDERATIONS

As pointed out in this review, many aspects of brain aromatase in teleost fishes are quite unique. It is possible that these unique features were acquired after the 3R duplication event that gave birth to the *cyp19a1a* and *cyp19a1b* genes, generating a gonadal aromatase and a brain aromatase, respectively (Tong et al., 2001). In agreement with this hypothesis, these two genes are found only in teleosts. The two copies may then have undergone subfunctionalization of the ancestral function (Ohno, 1999), the *cyp19a1b* gene evolving to primarily support brain functions and *cyp19a1a* being restricted mainly to gonadal expression and functions. In the case of *cyp19a1b*, subfunctionalization was possibly accompanied by an exaggeration of an ancestral function. This is supported by the finding of aromatase immunoreactivity in radial glial cells during embryogenesis in mice and quail (Martinez-Cerdeno et al., 2006; Aste et al., 2010) or following brain lesion in birds (Peterson et al., 2004), possibly indicating that aromatase expression in RGCs also occurs in the tetrapod lineage, but in a more discrete fashion. This then would mean that aromatase expression in RGCs could be an ancestral feature already existing in basal Vertebrates. As mentioned above, the Japanese and European eels have a unique *cyp19a1* gene expressed in both gonads and brains, notably in RGCs (S.R. Jeng et al., unpublished data). It would be highly interesting to investigate aromatase expression in the brain of early Actinopterygians (ray-finned fishes) and Sarcopterygians, the lobe-finned fishes that are at the origin of tetrapods.

On the other hand, there is some evidence to suggest that bona fide astrocytes are lacking in the forebrain of fishes at least in zebrafish. This is suggested by the fact that astroglial markers such as BLBP, protein S100ß, nestin, and GFAP all stain only radial glia (März et al., 2010). Because most steroidogenic enzymes are expressed in astrocytes in mammals (Zwain et al., 1997Zwain & Yen, 1999), it is not surprising that RGCs also express most of these steroidogenic enzymes. The capacity of radial glial cells to produce neurosteroids could then be an ancient function of ancestral astroglial cells.

Despite the recent progress, our knowledge of RGCs in teleost fishes remains very limited. Looking at *cyp19a1b*-GFP positive cells using confocal microscopy on brain slices indicates that RGCs organization and morphology is much more complex than previously envisioned. Such cells are either smooth or spiny (Fig. 7–5). Radial processes, which are not always radial, branch extensively,

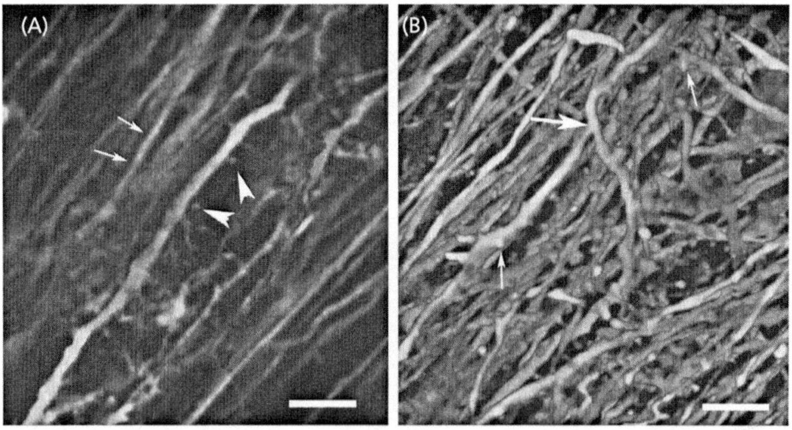

Figure 7–5. A. Confocal image of *cyp19a1b*-GFP expressing RGCs in the brain of zebrafish showing that parts of the radial processes are smooth (*arrows*), whereas others exhibit small spines (*arrowheads*). B. Three dimensional view obtained from a 100µm Z-stack showing the extreme abundance of the *cyp19a1b*-GFP radial processes containing aromatase B and the complexity of their organization. One can notably see that these processes divide (*large arrow*) and exhibit numerous small spines (*small arrows*). Bar = 10µm.

possibly surrounding or contacting neurons. They also entirely surround blood vessels similar to astrocytes, indicating that RGCs probably participate in the blood brain barrier.

CONCLUSIONS

Altogether, these data point to apparently unique features of the brain of fish with respect to brain aromatase. The high aromatase activity observed in mature fish is clearly due to upregulation of the *cyp19a1b* gene in the radial glial cells by estrogens and androgens. The fact that these cells are progenitors in adult fish suggests potential new roles of estrogens in neurogenesis that do not exclude their involvement in "classical" mechanisms. Aromatase is clearly involved in sexual behavior and there is increasing evidence for a role of aromatase in the sexualization of the brain. These new potential functions must also be considered in light of emerging data showing that the brain of fish de novo produces substrates for the aromatase enzyme. Recent studies in the brain of zebrafish indicated that key steroidogenic enzymes are active leading to the formation of aromatizable androgens, possibly in RGCs (Diotel, 2011; Diotel et al., 2011). Our recent data also show that blocking estrogen synthesis or receptors causes increased cell proliferation in the brain of adult zebrafish (Diotel et al., 2012), suggesting that estrogens would inhibit rather than stimulate neurogenesis in adult fish. However, these data will need to be reinforced and such effects need to be demonstrated in other fish species before they can be extrapolated.

Overall brain aromatase shows distinct features in fish compared to other well-studied vertebrate models like birds and rodents. It would be highly interesting to learn more about the situation in basal vertebrates, such as lamprey and sharks, in order to understand how aromatase expression, regulation, and functions have evolved in the teleost lineage compared to tetrapods. Apparent differences between fish and tetrapods are most likely due to the complexification of the brain, the fate of radial glial cells after the embryonic period and the increased importance of bona fide astrocytes in brain development and functioning. However, it is also possible that some ancestral functions of brain aromatase similar to what is observed in fishes remain to be uncovered during brain development in the tetrapod lineage.

ACKNOWLEDGMENTS

This work was supported by European project No. 222719 (LIFECYCLE to O.K.), and the Agence Nationale de la Recherche NEED (CES 2008–011 to O.K.).

REFERENCES

Andreassen, T. K., Skjoedt, K., Anglade, I., Kah, O., & Korsgaard, B. (2003). Molecular cloning, characterisation, and tissue distribution of oestrogen receptor alpha in eelpout (Zoarces viviparus). *Gen Comp Endocrinol, 132*(3), 356–368.

Anglade, I., Pakdel, F., Bailhache, T., Petit, F., Salbert, G., Jego, P., et al. (1994). Distribution of estrogen receptor-immunoreactive cells in the brain of the rainbow trout (Oncorhynchus mykiss). *J Neuroendocrinol, 6*(5), 573–583.

Aste, N., Watanabe, Y., Harada, N., & Saito, N. (2010). Distribution and sex differences in aromatase-producing neurons in the brain of Japanese quail embryos. *J Chem Neuroanat, 39*(4), 272–288.

Azcoitia, I., Sierra, A., Veiga, S., & Garcia-Segura, L. M. (2003). Aromatase expression by reactive astroglia is neuroprotective. *Ann N Y Acad Sci, 1007,* 298–305.

Azcoitia, I., Yague, J. G., & Garcia-Segura, L. M. (2011). Estradiol synthesis within the human brain. *Neuroscience, 191,* 139–147.

Balthazart, J., Baillien, M., Charlier, T. D., Cornil, C. A., & Ball, G. F. (2003). The neuroendocrinology of reproductive behavior in Japanese quail. *Domest Anim Endocrinol, 25*(1), 69–82.

Balthazart, J., & Ball, G. F. (1998). New insights into the regulation and function of brain estrogen synthase (aromatase). *Trends Neurosci, 21*(6), 243–249.

Balthazart, J., Cornil, C. A., Charlier, T. D., Taziaux, M., & Ball, G. F. (2009). Estradiol, a key endocrine signal in the sexual differentiation and activation of reproductive behavior in quail. *J Exp Zool A Ecol Genet Physiol, 311*(5), 323–345.

Black, M. P., Balthazart, J., Baillien, M., & Grober, M. S. (2005). Socially induced and rapid increases in aggression are inversely related to brain aromatase activity in a sex-changing fish, Lythrypnus dalli. *Proc Biol Sci, 272*(1579), 2435–2440.

Black, M. P., Balthazart, J., Baillien, M., & Grober, M. S. (2011). Rapid increase in aggressive behavior precedes the decrease in brain aromatase activity during socially mediated sex change in Lythrypnus dalli. *Gen Comp Endocrinol, 170*(1), 119–124.

Borg, B., Andersson, E., Mayer, I., & Lambert, J. G. (1989). Aromatase activity in the brain of the three-spined stickleback, Gasterosteus aculeatus. III. Effects of castration under different conditions and of replacement with different androgens. *Exp Biol, 48*(3), 149–152.

Borg, B., Timmers, R. J., & Lambert, J. G. (1987a). Aromatase activity in the brain of the three-spined stickleback, Gasterosteus aculeatus. I. Distribution and effects of season and photoperiod. *Exp Biol, 47*(2), 63–68.

Borg, B., Timmers, R. J., & Lambert, J. G. (1987b). Aromatase activity in the brain of the three-spined stickleback, Gasterosteus aculeatus. II. Effects of castration in winter. *Exp Biol, 47*(2), 69–71.

Breton, B., & Sambroni, E. (1996). Steroid activation of the brain-pituitary complex gonadotropic function in the triploid rainbow trout Oncorhynchus mykiss. *Gen Comp Endocrinol, 101*(2), 155–164.

Breton, B., Sambroni, E., Govoroun, M., & Weil, C. (1997). Effects of steroids on GTH I and GTH II secretion and pituitary concentration in the immature rainbow trout Oncorhynchus mykiss. *C R Acad Sci III, 320*(10), 783–789.

Callard, G. V., Petro, Z., & Ryan, K. J. (1978). Phylogenetic distribution of aromatase and other androgen-converting enzymes in the central nervous system. *Endocrinology, 103*(6), 2283–2290.

Chang, X., Kobayashi, T., Senthilkumaran, B., Kobayashi-Kajura, H., Sudhakumari, C. C., & Nagahama, Y. (2005). Two types of aromatase with different encoding genes, tissue distribution and developmental expression in Nile tilapia (Oreochromis niloticus). *Gen Comp Endocrinol, 141*(2), 101–115.

Charlier, T. D., Cornil, C. A., Ball, G. F., & Balthazart, J. (2010). Diversity of mechanisms involved in aromatase regulation and estrogen action in the brain. *Biochim Biophys Acta, 1800*(10), 1094–1105.
Chen, C. F., Wen, H. S., Wang, Z. P., He, F., Zhang, J. R., Chen, X. Y., et al. (2010). Cloning and expression of P450c17-I (17alpha-hydroxylase/17,20-lyase) in brain and ovary during gonad development in Cynoglossus semilaevis. *Fish Physiol Biochem, 36*(4), 1001–1012.
Cheshenko, K., Pakdel, F., Segner, H., Kah, O., & Eggen, R. I. (2008). Interference of endocrine disrupting chemicals with aromatase CYP19 expression or activity, and consequences for reproduction of teleost fish. *Gen Comp Endocrinol, 155*(1), 31–62.
Chiang, E. F., Yan, Y. L., Tong, S. K., Hsiao, P. H., Guiguen, Y., Postlethwait, J., et al. (2001). Characterization of duplicated zebrafish cyp19 genes. *J Exp Zool, 290*(7), 709–714.
Clotfelter, E. D. & Rodriguez, A. C. (2006). Behavioral changes in fish exposed to phytoestrogens. *Environ Pollut, 144*(3), 833–839.
Diotel, N., Vaillant, C., Gabbero, C., Mironov, S., Fostier, A., Anglade, I., Kah, O., Pellegrini, E. (2012). Effects of estradiol in adult neurogenesis and brain repair in zebrafish.*Horm Behav*,in press.
Diotel, N. (2011). Production et cibles des neurostéroïdes dans le cerveau du poisson zèbre adulte: Rôles potentiels dans la neurogenèse et la réparation cérébrale. *PhD dissertation, Université de Rennes 1*, 1–321.
Diotel, N., Do Rego, J. L., Anglade, I., Vaillant, C., Pellegrini, E., Gueguen, M. M., et al. (2011). Activity and expression of steroidogenic enzymes in the brain of adult zebrafish. *Eur J Neurosci, 34*(1), 45–56.
Diotel, N., Le Page, Y., Mouriec, K., Tong, S. K., Pellegrini, E., Vaillant, C., et al. (2010a). Aromatase in the brain of teleost fish: expression, regulation and putative functions. *Front Neuroendocrinol, 31*(2), 172–192.
Diotel, N., Vaillant, C., Gueguen, M. M., Mironov, S., Anglade, I., Servili, A., et al. (2010b). Cxcr4 and Cxcl12 expression in radial glial cells of the brain of adult zebrafish. *J Comp Neurol, 518*(24), 4855–4876.
Dubois, E. A., Slob, S., Zandbergen, M. A., Peute, J., & Goos, H. J. (2001). Gonadal steroids and the maturation of the species-specific gonadotropin-releasing hormone system in brain and pituitary of the male African catfish (Clarias gariepinus). *Comp Biochem Physiol B Biochem Mol Biol, 129*(2–3), 381–387.
Ekström, P., Johnsson, C. M., & Ohlin, L. M. (2001). Ventricular proliferation zones in the brain of an adult teleost fish and their relation to neuromeres and migration (secondary matrix) zones. *J Comp Neurol, 436*(1), 92–110.
Forlano, P. M., & Bass, A. H. (2005). Seasonal plasticity of brain aromatase mRNA expression in glia: divergence across sex and vocal phenotypes. *J Neurobiol, 65*(1), 37–49.
Forlano, P. M., Deitcher, D. L., Myers, D. A., & Bass, A. H. (2001). Anatomical distribution and cellular basis for high levels of aromatase activity in the brain of teleost fish: aromatase enzyme and mRNA expression identify glia as source. *J Neurosci, I*(22), 8943–8955.
Garcia-Ovejero, D., Azcoitia, I., Doncarlos, L. L., Melcangi, R. C., & Garcia-Segura, L. M. (2005) Glia-neuron crosstalk in the neuroprotective mechanisms of sex steroid hormones. *Brain Res Brain Res Rev, 48*(2), 273–286.
Gatson, J. W., Simpkins, J. W., Yi, K. D., Idris, A. H., Minei, J. P., & Wigginton, J. G. (2011). Aromatase is increased in astrocytes in the presence of elevated pressure. *Endocrinology, 152*(1), 207–213.

Gelinas, D., Pitoc, G. A., & Callard, G. V. (1998). Isolation of a goldfish brain cytochrome P450 aromatase cDNA: mRNA expression during the seasonal cycle and after steroid treatment. *Mol Cell Endocrinol, 138*(1–2), 81–93.

Gonzalez, A. & Piferrer, F. (2003). Aromatase activity in the European sea bass (Dicentrarchus labrax L.) brain. Distribution and changes in relation to age, sex, and the annual reproductive cycle. *Gen Comp Endocrinol, 132*(2), 223–230.

Goto-Kazeto, R., Kight, K. E., Zohar, Y., Place, A. R., & Trant, J. M. (2004). Localization and expression of aromatase mRNA in adult zebrafish. *Gen Comp Endocrinol, 139*(1), 72–84.

Guo, B., Gan, X., & He, S. (2010). Hox genes of the Japanese eel Anguilla japonica and Hox cluster evolution in teleosts. *J Exp Zool B Mol Dev Evol, 314*(2), 135–147.

Halm, S., Kwon, J. Y., Rand-Weaver, M., Sumpter, J. P., Pounds, N., Hutchinson, T. H., et al. (2003). Cloning and gene expression of P450 17alpha-hydroxylase,17,20-lyase cDNA in the gonads and brain of the fathead minnow Pimephales promelas. *Gen Comp Endocrinol, 130*(3), 256–266.

Hawkins, M. B., Godwin, J., Crews, D., & Thomas, P. (2005). The distributions of the duplicate oestrogen receptors ER-beta a and ER-beta b in the forebrain of the Atlantic croaker (Micropogonias undulatus): evidence for subfunctionalization after gene duplication. *Proc Biol Sci, 272*(1563), 633–641.

Hinfray, N., Palluel, O., Piccini, B., Sanchez, W., Ait-Aissa, S., Noury, P., et al. (2010). Endocrine disruption in wild populations of chub (Leuciscus cephalus) in contaminated French streams. *Sci Total Environ, 408*(9), 2146–2154.

Hinfray, N., Palluel, O., Turies, C., Cousin, C., Porcher, J. M., & Brion, F. (2006). Brain and gonadal aromatase as potential targets of endocrine disrupting chemicals in a model species, the zebrafish (Danio rerio). *Environ Toxicol, 21*(4), 332–337.

Hsu, H. J., Hsiao, P., Kuo, M. W., & Chung, B. C. (2002). Expression of zebrafish cyp11a1 as a maternal transcript and in yolk syncytial layer. *Gene Expr Patterns, 2*(3–4), 219–222.

Ijiri, S., Kazeto, Y., Lokman, P. M., Adachi, S., & Yamauchi, K. (2003). Characterization of a cDNA encoding P-450 aromatase (CYP19) from Japanese eel ovary and its expression in ovarian follicles during induced ovarian development. *Gen Comp Endocrinol, 130*(2), 193–203.

Jones-Villeneuve, E. M., McBurney, M. W., Rogers, K. A., & Kalnins, V. I. (1982). Retinoic acid induces embryonal carcinoma cells to differentiate into neurons and glial cells. *J Cell Biol, 94*(2), 253–262.

Kallivretaki, E., Eggen, R. I., Neuhauss, S. C., Kah, O., & Segner, H. (2007). The zebrafish, brain-specific, aromatase cyp19a2 is neither expressed nor distributed in a sexually dimorphic manner during sexual differentiation. *Dev Dyn, 236*(11), 3155–3166.

Kazeto, Y. & Trant, J. M. (2005). Molecular biology of channel catfish brain cytochrome P450 aromatase (CYP19A2): cloning, preovulatory induction of gene expression, hormonal gene regulation and analysis of promoter region. *J Mol Endocrinol, 35*(3), 571–583.

Kishida, M. & Callard, G. V. (2001). Distinct cytochrome P450 aromatase isoforms in zebrafish (Danio rerio) brain and ovary are differentially programmed and estrogen regulated during early development. *Endocrinology, 142*(2), 740–750.

Kishida, M., McLellan, M., Miranda, J. A., & Callard, G. V. (2001). Estrogen and xenoestrogens upregulate the brain aromatase isoform (P450aromB) and perturb markers of early development in zebrafish (Danio rerio). *Comp Biochem Physiol B Biochem Mol Biol, 129*(2–3), 261–268.

Kobayashi, M. & Nakanishi, T. (1999). 11-ketotestosterone induces male-type sexual behavior and gonadotropin secretion in gynogenetic crucian carp, Carassius auratus langsdorfii. *Gen Comp Endocrinol, 115*(2), 178–187.

Kriebel, R. M., Burke, J. D., & Meetz, G. D. (1979). Morphologic features of the caudal neurosecretory system in the blueback herring, Pomolobus aestivalis. *Anat Rec, 195*(3), 553–572.

Kriegstein, A., & Alvarez-Buylla, A. (2009). The glial nature of embryonic and adult neural stem cells. *Annu Rev Neurosci, 32,* 149–184.

Lam, C. S., Marz, M., & Strahle, U. (2009). Gfap and nestin reporter lines reveal characteristics of neural progenitors in the adult zebrafish brain. *Dev Dyn, 238*(2), 475–486.

Le Page, Y., Menuet, A., Kah, O., & Pakdel, F. (2008). Characterization of a cis-acting element involved in cell-specific expression of the zebrafish brain aromatase gene. *Mol Reprod Dev, 75*(10), 1549–1557.

Le Page, Y., Scholze, M., Kah, O., & Pakdel, F. (2006). Assessment of xenoestrogens using three distinct estrogen receptors and the zebrafish brain aromatase gene in a highly responsive glial cell system. *Environ Health Perspect, 114*(5), 752–758.

Leanos-Castaneda, O., van der Kraak, G., Rodriguez-Canul, R., & Gold, G. (2007). Endocrine disruption mechanism of o,p'-DDT in mature male tilapia (Oreochromis niloticus). *Toxicol Appl Pharmacol, 221*(2), 158–167.

Lepore, G., Gadau, S., Peruffo, A., Mura, A., Mura, E., Floris, A., et al. (2011). Aromatase expression in cultured fetal sheep astrocytes after nitrosative/oxidative damage. *Cell Tissue Res, 344*(3), 407–413.

Lester, N. P., Shuter, B. J., & Abrams, P. A. (2004). Interpreting the von Bertalanffy model of somatic growth in fishes: the cost of reproduction. *Proc Biol Sci, 271*(1548), 1625–1631.

Linard, B., Anglade, I., Corio, M., Navas, J. M., Pakdel, F., Saligaut, C., et al. (1996). Estrogen receptors are expressed in a subset of tyrosine hydroxylase-positive neurons of the anterior preoptic region in the rainbow trout. *Neuroendocrinology, 63*(2), 156–165.

Linard, B., Bennani, S., & Saligaut, C. (1995). Involvement of estradiol in a catecholamine inhibitory tone of gonadotropin release in the rainbow trout (Oncorhynchus mykiss). *Gen Comp Endocrinol, 99*(2), 192–196.

Lord, L. D., Bond, J., & Thompson, R. R. (2009). Rapid steroid influences on visually guided sexual behavior in male goldfish. *Horm Behav, 56*(5), 519–526.

Marsh, K. E., Creutz, L. M., Hawkins, M. B., & Godwin, J. (2006). Aromatase immunoreactivity in the bluehead wrasse brain, Thalassoma bifasciatum: immunolocalization and co-regionalization with arginine vasotocin and tyrosine hydroxylase. *Brain Res, 1126*(1), 91–101.

Martinez-Cerdeno, V., Noctor, S. C., & Kriegstein, A. R. (2006). Estradiol stimulates progenitor cell division in the ventricular and subventricular zones of the embryonic neocortex. *Eur J Neurosci, 24*(12), 3475–3488.

März, M., Chapouton, P., Diotel, N., Vaillant, C., Hesl, B., Takamiya, M., et al. (2010). Heterogeneity in progenitor cell subtypes in the ventricular zone of the zebrafish adult telencephalon. *Glia, 58*(7), 870–888.

Menuet, A., Anglade, I., Le Guevel, R., Pellegrini, E., Pakdel, F., & Kah, O. (2003). Distribution of aromatase mRNA and protein in the brain and pituitary of female rainbow trout: comparison with estrogen receptor alpha. *J Comp Neurol, 462*(2), 180–193.

Menuet, A., Pellegrini, E., Anglade, I., Blaise, O., Laudet, V., Kah, O., et al. (2002). Molecular characterization of three estrogen receptor forms in zebrafish: binding characteristics, transactivation properties, and tissue distributions. *Biol Reprod, 66*(6), 1881–1892.

Menuet, A., Pellegrini, E., Brion, F., Gueguen, M. M., Anglade, I., Pakdel, F., et al. (2005). Expression and estrogen-dependent regulation of the zebrafish brain aromatase gene. *J Comp Neurol, 485*(4), 304–320.

Mindnich, R. & Adamski, J. (2009). Zebrafish 17beta-hydroxysteroid dehydrogenases: an evolutionary perspective. *Mol Cell Endocrinol, 301*(1–2), 20–26.

Mouriec, K., Gueguen, M. M., Manuel, C., Percevault, F., Thieulant, M. L., Pakdel, F., et al. (2009). Androgens upregulate cyp19a1b (aromatase B) gene expression in the brain of zebrafish (Danio rerio) through estrogen receptors. *Biol Reprod, 80*(5), 889–896.

Munakata, A. & Kobayashi, M. (2010). Endocrine control of sexual behavior in teleost fish. *Gen Comp Endocrinol, 165*(3), 456–468.

Navas, J. M., Anglade, I., Bailhache, T., Pakdel, F., Breton, B., Jego, P., et al. (1995). Do gonadotrophin-releasing hormone neurons express estrogen receptors in the rainbow trout? A double immunohistochemical study. *J Comp Neurol, 363*(3), 461–474.

Ohmuro-Matsuyama, Y., Okubo, K., Matsuda, M., Ijiri, S., Wang, D., Guan, G., et al. (2007). Liver receptor homologue-1 (LRH-1) activates the promoter of brain aromatase (cyp19a2) in a teleost fish, the medaka, Oryzias latipes. *Mol Reprod Dev, 74*(9), 1065–1071.

Ohno, S. (1999). Gene duplication and the uniqueness of vertebrate genomes circa 1970–1999. *Semin Cell Dev Biol, 10*(5), 517–522.

Orrego, R., McMaster, M., Van Der Kraak, G., & Holdway, D. (2010). Effects of pulp and paper mill effluent extractives on aromatase CYP19a gene expression and sex steroid levels in juvenile triploid rainbow trout. *Aquat Toxicol, 97*(4), 353–360.

Pakdel, F., Petit, F., Anglade, I., Kah, O., Delaunay, F., Bailhache, T., et al. (1994). Overexpression of rainbow trout estrogen receptor domains in Escherichia coli: characterization and utilization in the production of antibodies for immunoblotting and immunocytochemistry. *Mol Cell Endocrinol, 104*(1), 81–93.

Parmentier, C., Taxi, J., Balment, R., Nicolas, G., & Calas, A. (2006). Caudal neurosecretory system of the zebrafish: ultrastructural organization and immunocytochemical detection of urotensins. *Cell Tissue Res, 325*(1), 111–124.

Pasmanik, M., & Callard, G. V. (1985). Aromatase and 5 alpha-reductase in the teleost brain, spinal cord, and pituitary gland. *Gen Comp Endocrinol, 60*(2), 244–251.

Pasmanik, M., & Callard, G. V. (1988a). Changes in brain aromatase and 5 alpha-reductase activities correlate significantly with seasonal reproductive cycles in goldfish (Carassius auratus). *Endocrinology, 122*(4), 1349–1356.

Pasmanik, M., & Callard, G. V. (1988b). A high abundance androgen receptor in goldfish brain: characteristics and seasonal changes. *Endocrinology, 123*(2), 1162–1171.

Pellegrini, E., Menuet, A., Lethimonier, C., Adrio, F., Gueguen, M. M., Tascon, C., et al. (2005). Relationships between aromatase and estrogen receptors in the brain of teleost fish. *Gen Comp Endocrinol, 142*(1–2), 60–66.

Pellegrini, E., Mouriec, K., Anglade, I., Menuet, A., Le Page, Y., Gueguen, M. M., et al. (2007). Identification of aromatase-positive radial glial cells as progenitor cells in the ventricular layer of the forebrain in zebrafish. *J Comp Neurol, 501*(1), 150–167.

Peterson, R. S., Lee, D. W., Fernando, G., & Schlinger, B. A. (2004). Radial glia express aromatase in the injured zebra finch brain. *J Comp Neurol, 475*(2), 261–269.

Pinto, L., & Gotz, M. (2007). Radial glial cell heterogeneity—the source of diverse progeny in the CNS. *Prog Neurobiol, 83*(1), 2–23.

Rakic, P. (1978). Neuronal migration and contact guidance in the primate telencephalon. *Postgrad Med J, 54*(Suppl 1), 25–40.

Remage-Healey, L. & Bass, A. H. (2005). Rapid elevations in both steroid hormones and vocal signaling during playback challenge: a field experiment in Gulf toadfish. *Horm Behav, 47*(3), 297–305.

Rotchell, J. M., & Ostrander, G. K. (2003). Molecular markers of endocrine disruption in aquatic organisms. *J Toxicol Environ Health B Crit Rev, 6*(5), 453–496.

Rothenaigner, I., Krecsmarik, M., Hayes, J. A., Bahn, B., Lepier, A., Fortin, G., et al. (2011a). Clonal analysis by distinct viral vectors identifies bona fide neural stem cells in the adult zebrafish telencephalon and characterizes their division properties and fate. *Development, 138*(8), 1459–1469.

Rothenaigner, I., Krecsmarik, M., Hayes, J. A., Bahn, B., Lepier, A., Fortin, G., et al. (2011b). Clonal analysis by distinct viral vectors identifies bona fide neural stem cells in the adult zebrafish telencephalon and characterizes their division properties and fate. *Development, 138,* 1459–1469.

Sakamoto, H., Ukena, K., & Tsutsui, K. (2001). Activity and localization of 3beta-hydroxysteroid dehydrogenase/Delta5-Delta4-isomerase in the zebrafish central nervous system. *J Comp Neurol, 439*(3), 291–305.

Saldanha, C. J., Duncan, K. A., & Walters, B. J. (2009). Neuroprotective actions of brain aromatase. *Front Neuroendocrinol, 30*(2), 106–118.

Schlinger, B. A., Amur-Umarjee, S., Shen, P., Campagnoni, A. T., & Arnold, A. P. (1994). Neuronal and non-neuronal aromatase in primary cultures of developing zebra finch telencephalon. *J Neurosci, 14*(12), 7541–7552.

Servili, A., Le Page, Y., Leprince, J., Caraty, A., Escobar, S., Parhar, I. S., et al. (2011). Organization of two independent kisspeptin systems derived from evolutionary-ancient kiss genes in the brain of zebrafish. *Endocrinology, 152*(4), 1527–1540.

Steinke, D., Hoegg, S., Brinkmann, H., & Meyer, A. (2006). Three rounds (1R/2R/3R) of genome duplications and the evolution of the glycolytic pathway in vertebrates. *BMC Biol, 4,* 16.

Strobl-Mazzulla, P. H., Lethimonier, C., Gueguen, M. M., Karube, M., Fernandino, J. I., Yoshizaki, G., et al. (2008). Brain aromatase (Cyp19A2) and estrogen receptors, in larvae and adult pejerrey fish Odontesthes bonariensis: neuroanatomical and functional relations. *Gen Comp Endocrinol, 158*(2), 191–201.

Strobl-Mazzulla, P. H., Moncaut, N. P., Lopez, G. C., Miranda, L. A., Canario, A. V., & Somoza, G. M. (2005). Brain aromatase from pejerrey fish (Odontesthes bonariensis): cDNA cloning, tissue expression, and immunohistochemical localization. *Gen Comp Endocrinol, 143*(1), 21–32.

Strobl-Mazzulla, P. H., Nunez, A., Pellegrini, E., Gueguen, M. M., Kah, O., & Somoza, G. M. (2010). Progenitor radial cells and neurogenesis in pejerrey fish forebrain. *Brain Behav Evol, 76*(1), 20–31.

Tang, B., Hu, W., Hao, J., & Zhu, Z. (2010). Developmental expression of steroidogenic factor-1, cyp19a1a and cyp19a1b from common carp (Cyprinus carpio). *Gen Comp Endocrinol, 167*(3), 408–416.

Tchoudakova, A., & Callard, G. V. (1998). Identification of multiple CYP19 genes encoding different cytochrome P450 aromatase isozymes in brain and ovary. *Endocrinology, 139*(4), 2179–2189.

Teitsma, C., Lethimonier, C., Tujague, M., Anglade, I., Saligaut, D., Bailhache, T., et al. (1998). Identification of potential sites of cortisol actions on the reproductive axis in rainbow trout. *Comp Biochem Physiol C Pharmacol Toxicol Endocrinol, 119*(3), 243–249.

Timmers, R. J., & Lambert, J. G. (1987). Measurement of aromatase activity in the brain of the African catfish, Clarias gariepinus—a comparison of two assay methods. *Comp Biochem Physiol B, 88*(2), 453–456.

Timmers, R. J., Lambert, J. G., Peute, J., Vullings, H. G., & van Oordt, P. G. (1987). Localization of aromatase in the brain of the male African catfish, Clarias gariepinus (Burchell), by microdissection and biochemical identification. *J Comp Neurol, 258*(3), 368–377.

Tong, S. K., Chiang, E. F., Hsiao, P. H., & Chung, B. (2001). Phylogeny, expression and enzyme activity of zebrafish cyp19 (P450 aromatase) genes. *J Steroid Biochem Mol Biol, 79*(1–5), 299–303.

Tong, S. K., Mouriec, K., Kuo, M. W., Pellegrini, E., Gueguen, M. M., Brion, F., et al. (2009). A cyp19a1b-gfp (aromatase B) transgenic zebrafish line that expresses GFP in radial glial cells. *Genesis, 47*(2), 67–73.

Tran, P. B., Banisadr, G., Ren, D., Chenn, A., & Miller, R. J. (2007). Chemokine receptor expression by neural progenitor cells in neurogenic regions of mouse brain. *J Comp Neurol, 500*(6), 1007–1033.

Tzchori, I., Degani, G., Hurvitz, A., & Moav, B. (2004). Cloning and developmental expression of the cytochrome P450 aromatase gene (CYP19) in the European eel (Anguilla anguilla). *Gen Comp Endocrinol, 138*(3), 271–280.

Vacher, C., Mananos, E. L., Breton, B., Marmignon, M. H., & Saligaut, C. (2000). Modulation of pituitary dopamine D1 or D2 receptors and secretion of follicle stimulating hormone and luteinizing hormone during the annual reproductive cycle of female rainbow trout. *J Neuroendocrinol, 12*(12), 1219–1226.

Valle, L. D., Ramina, A., Vianello, S., Belvedere, P., & Colombo, L. (2002). Cloning of two mRNA variants of brain aromatase cytochrome P450 in rainbow trout (Oncorhynchus mykiss Walbaum). *J Steroid Biochem Mol Biol, 82*(1), 19–32.

Vetillard, A., Atteke, C., Saligaut, C., Jego, P., & Bailhache, T. (2003). Differential regulation of tyrosine hydroxylase and estradiol receptor expression in the rainbow trout brain. *Mol Cell Endocrinol, 199*(1–2), 37–47.

Vosges, M., Le Page, Y., Chung, B. C., Combarnous, Y., Porcher, J. M., Kah, O., et al. (2010). 17alpha-ethinylestradiol disrupts the ontogeny of the forebrain GnRH system and the expression of brain aromatase during early development of zebrafish. *Aquat Toxicol, 99*(4), 479–491.

Wang, Y., & Ge, W. (2004). Cloning of zebrafish ovarian P450c17 (CYP17, 17alpha-hydroxylase/17, 20-lyase) and characterization of its expression in gonadal and extra-gonadal tissues. *Gen Comp Endocrinol, 135*(2), 241–249.

Yague, J. G., Azcoitia, I., DeFelipe, J., Garcia-Segura, L. M., & Munoz, A. (2010). Aromatase expression in the normal and epileptic human hippocampus. *Brain Res, 1315,* 41–52.

Yague, J. G., Munoz, A., de Monasterio-Schrader, P., Defelipe, J., Garcia-Segura, L. M., & Azcoitia, I. (2006). Aromatase expression in the human temporal cortex. *Neuroscience, 138*(2), 389–401.

Yague, J. G., Wang, A. C., Janssen, W. G., Hof, P. R., Garcia-Segura, L. M., Azcoitia, I., et al. (2008). Aromatase distribution in the monkey temporal neocortex and hippocampus. *Brain Res, 1209,* 115–127.

Zhang, Y., Zhang, S., Ye, X., Ge, W., Cheng, C. H. K., Lin, H., et al. (2012). Androgen rather than estrogen up-regulates brain-type cytochrome P450 aromatase (cyp19a1b) gene via tissue-specific promoters in the hermaphrodite teleost ricefield eel monopterus albus original research article. *Mol Cell Endocrinol,* 350, 125–135.

Zhao, C., Fujinaga, R., Tanaka, M., Yanai, A., Nakahama, K., & Shinoda, K. (2007). Region-specific expression and sex-steroidal regulation on aromatase and its mRNA in the male rat brain: immunohistochemical and in situ hybridization analyses. *J Comp Neurol, 500*(3), 557–573.

Zwain, I. H., & Yen, S. S. (1999). Neurosteroidogenesis in astrocytes, oligodendrocytes, and neurons of cerebral cortex of rat brain. *Endocrinology, 140*(8), 3843–3852.

Zwain, I. H., Yen, S. S., & Cheng, C. Y. (1997). Astrocytes cultured in vitro produce estradiol-17beta and express aromatase cytochrome P-450 (P-450 AROM) mRNA. *Biochim Biophys Acta, 1334*(2–3), 338–348.

Chapter 8

Molecular Mechanisms Controlling Brain Aromatase Expression

Nobuhiro Harada and Shin-ichiro Honda

INTRODUCTION

Aromatase, the *cyp19a* gene product, is a key enzyme that catalyzes the rate-limiting step of aromatization in the biosynthesis of estrogens from androgens. This enzyme plays a crucial role in reproduction because it mediates the production of estrogens in gonadal tissues. However, aromatase activity is also present in extragonadal tissues, including the brain (Naftolin et al., 1972), skin (Berkovitz et al., 1987), liver (Smuk & Schwers, 1977), bone (Nawata et al., 1995), blood vessels (Bayard et al., 1995), and adipose tissues (Schindler et al., 1972). These observations led to questions about the function and regulation of aromatase in extragonadal tissues. To address these questions, the aromatase enzyme was purified from human placenta (Nakajin et al., 1986; Hagerman 1987; Kellis & Vickery, 1987; Harada, 1988b), and an aromatase cDNA was isolated (Corbin et al., 1988; Harada, 1988a). Subsequently, it was revealed that (1) aromatase is a hemeprotein with a heme-sulfur configuration at the catalytic site (Harada, 1988b), (2) human aromatase is a microsomal transmembrane glycoprotein comprising 503 amino acid residues (Shimozawa et al., 1993), (3) aromatase can catalyze three successive hydroxylations in the aromatization process (Harada, 1988b), (4) aromatase is classified in a new family in the cytochrome P-450 superfamily (Harada, 1988b). Interestingly, human aromatase cDNAs isolated from various human tissues contain tissue-specific unique untranslated DNA sequences at the 5′-terminus (Means et al., 1991; Harada et al., 1993b), but all of these cDNAs had an identical open reading frame, indicating that the mRNAs encode aromatase enzymes with the same amino acid sequences. This chapter briefly reviews the current knowledge concerning the aromatase gene structure and the controls of aromatase expression in the mammalian and more specifically in the mouse brain. These data provide to a large extent a detailed explanation of the mechanisms that ensure a tissue-specific regulation of the expression of this critical enzyme.

AROMATASE GENE STRUCTURE AND AROMATASE EXPRESSION

This first section describes the unique gene structure of aromatase with multiple exons 1 and promoters, which makes it possible to express the enzyme in a tissue-specific manner. The concept that aromatase is tissue-specifically regulated and produces local estrogens in various tissues corresponds well with the new notion of intracrinology of estrogens.

The human aromatase gene was isolated from genomic libraries and subsequently characterized (Means et al., 1989; Harada et al., 1990; Toda et al., 1990). An analysis of total human genomic DNA indicated that the human aromatase gene exists in the haploid genome as a single-copy (Chen et al., 1988; Harada et al., 1990). The gene includes 10 exons and, unexpectedly, spans at least 120 kilobases (kb). Exons 2–10, located within approximately 35 kb of the 3′end of the gene, encode the complete sequence of the human aromatase protein. However, multiple exon 1 variants, encoding only the unique 5′-untranslated regions of tissue-specific aromatase mRNAs found in the various human tissues, are scattered in a region spanning over 85 kb upstream of exon 2. So far, at least 9 unique exon 1 variants, designated exons I.1 (1a), I.2 (1e), I.3 (1c), I.4 (1b), I.5, I.6, I.7, PII (1d), and 1f, have been isolated from the human genome and characterized (Fig. 8–1) (Harada, 1999; Harada et al., 1993b; Simpson et al., 1997).

The multiple exon 1 variants were incorporated into mRNA in a tissue-specific manner via alternative splicing. Exons I.1 (1a) and 1f were tissue-specifically expressed only in the placenta and brain, respectively; in contrast, exon I.4 (1b) was expressed widely, including in adipose tissue, skin fibroblasts, blood vessels, liver, bone, among other tissues. Exons I.3 (1c) and PII (1d) were mainly expressed in the gonadal tissues.

Sensitive analytical methods, including immunocytochemistry, reverse transcriptase-polymerase chain reaction, high performance liquid chromatography, and enzyme-linked immunosorbent assay, make it possible to detect trace levels of aromatase in various human tissues. Consequently, the expression of human aromatase in the gonadal and extragonadal tissues was shown to be regulated, induced or suppressed, by various tissue-specific factors (Table 8–1) (Harada, 1999).

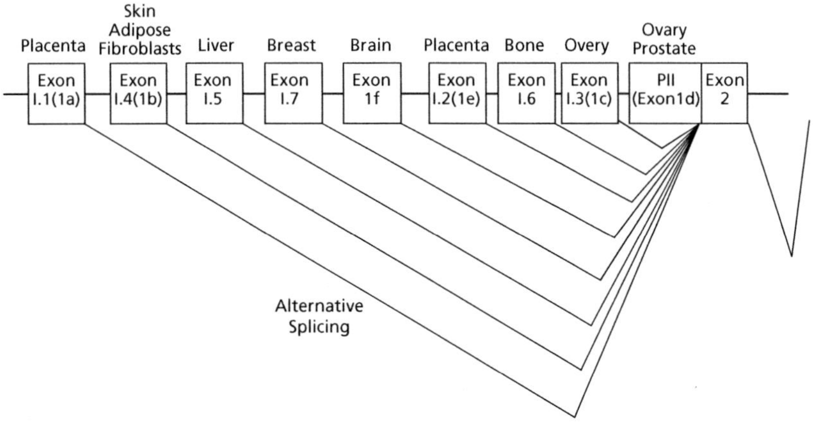

Figure 8–1. Structure of the Gene Encoding Human Aromatase. Multiple exon 1 variants in the human aromatase gene are alternatively spliced in a tissue-specific manner.

Table 8–1. Factors Regulating Aromatase in Various Tissues

Tissue	Cell Species	Regulatory Factors	
		Induction	Suppression
Placenta	Trophoblast cell	cAMP, TPA	IGF-I, Insulin
	Syncytiotrophoblast cell		GnRH, Glucose
Ovary	Granulosa cell	cAMP (FSH), Insulin	TPA, EGF, FGF
		TGF-β, IGF-I	Dexamethasone
		Activin-A	GnRH, Prolactin
Testis	Leydig cell	cAMP (FSH, LH, hCG)	EGF
		FGF, TPA	
Liver	Hepatocyte	cAMP, Dexamethasone	TGF-β
Adipose tissue	Stromal cell	cAMP, TPA	EGF, TGF-β
		Dexamethasone	
Skin	Fibroblast cell	Dexamethasone	
Bone	Resting osteoblast cell	TPA, Retinoic acd	
		Vitamin D_3	
Vascular tissue	Smooth muscle cell	cAMP, TPA	
		Dexamethasone	
Brain	Nerve cell	Androgen	

For example, glucocorticoids (dexamethasone) induce human aromatase in adipose stromal cells (Simpson et al., 1981) and suppress it in ovaries (Hsueh & Erickson, 1978). In contrast, transforming growth factor-β (TGF-β) induces human aromatase in ovaries (Ying et al., 1986) and suppresses it in adipose stromal cells (Harada, 1997). These complex mechanisms mediated by various factors regulate tissue-specific expression of aromatase. Among these factors, cyclic AMP (cAMP) induces human aromatase in a wide variety of tissues, but not in the brain. In rodents, only androgen apparently induces aromatase in the brain (Roselli et al., 1984, 1996; Yamada et al., 1993).Tissue-specific expression of human aromatase that is mediated by specific inducers and suppressors is dependent on tissue-specific exon 1 variants and on multiple promoters that have essential elements or binding sites in the proximal regions for regulatory factors and are involved in the tissue-specific transcriptional regulation of the aromatase gene (Fig. 8–2).

The placenta-specific promoter adjacent to exon I.1 (1a) has binding sites for activator protein 2γ (AP-2γ) and glial cell missing 1a (GCM1a) as

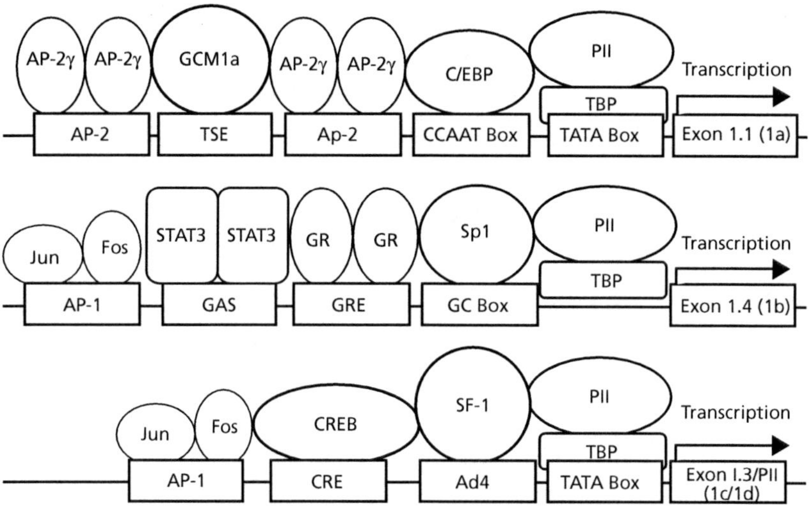

Figure 8–2. Schematic Diagram of the Promoter Structures Associated with Multiple Exon 1 Variants. Variants I.1 (1a), I.4 (1b), and I.3/PII (1c/1d) of the human aromatase gene are shown. C/EBP, CCAAT/enhancer binding protein; TBP, TATA binding protein; PII, RNA polymerase II; CREB, CRE binding protein.

trophoblast-specific elements (Yamada et al., 1995, 1999). Consequently, aromatase mRNA expressed in the human placenta is mainly transcribed from exon I.1 (1a). Kamat et al. also reported that the concerted interaction of placenta-specific enhancers and silencers within a 501 base pairs (bp) region of the placenta-specific promoter mediated the trophoblast-specific gene expression of human aromatase (Kamat & Mendelson, 2001; Kamat et al., 2005). In contrast, the gonad-specific promoter adjacent to exons I.3 (1c) and PII (exon 1d) has a cAMP-responsive element, an activating protein 1 (AP-1) binding site, and a steroidogenic factor 1 binding site (Ad4), indicating that, in gonadal tissues, aromatase transcription is regulated by cAMP-protein kinase A (PKA)-mediated and diacylglycerol-protein kinase C (PKC)-mediated signaling pathways and also by interaction with nuclear hormone receptors (Zhao et al., 1996; Harada, 1997; Zhou & Chen, 1999). The promoter region adjacent to exon I.4 (1b) is expressed in a wide variety of tissues and has a GC box, which is often found in housekeeping genes, rather than a TATA box; this promoter is regulated by glucocorticoids and cytokines via binding of glucocorticoid receptor to glucocorticoid responsive element and binding of signal transducers and activators of transcription protein to interferon-gamma activated sequence (Zhao et al., 1995, 1996; Harada, 1997), respectively.

The aromatase genes of many mammals, like the human aromatase gene, have multiple exon 1 variants and multiple promoters. The mammalian aromatase gene seems to be a single-copy gene and to be regulated through tissue-specific use of the promoters and exon 1 variants. In many animals, functional sexual behavior and normal reproductive-system activity are dependent on local aromatase-mediated production of estrogens in the brain and gonad, respectively. Therefore, it is not surprising that aromatase is found in the brain as well as the gonads of vertebrates from fish to mammals. In fish, two different aromatase transcripts were isolated, one from brain, the other from gonad. These transcripts are apparently products of different genes, indicating that two aromatase genes exist in these genomes (Kishida & Callard, 2001; Tchoudakova et al., 2001). Tissue-specific regulation of aromatase may be the result of the presence of two separate genes, a brain-specific and a gonad-specific gene, that occurred in an early stage of the molecular evolution of aromatase gene. However, aromatase also has another type of gene organization with multiple exon 1 variants, including brain-specific and gonad-specific variants (Golovine et al., 2003; S. Honda S. et al., 1994, 1996; Bulun et al., 2004; Corbin et al., 2009), as is true for later-emerging vertebrates, like mammals.

There are in addition multiple questions concerning aromatase regulation in the brain. These questions concern brain regions where and developmental stages when aromatase is expressed and what kinds of factors/hormones regulate this anatomical and temporal specificity. These questions are considered in the following section.

REGULATION OF THE NEONATAL PEAK OF AROMATASE EXPRESSION IN THE MAMMALIAN BRAIN

In vivo studies using rodents showed that aromatase in the brain is expressed mainly in the nerve cells of specific brain regions, with a transient peak at the perinatal period. This neonatal peak of aromatase expression was also observed in tissue slices and primary cultures of nerve cells prepared from fetal diencephala in the absence of any additional humoral factors, in a manner that closely mimicked the ontogeny of brain aromatase. Intracellular protein kinases of PKA, PKC, and PKG activated themselves by neurotransmitters were found to modify aromatase expression. Finally, studies using hypothalamic neural cell lines indicated that both estrogen receptor-α and progesterone receptor were also involved in the regulation of aromatase in the brain. We review here in vivo and in vitro studies on brain aromatase that were performed to elucidate the mechanisms controlling aromatase expression in the brain. The analysis of the expression of aromatase in the brain provides critical information on the mechanisms that control the organizing and activating effects of estrogens on sexual behaviors.

Aromatase in the mammalian brain is reportedly localized mainly in the medial preoptic area, hypothalamus, and amygdala (see chapter 3, this volume). In rodents, which have short gestation periods, brain aromatase first appears in the fetus, increases rapidly, and then decreases gradually to the adult level with a transient peak at the perinatal period (Harada & Yamada, 1992; Abe Dohmae et al., 1996b). In contrast, expression of brain hypothalamic aromatase in cows, which have a relatively long gestation period, was highest in the second quarter of gestation (Peruffo et al., 2008). Aromatase in mouse and human brains was transcribed from the brain-specific exon 1f (Honda et al. 1994, 1996). However, in the adult human brain except for the hypothalamic region, aromatase was also transcribed from exon I.4 (1b) utilized in many tissues (Sasano et al., 1998).

Immunocytochemical studies indicate that neuronal cells are aromatase-positive in the fetal and adult brain, but that glial cells or microglial cells become aromatase-positive in the adult brain after lesions (Azcoitia et al., 2003; Carswell et al., 2005; Saldanha et al., 2009). Several human and rat glioblastoma cell lines express aromatase which may participate in neuroprotection via autocrine or paracrine production of estrogen (Yague et al., 2004).

To analyze in detail the mechanisms regulating aromatase expression in the brain, diencephalic tissues were prepared from fetal mouse brains at various developmental stages and cultured in vitro as tissue slices or as monolayers of primary cells (Abe Dohmae et al., 1996b). Aromatase mRNA in the diencephalic region of the fetal mouse brain appears at embryonic day 13 (E13), increases for 3 to 4 days, and decreases after a transient peak at E16 to 17 (Harada & Yamada, 1992; Abe Dohmae et al., 1996b). Both tissue slices and primary cells grown in monolayers from the diencephalic region of E12 to 13 exhibited patterns of aromatase expression similar to those observed in vivo (Abe Dohmae et al., 1996b). The levels of aromatase mRNA in cultured diencephalic neurons of E13 mice increase for 3 days and then decrease, whereas no significant increase of the aromatase mRNA levels occurs in diencephalic neurons of E10 to 11 mice (Abe Dohmae et al., 1996b). Aromatase mRNA cannot be detected in cultured neurons from the cerebral cortex of E15 mice or in cultured astrocytes from the diencephalon or cerebral cortex of E15 mice, indicating that there is cell type-specific and region-specific expression of aromatase mRNA in the mouse brain (Abe Dohmae et al., 1996b). These results also indicate that developmental increases in aromatase mRNA expression in the diencephalic region are an autonomous event that is largely regulated endogenously within specific neurons themselves and probably genetically determined after E12 in fetal mice. Although the neuronal cells that were cultured in vitro in a chemically defined serum-free medium showed spontaneous changes in aromatase mRNA, in a manner that closely mimicked the ontogeny of mouse brain aromatase, the expression of aromatase in these cultured neurons is suppressed significantly when serum is added to the medium (Abe Dohmae et al., 1996a). This result may indicate that humoral factors in the serum affect, directly or indirectly, the expression of aromatase in diencephalic neurons.

The effect of activators of intracellular signal transduction pathways and neurotransmitters/neuropeptides on the expression of diencephalic aromatase was also examined in the cultured diencephalic neurons (Fig. 8–3) (Abe Dohmae et al., 1996a).

Aromatase expression increases in response to phorbol 12-myristate 13-acetate (PMA) and dibutyryl-cyclic GMP, which are activators of PKC and protein kinase G (PKG), respectively, and decreases in response to dibutyryl-cyclic AMP, a PKA activator (Abe Dohmae et al., 1996a). The possibility that neurotransmitter ligands of receptors that activate PKC or PKG function as inducers of

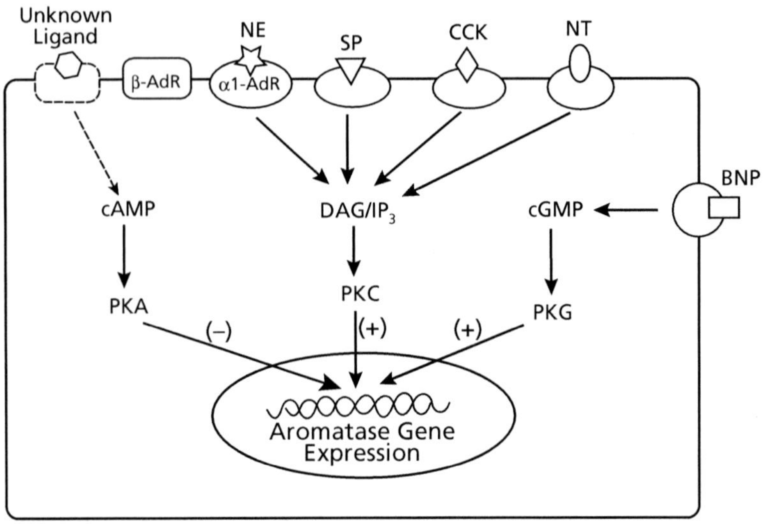

Figure 8–3. A Tentative Model for the Regulation of Aromatase Gene Expression in Diencephalic Neurons. Intracellular signal transduction systems induced by neurotransmitters may regulate aromatase expression. Symbols, (+) and (−) indicate inductive and suppressive effects, respectively. Abbreviations: NE, norepinephrine; EPI, epinephrine; SP, substance P; CCK, cholecystokinin; NT, neurotensin; BNP, brain natriuretic peptide; cAMP, cyclic AMP; DAG, diacylglycerol; IP3, inositol 1,4,5-triphosphate; cGMP, cyclic GMP; PKA, protein kinase A; PKC, protein kinase C; PKG, protein kinase G.

aromatase in diencephalic neurons was also examined. Neuronal aromatase is affected significantly by α1-adrenergic receptor ligands (Abe Dohmae et al., 1996a). Aromatase expression increases in neurons treated with an α1 agonist, phenylephrine, but an α1 antagonist, prazosin, suppresses this increase; however, agonists of α2- or β-adrenergic receptors do not exert any influence. The developmental expression profile of α1-adrenergic receptor subtypes indicates that the α1B-subtype is responsible for PKC-activating signal transduction in diencephalic neurons (Abe Dohmae et al., 1996a). Several neuropeptides including brain natriuretic peptide (a PKG activator) and substance P, cholecystokinin, and neurotensin (PKC activators) increase expression of neuronal aromatase (Abe Dohmae et al., 1996a). β-Adrenergic receptors are not detectable in the prenatal period; therefore, β-adrenergic receptors cannot affect expression of aromatase in diencephalic neurons at this stage. Nevertheless, cAMP does suppress expression of aromatase in diencephalic neurons at this stage (Abe Dohmae et al., 1996a).

The expression of aromatase in diencephalic neurons may be autonomously regulated by endogenous factors according to an internal genetic program and modified by neurotransmitters or neuropeptides secreted from projecting neurons or neighboring glia. The expression in the cultured neurons is not affected by androgens, a known inducer of brain aromatase in adult mouse (Yamada et al., 1993) and in quail (Harada et al., 1993a). Induction of brain aromatase by androgens may be mediated via several steps within a neural network. Aromatase in human glioma cells is transcriptionally regulated by vitamin D, dexamethasone, and mifepristone (Yague, et al., 2009).

MOLECULAR MECHANISMS REGULATING TRANSCRIPTION OF AROMATASE IN THE MOUSE BRAIN

After considering the regulation of the neonatal peak of aromatase expression in the developing brain by multiple neural factors or hormones, we will now review the transcriptional regulation of aromatase in the adult brain, mediated by binding of transcription factors to the corresponding DNA responsive elements on the brain-specific promoter.

The use of DNA footprinting assays indicated that there are several DNA elements for nuclear DNA binding factors on the proximal promoter of the aromatase gene. Both ARP1 and Lhx2 were characterized as DNA binding proteins to the identified elements, Arom-Aβ and Arom-B, respectively, and as crucial transcription factors of aromatase in the brain. Furthermore, promoter assays for a wide range of promoters using transgenic mice demonstrated that both proximal and distal promoter regions are necessary and essential for spatiotemporal expression of aromatase in the brain.

In this section, we further review the molecular mechanisms on the transcriptional regulation of aromatase in the brain. This regulation is mediated by binding of transcription factors to DNA elements on the promoter and their interaction with transcriptional machinery consisting of an RNA polymerase complex. Most studies have been performed using primary cultures of mouse diencephalic nerve cells, subcultures of mouse hypothalamic nerve cell lines, and transgenic mice. So, we will focus the review on the transcriptional regulation of brain aromatase in mouse specifically.

The brain-specific exon 1f and its promoter region have been isolated from the aromatase genes of several species (S. Honda et al., 1994, 1996; Yamada-Mouri et al., 1995, 1997; Ramachandran et al., 1999; Galmiche et al., 2006; Iwabuchi et al., 2007). The brain-specific promoter regions that mediate the expression of aromatase in the preoptic area, hypothalamus, and amygdala have DNA sequences that are highly conserved among mammals. The proximal promoter region adjacent to the brain-specific exon 1f was analyzed to elucidate the molecular mechanism(s) responsible for region-specific and developmental stage-specific expression of aromatase in the brain. A promoter activity assay that employed the chloramphenicol acetyltransferase (CAT) gene as a reporter was performed using segments of various lengths (57–954 base pairs) of the proximal promoter region (S. Honda et al., 1999). Expression from the reporter constructs transfected into cultured neurons from E13 fetal mouse showed that the promoter region 202 bp upstream from transcriptional initiation site was the minimal essential region for brain-specific promoter and resulted in the highest expression. Binding of *trans*-acting factors in a nuclear extract from the diencephalon to the -202 bp promoter region was analyzed using a DNase footprinting assay (S. Honda et al., 1999). Multiple areas, designated as Arom-Aα, -Aβ, -Aγ, -B, and -C, on this promoter segment were protected against DNase digestion. Gel-shift assay confirmed specific binding of the nuclear factors to the protected sites. Southwestern blotting analysis identified the single nuclear factor that binds to the Arom-Aα, -Aγ, and -C sites; it is a 42 kilodaltons protein, which is widely distributed in the nuclei of various tissues. However, unique nuclear factors bind to Arom-Aβ and -B sites; these were identified using a gel-shift assay involving competition with oligonucleotide probes. The Arom-Aβ-binding factor localizes in nuclei in brain, liver, and kidney cells, but not in cells of steroid-synthesizing tissues of testis, ovary, or adrenal (S. Honda et al., 1999). Furthermore, the nuclear factors isolated from the diencephalon and limbic region are of different molecular sizes; those isolated from other tissues are also different. These findings may suggest the presence of tissue-specific isoforms of these factors. Interestingly, the Arom-Aβ-binding factor in brain nuclei appears at E11, increases as development proceeds, decreases after a transient peak at E17, and reaches extremely low levels in adult mice (S. I. Honda et al., 2001). The pattern of developmental change of the factor is very consistent with that of aromatase in the brain. The Arom-B probe identified a few bands in the nuclear extracts prepared from fetal mouse diencephalic and limbic regions and from adult mouse liver tissue in the gel-shift assay (S. Honda et al., 1999). The intensity of each band was dependent on the brain region or tissues used in the assay. Mutations introduced into Arom-Aα, -Aβ, or -B sites greatly decreased the promoter activity of the -202 bp minimal essential promoter region in the CAT reporter assay (S. Honda et al., 1999); these findings suggested that the Arom-Aα, -Aβ, and -B sites in the brain-specific promoter have crucial roles in regulating transcription.

To isolate and characterize the candidate factors that may participate in the regulation of brain-specific expression of mouse aromatase, cDNA library derived from fetal mouse brains was subjected to a yeast one-hybrid screen using Arom-Aβ and Arom-B probes as baits. In this screen, ARP1 (also called COUP-TFII) was identified as an Arom-Aβ-binding factor, and Lhx2 as an Arom-B-binding factor (Fig. 8–4) (Harada & Honda, 2009; S. Honda et al., 2012).

Apolipoprotein Ai regulatory protein 1 (ARP1) is a transcription factor in the steroid hormone receptor family; it is an orphan-type receptor (Sagami et al., 1986). It is a transcriptional repressor of apolipoprotein A1 (apoAI) in the liver

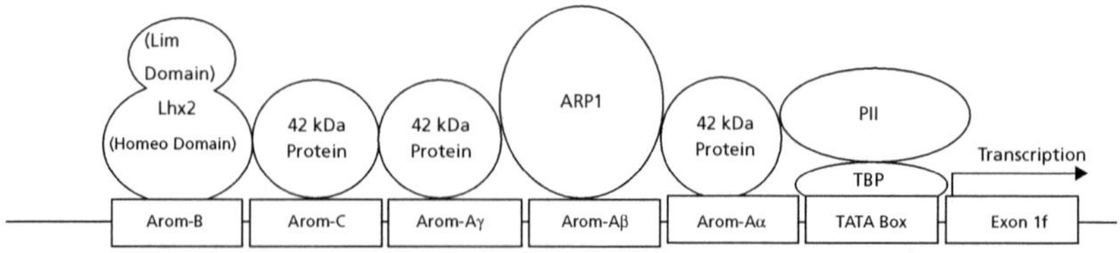

Figure 8–4. Schematic Diagram of the Promoter Structure Adjacent to the Brain-Specific Exon 1f of the Mouse Aromatase Gene

(Ladias & Karathanasis, 1991). Moreover, it is expressed in the developing fetal brain and participates in neurogenesis and differentiation of nerve cells (Qiu et al., 1994; Lopes da Silva et al., 1995). The identification of Arom-Aβ-binding factor as ARP1 was confirmed by competitive inhibition with ARP1 and apoAI probes and super-shift with anti-ARP1 antibody in gel-shift assays (Harada & Honda, 2009).

Lhx2 is a transcription factor with a Lim domain and a homeodomain (Xu et al., 1993). Reportedly, it regulates the glycoprotein hormone α-subunit gene and is expressed in the developing eye, forebrain, and erythrocytes (Porter et al., 1997). The identification of Arom-B-binding factor as Lhx2 was confirmed by competitive inhibition with a Lhx2 probe and super-shift with anti-Lhx2 antibody in the gel-shift assay (Harada & Honda, 2009; S. Honda et al., 2012). Furthermore, Lhx2 siRNA efficiently inhibited developmental increases in aromatase mRNA levels in cultured diencephalic neurons, and chromatin immunoprecipitation assays revealed that Lhx2 actually binds to the Arom-B site on the chromosomal DNA from E16 neurons in the fetal mouse brain (Harada & Honda, 2009; S. Honda et al., 2012). These results indicate that ARP1 and Lhx2 are actually involved in the ontogenesis of aromatase expression in mouse diencephalic neurons.

Whether the -202 bp minimal essential promoter region that contains the Arom-Aα, -Aβ, and -B sites is the necessary and sufficient region for neuron-specific, brain region-specific, and development stage-specific expression of aromatase in the mouse brain is difficult to determine. Experiments employing only in vitro assays or cultured cells cannot be used to definitively identify the essential promoter region for spatiotemporal regulation of brain aromatase. Therefore, the spatiotemporal regulation of aromatase in the brain was analyzed in vivo using transgenic animals carrying β-galactosidase reporter gene constructs containing various segments of the brain-specific promoter of the mouse aromatase gene (Harada & Honda, 2005). The reporter transgene carrying a 6.5 kb upstream region of the brain-specific promoter showed a transient expression peak of β-galactosidase reporter activity at the perinatal period in the preoptic area, medial hypothalamus, and amygdala of the transgenic mice, but not in other brain regions or other tissues, whereas transgenes carrying shorter fragments (the -4.0 or -2.0 kb promoter region) often showed ambiguous or inconsistent expression patterns (Harada & Honda, 2005). These results demonstrate that the -6.5 kb upstream region of the brain-specific promoter virtually contains the essential elements for spatiotemporal expression of aromatase in the mouse brain. Nausch et al. also assessed the activity of the brain-specific promoter region using transgenic mice carrying β-galactosidase reporter constructs with various lengths of this promoter. The data demonstrated that a -0.2 kb segment results in reporter gene expression specific to brain areas, and a -0.2 to -1.7 kb promoter region is necessary for expression in additional areas, but not sufficient for consistent replication of region-specific expression in mouse brain (Nausch, et al., 2007).

Recently, aromatase in the mouse hypothalamic neuronal cell lines was shown to be regulated by the proximal region of the brain-specific promoter via both the estrogen receptor-α and progesterone receptor (Yilmaz et al., 2009, 2011). The -1 to -200 bp region of the promoter confers estrogen responsiveness via binding of estrogen receptor-α and c-jun to two AP-1 elements in this region (Yilmaz et al., 2009), and the -600 to -800 bp region of the promoter is, by way of the progesterone receptor, involved in the regulation of aromatase in cultured hypothalamic neuronal cells (Yilmaz et al., 2011).

Moreover, the -202 bp minimal essential promoter region containing Arom-Aα, -Aβ, and -B sites is not sufficient for spatiotemporal expression of aromatase in the mouse brain (Harada & Honda, 2005). To determine whether the -202 bp promoter region is the region necessary for spatiotemporal expression of aromatase in the mouse brain, a point mutation was introduced into the Arom-Aβ sequence on the -6.5 kb upstream region of the brain-specific promoter, and a β-galactosidase reporter transgene, which was controlled by the mutated -6.5 kb upstream region, was constructed and introduced into mice (Harada & Honda, 2005). The transgenic mice exhibit disappearance of reporter gene expression in the diencephalon and appearance of the ectopic expression in the testis. This fact indicates that the -202 bp promoter region containing Arom-Aα, -Aβ, and -B sites is absolutely necessary for spatiotemporal expression of aromatase in the mouse brain. Presumably, there may be a number of essential elements in the -6.5 kb upstream region of the brain-specific promoter of mouse aromatase gene, and the strict spatiotemporal expression may be the result of multiple interactions among these elements.

CONCLUSIONS

Although the genetic and epigenetic regulatory mechanisms of aromatase in the brain remain somewhat unclear, substantial progress has been made during the past decades. A comprehensive understanding of these mechanisms may give us therapeutic clues to some estrogen-related neurodegenerative diseases as well as useful information on mechanisms mediating the organization and activation of sexual/reproductive behaviors.

REFERENCES

Abe Dohmae, S., Takagi, Y., & Harada, N. (1996). Neurotransmitter-mediated regulation of brain aromatase: protein kinase C- and G-dependent induction. *J Neurochem, 67*(5), 2087–2095.

Abe Dohmae, S., Tanaka, R., Takagi, Y., & Harada, N. (1996). In vitro increase of aromatase mRNA in diencephalic neurons. *Neuroendocrinology, 63*(1), 46–52.

Azcoitia, I., Sierra, A., Veiga, S., & Garcia-Segura, L. M. (2003). Aromatase expression by reactive astroglia is neuroprotective. *Ann N Y Acad Sci, 1007,* 298–305.

Bayard, F., Clamens, S., Delsol, G., Blaes, N., Maret, A., & Faye, J. C. (1995). Oestrogen synthesis, oestrogen metabolism and functional oestrogen receptors in bovine aortic endothelial cells. *Ciba Found Symp, 191,* 122–132; discussion 132–128.

Berkovitz, G. D., Brown, T. R., & Fujimoto, M. (1987). Aromatase activity in human skin fibroblasts grown in cell culture. *Steroids, 50*(1–3), 281–295.

Bulun, S. E., Takayama, K., Suzuki, T., Sasano, H., Yilmaz, B., & Sebastian, S. (2004). Organization of the human aromatase p450 (CYP19) gene. *Semin Reprod Med, 22*(1), 5–9.

Carswell, H. V., Dominiczak, A. F., Garcia-Segura, L. M., Harada, N., Hutchison, J. B., & Macrae, I. M. (2005). Brain aromatase expression after experimental stroke: topography and time course. *J Steroid Biochem Mol Biol, 96*(1), 89–91.

Chen, S. A., Besman, M. J., Sparkes, R. S., Zollman, S., Klisak, I., Mohandas, T., et al. (1988). Human aromatase: cDNA cloning, Southern blot analysis, and assignment of the gene to chromosome 15. *DNA, 7*(1), 27–38.

Corbin, C. J., Berger, T., Ford, J. J., Roselli, C. E., Sienkiewicz, W., Trainor, B. C., et al. (2009). Porcine hypothalamic aromatase cytochrome P450: isoform characterization, sex-dependent activity, regional expression, and regulation by enzyme inhibition in neonatal boars. *Biol Reprod, 81*(2), 388–395.

Corbin, C. J., Graham-Lorence, S., McPhaul, M., Mason, I., Mendelson, C. R., & Simpson, E. R. (1988). Isolation of a full-length cDNA insert encoding human aromatase system cytochrome P-450 and its expression in nonsteroidogenic cells. *Proc Natl Acad Sci U S A, 85*, 8948–8952.

Galmiche, G., Corvaisier, S., & Kottler, M. L. (2006). Aromatase gene expression and regulation in the female rat pituitary. *Ann N Y Acad Sci, 1070*, 286–292.

Golovine, K., Schwerin, M., & Vanselow, J. (2003). Three different promoters control expression of the aromatase cytochrome p450 gene (cyp19) in mouse gonads and brain. *Biol Reprod, 68*(3), 978–984.

Hagerman, D. D. (1987). Human placenta estrogen synthetase (aromatase) purified by affinity chromatography. *J Biol Chem, 262*(5), 2398–2400.

Harada, N. (1988a). Cloning of a complete cDNA encoding human aromatase; immunochemical identification and sequence analysis. *Biochem Biophys Res Commun, 156*, 725–732.

Harada, N. (1988b). Novel properties of human placental aromatase as cytochrome P-450; purification and characterization of a unique form of aromatase. *J Biochem, 103*, 106–113.

Harada, N. (1997). Aberrant expression of aromatase in breast cancer tissues. *J Steroid Biochem Mol Biol, 61*(3–6), 175–184.

Harada, N. (1999). Aromatase and intracrinology of estrogen in hormone-dependent tumors. *Oncology, 57*(Suppl S2), 7–16.

Harada, N., Abe-Dohmae, S., Loeffen, R., Foidart, A., & Balthazart, J. (1993a). Synergism between androgens and estrogens in the induction of aromatase and its messenger RNA in the brain. *Brain Res., 622*, 243–256.

Harada, N., & Honda, S. (2005). Analysis of spatiotemporal regulation of aromatase in the brain using transgenic mice. *J Steroid Biochem Mol Biol, 95*(1–5), 49–55.

Harada, N., & Honda, S.-I. (2009). Multiple transcription factors regulate a brain-specific promoter activity of mouse aromatase (CYP19A1) gene. In: Shoun, H. & Ohkawa H. (eds.), *16th International Conference on Cytochrome P450*. Bologna, Italy: Medimond S.r.l.

Harada, N., Utsumi, T., & Takagi, Y. (1993b). Tissue-specific expression of the human aromatase cytochrome P-450 gene by alternative use of multiple exons 1 and promoters, and switching of tissue-specific exons 1 in carcinogenesis. *Proc Natl Acad Sci U S A*, 90, 11312–11316.

Harada, N. & Yamada, K. (1992). Ontogeny of aromatase messenger ribonucleic acid in mouse brain: fluorometrical quantitation by polymerase chain reaction. *Endocrinology, 131*(5), 2306–2312.

Harada, N., Yamada, K., Saito, K., Kibe, N., Dohmae, S., & Takagi, Y. (1990). Structural characterization of the human estrogen synthetase (aromatase) gene. *Biochem Biophys Res Commun, 166*(1), 365–372.

Honda, S., Harada, N., Abe-Dohmae, S., & Takagi, Y. (1999). Identification of cis-acting elements in the proximal promoter region for brain-specific exon 1 of the mouse aromatase gene. *Brain Res Mol Brain Res, 66*(1–2), 122–132.

Honda, S., Harada, N., & Takagi, Y. (1994). Novel exon 1 of the aromatase gene specific for aromatase transcripts in human brain. *Biochem Biophys Res Commun, 198*(3), 1153–1160.

Honda, S., Harada, N., & Takagi, Y. (1996). The alternative exons 1 of the mouse aromatase cytochrome P-450 gene. *Biochim Biophys Acta, 1305*(3), 145–150.

Honda, S. I., Matsumoto, T., & Harada, N. (2001). Characterization and purification of a protein binding to the cis-acting element for brain-specific exon 1 of the mouse aromatase gene. *J Steroid Biochem Mol Biol, 79*(1–5), 255–260.

Honda, S., Kozako, T., Shimeno, H., Soeda, S., & Harada, N. (2012). LIM-homeodomain transcription factor, Lhx2, is involved in transcriptional control of brain-specific promoter/exon 1f of the mouse aromatase gene. *J Neuroendocrinol*, in press.

Hsueh, A. J., & Erickson, G. F. (1978). Glucocorticoid inhibition of FSH-induced estrogen production in cultured rat granulosa cells. *Steroids, 32*(5), 639–648.

Iwabuchi, J., Wako, S., Tanaka, T., Ishikawa, A., Yoshida, Y., & Miyata, S. (2007). Analysis of the p450 aromatase gene expression in the Xenopus brain and gonad. *J Steroid Biochem Mol Biol, 107*(3–5), 149–155.

Kamat, A., & Mendelson, C. R. (2001). Identification of the regulatory regions of the human aromatase P450 (CYP19) gene involved in placenta-specific expression. *J Steroid Biochem Mol Biol, 79*(1–5), 173–180.

Kamat, A., Smith, M. E., Shelton, J. M., Richardson, J. A., & Mendelson, C. R. (2005). Genomic regions that mediate placental cell-specific and developmental regulation of human cyp19 (aromatase) gene expression in transgenic mice. *Endocrinology, 146*(5), 2481–2488.

Kellis, J. T., Jr., & Vickery, L. E. (1987). Purification and characterization of human placental aromatase cytochrome P-450. *J Biol Chem, 262*(9), 4413–4420.

Kishida, M., & Callard, G. V. (2001). Distinct cytochrome P450 aromatase isoforms in zebrafish (Danio rerio) brain and ovary are differentially programmed and estrogen regulated during early development. *Endocrinology, 142*(2), 740–750.

Ladias, J. A., & Karathanasis, S. K. (1991). Regulation of the apolipoprotein AI gene by ARP-1, a novel member of the steroid receptor superfamily. *Science, 251*(4993), 561–565.

Lopes da Silva, S., Cox, J. J., Jonk, L. J., Kruijer, W., & Burbach, J. P. (1995). Localization of transcripts of the related nuclear orphan receptors COUP-TF I and ARP-1 in the adult mouse brain. *Brain Res Mol Brain Res, 30*(1), 131–136.

Means, G. D., Kilgore, M. W., Mahendroo, M. S., Mendelson, C. R., & Simpson, E. R. (1991). Tissue-specific promoters regulate aromatase cytochrome P450 gene expression in human ovary and fetal tissues. *Mol Endocrinol, 5*(12), 2005–2013.

Means, G. D., Mahendroo, M. S., Corbin, C. J., Mathis, J. M., Powell, F. E., Mendelson, C. R., et al. (1989). Structural analysis of the gene encoding human aromatase cytochrome P-450, the enzyme responsible for estrogen biosynthesis. *J Biol Chem, 264*(32), 19385–19391.

Naftolin, F., Ryan, K. J., & Petro, Z. (1972). Aromatization of androstenedione by the anterior hypothalamus of adult male and female rats. *Endocrinology, 90,* 295–298.

Nakajin, S., Shinoda, M., & Hall, P. F. (1986). Purification to homogeneity of aromatase from human placenta. *Biochem Biophys Res Commun, 134*(2), 704–710.

Nausch, N., Manteuffel, G., & Vanselow, J. (2007). 0.2kb promoter sequence of the murine Cyp19 gene target beta-galactosidase expression to specific brain areas of transgenic mice. *J Steroid Biochem Mol Biol, 103*(2), 119–128.

Nawata, H., Tanaka, S., Takayanagi, R., Sakai, Y., Yanase, T., Ikuyama, S., et al. (1995). Aromatase in bone cell: association with osteoporosis in postmenopausal women. *J Steroid Biochem Mol Biol, 53*(1–6), 165–174.

Peruffo, A., Cozzi, B., & Ballarin, C. (2008). Ontogenesis of brain aromatase P450 expression in the bovine hypothalamus. *Brain Res Bull, 75*(1), 60–65.

Porter, F. D., Drago, J., Xu, Y., Cheema, S. S., Wassif, C., Huang, S. P., et al. (1997). Lhx2, a LIM homeobox gene, is required for eye, forebrain, and definitive erythrocyte development. *Development, 124*(15), 2935–2944.

Qiu, Y., Cooney, A. J., Kuratani, S., DeMayo, F. J., Tsai, S. Y., & Tsai, M. J. (1994). Spatiotemporal expression patterns of chicken ovalbumin upstream promoter-transcription factors in the developing mouse central nervous system: evidence for a role in segmental patterning of the diencephalon. *Proc Natl Acad Sci U S A, 91*(10), 4451–4455.

Ramachandran, B., Schlinger, B. A., Arnold, A. P., & Campagnoni, A. T. (1999). Zebra finch aromatase gene expression is regulated in the brain through an alternate promoter. *Gene, 240*(1), 209–216.

Roselli, C. E., Abdelgadir, S. E., Jorgensen, E., & Resko, J. A. (1996). Sex differences in androgen-regulated cytochrome P450 aromatase mRNA in the rat brain. *Endocrine, 5*(1), 59–65.

Roselli, C. E., Ellinwood, W. E., & Resko, J. A. (1984). Regulation of brain aromatase activity in rats. *Endocrinology, 114*(1), 192–200.

Sagami, I., Tsai, S. Y., Wang, H., Tsai, M. J., & O'Malley, B. W. (1986). Identification of two factors required for transcription of the ovalbumin gene. *Mol Cell Biol, 6*(12), 4259–4267.

Saldanha, C. J., Duncan, K. A., & Walters, B. J. (2009). Neuroprotective actions of brain aromatase. *Front Neuroendocrinol, 30*(2), 106–118.

Sasano, H., Takashashi, K., Satoh, F., Nagura, H., & Harada, N. (1998). Aromatase in the human central nervous system. *Clin Endocrinol, 48*(3), 325–329.

Schindler, A. E., Enert, A., & Friedrich, E. (1972). Conversion of androstenedione to estrone by human fat tissue. *J Clin Endocrinol Metab, 35,* 627–630.

Shimozawa, O., Sakaguchi, M., Ogawa, H., Harada, N., Mihara, K., & Omura, T. (1993). Core glycosylation of cytochrome P-450(arom). Evidence for localization of N terminus of microsomal cytochrome P-450 in the lumen. *J Biol Chem, 268*(28), 21399–21402.

Simpson, E. R., Ackerman, G. E., Smith, M. E., & Mendelson, C. R. (1981). Estrogen formation in stromal cells of adipose tissue of women: induction by glucocorticosteroids. *Proc Natl Acad Sci U S A, 78*(9), 5690–5694.

Simpson, E. R., Michael, M. D., Agarwal, V. R., Hinshelwood, M. M., Bulun, S. E., & Zhao, Y. (1997). Cytochromes P450 11: expression of the CYP19 (aromatase) gene: an unusual case of alternative promoter usage. *FASEB Journal, 11*(1), 29–36.

Smuk, M., & Schwers, J. (1977). Aromatization of androstenedione by human adult liver in vitro. *J Clin Endocrinol Metab, 45*(5), 1009–1012.

Tchoudakova, A., Kishida, M., Wood, E., & Callard, G. V. (2001). Promoter characteristics of two cyp19 genes differentially expressed in the brain and ovary of teleost fish. *J Steroid Biochem Mol Biol, 78*(5), 427–439.

Toda, K., Terashima, M., Kawamoto, T., Sumimoto, H., Yokoyama, Y., Kuribayashi, I., et al. (1990). Structural and functional characterization of human aromatase P-450 gene. *Eur J Biochem, 193*(2), 559–565.

Xu, Y., Baldassare, M., Fisher, P., Rathbun, G., Oltz, E. M., Yancopoulos, G. D., et al. (1993). LH-2: a LIM/homeodomain gene expressed in developing lymphocytes and neural cells. *Proc Natl Acad Sci U S A, 90*(1), 227–231.

Yague, J. G., Garcia-Segura, L. M., & Azcoitia, I. (2009). Selective transcriptional regulation of aromatase gene by vitamin D, dexamethasone, and mifepristone in human glioma cells. *Endocrine, 35*(2), 252–261.

Yague, J. G., Lavaque, E., Carretero, J., Azcoitia, I., & Garcia-Segura, L. M. (2004). Aromatase, the enzyme responsible for estrogen biosynthesis, is expressed by human and rat glioblastomas. *Neurosci Lett, 368*(3), 279–284.

Yamada, K., Harada, N., Honda, S., & Takagi, Y. (1995). Regulation of placenta-specific expression of the aromatase cytochrome P-450 gene. Involvement of the trophoblast-specific element binding protein. *J Biol Chem, 270*(42), 25064–25069.

Yamada, K., Harada, N., Tamaru, M., & Takagi, Y. (1993). Effects of changes in gonadal hormones on the amount of aromatase messenger RNA in mouse brain diencephalon. *Biochem Biophys Res Commun, 195*(1), 462–468.

Yamada, K., Ogawa, H., Honda, S., Harada, N., & Okazaki, T. (1999). A GCM motif protein is involved in placenta-specific expression of human aromatase gene. *J Biol Chem, 274*(45), 32279–32286.

Yamada-Mouri, N., Hirata, S., Hayashi, M., & Kato, J. (1995). Analysis of the expression and the first exon of aromatase mRNA in monkey brain. *J Steroid Biochem Mol Biol, 55*(1), 17–23.

Yamada-Mouri, N., Hirata, S., Kato, J., & Hoshi, K. (1997). Expression and distribution of cortical type aromatase mRNA variant in the adult rat brain. *J Steroid Biochem Mol Biol, 60*(5–6), 325–329.

Yilmaz, M. B., Wolfe, A., Cheng, Y. H., Glidewell-Kenney, C., Jameson, J. L., & Bulun, S. E. (2009). Aromatase promoter I.f is regulated by estrogen receptor alpha (ESR1) in mouse hypothalamic neuronal cell lines. *Biol Reprod, 81*(5), 956–965.

Yilmaz, M. B., Wolfe, A., Zhao, H., Brooks, D. C., & Bulun, S. E. (2011). Aromatase promoter I.f is regulated by progesterone receptor in mouse hypothalamic neuronal cell lines. *J Mol Endocrinol, 47*(1), 69–80.

Ying, S. Y., Becker, A., Ling, N., Ueno, N., & Guillemin, R. (1986). Inhibin and beta type transforming growth factor (TGF beta) have opposite modulating effects on the follicle stimulating hormone (FSH)-induced aromatase activity of cultured rat granulosa cells. *Biochem Biophys Res Commun, 136*(3), 969–975.

Zhao, Y., Agarwal, V. R., Mendelson, C. R., & Simpson, E. R. (1996). Estrogen biosynthesis proximal to a breast tumor is stimulated by PGE2 via cyclic AMP, leading to activation of promoter II of the CYP19 (aromatase) gene. *Endocrinology, 137*(12), 5739–5742.

Zhao, Y., Nichols, J. E., Bulun, S. E., Mendelson, C. R., & Simpson, E. R. (1995). Aromatase P450 gene expression in human adipose tissue. Role of a Jak/STAT pathway in regulation of the adipose-specific promoter. *J Biol Chem, 270*(27), 16449–16457.

Zhao, Y., Nichols, J. E., Valdez, R., Mendelson, C. R., & Simpson, E. R. (1996). Tumor necrosis factor-alpha stimulates aromatase gene expression in human adipose stromal cells through use of an activating protein-1 binding site upstream of promoter 1.4. *Mol Endocrinol, 10*(11), 1350–1357.

Zhou, D., & Chen, S. (1999). Identification and characterization of a cAMP-responsive element in the region upstream from promoter 1.3 of the human aromatase gene. *Arch Biochem Biophys, 371*(2), 179–190.

Part III

Behavioral (Genomic) Effects of Locally Produced Estrogen in the Brain

Chapter 9

The Aromatization Hypothesis, 1970–1990

Lynwood G. Clemens

> "The growth of our knowledge is the result of a process closely resembling what Darwin called "natural selection"; that is, *the natural selection of hypotheses*: our knowledge consists, at every moment, of those hypotheses which have shown their (comparative) fitness by surviving so far in their struggle for existence, a competitive struggle which eliminates those hypotheses which are unfit."
> —*Karl Popper* (1971)

> In so far as a scientific statement speaks about reality, it must be falsifiable; and in so far as it is not falsifiable, it does not speak about reality.
> Karl Popper (2002)

INTRODUCTION

Based on castration studies in the nineteenth and early twentieth century, the relationship between gonadal hormones and sexual behavior began with the simple idea that androgens control male behavior and estrogens activate feminine behavior (see Young, 1961 for review). However, by the end of the 1950s it was clear that this idea of a simple specificity of one hormone to one behavior profile was not going to tell the whole story; males of numerous species showed some masculine responses to estrogens, and females showed feminine responses to androgens, though by no means were androgens and estrogens an adequate substitute for the other.

The search for an explanation for this phenomenon led to the proposal in 1970 of what we now refer to as the aromatization hypothesis: the effect of testosterone on sexual behavior is mediated by estrogen metabolites (Beyer et al., 1970a, 1970b; McDonald et al., 1970). Experiments designed to test the predictions derived from this hypothesis eventually led to a more general conceptualization of testosterone as a "prohormone" whereby multiple metabolites of testosterone were needed concurrently to replicate, in full, its effects on male sexual behavior.

The objective of this review is to trace the development of this aromatization hypothesis through its early stages and examine how it served as a guide in developing predictions about how testosterone affected the nervous system and more specifically male sexual behavior. It is a classic example of deductive scientific investigation: identify an unexplained phenomenon, develop a hypothesis or competing hypotheses to explain such a phenomenon, then derive predictions from that hypothesis to test its validity, or to use Popper's words, "its fitness" (Popper, 1971). This review is not intended as an exhaustive, historical review of the literature on male sexual behavior, or even aromatization studies; a number of outstanding reviews of this nature already exist on those topics (Lephart, 1996; Hull et al., 2002; Baum 2006). Instead, this introductory review is intended to capture the way in which an idea was systematically examined and found to be extremely fruitful in explaining some of the neuroendocrine mechanisms that underlie an important behavior. In the interest of brevity this review focuses primarily on steroid control of adult male sexual behavior; it leaves out those studies that tested predictions from the aromatization hypothesis concerning its relevance to masculinization of the developing brain—studies that were conducted during the same period as those on male sexual behavior reviewed here.

For this review a formal distinction is made between hypotheses and predictions (McPherson, 2001). Hypotheses are those statements that offer a candidate mechanism or explanation for some effect or relationship. A prediction, on the other hand, is generally a factual statement that can be verified by observation or experiment. Additional readings on the role hypotheses play in the development of science can be found in (Platt, 1964; Popper, 1981; Helfenbein & DeSalle, 2005).

EARLY STUDIES OF GONADAL HORMONES ON SEXUAL BEHAVIOR

In 1939 Stone showed that male sexual behavior could be restored in castrated male rats by treating them with testosterone propionate (Stone, 1939). Shortly after, Beach (1942a), following up on work reported by Ball (1939), found that estrogen also facilitated male sexual behavior but was not as effective as testosterone (T). He also found that T appeared to increase feminine sexual responses in female rats (Beach, 1942b). In trying to account for these effects, we see the first glimmer of a "metabolite explanation" for the similar effects of estradiol (E) and T on sexual behavior: "The appearance of sexually receptive reactions in the castrated females following testosterone propionate administration might be tentatively explained by assuming that the organism transforms some fraction of the injected androgen into estrogen (Geist 1941; Salmon 1941)." This interpretation was based on clinical and animal studies showing that T treatment of males and females often resulted in an increase in urinary estrogens, thus leading to the suggestion that T was being metabolized to estrogen, (Steinach, 1937; Hoskins et al., 1939; Nathanson et al., 1939; Geist 1941; Salmon, 1941).

By the mid-1950s the steroid synthetic pathways were pretty well established, showing that E is a metabolite of T (Dorfman & Ungar, 1953). But the idea that E effects might account for some of T's activational actions on sexual behavior did not attract a lot of attention. In his 1961 review, William Young (1961) examined the literature showing facilitation of male sexual behavior by E and female sex responses facilitated by androgens. At the end of a section in which he reviewed studies of the enhancement of feminine sexual behavior following androgen treatment, he acknowledged the possibility that metabolism of T to an estrogen might explain T's enhancing effects. After cautioning readers against ready acceptance of the idea that T is simply a substitute for E he noted: "…before the possibility of such an action is given complete acceptance, we should be able to exclude the *remote possibility* [italics mine] that the effective substance is the androgen rather than a metabolite with estrogenic action." p. 1196. While grammatically tortured, the statement makes it clear that Young saw metabolism of T to E (aromatization) as a possible, albeit remote, explanation for T's estrogen-like effects on sexual behavior. However, 9 years later Carlos Beyer and his group of collaborators saw it as more than a remote possibility and, citing Young's cautionary note, initiated a series of investigations to test the possibility that aromatization was crucial to understanding T's effect on sexual behavior. Their strategy was straightforward, treat castrated animals with androgens like T or androstenedione that can be aromatized to E, and compare their effects on sexual behavior with androgens that cannot be aromatized such as dihydrotestosterone (DHT) or androsterone. In 1970 they reported that whereas T restored both sexual behavior and penile morphology in castrated male rats, DHT had no effect on sexual behavior but did restore penile physiology (McDonald et al. 1970).[1] They were the first to use the aromatization hypothesis to account for the differential effects of T and DHT, pointing out that, whereas DHT cannot be metabolized to E: "Testosterone can be aromatized *in vivo* to give rise to oestrogenic metabolites…and this has been suggested as a possible mechanism by which testosterone could induce sexual behavior (Young, 1961)" (McDonald et al. 1970) p964.

As a further test of this "remote possibility," Beyer and his colleagues tested ovariectomized rabbits and found that T but not DHT induced estrous behavior as well as some mounting behavior (Beyer et al., 1970a, 1970b). Tests of other nonaromatizable androgens such as androsterone and chlorotestosterone acetate also failed to affect estrous behavior in rabbits:

1 Peter McDonald was a young scientist at Cambridge University in the UK. Tragically, he and his wife died in an airplane crash in France in 1974. His untimely death was a terrible loss for all those who knew him and for the field of behavioral endocrinology.

"In conclusion, our observations suggest that the induction of estrous behavior by androgens is mainly due to androgen aromatization." Some measure of the group's growing confidence in the aromatization idea may be reflected by the fact that this report used the term "aromatization" in the title (Beyer et al., 1970b).

It is difficult to overestimate the effect this set of three papers had on the next two decades of research into the mechanism of T action on sexual behavior as well as other behaviors. Although these studies with diverse androgens by no means proved that aromatization of T to E was an obligatory pathway for its effect on sexual behavior, they did form the first challenge that attempted to falsify the "aromatization hypothesis": *Testosterone effects on male and female sexual behavior are mediated by its estrogenic metabolites.* It was a close call in some ways. Some nonaromatizable androgens were later found to activate estrogen receptors and male sexual behavior (Yahr and Gerling, 1978; Nordeen and Yahr, 1981); had these reduced androgens been used in the initial tests of the aromatization hypothesis they might have falsely "falsified" it early on.

Following Beyer's publications, the aromatization hypothesis quickly garnered support from studies showing that incubation of diencephalic tissues from human fetal males containing median eminence and hypothalamus showed conversion of androstenedione to estrone and E (Naftolin et al., 1971). Subsequently, Naftolin (Naftolin et al., 1972; Reddy et al., 1974) reported that incubation of fresh tissue from the hypothalamus of male and female rats also contained estrone following treatment with androstenedione. Although it had been known for at least 30 years that treatment with T resulted in an increase in urinary estrogens, and it was well established that aromatization was the main metabolic pathway for synthesis of E, (Dorfman & Ungar, 1953, 1965), these in vitro studies were critical in showing that brain tissues associated with sexual behavior had the capability of converting T to E.

In the decade following its proposal, there was a flurry of experimental activity to further test the aromatization hypothesis. In this work, the aromatization hypothesis was used to derive predictions about how T (and E) affect male sexual behavior. With these predictions came experiments that presented strong, potentially falsifying challenges to the validity of the hypothesis. For purposes of discussion, it is convenient to divide these challenges into three groups.

Predictions Derived from the Aromatization Hypothesis

1. Diverse androgen prediction. If estrogen metabolites of T are necessary for it to facilitate sexual behavior, then other androgens that are not aromatized to estrogens will not sustain male or female sexual behavior. This was the basis for the initial and perhaps the most critical test of the aromatization hypothesis reported by Beyer and his colleagues. Following their initial reports, other investigators soon expanded the experimental base to include

other steroids and different species leading to limitations on the generality of the hypothesis across species.

2. Antiestrogen prediction. If estrogen metabolites of T are necessary to facilitate sexual behavior, antiestrogenic compounds should block T facilitation of sexual behavior.
3. The aromatase prediction. If estrogen metabolites of T are necessary for it to facilitate sexual behavior, compounds that inhibit the aromatase enzyme should block T facilitation of sexual behavior.

Diverse Androgen Predictions

Replications of Beyer's studies soon followed his 1970 publications. A year later, two reports confirmed the lack of DHT effects in castrated rats even if the longer lasting propionate form was used (Feder, 1971; Whalen et al., 1971). To exclude the possibility that DHT was not reaching the appropriate brain sites needed to activate male sexual behavior, Johnston and Davidson implanted T, DHT, and DHTP directly into the hypothalamus but found that same differences between T and DHT (Johnston & Davidson 1972).

Some nonaromatizable androgens, such as, 3β-androstanediol and 6α androstanediol were reported to facilitate male sexual behavior in the rat (Baum & Vreeburg, 1976; Yahr & Gerling, 1978), leading to the suggestion that some aspect of androgens other than estrogen metabolites was responsible for the differences between androgens in maintenance of male sexual behavior (Yahr & Gerling, 1978). In both cases, however, it was found that these androgens had an affinity for the estrogen receptor (Vreeburg et al., 1975; Nordeen & Yahr, 1981) and their facilitative effects most likely resulted from their interaction with the estrogen receptor.

Other Species Comparative studies looking at the effects of E and DHT in terms of their effects on male sexual behavior significantly expanded the generality of the aromatization hypothesis across numerous species. Early studies of the hamster were contradictory with regard to the effectiveness of DHT in activating or maintaining male sexual behavior in this species. High doses of DHT, 1000 μg/day maintained male sexual behavior, (Whalen & Debold, 1974), whereas 500 μg/day did not, (Christensen et al., 1973). Of four androgens compared for their effectiveness in maintaining sexual behavior of castrated male golden hamsters, T and androstenedione were more effective than the reduced androgens androsterone and dihydrotestosterone in maintaining intromission behavior over a 10-week period. However, during this time androsterone but not DHT was as effective as T and androsterone in maintaining mounting behavior. No differences were detected between these 4 androgens in their maintenance of penile morphology. The finding that E also facilitated mounting behavior in the castrated hamster strengthened the idea that aromatization was important for male sexual behavior in this species (Debold & Clemens, 1978), in spite of the fact that high doses of DHT maintained copulation.

Variations in the response to DHT were also seen in mice, wherein the DHT restored male sexual behavior in SW mice but had no effect on the CD-1 strain of males (Luttge & Hall, 1973). Other reduced androgens were also shown to facilitate male sexual behavior in SW males (Luttge et al., 1974). Again, the finding that E also facilitated some aspects of male sexual behavior in both strains (Edwards & Burge, 1971) made it difficult to reject the aromatization hypothesis for the SW strain based entirely on its responses to DHT.

Experiments similar to those for the hamster and mouse also implicated the process of aromatization in the control of male sexual behavior in quail (Adkins et al., 1980), ferrets (Baum, 1976; Baum et al., 1983), deer (Fletcher & Short, 1974), pigs (Joshi & Raeside, 1973), rams (Docchio & Brooks 1980), and Hereford cattle (Dykeman et al., 1982). In contrast, there was little evidence to suggest that aromatization was important for the guinea pig or the rhesus macaque. Estrogens failed to facilitate any aspect of male sexual behavior in either species (Alsum & Goy, 1974; Michael et al., 1990), while "dihydrotestosterone propionate and T propionate were both effective in establishing the complete male sexual behavior pattern." (Alsum & Goy, 1974; Phoenix, 1974). The effectiveness of DHTP in the absence of an effect of E was seen as a falsification of the aromatization hypothesis for these species.

In retrospect, most of these early reports on diverse androgens lacked critical control data to warrant many of the conclusions that were drawn about the role of aromatization. At the time these studies were conducted, it was a common assumption that castration eliminated all T as well as E from systemic circulation. In light of that assumption, facilitation of male sexual behavior by a reduced androgen was seen as evidence against the aromatization hypothesis. However, we now know that castration does not eliminate endogenous E in all rodents (Sinchak et al., 1996) and that DHT in combination with low dosages of E2 is very effective in facilitating male sexual behavior. Consequently, there were alternative interpretations for the DHT effects that were not apparent at the time. The availability and use of hormone assays in conjunction with treatments and behavioral testing in more recent studies have proven to be invaluable controls that are essential to better understand how these steroids may be interacting e.g.(Arteaga-Silva et al., 2005; Piekarski et al., 2010).

The Antiestrogen Prediction

A second prediction derived from the aromatization hypothesis was that if the facilitative effect of T on male sexual behavior was mediated by estrogenic metabolites, then blocking estrogenic receptors should block the effect of T. Early tests of this prediction used four different estrogen receptor antagonists, CI-628 and MER 25, ICI-46474 and clomiphene. While the overall assessment based on these studies was that the aromatization hypothesis could not be falsified, this line of investigation had two weaknesses: first, the antagonists often appeared to have both agonist and antagonist actions and second, the experimental designs used to test the drugs often had serious limitations. Few of the studies began with a clear dose–response curve to the antagonist, thereby making it difficult to know whether failure to see an effect was caused by the dosage level being too high or too low, or that blocking estrogen receptors was without effect.

CI-628 One of the first studies to look at the effects of CI-628 on T induced male sexual behavior reported that treatment with 1 mg CI-628 failed to block T-maintained sexual behavior in male rats (Whalen et al., 1972). However, CI-628 did block T induced lordosis in female rats. Although Yahr and Gerling (1978) failed to see any decrease in T-induced male sexual behavior using a dose of 2.5mg, Luttge (1975) found significant reductions for both 2.5- and 4-mg dosages.

MER 25 The mixed agonist/antagonist actions of MER-25 proved particularly troublesome in trying to test predictions about estrogenic actions, but that certainly did not inhibit investigators from its use. Södersten (1974) reported that although four daily injections of 20 mg of MER-25 inhibited the display of lordosis in intact female rats, it did not inhibit mounting in the same females. Ten mg/day for 9 days partially antagonized the activation of lordosis by E benzoate in ovariectomized female rats but had no inhibitory effect on mounting behavior. When given without other hormone treatments, MER-25 (10 mg/day for 9 days) actually resulted in the display of mounting behavior in ovariectomized rats.

The results of Sodersten (1974) on male-like behavior in female rats were confirmed and extended using male rats (Baum & Vreeburg, 1976). Treatment with 10 mg MER 25 blocked or suppressed lordosis in female rats treated with either EB or TP, but failed to disrupt TP-induced mounting behavior in males (Baum & Vreeburg, 1976). These same investigators also reported that MER-25 treatment activated male sexual behavior in males treated simultaneously with DHT, again suggesting that MER-25 was acting as an estrogenic agonist. In spite of MER-25 not blocking male responses to T treatment, the authors concluded, based on the known affinity of MER-25 for estrogen receptors, (Vreeburg et al., 1975), that these results do "not argue against the aromatization hypothesis for masculine sexual behavior."

Others also reported a failure of MER25 to block androgen induced male sexual behavior in rats (Beyer et al., 1976). However, treatment with high doses of the antiestrogen, ICI-46474 (1mg/kg) did significantly reduce T induced mounting behavior in 56% of male rats.

The picture presented by these early estrogen receptor blocking studies was a mixed bag; while some found blocking of T effects with high dosages of some antiestrogens such as CI-628 and ICI-46474, others saw little or no effect. One is left with an odd state of affairs: in those instances where the antiestrogen blocked T induction of male sexual behavior, the conclusion was drawn that it blocked the effects of the estrogen metabolites of T and therefore the results were seen to support the aromatization hypothesis; in the case of the antiestrogen, MER-25, its failure to block T effects and its facilitation of male sexual behavior were "blamed" on its ability to act as an estrogen agonist and those results were also used to support the aromatization hypothesis. Not the kind of situation that Popper would have enjoyed, I am sure.

The aromatase prediction

If estrogen metabolites of T are necessary to facilitate sexual behavior, compounds that inhibit or block the aromatase enzyme should block T facilitation of sexual behavior.

Following an experiment in which we showed that intracranial implants of E into the medial preoptic area (MPOA) of the hypothalamus were effective in stimulating mounting behavior in castrated male rats, (Christensen & Clemens, 1974), we were intrigued by the recent finding that a compound, androst-1,4,6,-triene-3,17-dione (ATD) was found to inhibit aromatization of T to E in placental cultures (Schwarzel et al., 1973). Because it had been demonstrated that T was most likely being aromatized in or near the tissues that mediate male sexual behavior (Naftolin et al., 1972), my graduate student, Larry Christensen, decided to test the prediction that blocking aromatase enzymes in the medial preoptic area should block T facilitation of male sexual behavior. When he implanted ATD into the MPOA he found that while E implants continued to maintain male mounting behavior, T implants were ineffective (Christensen & Clemens, 1975). Further confirmation of the aromatase prediction was reported by Beyer et al. (1976), who found that systemic administration of the aromatase inhibitor, aminoglutethimide (5 or 15 mg every 12 hr for 96 hr) also abolished mounting responses following administration of TP to castrated male rats. The aromatase inhibitors, metopirone or 5α-androstanedione did not affect TP facilitated mounting. Many studies of the role of the aromatase enzyme in reproductive function followed these early observations and the reader can find these discussed in a number of excellent reviews: (Callard et al., 1978; Lephart, 1996; Roselli, 2007).

By the late 1980s it was clear that the aromatase enzyme was present in the brain of every major vertebrate group (Callard et al., 1978). In both in vivo and in vitro studies, the greatest activity was located in areas associated with reproductive function, that is, the hypothalamus and the limbic system (Weisz & Gibbs, 1974; Lieberburg & McEwen, 1977; Selmanoff et al., 1977; Callard et al., 1979). The situation was summarized by Roselli in 1993 (Roselli & Resko 1993): "Studies using micropunch dissections revealed that the highest levels of aromatase activity are found in an interconnected group of sexually dimorphic nuclei which constitutes a neural circuit important in the control of male sexual behavior."

Refinement of the Aromatization Hypothesis: T as a "Prohormone"

Multiple metabolite prediction: Although there was clear support for the idea that estrogenic metabolites of T were involved in male rat sexual behavior, it was also clear that E alone could not fully substitute for T in the restoration of male sexual behavior in castrated males rats because male rats treated with E failed to show normal levels of intromissions and ejaculation and E did not maintain the males' sexual accessory structures (Davidson, 1969; Pfaff, 1970; Södersten, 1974). On the other hand, DHT maintained the male's peripheral sexual tissues but did not facilitate male sexual behavior. These findings along with the report that DHT, an androgenic metabolite of T, can be formed in the hypothalamus of the male rat (Jaffe, 1969; Rommerts & Vandermo, 1971), led several groups to propose a refinement of the aromatization hypothesis suggesting that both estrogenic and androgenic metabolites of T were necessary to mediate male sexual behavior (Baum & Vreeburg 1973; Larsson et al., 1973; Feder et al., 1974). This led to the prediction that a combination of E and DHT treatments should be more effective than E or DHT alone; and indeed such treatments resulted in male sexual behavior in castrated rats that was not distinguishable from normal males. Similar findings were soon reported for the mouse as well (Wallis & Luttge, 1975).

Whether DHT was acting centrally remained unclear, but in their closing arguments Baum and Vreeburg (1973) suggested that although the DHT is most likely functioning to maintain the male's peripheral sex tissues so that he can achieve the stimulation needed for ejaculation, they also argued, based on their behavioral data, that DHT might also be having some effects on central neural tissues as well, an argument that received substantial support several decades later. Vagel and McGinnis (1997) found that the aromatase inhibitor, Fadrozole, was quite effective in blocking male sexual behavior in castrated males that were implanted with two 10-mm Silastic capsules filled with T. However males receiving Fadrozole while implanted with two T capsules and one 5-mm 1% E capsule copulated normally, indicating that Fadrozole's effect was due specifically to aromatase inhibition. Interestingly, animals that received only one E capsule failed to exhibit male sexual behavior, leading the authors to suggest that both E and androgen receptor activation in the brain are necessary for male sexual behavior in rats; a conclusion suggested by Baum and Vreeburg (1973) 20 years earlier but based on very different evidence.

These cocktail studies using a combination of E and DHT supported the modification of the aromatization hypothesis, which now viewed T as a prohormone: *T effects on male sexual behavior are mediated by estrogenic and androgenic metabolites*. An example of comparative support for this hypothesis was found using the male deermouse, *Peromyscus maniculatus bairdi* (Clemens & Pomerantz, 1982). Activation of male sexual behavior in castrated deermice treated with TP was blocked when males were treated simultaneously with a 5α -reductase inhibitor (17βC). This treatment prevented androgenic stimulation of peripheral sex tissues as well. In a separate experiment male sexual behavior was also prevented in TP treated castrates by administration of an aromatase inhibitor, ATD.

CONCLUSIONS

Karl Popper speaks of hypothesis fitness and the ability of hypotheses to survive the ongoing "natural selection of hypotheses," which results in elimination of those that are unfit. Viewed from this metaphor, the aromatization hypothesis had a very short life. It appeared in 1970 but quickly started to lose the struggle based on its highly specialized phenotype: estrogens were not a complete replacement for testosterone. In 1973 it underwent a mutation and a new phenotype appeared: the *testosterone prohormone hypothesis*. How well this hypothesis will fare in light of continuing selection pressure remains to be seen, but each of these hypotheses has provided important insights into how androgens affect sexual behavior, and each provides the seeds for its own survival or elimination; outcomes that speak volumes about scientific reality.

ACKNOWLEDGMENT

The author wishes to thank Dr. Antonio Nunez for his thoughtful comments and discussions of earlier drafts of this manuscript.

REFERENCES

Adkins, E. K., & Boop, J. J., et al. (1980). Further evidence that androgen aromatization is essential for the activation of copulation in male quail. *Physiol Behav, 24*(3), 441–446.

Alsum, P., & Goym R. W. (1974). Actions of esters of testosterone, dihydrotestosterone, or estradiol on sexual behavior in castrated males guinea pigs. *Horm Behav, 5*(3), 207–217.

Arteaga-Silva, Marquez-Villanueva, Y., et al. (2005). Effects of hormonal replacement with androgens and estrogens on male sexual behavior and plasma levels of these steroids in gonadectomized golden hamsters (Mesocricetus auratus). *Physiol Behav, 85*(5), 571–580.

Ball, J. (1939). Male and female mating behavior in prepubertally castrated male rats receiving estrogens." *J Comp Physiol Psychol, 28,* 273–283.

Baum, M. J. (1976). Effects of testosterone propionate administered perinatally on sexual behavior of female ferrets. *J Comp Physiol Psychol, 90*(4), 399–410.

Baum, M. J. (2006). Mammalian animal models of psychosexual differentiation: when is 'translation' to the human situation possible? *Horm Behav, 50*(4): 579–588.

Baum, M. J., Canick, J. A., et al. (1983). Normal differentiation of masculine sexual behavior in male ferrets despite neonatal inhibition of brain aromatase or 5-alpha-reductase activity. *Neuroendocrinology, 36*(4), 277–284.

Baum, M. J., & Vreeburg, J. T. (1973). Copulation in castrated male rats following combined treatment with estradiol and dihydrotestosterone. *Science, 182*(4109), 283–285.

Baum, M. J., & Vreeburgm J. T. M. (1976). Differential effects of anti-estrogen MER-25 and of 3 5=lpha reduced androgens on mounting and lordosis behavior in rat. *Horm Behav, 7*(1), 87–104.

Beach, F. (1942a). Copulatory behavior in prepuberally castrated male rats and its modification by estrogen administration. *Endocrinology, 31*(6), 679–683.

Beach, F. (1942b). Male and female mating behavior in prepuberally castrated female rats treated with androgens. *Endocrinology, 31*(6), 673–678.

Beyer, C., McDonald, P., et al. (1970a). Failure of 5 alpha-dihydrotestosterone to elicit estrous behavior in ovariectomized rabbit. *Endocrinology, 86*(4), 939–941.

Beyer, C., Morali, G., et al. (1976). Effect of some antiestrogens and aromatase inhibitors on androgen induced sexual behavior in castrated male rats. *Horm Behav, 7*(3), 353–363.

Beyer, C., Vidall, N., et al. (1970b). Probable role of aromatization in the Induction of estrous behavior by androgens in the ovariectomized rabbit. *Endocrinology, 87,* 1386–1389.

Callard, G. V., Hoffman, R. A., et al. (1979). IIn vitro aromatization and other androgen transformations in the brain of the hamster (Mesocricetus auratus). *Biol Reprod, 21*(1), 33–38.

Callard, G. V., Petro, Z., et al. (1978). Phylogenetic distribution of aromatase and other androgen-converting enzymes in central nervous system. *Endocrinology, 103*(6), 2283–2290.

Christensen, L., & Clemens, L. G. (1974). Intrahypothalamic implants of testosterone or estradiol and resumption of masculine sexual behavior in long-term castrated male rats. *Endocrinology, 95*(4), 984–990.

Christensen, L. W., & Clemens, L. G. (1975). Blockade of testosterone-induced mounting behavior in male rat with intracranial application of aromatization inhibitor, androst-1,4,6-triene-3, 17-dione. *Endocrinology, 97*(6), 1545–1551.

Christensen, L., Coniglio, L. P., et al. (1973). Sexual behavior of male golden hamsters receiving diverse androgen treatments. *Horm Behav, 4*(3), 223–229.

Clemens, L. G., & Pomerantz, S. M. (1982). Testosterone acts as a prohormone to stimulate male copulatory behavior in male deer mice (Peromyscus maniculatus bairdi). *J Comp Physiol Psychol, 96*(1), 114–122.

Davidson, J. M. (1969). Effects of estrogen on the sexual behavior of male rats. *Endocrinology, 84,* 1365–1372.

Debold, J. F., & Clemen, L. G. (1978). Aromatization and the induction of male sexual behavior in male, female, and androgenized female hamsters. *Horm Behav, 11*(3), 401–413.

Docchio, M. J., & Brooks, D. E. (1980). Effects of androgenic and estrogenic hormones on mating behavior in rams castrated before or after puberty. *J Endocrinol, 86*(3), 403–411.

Dorfman, R., & Ungar, F. (1953). *Metabolism of Steroid Hormones.* Minneapolis, MN: Burgess.

Dorfman, R., & Ungar, F. (1965). *Metabolism of Steroid Hormones.* New York: Academic.

Dykeman, D. A., Katz, L. S., et al. (1982). Behavioral characteristics of beef steers administered estradiol, testosterone and dihydrotestosterone. *J Anim Sci, 55*(6), 1303–1309.

Edwards, D. A., & Burge, K. G. (1971). Estrogenic arousal of aggressive behavior and masculine sexual behavior in male and female mice. *Horm Behav, 2*(3), 239–245.

Feder, H. H. (1971). Comparative actions of testosterone propionate and 5 alpha-androstan-17 beta-OL-3 ONE propionate on reproductive behavioural physiology and morphology of male rats. *J Endocrinol, 51*(2), 241–252.

Feder, H. H., Naftolin, F., et al. (1974). Male and female sexual responses in male rats given estradiol benzoate and 5α-androstan-17β-ol-3-one propionate. *Endocrinology, 94*(1), 136–141.

Fletcher, T. J., & Short, R. V. (1974). Restoration of libido in castrated red deer stag (Cervys ekaogys) with estradiol-17 beta. *Nature, 248*(5449), 617–618.

Geist, S. (1941). Androgen therapy in human females. *J Clin Endocrinol Metab, 1,* 154–161.

Helfenbein, K., & DeSalle, R. (2005). Falsifications and corroborations: Karl Popper's influence on systematics. *Mol Phylogenet Evol, 35,* 271–280.

Hoskins, W. H., Coffman, J. R., et al. (1939). The effect of testosterone propionate on the urinary excretion of androgens and estrogens in eunuchoidism. *Endocrinology, 24*(5), 702–710.

Hull, E., Meisel, R., et al. (2002). Male sexual behavior. In: Pfaff, D., Arnold A., Etgen, A., Fahrbach, S., & Rubin, R. (eds.), *Hormones, Brain and Behavior* (pp. 1–138). New York: Academic Press.

Jaffe, R. B. (1969). Testosterone metabolism in target tillues. Hypothalamic and pitutitary tissues of the adult rat and human fetus, and the immature rat epiphysis. *Steroids, 14,* 483–498.

Johnston, P., & Davidson, J. M. (1972). Intracerebral androgens and sexual behavior in male rat. *Horm Behav, 3*(4), 345–357.

Joshi, H. S., & Raeside, J. I. (1973). Synergistic effects of testosterone and estrogens on sex glands and sexual behavior of boar. *J Reprod Fertil, 33*(3), 411–423.

Larsson, K., Sodersten, P., et al. (1973). Sexual behavior in male rats treated with estrogen in combination with dihydrotestosterone. *Horm Behav, 4*(4), 289–299.

Lephart, E. D. (1996). A review of brain aromatase cytochrome P450. *Brain Res Rev 22*(1), 1–26.

Lieberburg, I., & McEwen, B. S. (1977). Brain cell nuclear retintion of testosterone metabolites, 5 alpha-dihydrotestosterone and estradiol-17 beta, in adult rats. *Endocrinology, 100*(2), 588–597.

Luttge, W. G. (1975). Effects of anti-estrogens on testosterone stimulated male sexual behavior and peripheral target tissues in castrate male rat. *Physiol Behav, 14*(6), 839–846.

Luttge, W. G., & Hall, N. R. (1973). Differential effectiveness of testosterone and its metabolites in induction of male sexual behavior in 2 strain of albino mice. *Horm Behav, 4*(1–2), 31–43.

Luttge, W. G., Hall, N. R., et al. (1974). Studies on neuroendocrine, somatic and behavioral effectiveness of testosterone and its 5-alpha reduced metabolites in swiss-webster mice. *Physiol Behav, 13*(4), 553–561.

McDonald, P., Beyer C., et al. (1970). Failure of 5 alpha-dihydrotestosterone to initiated sexual behaviour in castrated male rat. *Nature 227*(5261), 964–965.

McPherson, G. (2001). Teaching & learning the scientific method. *Am Biol Teach, 63*(4), 242–245.

Michael, R. P., Zumpe, D., et al. (1990). Estradiol administration and the sexual activity of castrated male rhesus mondeys (Macaca mulatta). *Horm Behav, 24*(1), 71–88.

Naftolin, F., Ryan, K. J., et al. (1971). Aromatization of androstenedione by diencephalon. *J Clin Endocrinol Metab, 33*(2), 368–370.

Naftolin, F., K. J. Ryan, et al. (1972). Aromatization of androstenedione by anterior hypothalamus of adult male and female rat. *Endocrinology, 90*(1), 295–298.

Nathanson, I., J. Jones, et al. (1939). The urinary excretion of estrogens, androgens and FSH following the administration of testosterone to human female castrates. *Endocrinology, 25*(5), 754–758.

Nordeen, E. J., and P. Yahr (1981). Activation and differentiation of sexual behavior and translocation of hypothalamic estrogen receptors in rats by 6-alpha fluorotestosterone. *Horm Behav, 15*(2), 123–140.

Pfaff, D. (1970). Nature of sex hormone effects on rat sex behavior: specificity of effects and individual patterns of response. *J Comp Physiol Psychol, 73*(3), 349–358.

Phoenix, C. H. (1974). Effects of dihydrotestosterone on sexual behavior of castrated male rhesus monkeys. *Physiol Behav, 12*(6), 1045–1055.

Piekarski, D. J., Place, N. J., & Zucker, I. (2010). Facilitation of male sexual behavior in syrian hamsters by the combined action of dihydrotestosterone and testosterone. *Plos One, 5*(9), e12749.

Platt, J. R. (1964). Strong inference. *Science, 146*(364), 347–353.

Popper, K. (1971). Objective Knowledge: An Evolutionary Approach, In: Simonton, D. K. (ed.),*Origins of Genius: Darwinian Perspectives on Creativity*, 1999 (pp 26). New York: Oxford University Press.

Popper, K. (1981). Science, pseudo-science, and falsifiability. In: Tweney, R., Doherty, M., & Mynatt, C. (eds.), *On Scientific Thinking* (pp. 92–99). New York: Columbia University Press.

Popper, K. (2002) The Logic of Scienctific Discovery. Routledge Press, London

Reddy, V. V. R., Naftolin, F., et al. (1974). Conversion of androstenedione to estrone by neural tissues from fetal and neonatal rats. *Endocrinology, 94*(1), 117–121.

Rommerts, F. E., & Vandermo. H (1971). Occurrence and localization of 5-alpha-steroid reductase, 3-alpha-hydroxysteroied and 17-beta-hydroxysteroied dehydrogenases in hypothalamus and other brain tissues of male rat. *Biochimica Et Biophysica Acta, 248*(3), 489.

Roselli, C. E., & Resko, J. A. (1993). Aromatase activity in the rat brain—hormonal regulation and sex differences. *J Steroid Biochem Mol Biol, 44*(4–6), 499–508.

Roselli, C. F. (2007). Brain aromatase: roles in reproduction and neuroprotection. *J Steroid Biochem Mol Biol, 106*(1–5), 143–150.

Salmon, U. (1941). Rationale for androgen threrapy in gynecology. *J Clin Endocrinol Metab, 1,* 162–179.

Schwarzel, W., W. Kruggel, et al. (1973). Studies on the mechanism of estrogen biosynthesis. VIII. The development of inhibitors of the enzyme system in human placenta. *Endocrinology, 92*(3), 866–880.

Selmanoff, M. K., Brodkin, L. D., et al. (1977). Aromatization and 5-alpha-reduction of androgens in discrete hypothalamic and limbic regions of male and female rat. *Endocrinology, 101*(3), 841–848.

Sinchak, K., Roselli, C. E., et al. (1996). Levels of serum steroids, aromatase activity, and estrogen receptors in preoptic area, hypothalamus, and amygdala of B6D2F1 male house mice that differ in the display of copulatory behavior after castration. *Behav Neurosci, 110*(3), 593–602.

Södersten, P. (1974). Effects of an estrogen antagonist, MER-25, on mounting behavior and lordosis behavior in the female rat. *Horm Behav, 5*(1), 111–121.

Steinach, E. (1937). Transformation of male sex hormones into a substance with action of a female hormone. *Lancet, 2,* 845.

Stone, C. (1939). Copulatory activity in adult male rats following castration and injections of testosterone propionate *Endocrinology, 24*(2), 165–174.

Vagell, M. E., & McGinnis, M. Y. (1997). The role of aromatization in the restoration of male rat reproductive behavior. *J Neuroendocrinol, 9*(6), 415–421.

Vreeburg, J. T. M., Schretlen, P. J. M., et al. (1975). Specific, high affinity binding of 17-beta-estradiol in cytosols from several brain regions and pituitary of intact and castrated adult male rats. *Endocrinology, 97*(4), 969–977.

Wallis, C. J., & Luttge, W. G. (1975). Maintenance of male sexual behavior by combined teatment with estrogen and dihydrotestosterone in CD-1 mice. *J Endocrinol, 66*(2), 257–262.

Weisz, J., & Gibbs, C. (1974). Conversion of testosterone and androstenedione to estrogens in-vitro by brain of female rats. *Endocrinology, 94*(2), 616–620.

Whalen, R. E., Battie, C., et al. (1971). Androgen induced receptivity. *Horm Behav* *2*(4), 355.

Whalen, R. E., Battie, C., et al. (1972). Anti-estrogen inhibition of androgen induced sexual receptivity in rats. *Behav Biol, 7*(3), 311–320.

Whalen, R. E., & Debold, J. F. (1974). Comparative effectiveness of testosterone, androstenedione and dihydrotestosterone in maintainning mating behavior in castrated male hamster. *Endocrinology, 95*(6), 1674–1679.

Yahr, P., & Gerling, S. A. (1978). Aromatization and androgen stimulation of sexual behavior in male and female rats. *Horm Behav, 10*(2), 128–142.

Young, W. (1961). The hormones and mating behavior. In: Young, W. (ed.), *Sex and Internal Secretions (pp. 1173–1239).* Baltimore: Williams & Wilkins.

Chapter 10

Aromatase and Behavior: Concepts Gained from Studies of Aromatase in the Avian Brain

Barney Schlinger and Jacques Balthazart

INTRODUCTION

Birds have long fascinated both amateur and professional scientists alike because they are taxonomically diverse, relatively easy to observe, and exhibit complex behaviors and social systems. Studies aimed at unraveling avian biology have served as the basis for the emergence of many new concepts in biology in general (Konishi et al., 1989). This has been particularly the case in the field of behavioral neuroendocrinology (Goodson et al., 2005; Wingfield, 2005), in which birds were at the origin, among other things, of (1) the renewed interest in sex differences in brain structures and behavior (Arnold, 2002); (2) the rediscovery of adult neurogenesis (Goldman & Nottebohm, 1983; Paton & Nottebohm, 1984) and the associated seasonal plasticity in neural structures (Tramontin & Brenowitz, 2000); and (3) the refined understanding of bidirectional causal relationships linking hormone secretion and behavior (Lehrman, 1965; Wingfield & Marler, 1988; Ball & Balthazart, 2009). Not only are estrogens vital for numerous organizational and activational effects on the avian brain, but investigation of avian brain and behavior has also identified a surprisingly rich diversity of mechanisms for neuroestrogen production and action (Ball & Balthazart, 2009; Schlinger & Brenowitz, 2009). This chapter describes advances in neuroestrogen biology that arose from studies of birds such as Japanese quail and ring doves (nonsongbirds) as well as zebra finches (songbirds). From this foundation it introduces a series of newly emerging concepts regarding aromatase in the avian brain, including the role of neuroestrogens in nonreproductive functions, the local fluctuation of estrogens in brain arising from rapid regulation of aromatase catalytic activity, the role of aromatase as a terminal reaction in the broader process of neurosteroidogenesis, and the synthesis and actions of estrogens at the synapse. As the reader will see, many of these advances have gained considerable traction and widespread importance, thus warranting more detailed chapters of their own in this volume.

Nonsongbirds

In the 1950s and 1960s, Daniel Lehrman, Robert Hinde, and their collaborators initiated studies on ring doves (*Streptopelia risoria*), canaries (*Serinus canaria*), and other species that not only identified in a precise manner the endocrine controls of reproductive behaviors, but also demonstrated the subtle retroaction of behavioral performance, in particular sociosexual interactions, on the control of the endocrine state of an individual (Hinde, 1965; Lehrman, 1965). Lehrman's colleague, Mei Cheng later provided conclusive evidence indicating that endocrine secretions are affected not only by the behaviors of the sexual partner, but also by the behavioral activity of the focal subject itself (the "self-stimulation" idea) (Cheng, 1992, 2003). Subsequent studies that focused on the Japanese quail (*Coturnix japonica*) provided insight into hormonal control of the developing neural circuits that would later control reproduction and solidified our appreciation for the ways in which hormones activate neural circuits of reproductively mature individuals (Adkins, 1981; Balthazart, 1983; Balthazart et al., 2009a,b).

Because they readily exhibit high rates of sexual behavior under standardized captive conditions, Japanese quail have turned out to be outstanding subjects for the study of the endocrine basis of behavior (Ball & Balthazart, 2011). Even in relatively impoverished laboratory conditions, quail readily display a variety of reproductive behaviors. In the absence of a female, sexually motivated male quail produce a loud, broadband, two- to three-syllable vocalization, referred to as the "crow," that is attractive to female quail (Goodson & Adkins-Regan, 1997). When a male comes in contact with a female, he very reliably engages in a stereotyped sequence of copulatory behaviors that includes a precopulatory (and also often postcopulatory) display called *strutting,* as well as the copulatory sequence *sensu stricto,* during which he grabs the female's neck feathers (neck grab, NG), attempts to mount her (mount attempts, MA), eventually succeeds in mounting (mount, M), and finally apposes his cloaca to the female's cloaca so that sperm transfer can occur (Adkins & Adler, 1972; Hutchison, 1978b). All these behaviors are exquisitely sensitive to testosterone (T): they rapidly disappear after castration and are restored within 2 to 5 days after treatment with exogenous T (Beach & Inman, 1965; Sachs, 1969). The relationship between sexual behavior and exposure to high circulating levels of T is thus tighter (almost deterministic) in quail than in many other vertebrate species, which greatly facilitates the experimental analysis of the behavioral effects of sex steroids.

Sexual behavior is often divided, in a somewhat arbitrary manner, into a consummatory and an appetitive component. Consummatory sexual behavior (CSB) refers that the copulatory sequence itself (as described for quail), whereas appetitive sexual behavior (ASB) consists of more variable behaviors associated with the search and initial approach of a sexual partner (see Beach, 1956; Pfaus, 1996; for a more recent discussion see Sachs, 2007; Ball & Balthazart, 2008). Most of the research on the activation of sexual behavior by sex steroids has focused on CSB, but during the last decade more attention has also devoted to the endocrine control of ASB, particularly in quail.

A large part of these studies on ASB were based on the analysis of a behavior called *learned social proximity response* initially identified by Michael Domjan and colleagues at the University of Texas, Austin (Domjan & Hall, 1986a,b). The response involves a remarkable change in a male's behavior in that after a single copulation, males spend the majority of their time (up to 14 hours per day) standing in front of and looking through a window that provides visual access to the female (Domjan & Hall, 1986a,b). This response is acquired by sexually mature male quail but not castrated subjects (Balthazart et al., 1995, 1997). Treatment with exogenous T restores in castrates the ability to acquire this proximity response that provides a sensitive measure of both male appetitive sexual behavior and its sexual motivation. A faster version of this test that was developed in the Balthazart laboratory allows quantification of the response during a 5-minute period (Balthazart et al., 1995) and thus testing of a large number of subjects exposed to various endocrine conditions.

Another behavioral procedure that allows an easy assessment of the male ASB is based on the quantification of rhythmic contractions of the cloacal sphincter muscles (RCSM) in response to the simple view of a female (Seiwert & Adkins-Regan, 1998; Thompson et al., 1998). These contractions produce meringue-like foam that is transferred to the female cloaca during copulation and enhances the probability that the sperm will fertilize the egg (Seiwert & Adkins-Regan, 1998). Although the frequency of contractions is modulated by the bird's past experience (previous visual exposure to either a male or female (Cornil & Ball, 2010), they are also produced by naïve males. The frequency of RCSM is exquisitely T dependent: In response to the view of a female, the RCSM frequency increases nearly 20-fold in castrated males treated with T as compared with castrates that received no T (Balthazart et al., 1998a).

The neural sites of T action on quail sexual behavior are well characterized. The preoptic area (POA) contains a dense cluster of neurons called the medial preoptic nucleus (POM). The volume of this nucleus is significantly larger in males than in females (Viglietti-Panzica et al., 1986) and this difference reflects the higher circulating T concentrations in males (Panzica et al., 1987). Electrolytic lesions of the POM, but not of the surrounding POA, suppress copulatory behavior (CSB) activated by exogenous T in castrated males (Balthazart & Surlemont, 1990b). Conversely, stereotaxic implants filled with T reliably activate all aspects of copulatory behavior in castrated males if their tip is located within the cytoarchitectonic boundaries of the POM, but not if it is located in the adjacent POA (Balthazart & Surlemont, 1990b; Balthazart et al., 1992b). Testosterone action in the POM is thus sufficient to activate copulatory behavior in adult male quail (provided adequate stimuli are present).

Similar experiments suggest that the two aspects of appetitive sexual behavior are also activated by T action in the POM. Lesions of the POM that abolished CSB also markedly decreased the expression of RCSM in response to the view of a sexually mature female as well as measures of the learned social proximity response (Balthazart et al., 1998a). Note, however, that the two components of male sexual behavior were differentially affected by lesions of different subregions of the POM, thus suggesting the presence of a subtle anatomical specificity in these control mechanisms (Balthazart et al., 1998a; Balthazart & Ball, 2007). Together with studies analyzing the expression of the immediate early gene c-fos in the POA of subjects after they expressed ASB or CSB (Taziaux et al., 2006), these lesion and implant studies clearly suggest that the preoptic region is important in the regulation of motivational as well as performance aspects of male sexual behavior, but that there is at least a partial anatomical dissociation in the POM between structures involved in the control of both aspects of the behavior (Balthazart et al., 1998a; Balthazart & Ball, 2007). The brain circuitry controlling ASB and CSB is, however, not limited to POM and includes afferent sensory inputs, modulatory inputs originating from various neurotransmitter systems, and outputs to premotor and motor nuclei (Balthazart & Ball, 2007). The medial preoptic nucleus is nevertheless a critical node in this circuit and a central site of T action. Most importantly for the topic of the present chapter, most if not all behavioral effects of T in POM can be mimicked by estrogens and blocked by the concurrent administration of aromatase inhibitors. The experimental evidence supporting this notion is reviewed in the following.

Songbirds

Alongside this work on captive doves and quail there has emerged a rapidly growing field of investigation concerning the neurohormonal control of singing behavior in songbirds (Schlinger & Brenowitz, 2009). The oscine songbirds (suborder Passeri) of the avian order Passeriformes possess extraordinary control of their muscular syrinx, or vocal organ, which enables the production of especially complex songs. These birds also possess the innate capacity to learn their songs from parents or adult tutors. The sensorimotor requirements for singing and song learning are subserved by a remarkably complex neural circuitry, the neural song system (Nottebohm et al., 1976). These circuits are described in detail elsewhere (Zeigler & Marler, 2008). For our purposes, we can consider three overlapping elements composed of multiple discrete nuclei that collectively enable songbird singing behavior. The bulk of these nuclei are located in the songbird telencephalon that includes the basal ganglia and cortical-like structures. Auditory input to these circuits is required for song learning, in which a song template is acquired developmentally, but also for song stabilization and recognition by adults. An area of key interest because of its elevated expression of the aromatase enzyme is the caudomedial nidopallium (NCM), a higher-order auditory processing region. Auditory information is communicated to motor pathways that include nucleus HVC that sends afferents to nucleus RA that, in turn, projects to motoneurons of the avian syrinx located in the trachesyringeal (XIIth) nucleus. Song learning requires multiple telencephalic and thalamic nuclei termed the anterior forebrain pathway which overlaps with the motor system via nucleus HVC.

A large body of work has established sex steroids as crucial for appropriate growth and function of this neural song circuitry across the lifespan of diverse songbird species. Notable examples are described below. First, the gonadal hormone T has been shown to stimulate singing behavior in songbirds. Males sing more than females and they do so mostly at the height of their reproductive season when gonadal T was elevated (Armstrong, 1973; Catchpole & Slater, 2008). Castration of males reduces or eliminates song production, which is then restored by T replacement (Nottebohm, 1969; Prove, 1974; Arnold, 1975). In some species, such as canaries (*Serinus canaria*), females who do not ordinarily sing will do so after treatment with T (Shoemaker, 1939; Baldwin et al., 1940).

A second line of work suggested that sex steroids participate in the development of the neural song system. In some species, notably zebra finches (*Taeniopygia guttata*), adult males sing but females do not, even if treated with T. Examination of the brains of these birds revealed a striking sexual dimorphic song system with nuclei, such as HVC, substantially larger and more complex in males than females (Nottebohm & Arnold, 1976). Subsequent studies showed that sex steroids could functionally masculinize the developing female zebra finch brain, with estrogens playing a pivotal role (Gurney & Konishi, 1980). Not only were sex steroids shown to masculinize the developing neural song system, but androgens and estrogens were also seen to help define developmental periods of song learning (Marler et al., 1988). In birds such as zebra finches that learn adult song from their father or a tutor during critical phases of development, estrogens were viewed as maintaining the neural plasticity required for song learning, whereas androgens were shown to terminate this plasticity and crystallize adult song (Bottjer & Johnson, 1997).

In addition to these developmental effects, sex steroids are now known to promote plasticity in the adult songbird brain. In seasonally breeding birds with the yearly waxing and waning of singing behavior, song nuclei are seen to expand and contract across seasons in parallel with song (Nottebohm, 1981). For example, exposure to long day photoperiods that stimulate gonadal growth and increase circulating T induce significant growth of the song control system of male white-crowned sparrows (Tramontin et al., 2003) and removal of T induces a rapid and profound regression of the song system (Thompson et al., 2007). Thus, sex-steroids exert organizational and activational effects on the song circuitry and on singing behavior as well facilitating significant neuroplasticity.

Autoradiographic studies of nonsongbirds and mammals had shown that systemic injection of radiolabeled sex steroids led to cellular accumulation of radioactivity in discrete regions of the brain as the steroids bound to intranuclear receptors (Pfaff, 1968). These cells were found in highly conserved sites in the hypothalamus and limbic brain regions that corresponded with regions known to participate in sex steroid control of reproductive behavior and physiology. Studies of the songbird brain, however, demonstrated widespread accumulation of sex steroids, including in many regions in and near the neural song system (Zigmond et al., 1973; Arnold et al., 1976). We now know that receptors for androgens (AR) and estrogen (ERα and ERβ) are expressed widely in the songbird brain (Brenowitz et al., 1991; Balthazart et al., 1992a; Gahr et al., 1993; Bernard et al., 1999; Metzdorf et al., 1999) in and around song control circuits, making these birds exceptional models for assessing sex steroid actions on diverse neural end points.

Birds and Brain Aromatase

Thus, studies of both songbirds and of nonsongbirds have linked behavior to endocrine state, but as it turns out, not only with gonadal steroid secretion, but also with a number of brain enzymatic activities that either produce new steroids de novo (see chapter 11) or transform steroids coming from the periphery into more or into less active molecules. From a functional point of view, the most important of these enzymatic activities, and without a doubt the best studied to date, is aromatase or estrogen synthase. Aromatase activity (AA) catalyzes the transformation of aromatizable androgens such as T or androstenedione (A4) into estrogens such as estradiol (E2) or estrone (E1). Since the discovery of brain aromatase more than 40 years ago (Naftolin et al., 1971, 1975), this process has been shown to play a critical limiting role in the sexual differentiation during ontogeny and activation during adulthood of male sexual behavior in many vertebrate species.

The present chapter summarizes (1) The research performed during the last few decades that demonstrates beyond a doubt that brain aromatization of T is a critical step limiting the activation by T of many aspects of behavior in a variety of bird species, and (2) what is known about the brain expression of aromatase and control of AA activity in the avian brain. A third brief section then highlights recent discoveries concerning aromatase in the avian brain that are the focus of specific chapter in later sections of this book.

THE KEY ROLE OF AROMATIZATION PRODUCTS

Aromatase and Copulatory Behavior in Quail and Other Birds

Scientists analyzing the endocrine bases of reproductive behavior were initially surprised to discover that estrogens like E2 were activating a variety of behavior patterns in males that were also activated by T (Beach, 1948). The discovery of AA in the brain provided a potential mechanism to explain this astonishing result, and therefore intense research was carried out to investigate whether the similarity of behavioral effects of T and E2 was explained by the metabolism of T into E2 in the brain. Converging evidence was collected not only in mammals, but also in a variety of avian species. Three groups of facts essentially were established in birds. A large part of these studies were carried out in Japanese quail, but data are available suggesting that principles identified in quail can be generalized. Specifically (Fig. 10–1):

1. In multiple bird species, estrogens restore in a manner similar to T most aspects of reproductive behaviors in males and in particular copulatory behavior sensu stricto (Hutchison, 1971; Cheng &

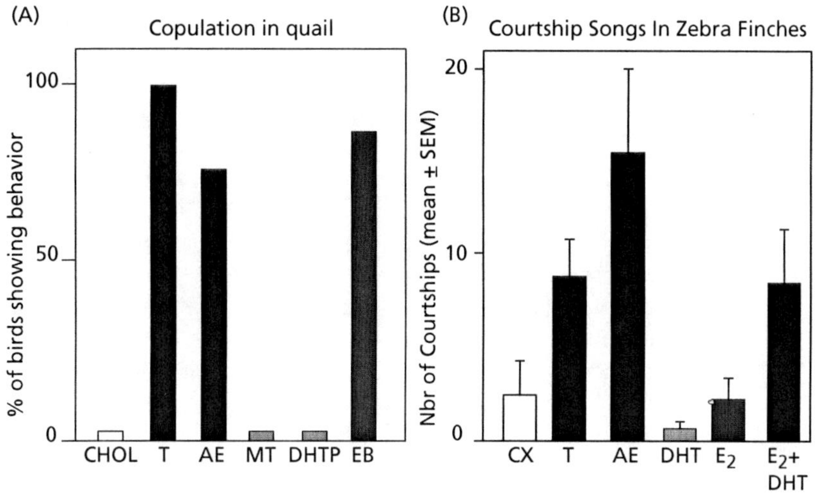

Figure 10–1. A. Percentage of castrated male quail implanted with steroid-filled Silastic capsules that copulated (exhibited at least one mount). Abbreviations: CHOL, cholesterol; T, testosterone; AE, androstenedione; MT, 17α-methyltestosterone; DHTP, 5α-dihydrotestosterone propionate; EB, estradiol benzoate. T and AE are aromatizable androgen; EB is an estrogen; MT and DHTP are nonaromatizable androgens. B. Number of courtship songs produced by castrated male zebra finches exposed to a female and treated with exogenous steroids that provide androgenic or estrogenic stimulation or both. T and AE are aromatizable ansrogens, 5α -dihydrotestosterone (DHT) is a nonaromatizable androgen. E2 = estradiol-17β. (*A*) Redrawn from data in Adkins, E. K., Boop, J. J., Koutnik, D. L., Morris, J. B., & Pnieswski, E. E. (1980). Further evidence that androgen aromatization is essential for the activation of copulation in male quail. *Physiol Behav, 24,* 441–446. (*B*) Redrawn from data in Harding, C. F., Sheridan, K., & Walters, M. J. (1983). Hormonal specificity and activation of sexual behavior in male zebra finches. *Horm Behav, 17,* 111–133.

Lehrman, 1975; Hutchison, 1978a; Adkins et al., 1980; Harding et al., 1983; Schumacher & Balthazart, 1983). Note, however, that various aspects of precopulatory behavior usually known as social or sexual displays are often strictly androgen dependent. They are activated by androgens acting directly in the brain as such (i.e., through interactions with androgen receptors) and are not activated or even sometimes inhibited by estrogens. This is, for example, the case for the bow-cooing display in doves (Cheng & Lehrman, 1975; Adkins-Regan, 1981) and the strutting display in quail (Adkins, 1977; Adkins et al., 1980). Such behaviors offer excellent internal controls for steroid specificity of behavior activation in these species.

2. Androgens activate copulatory behavior if and only if they are aromatizable. This is namely the case for T and A4 that are active

in a broad range of species, but nonaromatizable androgens such as 5α-dihydrotestosterone (DHT) or the synthethic androgen R1881 (methyltrienolone) have little or no effect on these behaviors (Adkins, 1977; Adkins et al., 1980; Harding et al., 1983; Alexandre & Balthazart, 1986). They are nevertheless fully effective in activating the purely androgen depend precopulatory displays mentioned before (Cheng & Lehrman, 1975; Adkins, 1977; Adkins et al., 1980; Schumacher & Balthazart, 1983), thus indicating that their failure to activate copulation is not related to a general technical problem such as the lack of access to the brain or a rapid peripheral catabolism.

3. Behavioral effects of T and other aromatizable androgens can be completely blocked or at least markedly attenuated by a concurrent treatment with aromatase inhibitors. This conclusion has been reached with quite a variety of species and with multiple aromatase inhibitors that have a steroidal (e.g., androstatrienedione, 4OH-A) or nonsteroidal (Fadrozole, Vorozole, or R76713) chemical nature (Adkins et al., 1980; Walters & Harding, 1988; Balthazart et al., 1990a; Schlinger & Callard, 1990; Soma et al., 1999a). Several studies have additionally shown that this behavioral inhibition can be reversed by a concurrent treatment with exogenous estrogens, thus demonstrating that the inhibition is the result of the depletion in estrogen production rather than being caused by some general toxic effects of the inhibitors.

Taken together, these data strongly support what came to be known as the *aromatization hypothesis,* which states that effects of T and other aromatizable androgens on behavior and many physiological processes are the results of the action of locally produced estrogens. A more restrictive theory, called the *estrogen receptor hypothesis* (Yahr, 1979), further states that these locally produced estrogens (largely) exert their action by interacting with nuclear ER. Several studies have demonstrated the validity of this concept in birds by showing that behavioral effects of T can be blocked by a concurrent treatment with an antiestrogen such as tamoxifen or nitromifene citrate (CI-628) (Adkins & Nock, 1976; Alexandre & Balthazart, 1986).

Studies in quail have demonstrated that the aromatization hypothesis, as well as the estrogen receptor hypothesis, also applies to the control of appetitive sexual behavior. Indeed the learned social proximity response is activated in castrated males by T, but this androgen can be substituted with E2 or with the synthetic estrogen diethylestradiol (DES) (Balthazart et al., 1995). In addition, when the proximity response has been acquired by castrated males treated with T, it will be blocked if the treatment with T is associated with injections of the aromatase inhibitor vorozole or the antiestrogen tamoxifen (Balthazart et al., 1995, 1997). Similarly, the rhythmic contractions of the cloacal sphincter musles (RCSM) in response to the view of a female that are activated by T progressively disappear if subjects are simultaneously treated with an aromatase inhibitor (Taziaux et al., 2003; Seredynski et al., unpublished). Together these data strongly suggest that aromatization of T is required for the activation of ASB as well as CSB, even if the evidence for ASB is more limited and only concerns a single avian species. Note, however, that singing in songbirds can also be considered as a form of ASB (Ball et al., 2002) and that this behavior also critically depends on T aromatization (see next section).

Although estrogens derived from brain aromatization of T play a critical role in the activation of male sexual behavior in many avian species, some contribution of androgens to this process also needs to be recognized because first, the treatment of castrates with 5α-DHT or other nonaromatizable androgens such as methyltrienolone partially restores a weak degree of sexual activity (Adkins, 1977; Balthazart et al., 1985; Alexandre & Balthazart, 1986; Schumacher et al., 1987), and second, the antiandrogen flutamide slightly reduces T-induced copulatory behavior in quail, indicating that interactions of T or its androgenic metabolites (e.g., 5α-DHT) with androgen receptors also play some role in the activation of this behavior (Balthazart et al., 1985; Balthazart & Surlemont, 1990a).

It should also be noted that castrated males must often be treated with very high doses of exogenous estrogens (producing circulating levels far above physiological concentrations) so as to activate sexual behavior. This requirement could reflect the need for local production of E_2 in high concentration at its site of action, but it might also be explained by the fact that under physiological conditions, T metabolism in the brain results in the availability of both estrogens (by aromatization) and androgens (T itself or its 5α-reduced metabolites) and that both types of steroids must interact to activate behavior. Multiple experiments support this interpretation by demonstrating the existence of a synergism between estrogens and androgens in the activation of behavior: The effects of low doses of E_2 that are behaviorally ineffective by themselves are markedly increased by a concurrent treatment with 5α-DHT (Adkins & Pniewski, 1978; Harding et al., 1983; Balthazart et al., 1985).

Multiple mechanisms have been proposed to explain this synergism: action of estrogens in the brain and androgens on sex organs in the periphery (penis in mammals, cloacal structures in birds) required for copulation, action of both steroids on the same neurons or different neurons whose outputs converge on the same functional targets, activation of estrogen receptors by 5α-DHT via its metabolism into weakly estrogenic diols, inhibition by 5α-DHT of E2 peripheral metabolism. However, despite much research on this topic, it remains impossible at this time to discriminate between these ideas. (See Ball & Balthazart, 2002, for more extensive discussion of this topic, which is beyond the scope of the present chapter.)

Singing Behavior and Brain Aromatase

In songbirds, estrogens have been found to play a pivotal role in neural development, the activation of neural circuits in seasonal neuroplasticity, and auditory perception (described in more detail in chapters 24 and 25).

With respect to neural development, E2 treatment of developing female zebra finches partially masculinizes song control circuits and gives adult females the capacity to sing (Gurney & Konishi, 1980; Gurney, 1982). Although estrogens have this potent developmental effect, there is convincing evidence that the masculine growth of the song circuitry has a considerable genetic basis, that is that the chromosomal complement of the zebra finch (ZZ male/ZW female) directs sex-specific gene expression patterns in brain that induce sex-specific growth of the neural song system (Arnold, 2000). It is likely, however, that the full masculine phenotype is acquired when E2 acts on a genetically male neural background (Gahr & Metzdorf, 1999). Evidence for de novo estrogen synthesis and estrogen-dependent masculine actions in vitro lends strong support for the conclusion that neural aromatization is crucial for this ontogeny of this sexual dimorphism (Holloway & Clayton, 2001).

Estradiol also participates in the activation of adult song (Harding, 1981; Harding et al., 1983) with HVC serving as a key nucleus for integrating estrogen-dependent signaling with motor control of song output (Meitzen et al., 2007). A final point concerns the role of estrogens in the prominent plasticity that is seen to occur in the songbird brain. Seasonal breeding songbirds sing maximally during the breeding season and little or not at all during the nonbreeding season. Various anatomical attributes of their neural song systems expand and contract in parallel with song behavior (Nottebohm, 1981). Facets of this adult seasonal neuroplasticity seem to depend on estrogens. Estrogens and androgens (probably metabolites of circulating T) stimulate the growth of nucleus HVC (DeVoogd & Nottebohm, 1981; Brenowitz & Lent, 2002; Soma et al., 2004b; Meitzen et al., 2007; Yamamura et al., 2011). Some of this growth involves the incorporation of new neurons that migrate into HVC from the subventricular zone of the nearby ipsilateral lateral ventricle (Paton & Nottebohm, 1984). Estrogens seem to promote the survival and appropriate differentiation of these newly born neurons as they migrate through a population of estrogen-receptor rich cells on their migration towards HVC (Hidalgo et al., 1995). When considered together, these observations spotlight E2's fundamental role in numerous core properties of the songbird brain emphasizing the importance of neural aromatization in this avian clade.

DISTRIBUTION OF AROMATASE IN THE AVIAN BRAIN

Soon after AA was discovered in the mammalian brain, comparative studies were initiated to research how broadly this enzymatic activity is distributed in vertebrates. Gloria Callard and her collaborators at Boston University quickly realized that conversion of radioactive A4 to E1 in the brain of representative species from all vertebrate classes, from fishes to mammals (Callard et al., 1978). Interestingly, the neuroanatomical distribution was broader and absolute enzymatic activity was higher in ectothermic vertebrates (e.g., fishes, amphibians, reptiles) than in mammals; birds were apparently intermediate in this respect (Callard, 1983).

More detailed analyses were then undertaken in ring doves (*Streptopelia risoria*) by Huchison et al. By assaying AA in multiple brain regions, they demonstrated that AA is higher in the anterior hypothalamus-preoptic area (AHPOA) than in all adjacent brain regions that had been tested (Steimer & Hutchison, 1980). Studies in Japanese quail (*Coturnix japonica*) confirmed this conclusion and showed that the highest enzymatic activity is actually present in the preoptic area (Schumacher & Balthazart, 1986; Schlinger & Callard, 1987).

This analysis of the neuroanatomical distribution of AA was refined by assaying this activity in specific nuclei of the quail brain that had been dissected by the Palkovits punch method (Palkovits, 1973). It was demonstrated that AA is significantly higher in the sexually dimorphic (volume larger in males than in females) (Viglietti-Panzica et al., 1986; Panzica et al., 1987) medial preoptic nucleus than in other hypothalamic or limbic nuclei that were studied. Little or no enzymatic activity was detected in telencephalic structures that were analyzed (Schumacher & Balthazart, 1987).

The same methodology was then used to study the distribution of AA in the zebra finch. This study confirmed the existence of a high AA in the preoptic area and ventromedial hypothalamus of male zebra finches and low or no activity in ventral parts of the telencephalon but quite surprisingly, a very high level of AA was detected in this species on the dorso-medial telencephalon, including the hippocampal area and the song control nucleus HVC (Vockel et al., 1990b). One surprise from these studies was the very high level of AA detected in the dorsal part of the telencephalon, a result that stood in contrast to that seen in brains of doves and quail, in which telencephalic AA was distinctively lower than in the hypothalamus. The pattern of elevated telencephalic aromatase was subsequently found in other species (albeit activity in the zebra finch brain was unusually high) indicating that elevated brain aromatase was likely an attribute of all oscine songbirds (Saldanha et al., 1999; Soma et al., 1999b; Fusani et al., 2000; Silverin et al., 2000; Riters et al., 2001; Soma et al., 2003).

Whereas measures of activity provided excellent information on the functional aromatase protein in brain, it was only when techniques became available to study the distribution of the aromatase protein and of the corresponding messenger RNA (by immunohistochemistry and in situ hybridization, respectively) that the precise distribution of the enzyme could be determined. An antibody raised against human placental aromatase (Harada, 1988) originally provided a first reliable visualization of the aromatase protein in the quail, ring dove, and zebra finch brain (Balthazart et al., 1990b,d). In all species, aromatase-immunoreactive cells were found only in the preoptic area and hypothalamus, with a high density of positive cells present in the medial preoptic area, the medial part of the bed nucleus of the stria terminalis (originally identified as septal area) above and caudal to the anterior commissure, the ventromedial nucleus of the hypothalamus, and the tuberal hypothalamus (see Fig. 10–2A for results in quail). Immunoreactivity was weaker in zebra finches, so no signal could be detected in the ventromedial and tuberal hypothalamus. The positive material was localized in the perikarya and in adjacent cytoplasmic processes, including the full length of axons always leaving a clear unstained cell nucleus.

Interestingly, the localization of labelled neurons coincided with the distribution of AA as previously studied by in vitro radioenzyme assays on brain nuclei dissected by the Palkovits punch method. There was, however, one striking exception to this rule: No immunoreactivity was detected in the zebra finch telencephalon, whereas assays had shown the presence of an active enzyme in broad parts of the dorsal telencephalon. Subsequently, antibodies raised against avian specific antigens: quail (Foidart et al., 1995), zebra finch (Saldanha et al., 2000) allowed refinement of these anatomical data and identified additional populations of labeled cells in zebra finches resolving earlier discrepancies. Very large populations of aromatase-immunoreactive cells were observed in the dorsal telencephalon of zebra finches (Fig. 10–2B) (Balthazart et al., 1996; Saldanha et al., 2000) and later on of other songbird species such as the pied flycatcher, *Ficedula hypoleuca* (Fig. 10–2C) (Foidart et al., 1998).

In general, in situ hybridization studies confirmed that the brain distribution of aromatase mRNA matches the distribution of the enzymatic protein: quail (Aste et al., 1998; Voigt et al., 2007), zebra finches (Shen et al., 1995), and coincides with the distribution of enzymatic activity. In the zebra finch in particular, aromatase mRNA was found throughout the mediocaudal telencephalon, which extended laterally through the hyperpallium and nidopallium (Shen et al., 1994, 1995).

A few key features emerged from the analysis of aromatase immunostaining in zebra finches. First, as had been observed in other species, neurons were the only cells with significant aromatase protein in the intact brain (i.e., in the absence of any damage) (see chapter 19). In the preoptic area, hippocampus, and NCM, aromatase was found throughout fibers, creating decidedly rich estrogen synthetic networks. This observation possibly justifies the surprising result of earlier in situ hybridization studies indicating that nucleus HVC, an estrogen target in some species that had been shown to possess AA (Vockel et al., 1990b), had little to no aromatase mRNA. Aromatase-immunoreactive fibers were indeed present in HVC (Saldanha et al., 2000) and possibly explain why enzymatic activity had been detected in this nucleus in the absence of the corresponding mRNA. It is also possible that the enzymatic activity originally reported in the HVC originated from a contamination by the adjacent nidopallium, which is one of the most active sites of the AA.

Moreover, quantitative analysis of these aromatase-positive fibers identified greater fiber density in males compared with females, providing the first evidence for a mechanism to provide estrogens in larger amounts to the telencephalon of male zebra finches (Saldanha et al., 2000)

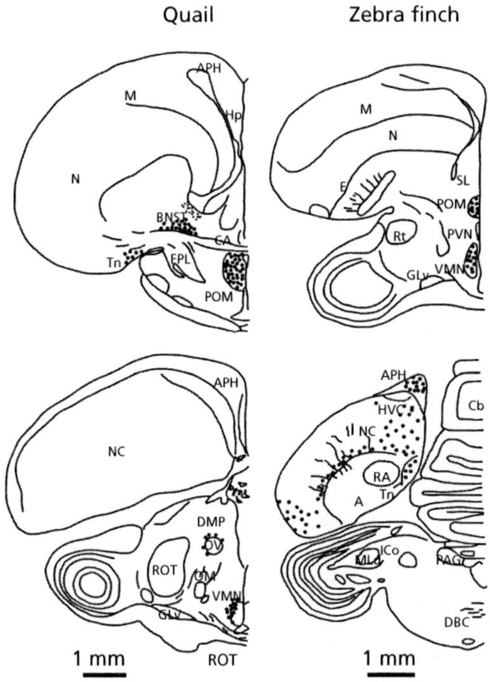

Figure 10–2. Schematic Representation of the Distribution of Aromatase Immunoreactive Cells in the Brain of Male Quail, Zebra Finches, and Pied Flycatchers Two rostral to caudal levels are shown from top to bottom. Darkly labeled cells are indicated by large dots; cells showing a weaker immunoreactivity are indicated by smaller dots. Abbreviations: A, arcopallium; APH, area parahippocampalis; BNST: bed nucleus of the stria terminalis; CA commissura anterior; Cb, cerebellum; DBC, decussatio brachiorum conjunctivorum; E, entopallium; DMP, nucleus dorsomedialis posterior thalami; FLM, fasciculus longitudinalis medialis; FPL, fasciculus prosencephali lateralis; GLv, nucleus geniculatus lateralis pars ventralis; Hp, hippocampus; HVC, used as a noun (formally high vocal center); ICo, nucleus intercollicularis; LAD, lamina arcopallialis dorsalis; M, mesopallium; MLd, nucleus mesencephalicus lateralis pars dorsalis; N, nidopallium; NC, nidopallium caudale caudale; OM, tractus occipitomesencephalicus; OV, nucleus ovoidalis; PAG, periaqueductal gray; POM, nucleus preopticus medialis; PVN, nucleus paraventricularis; RA, nucleus robustus arcopalliamis; Rt, nucleus rotundus; SL, nucleus septalis lateralis; Tn, nucleus taeniae of the amygdala; VMN, nucleus ventromedialis hypothalami. Redrawn from (Naftolin et al., 1996; quail), (Balthazart et al., 1996; Saldanha et al., 2000; zebra finch), and (Foidart et al., 1998; flycatcher).

This abundance of aromatase in the songbird brain, particularly in zebra finches, had interesting implications for estrogen physiology in songbirds. The suspected role of estrogens in development of the neural song system prompted investigation into the source of that estrogen. Surprisingly, little to no AA was detectable in testes or adrenals of male zebra finches or in other peripheral tissues of these birds. By contrast, the brain as a whole demonstrated considerable AA (Schlinger & Arnold, 1991). These results raised the possibility that the E2 that was detected in the blood of male zebra finches might derive from the brain. Studies involving injection of radiolabeled androgens into male finches with measures of carotid and jugular radiolabeled estrogens confirmed this idea in male zebra finches (Schlinger & Arnold, 1992, 1993) as well as in male brown-headed cowbirds (*Molothrus ater*) (Saldanha & Schlinger, 1997). Although there is no evidence that estrogens produced in brain exert systemic effects, these studies served to highlight the magnitude and importance of neural aromatization in the control of brain and behavior in this group of birds with implications for the role of neurosteroidogenesis in other vertebrates.

The Cellular Basis of Brain Aromatase

Neurons had been shown to be the principle if not only cell type expressing aromatase in the mammalian brain (Canick et al., 1986), a finding confirmed in birds and mammals by the neuronal profiles immunostained using aromatase-specific antibodies (Balthazart et al., 1996; Saldanha et al., 2000). Nevertheless, cell culture proved to be a useful method to investigate the neurochemistry of isolated cell populations and was applied to the study of aromatase in the songbird telencephalon. As expected, AA was detected in purified neuronal cultures of the developing zebra finch brain, in which neurons and their processes were also clearly immunostained for aromatase. What was surprising was that activity could also be detected in cultures enriched in glia and cells with the morphology of astrocytes showed obvious aromatase-positive immunostaining (Schlinger et al., 1994). Similar results were subsequently obtained from cells cultured from the mammalian brain (Zwain et al., 1997). Why cellular aromatase expression differs in vitro from what is seen in vivo might be explained by the fact that in both mammals (Garcia-Segura et al., 1999) and birds (Peterson et al., 2001; Spence et al., 2009), aromatase is induced in astrocytes in proximity to sites of neural injury. Cell culture may better reflect the injured neural state. Together with evidence that estrogens are neuroprotective and promote neural repair (Wise, 2002), the upregulation of aromatase in astrocytes first seen in birds appears to be an evolutionarily conserved response to neural injury. Details of aromatase and neural injury can be found in chapters 19 and 20.

Subcellular

Like many other steroidogenic enzymes, aromatase is bound to endoplasmic reticulum and thus is distributed throughout the cytoplasm of aromatase-expressing cells. The fact that neurons were known to possess aromatase in vivo raised questions as to whether the enzyme was restricted to somata, and thus located near intranuclear estrogen receptors, or was distributed more widely throughout the cell. Before the advent of ICC with aromatase-specific antibodies, biochemical procedures were the only mechanism available to investigate aromatase. Fueled by early results of the fish brain (Mak et al., 1985), detailed cellular subfractionation experiments of the quail hypothalamus demonstrate that, indeed, aromatase in brain was enriched in microsomes (tiny bits of endoplasmic reticulum) but was also found in synaptosomes (presynaptic nerve terminals that pinch-off and reseal after homogenization) (Schlinger & Callard, 1989). These results gave rise to the idea that aromatase-positive neurons project to and synapse on estrogen-sensitive cells to provide a form of targeted estrogen delivery. The synaptic localization of aromatase was confirmed by immunocytochemical studies of quail and mammalian species (Naftolin et al., 1996) and later in songbirds (Peterson et al., 2005). The presence of aromatase in terminals is the basis of new ideas regarding spatiotemporal properties of steroidal signaling in the vertebrate brain, topics that are discussed in the following and in detail in chapter 6.

CONTROLS OF AROMATASE ACTIVITY IN BIRDS

Given the prominent role of aromatase in the control of reproduction, many studies have been dedicated to the experimental analysis of the mechanisms that control this critical brain enzyme. In seasonal breeding male birds, aromatase in the preoptic area is elevated during the breeding season compared with the nonbreeding season or castrated breeding birds (Lea & Armstrong, 1986; Sharp et al., 1986; Riters et al., 2001; Soma et al., 2003). Studies in ring doves initially demonstrated that AA in the preoptic area-hypothalamus is markedly (fourfold to fivefold) increased in castrated birds that are treated for a few days with exogenous T (Steimer & Hutchison, 1981). This conclusion was later extended to a wide variety of vertebrates, including rodents (Roselli & Resko, 1984; Roselli et al., 1987) and multiple avian species; for example, quail (Schumacher & Balthazart, 1986) and zebra finches (Vockel et al., 1990a). This increase in activity is first detected in quail 8 hours after the initiation of T treatment and reaches its asymptote after 24 to 48 hours (Balthazart et al., 1990c), which is compatible with the idea that the increased activity results from a stimulation of the synthesis of the enzymatic protein. Subsequent studies indeed confirmed that the T-induced increase in preoptic AA is paralleled by a similar increase in aromatase mRNA concentration as measured by RT-PCR and in the number of aromatase-immunoreactive neurons in the POM (Fig. 10–3) (Balthazart & Foidart, 1993).

The effects of T on aromatase transcription appear to be largely mediated by an activation of androgen receptors in rats (Roselli & Resko, 1984), but in contrast by an action of locally produced estrogens in quail or ring doves (Hutchison & Steimer, 1986; Absil et al., 2001). In all species, however, there is a clear synergism between nonaromatizable androgens and estrogens in the mechanism that regulates aromatase, and this synergism has been observed

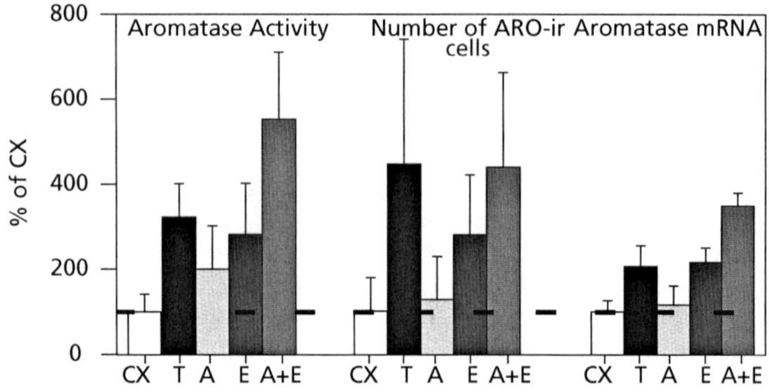

Figure 10–3. Effects of testosterone (T), nonaromatizable androgens (A) and estrogens (E) alone or in combination (A + E) on aromatase activity (*left*), on the number of aromatase-immunoreactive cells in the POM (*middle*), and on the aromatase mRNA concentration (*right*) in the POA in male quail. The nonaromatizable androgen was methyltrieneolone in the left panel and 5α-dihydrotestosterone in the other two panels. The estrogen used in left panel was diethylstilbestrol. Estradiol-17β was used in the other panels. All data (means ± SD) were expressed as percent of corresponding values in castrates (CX) to permit direct comparisons. Redrawn from data in Absil, P., Baillien, M., Ball, G. F., Panzica, G., & Balthazart, J. (2001) The control of preoptic aromatase activity by afferent inputs in Japanese quail. *Brain Res Rev,* 37, 38–58.

at the three different levels at which aromatase has been studied: the mRNA concentration, the protein as assessed semiquantitatively by immunocytochemistry, and the enzyme activity (Balthazart et al., 2004). The cellular mechanisms mediating this synergistic induction of brain AA by androgen and estrogens have been discussed in detail elsewhere (Balthazart & Foidart, 1993; Absil et al., 2001) and remain partially unexplained. Indeed, although androgen receptors are broadly expressed in aromatase cells (Balthazart et al., 1998b), only a small proportion of aromatase-immunoreactive cells in POM appear to express estrogen receptor alpha in quail (Balthazart et al., 1991) as well as in zebra finches (Saldanha & Coomaralingam, 2005). Furthermore, the aromatase gene does not appears to be clearly associated with both androgen-*and* estrogen-responsive elements (ARE; ERE) in the AHPOA of birds and mammals (Balthazart & Ball, 1998; Ramachandran et al., 1999) even if an ERE associated with brain aromatase has been detected in fishes (Diotel et al., 2010). A direct transcriptional control of avian brain aromatase by occupied androgen or estrogen receptors is at present unclear. Therefore, transsynaptic controls of aromatase expression have been considered (Balthazart & Ball, 1998; Absil et al., 2001), but have not been formally demonstrated to date.

The control of AA by sex steroids appears, to some extent, to be specific to the preoptic area and anterior hypothalamus. Several studies in a variety of species have indeed failed to detect an effect of castration and/or T treatment on this enzymatic activity in the area basalis of doves (Steimer & Hutchison,

1981), the area parahippocampalis or nidopallium of zebra finches (Vockel et al., 1990a), and the nucleus taeniae of the amygdala in quail (Balthazart et al., 1990e). Some of these negative results relate to the low enzymatic activity present in these structures that is obviously associated with a lack of assay sensitivity, but this explanation cannot be true for the zebra finch telencephalon, which expresses the highest level of enzymatic activity per mg protein ever identified in birds. Aromatase activity in these brain regions is, however, not necessarily stable. For example, prominent significant variations of this enzymatic activity were observed during the annual cycle in the telencephalon of starlings (Riters et al., 2001) as well as in the ventral telencephalon and NCM of song sparrows, *Melospiza melodia* (Soma et al., 2003). The factor(s) regulating AA in these regions remain(s) unidentified.

Although T-induced increases in preoptic AA are paralleled by changes in enzyme concentration that are largely mediated by the transcriptional action of the steroid, the apparent magnitude of this effect increases with the progression from DNA transcription, to protein content (number of immunoreactive cells) to enzyme activity (Fig. 10–4), suggesting that T may also regulate AA by other means. Indeed, work during the last decade has identified changes in AA that occur within minutes and are thus much too fast to possibly result from a change in enzyme concentration. A suite of experiments performed on quail preoptic homogenates or hypothalamic explants and finally in vivo has identified mechanisms based namely on protein phosphorylations that exert a powerful and rapid control on AA (see chapters 22 and 23). These rapid changes in estrogen production and therefore presumably availability have now been shown to affect not only cellular activity but also the expression of sexual behavior and physiology of the auditory system in birds (see chapters 23, 24 and 25).

AROMATASE AS A TERMINAL NEUROSTEROIDOGENIC REACTION

As described, the discovery of aromatase in the CNS shifted our attention from the gonads and adrenals as the sole site of synthesis of steroidal hormones to include the brain itself. The persistence of steroids extracted from the rodent brain long after gonadectomy and adrenalectomy gave new support for the idea that the brain might synthesize steroids de novo from cholesterol (Baulieu, 1997). This neurosteroidogenic potential has been confirmed in several vertebrate species with studies of both quail (Tsutsui et al., 2006) and songbirds (London et al., 2009), confirming neurosteroidogenesis in birds. Moreover, birds have provided new insights into the ways in which aromatase couples with other steroidogenic enzymes. For example, the enzyme 3β-hydroxysteroid dehydrogenase/isomerase (3β-HSD) that converts pregnenolone into progesterone and dehydroepiandrosterone (DHEA) into androstenedione (an aromatizable androgen) is expressed and active in the songbird brain. There is evidence both in vivo and in vitro that 3β-HSD readily couples with aromatase in using DHEA as a substrate for the formation of biologically active neutroestrogens (Vanson et al., 1996; Soma et al., 2004a; Tam & Schlinger, 2007). The source of that DHEA could be the brain, because cholesterol transporters and other steroidogenic enzymes are expressed in the songbird brain (London et al., 2006), or the DHEA could be derived from systemic (likely adrenal) synthesis (Soma & Wingfield, 2001). An in-depth discussion of this latter process can be found in chapter 11.

A ROLE FOR BRAIN AROMATASE IN NONREPRODUCTIVE CONTEXTS

Whereas we tend to think of sex steroids acting to participate in reproduction, there are a growing number of instances in which neuroestrogens participate more generally in primary neural functioning. Studies of songbirds have identified several estrogen-dependent neural attributes that have little or nothing to do with reproduction. We have already alluded to one of these, namely, the role of estrogens in neuroprotection and neural repair (see chapters 19 and 20). A second case involves the likelihood that peripheral or central DHEA serves as the substrate for neuroestrogen production to activate aggressive behavior in at least one songbird during the nonbreeding season (see chapter 11). The songbird hippocampus is especially rich in aromatase (Vockel et al., 1990b; Saldanha et al., 1998) and there is evidence that estrogens act in the hippocampus to strengthen songbird spatial learning and memory capabilities (Oberlander et al., 2004; Schlinger & Saldanha, 2006). Of course, song behavior itself is a behavior with reproductive and nonreproductive functions (Catchpole & Slater, 2008), so some role for estrogens in song system development, song learning, and/or song performance may be considered as occurring in a nonreproductive context.

A core feature of singing behavior lies in the capacity for songbirds to identify their own songs and that of conspecifics. The caudal telencephalon of songbirds includes the caudo-medial nidopallium or NCM, an area especially rich with aromatase (Vockel et al., 1990b; Saldanha et al., 2000) that serves as a high-order auditory processing center (Bolhuis & Gahr, 2006). Studies show that levels of E2 rapidly fluctuate in NCM in response to song playback (Remage-Healey et al., 2008). These estrogens in turn rapidly strengthen electrophysiological activity of NCM neurons and appear to enable song discrimination (Tremere et al., 2009; Remage-Healey et al., 2010; Tremere & Pinaud, 2011). These studies identify a new role for neuroestrogens in facilitating auditory discrimination and are developed more fully in chapters 24 and 25.

CONVERGENCE OF STUDIES ON SONGBIRDS AND NONSONGBIRDS

Although we have somewhat independently highlighted the strengths of studies on copulatory behavior of Japanese quail (nonsongbirds) and singing behavior of the zebra finch (songbirds), these lines of research are witnessing

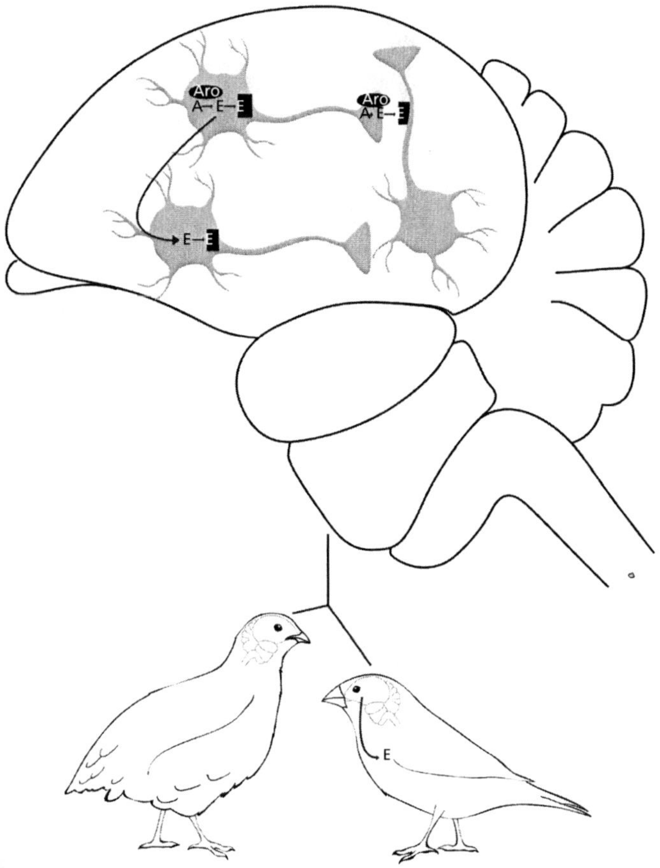

Figure 10–4. Estradiol Signaling in the Intact Male Avian Brain Studies of aromatase in the hypothalamus of the nonsongbird Japanese quail (*left*) and the telencephalon of the zebra finch songbird (*right*) have independently identified several mechanisms of neuroestrogen signaling. These include (1) autocrine supply of estradiol (E) to estrogen receptors (ER; *black rectangles with a white E sign*) from the neuronal colocalization of aromatase and ER; (2) paracrine supply of E to ER via diffusion of E from aromatase-expressing neurons to adjacent ER-expressing neurons; and (3) neuromodulatory (synaptocrine) release of E from synaptic terminals to postsynaptic neuronal ER. In songbirds, neural aromatase is sufficiently abundant to enrich the systemic vascular E content. A, androgen, Aro, aromatase.

increasing convergence (Fig. 10–4). Future research on the role of central aromatization on neuroestrogens function will likely be guided by these areas of overlap. Two crucial areas of convergence involve the discovery and impact of rapid regulation of aromatase protein and rapid (nongenomic) actions of estrogen on neurophysiology.

As briefly alluded to in the preceding and more fully discussed in other chapters (see chapters 22 and 23), studies in quail initially identified mechanisms based on protein phosphorylation that are able to profoundly alter brain AA within minutes. In vivo microdialysis studies recently demonstrated the existence of rapid variations in local E2 concentration within the NCM of zebra finches in response to social or auditory stimuli, thus indirectly confirming the existence of rapid control of AA in the songbird brain. In addition, NCM E2 levels rapidly changed on retrodialysis of the neurotransmitter glutamate (see chapter 25) similar to glutamatergic regulation of AA seen in quail (Balthazart & Ball, 2006). Thus, rapid changes in estrogen synthase activity appear to occur in multiple avian species and are likely to exist more broadly in vertebrates.

In parallel, functional studies have discovered that the rapid changes in estrogens bioavailability in the brain have a major physiological and behavioral impact on both quail and zebra finches. In quail, a suite of studies now demonstrates that rapid up- and downregulations of E2 availability respectively increase or decrease in parallel the expression of male sexual behavior (see chapter 23). In zebra finches, independent studies from two laboratories have demonstrated that changes in estrogen availability alters within minutes the electrophysiological activity and the processing of auditory signal in the aromatase-rich area of the nidopallium, NCM (see chapters 24 and 25). Thus, in two completely unrelated avian species, a galliform and a passeriform, similar mechanisms of control for AA appear to exist in brain areas as diverse as the preoptic area and the dorsal telencephalon. The resulting rapid changes in estrogen bioavailability have in parallel been shown to affect within minutes a classical "instinctive" (i.e., not learned) behavior (copulation) and the acoustic physiology that is the foundation for discrimination of learned songs produced by the male songbird and his congener. In that the brain areas and behaviors shown to be affected by the local rapid modulation of E2 levels are diverse, these studies suggest that other neurobiological systems may be similarly affected. Indeed, growing evidence suggests that these rapid changes in estrogens production and action are not limited to birds with data on these topics now available from fishes to mammals (see chapters 21, 25 and 26). These phenomena discovered mostly in birds are now seen to have vertebrate-wide significance. Undoubtedly, studies of avian brain aromatase will continue to bear fruit for years to come.

ACKNOWLEDGMENTS

The work described in this chapter was supported by grants from NIMH to BAS (MH061994) and JB (MH50388), the Belgian Science Foundation (FRFC2.4537.09), and the University of Liège (Fonds spéciaux) to JB. The authors thank their many students, postdoctoral fellows, and collaborators for their invaluable contributions to the research carried out during the past decades.

REFERENCES

Absil, P., Baillien, M., Ball, G. F., Panzica, G., & Balthazart, J. (2001). The control of preoptic aromatase activity by afferent inputs in Japanese quail. *Brain Res Rev, 37,* 38–58.

Adkins, E. K. (1977). Effects of diverse androgens on the sexual behavior and morphology of castrated male quail. *Horm Behav, 8,* 201–207.

Adkins, E. K. (1981). Early organization effects hormones. In: Adler, N. T. (ed.), *Neuroendocrinology of Reproduction: Physiology and Behavior.* New York: Plenum, pp. 159–228.

Adkins, E. K., & Adler, N. T. (1972). Hormonal control of behavior in the Japanese quail. *J Comp Physiol Psychol, 81,* 27–36.

Adkins, E. K., Boop, J. J., Koutnik, D. L., Morris, J. B., & Pnieswski, E. E. (1980). Further evidence that androgen aromatization is essential for the activation of copulation in male quail. *Physiol Behav, 24,* 441–446.

Adkins, E. K., & Nock, B. L. (1976). The effects of the antiestrogen CI-628 on sexual behavior activated by androgen and estrogen in quail. *Horm Behav, 7,* 417–429.

Adkins, E. K.,& Pniewski, E. E. (1978). Control of reproductive behavior by sex steroids in male quail. *J Comp Physiol Psychol, 92,* 1169–1178.

Adkins-Regan, E. (1981). Effect of sex steroids on the reproductive behavior of castrated male ring doves (*Streptopelia* sp.). *Physiol Behav, 26,* 561–565.

Alexandre, C., & Balthazart, J. (1986). Effects of metabolism inhibitors, antiestrogens and antiandrogens on the androgen and estrogen induced sexual behavior in Japanese quail. *Physiol Behav, 38,* 581–591.

Armstrong, E. A. (1973). *A Study of Bird Song.* New York: Dover.

Arnold, A. (2002). Concepts of genetic and hormonal induction of vertebrate sexual differentiation in the twentieth century, with special reference to the brain. In: Pfaff, D. W., Etgen, A. N., Fahrbach, S. E., & Rubin, R. T. (eds.), *Hormones, Brain and Behavior.* Amsterdam: Academic Press, pp. 105–135.

Arnold, A. P. (1975). The effects of castration and androgen replacement on song, courtship, and aggression in zebra finches (*Poephila guttata.*). *J Exp Zool, 191,* 309–326.

Arnold, A. P. (2000). Hormonal and non-hormonal mechanisms of sexual differentiation of the zebra finch brain:embracing the null hypothesis. In: Matsumoto, A. (ed.), *Sexual Differentiation of the Brain.* New York: CRC Press, pp. 131–148.

Arnold, A. P., Nottebohm, F., & Pfaff, D. W. (1976). Hormone concentrating cells in vocal control and other areas of the brain of the zebra finch (*Poephila guttata*). *J Comp Neurol, 165,* 487–511.

Aste, N., Panzica, G. C., Viglietti-Panzica, C., Harada, N., & Balthazart, J. (1998). Distribution and effects of testosterone on aromatase mRNA in the quail forebrain: a non-radioactive in situ hybridization study. *J Chem Neuroanat, 14,* 103–115.

Baldwin, F. M., Goldin, H. S., & Metfessel, M. (1940). Effects of testosterone propoinate on female roller canaries under complete song isolation. *Proc Soc Exp Biol Med, 44,* 373–375.

Ball, G. F., & Balthazart, J. (2002). Neuroendocrine mechanisms regulating reproductive cycles and reproductive behavior in birds. In: Pfaff, D. W., Arnold, A. P., Etgen, A. M., Fahrbach, S. E., & Rubin, R. T. (eds.), *Hormones, Brain and Behavior.* San Diego: Academic Press, pp. 649–798.

Ball, G. F., & Balthazart, J. (2008). How useful is the appetitive and consummatory distinction for our understanding of the neuroendocrine control of sexual behavior? *Horm Behav, 53,* 307–311; author reply 315–308.

Ball, G. F., & Balthazart, J. (2009). Neuroendocrine regulation of reproductive behavior in birds. In: Pfaff, D. W., Arnold, A. P., Etgen, A. M., Fahrbach, S. E., & Rubin, R. T. (eds.), *Hormones, Brain and Behavior.* San Diego: Academic Press, pp. 855–895.

Ball, G. F., & Balthazart, J. (2011). Japanese quail as a model system for studying the neuroendocrine control of reproductive and social behaviors. *ILAR J, 51,* 310–325.

Ball, G. F., Riters, L. V., & Balthazart, J. (2002). Neuroendocrinology of song behavior and avian brain plasticity: multiple sites of action of sex steroid hormones. *Front Neuroendocrinol, 23,* 137–178.

Balthazart, J. (1983). Hormonal corrleates of behavior. In: Farner, D. S., King, J. R., & Parkes, K. C. (eds.), *Avian Biology.* New York: Academic Press, pp. 221–365.

Balthazart, J., Absil, P., Foidart, A., Houbart, M., Harada, N., & Ball, G. F. (1996). Distribution of aromatase-immunoreactive cells in the forebrain of zebra finches (*Taeniopygia guttata*): implications for the neural action of steroids and nuclear definition in the avian hypothalamus. *J Neurobiol, 31,* 129–148.

Balthazart, J., Absil, P., Gérard, M., Appeltants, D., & Ball, G. F. (1998a). Appetitive and consummatory male sexual behavior in Japanese quail are differentially regulated by subregions of the preoptic medial nucleus. *J Neurosci, 18,* 6512–6527.

Balthazart, J., Baillien, M., Cornil, C. A., & Ball, G. F. (2004). Preoptic aromatase modulates male sexual behavior: slow and fast mechanisms of action. *Physiol Behav, 83,* 247–270.

Balthazart, J., & Ball, G. F. (1998). New insights into the regulation and function of brain estrogen synthase (aromatase). *TINS, 21,* 243–249.

Balthazart, J., & Ball, G. F. (2006). Is brain estradiol a hormone or a neurotransmitter? *Trends Neurosci, 29,* 241–249.

Balthazart, J., & Ball, G. F. (2007). Topography in the preoptic region: differential regulation of appetitive and consummatory male sexual behaviors. *Front Neuroendocrinol, 28,* 161–178.

Balthazart, J., Castagna, C., & Ball, G. F. (1997). Aromatase inhibition blocks the activation and sexual differentiation of appetitive male sexual behavior in Japanese quail. *Behav Neurosci, 111,* 381–397.

Balthazart, J., Cornil, C. A., Charlier, T. D., Taziaux, M., & Ball, G. F. (2009a). Estradiol, a key endocrine signal in the sexual differentiation and activation of reproductive behavior in quail. *J Exp Zool A Ecol Genet Physiol, 311,* 323–345.

Balthazart, J., Evrard, L., & Surlemont, C. (1990a). Effects of the non-steroidal aromatase inhibitor, R76713 on testosterone-induced sexual behavior in the Japanese quail (*Coturnix coturnix japonica*). *Horm Behav, 24,* 510–531.

Balthazart, J., & Foidart, A. (1993). Brain aromatase and the control of male sexual behavior. *J Steroid Biochem Mol Biol, 44,* 521–540.

Balthazart, J., Foidart, A., & Harada, N. (1990b). Immunocytochemical localization of aromatase in the brain. *Brain Res, 514,* 327–333.

Balthazart, J., Foidart, A., & Hendrick, J. C. (1990c). The induction by testosterone of aromatase activity in the preoptic area and activation of copulatory behavior. *Physiol Behav, 47,* 83–94.

Balthazart, J., Foidart, A., Houbart, M., Prins, G. S., & Ball, G. F. (1998b). Distribution of androgen receptor-immunoreactive cells in the quail forebrain and their relationship with aromatase immunoreactivity. *J Neurobiol, 35,* 323–340.

Balthazart, J., Foidart, A., Surlemont, C., & Harada, N. (1991). Neuroanatomical specificity in the co-localization of aromatase and estrogen receptors. *J Neurobiol, 22,* 143–157.

Balthazart, J., Foidart, A., Surlemont, C., Vockel, A., & Harada, N. (1990d). Distribution of aromatase in the brain of the Japanese quail, ring dove, and zebra finch: an immunocytochemical study. *J Comp Neurol, 301,* 276–288.

Balthazart, J., Foidart, A., Wilson, E. M., & Ball, G. F. (1992a). Immunocytochemical localization of androgen receptors in the male songbird and quail brain. *J Comp Neurol, 317,* 407–420.

Balthazart, J., Reid, J., Absil, P., Foidart, A., & Ball, G. F. (1995). Appetitive as well as consummatory aspects of male sexual behavior in quail are activated by androgens and estrogens. *Behav Neurosci, 109,* 485–501.

Balthazart, J., Schumacher, M., & Evrard, L. (1990e). Sex differences and steroid control of testosterone-metabolizing enzyme activity in the quail brain. *J Neuroendocrinol, 2,* 675–683.

Balthazart, J., Schumacher, M., & Malacarne, G. (1985). Interaction of androgens and estrogens in the control of sexual behavior in male Japanese quail. *Physiol Behav, 35,* 157–166.

Balthazart, J., & Surlemont, C. (1990a). Androgen and estrogen action in the preoptic area and activation of copulatory behavior in quail. *Physiol Behav, 48,* 599–609.

Balthazart, J., & Surlemont, C. (1990b). Copulatory behavior is controlled by the sexually dimorphic nucleus of the quail POA. *Brain Res Bull, 25,* 7–14.

Balthazart, J., Surlemont, C., & Harada, N. (1992b). Aromatase as a cellular marker of testosterone action in the preoptic area. *Physiol Behav, 51,* 395–409.

Balthazart, J., Taziaux, M., Holloway, K., Ball, G. F., & Cornil, C. A. (2009b). Behavioral effects of brain-derived estrogens in birds. *Ann N Y Acad Sci, 1163,* 31–48.

Baulieu, E. E. (1997). Neurosteroids: of the nervous system, by the nervous system, for the nervous system. *Rec Prog Horm Res, 52,* 1–32.

Beach, F. A. (1948). *Hormones and Behavior.* New York: Paul B. Hoeber.

Beach, F. A. (1956). Characteristics of masculine "sex drive." *Nebraska Symposium on Motivation, 4,* 1–32.

Beach, F. A., & Inman, N. G. (1965). Effects of castration and androgen replacement on mating in male quail. *Proc Natl Acad Sci U S A, 54,* 1426–1431.

Bernard, D. J., Bentley, G. E., Balthazart, J., Turek, F. W., & Ball, G. F. (1999). Androgen receptor, estrogen receptor alpha, and estrogen receptor beta show distinct patterns of expression in forebrain song control nuclei of European starlings. *Endocrinology, 140,* 4633–4643.

Bolhuis, J. J., & Gahr, M. (2006). Neural mechanisms of birdsong memory. *Nat Rev Neurosci, 7,* 347–357.

Bottjer, S. W., & Johnson, F. (1997). Circuits, hormones, and learning: vocal behavior in songbirds. *J Neurobiol, 33,* 602–618.

Brenowitz, E. A., & Lent, K. (2002). Act locally and think globally: intracerebral testosterone implants induce seasonal-like growth of adult avian song control circuits. *Proc Natl Acad Sci U S A, 99,* 12421–12426.

Brenowitz, E. A., Nalls, B., & Smith, G. T. (1991). Immunoreactive estrogen receptors in a song nucleus in female canaries. *Soc Neurosci Abst, 17,* 1052.

Callard, G. V. (1983). Androgen and estrogen action in the vertebrate brain. *Am Zool, 23,* 607–620.

Callard, G. V., Petro, Z., & Ryan, K. J. (1978). Phylogenetic distribution of aromatase and other androgen-converting enzymes in the central nervous system. *Endocrinology, 103,* 2283–2290.

Canick, J. A., Vaccaro, D. E., Livingston, E. M., Leeman, S. E., Ryan, K. J., & Fox, T. O. (1986). Localization of aromatase and 5 alpha-reductase to neuronal and non-neuronal cells in the fetal rat hypothalamus. *Brain Res, 372,* 277–282.

Catchpole, C. K., & Slater, P. J. B. (2008). *Bird Song. Biological themes and Variations.* Cambridge, UK: Cambridge University Press.

Cheng, M. F. (1992). For whom does the female dove coo? A case for the role of vocal self-stimulation. *Anim Behav, 43,* 1035–1044.

Cheng, M. F. (2003). Vocal self stimulation: from the ring dove story to emotion-based vocal communication. *Adv Study Behav, 33,* 309–353.

Cheng, M. F., & Lehrman, D. (1975). Gonadal hormone specificity in the sexual behavior of ring doves. *Psychoneuroendocrinology, 1,* 95–102.

Cornil, C. A., & Ball, G. F. (2010). Effects of social experience on subsequent sexual performance in naive male Japanese quail (Coturnix japonica). *Horm Behav, 57,* 515–522.

DeVoogd, T., & Nottebohm, F. (1981). Gonadal hormones induce dendritic growth in the adult avian brain. *Science, 214,* 202–204.

Diotel, N., Le Page, Y., Mouriec, K., Tong, S. K., Pellegrini, E., Vaillant, C., et al. (2010). Aromatase in the brain of teleost fish: expression, regulation and putative functions. *Front Neuroendocrinol, 31,* 172–192.

Domjan, M., & Hall, S. (1986a). Determinants of social proximity in Japanese quail (*Coturnix coturnix japonica*): male behavior. *J Comp Psychol, 100,* 59–67.

Domjan, M., & Hall, S. (1986b). Sexual dimorphism in the social proximity behavior of Japanese quail (Coturnix coturnix japonica). *J Comp Psychol, 100,* 68–71.

Foidart, A., Reid, J., Absil, P., Yoshimura, N., Harada, N., & Balthazart, J. (1995). Critical re-examination of the distribution of aromatase-immunoreactive cells in the quail forebrain using antibodies raised against human placental aromatase and against the recombinant quail, mouse or human enzyme. *J Chem Neuroanat, 8,* 267–282.

Foidart, A., Silverin, B., Baillien, M., Harada, N., & Balthazart, J. (1998). Neuroanatomical distribution and variations across the reproductive cycle of aromatase activity and aromatase-immunoreactive cells in the pied flycatcher (Ficedula hypoleuca). *Horm Behav, 33,* 180–196.

Fusani, L., Van't Hof, T., Hutchison, J. B., & Gahr, M. (2000). Seasonal expression of androgen receptors, estrogen receptors, and aromatase in the canary brain in relation to circulating androgens and estrogens. *J Neurobiol, 43,* 254–268.

Gahr, M., Geuttinger, H. R., & Kroodsma, D. E. (1993). Estrogen receptors in the avian brain: survey reveals general distribution and forebrain areas unique to songbirds. *J Comp Neurol, 327,* 112–122.

Gahr, M., & Metzdorf, R. (1999). The sexually dimorphic expression of androgen receptors in the song nucleus hyperstriatilis ventrale pars caudale of the zebra finch develops independently of gonadal steroids. *J Neurosci, 19,* 2628–2636.

Garcia-Segura, L. M., Wozniak, A., Azcoitia, I., Rodriguez, J. R., Hutchison, R. E., & Hutchison, J. B. (1999). Aromatase expression by astrocytes after brain injury: implications for local estrogen formation in brain repair. *Neuroscience, 89,* 567–578.

Goldman, S. A., & Nottebohm, F. (1983). Neuronal production, migration, and differentiation in a vocal control nucleus of the adult female canary brain. *Proc Natl Acad Sci U S A, 80,* 2390–2394.

Goodson, J., Saldanha, C., Hahn, T., & Soma, K. (2005). Recent advances in behavioral neuroendocrinology: insights from studies on birds. *Horm Behav, 37,* 31–39.

Goodson, J. L., & Adkins-Regan, E. (1997). Playback of crows of male Japanese quail elicits female phonotaxis. *The Condor, 99,* 990–993.

Gurney, M., & Konishi, M. (1980). Hormone-induced sexual differentiation of brain and behavior in zebra finches. *Science, 208,* 1380–1383.

Gurney, M. E. (1982). Behavioral correlates of sexual differentiation in the zebra finch song system. *Brain Res, 231,* 153–172.

Harada, N. (1988). Novel properties of human placental aromatase as cytochrome P-450: purification and characterization of a unique form of aromatase. *J Biochem, 103,* 106–113.

Harding, C. F. (1981). Social modulation of circulating hormone levels in the male. *Am Zool, 21,* 223–231.

Harding, C. F., Sheridan, K., & Walters, M. J. (1983). Hormonal specificity and activation of sexual behavior in male zebra finches. *Horm Behav, 17,* 111–133.

Hidalgo, A., Barami, K., Iversen, K., & Goldman, S. A. (1995). Estrogens and non-estrogenic ovarian influences combine to promote the recruitment and decrease the turnover of new neurons in the adult female canary brain. *J Neurobiol, 27,* 470–487.

Hinde, R. A. (1965) Interaction of internal and external factors in integration of canary reproduction. In: Beach, F. A. (ed.), *Sex and Behavior.* New York: Wiley, pp. 381–415.

Holloway, C. C., & Clayton, D. E. (2001). Estrogen synthesis in the male brain triggers development of the avian song control pathway in vitro. *Nat Neurosci, 4,* 170–175.

Hutchison, J. B. (1971). Effects of hypothalamic implants of gonadal steroids on courtship behaviour in Barbary doves (*Streptopelia risoria*). *J Endocrinol, 50,* 97–113.

Hutchison, J. B. (1978a). Hypothalamic regulation of male sexual responsiveness to androgen. In: Hutchison, J. B. (ed.), *Biological Determinants of Sexual Behaviour.* Chichester, UK: Wiley, pp. 277–319.

Hutchison, J. B., & Steimer, T. H. (1986). Formation of behaviorally effective 17beta-estradiol in the dove brain: steroid control of preoptic aromatase. *Endocrinology, 118,* 2180–2187.

Hutchison, R. E. (1978b). Hormonal differentiation of sexual behavior in Japanese quail. *Horm Behav, 11,* 363–387.

Konishi, M., Emlen, S., Ricklefs, R., & Wingfield, J. C. (1989). Contributions of bird studies to biology. *Science, 246,* 465–472.

Lea, R. W., & Armstrong, D. G. (1986). Changes in aromatase activity in the brain of the male ring dove (Streptopelia risoria) during the breeding cycle. *Comp Biochem Physiol, 4,* 693–698.

Lehrman, D. S. (1965). Interaction between internal and external environments in the regulation of the reproductive cycle of the ring dove. In: Beach, F. A. (ed.), *Sex and Behavior.* New York: Wiley, pp. 355–380.

London, S., Monks, D. A., Wade, J., & Schlinger, B. A. (2006). Widespread capacity for steroid synthesis in the avian brain and song system. *Endocrinology, 147,* 5975–5987.

London, S. E., Remage-Healey, L., & Schlinger, B. A. (2009). Neurosteroid production in the songbird brain: a re-evaluation of core principles. *Front Neuroendocrinol, 30,* 302–314.

Mak, P., Zenn, R., & Callard, G. V. (1985). Subcellular distribution and reaction kinetics of aromatase in goldfish brain. *Proc 67th Ann Meet Endocrine Soc, Baltimore,* 265.

Marler, P., Peters, S., Ball, G. F., Dufty, A. M., & Wingfield, J. C. (1988). The role of sex steroids in the acquisition and production of birdsong. *Nature, 336,* 770–772.

Meitzen, J., Moore, I. T., Lent, K., Brenowitz, E. A., & Perkel, D. J. (2007). Steroid hormones act transsynaptically within the Forebrain to regulate neuronal phenotype and song stereotypy. *J Neurosci, 27,* 12045–12057.

Metzdorf, R., Gahr, M., & Fusani, L. (1999). Distribution of aromatase, estrogen receptor, and androgen receptor mRNA in the forebrain of songbirds and nonsongbirds. *J Comp Neurol, 407,* 115–129.

Naftolin, F., Horvath, T. L., Jakab, R. L., Leranth, C., Harada, N., & Balthazart, J. (1996). Aromatase immunoreactivity in axon terminals of the vertebrate brain. An immunocytochemical study on quail, rat, monkey and human tissues. *Neuroendocrinology, 63,* 149–155.

Naftolin, F., Ryan, K. J., Davies, I. J., Reddy, V. V., Flores, F., Petro, Z., et al. (1975). The formation of estrogens by central neuroendocrine tissues. *Rec Prog Horm Res, 31,* 295–319.

Naftolin, F., Ryan, K. J., & Petro, Z. (1971). Aromatization of androstenedione by limbic system tissue from human foetuses. *J Endocrinol, 51,* 795–796.

Nottebohm, F. (1969). The critical period for song learning. *Ibis, 111,* 386–387.

Nottebohm, F. (1981). A brain for all seasons: cyclical anatomical changes in song control nuclei of the canary brain. *Science, 214,* 1368–1370.

Nottebohm, F. & Arnold, A. P. (1976). Sexual dimorphism in vocal control areas of the songbird brain. *Science, 194,* 211–213.

Nottebohm, F., Stokes, T. M., & Leonard, C. M. (1976). Central control of song in the canary, Serinus canarius. *J Comp Neurol, 165,* 457–486.

Oberlander, J. G., Schlinger, B. A., Clayton, N. S., & Saldanha, C. J. (2004). Neural aromatization accelerates the acquisition of spatial memory via an influence on the songbird hippocampus. *Horm Behav, 45,* 250–258.

Palkovits, M. (1973). Isolated removal of hypothalamic or other brain nuclei of the rat. *Brain Res, 59,* 449–450.

Panzica, G. C., Viglietti-Panzica, C., Calcagni, M., Anselmetti, G. C., Schumacher, M., & Balthazart, J. (1987). Sexual differentiation and hormonal control of the sexually dimorphic preoptic medial nucleus in quail. *Brain Res, 416,* 59–68.

Paton, J. A., & Nottebohm, F. N. (1984). Neurons generated in the adult brain are recruited into functional circuits. *Science, 225,* 1046–1048.

Peterson, R. S., Saldanha, C. J., & Schlinger, B. A. (2001). Rapid upregulation of aromatase mRNA and protein following neural injury in the zebra finch (*Taeniopygia guttata*). *J Neuroendocrinol, 13,* 317–323.

Peterson, R. S., Yarram, L., Schlinger, B. A., & Saldanha, C. J. (2005). Aromatase is pre-synaptic and sexually dimorphic in the adult zebra finch brain. *Proc R Soc B-Biol Sci, 272,* 2089–2096.

Pfaff, D. W. (1968). Autoradiographic localization of radioactivity in rat brain after injection of tritiated sex hormones. *Science, 161,* 1355–1356.

Pfaus, J. G. (1996). Frank A. Beach Award—Homologies of animal and human sexual behaviors. *Horm Behav, 30,* 187–200.

Prove, E. (1974). Der Einfluss von Kastration und Testosteronsubstitution auf das sexualverhalten mannlicher Zabrafinken (*Taeniophygia guttata castanotis* Gould). *J Ornithol, 115,* 338–347.

Ramachandran, B., Schlinger, B. A., Arnold, A. P., & Campagnoni, A. T. (1999). Zebra finch aromatase gene expression is regulated in the brain through an alternate promoter. *Gene, 240,* 209–216.

Remage-Healey, L., Coleman, M. J., Oyama, R. K., & Schlinger, B. A. (2010). Brain estrogens rapidly strengthen auditory encoding and guide song preference in a songbird. *Proc Natl Acad Sci U S A, 107,* 3852–3857.

Remage-Healey, L., Maidment N. T., & Schlinger B.A. 2008. Forebrain steroid levels fluctuate rapidly during social interactions. *Nat Neurosci, 11*:1327-1334.

Riters, L., Baillien, M., Eens, M., Pinxten, R., Foidart, A., Ball, G., et al. (2001). Seasonal variation in androgen-metabolizing enzymes in the diencephalon and telencephalon of the male European starling (*Sturnus vulgaris*). *J Neuroendocrinol, 13,* 985–997.

Roselli, C. E., Horton, L. E., & Resko, J. A. (1987). Time-course and steroid specificity of aromatase induction in rat hypothalamus-preoptic area. *Biol Reprod, 37,* 628–633.

Roselli, C. E. & Resko, J. A. (1984). Androgens regulate brain aromatase activity in adult male rats through a receptor mechanism. *Endocrinology, 114,* 2183–2189.

Sachs, B. D. (1969). Photoperiodic control of reproductive behavior and physiology of the male Japanese quail (*Coturnix coturnix japonica*). *Horm Behav, 1,* 7–24.

Sachs, B. D. (2007). A contextual definition of male sexual arousal. *Horm Behav, 51,* 569–578.

Saldanha, C. J., Clayton, N. S., & Schlinger, B. A. (1999). Androgen metabolism in the juvenile oscine forebrain: a cross-species analysis at neural sites implicated in memory function. *J Neurobiol, 40,* 397–406.

Saldanha, C. J. & Coomaralingam, L. (2005). Overlap and co-expression of estrogen synthetic and responsive neurons in the songbird brain—a double-label immunocytochemical study. *Gen Comp Endocrinol, 141,* 66–75.

Saldanha, C. J., Popper, P., Micevych, P. E., & Schlinger, B. A. (1998). The passerine hippocampus is a site of high aromatase: inter- and intraspecies comparisons. *Horm Behav, 34,* 85–97.

Saldanha, C. J., & Schlinger, B. A. (1997). Estrogen synthesis and secretion in the brown-headed cowbird (*Molothrus ater*). *Gen Comp Endocrinol, 105,* 390–401.

Saldanha, C. J., Tuerk, M. J., Kim, Y. H., Fernandes, A. O., Arnold, A. P., & Schlinger, B. A. (2000). Distribution and regulation of telencephalic aromatase expression in the zebra finch revealed with a specific antibody. *J Comp Neurol, 423,* 619–630.

Schlinger, B. A., Amur-Umarjee, S., Shen, P., Campagnoni, A. T., & Arnold, A. P. (1994). Neuronal and non-neuronal aromatase in primary cultures of developing zebra finch telencephalon. *J Neurosci, 14,* 7541–7552.

Schlinger, B. A., & Arnold, A. P. (1991). Brain is the major site of estrogen synthesis in a male songbird. *Proc Natl Acad Sci U S A, 88,* 4191–4194.

Schlinger, B. A., & Arnold, A. P. (1992). Circulating estrogens in a male songbird originate in the brain. *Proc Natl Acad Sci U S A, 89,* 7650–7653.

Schlinger, B. A., & Arnold, A. P. (1993). Estrogen synthesis in vivo in the adult zebra finch: additional evidence that circulating estrogens can originate in brain. *Endocrinology, 133,* 2610–2616.
Schlinger, B. A., & Callard, G. V. (1987). A comparison of aromatase, 5alpha-, and 5 beta- reductase activities in the brain and pituitary of male and female quail (C. c. japonica). *J Exp Zool, 242,* 171–180.
Schlinger, B. A., & Callard, G. V. (1989). Localization of aromatase in synaptosomal and microsomal subfractions of quail (Coturnix coturnix japonica) brain. *Neuroendocrinology, 49,* 434–441.
Schlinger, B. A., & Callard, G. V. (1990). Aromatization mediates aggressive behavior in quail. *Gen Comp Endocrinol, 79,* 39–53.
Schlinger, B. A., & Saldanha, C. J. (2006). Songbirds: a novel perspective on estrogens and the aging brain. *Age, 27,* 287–296.
Schlinger, B. A. & Brenowitz, E. A. (2009). Neuroendocrine regulation of reproductive behavior in birds. In: Pfaff , D. W., Arnold, A. P., Etgen, A. M., Fahrbach, S. E., & Rubin, R. T. (eds.), *Hormones, Brain and Behavior.* San Diego: Academic Press, pp. 897–941.
Schumacher, M., Alexandre, C., & Balthazart, J. (1987). Interactions des androgènes et des oestrogènes dans le contrôle de la reproduction. *CR Acad Sci Paris, Sér III, 305,* 569–574.
Schumacher, M., & Balthazart, J. (1983). The effects of testosterone and its metabolites on sexual behavior and morphology in male and female Japanese quail. *Physiol Behav, 30,* 335–339.
Schumacher, M., & Balthazart, J. (1986). Testosterone-induced brain aromatase is sexually dimorphic. *Brain Res, 370,* 285–293.
Schumacher, M., & Balthazart, J. (1987). Neuroanatomical distribution of testosterone-metabolizing enzymes in the Japanese quail. *Brain Res, 422,* 137–148.
Seiwert, C. M., & Adkins-Regan, E. (1998). The foam production system of the male Japanese quail: characterization of structure and function. *Brain Behav Evol, 52,* 61–80.
Sharp, P. J., Armstrong, D. G., & Moss, R. (1986). Changes in aromatase activity in the neuroendocrine tissues of red grouse (Lagopus lagopus scoticus) in relation to the development of long-day refractoriness. *J Endocrinol, 108,* 129–135.
Shen, P., Campagnoni, C. W., Kampf, K., Schlinger, B. A., Arnold, A. P., & Campagnoni, A. T. (1994). Isolation and characterization of a zebra finch aromatase cDNA: in situ hybridization reveals high aromatase expression in brain. *Mol Brain Res, 24,* 227–237.
Shen, P., Schlinger, B. A., Campagnoni, A. T., & Arnold, A. P. (1995). An atlas of aromatase mRNA expression in the zebra finch brain. *J Comp Neurol, 360,* 172–184.
Shoemaker, H. H. (1939). Effect of testosterone propionate on the behavior of the female canary. *Proc Soc Exp Biol Med, 41,* 299–302.
Silverin, B., Baillien, M., Foidart, A., & Balthazart, J. (2000). Distribution of aromatase activity in the brain and peripheral tissues of passerine and nonpasserine avian species. *Gen Comp Endocrinol, 117,* 34–53.
Soma, K. K., Alday, N. A., Hau, M., & Schlinger, B. A. (2004a). Dehydroepiandrosterone metabolism by 3 beta-hydroxysteroid dehydrogenase/ Delta 5-Delta 4 isomerase in adult zebra finch brain: Sex difference and rapid effect of stress. *Endocrinology, 145,* 1668–1677.

Soma, K. K,. Bindra, R. K., Gee, J., Wingfield, J., & Schlinger, B. A. (1999b). Androgen metabolizing enzymes show region-specific changes across the breeding season in the brain of a wild breeding song bird. *J Neurobiol, 41,* 176–188.

Soma, K. K., Schlinger, B. A., Wingfield, J. C., & Saldanha, C. J. (2003). Brain aromatase, 5 alpha-reductase, and 5 beta-reductase change seasonally in wild male song sparrows: relationship to aggressive and sexual behavior. *J Neurobiol, 56,* 209–221.

Soma, K. K., Sullivan, K., & Wingfield, J. (1999a). Combined aromatase inhibitor and antiandrogen treatment decreases territorial aggression in a wild songbird during the nonbreeding season. *Gen Comp Endocrinol, 115,* 442–453.

Soma, K. K., Tramontin, A. D., Featherstone, J., & Brenowitz, E. A. (2004b). Estrogen contributes to seasonal plasticity of the adult avian song control system. *J Neurobiol, 58,* 413–422.

Soma, K. K., & Wingfield, J. C. (2001). Dehydroepiandrosterone in songbird plasma: seasonal regulation and relationship to territorial aggression. *Gen Comp Endocrinol, 123,* 144–155.

Spence, R. D., Zhen, Y., White, S., Schlinger, B. A., & Day, L. B. (2009). Recovery of motor and cognitive function after cerebellar lesions in a songbird: role of estrogens. *Eur J Neurosci, 29,* 1225–1234.

Steimer, T., & Hutchison, J. B. (1980) Aromatization of testosterone within a discrete hypothalamic area associated with the behavioral action of androgen in the male dove. *Brain Res, 192,* 586–591.

Steimer, T., & Hutchison, J. B. (1981). Androgen increases formation of behaviourally effective oestrogen in dove brain. *Nature, 292,* 345–347.

Tam, H., & Schlinger, B. A. (2007). Activities of 3beta-HSD and aromatase in slices of developing and adult zebra finch brain. *Gen Comp Endocrinol, 150,* 26–33.

Taziaux, M., Cornil, C. A., & Balthazart, J. (2003). Testosterone aromatization modulates the expression of sexually-motivated cloacal gland movements in male quail. *Abst Winter Meeting Assoc Study Anim Behav(London), 83,* 329–346.

Taziaux, M., Cornil, C. A., Dejace, C., Arckens, L., Ball, G. F., & Balthazart, J. (2006). Neuroanatomical specificity in the expression of the immediate early gene c-fos following expression of appetitive and consummatory male sexual behaviour in Japanese quail. *Eur J Neurosci, 23,* 1869–1887.

Thompson, C. K., Bentley, G. E., & Brenowitz, E. A. (2007). Rapid seasonal-like regression of the adult avian song control system. *Proc Natl Acad Sci U S A, 104,* 15520–15525.

Thompson, R. R., Goodson, J. L., Ruscio, M. G., & Adkins-Regan, E. (1998). Role of the archistriatal nucleus taeniae in the sexual behavior of male Japanese quail (Coturnix japonica): a comparison of function with the medial nucleus of the amygdala in mammals. *Brain Behav Evol, 51,* 215–229.

Tramontin, A. D., & Brenowitz, E. A. (2000). Seasonal plasticity in the adult brain. *Trends Neurosci, 23,* 251–258.

Tramontin, A. D., Wingfield, J. C., & Brenowitz, E. A. (2003). Androgens and estrogens induce seasonal-like growth of song nuclei in the adult songbird brain. *J Neurobiol, 57,* 130–140.

Tremere, L. A., Jeong, J. K., & Pinaud, R. (2009). Estradiol shapes auditory processing in the adult brain by regulating inhibitory transmission and plasticity-associated gene expression. *J Neurosci, 29,* 5949–5963.

Tremere, L. A., & Pinaud, R. (2011). Brain-generated estradiol drives long-term optimization of auditory coding to enhance the discrimination of communication signals. *J Neurosci, 31,* 3271–3289.

Tsutsui, K., Matsunaga, M., Miyabara, H., & Ukena, K. (2006). Neurosteroid biosynthesis in the quail brain: a review. *J Exp Zool, 305A,* 733–742.

Vanson, A., Arnold, A. P., & Schlinger, B. A. (1996). 3 beta-hydroxysteroid dehydrogenase/isomerase and aromatase activity in primary cultures of developing zebra finch telencephalon: dehydroepiandrosterone as substrate for synthesis of androstenedione and estrogens. *Gen Comp Endocrinol, 102,* 342–350.

Viglietti-Panzica, C., Panzica, G. C., Fiori, M. G., Calcagni, M., Anselmetti, G. C., & Balthazart, J. (1986). A sexually dimorphic nucleus in the quail preoptic area. *Neurosci Lett, 64,* 129–134.

Vockel, A., Pröve, E., & Balthazart, J. (1990a). Effects of castration and testosterone treatment on the activity of testosterone-metabolizing enzymes in the brain of male and female zebra finches. *J Neurobiol, 21,* 808–825.

Vockel, A., Pröve, E., & Balthazart, J. (1990b). Sex- and age-related differences in the activity of testosterone-metabolizing enzymes in microdissected nuclei of the zebra finch brain. *Brain Res, 511,* 291–302.

Voigt, C., Ball, G. F., & Balthazart, J. (2007). Neuroanatomical specificity of sex differences in expression of aromatase mRNA in the quail brain. *J Chem Neuroanat, 33,* 75–86.

Walters, M. J., & Harding, C. F. (1988). The effects of an aromatization inhibitor on the reproductive behavior of male zebra finches. *Horm Behav, 22,* 207–218.

Wingfield, J. C. (2005). Historical contributions of research on birds to behavioral neuroendocrinology. *Horm Behav, 48,* 395–402.

Wingfield, J. C., & Marler, P. (1988). Endocrine basis of communication in reproduction and aggression. In: Knobil, E. & Neill, J. D. (eds.), *The Physiology of Reproduction.* New York: Raven Press, pp. 1647–1677.

Wise, P. M. (2002). Estrogens and neuroprotection. *Trends Endocrinol Metab, 13,* 229–230.

Yahr, P. (1979). Data and hypotheses in tales of dihydrotestosterone. *Horm Behav, 13,* 92–96.

Yamamura, T., Barker, J. M., Balthazart, J., & Ball, G. F. (2011). Androgens and estrogens synergistically regulate the expression of doublecortin and enhance neuronal recruitment in the song system of adult female canaries. *J Neurosci, 31,* 9649–9657.

Zeigler, H. P., & Marler, P. (2008). *Neuroscience of Birdsong.* Cambridge, UK: Cambridge University Press.

Zigmond, R. E., Notteboh, F., & Pfaff, D. W. (1973). Androgen-concentrating cells in midbrain of a songbird. *Science, 179,* 1005–1007.

Zwain, I. H., Yen, S. S. C., & Cheng, C. Y. (1997). Astrocytes cultured in vitro produce estradiol-17 beta and express aromatase cytochrome P-450 (P-450 AROM) mRNA. *Biochim Biophys Acta, 1334,* 338–348.

Chapter 11

Brain Aromatase and Territorial Aggression Across the Seasons in Male Song Sparrows

Sarah A. Heimovics, H. Bobby Fokidis and Kiran K. Soma

INTRODUCTION

Estrogens regulate a variety of social behaviors, including appetitive and consummatory sexual behavior, social affiliation, and aggression. Much of our understanding of the roles of estrogens in social behavior comes from laboratory studies of domesticated mice and rats (Blumstein et al., 2010; Soma et al., 2008). In addition to this work, field studies of free-living animals offer the opportunity to investigate how estrogens regulate naturally occurring social behaviors within an ecologically relevant context. Such work is important because it has provided novel insights and perspectives into the estrogenic regulation of complex social behaviors and has been critical to developing a comprehensive understanding of the variety of contexts in which estrogens regulate social behavior. One such example is research on the role of 17β-estradiol (E_2) in regulating male song sparrow (*Melospiza melodia*) territorial aggression in different seasons.

Our understanding of songbird natural history, behavior, social patterning, and seasonal endocrine physiology is far greater than for most vertebrates. Song sparrows played a critical role in the formulation of principles about testosterone and aggression, such as the Challenge Hypothesis described by Wingfield et al. (1990). Further, the songbird brain expresses the estrogen-synthetic enzyme aromatase more widely and in higher abundance than nonsongbird species, including rodents (Forlano et al., 2006). Moreover, songbird behavior and adult neuroplasticity are highly estrogen sensitive and seasonally variable (Ball et al., 2004; Soma, 2006; Schlinger et al., 2008). Together, these factors make songbirds outstanding model systems for studying estrogenic regulation of complex social behaviors.

ECOLOGICAL, SOCIAL, AND BEHAVIORAL CONTEXT OF TERRITORIALITY

Song sparrows are common and have a widespread distribution throughout North America. Populations in northern and eastern North America are typically long distance migrants to avoid the severe winter conditions in these regions (Arcese et al., 2002). In contrast, western populations along the Pacific

Table 11–1. Seasonal Changes in Daylength, Temperature, and Precipitation Near Vancouver, BC (49°12′N, 123°01′W)

	Jan	Feb	Mar	Apr	May	Jun	Jul	Aug	Sep	Oct	Nov	Dec
Daylength (hr:min)*	9:18	10:31	11:53	13:26	14:43	15:25	15:07	13:59	12:30	11:02	9:40	8:57
Mean temperature (°C)	3.6	4.9	6.6	9.1	12.3	14.7	16.9	17.1	14.5	10.3	6.1	3.8
Max. temperature (°C)	6	7.6	9.5	12.3	15.7	18	20.4	20.5	17.9	13.1	8.5	6.1
Min. temperature (°C)	1.2	2.3	3.6	5.7	8.8	11.4	13.2	13.5	11.1	7.5	3.7	1.5
Rainfall (mm)	146.5	125.2	118.7	89	68.3	55.5	39.3	48.1	58.6	113.3	196.1	167.9

*Daylength for day in middle of each month.

1971–2010 Canadian Weather Service (Environment Canada), Delta, British Columbia, Canada.

coast experience less severe winter climates with relatively moderate temperatures and little to no persistent snow cover (Table 11–1) (Arcese et al., 2002). As a result, song sparrows endemic to this region typically do not migrate or flock, but rather are sedentary and exhibit year-round territoriality (Arcese, 1989; Wingfield & Monk, 1992). For the western populations, territorial behavior in the autumn and winter (nonbreeding season) plays an important role in overwinter survival (Smith et al., 1980; Rogers et al., 1991).

Despite the relatively moderate winter conditions, low nightly temperatures have large metabolic costs for these small songbirds (23–25 g body mass) with a high body temperature (41°C) (Rogers et al., 1991; Rogers, 1995). For example, in captive male western song sparrows exposed to a short photoperiod (8L:16D), subjects exposed to an 8°C overnight temperature exhibit greater declines in body mass and fat stores than subjects exposed to a 15°C overnight temperature (Fig. 11–1) (Soma & Wingfield, unpublished data). Subjects exposed to an 8°C overnight temperature lost nearly 9% of their body mass in one night, and nightly temperatures in winter are often below 8°C. Thus, a central function of nonbreeding territoriality is likely resource (food) defense. It is critical to secure access to food so as to meet the large energetic costs of over-wintering at northern latitudes (Wingfield & Soma, 2002).

Breeding territories are defended by a male–female pair. In contrast, at some locations, only 30% of nonbreeding territories are defended by a male–female pair (Wingfield & Soma, 2002). Nonbreeding territories can also be defended by single birds, same-sex pairs, or associations of three or more birds. The majority of nonbreeding territory holders are adults that previously bred, but

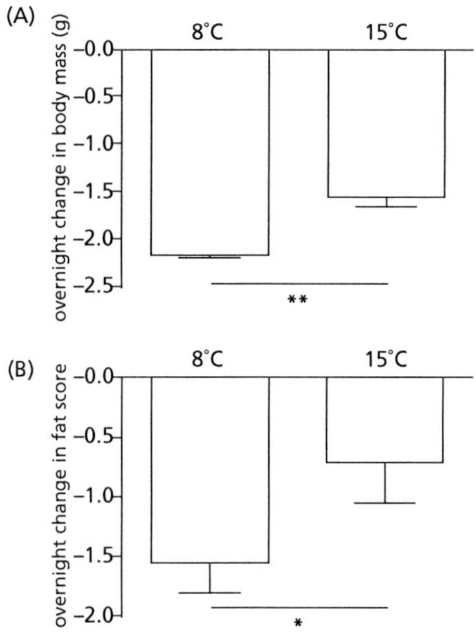

Figure 11–1. Overnight Change in (*A*) Body Mass and (*B*) Fat Score in Captive Adult Male Song Sparrows Housed at 8°C or 15°C. Subjects were in nonbreeding condition and were housed under a 8L:16D photoperiod. Abdominal and furcular fat reserves were each scored on a five-level visual fat index (Helms & Drury, 1960) and then these two scores were summed for each individual. Note that over a single night, subjects at 8°C lost approximately 8.9% of their body mass, and subjects at 15°C lost approximately 6.4% of their body mass. K. Soma and J. Wingfield (unpublished data). $^{*}p \leq .05$; $^{**}p \leq .01$.

many juveniles also establish territories in autumn (Arcese, 1987). In some cases, there is little spatial overlap between breeding and nonbreeding territories (Wingfield & Monk, 1992) (but see Arcese, 1987, 1989).

Song sparrows have been extensively used to study aggression during territorial disputes (Nowicki et al., 1998, 2002; Anderson et al., 2008a,b). When confronted with an intruder, a territorial male exhibits stereotypical vocal and display behaviors (Nice, 1943; Elekonich, 2000). These behaviors are easily elicited and quantified in the field in response to a simulated territorial intrusion (STI). An STI consists of placing a caged live decoy (or mount) in the approximate center of a territory along with tape-recorded conspecific song playback (Wingfield, 1985). The latency to respond, latency to sing, number of songs, number of flights (typically directed toward the decoy), time spent within 5 meters of the decoy, and closest approach to the decoy are then recorded. These behaviors are considered indicators of aggressiveness in this context. An STI typically lasts 10 to 30 minutes, depending on the experiment. Behavior can also be monitored after the STI (with the decoy removed and song playback turned off), to assess the persistence of the territorial response. This behavioral testing paradigm has provided much information about songbird territorial behavior and the roles of hormones, particularly steroids, in territorial aggression (Wingfield & Monk, 1992; Wingfield, 1994c).

In sedentary song sparrow populations, aggressive behavior during an STI is qualitatively and quantitatively similar in the breeding and nonbreeding seasons (Wingfield & Hahn, 1994; Mukai et al., 2009; Newman & Soma, 2011). *During* an STI, breeding and nonbreeding males exhibit similar levels of singing behavior, flights directed at the decoy, time spent within 5 meters, and closest approach to the decoy (Fig. 11–2) (Newman & Soma, 2011). However, *before* and *after* an STI, there are some seasonal differences in aggressive behavior. Breeding males exhibit higher baseline territorial singing behavior before an STI (see Fig. 11–2) and higher persistence of singing behavior after an STI

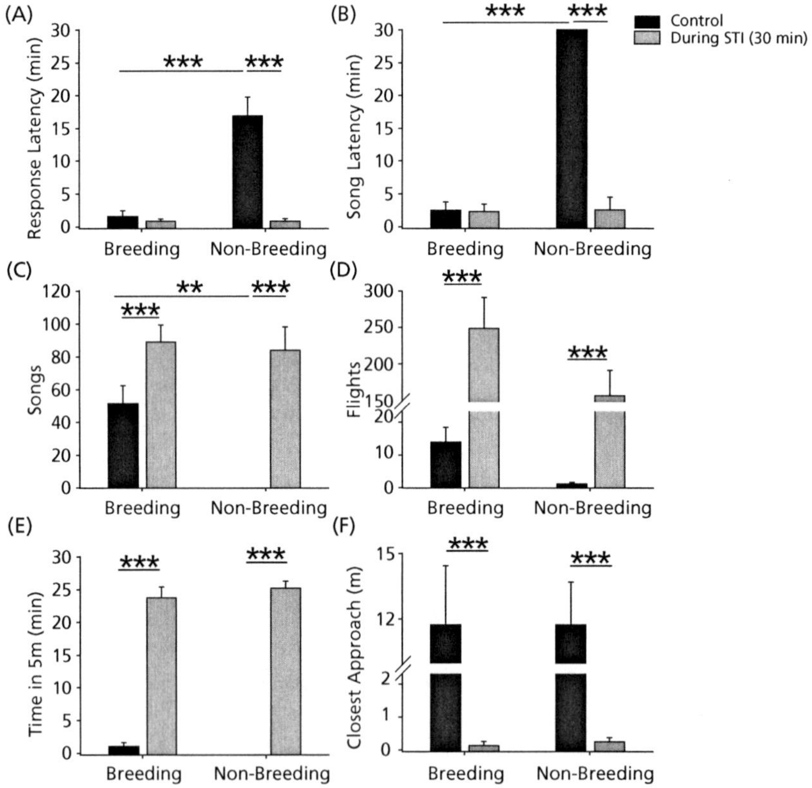

Figure 11–2. Effect of a Simulated Territorial Intrusion for 30 Minutes on Aggressive Responses of Wild Adult Male Song Sparrows During the Breeding and Nonbreeding Seasons. $^{**}p \leq .01$, $^{***}p \leq .0001$. Reprinted with permission from Newman, A. E. & Soma, K. K. (2011) Aggressive interactions differentially modulate local and systemic levels of corticosterone and DHEA in a wild songbird. *Horm Behav Sep, 60*(4), 389-396.

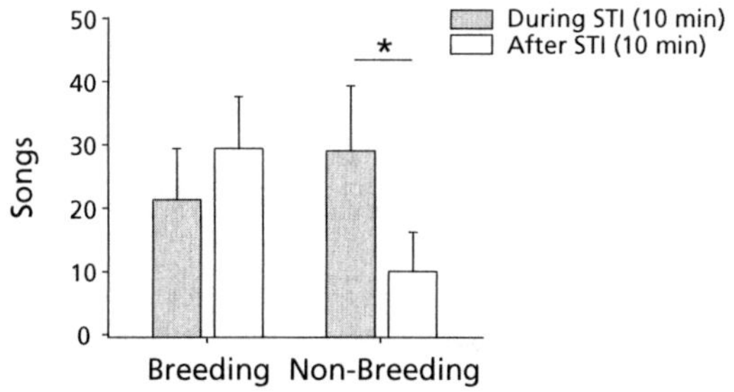

Figure 11–3. Number of Songs Given by Wild Adult Male Song Sparrows During (10 Minutes) and After (10 Minutes) a Simulated Territorial Intrusion in the Breeding and Nonbreeding Seasons. *$p < .05$. Redrawn from data published in Soma, K. K., Sullivan, K. A., Tramontin, A. D., Saldanha, C. J., Schlinger, B. A., & Wingfield, J. C. (2000a). Acute and chronic effects of an aromatase inhibitor on territorial aggression in breeding and nonbreeding male song sparrows. *J Comp Physiol A, 186*(7–8), 759–769; Soma, K. K., Tramontin, A. D., Wingfield, J. C. (2000). Oestrogen regulates male aggression in the non-breeding season. *Proc Biol Sci, 267*, 1089–1096.

is terminated (Fig. 11–3) (Wingfield, 1994b; Newman & Soma, 2011). One explanation for this seasonal plasticity in the persistence of aggression is that the function and context of aggressive behavior varies across different seasons. Breeding aggression primarily involves monopolizing access to a female mate to avoid cuckoldry, which may require continued vigilance even after an intruder has apparently disappeared (Wingfield, 1994b). Nonbreeding aggression primarily involves monopolizing food resources. Persistent nonbreeding aggression would be disadvantageous because it would reduce the amount of time available for foraging, which is already difficult because of shorter day lengths, inclement weather, and decreased food availability (Wingfield, 1994b).

Also, note that territorial aggression is dramatically reduced during the annual molt (August to September), when song sparrows lose their old feathers and grow new ones. Aggression might be reduced during the molt because agonistic encounters that escalate to fights might damage the fragile new feathers (Wingfield, 1994a). Alternatively, molting song sparrows cannot fly and evade predators very well, and they might reduce conspicuous behaviors (e.g., singing and other territorial behaviors) to avoid detection by predators.

SEASONAL CHANGES IN PHYSIOLOGY AND MORPHOLOGY

Although territorial aggression during an STI is similar in the breeding and nonbreeding seasons, the birds are in very different physiological conditions in each season. As with most songbirds living at temperate latitudes, the increasing daylengths typical of the early breeding season (March and April) (see Table 11–1) initiates gonadal recrudescence in song sparrows. In males, this increase in testes size is accompanied by elevated levels of circulating testosterone, which dramatically affect peripheral morphology, including decreased body mass and fat stores and increased cloacal protuberance length (used for sperm storage and delivery) (Table 11–2). After the breeding season ends, the gonads regress, circulating testosterone levels fall, body mass and fat stores increase, and the cloacal protuberance length decreases (see Table 11–2). Some of these seasonal changes in reproductive physiology (gonad size, plasma testosterone levels) are very pronounced, yet the territorial behavior remains similar.

Plasma testosterone levels also affect the brain, including various aspects of the chemical neuroanatomy of the song control system. This is a circuit of discrete brain nuclei dedicated to song learning and song production. For example, the volumes of HVC (abbreviation used as a proper name) and robust nucleus of the arcopallium (RA) are significantly smaller in nonbreeding than breeding male song sparrows (Smith et al., 1997). Both of these song nuclei contain androgen receptors, and HVC also expresses estrogen receptors (Bernard et al., 1999; Wacker et al., 2010). This is one of the best examples of naturally

Table 11–2. Seasonal Changes in Morphology and Plasma Steroid Hormones in Male Song Sparrows

	Breeding	Nonbreeding	Reference
Body mass (g)	23.32 ± 1.22	25.75 ± 1.86	Newman & Soma, unpublished
Fat score (furcular + abdominal)*	0.61 ± 0.18	5.22 ± 0.21	Newman & Soma (2011)
Left testis volume (mm^3)	285.2 ± 30.0	0.5 ± 0.3	Soma et al. (2003b)
Paired testes mass (g)	0.36 ± 0.02	0.01 ± 0.01	Tramontin & Brenowitz (1999)
Cloacal protuberance length (mm)	7.9 ± 0.3	3.5 ± 0.5	Soma et al. (2003b)
Plasma dehydroepiandrosterone (ng/ml)	0.80 ± 0.11	0.73 ± 0.09	Soma & Wingfield (2001)
Plasma androstenedione (ng/ml)	<0.1	< 0.1	Soma et al., unpublished results
Plasma testosterone (ng/ml)	4.18 ± 0.79	0.16 ± 0.01	Soma et al. (2003b)
Plasma estrone (ng/ml)	<0.1	<0.1	Soma et al., unpublished results
Plasma 17β-estradiol (ng/ml)	<0.1	<0.1	Soma & Wingfield (2001)

Sedentary song sparrow populations (WA, BC).

*A five-level visual fat index (Helms & Drury, 1960) was used to score furcular and abdominal fat which were then summed for each individual for analysis.

occurring adult neuroplasticity that is regulated by steroids (Tramontin & Brenowitz, 2000).

ESTROGEN EFFECTS ON SONG SPARROW BEHAVIOR AND BRAIN

Given that the testes are fully regressed and plasma sex steroid levels are basal in the nonbreeding season (Wingfield, 1994c; Soma & Wingfield, 2001; Soma et al., 2003), it was initially hypothesized that nonbreeding territorial aggression is not regulated by sex steroids in this species. Wingfield et al. made several additional observations in nonbreeding song sparrows in support of this idea, including: (1) plasma levels of testosterone do not rise in response to an STI during the nonbreeding season (Wingfield & Hahn, 1994); (2) removal of the regressed testes does not reduce aggression during an STI or impair the ability to maintain a nonbreeding territory (Wingfield, 1994b); (3) juvenile males with immature gonads successfully establish and defend territories in the nonbreeding season (Wingfield & Hahn, 1994; Nordby et al., 1999); and (4) increasing female sexual receptivity (via E_2 implants) does not affect sexual or aggressive behavior or plasma testosterone levels in nonbreeding males (Wingfield & Monk, 1994). Together, these data argued against a role for gonadal sex steroids in promoting nonbreeding aggression, especially because similar experimental manipulations in the breeding season do affect plasma testosterone levels and/or aggressive behavior (Runfeldt & Wingfield, 1985; Wingfield et al., 1989, 1990; Wingfield, 1994a).

Despite nondetectable concentrations of sex steroids in the general circulation during the nonbreeding season, the nonbreeding song sparrow brain expresses aromatase mRNA that is translated into active protein. Aromatase mRNA levels do not differ seasonally in brain areas strongly implicated in the regulation of aggression: nucleus taeniae (the avian homolog of the mammalian medial amygdala) and the ventromedial hypothalamus (Wacker et al., 2010). Also, aromatase activity in the ventromedial telencephalon (which contains nucleus taeniae) is high in both the breeding and nonbreeding season, but significantly reduced at molt, which parallels seasonal patterns of aggressive behavior (Soma et al., 2003) (Fig. 11–4). Furthermore, neither estrogen receptor alpha nor estrogen receptor beta mRNA levels differ across seasons in these brain regions (Wacker et al., 2010). Taken together, these studies raise the hypothesis that brain-generated estrogens are important for nonbreeding aggression, but the androgen substrate for brain aromatase might be of nongonadal origin during this season.

The first study to investigate the role of sex steroids in nonbreeding territorial aggression, treated free-living male song sparrows with the steroidal aromatase inhibitor 1,4,6-Androstatrien-3,17-dione (ATD) in combination with the androgen receptor antagonist flutamide (via subcutaneous Silastic implants) (Soma et al., 1999). Seven days of ATD + flutamide treatment had no effect on territorial aggression. However, 30 days of ATD + flutamide treatment significantly decreased aggressive behaviors both during and after STI.

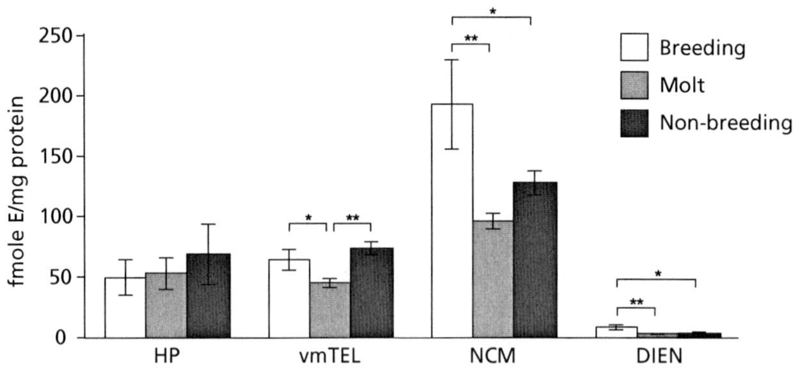

Figure 11–4. Seasonal Changes in Aromatase Activity in Wild Male Song Sparrow Brain. Free-ranging subjects were captured during the breeding season, molt, and the nonbreeding season. Aromatase activity was measured in hippocampus (HP), ventromedial telencephalon (vmTEL) including nucleus taeniae, caudomedial neostriatum (NCM), and diencephalon (DIEN). Brain tissue homogenates were incubated with tritiated androstenedione at 41°C, and the amounts of tritiated estrone and tritiated estradiol were measured. $^{*}p < .05$, $^{**}p < .01$. Reprinted with permission from Soma, K. K., Schlinger, B. A., Wingfield, J. C., & Saldanha, C. J. (2003). Brain aromatase, 5 alpha-reductase, and 5 beta-reductase change seasonally in wild male song sparrows: relationship to aggressive and sexual behavior. *J Neurobiol, 56*(3), 209–221.

There were no effects of treatment on body mass or fat score, suggesting that treated subjects were healthy and foraging normally. When administered alone, flutamide has no significant effect on nonbreeding aggression in song sparrows (Sperry et al., 2010) or another songbird species, the European robin (Schwabl & Kriner, 1991). Taken together, these data suggest that nongonadal sex steroids, particularly estrogens, regulate nonbreeding territorial aggression in male song sparrows.

Importantly, the preceding study employed ATD as an aromatase inhibitor, which is not as potent as other aromatase inhibitors (Soma et al., 1999). To demonstrate more clearly a role for estrogens in nonbreeding aggression, a series of follow-up studies examined the behavioral effects of a more potent and specific nonsteroidal aromatase inhibitor, fadrozole hydrochloride (FAD) (Wade et al., 1994). Fadrozole was administered acutely or chronically to free-living male song sparrows in both the breeding and nonbreeding seasons (Soma et al., 2000a,b). For acute treatment, fadrozole was administered via intramuscular injection, and behavior was assessed 24 hours later. For chronic treatment, fadrozole was administered via subcutaneous miniosmotic pump, and behavior was assessed approximately 10 days later. In the nonbreeding season, both acute and chronic fadrozole treatments significantly reduced multiple measures of territorial aggression *during* an STI (Figs. 11–5 and 11–6) (Soma et al.,

effects of an aromatase inhibitor on territorial aggression in breeding and nonbreeding male song sparrows. *J Comp Physiol A, 186*(7–8), 759–769.

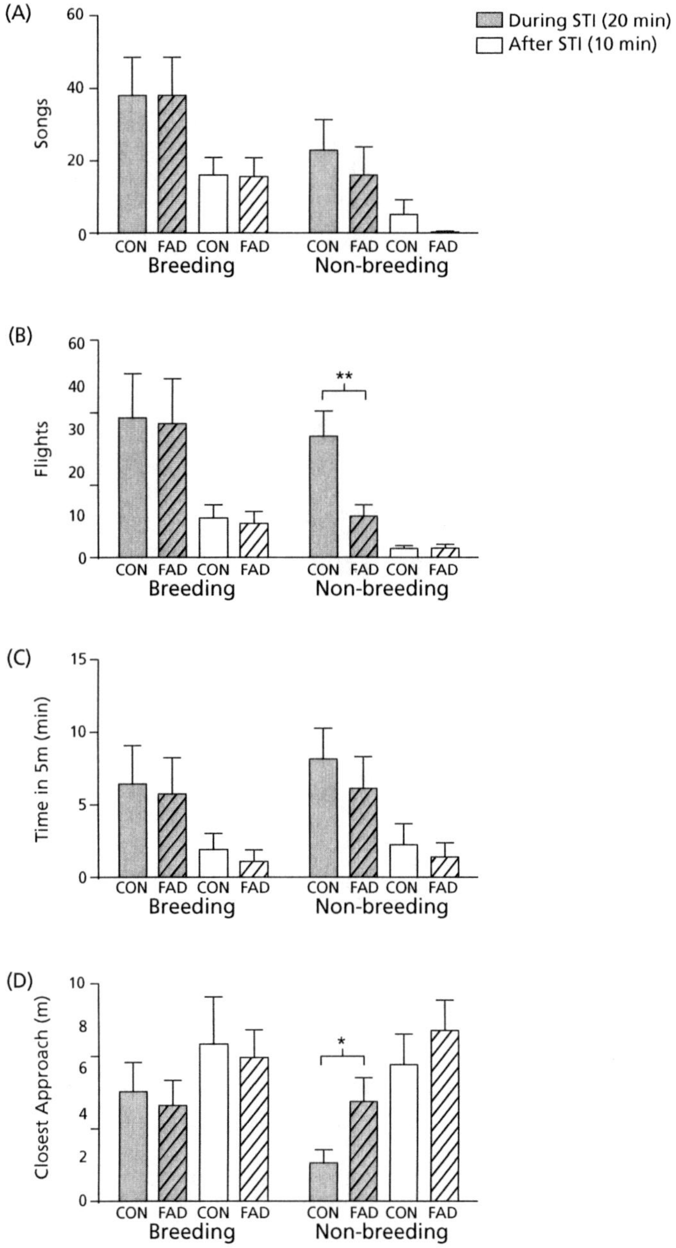

Figure 11–5. Acute Effects of the Aromatase Inhibitor Fadrozole or Vehicle on Territorial Behavior During (20 Minutes) and After (10 Minutes) a Simulated Territorial Intrusion. Subjects were wild adult male song sparrows during the breeding and non-breeding seasons. Subjects were tested 24 hours after FAD treatment (intramuscular injection). $*p < .05$, $**p < .01$. Redrawn from data published in Soma, K. K., Sullivan, K. A., Tramontin, A. D., Saldanha, C. J., Schlinger, B. A., & Wingfield, J. C. (2000a). chronic

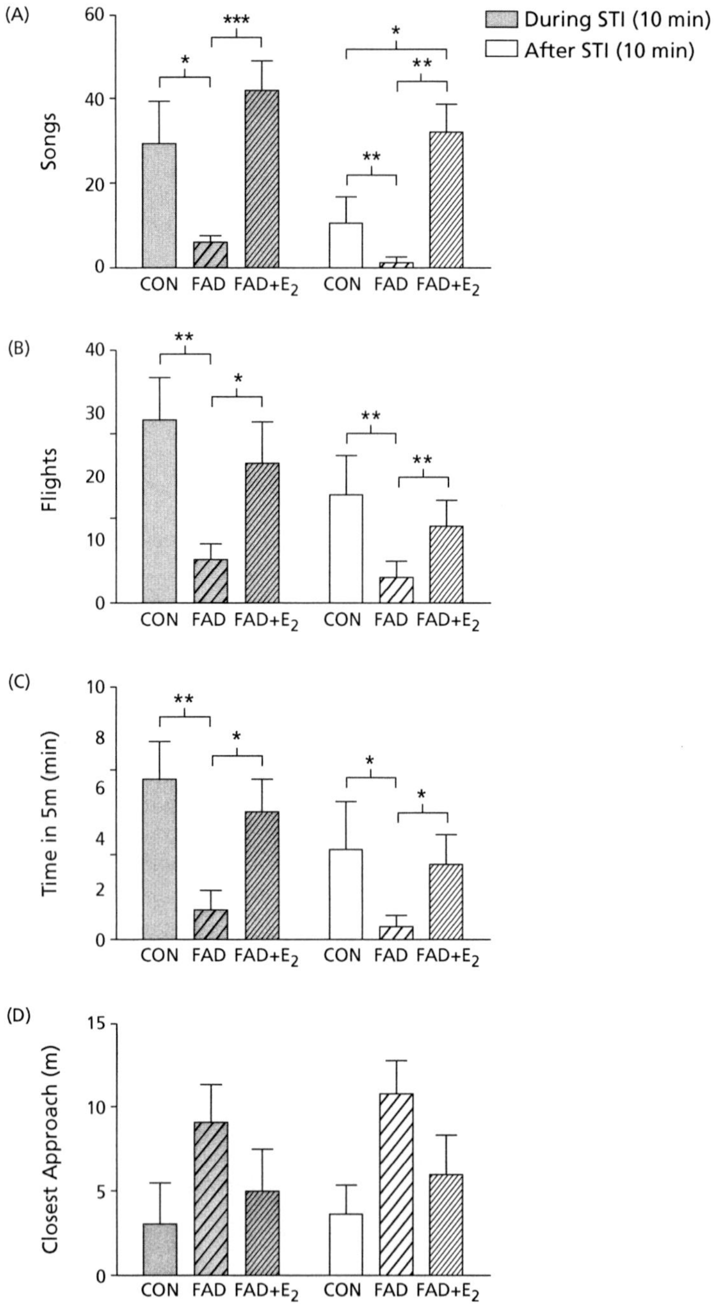

Figure 11–6. Chronic Effects of the Aromatase Inhibitor Fadrozole or Vehicle on Territorial Behavior During (10 Minutes) and After (10 Minutes) a Simulated Territorial Intrusion. Subjects were wild adult male song sparrows during the nonbreeding season.

2000a,b). Further, the effects of chronic fadrozole treatment on nonbreeding aggression were rescued by concurrent E_2 replacement (see Fig. 11–6) (Soma et al., 2000b). In contrast, neither acute nor chronic fadrozole treatment significantly affected territorial aggression in the breeding season (see Fig. 11–5) (Soma et al., 2000a). This lack of an effect of fadrozole on breeding aggression was likely owing to the fact that fadrozole causes a large increase in circulating testosterone levels in breeding subjects, which may have overcome the inhibition of aromatase activity and/or stimulated aggression via androgen receptors (Schlinger et al., 1999; Soma et al., 2000a). Fadrozole treatments did not significantly affect body mass or fat score, indicating that fadrozole did not reduce feeding behavior or debilitate subjects. Taken together, these studies show that estrogens are clearly involved in nonbreeding territorial aggression.

Interestingly, nonbreeding aggression *after* an STI was only affected by chronic fadrozole administration (Soma et al., 2000a,b) (see Figs. 11–5 and 11–6). Remarkably, E_2 replacement restored some features of post-STI aggression to control levels (flights and time in 5 meters), but elevated post-STI singing behavior above control levels to those typically seen during the breeding season (Soma et al., 2000a,b) (see Fig. 11–6). Taken together, these data suggest that persistence of territorial singing behavior after STI in the nonbreeding season may be tightly linked to long-lasting, genomic effects of E_2.

The effects of chronic fadrozole and E_2 treatments on the morphology of nuclei in the song control system were also investigated in nonbreeding song sparrows. Despite pronounced effects on territorial singing behavior, chronic fadrozole treatment did not significantly reduce HVC or RA volumes in nonbreeding song sparrows (Soma et al., 2004), perhaps because the song control system is already regressed at this time. In contrast, E_2 treatment significantly increased the volume of HVC to a size typical of the breeding season, indicating that E_2 can have profound neurotrophic and neuroanatomical effects during the nonbreeding season (Soma et al., 2004).

ANDROGEN SUBSTRATE FOR BRAIN AROMATASE IN THE NONBREEDING SEASON

These aromatase inhibitor studies clearly indicate that estrogens regulate nonbreeding territorial aggression as well as morphological aspects of behaviorally relevant brain regions. The androgen substrate for brain aromatase in the nonbreeding season was initially unclear, because circulating levels of testosterone and androstenedione are undetectable at this time (see Table 11–2). It was hypothesized that aromatizable androgens may be derived from the inactive

Subjects were tested approximately 10 to 12 days after FAD treatment (subcutaneous Alzet pump). $^*p < .05$, $^{**}p < .01$, $^{***}p < .001$. Redrawn from data published in Soma, K. K., Tramontin, A. D., & Wingfield, J. C. (2000b). Oestrogen regulates male aggression in the non-breeding season. *Proc Biol Sci, 267*(1448), 1089–1096.

prohormone dehydroepiandrosterone (DHEA). Dehydroepiandrosterone was measured in plasma from wild subjects using a specific radioimmunoassay, following dichloromethane extraction and Celite column chromatography (Soma & Wingfield, 2001). In the nonbreeding season, plasma levels of DHEA (0.8–1.75 ng/ml) were several times higher than plasma levels of testosterone and androstenedione. Moreover, the seasonal patterns of plasma DHEA levels and territorial behavior were similar: both are reduced during molt (Soma & Wingfield, 2001; Newman et al., 2008). The source of DHEA in the general circulation of nonbreeding birds may be either the regressed testes, the adrenals or the liver, which all have high DHEA levels (Soma & Wingfield, 2001; Newman & Soma, 2009, 2011). How DHEA levels in the general circulation are regulated remains unclear, as neither acute restraint stress nor GnRH administration affects DHEA levels in the brachial vein (Newman et al., 2008; Soma & Wingfield, 2001).

Next, wild nonbreeding male song sparrows were treated with DHEA to assess the effects on behavior and the brain (Soma et al., 2002). Importantly, this study used a dose of DHEA that elevated plasma DHEA levels within the physiological range. Chronic administration (>2 weeks) of DHEA (via subcutaneous Silastic implants) increased territorial singing behavior, including reduced song latency and increased persistence of singing after the STI. Surprisingly, other territorial behaviors (flights, closest approach, time in 5 meters) were not affected. Consistent with the effect on song, DHEA treatment also increased HVC volume by approximately 50% (to a size typical of the breeding season) (Soma et al., 2002; Newman et al., 2010). These effects of DHEA were similar to the chronic effects of E_2 on behavior (Soma et al., 2000b) and song system neuroanatomy (Soma et al., 2004) in the nonbreeding season. Thus, the behavioral and neuroanatomical effects of DHEA may occur through its conversion into active sex steroids, such as estradiol, within the brain.

Dehydroepiandrosterone cannot be directly aromatized by the aromatase enzyme and does not appear to bind to a classical intracellular steroid receptor with high affinity (Labrie et al., 2005). Rather, the enzyme 3β-hydroxysteroid dehydrogenase/isomerase (3β-HSD) converts DHEA into androstenedione, which can then be aromatized or converted by 17β-HSD into testosterone. 3β-HSD is abundant and active in the songbird brain, and 3β-HSD activity can be measured in vitro, using brain homogenates and tritiated DHEA as the substrate (Soma et al., 2004; Schlinger et al., 2008). Baseline 3β-HSD activity in brain areas implicated in song and aggression was assessed in wild male song sparrows across seasons. 3β-HSD activity was upregulated during the nonbreeding season in the central medial telencephalon (contains lateral septum and bed nucleus of the stria terminalis), ventromedial telencephalon (contains nucleus taeniae of the amygdala), and caudal telencephalon (contains RA) (Pradhan et al., 2010a). As mentioned, aromatase activity is also elevated in the ventromedial telencephalon during the nonbreeding season (Soma et al., 2003). Thus, the nonbreeding song sparrow brain can use circulating DHEA to produce high local levels of aromatizable androgens and estrogens to activate territorial aggression, without increasing sex steroid levels in the general circulation (Schmidt et al., 2008; Pradhan et al., 2010b) (Fig. 11–7).

To gain further insight into the role of DHEA metabolism in the regulation of nonbreeding aggression, the effect of an STI on brain 3β-HSD activity was examined using in vitro enzyme activity assays to quantify the conversion of tritiated DHEA to its androgenic and estrogenic metabolites. Free-living male song sparrows were exposed to an STI or control (empty cage and silence) for 30 minutes. The STI rapidly increased 3β-HSD activity in the central medial telencephalon and caudal telencephalon in nonbreeding song sparrows (Pradhan et al., 2010b). Further, 3β-HSD activity in these regions was positively correlated with the amount of time the subject spent near the decoy (Pradhan et al., 2010b). These data suggest that the local production of aromatizable androgens increases in response to an aggressive encounter in the nonbreeding season. Importantly, the effect of STI on 3β-HSD activity was only observed when exogenous NAD^+ (cofactor for 3β-HSD) was not added to the enzyme activity assay (Pradhan et al., 2010b). When exogenous NAD^+ was added to the assay (cofactors are typically added at saturating levels to maximize the reaction rate), no effect of STI on 3β-HSD activity was detected. The role of cofactors

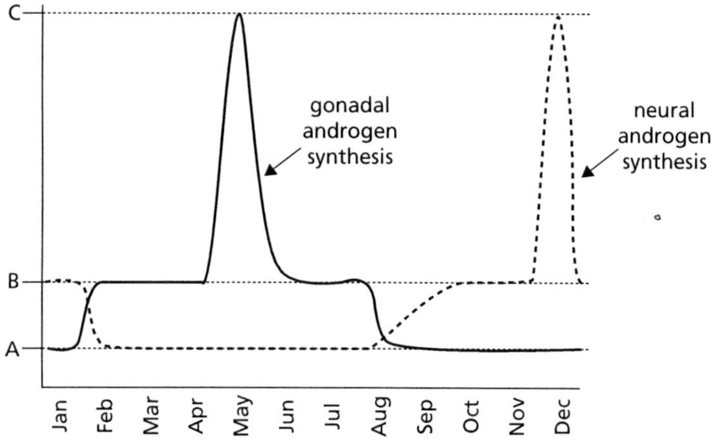

Figure 11–7. Hypothesized Patterns of Gonadal Androgen Synthesis (*Solid Line*) and Neural Androgen Synthesis (*Broken Line*) in Male Song Sparrows. Level A, constitutive levels of androgen synthesis for homeostatic functions; Level B, seasonal increase in androgen synthesis (regulated by predictable environmental cues); Level C, facultative and transient increase in androgen synthesis (regulated by unpredictable social cues). Reprinted with permission from Pradhan, D. S., Newman, A. E., Wacker, D. W., Wingfield, J. C., Schlinger, B. A., & Soma, K. K. (2010b). Aggressive interactions rapidly increase androgen synthesis in the brain during the non-breeding season. *Horm Behav,* *57*(4–5), 381–389.

in neurosteroid production is not well understood, but these data suggest that brain steroidogenic enzyme activity may be rapidly modulated by local bioavailability of cofactors. Alternatively, the social interaction might specifically affect 3β-HSD in a subcellular compartment that is NAD-rich (e.g., mitochondria) (Pradhan et al., 2010a,b).

Surprisingly, in this study STI had no effect on the absolute amounts of formed tritiated estrogens in the central medial telencephalon and caudal telencephalon (Pradhan et al., 2010b). However, STI did decrease the amount of formed tritiated estrogens relative to total 3β-HSD metabolites in these regions, indirectly suggesting that aromatase was inhibited by sustained aggressive behavior (Pradhan et al., 2010b). Research in Japanese quail has shown that dynamic social interactions lead to rapid and transient changes in brain aromatase activity. Specifically, brain aromatase activity in male quail is reduced after a 5-minute sexual interaction with a female, but is not significantly different from controls after 1- or 15-minute interactions (Cornil et al., 2005). Thus, in nonbreeding song sparrows, a reduction in brain aromatase activity after a 30-minute STI might reflect transient inhibition of aromatase activity. Given that sustained inhibition of aromatase activity inhibits nonbreeding aggression (Soma et al., 2000b), these data suggest that estrogenic regulation of aggressive behavior in the nonbreeding season may involve short pulses of local E_2 production. Future studies examining the effects of shorter STIs (e.g., 5 minutes) on 3β-HSD and aromatase activities and on androstenedione, testosterone, and E_2 levels in the brain are needed to test this hypothesis.

LOCAL VERSUS SYSTEMIC PRODUCTION OF STEROIDS

Although neural metabolism of DHEA appears to be important for nonbreeding aggression in song sparrows, the source of DHEA substrate for brain 3β-HSD remains enigmatic. During the nonbreeding season, levels of DHEA are high in the general circulation and are even higher in the regressed testes, adrenals and liver (Soma & Wingfield, 2001; Newman et al., 2008; Newman & Soma, 2009). Thus, DHEA may be secreted into the bloodstream by the gonads, adrenals, or liver and subsequently converted into active sex steroids in the brain (Labrie et al., 2005; Schlinger et al., 2008).

This hypothesis was tested by examining the effect of a 30-minute STI on circulating levels of DHEA. Simulated territorial intrusion had no effects on DHEA levels in the brachial plasma, gonads, or adrenals in the nonbreeding season (Soma & Wingfield, 2001; Newman & Soma, 2011) (Fig. 11–8). These data do not rule out the possibility of a peripheral source of DHEA substrate for brain 3β-HSD. However, STI did elevate DHEA levels in the jugular plasma (Newman & Soma, 2011) (see Fig. 11–8), which in songbirds is enriched with brain-derived steroids (Schlinger & Arnold, 1992; Taves et al., 2011). Taken together, these data suggest that in addition to being synthesized

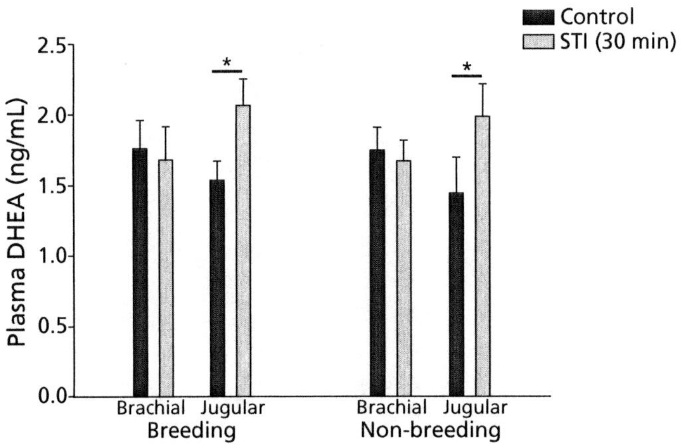

Figure 11–8. Effect of Simulated Territorial Intrusion for 30 Minutes on Brachial and Jugular Plasma Levels of Dehydroepiandrosterone During the Breeding Season and Nonbreeding Season $*p \leq .05$. Reprinted with permission from Newman, A. E., & Soma, K. K. (2011). Aggressive interactions differentially modulate local and systemic levels of corticosterone and DHEA in a wild songbird. *Horm Behav, 60*(4):389–396.

in the periphery, DHEA is synthesized de novo from cholesterol in the brain as a neurosteroid. Local production of DHEA is plausible given that the avian brain expresses the enzymes required for neurosteroid production, including cytochrome P450 side-chain cleavage (P450 scc) and cytochrome P450 17α-hydroxylase/17,20-lyase (P450c17) (Tsutsui et al., 2006; Schlinger & Remage-Healey, 2011).

Critically, these data suggest a shift from systemic to local sex steroid production and signaling in the brains of nonbreeding song sparrows (Soma, 2006; Schmidt et al., 2008) (Fig. 11–9). This shift may have evolved to avoid the costs of chronically elevated steroid hormones in the blood (Wingfield et al., 2001) (Fig. 11–10). For example, systemic testosterone treatment in the nonbreeding season suppresses cell-mediated and humoral immune responses in song sparrows (Owen-Ashley et al., 2004). Elevated circulating testosterone also decreases body mass and fat stores (Ketterson et al., 1991; Wikelski et al., 1999), which are essential for surviving snow storms and the low overnight temperatures during the nonbreeding season (see Fig. 11–1). Systemic testosterone also stimulates the development of secondary sexual characteristics (e.g., the cloacal protuberance) and activates reproductive behaviors, which are energetically expensive and inappropriate in the nonbreeding season. From an evolutionary perspective, the estrogenic regulation of male nonbreeding aggression may have become dissociated from systemic steroid signaling to reduce the exposure of peripheral tissues and other brain areas to testosterone and E_2. Importantly, this insight was made possible only by field studies of wild animals in their natural habitat.

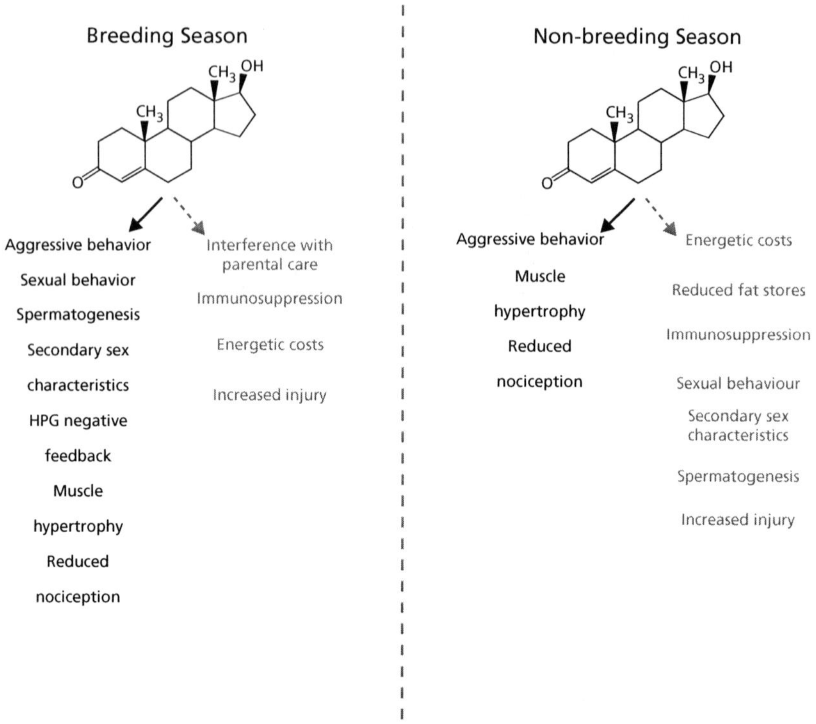

Figure 11–9. Actions and Costs of High Circulating Testosterone in the Breeding and Nonbreeding Season. Actions are in *black*, and potential costs are in *gray*. In the breeding season, a major cost of high circulating testosterone is interference with parental care. In the nonbreeding season, interference with parental care is not an issue, but a major cost of high circulating testosterone is increased energy expenditure. Modified from Soma, K. K. (2006). Testosterone and aggression: Berthold, birds and beyond. *J Neuroendocrinol, 18*(7), 543–551.

CONCLUSIONS

Local production of steroids within the brain occurs across vertebrate species, including rodents and humans (Schlinger et al., 2001; Balthazart et al., 2006; Tsutsui et al., 2006; Taves et al., 2011). The data reviewed in this chapter suggest the intriguing hypothesis that local steroid synthesis may be especially important when circulating steroid levels are low, such as during the nonbreeding season, and perhaps also during early development, postmenopause, and various endocrine disorders. Thus, understanding the roles of locally synthesized steroids such as E_2 has important and broad implications for neuroendocrinology, in which the focus historically has been on steroids that are in the general circulation. We still know relatively little about how neurosteroids influence physiology and behavior, and a comparative approach will be critical

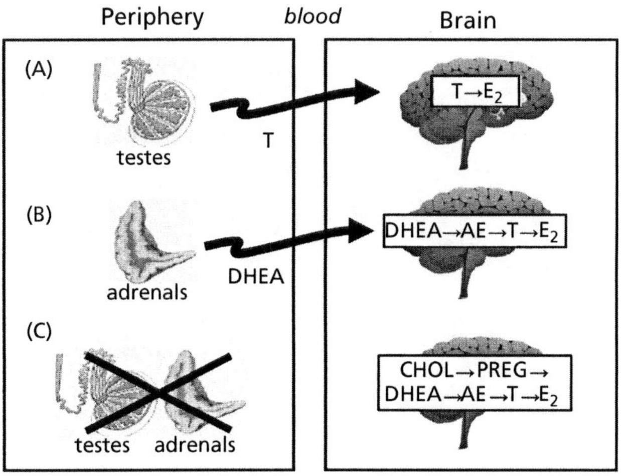

Figure 11–10. Potential Pathways for 17β-Estradiol Synthesis. A. Testosterone (T) can be synthesized in the gonads, released into the blood, travel to the brain, and be converted locally into E_2. B. The circulating prohormone dehydroepiandrosterone (DHEA) can be synthesized in the adrenals, released into the blood, travel to the brain, and be converted locally first into the aromatizable androgens T and androstenedione (AE) and then into E_2. C. 17β-Estradiol can be locally synthesized de novo from cholesterol in the brain. Modified from Schmidt, K. L., Pradhan, D. S., Shah, A. H., Charlier, T. D., Chin, E. H., & Soma, K. K. (2008). Neurosteroids, immunosteroids, and the Balkanization of endocrinology. *Gen Comp Endocrinol, 157*(3), 266–274.

for understanding the behavioral and functional consequences of neurosteroid production.

Future studies should further investigate neurosteroid production during times when circulating steroid levels are low. As one example, an important gap in our knowledge is the effect of STI on brain levels of testosterone, androstenedione, and E_2 and how these levels compare with plasma steroid levels. Local synthesis of steroids would be suggested if steroid levels were higher in brain tissue than in the circulation after STI. As a second example, the effects of STI on hypothalamic gene expression in the breeding and nonbreeding season were recently characterized by microarray (Mukai et al., 2009), but future research should also examine the effects of E_2 on gene expression in the songbird brain. Such studies have been greatly facilitated by the recent sequencing of the zebra finch genome and microarrays designed for use with songbird brain tissue (Replogle et al., 2008). Finally, it has been hypothesized that local steroid signals are more likely than systemic steroid signals to act via rapid, nongenomic mechanisms (Woolley, 2007; Schmidt et al., 2008; Heimovics et al., 2012). Thus, another important avenue of future research will be to explore the cellular and molecular mechanisms mediating the effects of E_2 on nonbreeding aggression.

ACKNOWLEDGMENTS

The authors gratefully acknowledge the support of grants from the Canadian Institutes of Health Research and the Michael Smith Foundation for Health Research to SAH and KKS, and from the Natural Sciences and Engineering Research Council of Canada to HBF. The authors also thank John Wingfield, Eliot Brenowitz, Barney Schlinger, Colin Saldanha, Scott MacDougall-Shackleton, Kim Sullivan, Michaela Hau, Danika Painter, Amy Newman, and Devaleena Pradhan for their assistance.

REFERENCES

Anderson, R. C., Searcy, W. A., & Nowicki, S. (2008a). Testing the function of song-matching in birds: responses of eastern male song sparrows Melospiza melodia to partial song-matching. *Behaviour, 145,* 347–363.

Anderson, R. C., Searcy, W. A., Peters, S., & Nowicki, S. (2008b). Soft song in song sparrows: Acoustic structure and implications for signal function. *Ethology,* 114(7), 662–676.

Arcese, P. (1987). Age, intrusion pressure and defense against floaters by territorial-male song sparrows. *Anim Behav, 35,* 773–784.

Arcese, P. (1989). Territory acquisition and loss in male song sparrows. *Anim Behav, 37,* 45–55.

Arcese, P., Sogge, M. K., Marr, A. B., & Patten, M. A. (2002). Song sparrow (*Melospiza melodia*). In: Poole, A. & Gill, F. (eds.), *The Birds of North America,* vol. 704 (pp. 1–40). Philadelphia: The Academy of Natural Sciences.

Ball, G. F., Auger, C. J., Bernard, D. J., Charlier, T. D., Sartor, J. J., Riters, L. V., et al. (2004). Seasonal plasticity in the song control system: multiple brain sites of steroid hormone action and the importance of variation in song behavior. *Ann N Y Acad Sci, 1016,* 586–610.

Balthazart, J., Cornil, C. A., Taziaux, M., Charlier, T. D., Baillien, M., & Ball, G. F. (2006). Rapid changes in production and behavioral action of estrogens. *Neuroscience, 138*(3), 783–791.

Bernard, D. J., Bentley, G. E., Balthazart, J., Turek, F. W., & Ball, G. F. (1999). Androgen receptor, estrogen receptor alpha, and estrogen receptor beta show distinct patterns of expression in forebrain song control nuclei of European starlings. *Endocrinology, 140*(10), 4633–4643.

Blumstein, D. T., Ebensperger, L. A., Hayes, L. D., Vasquez, R. A., Ahern, T. H., Burger, J. R., et al. (2010). Toward an integrative understanding of social behavior: new models and new opportunities. *Front Behav Neurosci, 4,* 34.

Cornil, C. A., Dalla, C., Papadopoulou-Daifoti, Z., Baillien, M., Dejace, C., Ball, G. F., et al. (2005). Rapid decreases in preoptic aromatase activity and brain monoamine concentrations after engaging in male sexual behavior. *Endocrinology, 146*(9), 3809–3820.

Elekonich, M. M. (2000). Female song sparrow, Melospiza melodia, response to simulated conspecific and heterospecific intrusion across three seasons. *Anim Behav, 59,* 551–557.

Forlano, P. M., Schlinger, B. A., & Bass, A. H. (2006). Brain aromatase: new lessons from non-mammalian model systems. *Front Neuroendocrinol, 27*(3), 247–274.

Heimovics, S. A., Prior, N. H., Maddison, C. J., Soma, K. K. (2012). Rapid and widespread effects of 17β-estradiol on intracellular signaling in the male songbird brain: a seasonal comparison. *Endocrinology, 153*(3):1364-1376.

Helms, C., & Drury, W. (1960). Winter and migratory weight and fat field studies on some north american buntings. *Bird Banding, 31*(1), 1–40.

Ketterson, E. D., Nolan, V., Wolf, L., Ziegenfus, C., Dufty, A. M., Ball, G. F., et al. (1991). Testosterone and avian life histories—the effect of experimentally elevated testosterone on corticosterone and body-mass in dark-eyed juncos. *Horm Behav, 25*(4), 489–503.

Labrie, F., Luu-The, V., Belanger, A., Lin, S. X., Simard, J., Pelletier, G., et al. (2005). Is dehydroepiandrosterone a hormone? *J Endocrinol, 187*(2), 169–196.

Mukai, M., Replogle, K., Drnevich, J., Wang, G., Wacker, D., Band, M., et al. (2009). Seasonal differences of gene expression profiles in song sparrow (*Melospiza melodia*) hypothalamus in relation to territorial aggression. *PLoS One, 4*(12), e8182.

Newman, A. E., MacDougall-Shackleton, S. A., An, Y. S., Kriengwatana, B., & Soma, K. K. (2010). Corticosterone and dehydroepiandrosterone have opposing effects on adult neuroplasticity in the avian song control system. *J Comp Neurol, 518*(18), 3662–3678.

Newman, A. E., Pradhan, D. S., & Soma, K. K. (2008). Dehydroepiandrosterone and corticosterone are regulated by season and acute stress in a wild songbird: jugular versus brachial plasma. *Endocrinology, 149*(5), 2537–2545.

Newman, A. E., & Soma, K. K. (2009). Corticosterone and dehydroepiandrosterone in songbird plasma and brain: effects of season and acute stress. *Eur J Neurosci, 29*(9), 1905–1914.

Newman, A. E., & Soma, K. K. (2011). Aggressive interactions differentially modulate local and systemic levels of corticosterone and DHEA in a wild songbird. *Horm Behav, 60*(4), 389–396.

Nice, M. M. (1943). Studies in the life history of the song sparrow. ii. the behavior of the song sparrow and other passerine birds. *Trans Linn Soc New York, 6,* 1–328.

Nordby, J. C., Campbell, S. E., & Beecher, M. D. (1999). Ecological correlates of song learning in song sparrows. *Behav Ecol, 10*(3), 287–297.

Nowicki, S., Searcy, W. A., & Hughes, M. (1998). The territory defense function of song in song sparrows: a test with the speaker occupation design. *Behaviour, 135,* 615–628.

Nowicki, S., Searcy, W. A., Krueger, T., & Hughes, M. (2002). Individual variation in response to simulated territorial challenge among territory-holding song sparrows. *J Avian Biol, 33*(3), 253–259.

Owen-Ashley, N. T., Hasselquist, D., & Wingfield, J. C. (2004). Androgens and the immunocompetence handicap hypothesis: Unraveling direct and indirect pathways of immunosuppression in song sparrows. *Am Natur, 164*(4), 490–505.

Pradhan, D. S., Lau, L. Y., Schmidt, K. L., & Soma, K. K. (2010a). 3beta-HSD in songbird brain: subcellular localization and rapid regulation by estradiol. *J Neurochem, 115*(3), 667–675.

Pradhan, D. S., Newman, A. E., Wacker, D. W., Wingfield, J. C., Schlinger, B. A., & Soma, K. K. (2010b). Aggressive interactions rapidly increase androgen synthesis in the brain during the non-breeding season. *Horm Behav, 57*(4–5), 381–389.

Replogle, K., Arnold, A. P., Ball, G. F., Band, M., Bensch, S., Brenowitz, E. A., et al. (2008). The Songbird Neurogenomics (SoNG) Initiative: community-based tools and strategies for study of brain gene function and evolution. *BMC Genomics, 9,* 131.

Rogers, C. M. (1995). Experimental-evidence for temperature-dependent winter lipid storage in the dark-eyed junco (junco-hyemalis-oreganus) and song sparrow (melospiza-melodia-morphna). *Physiol Zool, 68*(2), 277–289.
Rogers, C. M., Smith, J. N. M., Hochachka, W. M., Cassidy, A., Taitt, M. J., Arcese, P., et al. (1991). Spatial variation in winter survival of song sparrows Melospiza-melodia. *Ornis Scand, 22*(4), 387–395.
Runfeldt, S., & Wingfield, J. C. (1985). Experimentally prolonged sexual-activity in female sparrows delays termination of reproductive activity in their untreated mates. *Anim Behav, 33*(May), 403–410.
Schlinger, B. A., & Arnold, A. P. (1992). Circulating estrogens in a male songbird originate in the brain. *Proc Natl Acad Sci U S A, 89*(16), 7650–7653.
Schlinger, B. A., Lane, N. I., Grisham, W., & Thompson, L. (1999). Androgen synthesis in a songbird: a study of cyp17 (17alpha-hydroxylase/C17,20-lyase) activity in the zebra finch. *Gen Comp Endocrinol, 113*(1), 46–58.
Schlinger, B. A., Pradhan, D. S., & Soma, K. K. (2008). 3beta-HSD activates DHEA in the songbird brain. *Neurochem Int, 52*(4–5), 611–620.
Schlinger, B. A., & Remage-Healey, L. (2012). Neurosteroidogenesis: Insights from Studies of Songbirds. *J Neuroendocrinol, 24*(1):16–21.
Schlinger, B. A., Soma, K. K., & London, S. E. (2001). Neurosteroids and brain sexual differentiation. *Trends Neurosci, 24*(8), 429–431.
Schmidt, K. L., Pradhan, D. S., Shah, A. H., Charlier, T. D., Chin, E. H., & Soma, K. K. (2008). Neurosteroids, immunosteroids, and the Balkanization of endocrinology. *Gen Comp Endocrinol, 157*(3), 266–274.
Schwabl, H., & Kriner, E. (1991). Territorial aggression and song of male European robins (Erithacus rubecula) in autumn and spring: effects of antiandrogen treatment. *Horm Behav, 25*(2), 180–194.
Smith, G. T., Brenowitz, E. A., Beecher, M. D., & Wingfi eld, J. C. (1997). Seasonal changes in testosterone, neural attributes of song control nuclei, and song structure in wild songbirds. *J Neurosci, 17* (15), 6001–6010.
Smith, J. N. M., Montgomerie, R. D., Taitt, M. J., & Yomtov, Y. (1980). A winter feeding experiment on an island song sparrow population. *Oecologia, 47*(2), 164–170.
Soma, K. K. (2006). Testosterone and aggression: Berthold, birds and beyond. *J Neuroendocrinol, 18*(7), 543–551.
Soma, K. K., Alday, N. A., Hau, M., & Schlinger, B. A. (2004). Dehydroepiandrosterone metabolism by 3beta-hydroxysteroid dehydrogenase/Delta5-Delta4 isomerase in adult zebra finch brain: sex difference and rapid effect of stress. *Endocrinology, 145*(4), 1668–1677.
Soma, K. K., Schlinger, B. A., Wingfield, J. C., & Saldanha, C. J. (2003). Brain aromatase, 5 alpha-reductase, and 5 beta-reductase change seasonally in wild male song sparrows: relationship to aggressive and sexual behavior. *J Neurobiol, 56*(3), 209–221.
Soma, K. K., Scotti, M. A., Newman, A. E., Charlier, T. D., & Demas, G. E. (2008). Novel mechanisms for neuroendocrine regulation of aggression. *Front Neuroendocrinol, 29*(4), 476–489.
Soma, K. K., Sullivan, K., & Wingfield, J. (1999). Combined aromatase inhibitor and antiandrogen treatment decreases territorial aggression in a wild songbird during the nonbreeding season. *Gen Comp Endocrinol, 115*(3), 442–453.
Soma, K. K., Sullivan, K. A., Tramontin, A. D., Saldanha, C. J., Schlinger, B. A., & Wingfield, J. C. (2000a). Acute and chronic effects of an aromatase inhibitor on

territorial aggression in breeding and nonbreeding male song sparrows. *J Comp Physiol A, 186*(7–8), 759–769.

Soma, K. K., Tramontin, A. D., Featherstone, J., & Brenowitz, E. A. (2004). Estrogen contributes to seasonal plasticity of the adult avian song control system. *J Neurobiol, 58*(3), 413–422.

Soma, K. K., Tramontin, A. D., & Wingfield, J. C. (2000b). Oestrogen regulates male aggression in the non-breeding season. *Proc Biol Sci, 267*(1448), 1089–1096.

Soma, K. K., & Wingfield, J. C. (2001). Dehydroepiandrosterone in songbird plasma: seasonal regulation and relationship to territorial aggression. *Gen Comp Endocrinol, 123*(2), 144–155.

Soma, K. K., Wissman, A. M., Brenowitz, E. A., & Wingfield, J. C. (2002). Dehydroepiandrosterone (DHEA) increases territorial song and the size of an associated brain region in a male songbird. *Horm Behav, 41*(2), 203–212.

Sperry, T. S., Wacker, D. W., & Wingfield, J. C. (2010). The role of androgen receptors in regulating territorial aggression in male song sparrows. *Horm Behav, 57*(1), 86–95.

Taves, M. D., Gomez-Sanchez, C. E., & Soma, K. K. (2011). Extra-adrenal glucocorticoids and mineralocorticoids: evidence for local synthesis, regulation, and function. *Am J Physiol Endocrinol Metab, 301*(1), E11–E24.

Taves, M. D., Ma, C., Heimovics, S. A., Saldanha, C. J., & Soma, K. K. (2011). Measurement of steroid concentrations in brain tissue: methodological considerations. *Front Endocrinol, 2,* 1–13.

Tramontin, A. D., & Brenowitz, E. A. (1999). A field study of seasonal neuronal incorporation into the song control system of a songbird that lacks adult song learning.*J Neurobiol, 40,* 316 –326.

Tramontin, A. D., & Brenowitz, E. A. (2000). Seasonal plasticity in the adult brain. *Trends Neurosci, 23*(6), 251–258.

Tsutsui, K., Matsunaga, M., Miyabara, H., & Ukena, K. (2006). Neurosteroid biosynthesis in the quail brain: a review. *J Exp Zool A Comp Exp Biol, 305*(9), 733–742.

Wacker, D. W., Wingfield, J. C., Davis, J. E., & Meddle, S. L. (2010). Seasonal changes in aromatase and androgen receptor, but not estrogen receptor mRNA expression in the brain of the free-living male song sparrow, Melospiza melodia morphna. *J Comp Neurol, 518*(18), 3819–3835.

Wade, J., Schlinger, B. A., Hodges, L., & Arnold, A. P. (1994). Fadrozole: a potent and specific inhibitor of aromatase in the zebra finch brain. *Gen Comp Endocrinol, 94*(1), 53–61.

Wikelski, M., Lynn, S., Breuner, C., Wingfield, J. C., & Kenagy, G. J. (1999). Energy metabolism, testosterone and corticosterone in white-crowned sparrows. *J Comp Physiol, 185*(5), 463–470.

Wingfield, J. C. (1985). Short-term changes in plasma-levels of hormones during establishment and defense of a breeding territory in male song sparrows, Melospiza-melodia. *Horm Behav, 19*(2), 174–187.

Wingfield, J. C. (1994a). Control of territorial aggression in a changing environment. *Psychoneuroendocrinology, 19*(5–7), 709–721.

Wingfield, J. C. (1994b). Regulation of territorial behavior in the sedentary song sparrow, Melospiza-melodia morphna. *Horm Behav, 28*(1), 1–15.

Wingfield, J. C., & Hahn, T. P. (1994). Testosterone and territorial behavior in sedentary and migratory sparrows. *Anim Behav, 47*(1), 77–89.

Wingfield, J. C., Hegner, R. E., Dufty, A. M., & Ball, G. F. (1990). The challenge hypothesis—theoretical implications for patterns of testosterone secretion, mating systems, and breeding strategies. *Am Natur, 136*(6), 829–846.

Wingfield, J. C., Lynn, S., & Soma, K. K. (2001). Avoiding the "costs" of testosterone: ecological bases of hormone-behavior interactions. *Brain Behav Evol, 57*(5), 239–251.

Wingfield, J. C., & Monk, D. (1992). Control and context of year-round territorial aggression in the nonmigratory song sparrow Zonotrichia-melodia-morphna. *Ornis Scandinavica, 23*(3), 298–303.

Wingfield, J. C., & Monk, D. (1994). Behavioral and hormonal responses of male song sparrows to estradiol-treated females during the non-breeding season. *Horm Behav, 28*(2), 146–154.

Wingfield, J. C., Ronchi, E., Goldsmith, A. R., & Marler, C. (1989). Interactions of sex steroid-hormones and prolactin in male and female song sparrows, Melospiza-melodia. *Physiol Zool, 62*(1), 11–24.

Wingfield, J. C., & Soma, K. K. (2002). Spring and autumn territoriality in song sparrows: Same behavior, different mechanisms? *Int Comp Biol, 42*(1), 11–20.

Woolley, C. S. (2007). Acute effects of estrogen on neuronal physiology. *Annu Rev Pharmacol Toxicol, 47,* 657–690.

Chapter 12

Testosterone Metabolism in Reptiles

Rachel E. Cohen and Juli Wade

INTRODUCTION

Steroid hormones are critical to the control of reproductive behaviors across a wide variety of vertebrates (e.g., Baum & Vreeburg, 1973; Fabre-Nys & Martin, 1991; Fusani, 2008). These functions are commonly facilitated by the metabolism of these hormones locally within particular brain regions (e.g., Ball & Balthazart, 2002; Fabre-Nys & Martin, 1991; Fusani, 2008). Perhaps most important for the control of sexual behaviors is the metabolism of testosterone (T) into two other steroid hormones, 17β-estradiol (E2) and 5α-dihydrotestosterone (DHT) via the aromatase and 5α-reductase (5αR) enzymes, respectively (reviewed in Lephart et al., 2001). These molecules can act in both adult and developing animals to control the display of sexual behaviors. Hormonal effects that transiently elicit these displays in adults are classified as "activational," whereas those that induce permanent changes in the capacity to display sexually dimorphic behaviors during development are considered "organizational" (Arnold & Breedlove, 1985). Thus, T and its metabolites play pivotal roles in both the adult and developing brain.

The role of estrogen synthesis and the aromatase enzyme that catalyzes this reaction has been widely studied in birds and mammals (see other chapters in this volume). However, fewer studies exist on neural aromatase in reptiles. Its activity was first documented in the reptilian brain by Callard et al. (1977) in the painted turtle, *Chrysemys picta*, a native of North America. From this early study, many studies in reptiles have focused on the role of aromatase in gonadal differentiation (Pieau & Dorizzi, 2004; Lance, 2009; Nakamura 2010). The inhibition of aromatase activity during embryonic development results in testes formation in both a parthenogenetic lizard (triploid, XXX) and a turtle with temperature-dependent sexual determination (Wibbels & Crews, 1994).

Fewer studies have focused on aromatase in the reptilian nervous system. Activity of the enzyme has been detected in the developing brain of a few species, including alligators (*Alligator mississippiensis*) and red-eared slider turtles (*Trachemys scripta elegans*), and mRNA has been evaluated by RT-PCR in whole brains of diamondback terrapins (*Malaclemys terrapin*) (Jeyasuria & Place, 1997; Willingham et al., 2000; Milnes et al., 2002). In adults, neural aromatase mRNA has been examined in red-sided garter snakes (*Thamnophis sirtalis*) (Krohmer et al., 2002) and whiptail lizards (*Cnemidophorus uniparens*) (Dias et al., 2009), but the majority of data on reptilian neural aromatase has come from green anole lizards (*Anolis carolinensis*). 5α-reductase has not yet been examined in reptilian species other than the green anole. The remainder of this chapter will therefore focus on T metabolism in this species.

GREEN ANOLE REPRODUCTION

Green anole lizards are excellent models for the study of the control of reproductive behavior and have been studied since the 1940s, with their reproductive behaviors and neural anatomy well characterized. These lizards are found abundantly throughout the southeastern United States in forested habitats. They are sexually dimorphic such that, overall, males are larger than females. Males have a large, red throat fan called a dewlap (Fig. 12–1), which they use to court females and defend territories. Even controlling for body size, this structure is far smaller in females, who use it rarely (Lovern et al., 2004; Johnson et al., 2011). These lizards are seasonal breeders, and the breeding season lasts from approximately April through July (Wade, 2011). Females lay one egg (from alternating ovaries) about every 7 to 10 days and may mate multiple times throughout the season (Lovern & Wade, 2003). Developing anoles hatch approximately one month after oviposition. Prior to hatching, between embryonic days 16 and 24, gonads likely begin to produce T (Lovern & Wade, 2003). During embryonic

Figure 12–1. Male Green Anole Lizard. The dewlap is extended in the typical male courtship (shown here) and territorial display. Photo taken near St. Augustine, FL, by REC.

development, both sexes have bilateral copulatory organs (called hemipenes or hemipenises) and muscles associated with copulation; these completely regress in females by embryonic day 19 (Holmes & Wade, 2005). Circulating T is higher in male than in female juvenile anoles by approximately 30 days after hatching (Lovern et al., 2001). The cartilage and muscles that control dewlap become sexually dimorphic 2 to 3 months after hatching (O'Bryant & Wade, 2001). Juveniles become sexually mature by the next breeding season (Crews, 1980). Although juvenile anoles can likely breed by the next breeding season, recent evidence suggests that smaller (presumably younger) males may not be able to compete with larger males (Husak et al., 2007).

During the breeding season, adult male anoles defend large territories that overlap the territories of several females (Greenberg & Noble, 1944). Males have distinct courtship displays, which consist of extending the dewlap, headbobs, and pushups (Orrell & Jensson, 2003). If a female is sexually receptive, she will allow the male to mount, and he will grip her neck and insert one hemipenis into her cloaca. If the female is not receptive, she will run to escape (Greenberg & Noble, 1944).

Once the breeding season is over, gonads regress and these animals enter a refractory period during which time they do not display sexual behaviors, and their gonads cannot be stimulated to recrudesce (Licht, 1967b). Although it is unclear what causes this refractory period, once it is over (approximately 2 months after the end of the breeding season), both increased ambient temperatures and increased photoperiod stimulate gonad recrudescence (Licht, 1967a, 1969). Similar to other vertebrates, male green anoles have much higher levels of plasma T than females, and breeding males have approximately double the concentration of T compared with nonbreeding males (Lovern et al., 2001). Although females have less T than males, breeding females do have higher plasma T levels than do non-breeding females.

STEROID HORMONES INFLUENCE ANOLE REPRODUCTIVE BEHAVIOR

Castration inhibits the display of adult male sexual behaviors, and T activates them. Dihydrotestosterone can also activate sexual behaviors, but the effects are limited and inconsistent (Mason & Adkins, 1976; Adkins & Schlesinger, 1979; Crews, 1980; Rosen & Wade, 2000). Inhibition of 5αR in castrated males treated with T slightly diminishes courtship displays (Winkler & Wade, 1998; Rosen & Wade, 2000). Blocking aromatase activity (E2 synthesis) in castrated T-treated males does not reduce sexual behaviors (Winkler and Wade, 1998), although E2 administration in males increases motivation to copulate (mount attempts; Latham & Wade, 2010). Thus, sexual behaviors in adult male green anoles seem to be primarily regulated by T itself rather than its metabolites. However, both E2 and DHT play supporting roles (Wade, 2011). In addition, castrated males treated with T, E2, or DHT directly in the preoptic area (POA) can display normal sexual behavior, suggesting that direct neural application of high enough doses can facilitate male sexual behavior (Morgantaler & Crews, 1978; Crews & Morgentaler, 1979).

In females, E2 in the plasma is at moderate levels during periods of receptivity, but can increase during nonreceptive periods, perhaps to control oviposition and subsequent future ovulation (Jones et al., 1983). Receptivity in gonadectomized females is increased by E2, E2 plus progesterone, or T, and is decreased by antiestrogen treatments (Mason & Adkins, 1976; Crews, 1980; Tokarz & Crews, 1980). Additionally, inhibition of aromatase activity in T-treated gonadectomized females reduces receptivity, suggesting that neural E2 synthesis (either de novo in the brain, or from precursors supplied by the periphery) is likely important for this behavior (Winkler & Wade, 1998).

NEURAL AROMATASE AND 5α-REDUCTASE ACTIVITY

Activity of T-metabolizing enzymes can be readily determined by incubating homogenized brain tissue with 3H-T, separating androgens and estrogens with phenolic partition and thin layer chromatography, and measuring the resulting 3H-E2 and 3H-DHT produced by endogenous aromatase and 5αR (methods validated for anoles in Wade, 1997). Using pooled samples, Wade (1997) determined that aromatase activity is greatest in the hypothalamus and POA, whereas 5αR activity is highest in the brainstem. Additionally, the examination of enzyme activity in whole brains showed that breeding males have higher levels of neural aromatase activity than either breeding females or nonbreeding males, and aromatase activity is equivalent in males and females during the nonbreeding season (Rosen & Wade, 2001). In contrast, whole brain 5αR activity appears not to differ between the sexes or the seasons in gonadally intact adult green anoles.

Sex and seasonal differences in plasma T follow a similar pattern to the aromatase activity present in whole brains of breeding and nonbreeding anoles (see above). To test whether T actively regulates enzyme activity, we gonadectomized male and female lizards in both the breeding and nonbreeding seasons, treated them with T for one week, and determined the aromatase and 5αR activities present in the brain (Cohen & Wade, 2010b). We found that T treatment in males increases the activity of aromatase during the breeding season only, suggesting that aromatase is regulated by T specifically (Fig. 12–2). Testosterone treatment also increases 5αR activity in males regardless of season. No effect of treatment was detected on enzyme activity in females. The results suggest that the production of DHT and E2 in the brain are selectively controlled by T availability in a manner consistent with roles in male reproduction.

ANDROGEN METABOLIZING ENZYME EXPRESSION IN THE BRAIN

To determine the potential for aromatase and 5αR action in specific regions of the brain to influence behavior, we used in situ hybridization to evaluate where the enzymes are synthesized. The recent release of the green anole genome sequence (Alfoldi et al. 2011) facilitated the cloning of aromatase and two

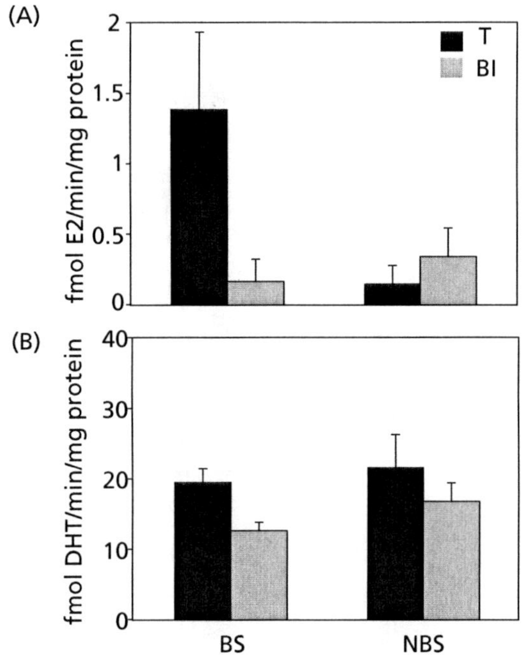

Figure 12–2. Aromatase (*A*) and 5αR (*B*) Activity in Gonadectomized and T- or Blank (Bl)-Treated Males. Aromatase activity was increased by T, but only in males and only during the breeding season (BS), not during the nonbreeding season (NBS). 5αR activity was increased by T regardless of season. Parallel effects were not detected in females. Modified from Cohen & Wade (2010b). Testosterone selectively affects aromatase and 5alpha-reductase activities in the green anole lizard. *Gen Comp Endocrinol.* 104: 437-445 © Elsevier 2010. Reprinted with permission.

isozymes of 5αR (5αR1 and 5αR2; Cohen & Wade, 2010a, 2011). In addition to identifying general locations of expression, we have begun to test the hypothesis that T regulates aromatase and 5αR expression in specific regions of the anole brain important for the display of reproductive behaviors. Three types of experiments addressed this idea. In adults, patterns of mRNA were analyzed in intact males and females in the breeding and nonbreeding seasons, as well as in T-manipulated animals across the two seasons. Sex differences were also evaluated at two stages of development—before and after males exhibit increased levels of plasma T compared to females.

Brain Regions Controlling Sexual Behavior in Anoles

As in other vertebrates, specific limbic brain regions are important for male- and female-specific reproductive behaviors in the green anole. The POA is important for green anole male sexual behaviors such that if this area is lesioned, these behaviors are absent (Wheeler & Crews, 1978). Similarly, if the ventromedial portion of the amygdala (AMY) is lesioned, courtship displays are decreased and neck grips are not achieved (Greenberg et al., 1984). Lesion studies have not been conducted in the ventromedial hypothalamus (VMH) in green anoles, but in other lizards, female sexual behaviors are absent after destruction of the VMH (Kendrick et al., 1995). Thus, these brain regions play important roles in anole reproductive behaviors and we hypothesized that aromatase and 5αR are expressed in the POA, AMY, and VMH in patterns consistent with their support of these functions.

Aromatase Expression in the Adult Brain: Intact Animals

Aromatase is expressed in specific regions throughout the lizard brain, and is similar to the distribution of the enzyme in red-sided garter snakes (Krohmer et al., 2002) and whiptail lizards (Dias et al., 2009). Aromatase is expressed in the reptilian telencephalon, as well as other parts of the brain. Its mRNA is present in all three reproductive nuclei we examined.

Using stereology, we estimated the total number and density of aromatase-positive cells in the three brain areas: POA, AMY, and VMH. Density was determined by dividing the number of aromatase-positive cells by the volume of the region. A male biased sex difference in the total number of cells expressing aromatase was detected in the POA, and female biased sex differences in the density of these cells was detected in both the AMY (Fig. 12–3) and VMH (Cohen & Wade, 2011). The biological function of the density of cells expressing aromatase mRNA may not be completely obvious, but it is possible that a greater density of aromatase-positive cells in specific regions could lead to increased local production of E2, much as a greater total number might across an entire region. Increased aromatase expression in the male POA suggests a potential for more E2 to be synthesized, and increased E2 production in the

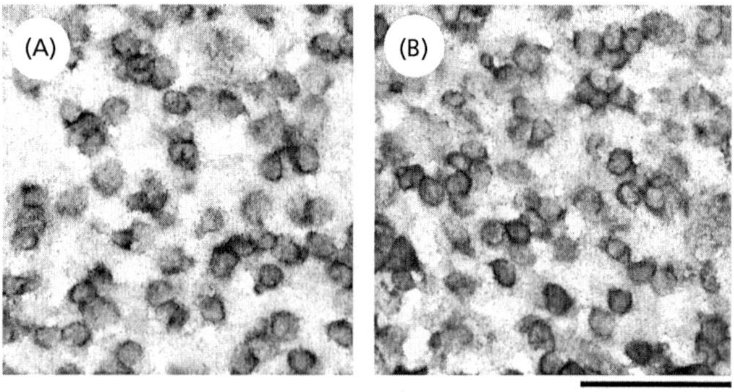

Figure 12–3. Aromatase mRNA Expression in the AMY Using Digoxigenin-Labeled in situ Hybridization. A male (*A*) and female (*B*) are depicted. Intact adult females have a greater density of aromatase expressing cells in this region than do males. Scale bar = 50 μm. From Cohen, R. E. & Wade, J. (2011). Aromatase mRNA in the brain of adult green anole lizards: effects of sex and season. *J Neuroendocrinol, 23*(3), 254–260.

male POA would be consistent with this hormone's facilitation of sexual motivation (Latham & Wade, 2010). These data are also consistent with those from other species including rats and birds, which show that many brain regions, including the POA, have higher aromatase activity and mRNA in males than in females (Roselli et al., 1985; Balthazart et al., 1990; Negri-Cesi et al., 1996; Wagner & Morrell, 1996).

Although the total number of cells in the anole AMY and VMH that express aromatase mRNA does not differ between the sexes, females have a higher density of aromatase positive cells in these regions than males, as calculated from the number of cells in which mRNA was detected divided by the volume of the region as defined by aromatase in situ hybridization. This idea is consistent with a role of aromatase in female receptivity in this species (Winkler & Wade, 1998). Although it is well established that the VMH regulates female sexual behaviors across vertebrates (La Vaque & Rodgers, 1975; Emery & Moss, 1984; Kendrick et al., 1995), the role of the AMY in female reproduction is less clear. This region is, however, important in sensory integration in general (Kostarczyk, 1986), which may allow the female anole to process male courtship signals.

The fact that sex differences in aromatase expression exist in the adult anole brain in regions associated with reproductive behavior is consistent with the hypothesis that T locally regulates enzyme availability, because males have higher levels of circulating plasma T than females (Lovern et al., 2001). However, if this mechanism is employed, it is region specific. For example, the greater overall number of aromatase expressing cells in the male compared with female POA could be caused by increased T, but this pattern was not detected in the VMH and AMY. The sex differences in aromatase expression that we detected could be controlled by T, such that the hormone activates expression in adulthood or organizes expression during development. To further examine this possibility, we measured aromatase expression during both development and in T- and blank-treated adults.

Aromatase Expression in the Developing Brain

To examine aromatase expression during development in the anole, we chose two ages: the day of hatching (P0), where circulating T is equivalent in the two sexes, and 50 days after hatching (P50), when males have higher levels of circulating T than females. We used in situ hybridization to quantify aromatase mRNA in the POA, AMY, and VMH (Cohen and Wade, 2012b).

Neural expression of this enzyme in development did not differ between the sexes or ages we examined, suggesting that T is not regulating mRNA distribution at those ages. It is possible that higher levels of T are required to produce a more robust effect on neural aromatase expression so that sex differences would be detected in older juvenile anoles (where the sex difference in circulating T is presumably larger). Testosterone treatment during development results in an increase in aromatase activity and the number of cells that contain the enzyme in other species, including mice and Japanese quail (Schumacher & Hutchison,

1986; Hutchison et al., 1997; Bardet et al., 2010). Thus, it is possible that increasing T during development also plays a role in anole aromatase expression, but if so, it largely happens at ages older than P50.

Aromatase Expression in the Adult Brain: Testosterone Manipulation

Using adult animals that were gonadectomized and treated with T in the breeding and nonbreeding seasons as in Cohen and Wade (2010b), and the same in situ hybridization procedures described above, we examined changes in expression of aromatase mRNA in the POA, AMY, and VMH. However, we did not detect effects of T treatment on the number or density of cells expressing aromatase mRNA in any of these three brain regions (Cohen and Wade, 2012a). Collectively, the data indicate that adult T upregulates whole brain aromatase activity, but this may not be apparent in cells expressing aromatase mRNA locally within the POA, AMY, or VMH. Work on Japanese quail is consistent with this pattern, as aromatase activity in the hypothalamus and POA is higher in males than females without sex differences in the number of cells that express aromatase protein or mRNA in similar regions (POA and bed nucleus of the stria terminalis [BNST]) (Voigt et al., 2007).

A few possible reasons exist for our observation of higher activity with increased T in breeding season males without an effect of T-treatment on the pattern of mRNA expression. For example, T simply may not regulate aromatase in the specific areas that were examined. Second, sexually dimorphic aromatase expression in the brain may be permanently organized during juvenile development and adult T is not necessary to maintain this pattern of expression. Third, T may increase aromatase activity without changing the number of detectable cells expressing aromatase mRNA. That is, the hormone may upregulate the amount per cell, which our procedure could not detect. Alternatively, the sensitivity of the technique may not have been great enough to allow visualization of cells that express very low levels of aromatase. It is also possible that the enzyme itself (either the amount of protein or its ability to react with substrate) may have been altered by hormone treatment without changes in mRNA. For example, increased translation of available mRNA, or posttranslational modifications of the aromatase protein are not detectable in an in situ hybridization analysis. Additionally, recent evidence has suggested that the availability of cofactors for the reaction can affect steroid metabolism (Pradhan et al., 2010). Calcium can also cause rapid changes in aromatase activity through phosphorylation (Charlier et al., 2011), an effect that differs between male and female Japanese quail (Konkle & Balthazart, 2011). Thus, many factors can lead to the modification of aromatase availability and activity, so the lack of an effect of T on mRNA as analyzed here does not necessarily mean that the hormone has no influence on the enzyme in the brain regions investigated. Further work should examine these possibilities in the anole brain as well. Collectively, the data from intact and treated adults suggest that aromatase is sexually dimorphic in brain areas controlling reproduction, although the number of cells expressing aromatase does not seem to be modulated by T. The time course of sexual differentiation of aromatase expression and the developmental role of T should be investigated more fully.

5α-Reductase Expression in the Brain

5α-reductase exists in two isozymes, 5αR1 and 5αR2 (Lephart et al., 2001). These isozymes have been primarily characterized in mammals, with no work conducted on them in reptiles. From data on humans, mice, and rats, it is clear that 5αR1 is expressed throughout the brain, whereas 5αR2 is found in very low concentrations in the adult brain and expression is greater overall for both isozymes in the brainstem (Celotti, Negri-Cesi, & Poletti, 1997). Isozyme 5αR2 is transiently expressed at high levels in the brain of developing rats, and this expression decreases to low adult levels (Poletti et al., 1998). Thus, these two isozymes may play different roles in the brain such that in mammals, 5αR2 may be important for development and 5αR1 may be important for physiological processes and maintenance throughout life.

We examined the distribution of both isozymes of 5αR in the adult anole brain using in situ hybridization (Cohen & Wade, 2010a). Anole 5αR1 had a higher sequence identity with sequences of 5αR1 from other species than with sequences for 5αR2, suggesting that anole 5αR1 is homologous to 5αR1 in other species. The same was true for 5αR2.

Isozyme 5αR2 is expressed in many regions of the adult anole brain, including portions of the forebrain, whereas 5αR1 is only expressed in the brainstem of adult lizards. Based on its expression pattern, 5αR2 may be important for either clearing excess T or maintaining local DHT production in adults. The pattern of expression of the two isozymes in anoles is opposite that seen in rodents, wherein 5αR1 is expressed in the adult brain and 5αR2 is expressed primarily during development (Melcangi et al., 1998; Poletti et al., 1998).

Similar to aromatase, we quantified expression of 5αR2 in the POA, AMY, and VMH in intact adults (5αR1 was not quantified because it was not detected in the adult forebrain). We found a sex difference in the density of 5αR2-expressing cells in the AMY, such that females have a denser population in this region than males (Cohen & Wade, 2010a). Isozyme 5αR2 in the AMY may function in females to facilitate the processing of male courtship behavior. The breakdown of T in females may serve to clear excess hormone that could be detrimental. It is also possible that DHT may play a role in female receptive behaviors, although it is difficult to speculate on exactly what that might be. 5αR is known to act on progesterone, converting it to 5α-dihydroprogesterone (DHP) (Lephart et al., 2001). Progesterone influences female reproductive behaviors across vertebrates (Rubin & Barfield, 1980; Steel, 1981; Fabre-Nys & Martin, 1991). In anoles, progesterone acts in synergy with E2 in females and T in males to facilitate sexual behaviors (McNicol & Crews, 1979; Wu, Whittier, & Crews, 1985; Young, Greenberg, & Crews, 1991). The data on 5αR2 expression in the anole suggest the possibility that metabolism of progesterone into DHP may be important and should be investigated.

5α-Reductase Expression in the Developing Brain

Using similar methods as described above, we examined 5αR1 and 5αR2 mRNA expression at two different ages in development: P0 and P50 (Cohen and Wade, 2012b). Unlike adult anoles, both isozymes of 5αR are expressed in the developing forebrain. Additionally, both the number of cells expressing 5αR1 in the POA and density per section in the AMY decrease between the day of and 50 days after hatching (Cohen and Wade, 2012b). The total number of cells in the POA and density of cells per section in the AMY do not differ between the ages we examined in Nissl stained tissue, suggesting that number of existing cells expressing 5αR1 decreases during this developmental period. This is consistent with the data from adults wherein 5αR1 is not expressed in the forebrain, and suggests that this isozyme may play a role in development. Because 5αR1 decreases between P0 and P50 while plasma T increases, the androgenic substrate may serve as a negative feedback signal in 5αR1 production. No differences in 5αR2 expression were detected at those ages, suggesting that 5αR2 expression has not yet differentiated.

Effects of Testosterone on Isozyme 5αR2 Expression in the Adult Brain

Testosterone treatment of females increases the density of 5αR2 mRNA expressing cells in the VMH (Cohen and Wade, 2012a). Increased density of 5αR2 expressing cells could lead to a greater capacity to locally synthesize DHT. Although there are no obvious behavioral or functional effects of DHT in females, perhaps increased 5αR2 expression in females is related to clearing excess T and/or decreasing the opportunity for higher than normal levels of E2 to be synthesized.

Additionally, within the breeding season, blank-treated males had a greater density of 5αR2 cells in the VMH than those that were treated with T (Cohen and Wade, 2012a). These data suggest that while T potentially increases whole brain 5αR activity in males regardless of season, it has opposite effects on the mRNA expression of 5αR2 in the VMH in these individuals. The relative absence of T (gonadectomy and treatment with a blank capsule) in males may cause more 5αR2 to be produced in order to increase production of DHT. Production of DHT is important for the full expression of male sexual behaviors (Rosen & Wade, 2000), and perhaps the VMH plays a role in this function. Immediate early gene expression (a measure of neural activation) is increased in the VMH of male hamsters after copulation (Kollack-Walker & Newman, 1995), suggesting a role of the VMH in male reproductive behaviors. This function should be more fully investigated in green anoles.

CONCLUSIONS

The examination of T metabolism across multiple taxa is important to fully understand the roles aromatase, 5αR1, and 5αR2 have in adult functions, such as the activation of sexual behavior, and brain development. This type of comparative research can also begin to address questions about the evolution of these mechanisms. Unlike avian and mammalian species, in the anole, T itself seems to be the more important hormone in regulating sexual behaviors (see previous) (Wade, 2011). However, specific sex and seasonal differences in whole brain aromatase activity (Rosen & Wade, 2001) and selective upregulation of this activity by T (Cohen & Wade, 2010b) suggest that key modulatory processes are in place. This idea and the facts that E2 facilitates male motivation and DHT is necessary for the full expression of male anole sexual behaviors (Rosen & Wade, 2000; Latham & Wade, 2010;) suggest that some basic mechanisms are in place, but the enzymes have been coopted for reproduction more fully in birds and mammals than in anoles, and perhaps reptiles in general. Testosterone metabolism should be examined in more species to determine whether the apparent role of these enzymes is the same for each vertebrate group. Additionally, 5αR1 and 2 are expressed in anoles in a pattern opposite that in rodents, which suggests that different enzymes are used for similar functions in these species: maintenance in the adult brain (5αR1 in rodents and 5αR2 in anoles) and masculinization and/or defeminization (5αR2 in rodents and 5αR1 in anoles) of the developing brain. Future work is needed to determine the roles of these enzymes in anole development. Overall, the experiments presented here highlight the importance of comparative work and lay important groundwork for future research that will ascertain the role of these enzymes in anoles and the extent of their regulation by T.

REFERENCES

Adkins, E., & Schlesinger, L. (1979). Androgens and the social behavior of male and female lizards (Anolis carolinensis). *Horm Behav, 13*(2), 139–152.

Alfoldi, J., Di Palma, F., Grabherr, M., Williams, C., Kong, L., Mauceli, E., et al. (2011). The genome of the green anole lizard and a comparative analysis with birds and mammals. *Nature, 477,* 587–591.

Arnold, A. P., & Breedlove, S. M. (1985). Organizational and activational effects of sex steroids on brain and behavior: a reanalysis. *Horm Behav, 19*(4), 469–498.

Ball, G. F., & Balthazart, J. (2002). Neuroendocrine mechanisms regulating reproductive cycles and reproductive behavior in birds. In: Pfaff, D. W., Arnold, A. P., Etgen, A. M., Fahrbach, S. E., & Rubin, R. T. (eds.), *Hormones, Brain and Behavior* (Vol. 2). San Diego, CA: Academic Press, pp. 649–798.

Balthazart, J., Schumacher, M., & Evrard, L. (1990). Sex differences and steroid control of testosterone-metabolizing enzyme activity in the quail brain. *J Neuroendocrinol, 2*(5), 675–683.

Bardet, S. M., Cornil, C. A., & Balthazart, J. (2010). Testosterone recruits new aromatase-imunoreactive cells in neonatal quail brain. *Neuroreport, 21*(5), 376–380.

Baum, M. J., & Vreeburg, J. T. (1973). Copulation in castrated male rats following combined treatment with estradiol and dihydrotestosterone. *Science, 182*(109), 283–285.

Callard, G. V., Z. Petro, & K. Ryan. (1977). Identification of Aromatase in the Reptilian Brain. *Endocrinology, 100,* 1214–1218.

Celotti, F., Negri-Cesi, P., & Poletti, A. (1997). Steroid metabolism in the mammalian brain: 5alpha-reduction and aromatization. *Brain Res Bull, 44*(4), 365–375.

Charlier, T. D., Harada, N., Balthazart, J., & Cornil, C. A. (2011). Human and quail aromatase activity is rapidly and reversibly inhibited by phosphorylating conditions. *Endocrinology, 152*(11), 4199–210.

Cohen, R. E., & Wade, J. (2010a). Distribution of two isozymes of 5alpha-reductase in the brains of adult male and female green anole lizards. *Brain Behav Evol, 76*(3–4), 279–288.

Cohen, R. E., & Wade, J. (2010b). Testosterone selectively affects aromatase and 5alpha-reductase activities in the green anole lizard brain. *Gen Comp Endocrinol, 166*(1), 128–133.

Cohen, R. E., & Wade, J. (2011). Aromatase mRNA in the brain of adult green anole lizards: effects of sex and season. *J Neuroendocrinol, 23*(3), 254–260.

Cohen, R. E., & Wade, J. (2012a). Aromatase and 5α-reductase type 2 mRNA in the green anole forebrain: an investigation into the effects of sex, season, and testosterone manipulation. *Gen Comp Endocrinol, 176*(3), 377–384.

Cohen, R. E., & Wade, J., (2012b). Expression of aromatase and two isozymes of 5α-reductase in the developing green anole forebrain. *J Neuroenocrinol,* April 9 [Epub ahead of print].

Crews, D. (1980). Interrelationships among ecological, behavioral, and neuroendocrine processes in the reproductive cycle of Anolis carolinensis and other reptiles. In Rosenblatt, J. S., Hinde, R. A., Beer, C. G. & Busnel, M.-C. (eds.), *Advances in the Study of Behavior* (Vol. 11). New York: Academic-Press, pp. 1–74.

Crews, D., & Morgentaler, A. (1979). Effects of intracranial implantation of oestradiol and dihydrotestosterone on the sexual behaviour of the lizard Anolis carolinensis. *J Endocrinol, 82*(3), 373–381.

Dias, B. G., Chin, S. G., & Crews, D. (2009). Steroidogenic enzyme gene expression in the brain of the parthenogenetic whiptail lizard, Cnemidophorus uniparens. *Brain Res, 1253,* 129–138.

Emery, D. E., & Moss, R. L. (1984). Lesions confined to the ventromedial hypothalamus decrease the frequency of coital contacts in female rats. *Horm Behav, 18*(3), 313–329.

Fabre-Nys, C., & Martin, G. B. (1991). Hormonal control of proceptive and receptive sexual behavior and the preovulatory LH surge in the ewe: reassessment of the respective roles of estradiol, testosterone, and progesterone. *Horm Behav,* i(3), 295–312.

Fusani, L. (2008). Testosterone control of male courtship in birds. *Horm Behav, 54*(2), 227–233.

Greenberg, B., & Noble, G. K. (1944). Social behavior of the American chameleon (Anolis carolinensis Voight). *Physiol Zool, 17,* 392–439.

Greenberg, N., Scott, M., & Crews, D. (1984). Role of the amygdala in the reproductive and aggressive behavior of the lizard, Anolis carolinensis. *Physiol Behav, 32*(1), 147–151.

Holmes, M. M., & Wade, J. (2005). Sexual differentiation of the copulatory neuromuscular system in green anoles (Anolis carolinensis): normal ontogeny and manipulation of steroid hormones. *J Comp Neurol, 489*(4), 480–490.

Husak, J. F., Irschick, D. J., Meyers, J. J., Lailvaux, S. P., & Moore, I. T. (2007). Hormones, sexual signals, and performance of green anole lizards (Anolis carolinensis). *Horm Behav, 52*(3), 360–367.

Hutchison, J. B., Beyer, C., Hutchison, R. E., & Wozniak, A. (1997). Sex differences in the regulation of embryonic brain aromatase. *J Steroid Biochem Mol Biol, 61*(3–6), 315–322.
Jeyasuria, P., & Place, A. R. (1997). Temperature-dependent aromatase expression in developing diamondback terrapin (Malaclemys terrapin) embryos. *J Steroid Biochem Mol Biol, 61*(3–6), 415–425.
Johnson, M. A., Cohen, R. E., Vandecar, J. R., & Wade, J. (2011). Relationships among testosterone, reproductive morphology and behavior in a natural population of green anole lizards. *Physiol Behav, 104,* 437–445.
Jones, R. E., Guillette, L. J., Jr., Summers, C. H., Tokarz, R. R., & Crews, D. (1983). The relationship among ovarian condition, steroid hormones, and estrous behavior in Anolis carolinensis. *J Exp Zool, 227*(1), 145–154.
Kendrick, A. M., Rand, M. S., & Crews, D. (1995). Electrolytic lesions to the ventromedial hypothalamus abolish receptivity in female whiptail lizards, Cnemidophorus uniparens. *Brain Res, 680*(1–2), 226–228.
Kollack-Walker, S., & Newman, S. W. (1995). Mating and agonistic behavior produce different patterns of Fos immunolabeling in the male Syrian hamster brain. *Neuroscience, 66*(3), 721–736.
Konkle, A. T., & Balthazart, J. (2011). Sex differences in the rapid control of aromatase activity in the quail preoptic area. *J Neuroendocrinol, 23*(5), 424–434.
Kostarczyk, E. M. (1986). The amygdala and male reproductive functions: I. Anatomical and endocrine bases. *Neurosci Biobehav Rev, 10*(2), 67–77.
Krohmer, R. W., Bieganski, G. J., Baleckaitis, D. D., Harada, N., & Balthazart, J. (2002). Distribution of aromatase immunoreactivity in the forebrain of red-sided garter snakes at the beginning of the winter dormancy. *J Chem Neuroanat, 23*(1), 59–71.
La Vaque, T. J., & Rodgers, C. H. (1975). Recovery of mating behavior in the female rat following VMH lesions. *Physiol Behav, 14*(1), 59–63.
Lance, V. A. (2009). Is regulation of aromatase expression in reptiles the key to understanding temperature-dependent sex determination? *J Exp Zool Part A Ecol Genet Physiol, 311*(5), 314–322.
Latham, S., & Wade, J. (2010). Estradiol facilitates mounting behavior in male green anole lizards. *Physiol Behav, 99*(1), 78–81.
Lephart, E. D., Lund, T. D., & Horvath, T. L. (2001). Brain androgen and progesterone metabolizing enzymes: biosynthesis, distribution and function. *Brain Res Brain Res Rev, 37*(1–3), 25–37.
Licht, P. (1967a). Environmental control of annual testicular cycles in the lizard Anolis carolinensis. I. Interaction of light and temperature in the initiation of testicular recrudescence. *J Exp Zool, 165*(3), 505–516.
Licht, P. (1967b). Environmental control of annual testicular cycles in the lizard Anolis carolinensis. II. Seasonal variations in the effects of photoperiod and temperature on testicular recrudescence. *J Exp Zool, 166*(2), 243–253.
Licht, P. (1969). Environmental control of annual testicular cycles in the lizard Anolis carolinensis. 3. Temperature thresholds for photoperiodism. *J Exp Zool, 172*(3), 311–322.
Lovern, M. B., Holmes, M. M., & Wade, J. (2004). The green anole (Anolis carolinensis): a reptilian model for laboratory studies of reproductive morphology and behavior. *ILAR J, 45*(1), 54–64.
Lovern, M. B., McNabb, F. M., & Jenssen, T. A. (2001). Developmental effects of testosterone on behavior in male and female green anoles (Anolis carolinensis). *Horm Behav, 39*(2), 131–143.

Lovern, M. B., & Wade, J. (2003). Sex steroids in green anoles (Anolis carolinensis): uncoupled maternal plasma and yolking follicle concentrations, potential embryonic steroidogenesis, and evolutionary implications. *Gen Comp Endocrinol, 134*(2), 109–115.

Mason, P., & Adkins, E. K. (1976). Hormones and social behavior in the lizard, Anolis carolinensis. *Horm Behav, 7*(1), 75–86.

McNicol, D., Jr., & Crews, D. (1979). Estrogen/progesterone synergy in the control of female sexual receptivity in the lizard, Anolis carolinensis. *Gen Comp Endocrinol, 38*(1), 68–74.

Melcangi, R. C., Poletti, A., Cavarretta, I., Celotti, F., Colciago, A., Magnaghi, V., et al. (1998). The 5alpha-reductase in the central nervous system: expression and modes of control. *J Steroid Biochem Mol Biol, 65*(1–6), 295–299.

Milnes, M. R., Roberts, R. N., & Guillette, L. J. (2002). Effects of incubation temperature and estrogen exposure on aromatase activity in the brain and gonads of embryonic alligators. *Environ Health Persp, 110,* 393–396.

Morgantaler, A., & Crews, D. (1978). Role of the anterior hypothalamus-preoptic area in the regulation of behavior in the lizard, Anolis carolinensis: implantation studies. *Horm Behav, 11,* 61–73.

Nakamura, M. (2010). The mechanism of sex determination in vertebrates-are sex steroids the key-factor? *J Exp Zool A Ecol Genet Physiol, 313*(7), 381–398.

Negri-Cesi, P., Poletti, A., & Celotti, F. (1996). Metabolism of steroids in the brain: a new insight into the role of 5alpha-reductase and aromatase in brain differentiation and functions. *J Steroid Biochem Mol Biol, 58*(5–6), 455–466.

O'Bryant, E. L., & Wade, J. (2001). Development of a sexually dimorphic neuromuscular system involved in green anole courtship behavior. *Brain Behav Evol, 58*(6), 362–369.

Orrell, K. S., & Jensson, T. A. (2003). Heterosexual signalling by lizard Anolis carolinensis, with intersexual comparisons across contexts. *Behaviour, 140,* 603–634.

Pieau, C., & Dorizzi, M. (2004). Oestrogens and temperature-dependent sex determination in reptiles: all is in the gonads. *J Endocrinol, 181*(3), 367–377.

Poletti, A., Negri-Cesi, P., Rabuffetti, M., Colciago, A., Celotti, F., & Martini, L. (1998). Transient expression of the 5 alpha-reductase type 2 isozyme in the rat brain in late fetal and early postnatal life. *Endocrinology, 139*(4), 2171–2178.

Pradhan, D. S., Newman, A. E., Wacker, D. W., Wingfield, J. C., Schlinger, B. A., & Soma, K. K. (2010). Aggressive interactions rapidly increase androgen synthesis in the brain during the non-breeding season. *Horm Behav, 57*(4–5), 381–389.

Roselli, C. E., Horton, L. E., & Resko, J. A. (1985). Distribution and regulation of aromatase-activity in the rat hypothalamus and limbic system. *Endocrinology, 117*(6), 2471–2477.

Rosen, G. J., & Wade, J. (2000). The role of 5alpha-reductase activity in sexual behaviors of the green anole lizard. *Physiol Behav, 69*(4–5), 487–498.

Rosen, G. J., & Wade, J. (2001). Androgen metabolism in the brain of the green anole lizard (Anolis carolinensis): effects of sex and season. *Gen Comp Endocrinol,* 122(1), 40–47.

Rubin, B. S., & Barfield, R. J. (1980). Priming of estrous responsiveness by implants of 17 beta-estradiol in the ventromedial hypothalamic nucleus of female rats. *Endocrinology, 106*(2), 504–509.

Schumacher, M& Hutchison, J. B. (1986). Testosterone induces hypothalamic aromatase during early development in quail. *Brain Res, 377*(1), 63–72.
Steel, E. (1981). Control of proceptive and receptive behavior by ovarian hormones in the Syrian hamster (Mesocricetus auratus). *Horm Behav, 15*(2), 141–156.
Tokarz, R. R., & Crews, D. (1980). Induction of sexual receptivity in the female lizard, Anolis carolinensis: effects of estrogen and the antiestrogen CL-628. *Horm Behav, 14*(1), 33–45.
Voigt, C., Ball, G. F., & Balthazart, J. (2007). Neuroanatomical specificity of sex differences in expression of aromatase mRNA in the quail brain. *J Chem Neuroanat, 33*(2), 75–86.
Wade, J. (1997). Androgen metabolism in the brain of the green anole lizard (Anolis carolinensis). *Gen Comp Endocrinol, 106*(1), 127–137.
Wade, J. (2011). Relationships among hormones, brain and motivated behaviors in lizards. *Horm Behav, 59*(5), 637–644.
Wagner, C. K. & Morrell, J. I. (1996). Distribution and steroid hormone regulation of aromatase mRNA expression in the forebrain of adult male and female rats: a cellular-level analysis using in situ hybridization. *J Comp Neurol, 370*(1), 71–84.
Wheeler, J. M., & Crews, D. (1978). The role of the anterior hypothalamus-preoptic area in the regulation of male reproductive behavior in the lizard, Anolis carolinensis: lesion studies. *Horm Behav, 11*(1), 42–60.
Wibbels, T., & Crews, D. (1994). Putative aromatase inhibitor induces male sex determination in a female unisexual lizard and in a turtle with temperature-dependent sex determination. *J Endocrinol, 141*(2), 295–299.
Willingham, E., Baldwin, R., Skipper, J. K., & Crews, D. (2000). Aromatase activity during embryogenesis in the brain and adrenal-kidney-gonad of the red-eared slider turtle, a species with temperature-dependent sex determination. *Gen Comp Endocrinol, 119*(2), 202–207.
Winkler, S. M., & Wade, J. (1998). Aromatase activity and regulation of sexual behaviors in the green anole lizard. *Physiol Behav, 64*(5), 723–731.
Wu, J., Whittier, J. M., & Crews, D. (1985). Role of progesterone in the control of female sexual receptivity in Anolis carolinensis. *Gen Comp Endocrinol, 58*(3), 402–406.
Young, L. J., Greenberg, N., & Crews, D. (1991). The effects of progesterone on sexual-behavior in male green anole lizards (Anolis carolinensis). *Horm Behav, 25*(4), 477–488.

Chapter 13

Aromatase, Low Temperature Dormancy, and Reproduction in the Red-Sided Garter Snake

Randolph W. Krohmer

The red-sided garter snake (*Thamnophis sirtalis parietalis;* RSGS) is the most extensively studied species exhibiting a dissociated reproductive pattern; that is, mating at a time when the gonads are quiescent and sex steroid hormone levels are purported to be basal (Licht, 1984; Woolley et al., 2004). Although studies have now shown that the circulating levels of sex steroid hormones remain elevated throughout winter dormancy and at the beginning of the spring breeding season (Krohmer et al., 1987; Moore et al., 2000, 2001), to date, the only cue identified that is responsible for the initiation of courtship behavior and mating in the RSGS is an extended period of low temperature dormancy (LTD) (Garstka et al., 1982).

Similar to species exhibiting an associated reproductive pattern, the RSGS forebrain contains sex steroid concentrating regions within pathways critical for the expression of reproductive behaviors (Friedman & Crews, 1985b; Krohmer & Crews, 1987a). In addition, these regions have been found to respond to both androgens and estrogens (Halpern et al., 1982). Therefore, the presence of sex steroid concentrating regions along with elevated levels of circulating sex steroid hormones during LTD suggests a possible role for sex steroid hormones in the initiation and control of reproductive behaviors in the RSGS. Recent investigations (Krohmer et al., 2002, 2010) suggest a mechanism by which estrogenic metabolites aromatized in the brain from circulating androgens during the prolonged dormancy period are able to accumulate in specific regions in the brain (Krohmer, in preparation), thus providing the hormonal stimulus required to initiate reproductive behaviors.

REPRODUCTIVE PATTERNS

Most seasonally breeding vertebrates mate at a time when gonadal activity and the elevation of circulating sex steroid hormones immediately precedes or coincides with the expression of courtship behavior and mating (Fig. 13–1). In species exhibiting this "associated reproductive pattern" castration eliminates

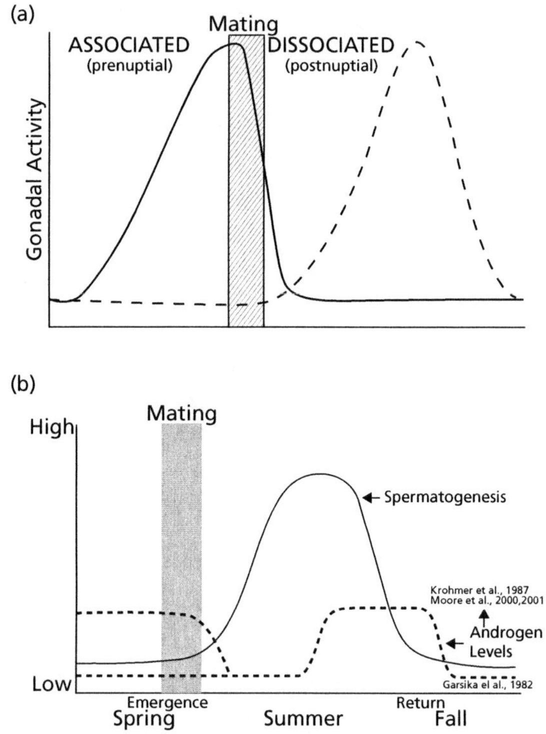

Figure 13–1. Temporal Differences in Gonadal Activity A. Comparison of the temporal differences in gonadal activity as related to the mating period in associated and dissociated reproductive patterns (Krohmer, 2004). B. Seasonal pattern of gonadal activity and reproductive behavior in the male red-sided garter snake. Courtship and mating is initiated on emergence from winter dormancy when the gonads are inactive. Early studies reported androgen levels to be low as animals entered low temperature dormancy and were basal at the initiation of the breeding season (Garstka et al., 1982), with spermatogenesis and steroidogenesis initiated at the end of the breeding season. Later studies demonstrated that androgen levels are elevated at the beginning of dormancy, and remain elevated throughout dormancy until the initiation of reproductive behaviors when a rapid drop in concentration is observed (Krohmer et al., 1987; Moore et al., 2000, 2001). (*A*) Reprinted with permission *ILAR Journal 45*(1), *Inst Lab Anim Res,* The Keck Center of the National Academies, 500 Fifth Street NW, Washington DC 20001 http://dels.nas.edu/ilar/ (*B*) Used with permission from Krohmer, R. W., & Lutterschmidt, D. I. (2011). Environmental and neuroendocrine control of reproduction in snakes. In: Aldridge, R. D., & Sever, D. M. (eds.), *Reproductive Biology and Phylogeny of Snakes.* Enfield, NH: Science Publishers, pp. 289–346.

mating behavior, whereas reimplantation or administration of exogenous sex steroid hormones reinstates reproductive behavior in castrated individuals and will initiate reproductive activity in nonbreeding individuals (Crews et al., 1984).

In contrast, a small number of vertebrate species, including bats (Gustafson & Shemesh, 1976; Gustafson, 1979), turtles (Lofts, 1977; Crews, 1984; Licht, 1984), and many temperate zone snakes (Saint-Girons, 1982; Crews & Garstka, 1982; Crews, 1983; Krohmer & Aldridge, 1985; Krohmer et al., 1987), display courtship behavior and mate at a time when the gonads are inactive and the level of circulating sex steroids have been reported to be low (Fig. 13–1A) (Crews, 1984; Licht, 1984). In species exhibiting this "dissociated reproductive pattern," castration does not eliminate sexual activity, nor does the administration of exogenous sex steroids initiate reproductive behavior in non-courting individuals (Woolley et al., 2004).

Over the past 35 years, the RSGS has become the most widely studied model system exhibiting a dissociated reproductive pattern (Crews, 1984; Crews et al., 1984; Krohmer & Lutterschmidt, 2011). In the adult male RSGS, initiation of courtship behavior and mating has been reported to be independent of testicular or pituitary hormone control (Camazine et al., 1980). Furthermore, systemic administration of sex steroid hormones, hypothalamic or pituitary hormones, or implantation of sex steroid hormones directly into the hypothalamic region fails to induce reproductive behavior in noncourting individuals (Camazine et al., 1980; Crews et al., 1984; Friedman & Crews, 1985a). Moreover, Crews (1991) reported that males continue to exhibit courtship behavior for up to 3 years following castration provided animals go through a period of LTD.

Although initial studies reported the level of circulating androgens in the male RSGS to be low or absent on emergence from winter dormancy (Garstka et al., 1982), later investigations found that circulating androgens, elevated in the fall before winter dormancy, remain elevated throughout the dormancy period and are not significantly reduced at emergence in the spring (Krohmer et al., 1987; Moore et al., 2000, 2001) (see Fig. 13–1B). However, Krohmer et al., (1987) found that the spring level of circulating androgens decrease rapidly, reaching basal levels 10 to 14 days following emergence.

LOW TEMPERATURE DORMANCY

Located in central Manitoba, Canada, the RSGS is found farther north than any other reptile in the western hemisphere. This extremely large population appears to be one of the most adaptable reptile populations to exist under climactic conditions that border on the subarctic (Fig. 13–2). Consequently, this species is able to survive in an environment in which winter temperatures can reach –40°C and snow cover can be continuous for 8 months or more. Aleksiuk and Stewart (1971) proposed that these extreme seasonal variations in temperature, photoperiod, and food availability may act as a major evolutionary force dictating the characteristics of the population.

It is interesting to note that although not all animals that undergo a period of LTD (hibernation or brumation) exhibit a dissociated reproductive pattern, all animals exhibiting the dissociated reproductive pattern do experience a period of LTD. In the RSGS, the only cue identified to date that will initiate courtship

Figure 13–2. Photograph Showing Spring Mating Activity in Newly Emerged Red-Sided Garter Snakes. Snakes shown are in a "mating ball," in which a number of males simultaneously attempt to mate with a single female. In the wild, mating will occur even when snow is present as long as there is sufficient sunlight available for the animals to warm. Photo by R. W. Krohmer.

behavior and mating is an extended period of LTD followed by exposure to warm temperatures (Camazine et al., 1980; Garstka et al., 1982; Bona-Gallo & Licht, 1983). In addition, the length of the dormancy period has been shown to have a profound effect on the intensity of the courtship behavior exhibited on emergence (Garstka et al., 1982; Krohmer, 2004). Therefore, seasonal activation of sexual behavior in the male RSGS appears to be dissociated from changes in gonadal development as well as maximal sex steroid hormone synthesis. For that reason, the majority of studies have proposed that rather than using the activational effects of sex steroid hormones to regulate sexual behavior, the RSGS reacts to physical cues by direct behavioral and physiological response (Crews, 1983).

LESIONS

As in species exhibiting an associated reproductive pattern, an intact anterior hypothalamus-preoptic area (AHPOA) has been shown to be essential for the display of reproductive behaviors in the male RSGS. Lesions placed in the AHPOA of actively courting males extinguished all courtship behavior

(Friedman & Crews, 1985b), whereas males receiving lesions in the AHPOA before entering winter dormancy failed to initiate courtship on emergence (Krohmer & Crews, 1987a). In contrast, bilateral lesions placed in the nucleus sphericus (NS) or septum (S) before LTD, reduced the time required to initiate intense courtship behavior (Krohmer & Crews, 1987b). The nucleus sphericus, a region analogous to the mammalian amygdaloid region, has been found to receive efferents from the accessory olfactory bulb, which in turn sends efferents to the bed nucleus of the stria terminalis, the preoptic area, and the ventromedial nucleus of the hypothalamus. In turn, the ventromedial nucleus of the hypothalamus and nucleus sphericus receives efferents from the accessory olfactory bulb (Halpern & Silfen, 1974; Halpern & Kubie, 1984). Krohmer and Crews (1987b) hypothesized that lesions of the nucleus sphericus affected the efferent outflow through the bed nucleus of the stria terminalis to the AHPOA. Also, lesions in the NS or S increased the total number of days that males exhibited moderate to intense courtship behavior suggesting that destruction of cells within the NS or S "disinhibit" sexual behaviors (Krohmer & Crews, 1987b).

SEX STEROID CONCENTRATING NUCLEI

Sex steroid concentrating nuclei have been identified in the brains of all vertebrate species studied to date. Although the species examined are quite diverse, there is a great deal of uniformity in their location (Morrell & Pfaff, 1978). These nuclei have been found to play a vital role in the regulation of species-typical behaviors (Ingle & Crews, 1985) and are often a component of the neural pathways that control reproductive behaviors (Crews & Silver, 1985). In many species, these nuclei may exhibit sex differences of various sorts and they may vary in size and other aspects of morphology depending on reproductive status (Gorski et al., 1978; Madeira et al., 1999; Simerly 2002). Although a large number of studies have examined the function of hormones acting in specific brain regions in mammals (Dohler et al., 1984; Tobet et al., 1986; Ulibarri & Yahr, 1988) and birds (Panzica et al., 1996; Ball & Balthazart 2009), only a relatively few studies have examined sex steroid concentrating regions in reptiles (lizards: Crews et al., 1993; Wade et al., 1993; O'Bryant & Wade, 2002; Beck et al., 2008; snakes: Halpern, 1980; Halpern et al., 1982; Krohmer, 2004).

Although the level of circulating sex steroid hormones in the male RSGS has now been shown to be elevated on emergence from LTD (Krohmer et al., 1987; Moore et al., 2000, 2001; Lutterschmidt & Mason, 2009), the majority of studies to date continue to support the perception that courtship behavior and mating in the RSGS are initiated independent of direct hormonal control. It is interesting to note, however, that the brains of RSGS exhibit a pattern of sex steroid concentrating nuclei similar to other vertebrate species (Halpern et al., 1982). And, although the garter snake brain was found to bind both testosterone and estradiol, the number of hormone-accumulating cells and density of labeling was greater in response to estrogen versus testosterone (Halpern et al., 1982).

In a recent study, it was found that the volume of the sex steroid–concentrating nucleus within the POA of the RSGS exhibits significant changes in response to both seasonal and hormonal manipulation (Baleckaitis & Krohmer, unpublished data; reviewed in Krohmer 2004). As in Halpern et al. (1982), this study found that the sex steroid–concentrating brain nucleus within the POA exhibited greater hypertrophy in animals treated with estradiol as compared to testosterone or a placebo. In addition, animals treated with either estradiol or testosterone maintained under conditions of LTD for 12 weeks exhibited greater hypertrophy of the sex steroid concentrating nucleus in the POA compared to their counterparts maintained under warm conditions (Baleckaitis & Krohmer, unpublished data; Krohmer 2004).

AROMATASE

Although aromatase activity (AA) has been documented in all classes of vertebrates (Naftolin et al., 1975; Callard et al., 1977, 1978a,b) and plays a key role in the activation of sexual behavior in many species (Vagell & McGinnis, 1997; Balthazart et al., 1990; Bonsall et al., 1992), relatively few studies have been conducted on reptiles. The majority of studies that examined AA in reptiles have been conducted on lizards, animals exhibiting an associated reproductive pattern (Gobbetti et al., 1994; Wade 1997). In the green anole (*Anolis carolinensis*), male courtship behavior was found to be partially mediated by 5α-reductase, the enzyme that converts testosterone to dihydrotestosterone (Rosen & Wade, 2001) and facilitated by testosterone (Winkler & Wade 1998; O'Bryant & Wade, 2002), whereas estrogen metabolites aromatized from androgens aided receptivity in females (Winkler & Wade, 1998). Although no sex or seasonal variation was found in 5α-reductase activity, AA was greater in males during the breeding season compared to nonbreeding males (Beck & Wade, 2008).

Callard et al. (1978a) identified AA in the forebrain of the female brown watersnake (*Natrix taxispilota*) but detected no activity in the mid- or hindbrain. This study however, did not examine specific regions within the forebrain or examine possible seasonal variations.

Employing an antibody raised against quail recombinant aromatase (Foidart et al., 1995a), Krohmer et al. (2002) were able to identify two types of aromatase-immunoreactive (ARO-ir) cells in the forebrain of the male RSGS. Large type I cells exhibited intense immunostaining in both the cell body and processes, whereas the smaller type II cells exhibited only weak to moderate immunostaining in the cell body with little or no staining in the processes (Fig. 13–3). The morphological characteristics of the ARO-ir cells identified in the RSGS (Krohmer et al., 2002) correspond to the descriptions of ARO-ir cells in both birds (Balthazart et al., 1996a,b) and mammals (Shinoda et al., 1994).

However, unlike the relatively restricted distribution reported in birds (Foidart et al., 1995; Balthazart et al., 2009) and mammals (Callard et al., 1983; Cherry et al., 1990), we found a relatively wide distribution of ARO-ir cells in the male RSGS (Krohmer et al., 2002). Correspondingly, Callard et al.

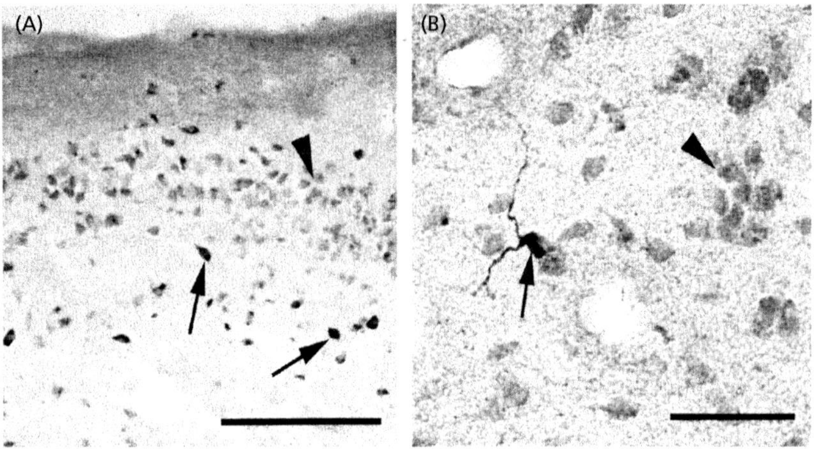

Figure 13–3. Photomicrographs of the Two Types of Aromatase Immunoreactive (ARO-ir) Cells in the Forebrain of the Male Red-Sided Garter Snake. Sections shown are from the preoptic area, which contains both types of immunoreactive cells. Type I ARO-ir cells (*arrows*) are larger darkly staining cells with visible processes. Type II ARO-ir cells (*arrowhead*) are smaller, lighter staining cells with few, if any, visible processes. Magnification bars: 200 μm (*A*), 50 μm (*B*). Used with permission from Figure 8.12, Krohmer and Lutterschmidt, 2011. Reprinted with permission of *ILAR Journal 45*(1), *Inst Lab Anim Res*, The Keck Center of the National Academies, 500 Fifth Street NW, Washington DC 20001 http://dels.nas.edu/ilar/.

(1978a,b) reported that AA is more broadly distributed in the brains of fish and reptiles than in homeotherm vertebrates (Callard et al., 1990). In addition to the wide distribution, AA was found to be 100 to 200 times greater in the RSGS (Krohmer et al., 2010) and 100 to 1,000 times greater in the goldfish compared with AA found in mammals (Pasmanik & Callard, 1985).

Nevertheless, the distribution of ARO-ir cells in the forebrain of the male RSGS is anatomically discrete and fluctuates seasonally (Krohmer et al., 2002). Small type II cells are prevalent throughout the forebrain of males collected in the fall as they return to the den before entering winter dormancy. However, a much smaller number of type I cells were concentrated primarily in regions identified as containing sex steroid concentrating neurons (Halpern et al., 1982) such as the AHPOA, S and NS that have been shown to be essential for the control and maintenance of reproductive behaviors (Friedman & Crews, 1985a; Krohmer & Crews, 1987a,b). Similar to fall-collected snakes, spring-collected, actively courting snakes had type II cells in all regions of the forebrain, but in significantly reduced numbers with the majority concentrated in those regions critical for the control of courtship and mating. Furthermore, type I cells, if present, were found primarily in the olfactory region (Krohmer, unpublished data). The distribution of ARO-ir cells found in the RSGS corresponds to the regions of elevated aromatase activity and/or areas of dense populations of ARO-ir cells in birds (Panzica et al., 1994; Foidart et al., 1995b; Balthazart et al., 1996a,b) and mammals (Roselli et al., 1985; Wagner & Morrell, 1996; Veney & Rissman, 1998). To date, no direct comparison of types and distribution of ARO-ir structures is available for any other reptile.

Aromatase activity in the forebrain of the male RSGS has also been found to vary regionally as well as seasonally (Krohmer et al., 2010). Animals collected in the fall before entering LTD had significantly greater AA in the AHPOA and S compared with the olfactory region (O) and NS. In contrast, actively courting males collected in the spring have significantly higher AA in the O compared with the AHPOA, S, or NS. These findings support earlier studies in other vertebrate species, suggesting that the function of target cells in the brain can be modified depending on hormonal status and/or environmental factors (Hutchison, 1974a,b; Hutchison et al., 1986). Consequently, regional and seasonal differences in AA in the brain of the RSGS may play a critical role in the priming of specific neural pathways that control stereotypical behavior patterns at the appropriate time.

The elevated AA observed in the S and AHPOA of male snakes collected in the fall (Krohmer et al., 2010), in combination with the known importance of the AHPOA in regulating reproduction (Friedman & Crews, 1985a; Krohmer & Crews, 1987a), suggests that the conversion of androgens to estrogens in the POA in fall animals might be essential to the expression of reproductive behaviors in the spring. Although mating behavior in the RSGS does not coincide with peak gonadal activity, this does not automatically rule out a role for sex steroid hormones in regulating reproductive behavior. Furthermore, because reproduction occurs immediately following spring emergence, the concomitant changes in neurophysiology and behavior that prepare and accompany reproduction are likely to occur during winter dormancy (Lutterschmidt & Mason, 2009).

What then is the importance of LTD in the organization and initiation of reproductive behavior? As described previously, the only cue found to initiate courtship behavior and mating in the RSGS is a prolonged period of LTD followed by exposure to warm temperatures. Moreover, Garstka et al. (1982) demonstrated that the duration of time in LTD plays an important role in the intensity at which courtship behavior is exhibited. Krohmer & Baleckaitis. (2000) demonstrated that the aromatase enzyme in the RSGS forebrain remained active under winter dormancy conditions. More recently, Krohmer (in preparation) conducted an aromatase assay under conditions of LTD. In this study, 32 adult male RSGS were collected in the fall from dens located in the Interlake Region north of Winnipeg, Manitoba, Canada and their brains were removed following the procedure detailed in Krohmer et al. (2010). Following removal, brains were frozen on dry ice and stored at –70°C until processed. Before the assay, brains were randomly assigned to one of four incubation groups (n = 8/group) and dissected into four regions: olfactory (O), nucleus sphericus (NS), septum/anterior hypothalamus preoptic area (S/AHPOA), and midbrain (Mb) (Fig. 13–4). The aromatase assay used in this study was a modification of the procedure developed for snakes and described in detail in Krohmer et al. (2010). Briefly, brain tissue homogenates were added to glass tubes containing 200 picomoles (pmol) 1,2,6,7-[^{3}H] androst-4-ene, 17-dione, adjusted to a specific activity of 5 Ci/mM (final concentration = 0.5 μM), 200 μl of 0.1 M potassium phosphate buffer (pH 7.4) and 2 mM NADPH cofactor. Tubes were covered and maintained under conditions of LTD (4.0 ± 2.0°C; 0L:24 D photic cycle) for 4, 8, 12, or 16 weeks. At termination, the supernatant was removed and evaluated for estrone production/accumulation. Assay results were expressed as pmol product per milligram (mg) protein per number of weeks maintained in LTD. This study (Krohmer, in preparation) identified a steady and significant increase in estrone accumulation in the S/AHPOA homogenate between 4, 8, 12, and 16 weeks (see Fig. 13–4). Although in vitro estrone accumulation may not mirror what occurs in vivo, the consistent increase of estrogenic metabolites, described in this study corresponds directly to the observed increase in courtship intensity as a function of LTD duration reported by Garstka et al. (1982) and Krohmer (2004) (see Fig. 13–4, *insert*).

In a continuation of the studies examining the role of LTD and neural aromatization on the initiation of reproductive behavior in the RSGS (Krohmer, in preparation), we implanted male RSGS subcutaneously with (1) an empty silastic tube (controls); (2) a tube containing the aromatase inhibitor, 1,4,6-androstatriene-3,17-dione (ATD); or (3) a tube containing ATD + a tube containing 17ß estradiol. Following implantation, animals were placed into LTD for a period of 16 weeks. On emergence, all animals were tested daily with attractive females. Control animals exhibited normal courtship behavior (Garstka et al., 1982), whereas animals implanted with only ATD failed to court attractive

Red-Sided Garter Snakes. Estrone accumulation increased significantly over the 16-week incubation period in assay tubes containing S/AHPOA tissue. In assay tubes containing nucleus sphericus tissue, estrone accumulation increased significantly over the first 12 weeks. Estrone accumulation in olfactory tissue was significantly higher in the first 4 weeks compared with weeks 8 to 16, whereas estrone accumulation in midbrain tissue was not significantly different between weeks 4 and 12, but then showed a significant drop in the 16-week samples. All statistical comparisons are from a two-way ANOVA followed by a Student-Newman-Keuls multiple comparisons test. Bar with different superscripts are significantly different for $p < .05$ (Krohmer, unpublished data). The insert shows the changes in the intensity of courtship generated by increased time maintained under conditions of LTD (redrawn from Krohmer, 2004) illustrating the similarity with estrone accumulation in the AHPOA/S during low-temperature dormancy.

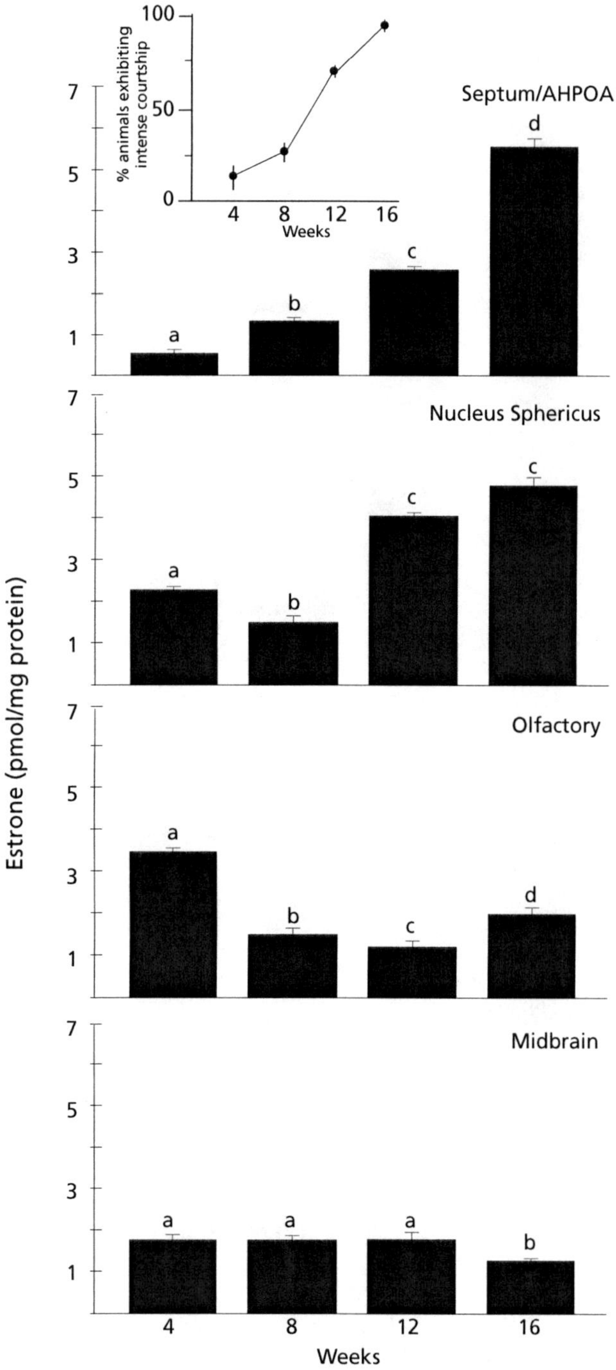

Figure 13–4. Aromatase Activity, as Measured by Estrone Accumulation (pmol/mg protein; mean + 1 SEM), of Regionally Dissected Brain Areas of Fall Collected Male

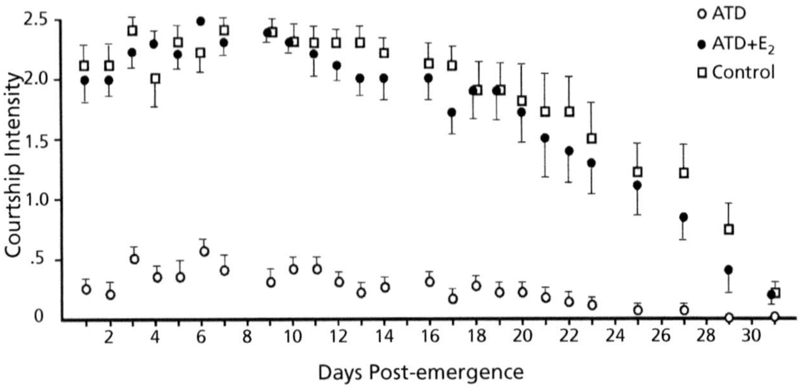

Figure 13–5. Intensity of Courtship Behavior Measured in Adult Male Red-Sided Garter Snakes Implanted with Silastic™ Tubes Containing ATD, an Aromatase Inhibitor, ATD + E2, or a Blank Tube Before a 16-Week Period of Low Temperature Dormancy. Intensity is defined as the mean courtship score (Camazine et al., 1980) of all individuals within each treatment group assessed during each testing period. On emergence, animals that received an ATD implant showed little, if any, courtship behavior during the testing period. Animals receiving implants of both ATD and E2 exhibited normal courtship behavior and mating and were indistinguishable from the blank implant controls (Krohmer, in preparation).

females. However, animals receiving implants of both ATD and 17β estradiol exhibited normal courtship behavior, mirroring control animals (Fig. 13–5).

CONCLUSIONS

The widespread presence of ARO-ir cells throughout the RSGS brain and persistence of enzymatic activity at low temperatures typical of LTD may represent an adaptation to cold. Under warm assay conditions, AA in S/AHPOA homogenates resulted in an accumulation of approximately 12 pMol estrone/mg protein in a 90-minute incubation (Krohmer et al., 2010), whereas under conditions of LTD an average of only 6 pMol estrone/mg protein was accumulated during a 16-week incubation (Krohmer, in preparation). Even though enzymatic activity is considerably reduced under conditions of LTD, the widespread presence of the aromatase enzyme at many levels of the RSGS forebrain, coupled with the long duration of the dormancy period would allow aromatization of T to progressively accumulate sufficient amounts of estrogenic metabolites during the winter that would then be available to initiate reproductive behaviors on emergence (Krohmer, in preparation).

In the RSGS, the decline of androgen levels during dormancy is the result of metabolic clearance and androgen levels reach basal levels significantly sooner in animals maintained at 10°C as compared with 5°C (Lutterschmidt & Mason, 2009). These findings appear to support the concept that the rapid clearance of androgens following spring emergence reported by Krohmer et al. (1987) is owing, at least in part, to a higher rate of clearance brought on by warm spring temperatures. It is also likely that this metabolic clearance induced by increasing temperature also applies to aromatized metabolites accumulated during LTD within the brain. In this way, estrogens would be present to initiate sexual behavior immediately after emergence from the den, but would also disappear very rapidly thereafter.

For the past 20 years, research in my lab has been looking beyond "LTD only" as the cue controlling the initiation of courtship behavior and mating in the RSGS. Current data have demonstrated that before entering winter dormancy, circulating

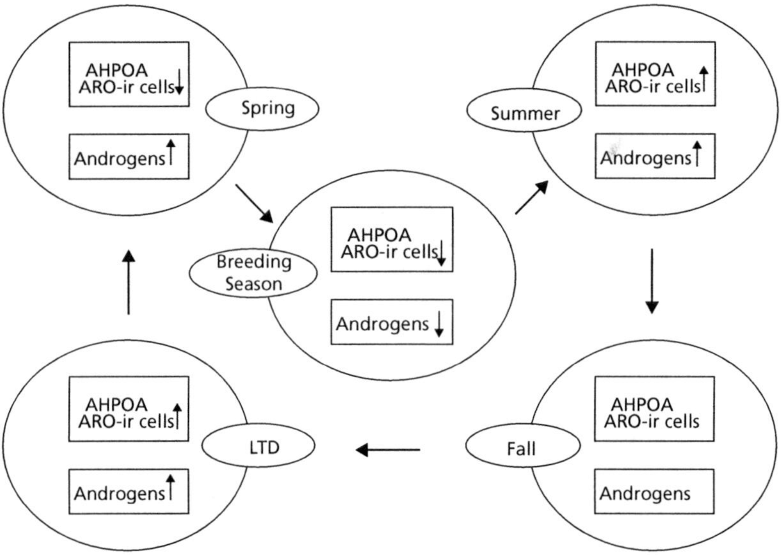

Figure 13–6. Conceptual Model Illustrating the Evolution of Hormonal and Neuroendocrine Events Regulating Reproduction in the Male RSGS; a Species Exhibiting the Dissociated Reproductive Pattern An intact AHPOA has been shown to be critical for the initiation and maintenance of courtship behavior and mating (Friedman & Crews, 1984; Krohmer & Crews, 1987) and blocking AA during LTD eliminates the display of mating behavior (Krohmer, in preparation). In the spring, reproductive behavior is initiated when circulating androgens are elevated, but the density of aromatase immunoreactive (ARO-ir) cells in the AHPOA is low (Krohmer, unpublished). Within 10 to 14 days postemergence androgen levels are basal, remaining low until spermatogenesis/steroidogenesis is initiated at the end of the breeding season. Androgen levels increase during the active season, reaching peak concentration by late summer. Concomitant with the rise in circulating androgens, the number of ARO-ir cells increase throughout the brain, with the majority of Type I cells concentrated in the AHPOA (Krohmer et al., 2002). As animals enter LTD, androgen levels remain elevated and ARO-ir cells are at their greatest density. Subsequently, aromatase activity continues throughout LTD accumulating estrogenic metabolites in pathways essential for the display of courtship behavior and mating on emergence.

androgens are elevated, ARO-ir cells are present in large numbers in the fall, specifically in areas shown to control courtship, the aromatase enzyme remains active at dormancy temperatures, and administration of an aromatase inhibitor during LTD will reduce or eliminate reproductive behavior. Together these data support our working hypothesis according to which the neural pathways controlling courtship and mating are "primed and activated" during LTD by the accumulation of estrogenic metabolites aromatized from circulating androgens (Fig. 13–6).

ACKNOWLEDGMENTS

The author thanks the following people for their technical assistance and critical feedback on various aspects of this review: Deborah I. Lutterschmidt, Robert T. Mason, Ignacio T. Moore, Michael Greene, Daniel D. Baleckaitis, and Jacques Balthazart. The author also thanks Christina Aviles for her help with the figures. RWK is funded in part by the Center for Engaged Learning and the Deans Fund, Saint Xavier University.

REFERENCES

Aleksiuk M., & Stewart, K. (1971). Seasonal changes in the body composition of the garter snake (*Thamnophis sirtalis parietalis*) at northern latitudes. *Ecology, 52*, 485–490.

Ball, G. F., & Balthazart, J. (2009). Neuroendocrine regulation of reproductive behavior in birds. In: Pfaff, D. W., Arnold, A. P., Etgen, A. M., Fahrbach, S. E., & Rubin, R. Y. (eds.), *Hormones Brain and Behavior*, vol 2. San Diego: Academic Press, pp. 855–896.

Balthazart, J., Absil, P., Foidart, A., Houbart, M., Harada, N., & Ball, G. F. (1996b). Distribution of aromatase immunoreactive cells in the forebrain of zebra finches (*Taeniopygia guttata*): implications for the neural action of steroids and nuclear definition in the avian hypothalamus. *J Neurobiol, 31*, 129–148.

Balthazart, J., Arnold, A. P., & Adkins-Regan, E. (2009). Sexual differentiation of brain and behavior in birds. In: Pfaff, D. W., Arnold, A. P., Etgen, A. M., Fahrbach, S. E., & Rubin, R. T. (eds.), *Hormones Brain and Behavior*, vol 3. San Diego: Academic Press, pp. 1745–1788.

Balthazart, J., Foidart, A., Surelmont, C., Vockel, A., & Harada, N. (1990). Distribution of aromatase in the brain of the Japanese quail, ringdove, and zebra finch: an immunocytochemical study. *J Comp Neurol, 301*, 276–288.

Balthazart, J., Tlemcani, O., & Harada, N. (1996a). Localization of testosterone sensitive and sexually dimorphic aromatase-immunoreactive cells in the quail preoptic area. *J Chem Neuroanat, 11*, 147–171.

Beck, L. A., O'Bryant, E. L., & Wade, J. (2008). Sex and seasonal differences in morphology of limbic forebrain nuclei in the green anole lizard. *Brain Res, 1227*, 68–75.

Beck, L. A. & Wade, J. (2008). Steroid receptor expression in the developing copulatory system of the green anole lizard (Anolis carolinensis). *Gen. Comp. Endocrin.* 157, 70-74.

Bona-Gallo, A., & Licht, P. (1983). Effect of temperature on sexual receptivity and ovarian recrudescence in the garter snake, *Thamnophis sirtalis parietalis*. *Herpetologica, 39,* 173–182.
Bonsall, R. W., Clancy, A. N., & Michael, R. P. (1992). Effects of the nonsteroidal aromatase inhibitor, fadrazole, on sexual behavior in male rats. *Hum Behav, 26,* 240–254.
Callard, G. V., Kunz, T. H., & Petro, Z. (1983). Identification of androgen metabolic pathways in the brain of little brown bats (*Myotis lucifugus*): sex and seasonal differences. *Biol Reprod, 28,* 1155–1161.
Callard, G. V., Petro, Z., & Ryan, K. J. (1977). Identification of aromatase in the reptilian brain. *Endocrinology, 100,* 1214–1218.
Callard, G. V., Petro, Z., & Ryan, K. J. (1978a). Phylogenetic distribution of aromatase and other androgen-converting enzymes in the central nervous system. *Endocrinology, 103,* 2283–2290.
Callard, G. V., Petro, Z., & Ryan, K. J. (1978b). Conversion of androgen to estrogen and other steroids in the vertebrate brain. *Am Zool, 18,* 511–523.
Callard, G. V., Schlinger, B. A., Pasmanik, M., & Corina, K. (1990). Aromatization and estrogen action in brain. *Progr Clin Biol Res, 342,* 105–111.
Camazine, B., Garstka, W., Tokarz, R., & Crews, D. (1980). Effects of castration and androgen replacement on male courtship behavior in the red-sided garter snake (*Thamnophis sirtalis parietalis*). *Hum Behav, 14,* 358–372.
Cherry, J. A., Basham, M. E., Weaver, C. E., Krohmer, R. W., & Baum, M. J. (1990). Ontogeny of the sexually dimorphic male nucleus in the preoptic/anterior hypothalamus of ferrets and its manipulation by gonadal steroids. *J Neurobiol, 21,* 844–847.
Crews, D. (1983). Control of male sexual behavior in the Canadian red-sided garter snake. In: Balthazart, J., Prove, E., & Gilles, R. (eds.), *Hormones and Behaviour in Higher Vertebrates.* Berlin: Springer-Verlag, pp. 398–405.
Crews, D. (1984). Gamete production, sex steroid secretion and mating behavior uncoupled. *Hum Behav, 18,* 21–28.
Crews, D. (1991). Trans-seasonal action of androgen in the control of spring courtship in male red-sided garter snakes. *Proc Natl Acad Sci U S A, 88,* 3545–3548.
Crews, D., Camazine, B., Diamond, M., Mason, R., Tokarz, R., & Gastka, W.R. (1984). Hormonal independence of courtship behavior in the male garter snake. *Hum Behav, 18,* 29–41.
Crews, D., & Garstka, W. R. (1982). The ecological physiology of a garter snake. *Sci Am, Nov,* 158–168.
Crews, D., Robker, R., & Mendonça, M. (1993). Seasonal fluctuations in brain nuclei in the red-sided garter snake and their hormonal control. *J Neurosci, 13,* 5356–5364.
Crews, D., & Silver, R. (1985). Reproductive physiology and behavior interactions in non-mammalian vertebrates. In: Adler, N., Pfaff, D., & Goy, R. W. (eds.), *Handbook of Behavioral Neurobiology,* vol 8. New York: Plenum, pp. 101–182.
Dohler, K. D., Coquelin, A., Davis, F., Hines, M., Shryne, J. E., & Gorski, R. A. (1984). Pre- and postnatal influences on testosterone propionate and diethylstilbestrol on differentiation of sexually dimorphic nucleus of the preoptic area of male and female rats. *Brain Res, 302,* 291–295.
Foidart, A., Reid, J., Absil, P., Yoshimura, N., Harada, N., & Balthazart, J. (1995a). Critical re-examination of the distribution of aromatase-immunoreactive cells in the quail

forebrain using antibodies raised against human placental aromatase and against the recombinant quail, mouse or human enzyme. *J Chem Neuroanat, 8,* 267–282.

Foidart, A., Reid, J., Absil, P., Yoshimura, N., Harada N., & Balthazart, J. (1995b). Critical re-examination of the distribution of aromatase-immunoreactive cells in the quail forebrain using antibodies raised against human placental aromatase and against the recombinant quail, mouse or human enzyme. *J Chem Neuroanat, 8,* 267–282.

Friedman, D., & Crews, D. (1985a). Role of the anterior hypothalamus-preoptic area in the regulation of courtship behavior in the male Canadian red-sided garter snake, *Thamnophis sirtalis parietalis:* intracranial implantation experiments. *Hum Behav, 19,* 122–136.

Friedman, D., & Crews. D. (1985b). Role of the anterior hypothalamus-preoptic area in the regulation of courtship behavior in the male Canadian red-sided garter snake (*Thamnophis sirtalis parietalis*): lesion experiments. *Behav Neurosci, 99,* 942–949.

Garstka, W. R., Camazine, B., & Crews, D. (1982). Interaction of behavior and physiology during the annual reproductive cycle of the red-sided garter snake (*Thamnophis sirtalis parietalis*). *Herpetologica, 38,* 104–123.

Gobbetti, A., Zerani, M., DiForce, M. M., & Botte, V. (1994). Relationships among GnRH, substance P, prostaglandins, sex steroids and aromatase activity in the brain of the male lizard *Podarcis sicula sicula* during reproduction. *J Reprod Fertil, 101,* 523–528.

Gorski, R. A., Shryne, J. H., & Southam, A. M. (1978). Evidence for a morphological sex difference within the medial preoptic area of the rat brain. *Brain Res, 143,* 333–346.

Gustafson, A. W. (1979). Male reproductive patterns in hibernating bats. *J Reprod Fertil, 56,* 317–331.

Gustafson, A. W., & Shemesh, M. (1976). Changes in plasma testosterone levels during the annual reproductive cycle of the hibernating bat, *Myotis lucifugus lucifugus* with a survey of plasma testosterone levels in adult male vertebrates. *Biol Reprod, 15,* 9–24.

Halpern, M. (1980). The telencephalon of snakes. In: Ebbesson, S. O. E. (ed.), *Comparative Neurology of the Telencephalon.* New York: Plenum, pp. 257–295.

Halpern, M., & Kubie, J. L. (1984). The role of the ophidian vomeronasal system in the species-typical behavior. *Trends Neurosci, 7,* 472–477.

Halpern, M., Morrell, J. I., & Pfaff, D. W. (1982). Cellular [^{3}H] testosterone localization in the brains of garter snakes: An autoradiographic study. *Gen Comp Endocrinol, 46,* 211–224.

Halpern, M., & Silfen, R. (1974). The efferent connections of the nucleus sphericus in the garter snake, *Thamnophis sirtalis. Anat Rec, 178,* 368.

Hutchison, J. B. (1974a). Effects of photoperiod on the decline in behavioral responsiveness to intra-hypothalamic androgen in doves (*Streptopelia risoria*). *J Endocrinol, 63,* 583–584.

Hutchison, J. B. (1974b). Post-castration decline in behavioral responsiveness to intra-hypothalamic androgen in doves. *Brain Res, 81,* 169–181.

Hutchison, J. B., Steimer, Th., & Jaggard, P. (1986). Effects of photoperiod on formation of oestradiol-17β in the dove brain. *J Endocrinol, 109,* 371–377.

Ingle, D., & Crews, D. (1985). Vertebrate neuroethology: definitions and paradigms. *Annu Rev Neurosci, 8,* 457–494.

Krohmer, R. W. (2004). The male red-sided garter snake (*Thamnophis sirtalis parietalis*): reproductive pattern and behavior. *Inst Lab Anim Res, 45,* 65–74.

Krohmer, R. W., & Aldridge, R. D. (1985). Male reproductive cycle of the lined snake (*Tropidoclonium lineatum*). *Herpetologica, 41,* 33–38.
Krohmer, R. W., & Baleckaitis, D. D. (2000). Does brain aromatase activity control courtship in a species with a dissociated reproductive cycle, the red-sided garter snake? *Trabajos del Instituto Cajal. Tomo, 77,* 102–105.
Krohmer, R. W., Bieganski, G.J., Baleckaitis, D. D., Harada, N., & Balthazart, J. (2002). Distribution of aromatase immunoreactivity in the forebrain of red-sided garter snakes at the beginning of the winter dormancy. *J Chem Neuroanat, 23,* 59–71.
Krohmer, R. W., Boyle, M. H., Lutterschmidt, D. I., & Mason, R. T. (2010). Seasonal aromatase activity in the brain of the red-sided garter snake. *Hum Behav, 58,* 485–492.
Krohmer, R. W., & Crews, D. (1987a). Temperature activation of courtship behavior in the male red-sided garter snake (*Thamnophis sirtalis parietalis*): role of the anterior hypothalamus-preoptic area. *Behav Neurosci, 101,* 228–236.
Krohmer, R. W., & Crews, D. (1987b). Facilitation of courtship behavior in the male red-sided garter snake (*Thamnophis sirtalis parietalis*) following lesions of the septum or nucleus sphericus. *Phsyiol Behav, 40,* 759–765.
Krohmer, R. W., Grassman, M., & Crews, D. (1987). Annual reproductive cycle in the male red-sided garter snake, *Thamnophis sirtalis parietalis:* field and laboratory studies. *Gen Comp Endocrinol, 68,* 64–75.
Krohmer, R. W., & Lutterschmidt, D. I. (2011). Environmental and neuroendocrine control of reproduction in snakes. In: Aldridge, R. D. & Sever, D. M. (eds.), *Reproductive Biology and Phylogeny of Snakes.* Enfield, NH: Science Publishers, pp. 289–346.
Licht, P. (1984). Reptiles. In: Lamming, G. E. (ed.), *Marshall's Physiology of Reproduction, Reproductive Cycles of Vertebrates,* vol. 1. Edinburgh: Livingstone, pp. 206–282.
Lofts, B. (1977). Patterns of spermatogenesis and steroidogenesis in male reptiles. In: Tyndale-Biscoe, C. H. (ed.), *Reproduction and Evolution.* Canberra: Australian Academy of Science, pp. 127–136.
Lutterschmidt, D. I., & Mason, R. T. (2009). Endocrine mechanisms mediating temperature-induced reproductive behavior in red-sided garter snakes (*Thamnophis sirtalis parietalis*). *J Exp Biol, 212,* 3108–3118.
Madeira, M. D., Leal, S., & Paula-Barbosa, M. M. (1999). Stereological evaluation and Golgi study of the sexual dimorphisms in volume, cell numbers, and cell size in the medial preoptic nucleus of the rat. *J Neurocytol, 28,* 131–148.
Moore, I. T., Green, M. J., & Mason, R. T. (2001). Environmental and seasonal adaptations of the adrenocortical and gonadal response to capture stress in two populations of the male garter snake, *Thamnophis sirtatis. J Exp Zool, 289,* 99–108.
Moore, I. T., LeMaster, M. P., & Mason, R. T. (2000). Behavioral and hormonal responses to capture stress in the male red-sided garter snake, *Thamnophis sirtalis parietalis. Anim Behav, 59,* 529–534.
Morrell, J. I., & Pfaff, D. W. (1978). A neuroendocrine approach to brain function: localization of sex steroid concentrating cells in vertebrate brains. *Am Zool, 18,* 447–460.
Naftolin, F., Ryan, K. J., Davies, I. J., Takaoka, Y., & Wolin, I. (1975). The formation of estrogens by central neuroendocrine tissues. *Rec Progr Horm Res, 31,* 295–319.
O'Bryant, E. L., & Wade, J. (2002). Seasonal and sexual dimorphisms in the green anole forebrain. *Hum Behav, 41,* 384–395.

Panzica, G. C., Arévalo, R., Sánchez, F., Alonso, J.R., Aste, N., Viglietti-Panzica, C., et al. (1994). Topographical distribution of reduced nicotinamide adenine dinucleotide phosphate-diaphorase in the brain of the Japanese quail. *J Comp Neurol, 342,* 97–114.

Panzica, G. C., Viglietti-Panzica, C., & Balthazart, J. (1996). The sexually dimorphic medial preoptic nucleus of quail: a key brain area mediating steroid action on male sexual behavior. *Front Neuroendocrinol, 17,* 51–125.

Pasmanik, M., & Callard, G.V. (1985). Aromatase and 5α-reductase in the teleost brain, spinal cord, and pituitary gland. *Gen Comp Endocrinol, 60,* 244–251.

Roselli, C. E., Horton, L. E., & Resko, J. A. (1985). Distribution and regulation of aromatase activity in the rat hypothalamus and limbic system. *Endocrinology, 117,* 2471–2477.

Rosen, G. J., & Wade, J. (2001). Androgen metabolism in the brain of the green anole lizard (*Anolis carolinensis*): effects of sex and season. *Gen Comp Endocrinol, 122,* 40–47.

Saint Girons, H. (1982). Reproductive cycles of male snakes and their relationships with climate and female reproductive cycles. *Herpetologica, 38,* 5–16.

Shinoda, K., Nagano, M., & Osawa, Y. (1994). Neural aromatase expression in preoptic, strial, and amygdaloid regions during late prenatal and early post natal development in the rat. *J Comp Neurol, 343,* 113–129.

Simerly, R. B. (2002). Wired for reproduction: Organization and development of sexually dimorphic circuits in the mammalian forebrain. *Annu Rev Neurosci, 25,* 507–536.

Tobet, S. A., Zahniser, D. J., & Baum, M. J. (1986). Differentiation in male ferrets of a sexually dimorphic nucleus in the preoptic/anterior hypothalamic area requires prenatal estrogen. *Neuroendocrinology, 44,* 299–308.

Ulibarri, C., & Yarh, P. (1988). Role of androgens in sexual differentiation of brain structure, scent marking, and gonadotropin secretion in gerbils. *Behav Neural Biol, 49,* 27–44.

Vagell, M. E., & McGinnis, M. Y. (1997). The role of aromatization in the restoration of male rat reproductive behavior. *J Neuroendocrinol, 9,* 415–421.

Veney, S. L., & Rissman, E. F. (1998). Co-localization of estrogen receptor and aromatase enzyme immunoreactivities in adult musk shrew brain. *Hum Behav, 33,* 151–162.

Wade, J. (1997). Androgen metabolism in the brain of the green anole lizard (*Anolis carolinensis*). *Gen Comp Endocrinol, 106,* 127–137.

Wade, J., Huang, J-M., & Crews, D. (1993). Hormonal control of sex differences in the brain, behavior and accessory sex structures of whiptail lizards (*Cnemidophorus* species). *J Neuroendocrinol, 5,* 81–93.

Wagner, C. K., & Morrell, J. L. (1996). Distribution and steroid hormone regulation of aromatase mRNA expression in the forebrain of adult male and female rats: a cellular-level analysis using in situ hybridization. *J Comp Neurol, 370,* 71–84.

Winkler, S. M., & Wade, J. (1998). Aromatase activity and regulation of sexual behaviors in the green anole lizard. *Phsyiol Behav, 64,* 723–731.

Woolley, S. C., Sakata, J. T., & Crews, D. (2004). Evolutionary insights into the regulation of courtship behavior in male amphibians and reptiles. *Phsyiol Behav, 83,* 347–360.

Chapter 14

Differential Actions of Estrogens in the Male and Female Brain: A Case for Sex-Specific Medicines

Glenda E. Gillies

INTRODUCTION

Estrogens are ubiquitous hormones with profound and complex actions in the mammalian brain in both sexes. However, many sex dimorphisms are seen in brain structure and function, including different, and even opposite, effects of estrogens in male and female subjects. With a main focus on the nigrostriatal dopaminergic (NSDA) system and Parkinson's disease, and a brief consideration of other brain networks, this chapter addresses the view that sex differences in brain organization underlie sex-specific effects of estrogens. It also considers the different modes of action of systemically and centrally generated estrogens that are relevant to unraveling their relative importance in brain physiopathology.

Estrogens Contribute to Sex Differences in Disease Susceptibility

The ovaries and placenta are the primary sources of circulating estrogens in mammalian species, and for much of the 20th century estradiol was viewed solely as "the woman's hormone" responsible for regulating reproduction. With the rise of hormone replacement therapy in peri/postmenopausal women, estrogens also became credited with abilities to protect against diseases of aging in women, such as cardiovascular disease, Alzheimer's disease (AD) and Parkinson's disease (PD) (Sherwin, 2002; Brinton, 2004; Craig & Murphy, 2007b; Wise et al., 2009; Gillies & McArthur, 2010b). These and the majority of complex diseases, including schizophrenia, attention-deficit hyperactivity disorder (ADHD), autism, substance abuse, and affective disorders, show sex differences in their incidence, severity, and/or manifestations (Aleman et al., 2003; Hennessy et al., 2004; Barnes et al., 2005; Baron-Cohen et al., 2005; Kaminsky et al., 2006; Shulman & Bhat, 2006; Becker & Hu, 2008; Gabory et al., 2009). Observational and retrospective studies of menopausal women with and without hormone replacement therapy suggest that estrogens are neuroprotective, neurotrophic, and psychotrophic (Fink et al., 1996; McEwen & Alves, 1999; Gillies et al., 2004; Craig et al., 2005; Cahill, 2006; Craig & Murphy, 2007a).

This led to the view that female ovarian steroids, and estrogens in particular, provide the female advantage and are responsible to a large degree for sex differences in disease susceptibility (Kaminsky et al., 2006; Gabory et al., 2009). This hypothesis is strongly supported by preclinical studies investigating the effects of estrogen treatment of ovariectomized laboratory animals (rodents, nonhuman primates) in experimental models of human disease (Dluzen & Horstink, 2003; Brinton, 2004; Gillies et al., 2004; Becker & Hu, 2008; Pike et al., 2009; Solomon & Herman, 2009; Gillies & McArthur, 2010b).

Estrogens Are Not Just Female Hormones

Several lines of evidence now suggest that the view of ovarian steroids in females as the only sex hormones that contribute to sex differences in central nervous system (CNS) disorders is too simplistic. Convergent evidence now reveals that estrogens have physiologically significant roles to play in males as well as females of all mammalian species studied. This concept is underpinned by the realization that estrogens can be synthesized locally in tissues throughout the body (including the brain) from circulating precursors (such as testis-derived testosterone that readily crosses the blood-brain barrier) by the aromatase enzyme (Sharpe, 1998; Jones et al., 2006; Roselli et al., 2009). A further advance came with the understanding that in vertebrates (including humans) and some insects, the brain itself expresses the full complement of steroid synthesizing enzymes capable of producing estrogens, progestagens, and androgens de novo from cholesterol, as well as from precursors (Tsutsui, 2008; Do Rego et al., 2009; Luchetti et al., 2010b). The relevance of these findings is supported by the cross-species, widespread distribution of the classical nuclear estrogen receptors (ERα and ERβ) in the body and brain of males and females, as well as the understanding that estrogens also exert rapid effects at the cell membrane of neuronal and non-neuronal cells (Toran-Allerand, 2005; Balthazart & Ball, 2006; Brann et al., 2007; Micevych & Dominguez, 2009). Together, these findings highlight the potential for systemically and/or centrally derived estrogens to influence processes ranging from effects on memory, mood, and behavior to neurodevelopment and neurodegeneration in both sexes (Fink et al., 1996; McEwen & Alves, 1999; Gillies et al., 2004; Belelli & Lambert, 2005; Craig et al., 2005; Cahill, 2006; Craig & Murphy, 2007a; Garcia-Segura, 2008; King, 2008). This offers great promise for estrogen-based therapies for brain disorders that affect both men and women, but that currently lack effective treatments (Rochira et al., 2002; Jones et al., 2006).

Brain Sex Dimorphisms

Although it appears incontrovertible that estrogens have potent actions in the male brain, the female brain has been the major focus for hormonal effects in experimental models of CNS disorders. It is noteworthy that this occurs despite the fact that males are more affected by certain CNS disorders, such as Parkinson's disease, schizophrenia, and autism-spectrum disorders, and yet the remaining vast body of neuroscience research has been carried out principally on the male brain (Cahill, 2006; Luine, 2007; Beery & Zucker, 2011). Of further significance is the emerging view that sex dimorphisms in regional brain structure, morphology, neurochemistry, connectivity, and function appear to be the norm rather than the exception. This includes regions in human and animals brains that are important for cognition, memory, and affect, such as the hippocampus, amygdala, and prefrontal cortex (Kelly et al., 1999; Allen et al., 2003; De Vries, 2004; Luders et al., 2004; Baron-Cohen et al., 2005; McCarthy & Konkle, 2005; Cahill, 2006; Cosgrove et al., 2007; Wilson & Davies, 2007; Swaab, 2008), and regions involved in sensorimotor control, reward, and motivation, especially the midbrain dopaminergic systems originating in the substantia nigra pars compacta (SNc) and the ventral tegmental area (VTA) (Becker, 1999; Dewing et al., 2006; Cantuti-Castelvetri et al., 2007; McArthur et al., 2007a). Furthermore, these differences may have significance for the increasing evidence that systemically administered estrogens can have different, sometimes opposite, effects on certain aspects of brain function in male and female subjects. This raises the possibility that evidence for an involvement of estrogens in the physiological or pathological processes in one sex may not translate to the opposite sex. Additionally, the knowledge that neurosteroids, including estrogens, are produced within the brain in both sexes adds a level of complexity to interpreting the role of systemic steroids in CNS function, and the relative contributions of steroids of systemic and central origin to brain health and disease remain to be fully established.

The foregoing discussions highlight two major areas that must be considered if we are to realize the translational potential of estrogens and their synthetic enzymes, namely the significance of sex dimorphisms in the brain, and the systemic versus central sources of sex hormones, which are the focus of this chapter. Specifically, clinical examples and experimental models of disease will be used to highlight the importance of understanding sex differences in brain organization if we are to understand sex-specific hormonal effects. Such knowledge will be critical for developing optimal therapies for many common brain disorders that differentially affect men and women.

PARKINSON'S DISEASE

For numerous reasons, Parkinson's disease (PD) and experimental parkinsonism represent excellent prototypes for analyzing sex differences and hormonal actions in the brain. Parkinson's disease is a debilitating, neurodegenerative, highly prevalent motor disorder, affecting approximately 1 in 500 people in the developed world, but in that number there at least twice as many men than women (Baldereschi et al., 2000; Swerdlow et al., 2001; Elbaz et al., 2002; Van Den Eeden et al., 2003; Pavon et al., 2010). The hallmark pathological lesion in PD centers on the degeneration of a discrete, midbrain neurotransmitter pathway, the nigrostriatal dopaminergic (NSDA) system, comprising cell bodies in the substantia nigra pars compacta (SNc) and their terminals in the dorsal striatum, which leads to a loss of striatal dopamine function. For all but a few percent of cases, the causes of PD are unknown, and the etiology of this multifactorial complex disease involves a gene–environment–aging interaction (Klein & Schlossmacher, 2007). Interestingly, molecular analyses of post-mortem brains and in vivo neuroimaging studies indicate marked sex differences in the mechanisms of disease processes (Cantuti-Castelvetri et al., 2007; Simunovic et al., 2011) and their clinical profiles (Shulman & Bhat, 2006; Haaxma et al., 2007; Miller & Cronin-Golomb, 2011). Even though the NSDA pathway performs the same function in normal men and women, real-time in vivo imaging techniques have also revealed notable sex differences in its functioning in the healthy brain. These include differences in amphetamine-stimulated striatal dopamine release (Munro et al., 2006), basal striatal dopamine dynamics (Pohjalainen et al., 1998; Lavalaye et al., 2000; Kaasinen et al., 2001; Mozley et al., 2001; Laakso et al., 2002; van Beilen et al., 2008), and in striatal responses while performing tasks that target cognitive and motor functioning associated with dopaminergic activity (Mozley et al., 2001). Although there is a growing appreciation that genetic factors, including factors linked to the sex chromosomes, as well as life-style could contribute to sex bias in disease (Federman, 2006; van Nas et al., 2009), sex hormones are considered by far to have the greatest influence on brain sex dimorphisms, whether in structure, function, or disease susceptibility (Kaminsky et al., 2006). The following sections consider what role estrogens may have in PD, first in humans, followed by experimental PD.

Clinical Studies

Clinical observations of parkinsonian symptoms in women over the menstrual cycle, at menopause and on withdrawal/reinstatement of hormone replacement therapy (all states of fluctuating circulating levels of estrogens) suggest that estrogens confer neuroprotection and are likely to contribute to the female advantage (Dluzen, 2000; Dluzen & Horstink, 2003; Shulman & Bhat, 2006). Although a large, randomized controlled study of estrogen treatment in postmenopausal women has revealed many, as yet unresolved, complexities in evaluating the risks and benefits of HRT and in optimizing treatment regimes and trial design (Rossouw et al., 2002; Wise et al., 2009), the potential for estrogenic compounds to act as neuroprotective agents in PD has been recognized (Ravina et al., 2003), and some studies have reported that the symptoms and risk of developing PD may be reduced by estradiol treatment (Saunders-Pullman et al., 1999; Tsang et al., 2000; Benedetti et al., 2001; Shulman & Bhat, 2006; Liu & Dluzen, 2007; Saunders-Pullman et al., 2009). However, clinical studies do not permit investigations into the nature of estradiol's actions, and cannot distinguish between the effects of gonadally derived, systemically administered or centrally generated steroids. This appears to be an increasingly important consideration, because, although there have been no specific reports of estradiol synthesis in the human NSDA system, expression (mRNA and protein) and activity of numerous steroid biosynthetic enzymes and their products have been located in nigrostriatal regions in human post-mortem brains, with reports of altered levels in neurodegenerative conditions, including PD (Bixo et al., 1997;

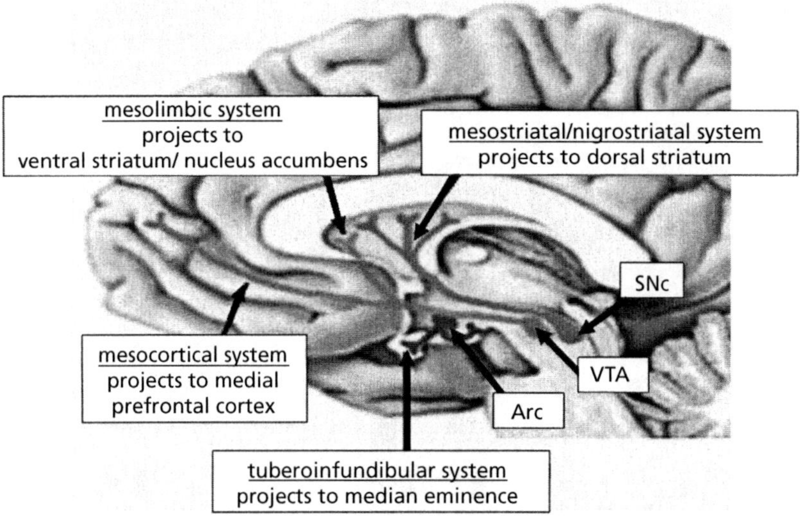

Figure 14–1. Schematic Diagram of the Midbrain Dopaminergic Systems. The dopaminergic neurons of the substantia nigra pars compacta (SNc) innervate the dorsal striatum and are involved in sensorimotor control. The dopaminergic neurons of the ventral tegmental area (VTA) innervate the prefrontal cortex (roles in regulating goal-directed behavior, cognition, attention/anticipation) and ventral striatum/nucleus accumbens (roles in regulating motivation, reward, and reinforcement). The dopaminergic neurons of the arcuate nucleus (Arc) of the hypothalamus terminate at the hypophysial portal vessels if the median eminence and inhibit prolactin secretion from the anterior pituitary gland.

di Michele et al., 2003; Luchetti et al., 2010a). Human studies also do not provide clues as to potential hormonal influences in men. However, experimental models of PD suggest that estrogens of systemic and central origins have distinct, differential influences, as discussed in the following sections.

Preclinical Studies

Neurotoxins such as 6-hydroxydopamine (6-OHDA) and 1-methyl-4-phenyl-1-,2,3,6-tetrahydropyridine (MPTP), selectively target dopaminergic neurons and are widely used to create experimental animal models that mimic many aspects of PD pathology (Schwarting & Huston, 1996b; Gillies et al., 2004; Fox & Brotchie, 2010).

Systemic sex steroids and susceptibility to parkinson's disease

Experimental models of PD have been used with great effect to investigate estrogenic influences on the loss of dopamine in the striatum or dopaminergic neurons in the SNc in rats, mice, and monkeys, using estradiol treatment of ovariectomized animals as the principal experimental paradigm. These studies have been unanimous in finding that physiological levels of circulating estradiol are neuroprotective in females and that striatal dopamine loss is significantly greater in males than females (Fig. 14–2) (Miller et al., 1998; Leranth et al., 2000; Yu & Liao, 2000; Datla et al., 2003; Dluzen & Horstink, 2003; Murray et al., 2003; Gillies et al., 2004; Ferraz et al., 2008; Morissette & Di Paolo, 2009; Gillies & McArthur, 2010b). Although fewer studies have been performed in

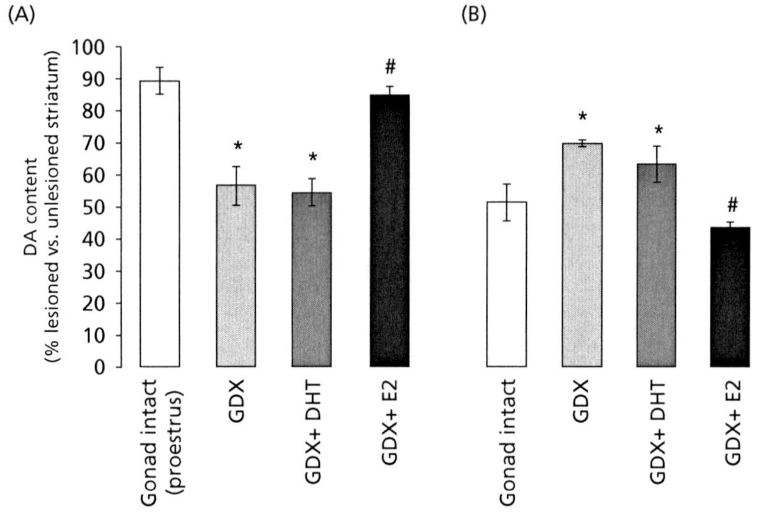

Figure 14–2. Manipulation of the Hormone Environment Exerts Sex-Specific Effects on Striatal Lesions Induced by 6-Hydroxydopamine (6-OHDA). Male and female rats underwent bilateral gonadectomy (GDX) or sham operation (gonad intact) followed immediately, while still under anesthesia, by subcutaneous implantation of a pellet, containing estradiol (replaced to proestrous levels), 5α-dihydrotestosterone (to mimic physiological androgen levels), or a placebo pellet (full details in Murray et al., 2003). Approximately 1 week later all animals received a unilateral infusion of 6-hydroxydopamine (6-OHDA; 1 μg) into the left medial forebrain bundle and, after killing 2 weeks later, brains were processed to assess lesion size. Results represent the percentage striatal dopamine (DA) remaining in the lesioned (left) side compared with the contralateral, unlesioned (right) side. Sham-operated animals (gonad intact) are taken as controls. A. Following ovariectomy levels of DA in the female striatum were significantly reduced compared with their sham-operated counterparts. This effect was unaltered by treatment with the nonaromatizable androgen, dihydrotestosterone (DHT,), but was reversed by estradiol (E2). B. Following orchidectomy, levels of DA in the male striatum were significantly greater than their sham-operated counterparts. This effect was unaffected by treatment with DHT, but was reversed by estradiol (E2). $^*p < .05$ versus gonad intact; $\#p < .05$ versus GDX; $\dagger p < .05$ for male versus female. Adapted from Murray, H. E., Pillai, A. V., Mcarthur, S. R., Razvi, N., Datla, K. P., Dexter, D. T., et al. (2003). Dose- and sex-dependent effects of the neurotoxin 6-hydroxydopamine on the nigrostriatal dopaminergic pathway of adult rats: differential actions of estrogen in males and females. *Neuroscience, 116*, 213–222.

males, the consensus of opinion suggests that gonadally derived testosterone is not neuroprotective, but may exacerbate striatal lesion size in male rodents (see Fig. 14–2) (Dluzen et al., 1994; Yu & Wagner, 1994; Gao & Dluzen, 2001; Murray et al., 2003; Lewis & Dluzen, 2008). Moreover, hormonal treatment of castrated male rats in the 6-OHDA model of PD showed that physiological levels of estradiol, but not the nonaromatizable androgen dihydrotestosterone (DHT), were capable of reversing the effects of gonadectomy (see Fig. 14–2). These findings suggest that any influences of circulating testosterone to promote degeneration in the male NSDA system operate only after its conversion to estradiol by aromatase enzymes that are prevalent in the striatum (Kuppers et al., 2000; Kipp et al., 2006). They therefore demonstrate that detrimental effects of gonadal factors in the male NSDA system, as well as protective effects of ovarian factors in females, have compounding influences to create sex differences in striatal susceptibility to injury. Additionally, they reveal that estradiol treatment has opposite effects on striatal dopamine loss in males (increased) and females (reduced). If recapitulation of these phenomena is found in the human striatum, these findings highlight the possibility that the reported clinical benefits of estrogens in the female sex may not simply translate to the male sex. Although data are not available for the injured human striatum, this proposal is supported by reports that, in healthy individuals, performance on a striatal-mediated sequential movement task was improved by endogenous estradiol levels in women (Jennings et al., 1998), but in men performance was not related to endogenous estradiol or testosterone (Siegel et al., 2008). Thus, sex hormone effects on nigrostriatal activity appear to be sexually dimorphic in humans as well as rodents.

Compensation versus neuron survival

The majority of investigations in experimental PD measure only striatal loss of dopamine, which is presumed to equate with the extent of dopaminergic neuron loss in the SNc. Accordingly, in gonad-intact rats we have shown that the significantly greater loss of striatal dopamine in males (48.4% ± 5.6%) as compared with females (17.0% ± 4.3%) is mirrored in the significantly greater loss of dopaminergic cells in the male SNc (35.9 ± 3.8) as compared with females (20.8% ± 3%) (Murray et al., 2003), establishing sex differences in both parameters. However, despite the substantial evidence that sex hormone levels in the peripheral circulation markedly affect striatal dopamine depletion in experimental PD (discussed above), gonadectomy with and without replacement with estradiol or DHT failed to affect the loss of dopaminergic neurons in the male or female rat SNc after 6-OHDA–induced lesions (Moroz et al., 2003; McArthur et al., 2007b; Ferraz et al., 2008). Although it might at first appear anomalous that hormonal influences on striatal dopamine depletion occur independently of dopaminergic cell survival, several lines of evidence support our proposal that physiological levels of circulating hormones influence the activity of the surviving NSDA neurons rather than neuron survival per se (McArthur et al., 2007b; Gillies & McArthur, 2010a,b). In clinical and experimental PD, it is well established that the surviving neurons and their associated circuitry have a remarkable capacity for compensation, such that overt symptoms appear only after 80% of striatal dopamine and 60% of SNc dopaminergic neurons are lost (Castaneda et al., 1990; Bezard et al., 2001; Bassilana et al., 2005). Compensatory processes include increased synthesis, metabolism, and release per impulse of dopamine to preserve extracellular levels of neurotransmitter and striatal functionality. Significantly, circulating estradiol stimulates all of these parameters in the normal NSDA system of female rats, as well as behavioral recovery in 6-OHDA lesioned females, but fails to do so in males (McDermott et al., 1994; Pasqualini et al., 1995; Becker, 1999; Ohtani et al., 2001; Thanky et al., 2003; Serova et al., 2004; Dluzen, 2005; Tamas et al., 2005). Together, these findings suggests a compelling explanation as to why estrogens appear to be neuroprotective at the levels of the striatal dopaminergic terminals, but not the SNc perikarya, in females, and indicate hormonal mechanisms that would render the female NSDA system more able to adapt to injury. These observations also provide further substantial evidence that the effects of estradiol on both the intact and degenerating NSDA are sexually dimorphic.

Centrally generated estrogens and parkinson's disease

The undamaged NSDA system expresses aromatase (Kuppers et al., 2000; Kipp et al., 2006). Upregulation of this enzyme in glial cells is emerging as an important mechanism for local estrogen synthesis and protection against excitotoxic and ischemic injury in various regions of male and female brains (Hurn & Macrae, 2000; Azcoitia et al., 2001; Garcia-Segura et al., 2001; Azcoitia et al., 2005; Carswell et al., 2005; Garcia-Segura, 2008). Recent evidence suggests that aromatase may also play an important role in the injured NSDA system (McArthur et al., 2007b). In the 6-OHDA rat model of PD, central administration of the aromatase inhibitor, anastrazole, into the male medial forebrain bundle (carrying the NSDA tract) just before infusion of the neurotoxin exacerbated striatal dopamine loss without affecting neuronal numbers in the male SNc (Fig.14–3). Therefore, these data extend the evidence that local estrogen synthesis can protect striatal dopaminergic innervation against injury in male rats. Investigations using aromatase knockout mice suggest a similar protective role for aromatase and local estrogen synthesis in females (Morale et al., 2008).

The ability of central aromatase inhibition to affect striatal, but not SNc, susceptibility to a mild neurotoxic insult supports the evidence discussed in the preceding section that prevailing physiological levels of estrogens primarily promote plastic/adaptive compensatory mechanisms within the surviving dopaminergic neurons, rather than affecting SNc cell survival. As these findings clearly implicate endogenous, centrally synthesized estrogens as a protective force in males, they appear at odds with the hypothesis that central aromatization of testosterone derived from the peripheral circulation could explain why the effect of castration to reduce striatal dopamine depletion in 6-OHDA–induced injury could be reversed by systemic treatment of male rats with physiological levels of estradiol, but not DHT (see Fig. 14–2). Thus, we need to reconcile the

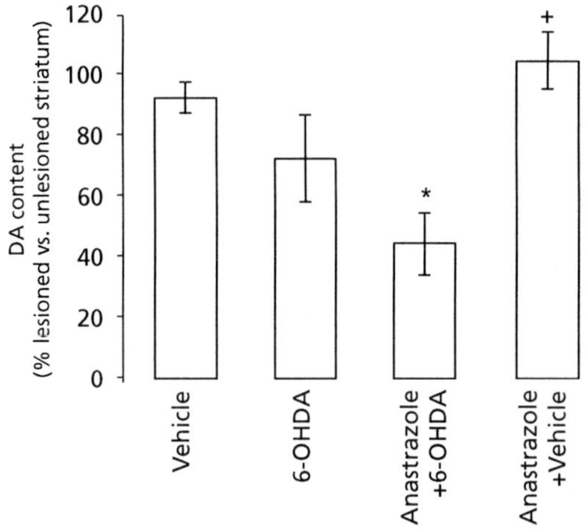

Figure 14–3. Inhibition of Striatal Aromatase Activity Augments the Lesion Size Following Administration of 6-Hydroxydopamine (6-OHDA). Administration of the selective aromatase inhibitor anastrazole (7 μg) to the left medial forebrain bundle of male rats concurrently with 1 μg 6-hydroxydopamine significantly increased the degree of striatal dopamine loss seen one week later. $^*p < .05$ versus vehicle treated group; $+p < .05$ versus anastrazole + 6-OHDA group. Adapted from McArthur, S., Murray, H. E., Dhankot, A., Dexter, D. T., & Gillies, G. E. (2007b). Striatal susceptibility to a dopaminergic neurotoxin is independent of sex hormone effects on cell survival and DAT expression but is exacerbated by central aromatase inhibition. *J Neurochem, 100,* 678–692.

findings that estrogens of peripheral or central origin have opposite effects on susceptibility to injury in the male, but not female, striatum. Several factors need to be considered, including the modes of action of systemic versus central steroids and their interactions with sexually dimorphic circuitry.

Systemic versus central steroids: modes of action and interactions with sexually dimorphic circuitry

Activation of neuronal aromatase or glial aromatase in the vicinity of neural injury, by definition, will cause a very local increase in estrogen synthesis. Concentrations of estradiol in male brains have been estimated to be at least six times greater than in the peripheral plasma (Mukai et al., 2006), and even considerably greater at certain locations, such as the synapse (Balthazart et al., 2002). Therefore, there could be considerable discrepancies between the concentrations achieved in the brain owing to estrogens of central or systemic origin, which in turn could dictate the neural response. Effects of centrally generated estrogens would also be rapid and restricted to the immediate neuronal environment (Balthazart & Ball, 2006; Prange-Kiel & Rune, 2006), with direct access to the NSDA system. In contrast, gonadal steroid levels change relatively slowly, and both gonadal and peripherally administered exogenous steroids can cross the blood-brain barrier, allowing them nonselective access to the whole brain. Therefore, systemic steroid hormones have the potential to act directly on the NSDA system as well as indirectly on networks that interact with these pathways, thereby controlling not only inherent NSDA activity, but also their plastic, compensatory responses to injury. Importantly, both the NSDA system and its regulatory input pathways exhibit sex dimorphisms, both inherently and in response to estradiol. Differences within the intact and degenerating NSDA pathway in men and women have already been highlighted, and this is supported and extended by abundant evidence in rodents. For example, sex differences are apparent in the numbers and topographical organization of SNc dopaminergic neurons (Dewing et al., 2006; McArthur et al., 2007a) that correlate with functional differences, such as an exaggerated response to cocaine, amphetamine, or apomorphine in female rats compared with males (Becker, 1999). It would also appear that similar extracellular striatal dopamine levels are maintained by a different balance of uptake and release mechanisms in males and females (Robinson et al., 1990; Walker et al., 2000). Moreover, these differences appear to affect responses to estradiol, which enhances amphetamine-stimulated release in vivo (Castner et al., 1993; Xiao & Becker, 1998), attenuates dopamine reuptake (Thompson, 1999) and promotes stereotypic behaviors involving the NSDA system (Becker, 1999) in females but not males. Sex dimorphisms are also thought to be present in the striatal medium spiny neurons providing critical GABAergic interneuron regulation of the basal ganglia, and furthermore, estradiol can suppress their negative influence on dopaminergic terminals to increases dopamine release in females, but not males (Hu et al., 2004, 2006; Schultz et al., 2009). Noradrenergic transmission from the locus coeruleus (LC) is also known to influence the NSDA system and its adaptive response to injury (Marien et al., 2004). Of particular interest, circulating estradiol upregulates the expression of tyrosine hydroxylase (TH, the rate-limiting enzyme for dopamine synthesis) in the female LC, whereas in males circulating testosterone downregulates LC TH expression only after conversion to estradiol (Thanky et al., 2002). The mesocortical dopaminergic pathways (Kritzer & Creutz, 2008) and the serotonergic system of the dorsal raphe (Klink et al., 2002) also represent sexually differentiated, estrogen-sensitive networks that could modify adaptations of the injured NSDA system to injury.

Together, these considerations illustrate how fundamental sex differences in the organization of NSDA-associated circuitry in the normal brain can account for the ability of gonadal factors and exogenous estradiol to preserve striatal dopamine in experimental PD in females, but not males, whereas estradiol generated locally may have direct effects to preserve striatal dopamine in both sexes. A hypothetical schema is summarized in Figure 14–4.

Origins of Sexually Dimorphic Responses to Estrogens

Our current understanding of sexual differentiation of the brain is based on concepts derived from a large number of studies of the hypothalamus. These focus on sexually dimorphic circuitries that are critical for ovulation in females and reproductive behaviors in both sexes and are excellently reviewed

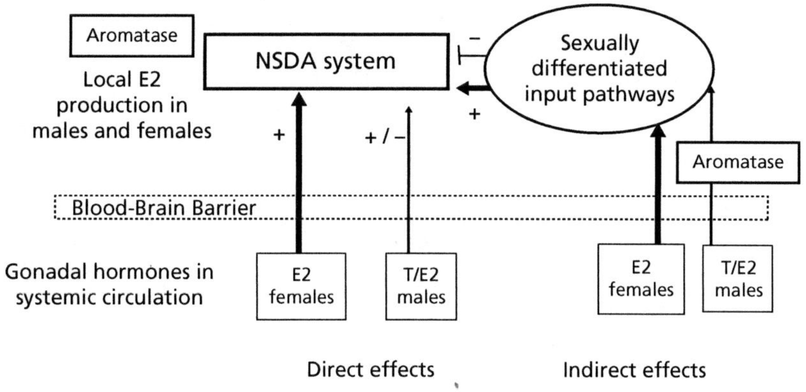

Figure 14–4. Summary of Potential Modes of Action of Peripheral and Central Sex Steroid Hormones on the Intact and Damaged Nigrostriatal Dopaminergic System in Male and Female Brains. Estradiol (E2) administered systemically (exogenous) or endogenous, peripherally derived E2 (ovarian production) in females, and its precursor in males (testicular production of testosterone), as well as centrally generated E2 (local upregulation of aromatase in response to NSDA injury) have the potential to act directly on the nigral dopaminergic neurons to upregulate activity/adaptive capacity in the surviving neurons of the partially injured NSDA system. However, peripherally derived E2 and its precursors also have the potential to act indirectly on sexually differentiated networks interacting with and regulating the NSDA system, resulting in an upregulation of NSDA activity in females, but a downregulation in males. These competing activities ultimately promote more effective adaptive responses in the partially injured NSDA system in females, whereas in males they offer no protection, and may even exacerbate striatal lesions. This hypothetical schema can explain why gonadal factors and exogenous E2 have the ability to preserve striatal dopamine levels in experimental Parkinson's disease only in females, but not males, whereas E2 generated in the region of the partially injured NSDA pathway reduces striatal dopamine loss in both sexes. Adapted from Gillies, G. E., & McArthur, S. (2010a). Estrogen actions in the brain and the basis for differential action in men and women: a case for sex-specific medicines. *Pharmacol Rev, 62,* 155–198.

elsewhere (Kelly et al., 1999; De Vries & Simerly, 2002; Arnold & Burgoyne, 2004; Morris et al., 2004; Simerly, 2005; Wilson & Davies, 2007; Arnold, 2009; McCarthy & Arnold, 2010). Although a genetic contribution, especially from the sex chromosomes (Arnold & Burgoyne, 2004), may instruct hypothalamic sexual differentiation, the classical view that this is driven principally by sex differences in the sex hormone environment at critical stages during development prevails. In summary, rodent studies have shown that the testes produce a transient rise in circulating testosterone just before and after birth, at a time when the ovaries are quiescent. This testosterone surge is thought to be instrumental in masculinizing/defeminizing the brain, but only after metabolism to estradiol by brain aromatase. In the neuroendocrine reproductive context, masculinization has been defined as the organization of the neural substrate required for male sexual behavior, which is distinct from defeminization, involving the loss of ability to display female sexual behavior. Most interestingly, defeminization has also been described as the loss of capacity to respond to the activational effects of estrogen to induce female sexual behavior (Schwarz & McCarthy, 2008). Underlying mechanisms include hard-wiring of sexually dimorphic neuronal networks (Simerly, 2005; Wilson & Davies, 2007), programming of astrocytic and synaptic plasticity (Naftolin et al., 2007), and differences in estrogen-stimulated intracellular signaling cascades (Abraham & Herbison, 2005). Thus, developmental exposure to testosterone (aromatized to estradiol) essentially suppresses responsiveness to estradiol in later life in specific hypothalamic processes that would otherwise characterize the female. We are still unclear as to the relevance these processes in extra-hypothalamic regions, but they are tantalizingly reminiscent of the evidence discussed in the preceding sections that the ability of estrogens to positively influence the female NSDA system is absent/suppressed in males. Although the origins of sex dimorphisms in the NSDA system are unclear, one study has reported that neonatal exposure to testosterone altered midbrain response to 6-OHDA–induced injury in adult rats and suppressed the neuroprotective effects of estradiol (Moroz et al., 2003).

Restrictions and Caveats

It must be emphasized that the foregoing discussions relate to experimental models of PD that generate partial lesions (<60%) of the NSDA pathway that are thought to represent the early preclinical stages of the disease (Schwarting & Huston, 1996a). Under such conditions, sex differences are widely detected, but with larger lesions (70%–90%) sex differences are no longer apparent, suggesting that the degree of intrinsic protection in the female brain can be overridden once degeneration reaches a certain extent (Gillies et al., 2004; Gillies & McArthur, 2010b). These observations are in accord with clinical findings suggesting that females may be more able to resist the onset and/or progression of PD, and highlight that only in cases in which submaximal, mild to moderate lesions are employed can factors that provide the intrinsic female advantage, such as sex hormones, be studied.

It must be emphasized also that the present discussions have focused on the influences of physiological levels of sex steroid hormones that, when combined with submaximal lesions of the NSDA system, primarily target the activity, not survival, of the remaining neurons (Ferraz et al., 2003; Moroz et al., 2003; Murray et al., 2003; McArthur et al., 2007b; Ferraz et al., 2008). In contrast, doses of estradiol that produce circulating levels that are severalfold greater than the highest, proestrous physiological levels have been reported to ameliorate SNc doapminergic cell loss in male and female rats (Quesada & Micevych, 2004; Schultz et al., 2009). Such pharmacological levels of estradiol have been shown to activate nonspecific process that prevent cell death in a variety of types of injury and brain regions, including antioxidant activity and upregulation of antiapoptotic genes and growth factors (Green & Simpkins, 2000; Behl, 2002; Garcia-Ovejero et al., 2005; Brann et al., 2007). This chapter has, therefore, focused on factors that are unique to a specific form of injury, rather than factors common to neuronal loss in general.

OTHER BRAIN NETWORKS

Evidence is rapidly emerging that networks in brain regions other than the NSDA system and hypothalamus are sexually dimorphic and respond to estradiol differentially in males and females, with sex dimorphisms in the underlying circuitry playing a prominent role (reviewed in Gillies & McArthur, 2010a). These include the hippocampus, amygdala, and prefrontal cortex. As already discussed for PD and the NSDA system, there are also interesting parallels emerging for sex differences in susceptibility to diseases involving these brain regions, in which sex steroid hormones, particularly estrogens, have a prominent role. This shall be illustrated briefly in the context of the hippocampus, with implications for learning, memory, and cognitive decline.

In humans, rodents, and monkeys sex differences are clearly present in certain behavioral and cognitive tasks, and this correlates with structural and functional sex differences in brain regions associated with cognition and emotions, such as the hippocampus, amygdala, striatum, and neocortex (McEwen, 1999; De Vries, 2004; Romeo et al., 2004; Piefke et al., 2005; Cahill, 2006; Cosgrove et al., 2007). A burgeoning literature documents potent effects of circulating sex hormones on learning, memory, behavior, and affect in humans and rodents, with a primary focus on the impact of estrogens as key contributors to the maintenance and/or generation of these differences (Dohanich, 2002; Sherwin, 2002; Gibbs & Gabor, 2003; Sherwin, 2003; Luine, 2008; Brinton, 2009; Pike et al., 2009; Rosario et al., 2009; McEwen, 2010). For example, in the hippocampal CA1 region estradiol treatment of ovariectomized female rats and monkeys rapidly (within seconds/minutes) increases the density of dendritic spines that typically represent sites of excitatory glutamatergic input and are the morphological correlates of long-term potentiation, learning, and memory (Woolley, 1999; Gibbs & Gabor, 2003; Leranth et al., 2003; Daniel, 2006; McEwen & Milner, 2007; Luine, 2008; Spencer et al., 2008). These dramatic influences of estradiol on synaptic plasticity may be owing in part to direct, rapid, nongenomic effects, as the hippocampus has a relative lack of classical estrogen receptors. However, other significant mechanisms involve indirect estradiol influences on transsynaptic pathways. These include activation of the basal forebrain (septohippocampal) cholinergic system and suppression of hippocampal GABAergic interneurons that mediate estrogenic influences on hippocampal memory and learning in females (McEwen & Alves, 1999; Leranth et al., 2003; Lee et al., 2004; Romeo et al., 2005b; Mitsushima et al., 2009). In contrast, in males estradiol fails to exert these effects on dendritic spine density and the septo-hippocampal system largely owing to estradiol exposure (masculinization/defeminization) in the neonatal period (Luine, 1985; Romeo et al., 2004; Cooke & Woolley, 2005; Romeo et al., 2005a; Mitsushima et al., 2009;

Gillies & McArthur, 2010a). These findings strongly suggest that masculinization/defeminization of the hippocampus occurs via much the same processes that prevail in the hypothalamus, and predict the formation of sex-specific circuitry that dictates long-term sex differences in responsiveness to estradiol.

As yet, there is little evidence from human studies to determine the extent to which sex differences in behavioral and cognitive strategies are caused by the organizational or activational effects of estradiol, and animal studies are only just beginning to address the contributions of centrally generated neurosteroids to these processes. Understandably, however, the preclinical evidence has generated considerable interest in elucidating the roles of sex hormones, and estrogens in particular, in mild cognitive impairment, age-associated dementia, and AD. Like PD, sex differences in the pathology and symptomatology have been documented for AD (Cahill, 2006; Bao et al., 2008) but, unlike PD, being female rather than male features among risk factors, after aging, for developing AD (Henderson et al., 2000; Pike et al., 2009). Interestingly, although the dramatic loss of estrogens at menopause is generally recognized as being a risk factor for women to develop AD (Sherwin, 2002; Brinton, 2004; Pike et al., 2009), there is no consensus that circulating levels of estrogens differ between control and AD women. However, post-mortem levels of estrogens have been reported to be lower than normal in brains of women who suffered from AD (Yue et al., 2005; Rosario et al., 2009). Furthermore deficiencies in aromatase expression have been reported in certain regions of AD brains (Ishunina et al., 2005; Hiltunen et al., 2006). Together, these observations raise the possibility that centrally generated estrogens may protect against cognitive decline. Fewer studies have been performed in men but, on balance, the evidence suggests that any estrogenic influences on cognition and behavior, whether the steroid is generated in the brain de novo or from circulating precursors, are different from those seen in women (Sherwin, 2003; Cherrier et al., 2005, 2007; Matousek & Sherwin, 2009). These observations are supported in experimental models in which, for example, estradiol treatment of gonadectomized females, but not males, reduced brain levels of soluble amyloid beta, which directly impairs synaptic plasticity and memory (Ramsden et al., 2003; Yue et al., 2005; Shankar et al., 2008; Pike et al., 2009).

On balance, the wealth of data relating to hippocampal function/dysfunction, memory, and AD indicate that the beneficial effects of physiological levels of estradiol seen in females is unlikely to have similar effects in males, in whom physiological (not pharmacological) levels of circulating testosterone, operating via androgenic mechanisms, may be much more important (reviewed in Pike et al., 2009; Gillies & McArthur, 2010a). These investigations further highlight the importance of understanding sex-specific mechanisms to target hormone-based therapies more effectively. However, as emphasized elsewhere for PD, there are other instances, such as experimentally induced ischemia (a model of stroke) or experimental autoimmune encephalomyelitis (a model of multiple sclerosis), in which estrogens have been reported to be protective in male and female brains (Murphy et al., 2003; Gold & Voskuhl, 2009; Suzuki et al., 2009).

This could be related, again, to issues of pharmacological versus physiological levels of hormones, and could also be influenced by the significant differences in the nature of the degenerating lesion in various neuropathologies, as well as the varied and complex potential mechanisms of action of estrogens (Garcia-Segura et al., 2001; Brann et al., 2007; Gillies & McArthur, 2010a).

SUMMARY AND IMPLICATIONS FOR SEX-SPECIFIC MEDICINES

Substantial evidence from preclinical studies, supported by clinical observations, suggests that estrogens have potential therapeutic benefits in the brain. Although the majority of investigations have employed female subjects, we now know that estrogens also have widespread actions in the male brain, raising the potential for estrogenic compounds, especially nonfeminizing estrogen mimics, to be of value in both sexes. However, accumulating evidence suggests that estrogens derived from the systemic circulation differentially influence male and female brains, whether healthy or diseased, probably because of underlying sex dimorphisms in certain neural networks that are programmed by gonadal steroids during development. Such observations, therefore, highlight that any potential clinically beneficial effects of estrogens in females may not be universally transferable to males. These concepts have been discussed in relation to the NSDA system, which exhibits inherent sex differences in humans and experimental animals in its normal and diseased state, as well as in its response to estrogens. Hence, in PD there is a long history suggesting that physiological levels of circulating estrogens can protect in women and female rodents subjected to experimental PD, whereas male rodent studies show that circulating estrogens or estrogen precursors (testosterone) cannot protect, and may even exacerbate striatal dopamine loss. To illustrate that sex dimorphisms are not unique to the NSDA and PD, this chapter has also briefly addressed the sex-specific sensitivites of the hippocampus to estrogens in terms of its structure, function, and cognitive behaviors that may have relevance to sex differences that are found in the pathology and manifestations of AD. These discussions illustrate the urgent need for a better understanding of brain sex dimorphisms and responses to estrogenic compounds if we are to reap their full translational potential to promote mental health, which clearly requires a sex-specific approach.

In contrast with systemic estrogens, this chapter has also highlighted that brain aromatase activity has the power to allow centrally generated estradiol to protect against striatal dopamine loss in both males and females subjected to experimental PD. This is probably because steroid synthesis is highly restricted to the region of damage, with negligible effect on sexually dimorphic NSDA regulatory pathways. Although the cells of origin of aromatase activity in experimental PD remain to be determined, it has been established that in other types of brain injury, including ischemia and excitotoxic injury, aromatase is upregulated in glial cells at the site of injury, and consequent estrogen synthesis appears to be protective (Azcoitia et al., 2003; Carswell et al., 2005; Azcoitia et al., 2011). As aromatase is regulated in a highly tissue- and brain region–specific manner (Simpson et al., 2002; Saldanha et al., 2009), the possibility of upregulating aromatase at sites of injury, or providing increased levels of aromatase substrates, such as DHEA, which has protective powers in experimental PD (Morissette et al., 2008), suggest interesting possibilities for future therapeutic exploitation. Thus, selective enhancers of aromatase transcription, translation, or activity would direct elevations in estradiol production to the required sites of damage, thereby avoiding the unwanted peripheral and central actions of systemically administered estrogens.

A better understanding of the local "neuroendocrine" mechanisms in the brain, and how they interrelate with steroids of systemic origin and their interacting pathways will be an important challenge for the future. However, our understanding of the pathophysiological role of neurosteroids in the human brain is in its infancy, and significant advances are required before we know whether brain aromatase and estrogens have analogous roles to those described in experimental species. Clinical conditions that could provide insight into such actions include cases of aromatase deficiency, which have helped unmask the biological significance of estrogens in men (Rochira & Carani, 2009). This is, however, an extremely rare condition that is treated soon after puberty with estradiol to combat bone and metabolic problems, with no reports of possible CNS effects. Thus, although the relevance of the findings in preclinical studies for clinical conditions remains to be established, the preclinical evidence is sufficiently compelling to suggest that the effective manipulation of estrogenic activity in the brain, whether via (brain) selective estrogen receptor modifying compounds or selective aromatase modifying agents, in males and females will have important clinical implications.

ACKNOWLEDGMENT

This work was supported by the Wellcome Trust (GEG grant ref 072634/Z/03/Z).

REFERENCES

Abraham, I. M., & Herbison, A. E. (2005). Major sex differences in non-genomic estrogen actions on intracellular signaling in mouse brain in vivo. *Neuroscience, 131,* 945–951.

Aleman, A., Kahn, R. S., & Selten, J.P. (2003). Sex differences in the risk of schizophrenia: evidence from meta-analysis. *Arch Gen Psychiatry, 60,* 565–571.

Allen, J. S., Damasio, H., Grabowski, T. J., Bruss, J., & Zhang, W. (2003). Sexual dimorphism and asymmetries in the gray-white composition of the human cerebrum. *Neuroimage, 18,* 880–894.

Arnold, A. P. (2009). The organizational-activational hypothesis as the foundation for a unified theory of sexual differentiation of all mammalian tissues. *Horm Behav, 55,* 570–578.

Arnold, A. P., & Burgoyne, P. S. (2004). Are XX and XY brain cells intrinsically different? *Trends Endocrinol Metab, 15,* 6–11.

Azcoitia, I., Arevalo, M. A., De Nicola, A. F., & Garcia-Segura, L. M. (2011). Neuroprotective actions of estradiol revisited. *Trends Endocrinol Metab, 22*(12), 467–473.

Azcoitia, I., Sierra, A., Veiga, S., & Garcia-Segura, L. M. (2003). Aromatase expression by reactive astroglia is neuroprotective. *Ann N Y Acad Sci, 1007,* 298–305.

Azcoitia, I., Sierra, A., Veiga, S., & Garcia-Segura, L. M. (2005). Brain steroidogenesis: emerging therapeutic strategies to prevent neurodegeneration. *J Neural Transm, 112,* 171–176.

Azcoitia, I., Sierra, A., Veiga, S., Honda, S., Harada, N., & Garcia-Segura, L. M. (2001). Brain aromatase is neuroprotective. *J Neurobiol, 47,* 318–329.

Baldereschi, M., Di Carlo, A., Rocca, W. A., Vanni, P., Maggi, S., Perissinotto, E., et al. (2000). Parkinson's disease and parkinsonism in a longitudinal study: two-fold higher incidence in men. ILSA Working Group. Italian Longitudinal Study on Aging. *Neurology, 55,* 1358–1363.

Balthazart, J., Baillien, M., & Ball, G. F. (2002). Interactions between aromatase (estrogen synthase) and dopamine in the control of male sexual behavior in quail. *Biochem Mol Biol, 132,* 37–55.

Balthazart, J., & Ball, G. F. (2006). Is brain estradiol a hormone or a neurotransmitter? *Trends Neurosci, 29,* 241–249.

Bao, A. M., Meynen, G., & Swaab, D. F. (2008). The stress system in depression and neurodegeneration: focus on the human hypothalamus. *Brain Res Rev, 57,* 531–553.

Barnes, L. L., Wilson, R. S., Bienias, J. L., Schneider, J. A., Evans, D. A., & Bennett, D. A. (2005). Sex differences in the clinical manifestations of Alzheimer disease pathology. *Arch Gen Psychiatry, 62,* 685–691.

Baron-Cohen, S., Knickmeyer, R. C., & Belmonte, M. K. (2005). Sex differences in the brain: implications for explaining autism. *Science, 310,* 819–823.

Bassilana, F., Mace, N., Li, Q., Stutzmann, J. M., Gross, C.E., Pradier, L., et al. (2005). Unraveling substantia nigra sequential gene expression in a progressive MPTP-lesioned macaque model of Parkinson's disease. *Neurobiol Dis, 20,* 93–103.

Becker, J. B. (1999). Gender differences in dopaminergic function in striatum and nucleus accumbens. *Pharmacol Biochem Behav, 64,* 803–812.

Becker, J. B., & Hu, M. (2008). Sex differences in drug abuse. *Front Neuroendocrinol, 29,* 36–47.

Beery, A. K., & Zucker, I. (2011). Sex bias in neuroscience and biomedical research. *Neurosci Biobehav Rev, 35,* 565–572.

Behl, C. (2002). Oestrogen as a neuroprotective hormone. *Nat Rev Neurosci, 3,* 433–442.

Belelli, D., & Lambert, J. J. (2005). Neurosteroids: endogenous regulators of the GABA(A) receptor. *Nat Rev Neurosci, 6,* 565–575.

Benedetti, M. D., Maraganore, D. M., Bower, J. H., Mcdonnell, S. K., Peterson, B. J., Ahlskog, J. E., et al. (2001). Hysterectomy, menopause, and estrogen use preceding Parkinson's disease: an exploratory case-control study. *Mov Disord, 16,* 830–837.

Bezard, E., Dovero, S., Prunier, C., Ravenscroft, P., Chalon, S., Guilloteau, D., et al. (2001). Relationship between the appearance of symptoms and the level of nigrostriatal degeneration in a progressive 1-methyl-4-phenyl-1,2,3,6-tetrahydropy ridine-lesioned macaque model of Parkinson's disease. *J Neurosci, 21,* 6853–6861.

Bixo, M., Andersson, A., Winblad, B., Purdy, R. H., & Backstrom, T. (1997). Progesterone, 5alpha-pregnane-3,20-dione and 3alpha-hydroxy-5alpha-pregnane-20-one in specific regions of the human female brain in different endocrine states. *Brain Res, 764,* 173–178.

Brann, D. W., Dhandapani, K., Wakade, C., Mahesh, V. B., & Khan, M. M. (2007). Neurotrophic and neuroprotective actions of estrogen: basic mechanisms and clinical implications. *Steroids, 72,* 381–405.

Brinton, R. D. (2004). Impact of estrogen therapy on Alzheimer's disease: a fork in the road? *CNS Drugs, 18,* 405–422.

Brinton, R. D. (2009). Estrogen-induced plasticity from cells to circuits: predictions for cognitive function. *Trends Pharmacol Sci, 30,* 212–222.

Cahill, L. (2006). Why sex matters for neuroscience. *Nat Rev Neurosci, 7,* 477–484.

Cantuti-Castelvetri, I., Keller-Mcgandy, C., Bouzou, B., Asteris, G., Clark, T. W., Frosch, M. P., et al. (2007). Effects of gender on nigral gene expression and parkinson disease. *Neurobiol Dis, 26,* 606–614.

Carswell, H. V., Dominiczak, A. F., Garcia-Segura, L. M., Harada, N., Hutchison, J. B., & Macrae, I. M. (2005). Brain aromatase expression after experimental stroke: topography and time course. *J Steroid Biochem Mol Biol, 96,* 89–91.

Castaneda, E., Whishaw, I. Q., & Robinson, T. E. (1990). Changes in striatal dopamine neurotransmission assessed with microdialysis following recovery from a bilateral 6-OHDA lesion: variation as a function of lesion size. *J Neurosci, 10,* 1847–1854.

Castner, S. A., Xiao, L., & Becker, J. B. (1993). Sex differences in striatal dopamine: in vivo microdialysis and behavioral studies. *Brain Res, 610,* 127–134.

Cherrier, M. M., Matsumoto, A. M., Amory, J. K., Ahmed, S., Bremner, W., Peskind, E. R., et al. (2005). The role of aromatization in testosterone supplementation: effects on cognition in older men. *Neurology, 64,* 290–296.

Cherrier, M. M., Matsumoto, A. M., Amory, J. K., Johnson, M., Craft, S., Peskind, E. R., et al. (2007). Characterization of verbal and spatial memory changes from moderate to supraphysiological increases in serum testosterone in healthy older men. *Psychoneuroendocrinology, 32,* 72–79.

Cooke, B. M., & Woolley, C. S. (2005). Gonadal hormone modulation of dendrites in the mammalian CNS. *J Neurobiol, 64,* 34–46.

Cosgrove, K. P., Mazure, C. M., & Staley, J. K. (2007). Evolving knowledge of sex differences in brain structure, function, and chemistry. *Biol Psychiatry, 62,* 847–855.

Craig, M. C., Maki, P. M., & Murphy, D. G. (2005). The Women's Health Initiative Memory Study: findings and implications for treatment. *Lancet Neurol, 4,* 190–194.

Craig, M. C., & Murphy, D. G. (2007a). Estrogen: effects on normal brain function and neuropsychiatric disorders. *Climacteric, 10*(Suppl 2), 97–104.

Craig, M. C., & Murphy, D. G. (2007b). Oestrogen, cognition and the maturing female brain. *J Neuroendocrinol, 19,* 1–6.

Daniel, J. M. (2006). Effects of oestrogen on cognition: what have we learned from basic research? *J Neuroendocrinol, 18,* 787–795.

Datla, K. P., Murray, H. E., Pillai, A. V., Gillies, G. E., & Dexter, D. T. (2003). Differences in dopaminergic neuroprotective effects of estrogen during estrous cycle. *Neuroreport, 14,* 47–50.

De Vries, G. J. (2004). Minireview: sex differences in adult and developing brains: compensation, compensation, compensation. *Endocrinology, 145,* 1063–1068.

De Vries, G. J., & Simerly, R. B. (2002). "Anatomy, development and function of sexually dimorphic neural circuits in the mammalian brain. In: Pfaff, D. W. E. A. (ed.), *Hormones, Brain and Behaviour.* San Diego: Elsevier, pp. 137–191.

Dewing, P., Chiang, C. W., Sinchak, K., Sim, H., Fernagut, P. O., Kelly, S., et al. (2006). Direct regulation of adult brain function by the male-specific factor SRY. *Curr Biol, 16,* 415–420.

Di Michele, F., Longone, P., Romeo, E., Lucchetti, S., Brusa, L., Pierantozzi, M., et al. (2003). Decreased plasma and cerebrospinal fluid content of neuroactive steroids in Parkinson's disease. *Neurol Sci, 24,* 172–173.

Dluzen, D., & Horstink, M. (2003). Estrogen as neuroprotectant of nigrostriatal dopaminergic system: laboratory and clinical studies. *Endocrine, 21,* 67–75.

Dluzen, D., Jain, R., & Liu, B. (1994). Modulatory effects of testosterone on 1-methyl-4-phenyl-1,2,3,6-tetrahydropyridine-induced neurotoxicity. *J Neurochem, 62,* 94–101.

Dluzen, D. E. (2000). Neuroprotective effects of estrogen upon the nigrostriatal dopaminergic system. *J Neurocytol, 29,* 387–399.

Dluzen, D. E. (2005). Unconventional effects of estrogen uncovered. *Trends Pharmacol Sci, 26,* 485–487.

Do Rego, J. L., Seong, J. Y., Burel, D., Leprince, J., Luu-the, V., Tsutsui, K., et al. (2009). Neurosteroid biosynthesis: enzymatic pathways and neuroendocrine regulation by neurotransmitters and neuropeptides. *Front Neuroendocrinol, 30,* 259–301.

Dohanich, G. (2002). *Gonadal Steroids, Learning and Memory.* San Diego: Academic Press.

Elbaz, A., Bower, J. H., Maraganore, D. M., Mcdonnell, S. K., Peterson, B. J., Ahlskog, J. E., et al. (2002). Risk tables for parkinsonism and Parkinson's disease. *J Clin Epidemiol, 55,* 25–31.

Federman, D. D. (2006). The biology of human sex differences. *N Engl J Med, 354,* 1507–1514.

Ferraz, A. C., Matheussi, F., Szawka, R. E., Rizelio, V., Delattre, A. M., Rigon, P., et al. (2008). Evaluation of estrogen neuroprotective effect on nigrostriatal dopaminergic neurons following 6-hydroxydopamine injection into the substantia nigra pars compacta or the medial forebrain bundle. *Neurochem Res, 33,* 1238–1246.

Ferraz, A. C., Xavier, L. L., Hernandes, S., Sulzbach, M., Viola, G. G., Anselmo-Franci, J. A., et al. (2003). Failure of estrogen to protect the substantia nigra pars compacta of female rats from lesion induced by 6-hydroxydopamine. *Brain Res, 986,* 200–205.

Fink, G., Sumner, B. E., Rosie, R., Grace, O., & Quinn, J. P. (1996). Estrogen control of central neurotransmission: effect on mood, mental state, and memory. *Cell Mol Neurobiol, 16,* 325–344.

Fox, S. H., & Brotchie, J. M. (2010). The MPTP-lesioned non-human primate models of Parkinson's disease. Past, present, and future. *Prog Brain Res, 184,* 133–157.

Gabory, A., Attig, L., & Junien, C. (2009). Sexual dimorphism in environmental epigenetic programming. *Mol Cell Endocrinol, 304,* 8–18.

Gao, X., & Dluzen, D. E. (2001). The effect of testosterone upon methamphetamine neurotoxicity of the nigrostriatal dopaminergic system. *Brain Res, 892,* 63–69.

Garcia-Ovejero, D., Azcoitia, I., Doncarlos, L. L., Melcangi, R. C., & Garcia-Segura, L. M. (2005). Glia-neuron crosstalk in the neuroprotective mechanisms of sex steroid hormones. *Brain Res, 48,* 273–286.

Garcia-Segura, L. M. (2008). Aromatase in the brain: not just for reproduction anymore. *J Neuroendocrinol, 20,* 705–712.

Garcia-Segura, L. M., Azcoitia, I., & Doncarlos, L. L. (2001). Neuroprotection by estradiol. *Prog Neurobiol, 63,* 29–60.

Gibbs, R. B., & Gabor, R. (2003). Estrogen and cognition: applying preclinical findings to clinical perspectives. *J Neurosci Res, 74,* 637–643.

Gillies, G. E., & McArthur, S. (2010a). Estrogen actions in the brain and the basis for differential action in men and women: a case for sex-specific medicines. *Pharmacol Rev, 62,* 155–198.

Gillies, G. E., & McArthur, S. (2010b). Independent influences of sex steroids of systemic and central origin in a rat model of Parkinson's disease: a contribution to sex-specific neuroprotection by estrogens. *Horm Behav, 57,* 23–34.

Gillies, G. E., Murray, H. E., Dexter, D., & Mcarthur, S. (2004). Sex dimorphisms in the neuroprotective effects of estrogen in an animal model of Parkinson's disease. *Pharmacol Biochem Behav, 78,* 513–522.

Gold, S. M., & Voskuhl, R. R. (2009). Estrogen and testosterone therapies in multiple sclerosis. *Prog Brain Res, 175,* 239–251.

Green, P. S., & Simpkins, J. W. (2000). Neuroprotective effects of estrogens: potential mechanisms of action. *Int J Dev Neurosci, 18,* 347–358.

Haaxma, C. A., Bloem, B. R., Borm, G. F., Oyen, W. J., Leenders, K. L., Eshuis, S., et al. (2007). Gender differences in Parkinson's disease. *J Neurol Neurosurg Psychiatry, 78,* 819–824.

Henderson, V. W., Paganini-Hill, A., Miller, B. L., Elble, R. J., Reyes, P. F., Shoupe, D., et al. (2000). Estrogen for Alzheimer's disease in women: randomized, double-blind, placebo-controlled trial. *Neurology, 54,* 295–301.

Hennessy, R. J., Lane, A., Kinsella, A., Larkin, C., O'Callaghan, E., & Waddington, J. L. (2004). 3D morphometrics of craniofacial dysmorphology reveals sex-specific asymmetries in schizophrenia. *Schizophr Res, 67,* 261–268.

Hiltunen, M., Iivonen, S., & Soininen, H. (2006). Aromatase enzyme and Alzheimer's disease. *Minerva Endocrinol, 31,* 61–73.

Hu, M., Crombag, H. S., Robinson, T. E., & Becker, J. B. (2004). Biological basis of sex differences in the propensity to self-administer cocaine. *Neuropsychopharmacology, 29,* 81–85.

Hu, M., Watson, C. J., Kennedy, R. T., & Becker, J. B. (2006). Estradiol attenuates the K+-induced increase in extracellular GABA in rat striatum. *Synapse, 59,* 122–124.

Hurn, P. D., & Macrae, I. M. (2000). Estrogen as a neuroprotectant in stroke. *J Cereb Blood Flow Metab, 20,* 631–652.

Ishunina, T. A., Van Beurden, D., Van Der Meulen, G., Unmehopa, U. A., Hol, E. M., Huitinga, I., et al. (2005). Diminished aromatase immunoreactivity in the hypothalamus, but not in the basal forebrain nuclei in Alzheimer's disease. *Neurobiol Aging, 26,* 173–194.

Jennings, P. J., Janowsky, J. S., & Orwoll, E. (1998). Estrogen and sequential movement. *Behav Neurosci, 112,* 154–159.

Jones, M. E., Chin Boon, W., Proietto, J., & Simpson, E. R. (2006). Of mice and men: the evolving phenotype of aromatase deficiency. *Trends Endocrinol Metab, 17,* 55–64.

Kaasinen, V., Nagren, K., Hietala, J., Farde, L., & Rinne, J. O. (2001). Sex differences in extrastriatal dopamine d(2)-like receptors in the human brain. *Am J Psychiatry, 158,* 308–311.

Kaminsky, Z., Wang, S. C., & Petronis, A. (2006). Complex disease, gender and epigenetics. *Ann Med, 38,* 530–544.

Kelly, S. J., Ostrowski, N. L., & Wilson, M. A. (1999). Gender differences in brain and behavior: hormonal and neural bases. *Pharmacol Biochem Behav, 64,* 655–664.

King, S. R. (2008). Emerging roles for neurosteroids in sexual behavior and function. *J Androl, 29,* 524–533.

Kipp, M., Karakaya, S., Pawlak, J., Araujo-Wright, G., Arnold, S., & Beyer, C. (2006). Estrogen and the development and protection of nigrostriatal dopaminergic

neurons: concerted action of a multitude of signals, protective molecules, and growth factors. *Front Neuroendocrinol, 27,* 376–390.
Klein, C., & Schlossmacher, M. G. (2007). Parkinson disease, 10 years after its genetic revolution: multiple clues to a complex disorder. *Neurology, 69,* 2093–2104.
Klink, R., Robichaud, M., & Debonnel, G. (2002). Gender and gonadal status modulation of dorsal raphe nucleus serotonergic neurons. Part II. Regulatory mechanisms. *Neuropharmacology, 43,* 1129–1138.
Kritzer, M. F., & Creutz, L. M. (2008). Region and sex differences in constituent dopamine neurons and immunoreactivity for intracellular estrogen and androgen receptors in mesocortical projections in rats. *J Neurosci, 28,* 9525–9535.
Kuppers, E., Ivanova, T., Karolczak, M., & Beyer, C. (2000). Estrogen: a multifunctional messenger to nigrostriatal dopaminergic neurons. *J Neurocytol, 29,* 375–385.
Laakso, A., Vilkman, H., Bergman, J., Haaparanta, M., Solin, O., Syvalahti, E., et al. (2002). Sex differences in striatal presynaptic dopamine synthesis capacity in healthy subjects. *Biol Psychiatry, 52,* 759–763.
Lavalaye, J., Booij, J., Reneman, L., Habraken, J. B., & Van Royen, E. A. (2000). Effect of age and gender on dopamine transporter imaging with [123I]FP-CIT SPET in healthy volunteers. *Eur J Nucl Med, 27,* 867–869.
Lee, S. J., Romeo, R. D., Svenningsson, P., Campomanes, C. R., Allen, P. B., Greengard, P., et al. (2004). Estradiol affects spinophilin protein differently in gonadectomized males and females. *Neuroscience, 127,* 983–988.
Leranth, C., Petnehazy, O., & Maclusky, N. J. (2003). Gonadal hormones affect spine synaptic density in the CA1 hippocampal subfield of male rats. *J Neurosci, 23,* 1588–1592.
Leranth, C., Roth, R. H., Elsworth, J. D., Naftolin, F., Horvath, T. L., & Redmond, D. E. Jr. (2000). Estrogen is essential for maintaining nigrostriatal dopamine neurons in primates: implications for Parkinson's disease and memory. *J Neurosci, 20,* 8604–8609.
Lewis, C., & Dluzen, D. E. (2008). Testosterone enhances dopamine depletion by methamphetamine in male, but not female, mice. *Neurosci Lett, 448,* 130–133.
Liu, B., & Dluzen, D. E. (2007). Oestrogen and nigrostriatal dopaminergic neurodegeneration: animal models and clinical reports of Parkinson's disease. *Clin Exp Pharmacol Physiol, 34,* 555–565.
Luchetti, S., Bossers, K., Frajese, G. V., & Swaab, D. F. (2010a). Neurosteroid biosynthetic pathway changes in substantia nigra and caudate nucleus in Parkinson's disease. *Brain Pathol, 20,* 945–951.
Luchetti, S., Huitinga, I., & Swaab, D. F. (2010b). Neurosteroid and GABA-A receptor alterations in Alzheimer's disease, Parkinson's disease and multiple sclerosis. *Neuroscience,* 191, 6–21.
Luders, E., Narr, K. L., Thompson, P. M., Rex, D. E., Jancke, L., Steinmetz, H., et al. (2004). Gender differences in cortical complexity. *Nat Neurosci, 7,* 799–800.
Luine, V. N. (1985). Estradiol increases choline acetyltransferase activity in specific basal forebrain nuclei and projection areas of female rats. *Exp Neurol, 89,* 484–490.
Luine, V. N. (2007). The prefrontal cortex, gonadal hormones and memory. *Horm Behav, 51,* 181–182.
Luine, V. N. (2008). Sex steroids and cognitive function. *J Neuroendocrinol, 20,* 866–872.
Marien, M. R., Colpaert, F. C., & Rosenquist, A. C. (2004). Noradrenergic mechanisms in neurodegenerative diseases: a theory. *Brain Res, 45,* 38–78.

Matousek, R. H., & Sherwin, B. B. (2009). A randomized controlled trial of add-back estrogen or placebo on cognition in men with prostate cancer receiving an antiandrogen and a gonadotropin-releasing hormone analog. *Psychoneuroendocrinology, 5*(11), 620–627.

McArthur, S., Mchale, E., & Gillies, G. E. (2007a). The size and distribution of midbrain dopaminergic populations are permanently altered by perinatal glucocorticoid exposure in a sex- region- and time-specific manner. *Neuropsychopharmacology, 32,* 1462–1476.

McArthur, S., Murray, H. E., Dhankot, A., Dexter, D. T., & Gillies, G. E. (2007b). Striatal susceptibility to a dopaminergic neurotoxin is independent of sex hormone effects on cell survival and DAT expression but is exacerbated by central aromatase inhibition. *J Neurochem, 100,* 678–692.

Mccarthy, M. M., & Arnold, A. P. (2010). Reframing sexual differentiation of the brain. *Nat Neurosci, 14,* 677–683.

Mccarthy, M. M., & Konkle, A. T. (2005). When is a sex difference not a sex difference? *Front Neuroendocrinol, 26,* 85–102.

Mcdermott, J. L., Liu, B., & Dluzen, D. E. (1994). Sex differences and effects of estrogen on dopamine and DOPAC release from the striatum of male and female CD-1 mice. *Exp Neurol, 125,* 306–311.

Mcewen, B. S. (1999). Permanence of brain sex differences and structural plasticity of the adult brain. *Proc Natl Acad Sci U S A, 96,* 7128–7130.

Mcewen, B. S. (2010). Stress, sex, and neural adaptation to a changing environment: mechanisms of neuronal remodeling. *Ann N Y Acad Sci,* 1204(Suppl), E38–E59.

Mcewen, B. S., & Alves, S. E. (1999). Estrogen actions in the central nervous system. *Endocr Rev, 20,* 279–307.

Mcewen, B. S., & Milner, T. A. (2007). Hippocampal formation: shedding light on the influence of sex and stress on the brain. *Brain Res, 55,* 343–355.

Micevych, P., & Dominguez, R. (2009). Membrane estradiol signaling in the brain. *Front Neuroendocrinol, 30*(3), 315–327.

Miller, D. B., Ali, S. F., O'callaghan, J. P., & Laws, S. C. (1998). The impact of gender and estrogen on striatal dopaminergic neurotoxicity. *Ann N Y Acad Sci, 844,* 153–165.

Miller, I. N., & Cronin-Golomb, A. (2011). Gender differences in Parkinson's disease: clinical characteristics and cognition. *Mov Disord, 25,* 2695–2703.

Mitsushima, D., Takase, K., Funabashi, T., & Kimura, F. (2009). Gonadal steroids maintain 24 h acetylcholine release in the hippocampus: organizational and activational effects in behaving rats. *J Neurosci, 29,* 3808–3815.

Morale, M. C., L'episcopo, F., Tirolo, C., Giaquinta, G., Caniglia, S., Testa, N., et al. (2008). Loss of aromatase cytochrome P450 function as a risk factor for Parkinson's disease? *Brain Res Rev, 57,* 431–443.

Morissette, M., Al Sweidi, S., Callier, S., & Di Paolo, T. (2008). Estrogen and SERM neuroprotection in animal models of Parkinson's disease. *Mol Cell Endocrinol, 290,* 60–69.

Morissette, M., & Di Paolo, T. (2009). Effect of estradiol on striatal dopamine activity of female hemiparkinsonian monkeys. *J Neurosci Res, 87,* 1634–1644.

Moroz, I. A., Rajabi, H., Rodaros, D., & Stewart, J. (2003). Effects of sex and hormonal status on astrocytic basic fibroblast growth factor-2 and tyrosine hydroxylase immunoreactivity after medial forebrain bundle 6-hydroxydopamine lesions of the midbrain dopamine neurons. *Neuroscience, 118,* 463–476.

Morris, J. A., Jordan, C. L., & Breedlove, S. M. (2004). Sexual differentiation of the vertebrate nervous system. *Nat Neurosci, 7,* 1034–1039.

Mozley, L. H., Gur, R. C., Mozley, P. D., & Gur, R. E. (2001). Striatal dopamine transporters and cognitive functioning in healthy men and women. *Am J Psychiatry, 158,* 1492–1499.

Mukai, H., Takata, N., Ishii, H. T., Tanabe, N., Hojo, Y., Furukawa, A., et al. (2006). Hippocampal synthesis of estrogens and androgens which are paracrine modulators of synaptic plasticity: synaptocrinology. *Neuroscience, 138,* 757–764.

Munro, C. A., Mccaul, M. E., Wong, D. F., Oswald, L. M., Zhou, Y., Brasic, J., et al. (2006). Sex differences in striatal dopamine release in healthy adults. *Biol Psychiatry, 59,* 966–974.

Murphy, S., Mccullough, L., Littleton-Kearney, M., & Hurn, P. (2003). Estrogen and selective estrogen receptor modulators: neuroprotection in the Women's Health Initiative era. *Endocrine, 21,* 17–26.

Murray, H. E., Pillai, A. V., Mcarthur, S. R., Razvi, N., Datla, K. P., Dexter, D. T., et al. (2003). Dose- and sex-dependent effects of the neurotoxin 6-hydroxydopamine on the nigrostriatal dopaminergic pathway of adult rats: differential actions of estrogen in males and females. *Neuroscience, 116,* 213–222.

Naftolin, F., Garcia-Segura, L. M., Horvath, T. L., Zsarnovszky, A., Demir, N., Fadiel, A., et al. (2007). Estrogen-induced hypothalamic synaptic plasticity and pituitary sensitization in the control of the estrogen-induced gonadotrophin surge. *Reprod Sci, 14,* 101–116.

Ohtani, H., Nomoto, M., & Douchi, T. (2001). Chronic estrogen treatment replaces striatal dopaminergic function in ovariectomized rats. *Brain Res, 900,* 163–168.

Pasqualini, C., Olivier, V., Guibert, B., Frain, O., & Leviel, V. (1995). Acute stimulatory effect of estradiol on striatal dopamine synthesis. *J Neurochem, 65,* 1651–1657.

Pavon, J. M., Whitson, H. E., & Okun, M. S. (2010). Parkinson's disease in women: a call for improved clinical studies and for comparative effectiveness research. *Maturitas, 65,* 352–358.

Piefke, M., Weiss, P. H., Markowitsch, H. J., & Fink, G. R. (2005). Gender differences in the functional neuroanatomy of emotional episodic autobiographical memory. *Hum Brain Mapp, 24,* 313–324.

Pike, C. J., Carroll, J. C., Rosario, E. R., & Barron, A. M. (2009). Protective actions of sex steroid hormones in Alzheimer's disease. *Front Neuroendocrinol, 30,* 239–258.

Pohjalainen, T., Rinne, J. O., Nagren, K., Syvalahti, E., & Hietala, J. (1998). Sex differences in the striatal dopamine D2 receptor binding characteristics in vivo. *Am J Psychiatry, 155,* 768–773.

Prange-Kiel, J., & Rune, G. M. (2006). Direct and indirect effects of estrogen on rat hippocampus. *Neuroscience, 138,* 765–772.

Quesada, A., & Micevych, P. E. (2004). Estrogen interacts with the IGF-1 system to protect nigrostriatal dopamine and maintain motoric behavior after 6-hydroxdopamine lesions. *J Neurosci Res, 75,* 107–116.

Ramsden, M., Nyborg, A. C., Murphy, M. P., Chang, L., Stanczyk, F. Z., Golde, T. E., et al. (2003). Androgens modulate beta-amyloid levels in male rat brain. *J Neurochem, 87,* 1052–1055.

Ravina, B. M., Fagan, S. C., Hart, R. G., Hovinga, C. A., Murphy, D. D., Dawson, T. M., et al. (2003). Neuroprotective agents for clinical trials in Parkinson's disease: a systematic assessment. *Neurology, 60,* 1234–1240.

Robinson, T. E., Castaneda, E., & Whishaw, I. Q. (1990). Compensatory changes in striatal dopamine neurons following recovery from injury induced by 6-OHDA or methamphetamine: a review of evidence from microdialysis studies. *Can J Psychol, 44,* 253–275.

Rochira, V., Balestrieri, A., Madeo, B., Spaggiari, A., & Carani, C. (2002). Congenital estrogen deficiency in men: a new syndrome with different phenotypes; clinical and therapeutic implications in men. *Mol Cell Endocrinol, 193,* 19–28.

Rochira, V., & Carani, C. (2009). Aromatase deficiency in men: a clinical perspective. *Nat Rev Endocrinol, 5,* 559–568.

Romeo, R. D., McCarthy, J. B., Wang, A., Milner, T. A., & McEwen, B. S. (2005a). Sex differences in hippocampal estradiol-induced N-methyl-D-aspartic acid binding and ultrastructural localization of estrogen receptor-alpha. *Neuroendocrinology, 81,* 391–399.

Romeo, R. D., Staub, D., Jasnow, A. M., Karatsoreos, I. N., Thornton, J. E., & McEwen, B. S. (2005b). Dihydrotestosterone increases hippocampal N-methyl-D-aspartate binding but does not affect choline acetyltransferase cell number in the forebrain or choline transporter levels in the CA1 region of adult male rats. *Endocrinology, 146,* 2091–2097.

Romeo, R. D., Waters, E. M., & McEwen, B. S. (2004). Steroid-induced hippocampal synaptic plasticity: sex differences and similarities. *Neuron Glia Biol, 1,* 219–229.

Rosario, E. R., Chang, L., Head, E. H., Stanczyk, F. Z., & Pike, C. J. (2009). Brain levels of sex steroid hormones in men and women during normal aging and in Alzheimer's disease. *Neurobiol Aging,* 30(2):239–258.

Roselli, C. E., Liu, M., & Hurn, P. D. (2009). Brain aromatization: classic roles and new perspectives. *Semin Reprod Med, 27,* 207–217.

Rossouw, J. E., Anderson, G. L., Prentice, R. L., Lacroix, A. Z., Kooperberg, C., Stefanick, M. L., et al. (2002). Risks and benefits of estrogen plus progestin in healthy postmenopausal women: principal results From the Women's Health Initiative randomized controlled trial. *JAMA, 288,* 321–333.

Saldanha, C. J., Duncan, K. A., & Walters, B. J. (2009). Neuroprotective actions of brain aromatase. *Front Neuroendocrinol, 30,* 106–118.

Saunders-Pullman, R., Derby, C., Santoro, N., Bressman, S., Chiu, B., Lipton, R. B., et al. (2012). Role of endogenous and exogenous hormone exposure on the risk of Parkinson's disease: results from the Women's Health Initiative Observational Study. *Am Acad Neurol,* S23.002.

Saunders-Pullman, R., Gordon-Elliott, J., Parides, M., Fahn, S., Saunders, H. R., & Bressman, S. (1999). The effect of estrogen replacement on early Parkinson's disease. *Neurology, 52,* 1417–1421.

Schultz, K. N., Von Esenwein, S. A., Hu, M., Bennett, A. L., Kennedy, R. T., Musatov, S., et al. (2009). Viral vector-mediated overexpression of estrogen receptor-alpha in striatum enhances the estradiol-induced motor activity in female rats and estradiol-modulated GABA release. *J Neurosci, 29,* 1897–1903.

Schwarting, R. K., & Huston, J. P. (1996a). The unilateral 6-hydroxydopamine lesion model in behavioral brain research. Analysis of functional deficits, recovery and treatments. *Prog Neurobiol, 50,* 275–331.

Schwarting, R. K. W., & Huston, J. P. (1996b). Unilateral 6-hydroxydopamine lesions of meso-striatal dopamine neurons and their physiological sequelae. *Prog Neurobiol, 49,* 215–266.

Schwarz, J. M., & McCarthy, M. M. (2008). Cellular mechanisms of estradiol-mediated masculinization of the brain. *J Steroid Biochem Mol Biol, 109,* 300–306.

Serova, L. I., Maharjan, S., Huang, A., Sun, D., Kaley, G., & Sabban, E. L. (2004). Response of tyrosine hydroxylase and GTP cyclohydrolase I gene expression to estrogen in brain catecholaminergic regions varies with mode of administration. *Brain Res, 1015,* 1–8.

Shankar, G. M., Li, S., Mehta, T. H., Garcia-Munoz, A., Shepardson, N. E., Smith, I., et al. (2008). Amyloid-beta protein dimers isolated directly from Alzheimer's brains impair synaptic plasticity and memory. *Nat Med, 14,* 837–842.

Sharpe, R. M. (1998). The roles of oestrogen in the male. *Trends Endocrinol Metab, 9,* 371–377.

Sherwin, B. B. (2002). Estrogen and cognitive aging in women. *Trends Pharmacol Sci, 23,* 527–534.

Sherwin, B. B. (2003). Steroid hormones and cognitive functioning in aging men: a mini-review. *J Mol Neurosci, 20,* 385–393.

Shulman, L. M., & Bhat, V. (2006). Gender disparities in Parkinson's disease. *Expert Rev Neurother, 6,* 407–416.

Siegel, J. A., Young, L. A., Neiss, M. B., Samuels, M. H., Roselli, C. E., & Janowsky, J. S. (2008). Estrogen, testosterone, and sequential movement in men. *Behav Neurosci, 122,* 955–962.

Simerly, R. B. (2005). Wired on hormones: endocrine regulation of hypothalamic development. *Curr Opin Neurobiol, 15,* 81–85.

Simpson, E. R., Clyne, C., Rubin, G., Boon, W. C., Robertson, K., Britt, K., et al. (2002). Aromatase—a brief overview. *Annu Rev Physiol, 64,* 93–127.

Simunovic, F., Yi, M., Wang, Y., Stephens, R., & Sonntag, K. C. (2011). Evidence for gender-specific transcriptional profiles of nigral dopamine neurons in Parkinson disease. *PLoS One, 5,* e8856.

Solomon, M. B., & Herman, J. P. (2009). Sex differences in psychopathology: of gonads, adrenals and mental illness. *Physiol Behav, 97,* 250–258.

Spencer, J. L., Waters, E. M., Romeo, R. D., Wood, G. E., Milner, T. A., & Mcewen, B. S. (2008). Uncovering the mechanisms of estrogen effects on hippocampal function. *Front Neuroendocrinol, 29,* 219–237.

Suzuki, S., Brown, C. M., & Wise, P. M. (2009). Neuroprotective effects of estrogens following ischemic stroke. *Front Neuroendocrinol, 30,* 201–211.

Swaab, D. F. (2008). Sexual orientation and its basis in brain structure and function. *Proc Natl Acad Sci U S A, 105,* 10273–10274.

Swerdlow, R. H., Parker, W. D. Jr, Currie, L. J., Bennett, J.P. Jr, Harrison, M. B., Trugman, J. M., et al. (2001). Gender ratio differences between Parkinson's disease patients and their affected relatives. *Parkinson Rel Disord, 7,* 129–133.

Tamas, A., Lubics, A., Szalontay, L., Lengvari, I., & Reglodi, D. (2005). Age and gender differences in behavioral and morphological outcome after 6-hydroxydopamine-induced lesion of the substantia nigra in rats. *Behav Brain Res, 158,* 221–229.

Thanky, N. R., Slater, R., & Herbison, A. E. (2003). Sex differences in estrogen-dependent transcription of gonadotropin-releasing hormone (GnRH) gene revealed in GnRH transgenic mice. *Endocrinology, 144,* 3351–3358.

Thanky, N. R., Son, J. H., & Herbison, A. E. (2002). Sex differences in the regulation of tyrosine hydroxylase gene transcription by estrogen in the locus coeruleus of TH9-LacZ transgenic mice. *Mol Brain Res, 104,* 220–226.

Thompson, T. L. (1999). Attenuation of dopamine uptake in vivo following priming with estradiol benzoate. *Brain Res, 834,* 164–167.

Toran-Allerand, C. D. (2005). Estrogen and the brain: beyond ER-alpha, ER-beta, and 17beta-estradiol. *Ann N Y Acad Sci, 1052,* 136–144.

Tsang, K. L., Ho, S. L., & Lo, S. K. (2000). Estrogen improves motor disability in parkinsonian postmenopausal women with motor fluctuations. *Neurology, 54,* 2292–2298.

Tsutsui, K. (2008). Neurosteroid synthesis and action in the cerebellum during development. *Cerebellum, 7,* 502–504.

Van Beilen, M., Portman, A. T., Kiers, H. A., Maguire, R. P., Kaasinen, V., Koning, M., et al. (2008). Striatal FDOPA uptake and cognition in advanced non-demented Parkinson's disease: a clinical and FDOPA-PET study. *Parkinsonism Relat Disord, 14,* 224–228.

Van Den Eeden, S. K., Tanner, C. M., Bernstein, A. L., Fross, R. D., Leimpeter, A., Bloch, D. A., et al. (2003). Incidence of Parkinson's disease: variation by age, gender, and race/ethnicity. *Am J Epidemiol, 157,* 1015–1022.

Van Nas, A., Guhathakurta, D., Wang, S. S., Yehya, N., Horvath, S., Zhang, B., et al. (2009). Elucidating the role of gonadal hormones in sexually dimorphic gene coexpression networks. *Endocrinology, 150,* 1235–1249.

Walker, Q. D., Rooney, M. B., Wightman, R. M., & Kuhn, C. M. (2000). Dopamine release and uptake are greater in female than male rat striatum as measured by fast cyclic voltammetry. *Neuroscience, 95,* 1061–1070.

Wilson, C. A., & Davies, D. C. (2007). The control of sexual differentiation of the reproductive system and brain. *Reproduction, 133,* 331–359.

Wise, P. M., Suzuki, S., & Brown, C. M. (2009). Estradiol: a hormone with diverse and contradictory neuroprotective actions. *Dialogues Clin Neurosci, 11,* 297–303.

Woolley, C. S. (1999). Effects of estrogen in the CNS. *Curr Opin Neurobiol, 9,* 349–354.

Xiao, L., & Becker, J. B. (1998). Effects of estrogen agonists on amphetamine-stimulated striatal dopamine release. *Synapse, 29,* 379–391.

Yu, L., & Liao, P. C. (2000). Sexual differences and estrous cycle in methamphetamine-induced dopamine and serotonin depletions in the striatum of mice. *J Neurol Transm, 107,* 419–427.

Yu, Y. L., & Wagner, G. C. (1994). Influence of gonadal hormones on sexual differences in sensitivity to methamphetamine-induced neurotoxicity. *J Neural Transm, Park Dis Dement Sect, 8,* 215–221.

Yue, X., Lu, M., Lancaster, T., Cao, P., Honda, S., Staufenbiel, M., et al. (2005). Brain estrogen deficiency accelerates Abeta plaque formation in an Alzheimer's disease animal model. *Proc Natl Acad Sci U S A, 102,* 19198–19203.

Part IV

Aromatase and Sexual Differentiation

Chapter 15

A Historical and Personal Perspective on the Aromatization Revolution: Steinach Confirmed[1]

Per Södersten

> Any mature science necessarily has inconsistencies; correcting these is a waste of time.
>
> —*Thomas Kuhn,* 1962

INTRODUCTION

The history of the endocrinology of sexual behavior is replete with exciting observations (Beach, 1948, 1950, 1971, 1975, 1976, 1978, 1981; Young, 1961), the ones that led to the aromatization hypothesis will be discussed in this essay.

Does taking an interest in the history of biology signal the end of creativity, and should thus be saved for retirement days (Gross, 1999)? This misconception perhaps appears more today than in the past, yet it still reflects more shallow scholarship than deep understanding. Indeed, those who do not know the history of their field are likely to be redundant, repeat mistakes already made, and even invent a few new ones. Sex behavior research has had its share of observations that did not stand the test of time. These were often presented dramatically, perhaps in an attempt not to yield to Victorian constraints: "rejuvenation after injection of testicular extracts" (Brown-Sequard, 1889), "hypersexuality" in male cats with rhinencephalic injury (Schreiner & Kling, 1953), "compulsive sexual activity" in male rats treated with serotonin synthesis inhibitors (Tagliamonte, Tagliamonte, Gessa, & Brodie, 1969; Ferguson et al., 1970; Gessa et al., 1971) and "full-blown male mating behaviours" in female mice deprived of pheromonal information (Kimchi, et al., 2007; Shah & Breedlove, 2007; Wu & Shah, 2011). When the rats, cats, and mice were appropriately tested with receptive females (rather than with toys, other males, or unreceptive females), a parameter of behavior or two changed a bit, the curtain fell and the show was over; less conspicuous, yet interesting effects had emerged (Moore, 1919; Michael, 1961; Whalen &

1 For Knut, Happy 90th Birthday.

Luttge, 1970; Zitrin et al., 1970; Ahlenius et al., 1971; Malmnäs & Meyerson, 1971; Salis & Dewsbury, 1971; Södersten et al., 1976a, 1976b; Martel & Baum, 2009).

How did all this start?

THE HISTORY OF THE TESTIS

> Let us take the testicle from a cock and put it into the belly of a gander- the inclination of the gander would not be towards the hen but towards the goose; for, although the testicles are the cause of the inclination, they do not direct these inclinations; the inclinations become an operation of the mind, after the mind is once stimulated by the testicle.
>
> —*John Hunter*, 1779

Clearly, Hunter suggested testicular control over sexual behavior; replace "mind" in the quote above with the "tissues mediating mating behavior" (Young, 1961) and it sounds familiar.

A surgeon working in the tradition of organ transplantation and reporting on the replacement of teeth, Hunter developed a method for introducing testes into the abdominal cavity of recipient animals. Indeed, when Hunter's transplants were examined 200 years later, the presence of normal tubular structures and spermatozoa was confirmed (Jørgensen, 1971). Although his method has been mentioned many times (e.g., Jørgensen, 1971; Setchell, 1990; Freeman et al., 2001; Wingfield, 2005), Hunter's observations of behavior and physiology have not. Remarkably, Hunter also noted the influence of photoperiod on the testes of the sparrow (*Passer domesticus*) (Fig. 15–1; Royal College of Surgeons of England, 2012) and observed ovarian compensatory hypertrophy after unilateral ovariectomy (Jørgensen, 1971).

Although these observations were not confirmed by controlled experiments, they were essentially correct and hundreds of years ahead of their time. However, because of conceptual constraints, Hunter never developed a concept of humoral signaling. Berthold (1849), of course, using Hunter's method, verified his behavioral hypothesis and concluded that the transplanted testes exerted their influence "durch deren Einwirkung des Blutes,"[2] (by acting via the blood) in the absence of neural support, and hence provided "the first proof of endocrine function" (Forbes, 1949). Harris (1955) was undoubtedly right in that Berthold's report is the "first paper of any importance in the field of endocrinology," but the idea was implicit in Hunter's work already.

Because Berthold did not replicate his results, and others failed (Wagner, 1851), his report went unnoticed for some time. However, interest in organ transplantation was recharged by Addison's description of patients suffering adrenal dysfunction (Addison, 1855) and by the subsequent demonstration by Brown-Sequard (1856) that replacing the adrenal gland reversed the effects of adrenalectomy (Tattersall & Turner, 2000). Optimism for the clinical use of organ transplantation for virtually any disorder surged (Hoberman & Yesalis, 1995; Sengoopta, 2003,

2 Quotes in German are used to encourage readers to take a look at papers written in other languages than English. Although it will become clear why this can be useful, translations will be provided in parenthesis immediately after the quotes, as needed.

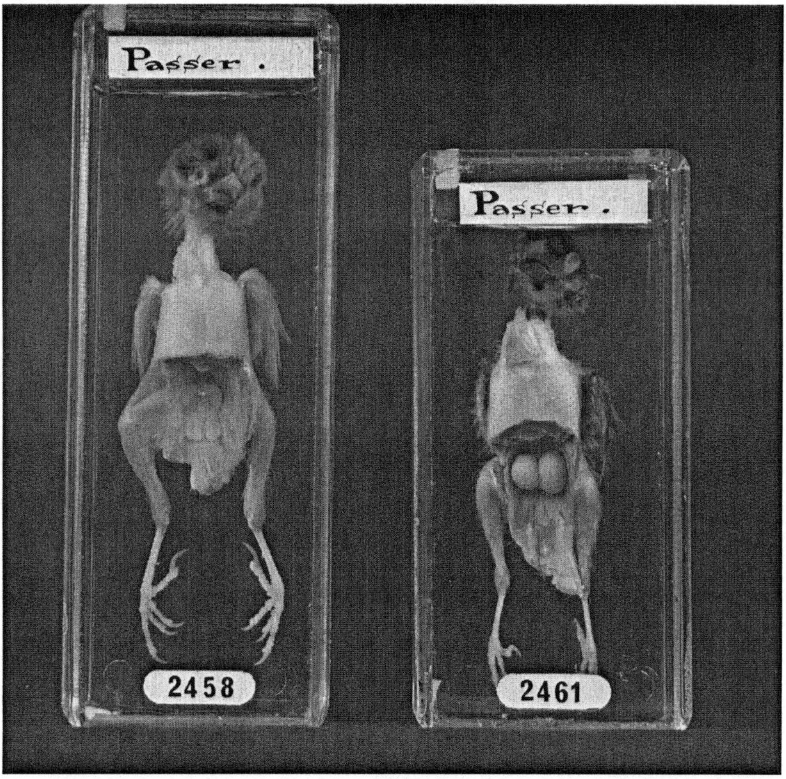

Figure 15–1. February (2458) and April (2461) Testicles of Sparrow (Passer domesticus) collected by John Hunter in 1767. Testicles from other months are found at http://bit.ly/nJaVGb. ©The Hunterian Museum at the Royal College of Surgeons. Reproduced with permission.

2006). When Brown-Sequard, already and rightly famous not only for his work on the adrenal gland but also for his contributions to the neurobiology of the spinal cord (Rengachary et al., 2008), reported the revitalizing effect of self administering extracts from testicles from "a dog or a guinea pig," testis transplantation was applied across the world on a surprisingly large scale (Brown-Sequard, 1889; Hoberman & Yesalis, 1995; Sengoopta, 2003, 2006).

The method was further simplified by Steinach (1920), who found that vasoligation increased testicular secretions in aged animals. The procedure was widely popular and between 1920 and 1940 men all over the world underwent the "Steinach operation" (reviewed by Steinach, 1940).

STEINACH'S THEORY

> Parallel mit der somatischen vollzieht sich die psychosexuelle Wandlung. Die maskulierten Weibchen erhalten ausgeprägt männlichen Sexualtrieb, sie underscheiden sofort ein nichtbrünstiges von einem brünstigen Weibchen. Sobald sie ein

solches agnoszieren, vervolgen sie es unaufhörich, umverben es leidenschaftlich und springen auf...Die Erotisierung des Zentralnervensystems ist also bei den maskulierten Weibchen in männlicher Richtung erfolgt.

—*Eugen Steinach*, 1913

(The psychosexual transformation occurs in parallel with the somatic change. The masculinized female displays a marked male sex drive, she immediately distinguishes a receptive females from a nonreceptive. As soon as she recognizes a receptive female, she will pursue her uninterruptedly, embrace her passionately and mount...Thus, the nervous system of the masculinized female has been "erotized" in male direction.)

Clearly, Steinach was a keen observer of sexual behavior, as testified by the quote above from an experiment with testicular grafts in female rats. Unsurprisingly, critical concerns were raised: "It is a drawback to the experimental method...that anyone may make a few random experiments and with the results lay some sort of claim to general attention" (the Lancet, 1917). However, the anonymous skeptic got the point: "Around these findings the theory has been constructed that the products of testicular and ovarian secretion—that is the specific reproductive hormone of the two sexes—are sharply antagonistic the one to the other. Their effect on the brain, from which the sexual impulse proceeds, is described as an 'erotising' one, in the direction of masculinity and femininity, respectively. The conclusions want more evidence to back them." That evidence was provided by Moore (1919), who replicated the effect of testis transplants on male sexual behavior in female rats: "These observations...support *very strongly (italics added)* his ideas of the transforming power of the gonad of one sex over...the psychic nature of the opposite sex."

Thus, in the first theory of the neuroendocrinology of sexual behavior, Steinach suggested that the behavioral sex is in the hormone, that testicular secretions ("male" hormones) antagonize the action ovarian of secretions ("female" hormones) and that these hormones act on the brain to cause behavior.

In parallel and with a long tradition, the influence of "gonadal secretions" was noted in a different context, discussed in the next paragraph.

400 MILLION LITERS OF BEER AND 25 THOUSAND LITERS OF URINE

Our soldiers complain bitterly that they don't get enough beer to maintain life.

—*Duke Johan of Sweden*, 1558[3]

In the 16th century, 400,000 Swedes drank 2.62 liters of beer every day, that is, about 400 million liters were consumed each year (Heckscher, 1939–1945; Morell, 1989; Tollin, 2011). As the quote indicates, the soldiers at the royal court considered the amount insufficient and demanded twice as much (Heckscher, 1939–1945). Domestic production could not possibly cover this impressive consumption; beer

3 Kungl. Majt:s vår käre herr faders och våre skytter sig fast hårt beklage, att the icke nödtorfteligen försörgde varde ved Kungl. Maj:t vår käre herr faders och våre slott och gårdar med så myckit öl som the sig till lifsens uppehälle behjälpe kunne (Heckscher, 1935-49). Readers will not be bothered by quotes in Swedish as very little of importance has been written in this language.

was imported from Prussia and Germany. Along with this trade, different kinds of hops (*Humulus lupus*) were also imported, in order to improve the taste and to preserve the beer for all subsequent days of drinking. Addition of hops to beer is perhaps the most important aspect of the brewing process and has been practiced since the 12th century; Without hops, beer loses all its taste pleasures; hopless beer is hopeless! (Verzele, 1986). Belgian monks, who brewed the best beer in the world, however, tried to prevent its introduction arguing that hops is the "fruit of the devil" and Hildegard of Bingen supported the use of the old spices, gruyt, by warning that hops causes melancholia and impotence (The Economist, 2011). These threats were motivated by local politics and had no effect on the thirsty Swedes. Millions of hop bars were grown in Sweden 500 years ago (Karlsson-Strese, 2008), and further than that, import of hops amounted for a fourth of the total foreign trade; the average Swede probably consumed 1 kg of hops each year (Heckscher, 1939–1945; Morell, 1989). As a side effect, the young women who came from their distant homes to pick the hops began to menstruate 2 days later (Koch & Heim, 1953), an interval conspicuously similar to the 48 hours often employed in studies on the induction of lordosis by estradiol in ovariectomized rats (Södersten, 1985), the commonly used bioassay of estrogen's effect on the brain!

Butenandt & Jacobi (1933) suggested the presence of "female" sex hormones (estrone) in plants, some disagreement emerged (Jacobsohn et al., 1965), but the presence of mammalian sex hormones in plants, including hops (Koch & Heim, 1953; Milligan et al., 1999), is now established (Janeczko & Skoczowski, 2005; Chadwick et al., 2006) and binding sites for estradiol and progesterone have been demonstrated in plants (Janeczko et al., 2008). Beer lovers, however, need not worry, they are not at risk of unwanted side effects; the concentration of estrogen in beer is low (Milligan et al., 1999). Interestingly, plant and mammalian 5α-reductases for steroids are structurally and functionally similar (Li et al., 1997), although receptors and intracellular signaling may differ (Wang & He, 2004). Steroid metabolic pathways are apparently evolutionarily conserved impacting on development and reproduction of individuals of a wide variety of species (Li et al., 1997).

The reported effects of testicular extracts and transplants in experimental animals and humans inspired the search for the active testicular component and in 1931 Butenandt isolated androsterone from tens of thousands of liters of policemen urine. Using testes as the source for androgenic activity, David et al. (1935) identified testosterone as the "male hormone" and testosterone was subsequently synthesized by Butenandt & Hanish (1935) and by Ruzicka & Wettstein (1935), who applied for a patent, very likely a sign of the expectation of financial rewards from clinical use by the supporting drug companies (Freeman et al., 2001). With the availability of the "male hormone" an important development of Steinach's theory emerged. But before that, the theory was "falsified" by Moore & Price (1932).

THE DISCOVERY OF NEUROENDOCRINE FEEDBACK

> Das geschlechtshormonische Gleichgewicht in individuellen Leben ist durch die Wechselwirkung von Hypophyse und Keimdrüse gesichert.
>
> —*Eugen Steinach & Heinrich Kun*, 1928

(The balance of the sex hormones is maintained throughout the life of the individual as the result of an interaction between the pituitary and the gonad.)

Steinach & Kun (1928) suggested testicular–pituitary feedback in 1928. In support, they had found that injection of pituitary extracts activated testicular function and sexual behavior in prepubertal rats, enuchoid rats, and in aged rats and referred to the effect as "reactivation." Administration of testicular hormones also caused reactivation and concomitantly influenced the pituitary gland as evidenced by the reversal of structural changes after castration (castration cells). Thus, Steinach & Kun (1928) suggested that the gonads exert a "Rückwirkung" (feedback) action on the pituitary gland.

Hohlweg & Junkmann (1932) expanded on this model by the observation that the appearance on castration cells in the pituitary required normal neural connections and suggested that a "Sexualzentrum" in the brain is involved in negative feedback action of gonadal hormones. Although it was suggested long ago that the hypothalamus has a role in the control of the pituitary (Anderson & Haymaker, 1974; Kreier & Swaab, 2010), only little attention was paid to the hypothalamo-pituitary-gonadal axis, very likely because the vascular control of the anterior pituitary was established relatively late (Harris, 1955; Anderson & Haymaker, 1974; Kreier & Swaab, 2010). It is perhaps for this reason that Hohlweg & Junkman (1932) suggested that a "nervös" link between the "Sexualzentrum" and the pituitary mediated in pituitary-gonadal feedback. Interestingly, Hohlweg & Junkman (1932) observed that gonadal hormones exert stronger negative feedback before than after puberty and they suggested therefore that puberty is caused by a decrease in the sensitivity of the Sexualzentrum to the hormones of the gonads, a hypothesis that was verified later 45 years later (Södersten, 1975b; Södersten, Damassa, & Smith, 1977).

Although commonly thought to be the originators of the feedback concept, Moore & Price (1932) were latecomers to this early neuroendocrine principle. Previously, Moore (1919) had studied gonadal grafts and was able, unlike Steinach, to maintain functional ovarian grafts in intact males and functional testicular grafts in intact females, thus challenging Steinach's concept of gonadal hormone antagonism. In 1927, a lipid extract of bull testes was found effective in a bioassay of the "testis hormone" (the term preferred by Moore) and an ovarian extract, estrin, that had been shown to cause vaginal cornification in ovariectomized rats, was already available. Administering estrin or the "testis hormone" to intact animals, Moore & Price (1932) discovered testicular shrinkage with both treatments and so launched the "Moore-Price negative feedback concept," or the "Moore-Price law" (Price, 1974; Sawin, 2002). Dorothy Price was the mastermind behind these elegant experiments but stereotyped sex roles, perhaps more prevalent at the time than today, prevented her from receiving the recognition she deserved (Sawin, 2002).

However, the confirmation of negative feedback by Moore & Price (1932) did not discourage Steinach's crew.

STEINACH'S HORMONE LIBERATION TEAM

> "das typische Bild der psychischen Feminierung" (the typical picture of behavioral feminization) in castrated male rats treated with estradiol.
>
> —*Heinrich Kun, 1934*

As administration of purified steroid hormones replaced grafting of gonads, a remarkable attempt was made to advance Steinach's theory. Kun (1934) reported that castrated male rats were "feminized" by just one injection of estradiol benzoate. Thus, 37 of 39 rats (95%), castrated between 4-6 months of age and tested 3-11 months later, displayed the "Lordosereflex" 48 hours after the injection; the two rats that were left behind in the first test were completely feminized a few days later without additional treatment. Kun (1934) also described "heftiges Zittern der Ohren" (earwiggling) and "springende Bewegungen" (hopping and darting), but it is not clear whether he observed these behaviors in the males. However, lordosis was readily elicited by mounts from a stimulus male, behaving equally "leidenschaftlich" (passionately) with an estradiol-treated male as with a female in estrus.

"Bertrand Russell once pointed out that experimental animals studied by American behaviorists behaved like Americans, running about in an almost random fashion, while those of Germans behaved like Germans, sitting and thinking" (Skinner, 1974). By contrast, rats studied by Steinach's team were hopping and darting rather than sitting and thinking! Clearly, "Ratten, Menschen und Meerschweinchen" (rats, humans, and guinea pigs) studied by followers of "das Fundament der Sexualhormonforschung" (the foundation of sex hormone research) behaved flexibly. To paraphrase a charismatic writer: "the Renschenmeer *(Iuravariabilis sexualis)* is a hybrid species unique in that all of the anatomical, physiological, and behavioral characteristics of one sex are completely reversible by treatment with the gonadal hormones of the other" (Beach, 1971). In fact, the Renschenmeer is the absolute opposite of the Ramstergig *(Iurasequens sexualis)* the virtual "Sexual follower of laws" rodent incarnation of the theory that sexual behavior and brain structure are irreversibly "organized" by androgens in early life (Beach, 1971).

Further defense of Steinach's hypothesis of hormone antagonism was Kun's (1934) observation that although intact male rats showed lordosis if treated with estradiol, a fourfold higher dose was needed to elicit the response and the latency increased to 4 to 13 days. Recognizing the contribution of the brain (Hohlweg & Junkmann, 1932), Kun (1934) suggested that in the presence of testicular secretions, the "Sexualzentrum" was less sensitive to estradiol and explained that Moore's reports mainly "confirm the results of earlier authors."

Kun's (1934) results have been ignored perhaps because they were considered contrary to a dogma that was already becoming solidified. However, Kun (1934) reported, as had Steinach (1894) already, effects of castration on mounting, intromitting, and ejaculation in the 39 males that he studied, which are identical to those reported in all other studies. Why, if Kun (1934) made valid observations on one behavior, would he report invalid results on the other?

AROMATIZATION DISCOVERED, AN OUTSTANDING STUDY BY STEINACH

> Der männliche Organismus imstande ist ... männliches Hormon in einen Stoff mit Östrogener Wirkung umzuwandlen.
>
> —*Eugen Steinach, Heinrich Kun, & Oskar Peczenik*, 1936
>
> (The male is able to convert male hormone into material with estrogenic action.)

In 1936, Steinach, Kun, & Peczenik (1936) reported results on the neuroendocrinology of sexual behavior, which were 37 years ahead of their time.

First, the effect of the synthetic male hormones on the brain was compared with that of testicular secretions. Steinach had found that vasoligation increases testicular secretions thereby increasing cerebral blood flow and used this finding as a model of hormone action on the brain. He referred to the behavioral effect as "revitalization", because vasoligation worked in old, but not in young rats. However, while daily injection of androgens had no effect, injection of estradiol benzoate markedly increased cerebral blood flow.

Because estradiol thus replicated the effect of testicular secretions on the brain, it was hypothesized that it might also affect "psychische Funktionen." Induction of ejaculation in young castrated, sexually quiescent male rats was chosen as the model of such a function. The threshold dose of a variety of androgens was first determined and because previous work had suggested that estradiol might enhance the effect of androgens on the seminal vesicles (Freud, 1933) (not that Freud!), a dose of estradiol benzoate, which had no behavioral effect in itself, was added to the androgens. Combined estradiol-androgen treatment reduced the threshold from between 2.5-fold with androsterone to 10-fold with nonesterified testosterone (0.1 mg/day for 9 days). Hence, Steinach's group suggested that estradiol and androgen act synergistically on the brain to induce of ejaculation in castrated rats, via a "Schrittmacherwirkung" similar to the "pacemaker" effect on the seminal vesicles demonstrated by Freud (1933). These two effects of combined estrogen-testosterone treatment were replicated by the present author, unaware that the exact same experiments had been done 40 years previously (Table 15–1, Södersten, 1975a).

Steinach et al. (1936) then noted that an estrogen had already been measured in blood and urine but because it might originate from ingested foods, an alternative hypothesis was tested, namely that male hormone can be converted into female hormone in vivo. Urine was collected from castrated and intact rats, which were treated with an extract of urine or androsterone, both of which with androgenic but no estrogenic activity. The urine was concentrated and injected into ovariectomized rats and observed to cause vaginal cornification. Hence, Steinach et al. (1936) concluded that the male is capable of converting male hormone into material with estrogenic action.

The presence of "female" hormones in males had been recognized for some time (see Sengoopta, 2006) and Zondek (1934a, 1934b; Zondek & von Euler, 1934) had suggested that males can metabolize testosterone to estrogen. Steinach verified this hypothesis with exceedingly elegant experiments and thus essentially discovered the role of estrogen in male behavioral neuroendocrinology in 1936.

Table 15–1 Synergistic effect of estradiol benzoate and testosterone propionate on seminal vesicles and sexual behavior.*

		Seminal Vesicle Weight	Percent Mounting	
Treatment	n	Percent of Intact	Before	After 6 TP inj
Oil + Oil	6	3	0	0
1 µg EB + Oil	8	34	25	25
Oil + 200 µg TP	9	141	0	0
1 µg EB + 200 µg TP	9	182	22	89
Intact	6	100	–	–

*Replication of Freud's (1933) "pace-maker" effect and Steinach et al.'s (1936) "Schrittwacherwirkung": "the two substances mutually reinforce each other's activity" (i.e., testosterone and estradiol) on seminal vesicle weight and sexual behavior in male rats. The rats were castrated at 30 days of age and treated with estradiol benzoate (EB) and testosterone propionate (TP) when 120 days old. Oil or EB were injected daily over 30 days and TP was then added over another 30 days.

From Södersten, P. (1975a). Mounting behavior and lordosis behavior in castrated male rats treated with testosterone propionate, or with estradiol benzoate or dihydrotestosterone in combination with testosterone propionate. *Horm Behav, 6,* 109–126.

STEINACH'S WORK RECOGNIZED

> Steinach's animal experiments were carefully conducted and his theoretical concepts were highly original, some of them so much so that their final test was delayed for half a century until necessary technical advances had taken place.
>
> —*Frank Beach*, 1981

In reviewing the history of research on hormones and behavior, Beach (1981) commended Steinach for his experimental skills and theoretical insights, noting that Steinach was ahead of his time. Beach was right, for Steinach's results have been confirmed in subsequent studies: increased testicular androgen secretion after vasectomy (Peng et al., 1987; Mo et al., 1995), androgen antagonism of estradiol-induced sexual receptivity (Davidson, 1969; Butler et al., 2001), induction of sexual receptivity, including "heftiges Zittern der Ohren" (vigorous shaking of the ears = earwiggling), by ovarian hormones in adult castrated male rats (Södersten et al., 1983), synergistic effect of estradiol and androgen on sexual behavior in castrated rats (Baum & Vreeburg, 1973; Larsson et al., 1973), ovarian hormones needed for the development of female sexual behavior (Brock et al. 2011), androgen and estrogen effects on cerebral blood flow (Gonzales et al., 2007; Krause et al., 2011), cerebral blood flow effects on sexual behavior (Georgiadiss et al., 2010), and, of course, conversion of androgens to estrogens by males, the topic of this book.

Virtually everything that has been written about Steinach concerns the clinical translation of his research (Hoberman & Yesalis, 1995; Sengoopta, 2003, 2006), his experimental work has been overlooked (Logan, 2010). The circumstances of his late life probably contributed to this neglect. Because Steinach was member of a Jewish academic family, his laboratory was closed by the national socialists just two years after his remarkable 1936 paper and he was prevented from returning to Vienna while on a lecture tour in Switzerland; after the second world war his work went into oblivion (Rudolf Soukup, 2010 and personal communication, 2011; Sonja Walch, personal communication, 2011).

One can only agree with Beach (1981) that: "it is regrettable that Steinach's imaginative and largely fruitful endeavors have never been accorded the importance they deserve." The two previous sections in the present chapter are an initial attempt to recognize the contributions made by Steinach in his extraordinary 1936 paper.

HORMONES AND BEHAVIOR, BEACH'S 1948 MASTERPIECE

> The specificity of the mating patterns for the two sexes, although probably inherited, is not rigidly dictated by the innately organized substratum... it is obvious that in a few individuals at least, there exists the innate organization essential to the mediation of the mating pattern of either sex.
>
> —*Frank Beach*, 1938

Many of the studies reviewed above relied on casual observation of sexual behavior (e.g., Moore, 1919; Green et al., 1957). Although this may seem surprising, statistical inference was a mature science long before that time (Hald, 2007), biological and physical scientists debated whether mathematical modeling was applicable to biology, including behavior in the early twentieth century (Coen, 2006). The situation changed very much because of the contributions of Beach (1948, 1950, 1971, 1975, 1976, 1981).

Reviewing the literature and listing the eight possible combinations of sex, hormone, and behavior, Beach (1948) pointed out that "none of the behavioral responses thus far investigated have been found to depend upon one and only one hormone," hence challenging Steinach's hormone-specificity theory. As indicated by the quote above, however, Beach realized that the "Perfect specimens of one single sex are in reality theoretical ideals; a complete man is as non-existent as a complete woman" (Steinach, 1940, p. 113).

Interestingly, in the context of the present book, intact male rats treated with 1 mg testosterone propionate daily had been reported to show not only "intensification of normal masculine mating behavior" but a few of these males also showed lordosis (Beach, 1941, 1945). In characteristic scholarly writing, Beach (1948) pointed out that Steinach et al. (1936) had suggested an explanation: males convert androgen to estrogen. However, rather than joining Steinach in taking an interest in the biochemistry of androgens, Beach (1948) introduced "the behavioral mechanism" as an intervening construct to describe hormone-behavior relationships. In discussing this relationship, Beach (1948) famously questioned Steinach's view that hormones can organize the behavioral mechanism, because rats castrated at various ages and subsequently treated with androgen readily displayed sexual behavior (Beach & Holz, 1946; Beach, 1971). In rats castrated on the day of birth and treated with androgen as adults there was a shift from intromitting and ejaculating to mounting, caused by reduced sensory feedback from a poorly developed penis. This observation and the subsequent report by Beach & Levinson (1950) that the sensory receptors on the glans penis are androgen dependent were crucial for the verification of the aromatization hypothesis.

Inspired by embryological studies the mid 1930s (Burns, 1961), the concepts of "organization" and "activation" of the behavioral mechanism were gaining ground as intervening variables summarizing the effects of gonadal hormones on sexual behavior. Thus, Ball (1937) noted that while the complete sex reversal in the estrogen-treated castrated rats reported by Kun (1934) suggested that the "gonadal hormone *organizes* [*italics added*] the mating behavior pattern," she found that the same "female hormone increased...the male activity, suggesting that its function in the adult animal is...merely to *activate* [*italics added*] a pattern already laid down either genetically or under the previous influence of the gonads." Add the observation of Dantchakoff (1938) that female guinea pigs treated prenatally with androgen displayed enhanced mounting as adult, and the "organization-activation" framework emerged in the magisterial review by Young (1961), which has permeated virtually all of the subsequent work on sex differences in sexual behavior and on sex differences in general (de Vries & Södersten, 2009; Crews, 2010).

ESTROGEN DISREGARDED

> To confirm the well-known inhibitory effects of estrogen on male sexual behavior.
>
> —*Julian Davidson*, 1969

Davidson (1969) correctly pointed that the quantitative methods introduced by Beach and others had not been used in studies on the effects of estrogen on sexual behavior in male rats. This was a major step ahead because studies discussed below had given estrogen a bad reputation in the context of male sexual behavior.

Once Phoenix, Gerall, Goy, & Young (1959) had verified Dantchakoff's (1938) hypothesis that sexual behavior is organized in a male direction by prenatal androgen stimulation, the possibility that female behavior is similarly organized by estrogen was tested. This is clearly a project in Steinach's spirit. However, rather than organizing anything, early estrogen treatment prevented the display of female sexual behavior in adult females (Harris & Levine, 1962; Whalen & Nadler, 1963; Feder & Whalen, 1964) and disorganized male sexual behavior in adult males (Harris & Levine, 1962; Levine & Mullins, 1964; Feder, 1967; Whalen, 1967). These observations probably discouraged further testing of the effects of estrogens in the male.

However, they did not discourage Davidson (1969). Using the lordosis quotient, Davidson (1969) replicated Kun's (1934) finding that the presence of the testes inhibited the induction of lordosis by estradiol and that high lordosis quotients could be induced by estradiol in the castrate. On the other hand, the experimental animals studied by Davidson (1969) did not behave like those studied by Kun (1934), they responded much less enthusiastically to estradiol.

Referring to Steinach (1940), Davidson (1969) then "*confirmed the well-known inhibitory effects of estrogen on male sexual behavior,*" [*italics added*] although Steinach et al. (1936) had shown the opposite. Paradoxically, Davidson (1969) found that estradiol inhibited ejaculation in intact rats and merely exerted a weak stimulatory effect on the sexual behavior of castrated rats.

In a comprehensive study, Pfaff (1970) tested all of the eight combinations of sex, hormone, and behavior listed by Beach (1948) and suggested that differences in "tissue sensitivity" explain individual differences in sexual behavior and that tissues are not "extremely hormone-specific." Interestingly, Pfaff (1970) pointed out that the biochemical pathways for conversion of testosterone to estradiol had been demonstrated, and, 34 years after Steinach et al. (1936), suggested that some of the behavioral effects of testosterone are mediated by estradiol, although this possibility was considered unlikely.

Steinach's concepts clearly exerted an influence in 1970, but the most imaginative ones developed in the 1936 paper, hibernated a few more years. Interestingly, however, the idea that testosterone is converted to estrogen, quickly influenced research in hypertension (Steinach et al., 1938; Walker, 1942) and cancer (Horning, 1958).

ESTROGEN REVITALIZED

> ...aromatization of testosterone may be an important step in the activation of sexual behaviour in both the male and female.
>
> —*Carlos Beyer*, 1970

As androsterone, used by Steinach as a potent androgen, is not present in sufficient amounts in the testes to explain the androgenic potency of testicular extracts, it was first supposed that testosterone is the main male hormone (Bruchovsky & Wilson, 1999). However, only 10% of the radioactivity of injection of ^{3}[H]testosterone was obtained from the nuclear fraction of the rat ventral prostate (Baulieu et al., 1968; Bruchovsky & Wilson, 1968a, 1968b). Subsequent chromatographic detective work identified 5α-dihydrotestosterone as the mediator of androgen action in the rat prostate (Bruchovsky & Wilson, 1968a, 1968b, 1999). Falsifying the hypothesis that 5α-dihydrotestosterone also activates sexual behavior in castrated rats, McDonald et al. (1970) suggested that aromatization to estrogenic metabolites is a mechanism whereby testosterone induces sexual behavior, as had Steinach 34 years earlier. However, Young (1961) had noted, as had Beach (1948) before him, and Davidson (1969) after him, that the evidence that estrogen stimulates masculine behavior in males was very weak. Two studies changed the situation.

In the first study, Baum (1972) hypothesized that treatment with estradiol should delay behavioral puberty in male rats by inhibiting gonadotropin secretion and therefore testicular testosterone secretion. Baum (1972) confirmed that estradiol inhibits testis development (Moore & Price, 1932), but found that it potently stimulates precocious sexual behavior. Prevention of behavioral puberty via the mechanism suggested by Baum (1972) was subsequently demonstrated by treatment with 5α-dihydrotestosterone (Södersten et al., 1977).

In the second study, Södersten (1973) found that estradiol induces all aspects of sexual behavior in castrated, sexually experienced males, confirmed Baum's (1972) finding that estradiol suppresses testis weight but has no effect on sexual behavior in intact rats and found that estradiol stimulates mounting and intromitting, but not ejaculation in prepubertally castrated rats.

Both Baum (1972) and Södersten (1973) used high doses of estradiol, but subsequent studies showed that daily treatment with just 1 μg estradiol benzoate stimulates all aspects of sexual behavior in castrated, sexually experienced male rats (Södersten, 1975a; Södersten & Larsson, 1975), that is, less than the 10 μg estradiol pellets that were found to activate sexual behavior in castrated rats from an intracerebral site (Christensen & Clemens, 1974).

The present writer had the privilege of examining the effect of sectioning of the dorsal penile nerve on the sexual behavior of male rats with the radiant Knut Larsson (Larsson & Södersten, 1973); we found that just as with neonatal castration (Beach & Holtz, 1946) or anesthetization of the glans penis (Carlsson & Larsson, 1964; Adler & Bermant, 1966), this maneuver caused a shift from intromitting and ejaculation to excessive mounting, that is, the same effect as that of estradiol in prepubertally castrated rats (Baum, 1972; Södersten, 1973). Hence the conclusion, confirmed by direct examination (Södersten, 1973), that estradiol does not stimulate the sensory receptors of the penis, which Beach & Levinson (1950) had demonstrated to be androgen dependent.

These findings opened the door for the reinvention of the wheel.

"SCHRITTMACHERWIRKUNG" (PACEMAKER ACTIVITY) OF ESTROGEN REDISCOVERED

> ... in the presence of estrogen, DHT acquires the ability to facilitate further the activity of neural tissues controlling mounts with pelvic thrusting.
>
> —*Michael Baum & Jan Vreeburg*, 1973

> ... the two hormones act synergistically in the brain in stimulating behavior.
>
> —*Knut Larsson, Per Södersten, & Carlos Beyer*, 1973

In the early 1970s, Michael Baum, having revitalized estrogen's role in male behavioral neuroendocrinology (Baum, 1972), worked with Jan Vreeburg who had examined the androgen dependency of the male accessory sexual organs (Aafjes & Vreeburg, 1972), and so they generated the obvious hypothesis that testosterone induces sexual behavior as a result of estrogenic stimulation of the brain and androgenic stimulation of the penis (first quote above).

At the same time, Larsson, on one of his infrequent visits to his laboratory in Göteborg, informed Södersten that neither estradiol nor 5α-dihydrotestosterone has an effect of sexual behavior in male rats and so noone knows how testosterone activates the behavior. This state of affairs left the ever enthusiastic Larsson clueless, but when informed that estrogen exerts a strong stimulatory effect on male sexual behavior (Södersten, 1973), Larsson entered a state of cerebral euphoria and so, with Södersten, he came up with the same hypothesis that Baum and Vreeburg had produced (second quote above).

Thus, daily treatment with low doses of estradiol induced all aspects of copulation in prepubertally castrated rats if combined with 5α-dihydrotestosterone; as little as 0.05 μg estradiol benzoate combined with 1 mg 5α-dihydrotestosterone induced ejaculation in 60% of the rats, with only occasional responses when either hormone was given alone (Larsson et al., 1973); virtually identical results were obtained by Baum & Vreeburg (1973) and similar results were also reported by Feder et al. (1974).

The reason this experiment was not performed by the group that had demonstrated that 5α-dihydrotestosterone is ineffective in activating sexual behavior in castrated rats and suggested that "aromatization of testosterone may be an important step in the activation of sexual behaviour in both the male and female" (McDonald et al., 1970) was, ironically, practical; the supply of 5α-dihydrotestosterone was limited (Carlos Beyer, personal communication, 2011).

While the synergistic effect of estradiol and 5α-dihydrotestosterone on sexual behavior in castrated male rats was greeted as breaking news, it was merely a replication of the "Schrittmacherwirkung" reported by Steinach et al. (1936) 37 years earlier. Recollecting the spirit of the 1970s and having written the first few pages of the present chapter before discovering the masterful paper from Steinach's group instilled a sentiment of "confessions of an imposter" in the mind of the present writer, feeling "forced to admit to myself that I should have acquired knowledge of Steinach's extraordinary work long ago" (Beach, 1978, paraphrased).

Verification of the aromatization hypothesis led to new discoveries: for example, 107 of 213 (50%) intact male rats showed lordosis in response to manual stimulation without treatment (Södersten et al., 1974; Södersten & Larsson, 1974, 1975), 34 of 35 (97%) day 30 castrated males showed lordosis after daily treatment with 1 to 15 μg estradiol benzoate (Södersten & Larsson, 1974, 1975; Södersten, 1975a), and male rats treated neonatally with antiestrogen had normal androgen levels during development and displayed both female and male sexual behavior as adults without treatment (Vreeburg et al., 1977; Södersten, 1978a, 1978b; Södersten & Hansen, 1978; Södersten & Eneroth, 1980; Södersten et al., 1981). In addition, the possibility emerged that rather than acting as a direct stimulator, estradiol acts indirectly by preventing androgen metabolism in the brain. Lacking a peripheral androgen binding protein, the male rat brain may need protection against exposure to excessive amounts of testosterone secreted by the testes; cerebral inactivation via 5α-reduction may provide that protection (Södersten & Gustafsson, 1980; Mode, et al., 1984; Södersten et al., 1983, 1986, 1988). Recent work has developed these ideas to an extraordinary level of sophistication (Balthazart, 2010).

Although these studies have been reviewed before (Södersten, 1984; de Vries & Södersten, 2009), they are listed again; if they are not, the present writer suspects that, just as in Steinach's papers, they might be forgotten only to be revitalized at a later point in time.

COMPLETELY FEMINIZED MALES DISREGARDED

> The lordosis response is not exhibited by intact males or even adult castrated males treated with ovarian hormones.
>
> -*Loretta Flanagan-Cato*, 2011

The quote above shows that it did not take long to forget the hundreds of male rats, reviewed previously, which showed lordosis. Memory is short; for

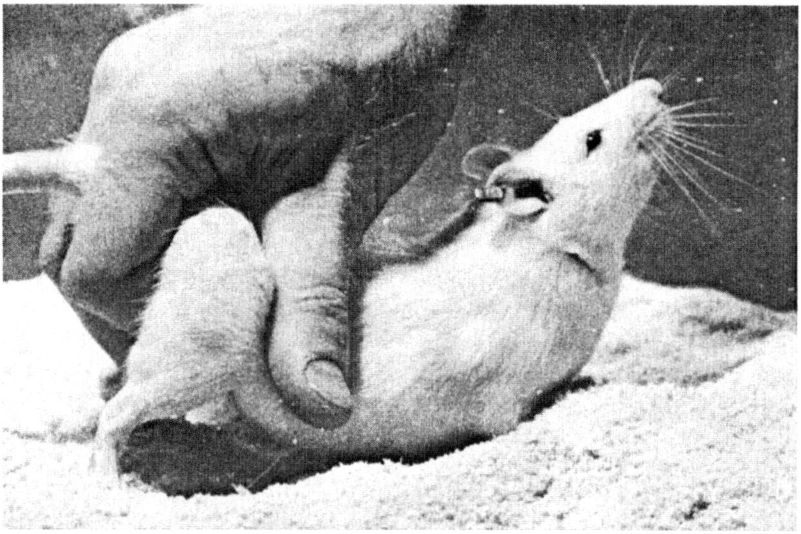

Figure 15–2. Lordosis Behavior in a Male Rat Castrated at 30 Days of Age and Treated with 1 mg Testosterone Propionate for 26 Days when 90 Days Old. Reproduced from Södersten & Larsson (1974) with permission from Elsevier Publishers.

example, carried away by enthusiasm after having induced lordosis in day 30 castrated male rats with testosterone propionate (Fig. 15–2), Beach reminded us that this had been done before and we therefore included reference to Beach (1941; 1945) in our publication (Södersten & Larsson, 1974).

Had he been around, Bertrand Russell would very likely have commented on the behavior of Danish Wistar rats (the strain used by the present writer) studied by Swedes, but he would not have disregarded these interesting animals. Their behavior emerged from the strategy of classical endocrinology; observing the effects of removal of an endocrine gland and then studying the effects of replacement with physiological doses of the hormones of that gland.

This strategy was adopted in a series of experiments on ovarian hormones and the display of sexual receptivity by female rats. The analysis was undertaken because it had been found that the presence of the ovaries during development masks the effect of neonatal androgen treatment on lordosis in adulthood in female rats (Södersten, 1976); suggesting that adult castrated male rats should be "completely feminized" (Kun, 1934) by the presence of ovaries during development. However, rather than using ovarian transplants, "artificial estrous cycles (i.e., a schedule of estradiol and progesterone injections that mimics the pattern of ovarian hormone secretion) were administered to adult castrated male rats. The results verified Steinach's hypothesis, to paraphrase Moore (1919): "The observations supported *very strongly* the idea of the transforming power of the hormones of one sex over the psychic nature of the opposite sex," and confirmed Kun's (1934) observation that adult castrated male rats can show all aspects of female sexual behavior, including "heftiges Zittern der Ohren" in response to physiological levels of ovarian hormones (Södersten et al., 1983). Conclusions about the behavioral capacity of male rats of other nationalities (Becker et al., 2005; Bonthuis et al., 2010; Flanagan-Cato, 2011) should not be drawn before a similar analysis has been undertaken. Findings that contradict prevailing theory should be explained rather than ignored; if they are not, endocrinology becomes an opinion.

The next two paragraphs depart therefore from the "personal perspectives" approach of this essay in that they review the details of the three successful and one unsuccessful attempt to obtain results comparable to those reported by Kun (1934).

COMPLETELY FEMINIZED MALES REVITALIZED

> We always interpreted our data as confirming your results.
>
> —*Jeff Blaustein*, 2012

Although the conditions used by Kun (1934) are difficult to replicate, the procedures of the four other studies can be compared and the differences in results can be explained. As indicated by the quote from Jeff Blaustein, three were successful (Södersten et al., 1983; Olster & Blaustein, 1988, 1990) another one was unsuccessful (Moreines et al., 1986).

The pattern of serum levels of ovarian hormones that precedes the onset and termination of sexual behavior during the estrous cycle has been determined in only one published study using Wistar rats (Södersten & Eneroth, 1981a). In that study, three main results were reported:

1. Cycling rats became receptive over a period of 3 to 4 hours after lights off in a 12/12 light:darkness (LD) cycle.
2. Serum levels of estradiol varied markedly starting about 2 hours before the L-phase of the LD cycle on the day before behavioral estrus (i.e., about 29 hours before), and continuing until the time when the rats were receptive.
3. Display of sexual receptivity occurred in close correlation with increasing levels of progesterone, which are required under physiological conditions.

These results were related to the details of the estradiol stimulus sufficient for induction of sexual receptivity in ovariectomized rats (Ydstebø & Södersten, 1977; Södersten & Hansen, 1977, 1979; Hansen et al., 1978, 1979; Södersten et al., 1981) with the following main results:

1. Rats must be exposed to at least 26 hours of estradiol stimulation to be prepared to respond to progesterone, increasing the strength of the estradiol stimulus does not compensate for a brief period of stimulation.
2. Continuous stimulation with estradiol can be replaced by 4-hour-long pulses in the beginning and end of the 16 hour period of estradiol stimulation.

3. If the two estradiol pulses are temporally desynchronized, the display of sexual receptive is reduced by 50%.
4. The effect of the estradiol stimulus depends on the phase of the LD cycle.
5. The lordosis quotients of gonadectomized rats treated with constant-release, estradiol-filled implants vary with the LD cycle in females, but not in males. Males show higher lordosis quotients than females in the L phase, but this sex difference is reversed 12 hours later.

The results can be summarized as follows: there is no evidence that the levels of ovarian hormones display the same pattern during the estrous cycle in different strains of rats (references in Södersten & Eneroth, 1981a). The period of variability in estradiol levels in cycling rats coincides with the period during which ovariectomized rats should be exposed to estradiol to be prepared to respond behaviorally to progesterone. The display of sexual receptivity is influenced by the LD cycle in the female, but not the male rat.

Thus, in order to replicate the condition of the estrous cycle in gonadectomized rats of this strain, estradiol should be given 29 and 17 hours before behavioral testing and progesterone should be given 3 hours before testing. The pulses of estradiol can be administered using constant-release implants or, less invasively, by subcutaneous injections. Tests for sexual receptivity should be done at least 3 hours after the onset of darkness during a 12/12 hour LD cycle. Using these procedures, injections of only 0.25 μg estradiol followed by injection of 0.5 mg progesterone induced lordosis quotients of 72±9 in 9 out of 14 ovariectomized rats. By contrast, if estradiol benzoate, rather than estradiol, was injected according to the same schedule, none out of 14 ovariectomized rats showed lordosis (Södersten et al., 1981).

To further validate these procedures physiologically, ovariectomized rats were injected with 1 μg estradiol and 0.5 mg progesterone and the levels of estradiol and progesterone were measured in blood samples obtained at hourly intervals and found to be within the range of those observed during the estrous cycle (Södersten et al., 1983). Administration of "artificial estrous cycles" thus defined, induced all aspects of female sexual behavior in adult castrated male rats and eliminated the sex difference (Södersten, et al., 1983).

Olster & Blaustein (1988) used similar procedures in gonadectomized Sprague-Dawley rats and essentially replicated the results, with the exception that they tested their rats 4, rather than 3 hours after the injection of progesterone and 2, rather than 3 hours into the D-phase of a 14/10 LD cycle (see Table 15–2). These procedural differences may seem minor, but considering the coupling of the display of sexual receptivity to the LD cycle, they may explain the small difference in the lordosis quotients of the female rats between the two studies (see Table 15–2). The difference between the lordosis quotients shown by the male rats in the two studies is more apparent than real. The quote from Jeff Blaustein: "We always interpreted our data as confirming your result," continues: "They may not have been the same as females, but repeated pulses brought them pretty close" (Blaustein, J. D., personal communication, January 2, 2012.).

Table 15–2 Lordosis quotients in gonadectomized rats treated with pulses of estradiol and progesterone.*

		Time of Injection						
Study	LD	D	E	E	P	E-P	Sex LQ	Sex LQ
1	12/12	−3	−29	−17	−3	26	Females: 90	Males: 90
2	14/10	−2	−30	−18	−4	26	Females: 80	Males: 50
3	12/12	?	−27	−15	−3	24	Females: 60	Males:5
E-filled implants								
4	12/12	−3	−24		−6	18	Females: 40	
4		−3	−28		−6	22	Females: 55	
4		−3	−32		−6	26	Females: 85	
4		−3	−48		-6	42	Females: 100	

*Lordosis quotients (LQ) in response to 10 mounts by stimulus male rats displayed by adult gonadectomized rats injected with 1 μg estradiol (E) twice and 0.5 mg progesterone (P). The hour of injection is related to behavioral testing and the hours of E-priming (E-P) are related to the time of the P injection. The hormones were injected at 4-day intervals and the results are from the fourth test. The hours of light-darkness (LD) are indicated as are the hours between the onset of darkness and behavioral testing (D). Study 4 shows LQs of female rats implanted with 5-mm long, E-filled constant-release capsules at various times before behavioral testing.

1. Södersten et al. (1983)
2. Olster & Blaustein (1988)
3. Moreines et al. (1986)
4. Södersten et al. (1981)

Moreines et al. (1986), who did not obtain these results, used different procedures than Södersten et al. (1983) and Olster & Blaustein (1988). Thus, the time period between the first injection of estradiol and the injection of progesterone, 24 hours, was too short (see Table 15–2). Insufficient estradiol priming probably explains why Moreines et al. (1986) obtained lordosis quotients of only 60 in their ovariectomized rats compared with 90 reported by Södersten et al. (1983) and 80 reported by Olster & Blaustein (1988). Note, however, that the rats studied by Södersten et al. (1983) showed lordosis quotients similar to those reported by Moreines et al. (1986) if treated with constant-release implants filled with estradiol for a period of time comparable to that of the rats studied by Moreines et al. (1986) (see Table 15–2). Moreines et al. (1986) did not state when during the LD cycle they tested their rats, but as they injected progesterone at 09.00 and tested their rats 3 hours later, it is possible that their rats were tested during a less than optimal phase of the LD cycle. And if they were not, the injections of estradiol were given at a less than optimal phase of the LD cycle.

In unpublished work, Moreines (1980) reported that ovariectomized Sprague-Dawley rats should be exposed to estradiol for 39 hours to be prepared to respond to progesterone by showing maximal sexual receptivity 6 hours later, and it is unclear, therefore, why Moreines et al. (1986) departed from procedures that they had themselves determined to be optimal. It is clear, however, that a 39 hour long estradiol priming period is too long because it yielded "concurrent inhibition" by progesterone, that is, by administering progesterone concurrently with estradiol, the rats did not respond to an injection of progesterone 39 hours later (Moreines, 1980). This effect is an artifact, because if ovariectomized rats are tested 32 hours after implantation of an estradiol-filled constant release capsule, concurrent implantation of a progesterone-filled constant release capsule *facilitates* sexual receptivity (Södersten & Eneroth, 1981b). If rats are tested 32 hours after estradiol implantation, progesterone facilitates sexual receptive independently of when it is administered (Södersten & Eneroth, 1981b) and progesterone can facilitate the behavioral response repeatedly simply by raising the dose (Hansen & Södersten, 1978, 1979). Progesterone is therefore not a behavioral antiestrogen in the rat. The hypothesis emerged from the observation that estradiol does not stimulate sexual receptivity in the presence of functional corpora lutea in the guinea pig (Goy et al., 1966). However, induction of functional corpora lutea in rats does not inhibit the induction of sexual receptivity by two properly timed pulses of 1 μg estradiol; quite the opposite, pulses of estradiol synergize with the already elevated endogenous progesterone levels to stimulate display of sexual receptivity in the pseudopregnant rat (Södersten & Eneroth, 1981b).

More interesting, Moreines et al. (1986) obtained lordosis quotients in their estradiol-treated males that were far below those reported by Södersten et al. (1983) and Olster & Blaustein (1988); essentially their males did not show lordosis (see Table 15–2). This is surprising because Olster & Blaustein (1988) and Moreines et al. (1986) studied Sprague-Dawley rats from the same breeder and one wonders why these rats behaved differently if tested in New York City than if tested in Amherst. Olster & Blaustein (1988) suggested that this difference is not likely due to the differences in hormonal treatment, and more likely caused by differences in the stimulus intensity used to elicit lordosis. This point was made long ago by Beach (1938): "The mating behavior to be displayed by a member of either sex may be in part or (in the cases reported), entirely predetermined by the behavior of the partner." In the studies by Kun (1934) and Södersten et al. (1983), this was not a problem, because, as noted above, the stimulus males behaved equally "leidenschaftlich" (passionately) with a castrated estradiol-treated male as with a female in estrus.

In the fourth study, which was done on guinea pigs and obtained results on female sexual behavior in males comparable to those reported by Kun (1934) and Södersten, et al. (1983), Olster & Blaustein (1990) kept the stimulus for eliciting lordosis constant using a procedure, which had been available for 53 years (Young et al., 1937). Two injections of 2 μg estradiol, 28 hours apart, followed by 0.5 mg progesterone 40 hours after the first injection of estradiol, induced lordosis in all animals and eliminated the sex difference in behavior and in hypothalamic progestin receptors concentrations. By contrast, the commonly observed, marked sex difference in lordosis was replicated by injection of 10 μg estradiol benzoate followed by 0.5 mg progesterone. Hence, Olster & Blaustein (1990) concluded that "the display of progesterone-facilitated lordosis is not inflexibly differentiated according to sex in guinea-pigs."

It is noteworthy that the high invididual variation in serum levels of estradiol observed in cycling rats (Moreines, 1980; Södersten & Eneroth, 1981a) had already been reported in the guinea pig (Croix & Franchimont, 1975). Considering the fact that luteinizing hormone (LH)-releasing hormone is secreted in a pulsatile manner by the hypothalamus in this species (McKibbin & Belchetz, 1986), it seems likely that guinea pig ovaries also secrete estradiol episodically. Despite possible species and strain differences in ovarian hormone secretary patterns and behavioral sensitivities (Goy & Young, 1956; Moreines, 1980; Södersten & Eneroth, 1981a; Spearow et al., 1999; Bonthuis et al., 2010) it is striking that display of sexual receptivity can be induced in adult castrated male rodents of different strains and species using pulsatile administraton of estradiol (Södersten, et al., 1983; Olster & Blaustein, 1988, 1990). The failure to obtain these results (Moreines et al., 1986) is not convincing.

A CHALLENGE OF THE PARADIGM OF THE NEUROENDOCRINOLOGY OF SEXUAL BEHAVIOR

> In 1959, a single publication by William C. Young and his colleagues changed the paradigm of behavioral endocrinology so much so there has been little work on the bisexuality of the brain since that time.
>
> —*David Crews, 2010*

As indicated by the quote from David Crews (2010), the report of the sex difference in lordosis behavior in guinea pigs by Phoenix et al. (1959) changed the focus of behavioral neuroendocrinology from 50 years of study of the similarities between the sexes to 50 years of study of the differences between the sexes. In that study, Phoenix et al. (1959) treated guinea pigs with lower doses of estradiol benzoate (1.66–6.64 versus 10 μg) and progesterone (0.2 versus 0.5 mg) than those used by Olster & Blaustein (1990) and injected the estrogen at a different point in time (36 versus 40 hours) before progesterone. However, both studies used the same method to elicit lordosis behavior (Young et al., 1937; Goy & Young, 1956) and obtained the same, marked sex difference. In the study by Olster & Blaustein (1990), however, the sex difference was eliminated by treatment with a smaller amount of estradiol administered as two pulses. These results challenge the conclusion that: "an androgen administered prenatally has an organizing action on the tissues mediating mating behavior in the sense of producing a responsiveness to exogenous hormones which differs from that of normal adult females" (Phoenix et al., 1959) and support the alternative interpretation: "Studies in which such effects have been demonstrated may merely be artifacts of pharmacological methods of hormone administration" (Södersten et al., 1983).

If Crews (2010) is right in that the study by Phoenix et al. (1959) "changed the paradigm of behavioral endocrinology," the foundation of the paradigm that has dominated the study of sexual behavior during the last 50 years is brought into question by the results reported by Olster & Blaustein (1990).

The time may be ripe to shift the paradigm for the next 50 years back to "das Fundament der Sexualhormonforschung" because most of what Steinach said was right. This essay now returns to the "personal perspective" approach and ends by revealing the discovery of another remarkable paper, which proves that he was right in all his long forgotten ideas.

OVULATION IN MALES

> Nobody has ever run samples from the pro-estrous LH surge and from the exogenous steroid-evoked surge in the same assay. Most of the work has involved different reference preparations (which differ enormously!). So there is no a priori basis for making any such comparisons from the literature.
>
> —*Clive Coen*, 2011

As the quote from Clive Coen makes clear, an experiment has still not been performed that precisely determines the engagement of ovarian hormones in the preovulatory surge of LH in the female rat. Therefore, and for reasons discussed in the following paragraphs, the conventional view that the LH-surge is permanently suppressed in male rats (e.g., Petersen et al., 2011) may be premature.

There is no sex difference in LH-secretion in gonadectomized rats (Gay & Sheth, 1972; Kalra & Kalra, 1983). Luteinizing hormone is secreted in pulses with a high frequency and low amplitude or with a low frequency and high amplitude with no sex difference (Fig. 15–3A); importantly, the pituitary of the male is capable of secreting large amounts of LH, well within the range of the

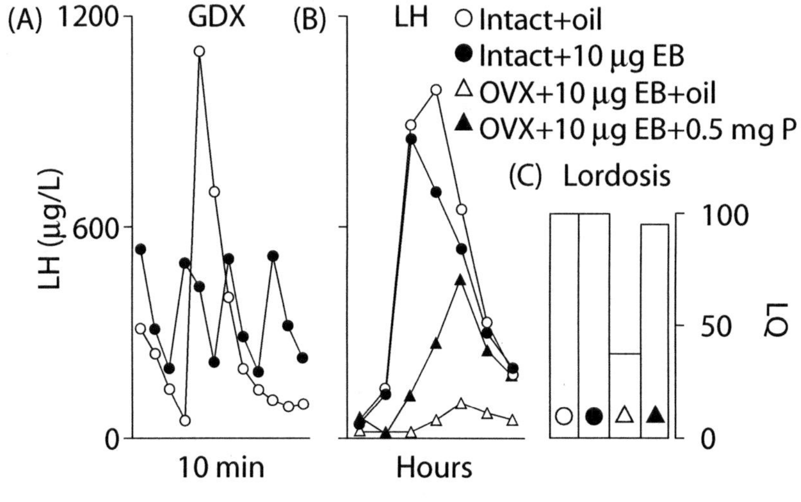

Figure 15–3. Levels of luteinizing hormone (LH) at ten minute intervals in a gonadectomized (GDX) male or female rat (*A*) and at hourly intervals in intact female rats and rats ovariectomized two days before behavioral estrus and injected with estradiol benzoate (EB) and progesterone (P) 42 hours later (*B*). The rats were tested for sexual receptivity after the last blood sample had been taken (*C*). Reproduced from Södersten & Eneroth (1983, 1987), with permission from Springer Verlag and the Journal of Endocrinology Ltd.

LH-surge in intact females (Fig. 15–3A,B) (Södersten & Eneroth, 1983). Thus, the failure to induce preovulatory LH-surges in castrated male rats (Petersen et al., 2011) is not caused by an inability in the capacity of the brain and the pituitary of the males. The reason for absence of LH-surges in castrated males is more likely the same as the reason for the failure to induce high levels of sexual receptivity in castrated males, that is, the use of nonphysiological schedules of hormone administration (Södersten et al., 1983). Coen, C. (personal communication, October 19, 2011). suggests that the crucial experiment has not been performed even in the female rat.

Treatment with various combinations of estradiol and progesterone induced LH-surges in ovariectomized rats, but, as opposed to display of sexual receptivity, the amplitude was never normal (Fig. 15–3B,C) (Södersten & Eneroth, 1987). The LH-level in sequential blood samples obtained from rats, ovariectomized 2 days before behavioral estrus and implanted with constant release capsules filled with crystalline estradiol or oil solutions of estradiol, injected with 1 or 10 μg estradiol at various frequencies, alone or in combination with 0.5 mg progesterone, reached a median of 43% (range: 25%–51%) of the LH-peak in cyclic rats (Södersten & Eneroth, 1987), confirming that LH-surges of normal amplitude are not inducible with estradiol and progesterone in ovariectomized rats (Kalra & Kalra, 1983). At the time, we concluded that "normal pre-ovulatory LH-surges require the action of additional, as yet unspecified, ovarian factors" (Södersten & Eneroth, 1987).

Although this conclusion was reached long ago, there appears to be no new information on this topic; commonly cited references to the role of the ovaries in the LH-surge are even older (listed in Petersen et al., 2011). The hormonal stimulus control of normal amplitude preovulatory LH-surges in female rats needs to be established before it is concluded that male rats are unable to show this neuroendocrine response.

In order to demonstrate that adult castrated male rats can maintain cyclic ovarian function, and therefore show ovulatory secretion of LH, we need to go backwards to the Hunter-Berthold-Steinach gonadal transplantation method. Harris (1955) cites three papers reporting that ovaries which are transplanted into the anterior chamber of the eye or to the ear of adult castrated male rats or mice fail to undergo cyclic variations (Deanesley, 1938; Goodman, 1934; Pfeiffer, 1936). The reason for this choice of site of transplantation was because the main aim of most of these early experiments was to study if the ovarian transplants secreted enough androgens to maintain the accessory sexual glands of the castrated male host. For this purpose, the temperature of the eye or ear is better suited than that of the abdominal cavity; testes that do not descend into their cool scrotal destination are kept at a temperature that is too high for normal functioning (Hill, 1937a,b).

Interestingly, ovaries transplanted into the abdominal cavity of adult castrated male rats were reported to develop corpora lutea (Moore, 1919; Pettinari, 1925; Wang, 1925). However, these studies relied on semiquantitative observations and also because the corpora lutea formed "seemed undersized and not quite normal," Pfeiffer (1936) disregarded the results.

Compelling evidence that prepubertal ovaries transplanted into the seminal vesicles at the time of castration of adult male rats develop normally was presented by Katsh (1950), who noted: "Follicles containing ova are found in all stages of growth. The normal complement of the mature follicle is well defined (see Fig. 15–4A). Fully formed and developing corpora lutea are seen (Fig. 15–4B,C)." Katsh (1950) suggested that "...the abdominal cavity is the normal site for ovarian histogenesis while the eye and ear sites present conditions abnormal to the gland."

Paradoxically, Harris (1955) was aware of Katsh's (1950) paper, but did not take it into consideration, although he had himself treated male rats with 100 μg estradiol benzoate on day 5 of life, castrated them as adults, transplanted ovarian tissue beneath the kidney capsule, and found that ovulation and formation of corpora lutea occurred; he even concluded that these males displayed the "female pattern of gonadotrophic release" (Harris & Levine, 1962).

Rusak & Zucker (1975), in another interesting version of Steinach's hypothesis, suggested that the capacity for surge secretion of LH is present in normal male rats but that it is uncoupled from the generator of circadian rhythms. Their hypothesis was verified for the development and display of sexual behavior

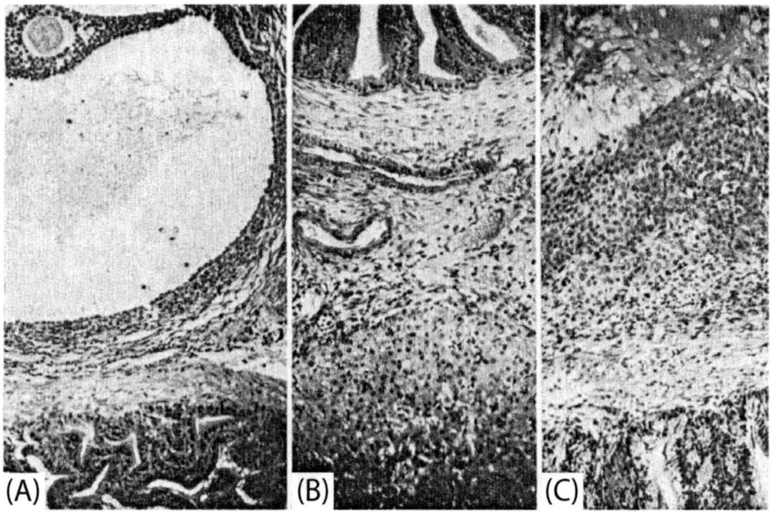

Figure 15–4. A. Seminal vesicle of castrated rat with ovarian graft showing complement of large Graafian follicle and tall vesicular epithelium, 25 days (×100) (A). B. Seminal vesicle with ovarian graft from castrate animal. A portion of a corpus luteum is shown (*lower half*); note the well-maintained secretory epithelium of vesicle (*upper quarter*) (x100) (*B*). C. A portion of a developing corpus luteum of an ovarian-grafted castrate rat. Note vesicular epithelium in upper part of photomicrograph (×100) (*C*). From Katsh, S. (1950). The androgenic activity of ovarian transplants to the seminal vesicles of castrated adult male rats. *Endocrinolgy, 47,* 370–383. Copyright 1950, The Endocrine Society, reproduced with permission.

(Hansen et al., 1978, 1979; Södersten & Eneroth, 1980; Södersten et al., 1980; Södersten et al., 1981a). Therefore, the reason the ovaries that Katsh (1950) transplanted to the seminal vesicle functioned in a normal female manner, may be because the circadian rhythm generator exerts its influence on these accessory sexual organs (Rui et al., 1985; Bebas, Goodall, Majewska et al., 2009). By contrast, an endocrine organ, such as the pineal, exhibits much less rhythmicity if transplanted into the anterior chamber of the eye (i.e., the site where ovarian transplants fail to function normally in a male), than if it remains in situ (Bäckström et al., 1976). Add the possible differences in sympathetic innervation of ovaries transplanted into the various bodily compartments and several new tests of Steinach's hypothesis appear.

REFLECTIONS

> One conclusion made clear is that although we may be able to describe it, we shall not understand the influence of the endocrines upon behavior until considerably more is known about the biochemistry of the central nervous system and the specific metabolic actions of each of the various hormones.
>
> —*John Brobeck*, 1949

This "recherche du temps perdu" brought back many memories of wonderfully exciting moments in the laboratory, but also sleepless nights of embarrassment of not previously having studied the work of Steinach. John Brobeck, in reviewing Beach's *Hormones and Behavior* (1948), accurately predicted the development of the field as shown by the quote. It appears that this development was started already by Steinach ten years before Beach's book was published and would probably have been more rapid had Steinach and his students been able to continue their work. It is clear from the present book, however, that the field has developed in a way that would have pleased Steinach.

ACKNOWLEDGMENTS

I thank Jacques Balthazart and Gregory Ball for encouraging me to write about the aromatization revolution in the endocrinology of sexual behavior, providing me the opportunity to revitalize experiments done at a time when I had the privilege to interact with the brilliant community of the neuroendocrinologists of sexual behavior. Sarah Pearson, Curator, Hunterian Museum, The Royal College of Surgeons of England, kindly allowed me to reproduce Hunter's testis preparations in Figure 15–1. Equally kindly, Elsevier permitted me to reproduce Figure 15–2, and the *Journal of Endocrinology* and the Endocrine Society agreed that I can reproduce Figures 15–3 and 15–4. I thank Stefan Arver, Michael Baum, Cecilia Bergh, Kent Berridge, Carlos Beyer, Jeff Blaustein, Peter Brzezinski, Ruud Buijs, Clive Coen, David Crews, Geert de Vries, Alonso Fernandez-Guasti, George Fink, Ove Hagelin, Stefan Hansen, Agneta Herlitz, Ioannis Ioakeimidis, Gertie Johansson, Cheryl Logan, Lars Olsson, Donald Pfaff, Martin Ritzén, Marianne Schultzberg, Jan Sjövall, Rudolf

Werner Soukup, Dick Swaab, Mona Tang, Sonja Walch, and Modjtaba Zandian for "out of the ordinary" discussions.

REFERENCES

Aafjes, J. H., & Vreeburg, J. T. (1972). Distribution of 5α-dihydrotestosterone in the epididymis of bull and boar, and its concentration in rat epididymis after ligation of efferent testicular ducts, castration and unilateral gonadectomy. *J Endocrinol, 53,* 85–93.

Addison, T. (1855). *On the Constitutional and Local Effects of Diseases of the Supra-Renal Capsules.* London: Warren and Son.

Adler, N., & Bermant, G. (1966). Sexual behavior of male rats: Effects of reduced sensory feedback. *J Comp Physiol Psychol, 61,* 240–243.

Ahlenius, S., Eriksson, H., Larsson, K., Modigh, K., & Södersten P. (1971). Mating behavior in the male rat treated with p-chlorophenylalanine methyl ester alone and in combination with pargyline. *Psychopharmacologia 20,* 383–388.

Anderson, E., & Haymaker, W. (1974). Breakthroughs in hypothalamic and pituitary research. *Prog Brain Res, 41,* 1–60.

Bäckström, M., Olson, L., & Seiger, Å. (1976). N-acetyltransferase and hydroxyindo le-O-methyltransferase activity in intraocular pineal transplants: diurnal rhythm as evidence for functional sympathetic adrenergic innervation. *Acta Physiologica Scandinavica, 96,* 64–71.

Ball, J. (1937). Sex activity of castrated male rats increased by estrin administration. *J Comp Psychol, 24,*136–146.

Balthazart, J. (2010). Behavioral implications of rapid changes in steroid production action in the brain [Commentary on Pradhan D. S., Newman A. E. M., Wacker D. W., Wingfield J. C., Schlinger B. A., and Soma K. K. Aggressive interactions rapidly increase androgen synthesis in the brain during the non-breeding season. *Horm Behav,* 2010]. *Horm Behav, 57,* 375–378.

Baulieu, E. E., Lasnizki, I., & Robel, P. (1968). Metabolism of testosterone and action of metabolites on prostate glands grown in organ culture. *Nature, 219,* 1155–1156.

Baum, M. J. (1972). Precocious mating in male rats following treatment with androgen or estrogen. *J Comp Physiol Psychol, 78,* 356–367.

Baum, M. J., & Vreeburg, J. T. (1973). Copulation in castrated male rats following combined treatment with estradiol and dihydrotestosterone. *Science, 182,* 283–285.

Beach, F. A. (1938). Sex reversals in the mating pattern of the rat. *J Genet Psychol, 53,* 329–334.

Beach, F. A. (1941). Female mating behavior shown by male rats after administration of testosterone propionate. *J Comp Psychol, 29,* 409–412.

Beach, F.A. (1945). Bisexual mating behavior in the male rat: effects of castration and hormone administration. *Physiol Zool, 18,* 390–402.

Beach, F. A. (1948). *Hormones and Behavior.* Harper: New York.

Beach, F. A. (1950). The snark was a boojum. *Am Psychol, 5,* 115–124.

Beach, F. A. (1971). Hormonal factors controlling the differentiation, development, and display of copulatory behavior in the ramstergig and related species. In: Tobach, E., Aronson, E. R., & Shaw, E. (eds.), *The Biopsychology of Development.* New York: Academic Press, pp. 249–296.

Beach, F. A. (1975). Behavioral endocrinology: an emerging discipline. *Am Sci, 63,* 178–187.

Beach, F. A. (1976). Sexual attractivity, proceptivity, and receptivity in female mammals. *Horm Behav, 7,* 105–138.

Beach, F. A. (1978). Confessions of an imposter. In: Meites, J., Donovan, B. T., & McCann, S. M. (eds.), *Pioneers in Neuroendocrinology*, vol. II. New York: Plenum Press, pp.19–35.

Beach, F. A. (1981). Historical origins of modern research on hormones and behavior. *Horm Behav, 15,* 325–376.

Beach, F. A., & Holtz, A. M. (1946). Mating behavior in male rats castrated at various ages and injected with androgen. *J Exp Zool, 101,* 91–142.

Beach, F. A., & Levinson, G. (1950). Effects of androgen on the glans penis and mating behavior of castrated male rats. *J Exp Zool, 114,* 159–171.

Bebas, P., Goodall, C. P., Majewska, M., Neumann, A., Giebultowicz, J. M., & Chappell, P.E. (2009). Circadian clock and output genes are rhythmically expressed in extratesticular ducts and accessory organs of mice. *FASEB J, 23,* 523–533.

Becker, J. B., Arnold, A. P., Berkley, K. J., Blaustein, J. D., Eckel, L. A., Hampson, E., et al. (2005). Strategies and methods for research on sex differences in brain and behavior. *Endocrinology 146,* 1650–1673.

Berthold, A. (1849). Transplantation der Hoden. *Archiv für Anatomie, Physiologie und wissenschaftliche Medicin 16,* 42–46.

Bonthuis, P. J., Cox, K. H., Searcy, B. T., Kumar, P., Tobet, S., & Rissman, E. F. (2010). Of mice and rats: key species variations in the sexual differentiation of brain and behavior. *Front Neuroendocrinol, 31,* 341–358.

Brobeck, J. R. (1949). Hormones and behavior. By Frank A. Beach. Paul B. Hoeber, Inc., New York, 1948. *Yale J Biol Med, 20,* 516.

Brock, O., Baum, M. J., & Bakker, J. (2011). The development of female sexual behavior requires prepubertal estradiol. *J Neurosci, 31,* 5574–5578.

Brown-Séquard, C. E. (1856). Recherches expérimentales sur la physiologie et la pathologie des capsules surrénales. *Comptes rendus de L'Académie des sciences 43,* 422–425.

Brown-Séquard, C. E. (1889). The effects produced on man by subcutaneous injections of liquid obtained from the testicles of animals. *Lancet* July 20,105–107.

Bruchovsky, N., & Wilson, J. D. (1968a). The conversion of testosterone to 5-alpha-androstan-17-beta-ol-3-one by rat prostate in vivo and in vitro. *J Biol Chem, 243,* 2012–2021.

Bruchovsky, N., & Wilson, J. D. (1968b). The intranuclear binding of testosterone and 5-alpha-androstan-17-beta-ol-3-one by rat prostate. *J Biol Chem, 243,* 5953–5960.

Bruchovsky, N., & Wilson, J. D. (1999). Discovery of the role of dihydrotestosterone in androgen action. *Steroids, 64,* 753–579.

Burns, R. K. (1961). Role of hormones in the differentiation of sex. In: .Young, W. C. (ed.), *Sex and Internal Secretions.* Baltimore, MD: Williams & Wilkins Co., pp. 76–158.

Butenandt, A. (1931). Uber die chemische Unterschung der sexual hormones. *Zeitschrift für Angew Chemie 44,* 905–908.

Butenandt, A., & Hanisch, G. (1935). Uber Testosterone. Umwandlung des Dehydro-Androsterones in Androstendiol und Testosteron; ein Weg zur Darstellung des Testosterons aus Cholestrin. *Hoppe-Seyler's Zeitschrift für physiologische Chemie 237,* 89–93.

Butenandt, A., & Jacobi, H. (1933). Über die Darstellung eines krystallisierten pflanzlichen Tokokinins (Thelykinins) und seine Identifizierung mit dem

α -Follikelhormon. Untersuchungen über das weibliche Sexualhormon, 10. Mitteilung. *Hoppe-Seyler's Zeitschrift für physiologische Chemie 218,* 104–112.
Butler, P.C., Mills, R.H., & Bloch, G.J. (2001). Inhibition of lordosis behavior in male and female rats by androgens and progesterone. *Horm Behav, 40,* 384–395.
Carlsson, S. G., & Larsson, K. (1964). Mating in male rats after local anesthetization of the glans penis. *Zeitschrift für Tierpsychologie, 21,* 854–856.
Chadwick, L. R., Pauli, G. F., & Farnsworth, N. R. (2006). The pharmacognosy of Humulus lupulus L. (hops) with an emphasis on estrogenic properties. *Phytomedicine, 13,* 119–131.
Christensen, L. W., & Clemens, L. G. (1974). Intrahypothalamic implants of testosterone or estradiol and resumption of masculine sexual behavior in long-term castrated male rats. *Endocrinology, 95,* 984–990.
Coen, D. R. (2006). Living precisely in fin-de-siècle Vienna. *Journal of the History of Biology 39,* 493–523.
Crews, D.A. (2010). Neural control of sexual behavior. In: Breed, M. D., & Moore, J. (eds.), *Encyclopedia of Animal Behavior,* vol. 2. Oxford: Academic Press, pp. 541–548.
Croix, D., & Franchimont, P. (1975). Changes in the serum levels of the gonadotrophins progesterone and estradiol during the estrous cycle of the guinea pig. *Neuroendocrinology, 19,* 1–11.
Dantchakoff, V. (1938). Rôle des hormones dans la manifestation des instincts sexuels. *Comptes rendus de L'Académie des sciences, 206,* 945–948.
David, K., Dingemanse, E., Freud, J., & Laqueur, E. (1935). Uber krystallinisches mannliches Hormon aus Hoden (Testosteron) wirksamer als aus harn oder aus Cholesterin bereitetes Androsterone. *Hoppe-Seyler's Zeitschrift für physiologische Chemie, 233,* 281–283.
Davidson, J. M. (1969). Effects of estrogen on the sexual behavior of male rats. *Endocrinology, 84,* 1365–1372.
Deanesly, R. (1938). The androgenic activity of ovarian grafts in castrated male rats. *Proc R Soc B 126,* 122–135.
de Vries, G. J., & Södersten, P. (2009). Sex differences in the brain: the relation between structure and function. *Horm Behav, 55,* 589–596.
The Economist. (2011). Belgian beer: Brewed force. *The Economist, December 17,* 128–129.
Feder, H. H. (1967). Specificity of testosterone and estradiol in the differentiating neonatal rat. *Anat Rec, 157,* 79–86.
Feder, H. H., Naftolin, F., & Ryan, K. J. (1974). Male and female sexual responses in male rats given estradiol benzoate and 5 alpha-androstan-17 beta-ol-3-one propionate. *Endocrinology, 94,* 136–141.
Feder, H. H., & Whalen, R. E. (1965). Feminine behavior in neonatally castrated and estrogen-treated male rats. *Science, 147,* 306–307.
Ferguson, J., Henriksen, S., Cohen, H., Mitchell, G., Barchas, J., & Dement, W. (1970). Hypersexuality and behavioral changes in cats caused by administration of p-chlorophenylalanine. *Science, 168,* 499–501.
Flanagan-Cato LM. (2011). Sex differences in the neural circuit that mediates female sexual receptivity. *Front Neuroendocrinol, 32,* 124–136
Forbes, T. R. (1949). A. A. Berthold and the first endocrine experiment: Some speculation as to its origin. *Bull Hist Med, 23,* 263–267.
Freeman, E. R., Bloom, D. A., & McGuire, E. J. (2001). A brief history of testosterone. *J Urol, 165,* 371–373.

Freud, J. (1933). CXCV. Conditions of hypertrophy of the seminal vesicles in rats. *Biochem J, 27,* 1438–1450.

Gay, V. L., & Sheth, N. A. (1972). Evidence for a periodic release of LH in castrated male and female rats. *Endocrinology, 90,* 158–162.

Georgiadis, J. R., Farrell, M. J., Boessen, R., Denton, D. A., Gavrilescu, M., Kortekaas, R., et al. (2010). Dynamic subcortical blood flow during male sexual activity with ecological validity: a perfusion fMRI study. *Neuroimage, 50,* 208–216.

Gessa, G. L., Tagliamonte, A., & Tagliamonte, P. (1971). Aphrodisiac effect of p-chlorophenylalanine. *Science, 171,* 706.

Gonzales, R. J., Ansar, S., Duckles, S. P., & Krause, D. N. (2007). Androgenic/estrogenic balance in the male rat cerebral circulation: metabolic enzymes and sex steroid receptors. *J Cereb Blood FlowMetab, 27,* 1841–1852.

Goodman, L. (1934). Observations on transplanted immature ovaries in the eye of adult male and female rats. *Anat Rec, 59,* 223–251.

Goy, R. W., Phoenix, C. H., & Young, W. C. (1966). Inhibitory actions of the corpus luteum on the hormonal induction of estrous behavior in the guinea pig. *Gen Comp Endocrinol, 6,* 267–275.

Goy, R. W., & Young, W. C. (1956). Strain differences in the behavioral responses of spayed female guinea pigs to alpha-estradiol benzoate and progesterone. *Behaviour, 10,* 340–354.

Green, J. D., Clemente, C.D., & De Groot, J. (1957). Rhinencephalic lesions and behavior in cats: an analysis of the Kluver-Bucy syndrome with particular reference to normal and abnormal sexual behavior. *J Comp Neurol, 108,* 505–545.

Gross, C. G. (1999). *Brain, Vision, Memory: Tales In The History Of Neuroscience.* Cambridge, MA: MIT Press.

Hald, A. (2007). A history of parametric statistical inference from Bernoulli to Fisher, 1713–1935. In: Buchwald, J. Z., Lützen, J., & Hogendijk, J. (eds.), *Sources and Studies in the History of Mathematics and Physical Sciences.* New York: Springer, pp.

Hansen, S., & Södersten, P. (1978). Effects of subcutaneous implants of progesterone on the induction and duration of sexual receptivity in ovariectomized rats. *J Endocrinol, 77,* 373–379.

Hansen, S., & Södersten, P. (1979). Reversal of progesterone inhibition of sexual behaviour in ovariectomized rats by high doses of progesterone. *J Endocrinol, 80,* 381–388.

Hansen, S., Södersten, P., Eneroth, P., Srebro, B., & Hole, K. (1979). A sexually dimorphic rhythm in oestradiol-activated lordosis behaviour in the rat. *J Endocrinol, 83,* 267–274.

Hansen, S., Södersten, P., & Srebro, B. (1978). A daily rhythm in the behavioural sensitivity of the female rat to oestradiol. *J Endocrinol, 77,* 381–388.

Harris, G.W. (1955). *Neural Control Of The Pituitary Gland.* London: Edward Arnold, Ltd.

Harris, G. W., & Levine, S. (1962). Sexual differentiation of the brain and its experimental control. *J Physiol, 163,* 42–43P.

Heckscher, E. F. (1935–49). *Sveriges ekonomiska historia från Gustav Vasa.* Stockholm: Albert Bonniers förlag.

Hill, R. T. (1937a). Ovaries secrete male hormone. I: Restoration of the castrate type of seminal vesicle and prostate glands to normal by grafts of ovaries in mice. *Endocrinology, 21,* 495–502.

Hill, R. T. (1937b). Ovaries secrete male hormone. III: Temperature control of male hormone output by grafted ovaries. *Endocrinology, 21,* 633–636.

Hoberman, J. M., & Yesalis, C.E. (1995). The history of synthetic testosterone. *Sci Am, 272,* 76–81.
Hohlweg, W., & Junkmann, K. (1932). Die hormonal-nervöse Regulierung der Fuktion des Hypophysenvorderlappens. *Klinische Wochenschrift, 11,* 1832–1835.
Horning, E. S. (1958). Carcinogenic action of androgens. *Brit J Cancer 12,* 414–418.
Hunter, J. (1779). Account of the Free Martin. *Philos T R Soc, 69,* 279–293.
Jacobsohn, G. M., Frey, M. J., & Hochberg, R. B. (1965). The absence of steroid estrogens in plants. *Steroids, 6,* 93–99.
Janeczko, A., Budziszewska, B., Skoczowski, A., & Dybałam, M. (2008). Specific binding sites for progesterone and 17beta-estradiol in cells of Triticum aestivum L. *Acta Biochimica Polonica, 55,* 707–711.
Janeczko, A., & Skoczowski, A. (2005). Mammalian sex hormones in plants. *Folia Histochemica et Cytobiologica, 43,* 71–79.
Jørgensen, C. B. (1971). John Hunter, A. A. Berthold, and the origins of endocrinology. *Acta Historica Scientiarum Naturalium et Medicinalium 24,* Editit Bibliotheca Universitatis Hauniensis, Odense University Press, Odense.
Kalra, P. S., & Kalra, S. P. (1983). Neural regulation of luteinizing hormone secretion in the rat. *Endocr Rev, 4,* 311–351.
Karlsson Strese, E.-M. (2008). Humlegårdar på kartor, växter i naturen ur 1600-talets jordbrukslandskap. In: Höglund, M. (ed.), *Skrifter utgivna av Riksarkivet,* nr 29. Stockholm: Riksarkivet, pp. 46–52.
Katsh, S. (1950). The androgenic activity of ovarian transplants to the seminal vesicles of the castrated adult male rat. *Endocrinology, 47,* 370–383.
Kimchi, T., Xu, J., & Dulac, C. (2007). A functional circuit underlying male sexual behaviour in the female mouse brain. *Nature, 448,* 1009–1014.
Koch, W., & Heim, G. (1953). Östrogene harmone in Hopfen und Bier. *Münchner Medizinische Wochenschrift 95,* 845.
Krause, D. N., Duckles, S. P., & Gonzales, R. J. (2011). Local oestrogenic/androgenic balance in the cerebral vasculature. *Acta Physiologica, 203,* 181–186.
Kreier, F., & Swaab, D. F. (2010). Chapter 23: history of neuroendocrinology "the spring of primitive existence." *Handb Clin Neurol, 95,* 335–360.
Kuhn, T. S. (1962). *The structure of scientific revolutions.* Chicago: University of Chicago Press.
Kun, H. (1934). Psychische Feminierung unde Hermaphrodisierung von Männchen durch weibliches Sexualhormon. *Endokrinologie, 13,* 323–337.
The Lancet. (1917). The internal secretions of the reproductive glands. *The Lancet, November 3,* 687.
Larsson, K., & Södersten, P. (1973). Mating in male rats after section of the dorsal penile nerve. *Physiol Behav, 10,* 567–571
Larsson, K., Södersten, P., & Beyer, C. (1973). Sexual behavior in male rats treated with estrogen in combination with dihydrotestosterone. *Horm Behav, 4,* 289–299.
Levine, S., & Mullins, R. (1964). Estrogen administration neonatally affects adult sexual behavior in male and female rats. *Science, 144,* 185–187.
Li, J., Biswas, M.G., Chao, A., Russell, D.W., & Cory, J. (1997). Conservation of function between mammalian and plant steroid 5α-reductases. *Proc Natl Acad Sci U S A, 94,* 3553–3559.
Logan, C. A. (2010). Reproduction versus endocrine bisexuality: Eugen Steinach and the mammalian puberty gland. In: Soukup, R. W., & Noe, C. (eds.), *Pioneere der Sexualforschung.* Gumpoldskirchen: Ignaz-Lieben-Gesellschaft Austria, pp. 35–54.

Malmnäs, C. O., & Meyerson, B. J. (1971). P-chlorophenylalanine and copulatory behaviour in the male rat. *Nature, 232,* 398–400.

Martel, K. L., & Baum, M. J. (2009). Adult testosterone treatment but not surgical disruption of vomeronasal function augments male-typical sexual behavior in female mice. *J Neurosci, 29,* 7658–7666.

McDonald, P., Beyer, C., Newton, F., Brien, B., Baker, R., Tan, H. S., et al. (1970). Failure of 5alpha-dihydrotestosterone to initiate sexual behaviour in the castrated male rat. *Nature, 227,* 964–965.

McKibbin, P. E., & Belchetz, P. E. (1986). Prolonged pulsatile release of gonadotropin-releasing hormone from the guinea pig hypothalamus in vitro. *Life Sci, 38,* 2145–2150.

Michael, R. P. (1961). Hypersexuality in male cats without brain damage. *Science, 134,* 553–554.

Milligan, S. R., Kalita, J. C., Heyerick, A., Rong, H., De Cooman, L., & De Keukeleire, D. (1999). Identification of a potent phytoestrogen in hops (Humulus lupulus L.) and beer. *J Clin Endocrinol Metab, 84,* 2249–2252.

Mo, Z. N., Huang, X., Zhang, S. C., & Yang, J. R. (1995). Early and late long-term effects of vasectomy on serum testosterone, dihydrotestosterone, luteinizing hormone and follicle-stimulating hormone levels. *J Urol, 154,* 2065–2069.

Mode, A., Gustafsson, J. Å, Södersten, P., & Eneroth, P. (1984). Sex differences in behavioural androgen sensitivity: possible role of androgen metabolism. *J Endocrinol, 100,* 245–248.

Moore, C. P. (1919). On the physiological properties of the gonads as controllers of somatic and psychical characteristics. I. The rat. *J Exp Zool, 28,* 137–160.

Moore, C. R., & Price, D. (1932). Gonad hormone functions, and the reciprocal influence between gonads and hypophysis with its bearing on the problem of sex hormone antagonism. *Am J Anat, 50,* 13–67.

Moreines, J. K. (1980). Ovarian estradiol and progesterone regulation of sexual receptivity of female rats. Unpublished doctoral thesis, University of Michigan.

Moreines, J., McEwen, B., & Pfaff, D. (1986). Sex differences in response to discrete estradiol injections. *Horm Behav, 20,* 445–451.

Morell, M. (1989). Studies in the history of Swedish food consumption: food consumption among institutionally supported paupers 1621–1872. University of Uppsala. Uppsala studies in economic history, 0346–6493; 29.

Olster, D. H., & Blaustein, J. D. (1988). Progesterone facilitation of lordosis in male and female Sprague-Dawley rats following priming with estradiol pulses. *Horm Behav, 22,* 294–304.

Olster, D. H., & Blaustein, J.D. (1990). Biochemical and immunocytochemical assessment of neural progestin receptors following estradiol treatments that eliminate the sex difference in progesterone facilitated lordosis in guinea pigs. *J Neuroendocrinol, 2,* 79–86.

Peng, X. S., Li, F. D., Miao, Z. R., et al. (1987). Plasma reproductive hormones in normal and vasectomized Chinese males. *Int J Androl, 10,* 471–479.

Petersen, S. L., Krishnan, S., Aggison, L. K., Intlekofer, K. A., & Moura, P. J. (2011). Sexual diffentiation of the gonadotropin surge release mechanism: A new role for the canonical NfκB signaling pathway. *Front Neuroendocrinol,* doi:10.1016/j.yfrne.2011.06.002.

Pettinari, V. (1925). Feminisation et hyperfeminisation par greffe ovarienne. *Compte Rendu de la Societe Biologique 92,* 1228–1229.

Pfaff, D. (1970). Nature of sex hormone effects on rat sex behavior: specificity of effects and individual patterns of response. *J Comp Physiol Psychol, 73,* 349–358.
Pfeiffer, C. A. (1936). Sexual differences of the hypophyses and their determination by the gonads. *Am J Anat, 58,* 195–125.
Phoenix, C. H., Goy, R. W., Gerall, A. A., & Young, W. C. (1959). Organizing action of prenatally administered testosterone propionate on teh tissues mediating mating behavior in the female guinea pig. *Endocrinology, 65,* 369–382.
Price, D. (1974). Carl Richard Moore. December 5, 1892–October 16, 1955. *Biogr Mem Natl Acad Sci, 45,* 385–412.
Rengachary, S. S., Colen, C., & Guthikonda, M. (2008). Charles-Edouard Brown-Séquard: an eccentric genius. *Neurosurgery, 62,* 954–964.
Royal College of Surgeons of England. (2012). http://bit.ly/nJaVGb
Rui, H., Gordeladze, J. O., Mevåg, B., Haug, E., Brekke, I., & Purvis, K. (1985). Circadian rhythms in accessory sex gland function in the male rat. *Uro Int, 40,* 331–336.
Rusak, B., & Zucker, I. (1975). Biological rhythms and animal behavior. *Ann Rev Psychol, 26,* 137–171.
Ruzicka, L., & Wettstein, A. (1935). Sexualhormone VII. Uber die kunstliche Herstellung des Testikelhormons. Testosteron (Androsten-3-on-17-ol.). *Helvetica Chimica Acta, 18,* 1264–1275.
Salis, P. J., & Dewsbury, D. A. (1971). p-Chlorophenylalanine facilitates copulatory behaviour in male rats. *Nature, 232,* 400–401.
Sawin, C. T. (2002). Dorothy Price, Carl Moore, and the regulation of gonadal hormone secretion. *Endocrinologist, 12,* 1–4.
Schreiner, L., & Kling, A. (1953). Behavioral changes following rhinenecephalic injury in cat. *J Neurophysiol, 16,* 643–659.
Sengoopta, C. (2003). 'Dr Steinach coming to make old young!': sex glands, vasectomy and the quest for rejuvenation in the roaring twenties. *Endeavour, 27,* 122–126.
Sengoopta, C. (2006). *The Most Secrete Quintessence of Life. Sex, glands, and hormones, 1850–1950.* Chicago and London: University of Chicago Press.
Setchell, B. P. (1990). The testis and tissue transplantation: historical aspects. *J Reprod Immunol, 18,* 1–8.
Shah, N. M., & Breedlove, S. M. (2007). Behavioural neurobiology: females can also be from Mars. *Nature, 448,* 999–1000.
Skinner, B. F. (1974). *About behaviorism.* New York: Alfred A. Kopf.
Södersten, P. (1973). Estrogen-activated sexual behavior in male rats. *Horm Behav, 4,* 247–256.
Södersten, P. (1975a). Mounting behavior and lordosis behavior in castrated male rats treated with testosterone propionate, or with estradiol benzoate or dihydrotestosterone in combination with testosterone propionate. *Horm Behav, 6,* 109–126.
Södersten, P. (1975b). Receptive behavior in developing female rats. *Horm Behav, 6,* 307–317.
Södersten, P. (1976). Lordosis behaviour in male, female and androgenized female rats. *J Endocrinol, 70,* 409–420.
Södersten, P. (1978a). Lordosis behaviour in immature male rats. *J Endocrinol, 76,* 233–240.
Södersten, P. (1978b). Effects of anti-oestrogen treatment of neonatal male rats on lordosis behaviour and mounting behaviour in the adult. *J Endocrinol, 76,* 241–249.

Södersten, P. (1984). Sexual diff erentiation: do males diff er from females in behavioral sensitivity to gonadal hormones? *Prog Brain Res, 61,* 257–270.
Södersten, P. (1985). Estradiol-progesterone inteactions in the reproductive behavior of female rats. In: Pfaff, D. W. & Danten, D. (eds.), *Current Topics in Neuroendocrinology,* vol. 5. Berlin-Heidelberg: Springer Verlag, pp. 141–174.
Södersten, P., Damassa, D.A., & Smith, E.R. (1977). Sexual behavior in developing male rats. *Horm Behav, 8,* 320–341.
Södersten, P., de Jong, F.H., Vreeburg, J.T., & Baum, M.J. (1974). Lordosis behavior in intact male rats: absence of correlation with mounting behavior or testicular secretion of estradiol-17 beta and testosterone. *Physiol Behav, 13,* 803–808.
Södersten, P., & Eneroth, P. (1980). Neonatal treatment with antioestrogen increases the diurnal rhythmicity in the sexual behaviour of adult male rats. *J Endocrinol, 85,* 331–339.
Södersten, P., & Eneroth, P. (1981a). Serum levels of oestradiol-17β and progesterone in relation to sexual receptivity in intact and ovariectomized rats. *J Endocrinol, 89,* 45–54.
Södersten, P., & Eneroth, P. (1981b). Evidence that progesterone does not inhibit the induction of sexual receptivity by oestradiol-17β in the rat. *J Endocrinol, 89,* 63–69.
Södersten, P., & Eneroth, P. (1983). Reproductive neuroendocrine rhythms. In: Balthazart, J., Pröve, E., & Gilles, R. (eds.), *Hormones and Behaviour in Higher Vertebrates.* Heidelberg: Springer-Verlag, pp. 178–193.
Södersten, P., & Eneroth, P. (1987). Dissociation between the ovarian factors controlling sexual receptivity and preovulatory secretion of LH in cyclic female rats. *J Endocrinol, 112,* 133–138.
Södersten, P., Eneroth, P., & Ekberg, P.H. (1980). Episodic fluctuations in concentrations of androgen in serum of male rats: possible relationship to sexual behaviour. *J Endocrinol, 87,* 463–471.
Södersten, P., Eneroth, P., Hansen, S. (1981). Induction of sexual receptivity in ovariectomized rats by pulse administration of oestradiol-17β. *J Endocrinol, 89,* 55–62.
Södersten, P., Eneroth, P., & Hansson, T. (1988). Oestradiol synergizes with 5 alpha-dihydrotestosterone or 3 alpha- but not 3 beta-androstanediol in inducing sexual behaviour in castrated rats. *J Endocrinol, 119,* 461–465.
Södersten, P., Eneroth, P., Hansson, T., Mode, A., Johansson, D., Näslund, B., et al. (1986). Activation of sexual behaviour in castrated rats: the role of oestradiol. *J Endocrinol, 111,* 455–462.
Södersten, P., Eneroth, P., & Pettersson, A. (1983). Episodic secretion of luteinizing hormone and androgen in male rats. *J Endocrinol, 97,* 145–153.
Södersten, P., & Gustafsson, J.Å. (1980). Activation of sexual behaviour in castrated rats with the synthetic androgen 17 beta-hydroxy-17 alpha-methyl-estra-4,9,11-triene-3-one (R 1881). *J Endocrinol, 87,* 279–283.
Södersten, P., & Hansen, S. (1977). Effects of oestradiol and progesterone on the induction and duration of sexual receptivity in cyclic female rats. *J Endocrinol, 74,* 477–485.
Södersten, P., & Hansen, S. (1978). Effects of castration and testosterone, dihydrotestosterone or oestradiol replacement treatment in neonatal rats on mounting behaviour in the adult. *J Endocrinol, 76,* 251–260.
Södersten, P., & Hansen, S. (1979). Induction of sexual receptivity by oestradiol benzoate in cyclic female rats: influence of ovarian secretions before injection of oestradiol benzoate. *J Endocrinol, 80,* 389–395.

Södersten, P., Hansen, S., & Srebro, B. (1981). Suprachiasmatic lesions disrupt the daily rhythmicity in the sexual behaviour of normal male rats and of male rats treated neonatally with antioestrogen. *J Endocrinol, 88,* 125–130.

Södersten, P., & Larsson, K. (1974). Lordosis behavior in castrated male rats treated with estradiol benzoate or testosterone propionate in combination with an estrogen antagonist, MER-25, and in intact male rats. *Horm Behav, 5,* 13–18.

Södersten, P., & Larsson, K. (1975). Lordosis behavior and mounting behavior in male rats: effects of castration and treatment with estradiol benzoate or testosterone propionate. *Physiol Behav, 14,* 159–164.

Södersten, P., Larsson, K., Ahlenius, S., & Engel, J. (1976a). Sexual behavior in castrated male rats treated with monoamine synthesis inhibitors and testosterone. *Pharmacol Biochem Behav, 5,* 319–327.

Södersten, P., Larsson, K., Ahlenius, S., & Engel, J. (1976b). Stimulation of mounting behavior but not lordosis behavior in ovariectomized female rats by p-chlorophenylalanine. *Pharmacol Biochem Behav, 5,* 329–333.

Södersten, P., Pettersson, A., & Eneroth, P. (1983). Pulse administration of estradiol-17 beta cancels sex difference in behavioral estrogen sensitivity. *Endocrinology, 112,* 1883–1885.

Soukup, R. W. (2010), Eugen Steinach: Spiritus rector der ersten zyklusregulierten Hormonpräparates. In: Soukup, R. W. & Noe, C. (eds.), *Pioneere der Sexualforschung,* Gumpoldskirchen: Ignaz-Lieben-Gesellschaft Austria, pp. 16–34..

Spearow, J. L., Doemeny, P., Sera, R., Leffler, R., & Barkley, M. (1999). Genetic variation in susceptibility to endocrine disruption by estrogen in mice. *Science, 285,* 1259–1261.

Steinach, E. (1894). Untersuchungen zur vergleichenden Physiologie der männlichen Geschlechtsorgane inbesondere der accessorischen Geschlechtsdrüsen. *Pflügers Archiv für die Gesamte Physiologie des Menschen und der Tiere, 56,* 304–338.

Steinach, E. (1913). Feminierung von Männlichen und Maskulierung von Weibchen. *Zentralblatt für Physiologie, 27,* 717–723.

Steinach, E. (1920). Verjüngung durch experimentelle Neubelung der alternden Pubertätsdrüse. *Archiv für Entwicklungsmechanik der Organismen, 46,* 553–619.

Steinach, E. (1940). *Sex and Life: Forty Years of Biological and Medical Experiments.* London: Faber and Faber.

Steinach, E., & Kun, H. (1928). Die entwicklungsmechanische Bedeutung der Hypophysis als Aktivator der Keimdrüseninkretion. *Medizinische Klinik, 24,* 524–529.

Steinach, E., Kun, H., & Peczenik, O. (1936). Beiträge zur Analyse der Sexualhormonwirkung. Tierexperimentelle und klinische Untersuchungen. *Wiener klinische Wochenschrift, 49,* 899–903.

Steinach, E., Peczenik, O., & Kun, H. (1938). Über hormonale Hyperämisierung inbesondere über den Einfluss der männlichen Sexualhormone und ihrer Kombination mit weiblichem Hormon auf erhöhten Blutdruch und Hypertonus. *Wiener klinische Wochenschrift, 51,* 65, 102, 134.

Talgliamonte, A., Tagliamonte, P., Gessa, G.L., & Brodie, B.B. (1969). Compulsive sexual activity induced by p-chlorophenylalanine in normal and pinealectomized male rats. *Science, 166,* 1433–1435.

Tattersall, R., & Turner, B. (2000). Brown-Séquard and his syndrome. *Lancet, 356,* 61–63.

Tollin, C. (2011). Den hemliga humlen. www.bok-bibliotek.se/sv/program/program-2011/medverkande/?person=90314.

Verzele, M. (1986). 100 years of hop chemistry and its relevance to brewing. *J Inst Brew, 92*, 32–48.

Vreeburg, J.T., van der Vaart, P.D., & der Schoot, P. (1977). Prevention of central defeminization but not masculinization in male rats by inhibition neonatally of oestrogen biosynthesis. *Journal of Endocrinology, 74*, 375-382.

Wagner, R. (1851). Mittheilung einer einfachen Methoden zu Versuchen über die Veränderungen thierischer Gewebe in morphologischer und chemicher Beziehung. *Archive für Physiologischen Heilkunde, 10*, 521–528.

Walker, T. C. (1942). Use of testosterone propionate and estrogenic substance in treatment of essential hypertension, angina pectoris and peripheral vascular disease. *J Clin Endocrinol, 2*, 560–568.

Wang, Z. Y., & He, J. X. (2004). Brassinosteroid signal transduction-choices of signals and receptors. *Trends Plant Sci, 9*, 91–96.

Wang, G. H., Richter, C. P., & Guttmacher, A. F. (1925). Activity studies on male castrated rats with ovarian transplants, and correlation of the activity with the histology of the grafts. *Am J Physiol, 73*, 581–599.

Whalen, R. E. (1964). Hormone-induced changes in the organization of sexual behavior in the male rat. *J Comp Physiol Psychol, 57*, 175–182.

Whalen, R. E., & Luttge, W. G. (1970). p-chlorophenylalanine methyl ester: an aphrodisiac? *Science, 169*, 1000–1001.

Whalen, R. E., & Nadler, R. D. (1963). Suppression of the development of female mating behavior by estrogen administered in infancy. *Science, 141*, 273–274.

Wingfield, J. C. (2005). Historical contributions of research on birds to behavioral neuroendocrinology. *Horm Behav, 48*, 395–402.

Wu, M. V., & Shah, N. M. (2011). Control of masculinization of the brain and behavior. *Curr Opin Neurobiol, 21*, 116–123.

Ydstebø, R., & Södersten, P. (1977). Induction of sexual receptivity in ovariectomized female rats by subcutaneous implants of estradiol-17β. *Horm Behav, 9*, 130–140.

Young, W. C. (1961). The hormones and mating behavior. In: Young, W. Y. (ed.), *Sex and Internal Secretions.* Baltimore, MD: Williams & Wilkins Co, pp. 1173–1239.

Young, W. C., Dempsey, E., Hagquist, C. W., & Boling, J. L. (1937). The determination of heat in the guinea pig. *J Lab Clin Med, 23*, 300–302.

Zondek, B. (1934a). Mass excretion of oestrogenic hormone in the urine of the stallion. *Nature, 133*, 209.

Zondek, B. (1934b). Oestrogenic hormone in the urine of the stallion. *Nature, 133*, 494.

Zondek, B., & von Euler, H. (1934). Follikulinausschiedung im Harn des Kindes, der Frau und des Mannes. *Skandinavisches Archiv für Physiologie, 67*, 259–264.

Zitrin, A., Beach, F. A., Barchas, J. D., & Dement, W. C. (1970). Sexual behavior of male cats after administration of parachlorophenylalanine. *Science, 170*, 868–870.

Chapter 16

Aromatase and Sexual Differentiation of the Rodent Brain: The Old, the New, and the Unexpected

Margaret M. McCarthy, Christopher L. Wright, and Anne T. M. Konkle

INTRODUCTION

Aromatase is the rate-limiting enzyme responsible for the biosynthesis of estrogens, mediating the conversion of testosterone to estradiol and androstenedione to estrone. This enzyme complex is part of the family of cytochrome P450 enzymes important in many metabolic oxydation-reduction reactions. Aromatase activity and/or expression is found in many peripheral body tissues, in those necessary for reproduction such as Sertoli and Leydig cells in the testis (Valladares & Payne, 1979) and granulosa cells in the ovary (Dorrington et al., 1987) as well as nonreproductively related tissues such as skin fibroblasts (Harada & Yamada, 1992) and adipose tissue (Simpson et al., 1989). Circulating androgens of gonadal or adrenal origin are frequently aromatized in these various tissues. Appreciation of aromatase activity as a major contributor to phenotypic variability in brain and behavior began with the discovery of its central role in hormonally mediated sexual differentiation of the rodent brain. In the more than 50 years since the articulation of the now iconic Organizational/Activational Hypothesis codifying that developmental steroid effects predispose adult brain steroid responsiveness (Phoenix et al., 1959), the addendum of the Aromatization Hypothesis in the early 1970s remains its most significant expansion (Naftolin et al., 1975); so much so that even today researchers interested in identifying the origins of a sex difference in the rat or mouse brain must first face the quixotic proposition that estradiol, the archetypal female hormone, mediates the establishment of male characteristics. We now know that estradiol produced within the brain mediates at least three distinct processes: masculinization, defeminization, and feminization. However, these processes occur at differing but sometimes overlapping time points and brain regions. What appears to be unique in each process is the myriad of cellular pathways regulated by estradiol that then mediate the attendant differentiation. The goal of this chapter is to review briefly the established and well-known historical antecedents to more recent discoveries regarding neural aromatase, including unexpected and as of yet unexplained findings that prompt a revised view of brain steroidogenesis and its importance to brain development.

HISTORICAL ANTECEDENTS

Estradiol Masculinizes the Developing Rodent Brain

Some 20 years after the Organizational/Activational Hypothesis, McEwen et al. treated newborn female rats with either testosterone or estradiol to not only confirm that testosterone is a masculinizing hormone but also expecting that estradiol would serve as a control steroid and thereby have no effect on brain masculinization (Naftolin, 1975; Naftolin et al., 1975; McEwen et al., 1977a,b). Unexpectedly, the females treated perinatally with estradiol were just as behaviorally masculinized as intact males or testosterone-treated females (Booth, 1977; Baum et al., 1982). This surprising finding combined with the discovery of estrogen receptors and aromatase activity in the brain, led to the development of the Aromatization Hypothesis, which postulates that in the neonatal brain testosterone does not act directly via its binding to the androgen receptor to organize male and suppress female sex behavior, but rather testosterone functions as a prohormone that is aromatized to estradiol in the brain (Naftolin et al., 1975a,b). Estradiol then binds to its cognate receptor to masculinize and defeminize brain and behavior (Pfaff & Keiner, 1973; MacLusky et al., 1979; McCarthy et al., 1993; Wersinger et al., 1997). As proof of principle, neonatal males treated with an aromatase inhibitor, express female but not male sexual behavior despite availability of the appropriate hormones in adulthood (Gladue & Clemens, 1980a,b; Roselli & Resko, 1984, 1993).

Alpha-Fetoprotein Protects the Developing Female Brain from Masculinization

The discovery that estradiol, not testosterone, was the masculinizing hormone in rodent brain presented a new puzzle to solve. Pregnancy in all mammals, including rodents, is characterized by exceedingly high estrogen production synthesized by the placenta and released into the maternal circulation and amniotic fluid. As a result, the developing fetus is exposed to very high levels of estrogens of nongonadal origin, raising the question of how do female fetuses avoid being masculinized? The answer lies in the steroid binding globulin, alpha-fetoprotein (AFP), a plasma protein elevated during embryonic development and that has a high affinity for estradiol (Raynaud, 1973; Andrews et al., 1982). By binding circulating estrogens of a maternal, placental, or male sibling origin, prenatal AFP prevents these estrogens from reaching the brain and influencing its development. In males, estrogens are synthesized by aromatization of circulating gonadal androgens within the brain, whereas no such substrate is available in females at this early age. Therefore, AFP protects from many estrogenic effects on the brain; that is, preventing masculinization and allowing it to be feminized (Mizejewski et al., 1980; Bakker et al., 2006).

Aromatization Contributes to Adult Male Sexual Behavior

The discovery that estradiol is a developmental masculinizing hormone led to an interest in whether local aromatization in the adult brain is an important contributor to activational hormone effects. Adult castrate male rats provided estradiol hormone replacement instead of testosterone can show an almost complete behavioral copulatory repertoire; mounting and thrusting as frequently as males treated with testosterone in adulthood (Putnam et al., 2003, 2005). On the other hand, males do not exhibit male sexual behavior if given only a nonaromatizable metabolite of testosterone called dihydrotestosterone (DHT) (McDonald et al., 1970; Baum & Vreeburg, 1973). The ability of estradiol to induce malelike behavior in normal males is presumed to be the result of associated increases in extracellular glutamate and dopamine in the medial preoptic area that are of the same magnitude as those seen in gonadally intact behaving male rats (Putnam et al., 2003, 2005). The signaling of these neurotransmitters in the medial preoptic area (POA) is essential for the expression of male behavior. Estradiol-replaced males do show impairments related to erections and ejaculations, which involve sympathetic and parasympathetic signaling and are likely caused by the effects of testosterone but not estradiol on the spinal cord (Baum & Vreeburg, 1973; Booth, 1977). It should also be noted that primate male sex behavior seems to be organized by androgen and not estrogen receptor activation developmentally (Resko & Roselli, 1997), but anecdotal evidence from men with dysfunctional aromatase and receiving hormone replacement therapy suggests a role for estrogens in activation of male primate sexual behavior.

Estrogen Receptors Are Discretely Distributed in the Brain

The local production of estrogens is pointless in the absence of a receptor to both transduce the biological signal and localize it to a particular brain region or cell type. Thus the precise mapping of estrogen receptor (ER) expression has been a topic of intense attention. Estrogen receptors are members of an extended family of nuclear transcription factors and exist in two isoforms referred to as ER-α (*Esr1*) and *Esr2* or ER-β (Toft & Gorski, 1966; Tremblay et al., 1997). Both ER-α and ER-β are found throughout the central nervous system but in varying degrees of concentration, and this concentration may change across development. The expression of both receptors is greatest in the preoptic area, hypothalamus, bed nucleus of the stria terminalis, periaqueductal grey, and medial and cortical amygdaloid nuclei; and moderate in the lateral habenula, parabrachial nucleus, locus coeruleus, nucleus of the solitary tract, spinal trigeminal nucleus, and superficial laminae of the spinal cord. Estrogen receptor-β is found in the olfactory bulb, supraoptic, paraventricular, suprachiasmatic, and tuberal hypothalamic nuclei, zona incerta, ventral tegmental area, cerebellum (Purkinje cells), laminae III to V, VIII, and IX of the spinal cord, and pineal gland. Estrogen receptor-β is also found in the cortex and hippocampus, but recently weak staining of ER-α has also been reported there. Estrogen receptor-α is also found in areas making extensive projection to other brain regions, such as lateral hypothalamus, intralaminar thalamic nuclei, and deep layers of the cerebral cortex (Simerly et al., 1990). Both receptors are highly expressed by neurons, but have also been found in nonneuronal cell types, including microglia and astrocytes (Jung-Testas et al., 1992; Garcia-Ovejero et al., 2002; Sierra et al., 2008). Dynamic changes across development are particularly notable in the prefrontal cortex and hippocampus, both of which show very high expression early in the postnatal period, which then declines with increasing maturity (O'Keefe & Handa, 1990; O'Keefe et al., 1995), suggesting that estradiol plays a particularly important role in the development of these structures, but precisely what that role is remains largely unknown.

REAFFIRMATION OF THE IMPORTANCE OF AROMATIZATION TO SEXUAL DIFFERENTIATION OF THE BRAIN IN THE 21ST CENTURY

Null Mutation Mice Confirm the Old, but Also Establish the New

Technological advances in the tardive 20th and early 21st centuries allowed a more thorough examination of the role of estradiol in differentiation of brain and behavior. Mapping of the mouse genome provided the impetus for its controlled manipulation, one example being the targeted deletions of estrogen receptor α (ERKO) (Lubahn et al., 1993), β (βERKO) (Krege et al., 1998), or mutations of both ER gene isoforms at once (αβERKO) (Couse et al., 1999; Dupont et al., 2000). Although still capable of estrogen synthesis, the resulting knockout (KO) mice were deficient in these receptors, thus, without estrogens' typical targets for action. Characterization of ERKOs generally confirmed the critical role of estradiol in brain masculinization (Ogawa et al., 2000). Studies with βERKO mice show that both males and females are fertile, but litters born to these females are typically smaller than those born to their wild-type counterparts (Krege et al., 1998) and behavioral defeminization of males is impaired (Kudwa et al., 2005).

Subsequent to characterization of mice lacking estrogen receptors was development of three separate models whereby targeted disruptions were made to the Cyp19 gene, referred to as ArKOs (Fisher et al., 1998; Honda et al., 1998; Toda et al., 2001). Although completely devoid of estradiol, ArKOs have elevated circulating levels of testosterone and luteinizing hormone but normal follicle-stimulating hormone (Fisher et al., 1998). In one model the young males are fertile (Fisher et al., 1998) but show age-dependent disrupted spermatogenesis and reduction in testis weight (Robertson et al., 1999). Male fertility is reduced in another ArKO model (Toda et al., 2001) and totally compromised in a third (Honda et al., 1998). Reproductive performance was severely affected in all three models, with males exhibiting early exploratory interest of the stimulus female, with a preference for estrus females versus intact males (Bakker et al., 2002) but reduced or no mounting attempts (Honda et al., 1998; Robertson et al., 1999; Toda et al., 2001). Administration of estrogens on the first day of life successfully restored appropriate mounting behavior in one study (Toda et al., 2001), whereas it was also restored by administration in adulthood in another (Bakker et al., 2004). These males showed altered aggressive behavior, with infanticide toward pups and increased aggression toward reproductively receptive females (Matsumoto et al., 2003a,b), but reduced aggressive behavior toward a male intruder, the latter reversed by estradiol administration after birth. In a transgenic ArKO model, with a human transgene regulated by the brain-specific promoter of the mouse aromatase, males no longer displayed impaired reproductive behavior and aggression, further supporting the importance of estradiol in key brain areas for the organization of behavioral masculinization (Harada et al., 2009).

Neural Aromatization Contributes to a Late-Phase Sensitive Period for Feminization of Brain and Behavior

Although development of the female brain phenotype has long been believed to occur in the absence of gonadal hormones, female ArKO mice point to the potential necessity of small amounts of estrogens for normal development. Female ArKO mice show reduced lordosis when treated in adulthood with appropriate ovarian hormones when compared with females with normal aromatase, and presumably estradiol levels, during development (Bakker et al., 2003), suggesting an organizational role for estrogens in females as well. A sex difference in the critical periods for these organizational effects has been proposed (Whalen & Edwards, 1967; Quadros et al., 2002), as circulating estrogens are typically bound to AFP only until the postnatal period in females. The ovaries are capable of estradiol synthesis around the first week of life (Weniger, 1993; Weniger et al., 1993) and the observation of a late increase in circulating estradiol, around PN6–10 (Konkle & McCarthy, 2011), is consistent with the notion of a later occurring sensitive period for estradiol-mediated feminization of the brain.

Alpha-Fetoprotein Protein Does More Than Prevent Behavioral Masculinization

Given that aromatization is important for the developing female brain, a theory present in the 1980s suggested an active role for AFP in carrying estrogens into select brain areas, thus playing a role in the development of the female rodent brain (Toran-Allerand, 1980, 1987). Dysfunction of the GnRH and luteinizing hormone (LH) systems in the AFP KO mouse demonstrate alpha-fetoprotein's importance in the protection of the female brain against defeminization of gonadotropin secretion and foreclosed on its active role in feminization. These female mice are sterile (De Mees et al., 2006), with infertility linked to impaired steroid-induced preovulatory GnRH and LH surges in adulthood (Gonzalez-Martinez et al., 2008). Fewer Kisspeptin-10 neurons were detected in the hypothalamus of these AFP KO females but of equal importance is the inability of these neurons to be activated by estrogens in adulthood, a necessity for the cascade of GnRH and LH surges in the cycling adult female rodent (Gonzalez-Martinez et al., 2008). Nonetheless, the possibility that AFP contributes to feminization remains open, particularly as we learn more about the role of estradiol in brain regions outside the reproductive axis and the contributions of estradiol to the sexual differentiation of the brain between PN15–25, outside the perinatal surges in testosterone.

Parental behavior, another highly sexually dimorphic behavior in many rodents, also appears to be influenced by early life hormone exposure, but the emphasis was on androgens (Gandelman, 1973; Gandelman & Vom Saal, 1975). That the critical hormone is estradiol was established using AFP KO mice as these transgenic females showed dramatically reduced maternal behavior, with pup directed caring activities at low and infanticide at high male-like levels (Keller et al., 2010). Note that these were observed in the absence of any differences in anxiety, recognition, or olfaction. These data support an important role for AFP in protecting the female brain against masculinization and defeminization during fetal development.

Thus the advent of knockout mice models have helped tremendously in understanding the role of estrogens in brain development, but also in teasing apart its role and time frame by which it acts to masculinize versus defeminize brain and behavior.

Membrane Estrogen Receptors Mediate Rapid Estradiol Effects

Historically the distribution of ER was considered important for demonstrating which brain regions responded to estradiol, the hormone, but investigators in the field never considered estradiol as having the characteristics of localized cell-to-cell signaling molecules. To contrast, epinephrine/norepinephrine, vasopressin/antidiuretic hormone, and oxytocin signaling were distinguished as having rapid characteristics because they were both hormones, released from an endocrine gland into the bloodstream, and neurotransmitters, released presynaptically. Either way the signaling molecule would act on receptors to induce the rapid cascade of second messengers. Even when the brain aromatized testosterone to estradiol, its regulation was viewed as limited by the relatively slow transcription of aromatase, thus requiring prolonged serum testosterone exposure. Then even after ER was activated, transcription and translation of estradiol-sensitive gene products were still required. The sum of the physiological response took days. In recent years a new appreciation has emerged for steroids as highly local and rapidly acting signaling molecules that are synthesized in a tightly controlled and regulated manner. The phosphorylation of aromatase by calmodulin can change its activity within minutes of intracellular calcium transients (Balthazart et al., 2005), thus estradiol synthesis and signaling may be coupled to neural activity and rapid enzymatic processes. Estradiol exposure can change the likelihood of glutamate release (Smejkalova & Woolley, 2010) and facilitate long-term potentiation within minutes (Dai et al., 2007; Grassi et al., 2011). Rapid actions are mediated at the membrane, by so-called mER, which appear to be the same nuclear receptor but possessing modifications that permit trafficking to and association with the bi-lipid membrane (Luoma et al., 2008; Mermelstein & Micevych, 2008; Micevych & Mermelstein, 2008; Keller et al., 2010). Moreover, the detection of ER in dendrites (Blaustein, 1992; Blaustein et al., 1992) and more recently in synaptic specializations of the hippocampus and cortex (Milner et al., 2005; Hart et al., 2007; Smejkalova & Woolley, 2010; Waters et al., 2011), contributed to the notion of finely tuned localized effects not unlike those of neurotransmitters (Balthazart & Ball, 2006).

Rapid Effects of Estradiol Mediate Sexual Differentiation of the Ventromedial Nucleus

Although this revolution in hormone action took relatively easy hold for considerations of activational effects in the adult brain, there was little consideration of its role in organizational hormonal effects. This was reasonable in light of the respective purposes of hormone action during the different phases of life, with adult effects providing agility in responsiveness to rapidly changing environmental or social circumstances, whereas developmental effects are meant to endure across the lifespan. Thus it was surprising to discover a rapid effect of estradiol mediating sexual differentiation of the ventromedial nucleus of the hypothalamus (Schwarz et al., 2008) a brain region critical to control of female reproductive behavior. Neurons of the VMN of males exhibit twofold more dendritic spines than female ventromedial hypothalamic nucleus (VMN) neurons because of an increased number of dendritic branches and dendrite length (Todd et al., 2007). During the neonatal sensitive period, estradiol activates PI3kinase to enhance glutamate release from medial basal hypothalamic neurons within an hour. This enhanced glutamate release is insensitive to protein synthesis inhibitors, revealing its independence from transcription and translation. The subsequent activation of postsynaptic NMDA receptors is followed by a calcium influx, which initiates dendritic branching and spine formation. It is only in the postsynaptic cell that protein synthesis is required and initiated through a MAP kinase pathway. Estrogen receptor activation in the postsynaptic neuron is not necessary; rather, NMDA receptor agonists alone suffice for the formation of dendritic spines (Schwarz et al., 2008). The importance of this sequelae to sexual differentiation of behavior is demonstrated by the effect of treating neonatal females with NMDA agonists, which precludes expression of female sexual behavior in adulthood; in other words, they have been defeminized by NMDA receptor activation during the perinatal critical period. Dendritic spine synapses are predominantly glutamatergic, therefore more spines are considered to reflect a greater excitatory input into this region in males compared to females. In the adult, female sexual receptivity requires sequential estradiol and progesterone, and one action of these hormones is to blunt the activation and signaling of glutamatergic neurons that inhibit the lordosis circuitry, thereby facilitating female sexual behavior (Georgescu & Pfaus, 2006a,b; Georgescu et al., 2009; Booth et al, 2010). Males also have more serotoninergic projections into the VMN than females, and removal of this input unmasks a latent lordosis behavior in that sex (Moreines et al., 1988). Serotoninergic projections typically synapse onto the dendritic shaft (Chan-Palay, 1975). Combining this evidence generates a working model in which neonatal aromatization within the VMN provides local and rapid activation of ER, enhancing glutamate release, which organizes dendritic morphology. In adulthood, the facilitated glutamatergic and serotoninergic neurotransmission preemptively clamps lordosis behavior such that the VNM cannot be suppressed by adult exposure to estradiol and progesterone and lordosis behavior cannot be disinhibited. This model has some pertinent advantages because the same area important for the control of feminization, or the expression of both appetitive and consummatory female sex behavior, is also organized during defeminization (Fig. 16–1).

Aromatization Regulates Neuronal Proliferation, Survival, and Death in a Region-Specific Manner

The sexually dimorphic nucleus of the preoptic area (SDN-POA) is arguably the most intensely studied sex difference in the mammalian brain, being three to five times larger in male rats compared with females (Gorski et al., 1980),

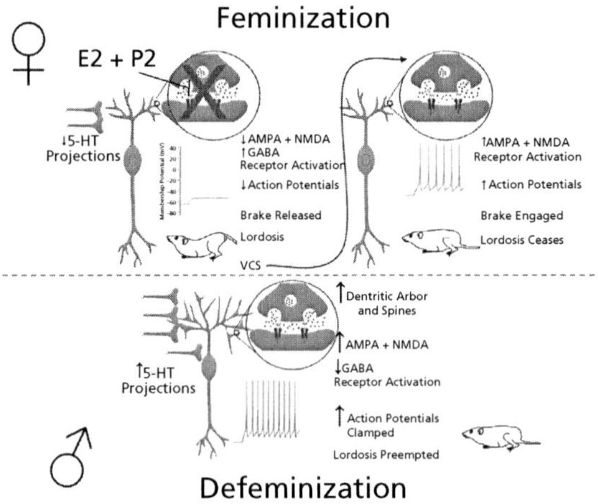

Figure 16–1. Sexually Dimorphic Serotonergic and Glutamatergic Synapses Organize the Ventromedial Hypothalamic Nucleus into a Brake on Lordosis in Females and a Clamp in Males. Male ventromedial hypothalamic nucleus (VMN) neurons have more 5-HT projections, which infers more axodendritic shaft synapses, and longer, dendrites with more branches and total spines, inferring more glutamatergic synapses. Increased glutamate neurotransmission correlates with the cessation of lordosis in females. Ablation of serotonergic projections unmasks a latent lordosis behavior in males. In females, estradiol and progesterone might reduce either serotonergic or glutamatergic neurotransmission or the frequency of action potentials (APs) to initiate lordosis behavior. The serotonergic and glutamatergic neurotransmission in the VMN may serve as a brake on lordosis behavior when appropriate behavioral and sensory cues such as vaginal cervical stimulation (VCS) increase glutamatergic or serotonergic neurotransmission to increase frequency of action potentials and stop lordosis. In males, the combined increases in serotonergic and glutamatergic neurotransmission may preemptively clamp lordosis behavior by overwhelmingly activating VMN neurons such that estradiol and progesterone cannot suppress VMN activity. See text for more information and references.

with homologs in humans and nonhuman primates (LeVay, 1991; Byne, 1998) and, to a lesser degree, in mice (Wolfe et al., 2005). The initial discovery of the SDN in the rat occurred in close proximity to the elucidation of the aromatization hypothesis, both in the early 1970s; thus, there was little delay in discerning that estradiol was the hormonal basis for the larger nucleus in males. Estradiol-mediated prevention of apoptotic cell death was determined to be the dominant mechanism by which males ultimately have a larger volumetric SDN, consisting of more neurons than the same nucleus in females (Davis et al., 1996). The mechanism mediating a sex difference in the putative SDN of the mouse appears to be different in that the same number of neurons are found in the nucleus, but a late-stage or radial scattering in males resulting in a larger overall volume (Orikasa et al., 2010). Nonetheless, the general principle that sex differences in cell death is a major source of phenotypic variance between male and female brains was built on observations of the SDN and extended to other brain regions and the spinal cord. An interesting twist on this generality is the anterior ventral periventricular area, or AVPV, which is larger in females because of estradiol-directed cell death in males. This nucleus is part of a circuit in the HPG axis that controls the LH surge in females and the pulsatile release of LH in males. In the developing male, estradiol binds to both the ER-α and ER-β receptors to initiate the cell death of AVPV neurons. Three unique and distinct populations of neurons are lost: (1) kisspeptin neurons that only express this neurotransmitter in adulthood, (2) a group of dopaminergic neurons, and (3) a neuron subclass that coreleases glutamate and GABA. In the glutamatergic/GABAergic neurons of females, tumor necrosis factor-α, an inflammatory molecule, and its affector NF-κβ are constitutively active and promote cell survival. Estradiol signaling in males activates the TNF receptor–associated factor 2–inhibiting protein (TRIP), which in turn suppresses the TNFα-NF-κβ pathway leading to cell death (Krishnan et al., 2009). This recent discovery created a paradigm shift in that potent inflammatory mediators previously thought to initiate cell death actually promoted cell survival, as indicated here. The dopaminergic neurons also die; here in response to the putative transition in mitochondrial permeability that then initiates the formation of reactive oxygen species, caspase signaling, and apoptosis (Waters & Simerly, 2009). Changes in bcl-2 (an antiapoptosis gene), BAX, and BAD (both are proapoptosis) seen in the AVPV could also be indicative of this mitochondrial mechanism unique to the dopaminergic cell types (Zup et al., 2003). These distinct proapoptotic cellular pathways have not been fully delineated or tied together, but the common theme is estradiol-mediated cell death, as opposed to cell survival. Again as with the SDN, there is also a later developmental phase of modification, but in this case involving addition of new neurons via proliferation to help maintain the sex difference across the lifespan (Ahmed et al., 2008).

Aromatase-mediated sex differences in neuronal proliferation have emerged as a major factor in hippocampal development, with males birthing almost twice as many new cells as females within the first few days of life. Treatment of newborn females with estradiol raises rates of cell proliferation to that of males, whereas treatment of males with either inhibitors of aromatase activity or ER antagonists lowers rates of proliferation to even below that of females (Zhang et al., 2008; Bowers et al., 2010). Of interest is the exceedingly low levels of estradiol detected in the hippocampus at this developmental time point (see discussion that follows), leaving two possible interpretations; either estradiol is synthesized in a narrow microdomain that provides sufficiently high local levels to regulate neural progenitor proliferation, or estradiol acting outside the hippocampus is impinging on neuronal proliferation within the hippocampus. Either way, this represents an important new source of sex differences in the brain that can be directly tied to aromatization.

Aromatization Promotes Expression of Inflammatory Mediators to Masculinize the Preoptic Nucleus of the Hypothalamus and Sexual Behavior

Males have twofold higher density of dendritic spines on medial POA neurons than females. Similarly, estradiol treatment on the day of birth and postnatal day one induces dendritic spine formation in females such that spine levels are equivalent to the male (Amateau & McCarthy, 2002). Similar to the VMN, increased dendritic spines in males results in increased glutamatergic neurotransmission in adulthood, which in turn facilitates the expression of male behavior under the appropriate adult hormonal milieu. Thus, when considering how estradiol mediates sexual differentiation of this brain area, it would be reasonable to assume that glutamate receptors or some other component of dendritic spines would be the primary target. But this is decidedly not the case, as the signal transduction cascade leading to increased dendritic spines in the male POA in response to estradiol begins with the increased expression of the prostaglandin synthetic enzymes, COX-1 and COX-2 (Amateau & McCarthy, 2004; Wright, 2009). The cyclooxygenase enzymes, along with prostaglandin E synthase, produce prostaglandin E2 (PGE2), a labile arachadonic acid metabolite associated with other endocrine and reproductive functions but best known for its role in inflammation and induction of fever. In response to locally synthesized estradiol, PGE2 levels rise substantially from previously undetectable. PGE2 then activates the EP2 and EP4 receptors and subsequently protein kinase A (PKA) to phosphorylate the AMPA/kainate receptor, thereby causing its insertion into the membrane and increased glutamatergic signaling (Wright et al., 2008; Wright & McCarthy, 2009; Lenz et al., 2011). Increased AMPA/Kainate and metabotropic but not NMDA glutamatergic signaling then initiates a twofold increase in the density of dendritic spines on POA neurons, which in turn permits facilitated glutamatergic signaling in adulthood (Wright & McCarthy, 2009). This adult glutamatergic neurotransmission, along with dopaminergic, activates NMDA glutamatergic and D1- and D2-like dopaminergic receptors to facilitate the learning and expression of male sexuality (Hull et al., 1992; Powell et al., 2003; Hull & Dominguez, 2006).

AROMATIZATION: THE UNEXPECTED AND UNEXPLAINED

The Immune System Is a Target for Regulation by Aromatization in the Developing Preoptic Nucleus of the Hypothalamus

The discovery that PGE2 is the downstream signaling pathway for estradiol-mediated masculinization is not entirely unexpected when one considers the importance of this molecule to reproductive physiology, including in the brain. What was unexpected, however, was that confirmation that PGE2 is the initiating signal for masculinization is found in its ability to wholly substitute for estradiol. In other words, treating neonatal females with PGE2 directly into the POA masculinizes the POA and male sexual behavior in a manner that is indistinguishable from females masculinized with estradiol or normal males (Amateau & McCarthy, 2004). Indeed, the system is so sensitive to PGE2 that a single infusion of this labile molecule into a neonatal females brain is sufficient to induce the full complement of masculinization in adulthood (Wright & McCarthy, 2009). This surprising finding suggested some type of feed forward mechanism and recent evidence in support of such a positive feedback is found in our discovery of a sex difference in microglia, the immune cells of the brain. There are almost twice as many microglia in the developing male POA, and inhibition of microglia with minocyclin during the sensitive period impairs estradiol-mediated masculinization (Lenz et al., 2010). That the immune system would be critical to aromatization-mediated organization of the brain most certainly falls in the realm of the unexpected, and details of how this occurs are currently unexplained.

Inflammatory Mediators Regulate Aromatase Activity in the Developing Cerebellum

Inflammation early in development is emerging as a major contributing variable to later life vulnerabilites (Bale et al., 2010), which prompted us to consider whether estradiol-induced prostaglandin synthesis is generalizable to brain areas other than the POA, in particular those brain areas previously shown to be vulnerable to the deleterious effects of early life inflammation, the cerebellum being one such area (Dean & McCarthy, 2008). Aromatase is expressed in the Purkinje cells of the rodent cerebellum, a brain region that also expresses many other enzymes essential for the de novo synthesis of steroids (Price & Handa, 2000; Abel et al., 2011). In fact, cerebellar E2 levels have been reported to be higher than serum levels during the first week of life. In the course of exploring whether this estradiol increases COX synthesis and prostaglandin production, we made the unexpected and opposite finding that PGE2 stimulates aromatase in the cerebellum to increase estradiol production (Fig. 16–2). There is precedence in the periphery for PGE2 stimulation of aromatase, in the breast and ovary (Cai et al., 2007; Subbaramaiah et al., 2008), but to our knowledge this is the first report of an inflammatory molecule mediating aromatase activity in the brain. Injury can also regulate the transcription and translation of enzymes

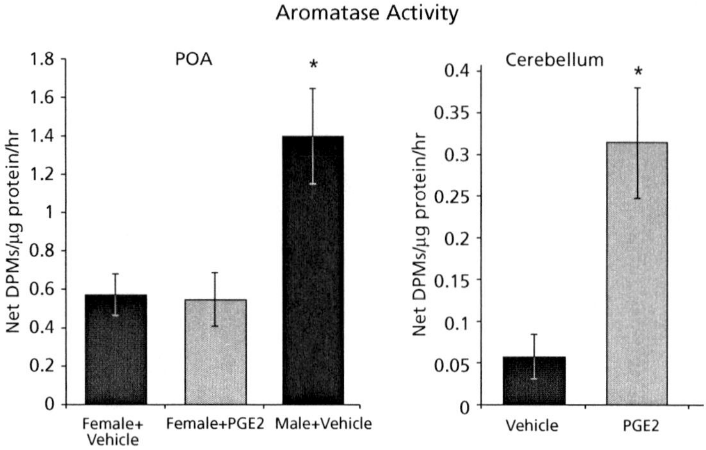

Figure 16–2. Prostaglandin E2 Stimulates Aromatase Activity in the Developing Cerebellum but Not the Preoptic Area. Aromatase activity by cultured cerebellar neurons was quantified with and without treatment with the prostaglandin PGE2 for 48 hours or in the in vivo POA of males and females and females receiving intracerebroventricular injections of PGE2 on the day of birth and postnatal day 1, a treatment paradigm known to induce masculinization of the brain. We found that PGE2 significantly increased aromatase activity in cerebellar neurons, but did not have the same effect in POA neurons. Detection of a significant sex difference in aromatase activity in the POA is consistent with published reports. Net disintegrations per minute (DPM) are the average of aromatase reaction replicates minus the DPM values from letrozole (40 µM) treated reactions. To our knowledge this is the first demonstration of PGE2 stimulated aromatase activity in the brain. Dean, Wright, Knutsen, & McCarthy, submitted for publication.

involved in de novo estrogen synthesis in the cerebellum of the zebra finch (Mirzatoniet al., 2010) and rat (Lavaque et al., 2006; Biamonte et al., 2009). Our recent observation that inflammatory mediators, which are up regulated after injury, induce aromatase activity in the brain may be a previously unappreciated component of this response.

Steroid Levels in the Neonatal Brain Do Not Abide by Predictions of the Organizational/Activational Hypothesis—Evidence for Neurosteroidogenesis?

The concept of neurosteroidogenesis was first put forth by Baulieu et al. (Corpechot et al., 1981, 1983) when they established that the mammalian brain is capable of steroidogenesis beyond aromatization of circulating androgens to estrogens. Over the years, functional enzymes important in steroidogenesis have been reported throughout the developing brain (Mellon et al., 2001). Others have shown induction of estradiol synthesis in hippocampal slices of either postnatal (Prange-Kiel et al., 2003) or adult rats (Hojo et al., 2004). Taken together, these findings further support a theory of neurosteroidogenesis that may differ across the lifespan in these various brain loci.

Until relatively recently, data regarding the actual concentrations of steroids in the developing rodent brain was either inferred from levels detected in the plasma or from a small number of studies that examined only select brain regions and used pooled samples from multiple animals. The advent of ultra-sensitive radioimmunoassay allowed for the opportunity to quantify exceedingly low levels of estradiol as well as testosterone and dihydrotestosterone in microdissected regions of the neonatal brain from individual animals. Although the use of RIA is not without its limitations (Konkle & McCarthy, 2011), a broad survey of steroid levels in multiple brain regions revealed a number of unexpected findings. Among these was the determination of relatively few sex differences, largely limited to higher estradiol concentrations in male hypothalamus on embryonic day 19, but at not other times examined, and as expected, higher estradiol in male POA within 2 hours of birth. The concordance in steroid levels between males and females was particularly notable in the hippocampus, which was also characterized by exceedingly low levels of all three steroids, with estradiol below 1.5 pg/mg protein (as opposed to >20 pg/mg protein for the male POA). The low estradiol levels of the hippocampus were paralleled by an equally low level of aromatase activity, but which was nonetheless subject to detectable inhibition by formestane and letrozol. In attempts to isolate the functional significance of this low but regulated estradiol synthesis, we assessed whether there were any changes in synaptogenesis, but found none (Konkle & McCarthy, unpublished data). However, we did find a potent effect of endogenous estradiol synthesis on neurogenesis in the developing male hippocampus (Bowers et al., 2010). This suggests that in addition to the high local production of estradiol postulated to modulate synaptic plasticity in the adult hippocampus (Woolley, 2007), there is also a role for local aromatization in the developing hippocampus that impacts on the proliferation of neural stem cells and possibly neuronal survival, selectively in males.

A third unexpected finding was the resistance of neonatal brain steroid levels to removal of the gonads and adrenals, with extremely high levels of estradiol maintained in the hypothalamus and even a significant increase in the female cortex, suggesting de novo steroidogenesis within the brain that is responsive to peripheral changes. Further evidence for de novo steroidogenesis is found in the unusual pattern of steroid levels seen in the cortex, with late peaks in estradiol appearing on PN4 for males and PN6 for females, resulting in sex differences at both those time points (Fig. 16–3). There is no evidence that gonadal activity would underlie these peaks, especially because they are unique to the cortex, with other regions remaining largely flat until PN20, when yet another unexplained rise in estradiol and testosterone is observed in the cortex and to a lesser degree, the hippocampus. The cortex expresses high levels of ERα neonatally, which then declines with advancing maturity (Wilson et al., 2011); thus, there is concordance between ligand and receptor, but the functional role of estradiol in cortical development is unknown.

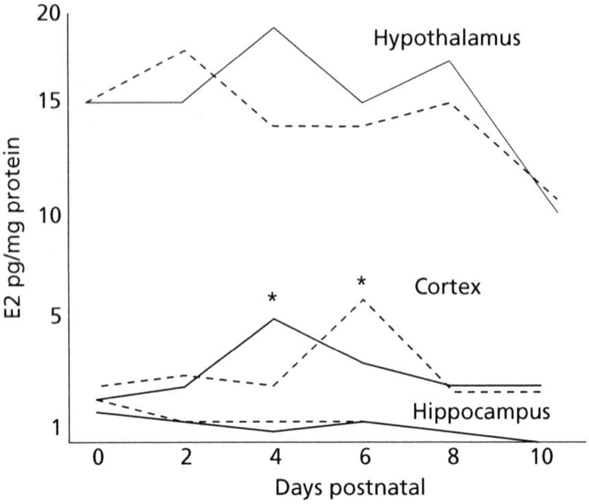

Figure 16–3. Estradiol Levels Are Dynamic in the Developing Brain. Estradiol content was measured by radioimmunoassay in microdissected brain regions of individual developing male (*solid lines*) and female (*dashed lines*) rat pups from the day of birth until adult, with the first 10 days of life shown here. Both the amount and the developmental pattern of estradiol are markedly different across the first 10 days of postnatal life. As expected, the highest levels are found in the hypothalamus and remain high relative to the other brain regions until at least postnatal day 10. The lowest levels are in the hippocampus, and both the pattern and amount are strikingly similar in males and females. Nonetheless, this low level of estradiol is paralleled by a low level of aromatase activity that is responsive to two different aromatase inhibitors, suggestive of biologically significant but highly local estrogen production. The most interesting pattern was seen in the cortex, in which a late surge in estradiol is seen in females on PN4 and males on PN6, in both cases resulting in a sex difference (*$p < .05$). (See text for details.) Redrawn from Konkle, A. T., & McCarthy, M. M. (2011). Developmental time course of estradiol, testosterone, and dihydrotestosterone levels in discrete regions of male and female rat brain. *Endocrinology, 152*(1), 223–235.

CONCLUSIONS

Steroid hormones remain one of the most powerful and pervasive nongenomic factors regulating brain development. The number of systems modulated by estradiol continues to grow, ranging from other steroid receptors, to neurotransmitter systems, and now on to lipid membrane signaling molecules, cytokines, and growth factors. Uncertainties in the precise amount and local synthesis of estradiol limit our ability to fully grasp the potential for hormonal modulation of the developing brain, but we are gaining good traction and progress is being made. The challenges ahead include creating an integrated view of how steroid synthesis, in particular estradiol, affects the brain both as a whole and also as a sum of its parts. The brain is one organ, but also a collection of many discrete units, and their appropriate connectedness and interaction is paramount to the generation of effective behavioral and reproductive output. Aromatization is central to this integration and worthy of our continued attention.

REFERENCES

Abel, J. M., Witt, D. M., & Rissman, E. F. (2011). Sex differences in the cerebellum and frontal cortex: roles of estrogen receptor alpha and sex chromosome genes. *Neuroendocrinology, 93*(4), 230–240.

Ahmed, E. I., Zehr, J. L., Schulz, K. M., Lorenz, B. H., Doncarlos, L. L., & Sisk, C. L. (2008). Pubertal hormones modulate the addition of new cells to sexually dimorphic brain regions. *Nat Neurosci, 55,* 597–604.

Amateau, S. K., & McCarthy, M. M. (2002). A novel mechanism of dendritic spine plasticity involving estradiol induction of prostaglandin-E2. *J Neurosci, 22*(19), 8586–8596.

Amateau, S. K., & McCarthy, M. M. (2004a). Induction of PGE2 by estradiol mediates developmental masculinization of sex behavior. *Nat Neurosci, 7*(6), 643–650.

Andrews, G. K., Dziadek, M., & Tamaoki, T. (1982). Expression and methylation of the mouse alpha-fetoprotein gene in embryonic, adult, and neoplastic tissues. *J Biol Chem, 257*(9), 5148–5153.

Bakker, J., De Mees, C., Douhard, Q., Balthazart, J., Gabant, P., Szpirer, J., et al. (2006). Alpha-fetoprotein protects the developing female mouse brain from masculinization and defeminization by estrogens. *Nat Neurosci, 9*(2), 220–226.

Bakker, J., Honda, S., Harada, N., & Balthazart, J. (2002). Sexual partner preference requires a functional aromatase (cyp19) gene in male mice. *Horm Behav, 42*(2), 158–171.

Bakker, J., Honda, S., Harada, N., & Balthazart, J. (2003). The aromatase knockout (ArKO) mouse provides new evidence that estrogens are required for the development of the female brain. *Ann N Y Acad Sci, 1007,* 251–262.

Bakker, J., Honda, S., Harada, N., & Balthazart, J. (2004). Restoration of male sexual behavior by adult exogenous estrogens in male aromatase knockout mice. *Horm Behav, 46*(1), 1–10.

Bale, T. L., Baram, T. Z., Brown, A. S., Goldstein, J. M., Insel, T. R., McCarthy, et al. (2010). Early life programming and neurodevelopmental disorders. *Biol Psychiatry, 68*(4), 314–319.

Balthazart, J., Baillien, M., Charlier, T. D., & Ball, G. F. (2005). Effects of calmodulin on aromatase activity in the preoptic area. *J Neuroendocrinol, 17*(10), 664–671.

Balthazart, J., & Ball, G. F. (2006). Is brain estradiol a hormone or a neurotransmitter? *Trends Neurosci, 29*(5), 241–249.

Baum, M. J., Gallagher, C. A., Martin, J. T., & Damassa, D. A. (1982). Effects of testosterone, dihydrotestosterone, or estradiol administered neonatally on sexual behavior of female ferrets. *Endocrinology, 111*(3), 773–780.

Baum, M. J., & Vreeburg, J. T. (1973). Copulation in castrated male rats following combined treatment with estradiol and dihydrotestosterone. *Science, 182*(109), 283–285.

Biamonte, F., Assenza, G., Marino, R., D'Amelio, M., Panteri, R., Caruso, D., et al. (2009). Interactions between neuroactive steroids and reelin haploinsufficiency in Purkinje cell survival. *Neurobiol Dis, 36*(1), 103–115.

Blaustein, J. D. (1992). Cytoplasmic estrogen receptors in rat brain: immunocytochemical evidence using three antibodies with distinct epitopes. *Endocrinology, 131*(3), 1336–1342.

Blaustein, J. D., Lehman, M. N., Turcotte, J. C., & Greene, G. (1992). Estrogen receptors in dendrites and axon terminals in the guinea pig hypothalamus. *Endocrinology, 131*(1), 281–290.

Booth, C., Wayman, C. P., & Jackson, V. M. (2010). An ex vivo multi-electrode approach to evaluate endogenous hormones and receptor subtype pharmacology on evoked and spontaneous neuronal activity within the ventromedial hypothalamus; translation from female receptivity. *J Sex Med, 7*(7), 2411–2423.

Booth, J. E. (1977). Sexual behaviour of neonatally castrated rats injected during infancy with oestrogen and dihydrotestosterone. *J Endocrinol, 72*(2), 135–141.

Bowers, J., Waddell, J., & McCarthy, M. (2010). A developmental sex difference in hippocampal neurogenesis is mediated by endogenous estradiol. *Biol Sex Diff, 1,* 8.

Byne, W. (1998). The medial preoptic and anterior hypothalamic regions of the rhesus monkey: cytoarchitectonic comparison with the human and evidence for sexual dimorphism. *Brain Res, 793*(1–2), 346–350.

Cai, Z., Kwintkiewicz, J., Young, M. E., & Stocco, C. (2007). Prostaglandin E2 increases cyp19 expression in rat granulosa cells: implication of GATA-4. *Mol Cell Endocrinol, 263*(1–2), 181–189.

Chan-Palay, V. (1975). Fine structure of labelled axons in the cerebellar cortex and nuclei of rodents and primates after intraventricular infusions with tritiated serotonin. *Anat Embryol (Berl), 148*(3), 235–265.

Corpechot, C., Robel, P., Axelson, M., Sjovall, J., & Baulieu, E. E. (1981). Characterization and measurement of dehydroepiandrosterone sulfate in rat brain. *Proc Natl Acad Sci U S A, 78*(8), 4704–4707.

Corpechot, C., Synguelakis, M., Talha, S., Axelson, M., Sjovall, J., Vihko, R., et al. (1983). Pregnenolone and its sulfate ester in the rat brain. *Brain Res, 270*(1), 119–125.

Couse, J. F., Hewitt, S. C., Bunch, D. O., Sar, M., Walker, V. R., Davis, B. J., et al. (1999). Postnatal sex reversal of the ovaries in mice lacking estrogen receptors alpha and beta. *Science, 286*(5448), 2328–2331.

Dai, X., Chen, L., & Sokabe, M. (2007). Neurosteroid estradiol rescues ischemia-induced deficit in the long-term potentiation of rat hippocampal CA1 neurons. *Neuropharmacology, 52*(4), 1124–1138.

Davis, E. C., Popper, P., & Gorski, R. A. (1996). The role of apoptosis in sexual differentiation of the rat sexually dimorphic nucleus of the preoptic area. *Brain Res., 734,* 10–18.

De Mees, C., Laes, J. F., Bakker, J., Smitz, J., Hennuy, B., Van Vooren, P., et al. (2006). Alpha-fetoprotein controls female fertility and prenatal development of the gonadotropin-releasing hormone pathway through an antiestrogenic action. *Mol Cell Biol, 26*(5), 2012–2018.

Dean, S. L., & McCarthy, M. M. (2008). Steroids, sex and the cerebellar cortex: implications for human disease. *Cerebellum 7*(1) 38-47.

Dorrington, J. H., Bendell, J. J., & Lobb, D. K. (1987). Aromatase activity in granulosa cells: regulation by growth factors. *Steroids, 50*(4–6), 411–421.

Dupont, S., Krust, A., Gansmuller, A., Dierich, A., Chambon, P., & Mark, M. (2000). Effect of single and compound knockouts of estrogen receptors alpha (ERalpha) and beta (ERbeta) on mouse reproductive phenotypes. *Development, 127*(19), 4277–4291.

Fisher, C. R., Graves, K. H., Parlow, A. F., & Simpson, E. R. (1998). Characterization of mice deficient in aromatase (ArKO) because of targeted disruption of the cyp19 gene. *Proc Natl Acad Sci U S A, 95*(12), 6965–6970.

Gandelman, R. (1973). The ontogeny of maternal responsiveness in female Rockland-Swiss albino mice. *Horm Behav, 4*(3), 257–268.

Gandelman, R., & Vom Saal, F. S. (1975). Pup-killing in mice: the effects of gonadectomy and testosterone administration. *Physiol Behav, 15*(6), 647–651.

Garcia-Ovejero, D., Veiga, S., Garcia-Segura, L. M., & Doncarlos, L. L. (2002). Glial expression of estrogen and androgen receptors after rat brain injury. *J Comp Neurol, 450*(3), 256–271.

Georgescu, M., & Pfaus, J. G. (2006a). Role of glutamate receptors in the ventromedial hypothalamus in the regulation of female rat sexual behaviors I. Behavioral effects of glutamate and its selective receptor agonists AMPA, NMDA and kainate. *Pharmacol Biochem Behav, 83*(2), 322–332.

Georgescu, M., & Pfaus, J. G. (2006b). Role of glutamate receptors in the ventromedial hypothalamus in the regulation of female rat sexual behaviors. II. Behavioral effects of selective glutamate receptor antagonists AP-5, CNQX, and DNQX. *Pharmacol Biochem Behav, 83*(2), 333–341.

Georgescu, M., Sabongui, C., Del Corpo, A., Marsan, L., & Pfaus, J. G. (2009). Vaginocervical stimulation induces Fos in glutamate neurons in the ventromedial hypothalamus: attenuation by estrogen and progesterone. *Horm Behav, 56*(4), 450–456.

Gladue, B. A., & Clemens, L. G. (1980a). Flutamide inhibits testosterone-induced masculine sexual behavior in male and female rats. *Endocrinology, 106*(6), 1917–1922.

Gladue, B. A., & Clemens, L. G. (1980b). Masculinization diminished by disruption of prenatal estrogen biosynthesis in male rats. *Physiol Behav, 25*(4), 589–593.

Gonzalez-Martinez, D., De Mees, C., Douhard, Q., Szpirer, C., & Bakker, J. (2008). Absence of gonadotropin-releasing hormone 1 and Kiss1 activation in alpha-fetoprotein knockout mice: prenatal estrogens defeminize the potential to show preovulatory luteinizing hormone surges. *Endocrinology, 149*(5), 2333–2340.

Gorski, R. A., Harlan, R. E., Jacobson, C. D., Shryne, J. E., & Southam, A. M. (1980). Evidence for the existence of a sexually dimorphic nucleus in the preoptic area of the rat. *J Comp Neurol, 193*, 529–539.

Grassi, S., Tozzi, A., Costa, C., Tantucci, M., Colcelli, E., Scarduzio, M., et al. (2011). Neural 17beta-estradiol facilitates long-term potentiation in the hippocampal CA1 region. *Neuroscience, 192*, 67–73.

Harada, N., Wakatsuki, T., Aste, N., Yoshimura, N., & Honda, S. I. (2009). Functional analysis of neurosteroidal oestrogen using gene-disrupted and transgenic mice. *J Neuroendocrinol, 21*(4), 365–369.

Harada, N., & Yamada, K. (1992). Ontogeny of aromatase messenger ribonucleic acid in mouse brain: fluorometrical quantitation by polymerase chain reaction. *Endocrinology, 131*(5), 2306–2312.

Hart, S. A., Snyder, M. A., Smejkalova, T., & Woolley, C. S. (2007). Estrogen mobilizes a subset of estrogen receptor-alpha-immunoreactive vesicles in inhibitory presynaptic boutons in hippocampal CA1. *J Neurosci, 27*(8), 2102–2111.

Hojo, Y., Hattori, T. A., Enami, T., Furukawa, A., Suzuki, K., Ishii, H. T., et al. (2004). Adult male rat hippocampus synthesizes estradiol from pregnenolone by cytochromes P45017alpha and P450 aromatase localized in neurons. *Proc Natl Acad Sci U S A, 101*(3), 865–870.

Honda, S., Harada, N., Ito, S., Takagi, Y., & Maeda, S. (1998). Disruption of sexual behavior in male aromatase-deficient mice lacking exons 1 and 2 of the cyp19 gene. *Biochem Biophys Res Commun, 252*(2), 445–449.

Hull, E. M., & Dominguez, J. M. (2006). Getting his act together: roles of glutamate, nitric oxide, and dopamine in the medial preoptic area. *Brain Res, 1126*(1), 66–75.

Hull, E. M., Eaton, R. C., Markowski, V. P., Moses, J., Lumley, L. A., & Loucks, J. A. (1992). Opposite influence of medial preoptic D1 and D2 receptors on genital reflexes: implications for copulation. *Life Sci, 51*(22), 1705–1713.

Jung-Testas, I., Renoir, M., Bugnard, H., Greene, G. L., & Baulieu, E. E. (1992). Demonstration of steroid hormone receptors and steroid action in primary cultures of rat glial cells. *J Steroid Biochem Mol Biol, 41*(3–8), 621–631.

Keller, M., Pawluski, J. L., Brock, O., Douhard, Q., & Bakker, J. (2010). The alpha-fetoprotein knock-out mouse model suggests that parental behavior is sexually differentiated under the influence of prenatal estradiol. *Horm Behav, 57*(4–5), 434–440.

Konkle, A. T., & McCarthy, M. M. (2011). Developmental time course of estradiol, testosterone, and dihydrotestosterone levels in discrete regions of male and female rat brain. *Endocrinology, 152*(1), 223–235.

Krege, J. H., Hodgin, J. B., Couse, J. F., Enmark, E., Warner, M., Mahler, J. F., et al. (1998). Generation and reproductive phenotypes of mice lacking estrogen receptor beta. *Proc Natl Acad Sci U S A, 95*(26), 15677–15682.

Krishnan, S., Intlekofer, K. A., Aggison, L. K., & Petersen, S. L. (2009). Central role of TRAF-interacting protein in a new model of brain sexual differentiation. *Proc Natl Acad Sci U S A, 106*(39), 16692–16697.

Kudwa, A. E., Bodo, C., Gustafsson, J. A., & Rissman, E. F. (2005). A previously uncharacterized role for estrogen receptor beta: defeminization of male brain and behavior. *Proc Natl Acad Sci U S A, 102*(12), 4608–4612.

Lavaque, E., Mayen, A., Azcoitia, I., Tena-Sempere, M., & Garcia-Segura, L. M. (2006). Sex differences, developmental changes, response to injury and cAMP regulation of the mRNA levels of steroidogenic acute regulatory protein, cytochrome p450scc, and aromatase in the olivocerebellar system. *J Neurobiol, 66*(3), 308–318.

Lenz, K. M., Nugent, B. M., & McCarthy, M. M. (2010). *Do microglia mediate sexual differentiation of the preoptic area and behavior?* Paper presented at the Society For Neuroscience Annual Meeting, Washington, DC Poster Presentation.

Lenz, K. M., Wright, C. L., Martin, R. C., & McCarthy, M. M. (2011). Prostaglandin E regulates AMPA receptor phosphorylation and promotes membrane insertion in preoptic area neurons and glia during sexual differentiation. *PLoS One, 6*(4), e18500.

LeVay, S. (1991). A difference in hypothalamic structure between heterosexual and homosexual men. *Science, 9*, 497–506.

Lubahn, D. B., Moyer, J. S., Golding, T. S., Couse, J. F., Korach, K. S., & Smithies, O. (1993). Alteration of reproductive function but not prenatal sexual development after insertional disruption of the mouse estrogen receptor gene. *Proc Natl Acad Sci U S A, 90*(23), 11162–11166.

Luoma, J. I., Boulware, M. I., & Mermelstein, P. G. (2008). Caveolin proteins and estrogen signaling in the brain. *Mol Cell Endocrinol, 290*(1–2), 8–13.

MacLusky, N. J., Chaptal, C., & McEwen, B. S. (1979). The development of estrogen receptor systems in the rat brain and pituitary: postnatal development. *Brain Res, 178*(1), 143–160.

Matsumoto, T., Honda, S., & Harada, N. (2003a). Alteration in sex-specific behaviors in male mice lacking the aromatase gene. *Neuroendocrin, 77*(6), 416–424.

Matsumoto, T., Honda, S., & Harada, N. (2003b). Neurological effects of aromatase deficiency in the mouse. *J Steroid Biochem Mol Biol, 86*(3–5), 357–365.

McCarthy, M. M., Schlenker, E. H., & Pfaff, D. W. (1993). Enduring consequences of neonatal treatment with antisense oligodeoxynucleotides to estrogen receptor messenger ribonucleic acid on sexual differentiation of rat brain. *Endocrinology, 133*(2), 433–439.

McDonald, P., Beyer, C., Newton, F., Brien, B., Baker, R., Tan, H. S., et al. (1970). Failure of 5alpha-dihydrotestosterone to initiate sexual behaviour in the castrated male rat. *Nature, 227*(5261), 964–965.

McEwen, B. S., Lieberburg, I., Chaptal, C., & Krey, L. C. (1977a). Aromatization: important for sexual differentiation of the neonatal rat brain. *Horm Behav, 9*(3), 249–263.

McEwen, B. S., Lieberburg, I., Maclusky, N., & Plapinger, L. (1977b). Do estrogen receptors play a role in the sexual differentiation of the rat brain? *J Steroid Biochem, 8*(5), 593–598.

Mellon, S. H., Griffin, L. D., & Compagnone, N. A. (2001). Biosynthesis and action of neurosteroids. *Brain Res Brain Res Rev, 37*(1–3), 3–12.

Mermelstein, P. G., & Micevych, P. E. (2008). Nervous system physiology regulated by membrane estrogen receptors. *Rev Neurosci, 19*(6), 413–424.

Micevych, P. E., & Mermelstein, P. G. (2008). Membrane estrogen receptors acting through metabotropic glutamate receptors: an emerging mechanism of estrogen action in brain. *Mol Neurobiol, 38*(1), 66–77.

Milner, T. A., Ayoola, K., Drake, C. T., Herrick, S. P., Tabori, N. E., McEwen, B. S., et al. (2005). Ultrastructural localization of estrogen receptor beta immunoreactivity in the rat hippocampal formation. *J Comp Neurol, 491*(2), 81–95.

Mirzatoni, A., Spence, R. D., Naranjo, K. C., Saldanha, C. J., & Schlinger, B. A. (2010). Injury-induced regulation of steroidogenic gene expression in the cerebellum. *J Neurotrauma, 27*(10), 1875–1882.

Mizejewski, G. J., Vonnegut, M., & Simon, R. (1980). Neonatal androgenization using antibodies to alpha-fetoprotein. *Brain Res, 188*(1), 273–277.

Moreines, J., Kelton, M., Luine, V. N., Pfaff, D. W., & McEwen, B. S. (1988). Hypothalamic serotonin lesions unmask hormone responsiveness of lordosis behavior in adult male rats. *Neuroendocrinology, 47*(5), 453–458.

Naftolin, F. (1975). Proceedings: Gonadotrophin rhythms. *J Endocrinol, 66*(2), 9P–10P.

Naftolin, F., Ryan, K. J., Davies, I. J., Petro, Z., & Kuhn, M. (1975a). The formation and metabolism of estrogens in brain tissues. *Adv Biosci, 15*, 105–121.

Naftolin, F., Ryan, K. J., Davies, I. J., Reddy, V. V., Flores, F., Petro, Z., et al.. (1975b). The formation of estrogens by central neuroendocrine tissues. *Recent Prog Horm Res, 31*, 295–319.

O'Keefe, J. A., & Handa, R. J. (1990). Transient elevation of estrogen receptors in the neonatal rat hippocampus. *Brain Res Dev Brain Res, 57*(1), 119–127.

O'Keefe, J. A., Li, Y., Burgess, L. H., & Handa, R. J. (1995). Estrogen receptor mRNA alterations in the developing rat hippocampus. *Brain Res Mol Brain Res, 30*(1), 115–124.

Ogawa, S., Chester, A. E., Hewitt, S. C., Walker, V. R., Gustafsson, J. A., Smithies, O., et al. (2000). Abolition of male sexual behaviors in mice lacking estrogen receptors alpha and beta (alpha beta ERKO). *Proc Natl Acad Sci U S A, 97*(26), 14737–14741.

Orikasa, C., Kondo, Y., Usui, S., & Sakuma, Y. (2010). Similar numbers of neurons are generated in the male and female rat preoptic area in utero. *Neurosci Res, 68*(1), 9–14.

Pfaff, D., & Keiner, M. (1973). Atlas of estradiol-concentrating cells in the central nervous system of the female rat. *J Comp Neurol, 151*(2), 121–158.

Phoenix, C. H., Goy, R. W., Gerall, A. A., & Young, W. C. (1959). Organizing action of prenatally administered testosterone proprionate on the tissues mediating mating behavior in the female guinea pig. *Endocrinology, 65*, 369–382.

Powell, W. S., Dominguez, J. M., & Hull, E. M. (2003). An NMDA antagonist impairs copulation and the experience-induced enhancement of male sexual behavior in the rat. *Behav Neurosci, 117*(1), 69–75.

Prange-Kiel, J., Wehrenberg, U., Jarry, H., & Rune, G. M. (2003). Para/Autocrine regulation of estrogen receptors in hippocampal neurons. *Hippocampus, 13*, 226–234.

Price, R. H., Jr., & Handa, R. J. (2000). Expression of estrogen receptor-beta protein and mRNA in the cerebellum of the rat. *Neurosci Lett, 288*(2), 115–118.

Putnam, S. K., Sato, S., & Hull, E. M. (2003). Effects of testosterone metabolites on copulation and medial preoptic dopamine release in castrated male rats. *Horm Behav, 44*(5), 419–426.

Putnam, S. K., Sato, S., Riolo, J. V., & Hull, E. M. (2005). Effects of testosterone metabolites on copulation, medial preoptic dopamine, and NOS-immunoreactivity in castrated male rats. *Horm Behav, 47*(5), 513–522.

Quadros, P. S., Pfau, J. L., Goldstein, A. Y., De Vries, G. J., & Wagner, C. K. (2002). Sex differences in progesterone receptor expression: a potential mechanism for estradiol-mediated sexual differentiation. *Endocrinology, 143*(10), 3727–3739.

Raynaud, J. P. (1973). Influence of rat estradiol binding plasma protein (EBP) on uterotrophic activity. *Steroids, 21*(2), 249–258.

Resko, J. A., & Roselli, C. E. (1997). Prenatal hormones organize sex differences of the neuroendocrine reproductive system: observations on guinea pigs and nonhuman primates. *Cell Mol Neurobiol, 17*(6), 627–648.

Robertson, K. M., O'Donnell, L., Jones, M. E., Meachem, S. J., Boon, W. C., Fisher, C. R., et al. (1999). Impairment of spermatogenesis in mice lacking a functional aromatase (cyp 19) gene. *Proc Natl Acad Sci U S A, 96*(14), 7986–7991.

Roselli, C. E., & Resko, J. A. (1984). Androgens regulate brain aromatase activity in adult male rats through a receptor mechanism. *Endocrinology, 114*(6), 2183–2189.

Roselli, C. E., & Resko, J. A. (1993). Aromatase activity in the rat brain: hormonal regulation and sex differences. *J Steroid Biochem Mol Biol, 44*(4–6), 499–508.

Schwarz, J. M., Liang, S.-L., Thompson, S. M., & McCarthy, M. M. (2008). Estradiol induces hypothalamic dendritic spines by enhancing glutamate release: A mechanism for organizational sex differences. *Neuron, 58*, 584–598.

Sierra, A., Gottfried-Blackmore, A., Milner, T. A., McEwen, B. S., & Bulloch, K. (2008). Steroid hormone receptor expression and function in microglia. *Glia, 56*(6), 659–674.

Simerly, R. B., Chang, C., Muramatsu, M., & Swanson, L. W. (1990). Distribution of androgen and estrogen receptor mRNA-containing cells in the rat brain: an in situ hybridization study. *J Comp Neurol, 294*(1), 76–95.

Simpson, E. R., Merrill, J. C., Hollub, A. J., Graham-Lorence, S., & Mendelson, C. R. (1989). Regulation of estrogen biosynthesis by human adipose cells. *Endocr Rev, 10*(2), 136–148.

Smejkalova, T., & Woolley, C. S. (2010). Estradiol acutely potentiates hippocampal excitatory synaptic transmission through a presynaptic mechanism. *J Neurosci, 30*(48), 16137–16148.

Subbaramaiah, K., Hudis, C., Chang, S. H., Hla, T., & Dannenberg, A. J. (2008). EP2 and EP4 receptors regulate aromatase expression in human adipocytes and breast cancer cells. Evidence of a BRCA1 and p300 exchange. *J Biol Chem, 283*(6), 3433–3444.

Toda, K., Okada, T., Takeda, K., Akira, S., Saibara, T., Shiraishi, M., et al. (2001). Oestrogen at the neonatal stage is critical for the reproductive ability of male mice as revealed by supplementation with 17beta-oestradiol to aromatase gene (Cyp19) knockout mice. *J Endocrinol, 168*(3), 455–463.

Toda, K., Saibara, T., Okada, T., Onishi, S., & Shizuta, Y. (2001). A loss of aggressive behaviour and its reinstatement by oestrogen in mice lacking the aromatase gene (Cyp19). *J Endocrinol, 168*(2), 217–220.

Todd, B. J., Schwarz, J. M., Mong, J. A., & McCarthy, M. M. (2007). Glutamate AMPA/kainate receptors, not GABAA receptors, mediate estradiol-induced sex differences in the hypothalamus. *Dev Neurobiol, 67*, 304–315.

Toft, D., & Gorski, J. (1966). A receptor molecule for estrogens: isolation from the rat uterus and preliminary characterization. *Proc Natl Acad Sci U S A, 55*(6), 1574–1581.

Toran-Allerand, C. D. (1980). Coexistence of alpha-fetoprotein, albumin and transferrin immunoreactivity in neurones of the developing mouse brain. *Nature, 286*(5774), 733–735.

Toran-Allerand, C. D. (1987). Neuronal uptake of alpha-fetoprotein (AFP) synthesized and secreted by hepatocytes in liver/brain co-cultures. *Neurosci Lett, 83*(1–2), 35–40.

Tremblay, G. B., Tremblay, A., Copeland, N. G., Gilbert, D. J., Jenkins, N. A., Labrie, F., et al. (1997). Cloning, chromosomal localization, and functional analysis of the murine estrogen receptor beta. *Mol Endocrinol, 11*(3), 353–365.

Valladares, L. E., & Payne, A. H. (1979). Acute stimulation of aromatization in Leydig cells by human chorionic gonadotropin in vitro. *Proc Natl Acad Sci U S A, 76*(9), 4460–4463.

Waters, E. M., & Simerly, R. B. (2009). Estrogen induces caspase-dependent cell death during hypothalamic development. *J Neurosci, 29*(31), 9714–9718.

Waters, E. M., Yildirim, M., Janssen, W. G., Lou, W. Y., McEwen, B. S., Morrison, J. H., et al. (2011). Estrogen and aging affect the synaptic distribution of estrogen receptor beta-immunoreactivity in the CA1 region of female rat hippocampus. *Brain Res, 1379*, 86–97.

Weniger, J. P. (1993). Estrogen production by fetal rat gonads. *J Steroid Biochem Mol Biol, 44*(4–6), 459–462.

Weniger, J. P., Zeis, A., & Chouraqui, J. (1993). Estrogen production by fetal and infantile rat ovaries. *Reprod Nutr Dev, 33*(2), 129–136.

Wersinger, S. R., Sannen, K., Villalba, C., Lubahn, D. B., Rissman, E. F., & De Vries, G. J. (1997). Masculine sexual behavior is disrupted in male and female mice lacking a functional estrogen receptor alpha gene. *Horm Behav, 32*(3), 176–183.

Whalen, R. E., & Edwards, D. A. (1967). Hormonal determinants of the development of masculine and feminine behavior in male and female rats. *Anat Rec, 157*(2), 173–180.

Wilson, M. E., Westberry, J. M., & Trout, A. L. (2011). Estrogen receptor-alpha gene expression in the cortex: sex differences during development and in adulthood. *Horm Behav, 59*(3), 353–357.

Wolfe, C. A., Van Doren, M., Walker, H. J., Seney, M. L., McClellan, K. M., & Tobet, S. A. (2005). Sex differences in the location of immunochemically defined cell populations in the mouse preoptic area/anterior hypothalamus. *Brain Res Dev Brain Res, 157*(1), 34–41.

Woolley, C. S. (2007). Acute effects of estrogen on neuronal physiology. *Annu Rev Pharmacol Toxicol, 47*, 657–680.

Wright, C. L. (2009). *Identification of the Prostaglandin E2 Receptors Mediating the Perinatal Organization of Male Sexual Behavior and Neuroanatomica Correlates.* Ph.D. Dissertation, University of Maryland, Baltimore.

Wright, C. L., Burks, S. R., & McCarthy, M. M. (2008). Identification of prostaglandin E2 receptors mediating perinatal masculinization of adult sex behavior and neuroanatomical correlates. *Dev Neurobiol, 68*(12), 1406–1419.

Wright, C. L., & McCarthy, M. M. (2009). Prostaglandin E2-induced masculinization of brain and behavior requires protein kinase A, AMPA/kainate, and metabotropic glutamate receptor signaling. *J Neurosci, 29*(42), 13274–13282.

Zhang, J.-M., Konkle, A. T. M., Zup, S. L., & McCarthy, M. M. (2008). Impact of sex and hormones on new cells in the developing rat hippocampus: A novel source of sex dimorphism? *Eur J Neurosci, 27*, 791–800.

Zup, S. L., Carrier, H., Waters, E. M., Tabor, A., Bengston, L., Rosen, G. J., et al. (2003). Overexpression of bcl-2 reduces sex differences in neuron number in the brain and spinal cord. *J Neurosci, 23*(6), 2357–2362.

Chapter 17

Aromatase and Sexual Differentiation: Lessons from the Aromatase and Alpha-Fetoprotein Knockout Mice

Julie Bakker

INTRODUCTION

The classic theory of brain sexual differentiation holds that male-typical behavioral and neural characteristics develop under the influence of testosterone secreted by the male's testes and that female-typical behavioral and neural characteristics develop in its absence (e.g., Baum, 1979; Schwarz & McCarthy, 2008). A corollary of this theory is that many of the organizational actions of testosterone on the male brain result from estradiol formed via neural aromatization of testosterone. Accordingly, McEwen et al. (1975) proposed that the female rodent brain needs to be protected from any estradiol produced by the placenta or by males lying in close approximation to a female fetus in utero, and that alpha-fetoprotein (AFP)—an important plasma protein present during fetal development in high concentrations—is the most likely candidate to achieve this protection because of its high estradiol-binding capacity.

This idea of a default organizational program for the female brain as well as a protective role for AFP has been challenged. First, there is some behavioral evidence that the normal sexual differentiation of the female brain actually requires estradiol because female sexual behavior was affected up on removal of the ovaries (Dunlap et al., 1973; Gerall et al., 1973). Second, the presence of AFP within neurons in the absence of any local AFP synthesis suggests that AFP may act as a carrier transporting estradiol into the brain and actively participates in the sexual differentiation of the female brain (Toran-Allerand, 1984). Thus two opposing hypotheses on the role of AFP and estradiol in the development of the female brain exist, but it has been difficult to test any of these hypotheses experimentally owing to the absence of a suitable animal model as well as technical difficulties associated with measuring estradiol concentrations in the brain.

The introduction of transgenic mouse models, that is, the aromatase knockout (ArKO) and the alpha-fetoprotein knockout (AFP-KO), has made it possible to solve this longstanding dilemma about the role of estradiol and AFP in brain sexual differentiation. Results obtained in female ArKO mice provided important evidence for a feminizing action for estradiol on the brain and behavior during prepubertal development (Bakker et al., 2002; Brock et al., 2011), whereas results obtained in female AFP-KO mice showed that the principal action of prenatal estrogen exposure, regardless of whether it occurs in female or male mice, is to defeminize, and to some extent, masculinize the brain and behavior (Bakker et al., 2006). Based on the results obtained in both knockout mouse models, it is proposed that the defeminizing actions of estradiol normally occur prenatally in males and are avoided in fetal females because of the protective actions of AFP, whereas the feminizing actions of estradiol normally occur prepubertally in genetic females, when the ovaries start to produce estrogens, that is, in the second week after birth, and when circulating concentrations of AFP have diminished so that it no longer plays a protective role.

ROLE OF ESTRADIOL IN FEMALE-TYPICAL BRAIN AND BEHAVIORAL SEXUAL DIFFERENTIATION: LESSONS FROM ArKO MICE

A possible contribution of estradiol to the development of female-typical behavioral characteristics was first suggested by some early behavioral studies showing that female rats ovariectomized on the day of birth had lower lordosis quotients following adult treatment with estradiol and progesterone than females that either kept their ovaries (Lisk, 1969) or were ovariectomized at birth and at the same time implanted with ovaries until postnatal day 60 (Gerall et al., 1973). In addition, Döhler et al. (1984) showed that neonatal treatment of female rats with tamoxifen, an estradiol receptor antagonist, decreased their capacity to show lordosis behavior in adulthood, whereas concurrent neonatal administration of a low dose of estradiol benzoate (EB) prevented this effect. Finally, Baum and Tobet (1986) found that female ferrets, which were treated prenatally with the aromatase inhibitor 1,4,6-androstatriene-3,17-dione (ATD), and then treated in adulthood with a low or moderate dose of EB, were less receptive to the mounts of a stimulus male.

These various behavioral results suggested that exposure to a low level of estradiol over a postnatal interval between birth and the age of puberty facilitated the later capacity to display female sexual behavior; however, these data did not provide incontrovertible evidence that estradiol normally contributes to the development of female sexual behavior in female mammals. First, the effects of neonatal ovariectomy on the potential to show lordosis behavior later in life were only transient because any differences in lordosis behavior disappeared following repeated testing (Gerall et al., 1973). Perhaps more important, in addition to its antiestrogenic actions, tamoxifen can exert estrogen-like agonist actions in the brain (Mathews et al., 1988). Therefore, the observed reduction in lordosis behavior induced by administering tamoxifen neonatally to female rats (Dohler et al., 1984) may actually have resulted from a partial defeminization of the brain by the estradiol-like actions of tamoxifen acting on neural estradiol receptors. Finally, prenatally ATD-treated female ferrets and control females displayed equivalent, high, acceptance quotients when tested after receiving a high dose of EB in adulthood (Baum & Tobet,

1986) indicating that ATD-treated females were capable of displaying normal female-typical receptive behavior. Thus, it has been difficult to provide conclusive evidence of a role for estradiol in the development of the female brain. As a result, the hypothesis has languished since the mid-1980s because of the absence of a suitable animal model in which to assess the possible contribution of estradiol more rigorously. The creation of the ArKO mouse (Fisher et al., 1998; Honda et al., 1998) provided an opportunity to reopen work on this hypothesis. Because ArKO mice cannot produce estradiol themselves owing to a targeted mutation in the aromatase (*Cyp19*) gene, but do have functional estradiol receptors, they can respond to exogenous estradiol at any time during their lifespan. This means that this model can be used to distinguish between organizational and activational effects of estradiol on brain and behavioral functions. Here a short overview is presented of results obtained in the ArKO mouse model that provide new evidence for a role of estradiol in female neural and behavioral differentiation.

Reduced Female Sexual Behavior in Female ArKO Mice

A first experiment determined whether lordosis behavior was affected in female ArKO mice (Bakker et al., 2002). If the female brain indeed develops in the absence of any estradiol, ArKO female mice should show normal, wild-type (WT) levels of lordosis behavior, provided that they are supplemented with estradiol and progesterone in adulthood, because they cannot produce estradiol themselves. Furthermore, progesterone levels might also be affected in ArKO females because they do not show any signs of ovulation. Thus, WT and ArKO female mice of three genotypes, that is, WT, heterozygous (HET), and ArKO were ovariectomized in adulthood and subsequently implanted with a Silastic™ capsule containing crystalline estradiol. All females were injected subcutaneously with 500 μg progesterone 2 to 4 hours before each lordosis test. ArKO females showed fewer lordotic responses when mounted by the stimulus male (Fig. 17–1A), but there were no genotype differences in the number of mounts received. In addition, this decrease in lordosis behavior did not disappear with repeated testing, as was observed in the early studies using neonatally ovariectomized female rats (Gerall et al., 1973), suggesting permanent effects of early estradiol deprivation on the later ability to show lordosis behavior.

The absence of lordosis behavior in ArKO females might have been caused by a partial masculinization of the brain because testosterone might be increased during early development. Fisher et al. (1998) reported that ArKO mice of both sexes had increased plasma levels of testosterone in adulthood compared with WT mice. These increased levels of testosterone probably result from the lack of negative feedback action of estradiol on the hypothalamic–pituitary–gonadal axis, as suggested by the increased levels of circulating luteinizing hormone and follicle-stimulating hormone in these mice (Fisher et al., 1998). Alternatively, the increase in plasma testosterone could be caused by an accumulation of the androgenic substrate, because it is no longer aromatized to estradiol owing to the targeted mutation in the aromatase gene.

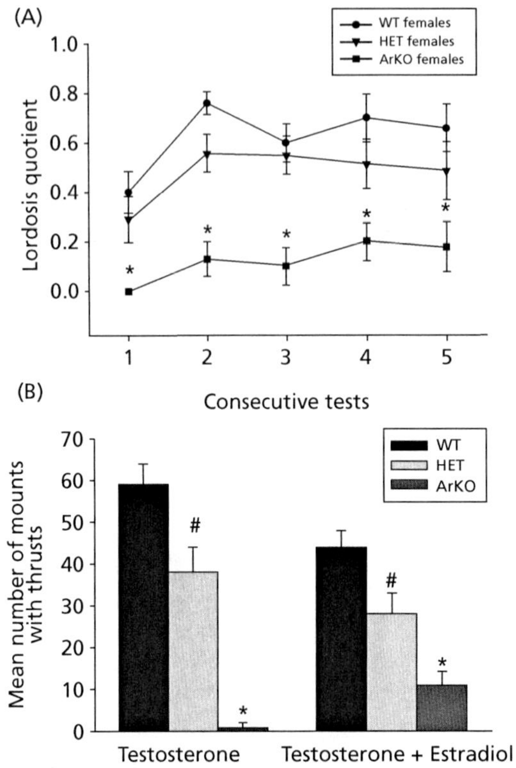

Figure 17–1. Sexual Behavior of Wild-Type (WT), Heterozygous (HT), and Aromatase Knockout (ArKO) Female Mice That Were Treated with Estradiol and Progesterone in Adulthood. A. Mean (± SEM) lordosis quotients calculated by dividing the number of lordotic responses shown by the number of mounts received from the stimulus male. $^{*}p < .05$ Compared with WT females. B. Mean (± SEM) number of mounts with thrusts displayed with an estrous female. Females were first tested with testosterone and then with testosterone and estradiol. $^{*}p < .05$ Compared with WT and HT females; $\#p < .05$ compared with WT and ArKO females. Adapted from Bakker, J., Honda, S., Harada, N., Balthazart, J. (2002). The aromatase knock-out mouse provides new evidence that estradiol is required during development in the female for the expression of sociosexual behaviors in adulthood. *J Neurosci, 22*(20), 9104–9112.

Because the ovaries do not start to secrete high levels of steroid hormones until the second week after birth (Mannan & O'Shaughnessy, 1991), it is unlikely that this increase in testosterone levels also takes place during fetal and early neonatal development, but because no data are available at present to evaluate this question, it could be speculated that increased levels of testosterone contribute to the development of the behavioral phenotype of ArKO female mice. Therefore, to evaluate whether the reduction in lordosis behavior in ArKO females reflects masculinizing and defeminizing actions of testosterone on their brains during early development, female WT, HET, and ArKO mice were tested for their ability to show male-typical mounting behavior with an estrous female (Bakker et al., 2002). Female mice normally display surprisingly high levels of male-like mounting behavior toward other females around the time of estrus; they also show high levels of mounting after adult ovariectomy and testosterone treatment (Edwards, 1971; Martel & Baum, 2009). Thus subjects were ovariectomized in adulthood and subsequently implanted with a Silastic™ capsule containing testosterone. Following 3 weeks of testosterone treatment, female ArKO mice showed very little mounting behavior, whereas WT and to some lesser extent HET females readily displayed mounting behavior (Bakker et al., 2002) (Fig. 17–1B). Adding EB to the testosterone treatment stimulated some mounting behavior in ArKO females, but not up to the levels observed in WT and HET females. Therefore, there is no clear behavioral evidence for brain masculinization in ArKO females.

Reduced Prepubertal Expression of Progesterone Receptor in the Hypothalamus of Female ArKO Mice

The observation of clear deficits in lordosis behavior in female ArKO mice resurrected the question of whether estradiol actively contributes to female-typical brain and behavioral sexual differentiation, perhaps beginning after postnatal day 7 when the ovary starts to produce estradiol (Mannan & O'Shaughnessy, 1991). One approach to this question would be to compare the expression of an estradiol-dependent gene in ArKO female mice, which produce no estradiol in any tissue, and in WT female mice. The expression of the progesterone receptor (PR) in the hypothalamus of adult female rodents is well known to be dramatically upregulated by estradiol (Quadros et al., 2002a,b). Likewise, Wagner et al. have conducted an extensive series of experiments showing that the number of PR-immunoreactive (ir) cells in the medial preoptic area (MPOA), the lateral part of the ventromedial hypothalamus (VMHvl), and the anteroventral periventricular area (AVPV) of both rat (Quadros & Wagner, 2008) and mouse (Wagner et al., 2001; Kudwa et al., 2009) is significantly greater in neonatal males than females. Additional work from this group (Wagner et al., 2001; Quadros et al., 2002b) showed that estradiol, formed via the neural aromatization of testosterone in male neonates, is responsible for the higher, male-typical levels of PR expression in the MPOA. Finally, another study from this group (Jahagirdar et al., 2008) showed that administration of the aromatase inhibitor ATD reduced PR-ir in the MPOA of male rats killed at the end of gestation. Therefore, the expression of PR-ir in three different hypothalamic regions (i.e., AVPV, MPOA, and VMHvl), was used as an index of estradiol's action in the female mouse brain across the first 25 days of postnatal development (Brock et al., 2010). The amount of PR-ir in the AVPV and MPOA was significantly lower in ArKO female mice than in WT females at several prepubertal ages, including postnatal day P15, P20, and P25, but not neonatally at P0, P5, or P10. Likewise, PR-ir was significantly lower in the VMHvl at P25 in ArKO versus WT females, but not at earlier postnatal ages (Brock et al., 2010). Furthermore, PR-ir was consistently higher in male than female WT mice in the AVPV and MPOA over P0 to P10 and in the VMHvl over P0 to P20. In addition, PR-ir in ArKO males was significantly lower than in WT males in these brain areas across these latter ages, and resembled the values observed in WT females, confirming previous reports that estradiol formed in the male hypothalamus by aromatization of testosterone is responsible for inducing male-typical levels of neural PR expression. Thus, these results support the view that estradiol contributes to at least one female-typical aspect of development in both the MPOA and AVPV, the expression of PR (Brock et al., 2010). Interestingly, these results also indicate that estradiol induces both female- and male-typical expression of PR in the mouse hypothalamus with the male-typical population developing prenatally and the female-typical population developing prepubertally.

Rescue of Lordosis Behavior in Female ArKO Mice by Estradiol Treatment over a Specific Prepubertal Period

The observation of an increase in PR expression over the prepubertal period in WT female mice suggests that there might be a specific prepubertal period in which estradiol feminizes the female brain and behavior. Therefore, in a follow-up experiment (Brock et al., 2011), it was determined whether treatment with EB over a specific early postnatal (P5–P15) or prepubertal (P15–P25) period would reverse later deficits in lordosis behavior in female ArKO mice tested following adult treatment with estradiol and progesterone. The ability of ArKO female mice to show WT-like levels of lordosis behavior in response to adult ovarian hormones was almost completely rescued by daily treatment with EB over the prepubertal period, P15 to P25, whereas EB given over an earlier postnatal period (P5–P15) had no such effect in ArKO females and actually completely suppressed lordosis behavior in WT females (Fig. 17–2) (Brock et al., 2011). These latter results are in line with the classic view of the sexual differentiation of brain and behavior that in male rodents perinatal testosterone exerts strong defeminizing effects on the neural mechanisms mediating sexual behavior, with these effects of testosterone being mediated by estradiol derived from local aromatization of testosterone. They also corroborate an early study showing that neonatal testosterone and estradiol were equally effective in suppressing sexual receptivity in female mice (Edwards & Thompson, 1970), and are in line with many additional studies conducted in other rodent species (Wallen & Baum, 2002). In addition, prepubertal EB treatment restored the motivation of ArKO females to approach a stimulus male as opposed to another female. These results thus specify a specific prepubertal period during which estradiol exerts a feminizing action on both appetitive and consummatory aspects of female sexual behavior in mice. It should be noted that prepubertal (P15-P25) EB treatment had no detrimental effect on the expression of lordosis behavior by WT females, whereas earlier neonatal (P5-P15) EB treatment strongly defeminized WT females, indicating that the defeminization of sexual behavior capacity that normally occurs in male mice is completed by the actions of estradiol before P15. Taken together, results obtained in the ArKO mouse model (Bakker et al., 2002;

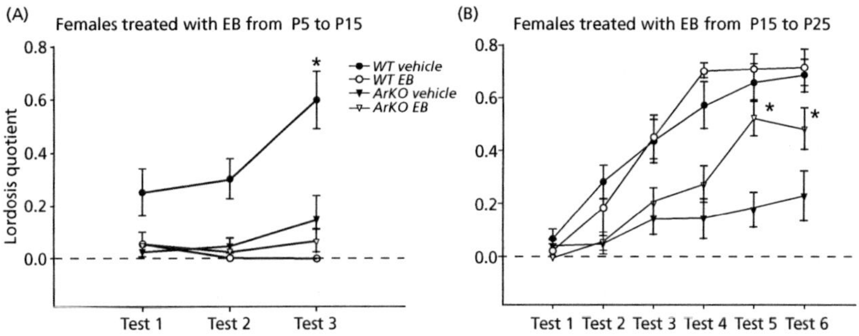

Figure 17–2. Lordosis Behavior of Wild-Type (WT) and Aromatase Knockout (ArKO) Female Mice Treated with Estradiol Benzoate (EB) Either Between P5 and P15 (0.5 μg every other day) (*A*) or Between P15 and P25 (0.05 μg/day) (*B*). Controls received oil vehicle. All females were treated with estradiol and progesterone in adulthood before behavioral testing. *A*. $*p < .05$ significantly different from all other groups using repeated measures ANOVA. B. $*p < .05$ significantly different from ArKO vehicle-injected controls. Data were analyzed using repeated measures ANOVA with genotype, and postnatal treatment as independent factors and test as the repeated factor. Adapted from Brock, O., Baum, M. J., Bakker, J. (2011). The development of female sexual behavior requires prepubertal estradiol. *J Neurosci, 31,* 5574–5578.

Brock et al., 2011) warrant a revision of the classic view of a default organization of female-typical neural and behavioral characteristics.

ROLE OF ALPHA-FETOPROTEIN IN THE SEXUAL DIFFERENTIATION OF THE BRAIN: LESSONS FROM AFP-KO MICE

Alpha-fetoprotein (AFP) is a plasma glycoprotein produced in great quantities during fetal life by the hepatocytes and the endodermal cells of the visceral yolk, and in lesser amounts by the gastrointestinal tract (Andrews et al., 1982; Tilghman & Belayew, 1982). The synthesis of AFP decreases rapidly after birth and only trace amounts are detected in adulthood. At present the physiological function of AFP during embryonic development remains largely unidentified, but the observation that AFP is able to bind estradiol with high affinity in rodent species has led to the hypothesis that AFP may play an important role in the sexual differentiation of the brain because estradiol has been shown to induce male-typical neural and behavioral characteristics during perinatal development. Accordingly, AFP was proposed to either protect the female brain from being defeminized by estradiol (McEwen et al., 1975) or act as a carrier to transport estradiol into the female brain (Toran-Allerand, 1984). It has proved difficult to test experimentally these two opposing hypotheses on the function of AFP in the sexual differentiation of the brain because there was no suitable animal model. Furthermore, it has also been very difficult to measure brain estradiol concentrations. Some indirect evidence for a protective role for AFP was obtained by adding neonatal serum, which was assumed to contain high levels of AFP, to [3H]-estradiol and subsequently to determine its uptake into the brain of adult female rats (McCall et al., 1981). However, the introduction of an AFP mutant mouse model (Gabant et al., 2002) finally made it possible to determine directly the role of AFP in the sexual differentiation of the brain. Male AFP-KO mice are fertile, but females are not, owing to a complete absence of ovulation (Gabant et al., 2002). Ovarian transplantations clearly demonstrated that anovulation in AFP-KO females is caused by an extragonadal factor probably localized at the level of the hypothalamus. Here a short overview is provided of the results obtained in the AFP-KO mouse model showing that AFP protects the female brain from masculinization and defeminization by estradiol (Bakker et al., 2006).

Absence of Female Sexual Behavior in AFP-KO Mice

Two different AFP-KO mice have been developed by Gabant et al.: AFP-KO1 (in which a portion of exon 1 and all of exons 2 and 3 were deleted) and KO2 (in which a portion of exon 2 and all of exon 3 were deleted; for more details see Gabant et al., 2002). The KO1 alleles were generated into two different genetic backgrounds, the outbred CD-1 and the inbred C57Bl6/j strain, the KO2 allele only in the CD-1 background. Heterozygous (HET) females were bred with homozygous and HET AFP-KO males to generate wild-type (WT), HET, and AFP-KO offspring. All AFP-KO mouse models were tested for their ability to show lordosis behavior with a sexually active male. The same experimental procedure as described for the studies on ArKO mice were used; that is, females of three genotypes (WT, HET, and AFP-KO) were ovariectomized in adulthood and at the same time implanted with a Silastic™ capsule containing estradiol. Females were then tested for lordosis behavior over three weekly trials following a subcutaneous injection with 500 μg progesterone 2 to 4 hours before testing. None of the AFP-KO mouse models showed any lordosis behavior when mounted by the male, whereas WT and HET females were clearly sexually receptive (Fig. 17–3, showing the results for the KO2 model). These results thus suggest that AFP-KO females have been defeminized with regard to lordosis behavior.

Increased Male-Typical Sexual Behavior in AFP-KO Mice

To determine whether AFP-KO females have been behaviorally masculinized as well, they were tested for their ability to show male-typical mounting behavior with an estrous female over three weekly trials. Thus, females of three genotypes (WT, HET, and AFP-KO) were ovariectomized in adulthood and implanted with a Silastic™ capsule containing estradiol. It was found that female AFP-KO mice showed more mounting behavior than WT females suggesting that they have been masculinized as well (Bakker et al., 2006).

Rescue of the Female Phenotype by Prenatal ATD Treatment

The observation that AFP-KO females do not show any lordosis behavior suggests that they are defeminized probably because during embryonic life they

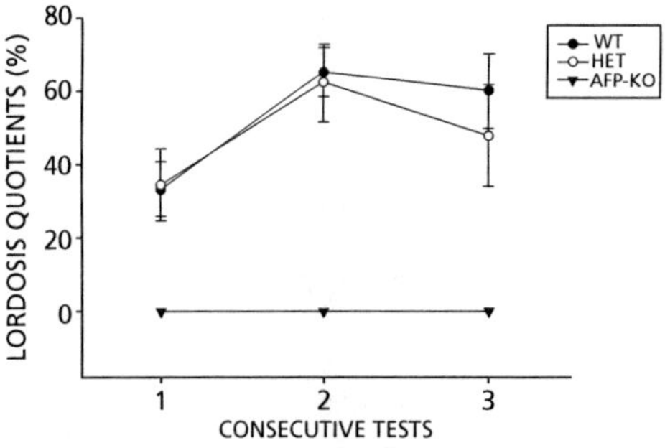

Figure 17–3. Lordosis Behavior of Wild-Type (WT), Heterozygous (HET), and Alpha-Fetoprotein Knockout (AFP-KO) Female Mice Carrying the KO2 Allele. For more details see Gabant et al., 2002. $^{*}p < .05$ compared with WT and HET females.

were no longer protected by AFP against any estradiol produced by their mother or male siblings. These data thus support the hypothesis that AFP protects the female brain from excessive exposure to estradiol (McEwen et al., 1975). However, it could also be argued that the brains of AFP-KO females were not feminized because they were lacking AFP to transport small amounts of estradiol to target brain areas and thus to promote the differentiation of these neural structures in a female direction. This argument favors the hypothesis that AFP is needed to transport estradiol into the brain and that it participates actively in the sexual differentiation of the female brain. To discriminate between these opposing hypotheses, pregnant female mice heterozygous for the AFP mutation were treated with the aromatase inhibitor ATD so as to inhibit estradiol synthesis during prenatal development. If AFP protects the female brain from being exposed to estradiol, then AFP-KO offspring of ATD-treated mothers should not be defeminized, because the ATD treatment would prevent the formation of high defeminizing levels of estradiol during prenatal development. However, if AFP actually acts as a carrier for estradiol and thus is necessary for the feminization of the brain, then AFP-KO offspring of ATD-treated mothers should not show normal levels of lordosis behavior because they lack both the steroid and the carrier and they will thus not be feminized. It was found that prenatal treatment with ATD completely rescued lordosis behavior in AFP-KO females (Bakker et al., 2006) (Fig. 17–4). Such females had similar lordosis quotients as WT females and HET females treated with ATD prenatally, whereas AFP-KO females, whose mothers were treated with vehicle, showed no lordosis behavior as was expected. These results thus indicate that prenatatal estradiol defeminizes the brain and that AFP protects the female brain from this action of estradiol.

CONCLUSIONS

Results obtained in AFP-KO and ArKO mice suggest that estradiol can have both defeminizing and feminizing actions on the brain, depending on when

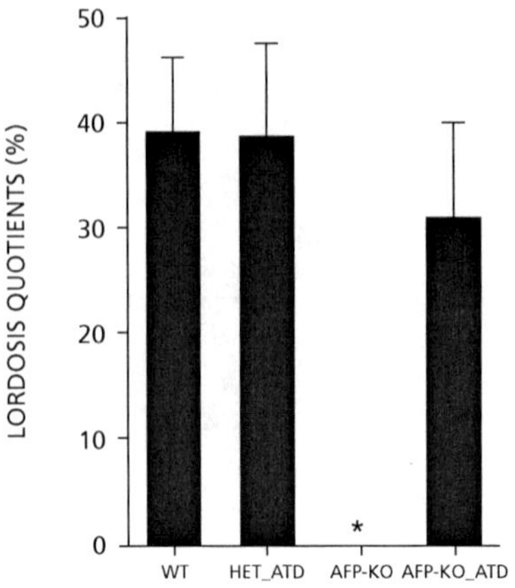

Figure 17–4. Lordosis Behavior of Wild-Type (WT) and Alpha-Fetoprotein Knockout (AFP-KO) Female Mice, and of Heterozygous (HET) and AFP-KO Female Mice Treated with the Aromatase Inhibitor ATD During Fetal Development. Values are means ± SEM of three behavioral tests. $^{*}p < .05$ compared with WT, HET_ATD, and AFP-KO_ATD females. Adapted from Bakker, J., De Mees, C., Douhard, Q., Balthazart, J., Gabant, P., Szpirer, J., et al. (2006). Alpha-fetoprotein protects the developing female mouse brain from masculinization and defeminization by estrogens. *Nat Neurosci, 9*(2), 220–226.

during development estradiol is available. The defeminizing actions of estradiol normally occur prenatally in males and are avoided in fetal females thanks to the protective actions of AFP, whereas the feminizing actions of estradiol normally occur prepubertally in genetic females, when the ovaries start to produce estrogens (i.e., in the second week after birth), and when circulating concentrations of AFP have diminished so that this protein no longer plays a protective role. These results also suggest a differential role for brain versus ovarian aromatase in the sexual differentiation of the brain. The ovary is most likely the source of estradiol in the female because the development of female-typical neural and behavioral characteristics coincides with increased secretion of estradiol by the ovary. By contrast, the brain is the most likely source of estradiol in the male because the testes do not show any significant aromatase activity and any estradiol produced by the testes will be bound to alpha-fetoprotein and thus not available for the brain. In future studies, it might be interesting to analyze the phenotype of brain-specific ArKO mice, because males might be as affected as in the general KO model, whereas females would not if the ovary is the only important source of estradiol in the female.

REFERENCES

Andrews, G. K., Dziadek, M., & Tamaoki, T. (1982). Expression and methylation of the mouse alpha-fetoprotein gene in embryonic, adult, and neoplastic tissues. *J Biol Chem, 257*(9), 5148–5153.

Bakker, J., De Mees, C., Douhard, Q., Balthazart, J., Gabant, P., Szpirer, J., et al. (2006). Alpha-fetoprotein protects the developing female mouse brain from masculinization and defeminization by estrogens. *Nat Neurosci, 9*(2), 220–226.

Bakker, J., Honda, S., Harada, N., Balthazart, J. (2002). The aromatase knock-out mouse provides new evidence that estradiol is required during development in the female for the expression of sociosexual behaviors in adulthood. *J Neurosci, 22*(20), 9104–9112.

Baum, M. J. (1979). Differentiation of coital behavior in mammals: a comparative analysis. *Neurosci Biobehav Rev, 3,* 265–284.

Baum, M. J., & Tobet, S. A. (1986). Effect of prenatal exposure to aromatase inhibitor, testosterone, or antiandrogen on the development of feminine sexual behavior in ferrets of both sexes. *Physiol Behav, 37*(1), 111–118.

Brock, O., Baum, M. J., Bakker, J. (2011). The development of female sexual behavior requires prepubertal estradiol. *J Neurosci, 31,* 5574–5578.

Brock, O., Douhard, Q., Baum, M. J., Bakker, J. (2010). Reduced prepubertal expression of progesterone receptor in the hypothalamus of female aromatase knockout mice. *Endocrinology, 151*(4), 1814–1821.

Dohler, K. D., Hancke, J. L., Srivastava, S. S., Hofmann, C., Shryne, J. E., Gorski, R. A. (1984). Participation of estrogens in female sexual differentiation of the brain; neuroanatomical, neuroendocrine and behavioral evidence. *Prog Brain Res, 61,* 99–117.

Dunlap, J. L., Gerall, A. A., McLean, L. D. (1973). Enhancement of female receptivity in neonatally castrated males by prepuberal ovarian transplants. *Physiol Behav, 10*(4), 701–705.

Edwards, D. A. (1971). Neonatal administration of androstenedione, testosterone or testosterone propionate: effects on ovulation, sexual receptivity and aggressive behavior in female mice. *Physiol Behav, 6,* 223–228.

Edwards, D. A., & Thompson, M. L. (1970). Neonatal androgenization and estrogenization and the hormonal induction of sexual receptivity in rats. *Physiol Behav, 5,* 1115–1119.

Fisher, C. R., Graves, K. H., Parlow, A. F., & Simpson, E. R. (1998). Characterization of mice deficient in aromatase (ArKO) because of targeted disruption of the cyp19 gene. *Proc Natl Acad Sci U S A, 95*(12): 6965–6970.

Gabant, P., Forrester, L., Nichols, J., Van Reeth, T., De Mees, C., Pajack, B., et al. (2002). Alpha-fetoprotein, the major fetal serum protein, is not essential for embryonic development but is required for female fertility. *Proc Natl Acad Sci U S A, 99*(20), 12865–12870.

Gerall, A. A., Dunlap, J. L., & Hendricks, S. E. (1973). Effect of ovarian secretions on female behavioral potentiality in the rat. *J Comp Physiol Psychol, 82*(3), 449–465.

Honda, S., Harada, N., Ito, S., Takagi, Y., & Maeda, S. (1998). Disruption of sexual behavior in male aromatase-deficient mice lacking exons 1 and 2 of the cyp19 gene. *Biochem Biophys Res Commun, 252*(2), 445–449.

Jahagirdar, V., Quadros, P. S., & Wagner, C. K. (2008). Endogenous oestradiol regulates progesterone receptor expression in the brain of female rat fetuses: what is the source of oestradiol? *J Neuroendocrinol, 20*(3), 359–365.

Kudwa, A. E., Harada, N., Honda, S. I., Rissman, E. F. (2009). Regulation of progestin receptors in medial amygdala: estradiol, phytoestrogens and sex. *Physiol Behav, 97*(2), 146–150.

Lisk, R. D. (1969). Progesterone: biphasic effects on the lordosis response in adult or neonatally gonadectomized rats. *Neuroendocrinology, 5*(3), 149–160.

Mannan, M. A., & O'Shaughnessy, P. J. (1991). Steroidogenesis during postnatal development in the mouse ovary. *J Endocrinol, 130,* 101–106.
Martel, K. L., & Baum, M. J. (2009). Adult testosterone treatment but not surgical disruption of vomeronasal function augments male-typical sexual behavior in female mice. *J Neurosci, 29,* 7658–7666.
Mathews, G. A., Brenowitz, E. A., & Arnold, A. P. (1988). Paradoxical hypermasculinization of the zebra finch song system by an antiestrogen. *Horm Behav, 22*(4), 540–551.
McCall, A. L., Han, S.-J., Millington, W. R., & Baum, M. J. (1981). Non-saturable transport of [^{3}H]-oestradiol across the blood-brain barrier in female rats is reduced by neonatal serum. *J Reprod Fertil, 61,* 103–108.
McEwen, B. S., Plapinger, L., Chaptal, C., Gerlach, J., & Wallach, G. (1975). Role of fetoneonatal estrogen binding proteins in the associations of estrogen with neonatal brain cell nuclear receptors. *Brain Res, 96*(2), 400–406.
Quadros, P. S., Goldstein, A. Y., De Vries, G. J., & Wagner, C. K. (2002a). Regulation of sex differences in progesterone receptor expression in the medial preoptic nucleus of postnatal rats. *J Neuroendocrinol, 14*(10), 761–767.
Quadros, P. S., Pfau, J. L., Goldstein, A. Y., De Vries, G. J., & Wagner, C. K. (2002b). Sex differences in progesterone receptor expression: a potential mechanism for estradiol-mediated sexual differentiation. *Endocrinology, 143*(10), 3727–3739.
Quadros, P. S., & Wagner, C. K. (2008). Regulation of progesterone receptor expression by estradiol is dependent on age, sex and region in the rat brain. *Endocrinology, 149*(6), 3054–3061.
Schwarz, J. M., & McCarthy, M. M. (2008). Steroid-induced sexual differentiation of the developing brain: multiple pathways, one goal. *J Neurochem, 105,* 1561–1572.
Tilghman, S. M., & Belayew, A. (1982). Transcriptional control of the murine albumin/alpha-fetoprotein locus during development. *Proc Natl Acad Sci U S A, 79*(17), 5254–5257.
Toran-Allerand, C. D. (1984). On the genesis of sexual differentiation of the central nervous system: morphogenetic consequences of steroidal exposure and possible role of alpha-fetoprotein. In: De Vries, G. J., De Bruin, J. P. C., Uylings, H. B. M., & Corner, M. A. (Eds). *Progress in Brain Research, Sex Differences in the Brain, the Relation Between Structure and Function*. Amsterdam: Elsevier, pp. 63–98.
Wagner, C. K., Pfau, J. L., De Vries, G. J., & Merchenthaler, I. J. (2001). Sex differences in progesterone receptor immunoreactivity in neonatal mouse brain depend on estrogen receptor alpha expression. *J Neurobiol, 47*(3), 176–182.
Wallen, K., & Baum, M. J. (2002). Masculinization and defeminization in altricial and precocial mammals: Comparative aspects of steroid hormone action. In: Pfaff, D. W. (ed.), *Hormones, Brain and Behavior*. New-York: Academic Press, pp. 385–423.

Chapter 18

Aromatase, Estrogens, and Differentiation of Sexual Behavior and Partner Preference in Birds

Elizabeth Adkins-Regan

INTRODUCTION

Aromatase appears to be present in at least some brain regions, such as the hypothalamus and preoptic area, in all vertebrates that have been examined (Forlano et al., 2006). Birds are one of two vertebrate groups, along with teleost fish, that have relatively high levels of aromatase and estrogen production in the brain. In the largest lineage of birds, the passeriform birds (oscine songbirds and sub-oscines), aromatase is found not only in the brainstem, but also at high levels in parts of the telencephalon, including areas such as the hippocampus that are not traditionally associated with reproduction (Saldanha et al., 2000). Birds are also one of two vertebrate groups, along with mammals, in which there is substantial evidence that estrogens are involved in sexual differentiation of behavior through actions on the brain (Balthazart et al., 2009).

What is the connection between these facts about birds? Is brain aromatase a mechanism for the development of sex differences in behavior (i.e., for the hormonal organization of sex differences)? What is the role of brain aromatase as opposed to gonadal aromatase (estrogen production by the gonads)? This chapter will discuss what is known (or not known) to try to answer these questions.

Nearly all of the pertinent literature comes from two species: the Japanese quail (*Coturnix japonica*) and the zebra finch (*Taeniopygia guttata*). The Japanese quail is in the same family (Phasianidae) as the chicken (*Gallus domesticus*), a species with a sequenced genome. The neural and neuroendocrine mechanisms for its mating (copulatory) behavior have been extensively investigated (Balthazart et al., 2009). The zebra finch is in the estrildid finch family, and has a sequenced genome (Warren et al., 2010). The zebra finch brain has proven to be exceptionally valuable for studying neuroendocrine mechanisms for the sexually dimorphic production of learned vocalizations, especially songs (Schlinger and Brenowitz, 2002). Both species have great potential for future research directions involving genomic and molecular mechanisms. They differ markedly in phylogenetic position, with phasianids belonging to a relatively basal lineage of living birds, and estrildids to a relatively apical lineage. The two species are also very different in their social lives. Zebra finches are gregarious, colonially breeding, socially monogamous, and biparental (Zann, 1996). Quail are a bit of a mystery, because they are so poorly known in the wild, but are clearly not like zebra finches (Mills et al., 1997). They also differ in developmental mode; quail have precocial young, whereas zebra finches have altricial young. Quail are highly photoperiodic seasonal breeders; zebra finches less so (Perfito et al., 2007). Because these are such different kinds of birds, any discoveries found in both species are likely to generalize to other birds. Any differences in mechanisms are more difficult to interpret, however, and their generality is unclear. Birds are a species-rich group of vertebrates, and two species is a very sparse sample of the many branches on the avian tree. In the discussion that follows, there will be some similarities but also some differences in how sexual differentiation of behavior takes place and in the contribution of estrogen and aromatase to the process.

QUAIL MATING BEHAVIOR

The mating behavior of male Japanese quail is easy to study and measure. Males housed individually are usually quite sexually motivated and tend to initiate copulation very quickly when presented with another quail. In a successful mating sequence with a receptive female, the male grabs the feathers of the head or neck, mounts, and performs a cloacal contact movement, leaning back with wings outstretched to press his cloacal opening against the female's to achieve insemination. Following a successful mating, the male often fluffs his feathers and struts. All these behavioral components are dependent on testicular hormones. They are never seen in males whose gonads have been removed or are regressed due to exposure to short daylengths. They are restored by treatment with testosterone. Females never show any of this behavior, even if their ovaries are regressed and they are treated with testosterone. Thus, the sex difference in male-typical copulatory behavior is not an activational hormone phenomenon (i.e., merely the result of sex differences in circulating testosterone concentrations in adulthood) (Balthazart et al., 2009).

Female sexual behavior (including receptivity) is more subtle and harder to measure. A receptive female squats and holds still while the male is trying to copulate. Males normally do not display these behaviors indicative of sexual receptivity. It is clear that the sex difference in receptivity is an activational phenomenon, for adult males with regressed testes treated with estradiol show female-typical levels of receptivity to mating attempts by another male (Adkins, 1975). Although males with developed testes have circulating estradiol, the amounts are not sufficient to reliably activate receptive behavior (Watson et al., 1990).

Therefore, it is male copulatory behavior that is of greatest interest for understanding the role of aromatase and estrogens in sexual differentiation of behavior in this species. It has been known since the 1970s that estradiol has a strongly demasculinizing effect on this behavior during early development (Adkins, 1975, 1979). A single injection of estradiol benzoate into eggs given roughly midway through incubation (during embryonic development) results in males that fail

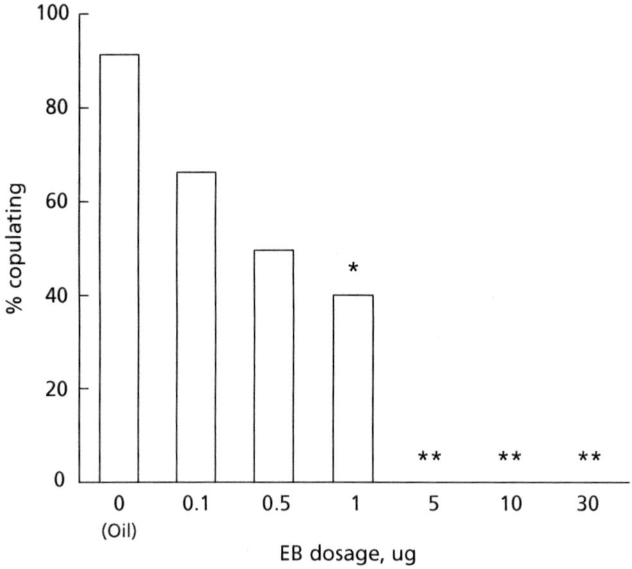

Figure 18–1. Male Japanese quail embryos are very sensitive to the demasculinizing effect of estrogen. Shown is the effect of a single injection of different dosages of estradiol benzoate given on day 10 of incubation on the percentage of adult males that copulated. * $p < 0.05$ compared with oil control; ** $p < 0.01$. Originally published in: Adkins, E.K. (1979). Effect of embryonic treatment with estradiol or testosterone on sexual differentiation of the quail brain: critical period and dose-response relationships. *Neuroendocrinology, 29,* 178–185. © 1979 S. Karger AG, Basel. Reprinted with permission.

to copulate as adults, even if given plenty of activational testosterone. Even a low dose of the estrogen produces a noticeable effect (Fig. 18–1).

Females are not affected behaviorally by being exposed to exogenous estradiol as embryos. The complementary experiment, in which eggs are injected with an estrogen synthesis inhibitor, results in females that show male-typical copulatory behavior as adults when given activational testosterone, which control females never do (Fig. 18–2) (Balthazart et al., 1992). These robust and highly replicable behavioral sex-reversing effects are paralleled by sex reversal of some sexually dimorphic features of brain regions implicated in reproductive behavior, most notably the vasotocinergic innervation of the preoptic area and bed nucleus of the stria terminalis (Panzica et al., 1998). Males have much more vasotocinergic innervation in those regions. Males from eggs injected with estradiol benzoate have female-typical levels, and females from eggs injected with an aromatase inhibitor have male-typical levels.

These results indicate a pattern of sexual differentiation in which male behavioral development (potential for the male-typical copulatory sequence to be activated in adulthood) occurs when embryonic levels of estrogens are low, and female behavioral development (no potential for male-typical copulation)

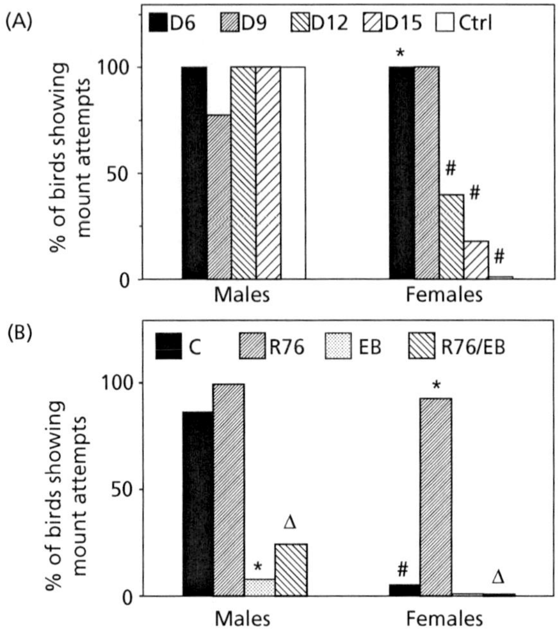

Figure 18–2. Female quail embryos treated with an aromatase inhibitor show male-typical copulatory behavior as adults, unlike control females. Males are not affected by the same treatment. Shown are the percentages of females and males that exhibited mount attempts in two experiments. A. Eggs were injected with the aromatase inhibitor R76713 on one of four incubation days (D) or with vehicle on D6 (Ctrl). Treatment on D6 or D9 (prior to or during the critical period for sexual differentiation) masculinized the females. * $p < 0.05$ compared with Ctrl females; # $p < 0.05$ compared with males given the same treatment. B. Eggs were injected on D9 with vehicle (C), R76713 (R76), estradiol benzoate (EB) or R76 + EB. Both treatments containing EB demasculinized males, and again R76713 masculinized females. * $p < 0.05$ compared to same-sex controls; # $p < 0.05$ compared to males given the same treatment; Δ $p < 0.05$ compared to birds given R76713 alone. Originally published in: Balthazart, J., & Adkins-Regan, E. (2002). Sexual differentiation of brain and behavior in birds. In: Pfaff, D. et al. (eds.), *Hormones, Brain and Behavior,* vol. 4. San Diego: Academic Press, pp. 223–301. © 2002 Academic Press (Elsevier). Reprinted with permission.

occurs when they are higher. Measurements of embryonic sex steroid levels during the critical period when hormone manipulations have their greatest effect also support this hypothesis. Both sexes of quail embryonic gonads produce both androgens and estrogens, but female ovaries express more aromatase, and female embryos produce quite a bit more estradiol (Schumacher et al., 1988; Ottinger et al., 2001; Koba et al., 2008).

Also consistent with a pattern of sexual differentiation in which estrogen is organizing the quail brain is that estrogen receptors (ERs) are present in the brains of embryos during the period when hormone treatments demasculinize males (Gahr and Balaban, 1996). More recent studies have addressed the question of whether those receptors are ERα or ERβ, and which is the mechanism for the organization of copulatory behavior. The evidence seems to be pointing to ERβ. A selective ERα agonist injected into eggs did not reduce copulatory behavior by males, and a selective ERα antagonist injected into eggs altered later copulatory behavior by females only at a high dose (Mattsson & Brunström, 2010). In addition, using in situ hybridization to localize the mRNAs, more ERβ than ERα was expressed in three embryonic brain regions known to be important for adult copulatory behavior: the medial preoptic nucleus, the medial bed nucleus of the stria terminalis, and nucleus taeniae of the amygdala (Fig. 18–3) (Axelsson et al., 2007).

Clearly aromatase and estrogens are key mechanisms for the differentiation of male-typical copulatory behavior potential during the embryonic development

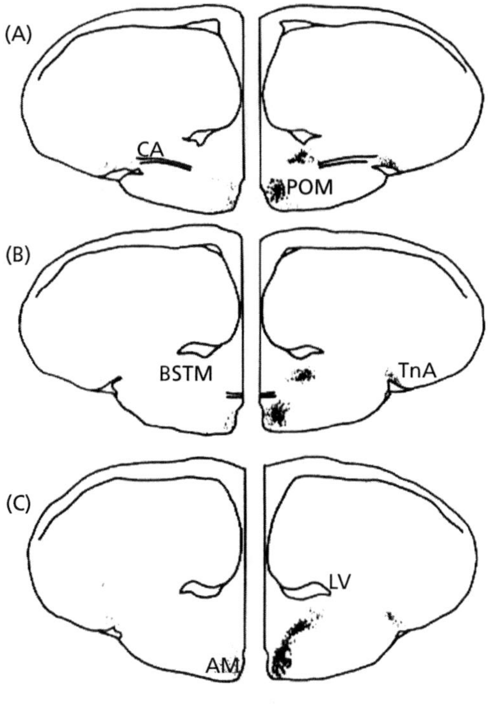

Figure 18–3. Quail embryos in late incubation have more mRNA expression for ERβ than for ERα in brain regions related to sexual behavior in adult males. Shown are schematics of embryonic brains from rostral (*A*) to caudal (*C*) with the distribution of ERα mRNA represented on the *left* and ERβ mRNA on the *right*. AM, nucleus anterior medialis hypothalami; BSTM, medial bed nucleus of the stria terminalis; CA, anterior commissure; LV, lateral ventricle; POM, medial preoptic nucleus; TnA, nucleus taeniae of the amygdala. Originally published in: Axelsson, J., Mattsson, A., Brunstrom, B., & Halldin, K. (2007). Expression of estrogen receptor-alpha and -beta mRNA in the brain of Japanese quail embryos. *Dev Neurobiol, 67,* 1742–1750. © 2007 Wiley-Blackwell. Reprinted with permission.

of Japanese quail. But aromatase and estrogen production where? What is the role of brain aromatase, as distinct from gonadal aromatase?

The sexual differentiation sequence of quail begins with differentiation of the gonads themselves, in which a gonadal primordium becomes a testis or an ovary. Based on experiments with chickens, and to a lesser extent quail, there is every reason to think that aromatase gene expression and production are both an important part of the molecular cascade responsible for ovarian differentiation and also a result of this process. The crucial piece of evidence for the former is that treatment of embryos with an estrogen synthesis inhibitor very early in incubation, before the gonads have differentiated, results in testicular development in genetic females (Elbrecht and Smith, 1992; Andrews et al., 1997; Koba et al., 2008). Once differentiated, ovaries continue to have higher aromatase expression throughout embryonic development (Koba et al., 2008; Nakamura et al., 2008).

Is this sex difference in embryonic gonadal aromatase and gonadal estrogen production sufficient by itself to explain the hormonal organization of male-typical copulatory behavior? What are the brain aromatase levels during embryonic development, and are neuroestrogens playing a role as well? It is not possible to remove the gonads during embryonic development, prior to the critical period for sexual differentiation, and have embryos survive to produce adult behavior. That technical limitation makes it difficult to know through a direct experimental approach what brain aromatase is doing for later behavior independently of the gonads.

Aste et al. (2010), using immunocytochemistry, found aromatase positive cells in the medial preoptic nucleus and medial and lateral bed nucleus of the stria terminalis in 9, 11, and 15 day old embryos. There were more aromatase positive cells in female than male brains at ages 9 and 11 days (corresponding to the heart of the critical period for sexual differentiation of copulatory potential), but no sex difference later on. Male embryos in eggs injected on day 9 with estradiol had female-typical numbers of aromatase positive cells two days later (Fig. 18–4). This suggests that the higher female aromatase cell numbers of untreated embryos are caused by their higher gonadal aromatase and estrogen production.

What might this brain aromatase be doing? Because it is upregulated by estrogen, one possibility is that it is contributing to female brain sexual differentiation by amplifying the estrogenic signal from the ovary to produce an even greater difference between the sexes in brain estrogens levels than would occur from gonadal estrogens production alone. In addition, because it is present in both sexes, it could be contributing to the establishment of mechanisms for future receptive behavior potential in both sexes. Evidence against this hypothesis is that adult receptivity is not affected by treatment of embryos with an aromatase inhibitor (Adkins-Regan, 1985). Alternatively, it could be having more general neurotrophic effects unrelated to the sexual differentiation of behavior.

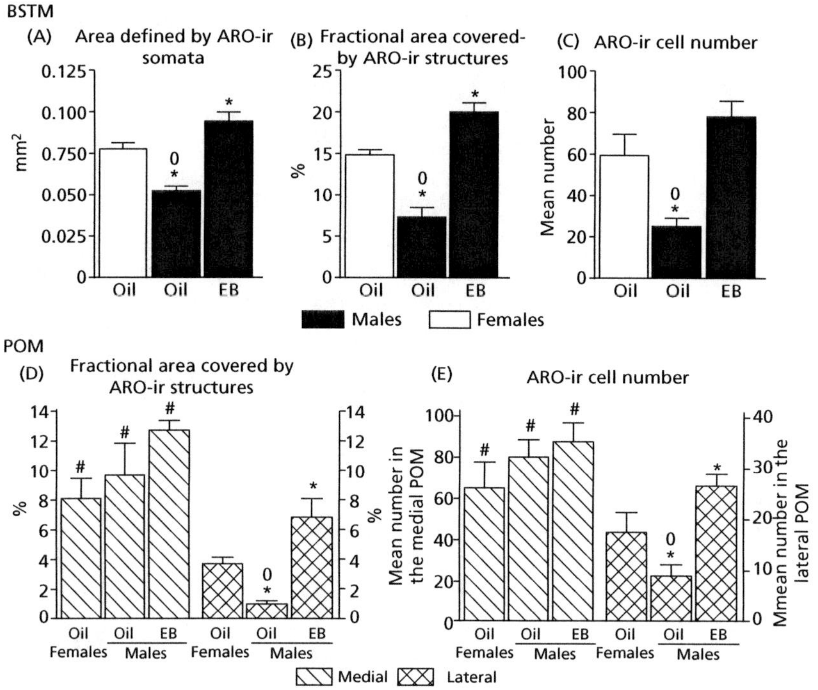

Figure 18–4. Aromatase in quail embryos is upregulated by estradiol during the critical period for behavioral sexual differentiation. Shown are measures of aromatase-immunoreactivity (ARO-ir) in two brain regions, the bed nucleus of the stria terminalis (BST) and the medial preoptic nucleus (POM; medial and lateral parts of the nucleus considered separately), on embryonic day 11 in male and female control embryos from eggs injected with oil on day 9, and male embryos from eggs injected with estradiol benzoate (EB) on day 9. * $p < 0.05$ compared with oil females; 0 $p < 0.01$ compared with EB treated males; # $p < 0.01$ compared with lateral POM in embryos of the same sex and treatment. Originally published in: Aste, N., Watanabe, Y., Harada, N., & Saito, N. (2010). Distribution and sex differences in aromatase-producing neurons in the brain of Japanese quail embryos. *J Chem Neuroanat, 39,* 272–288. © 2010 Elsevier. Reprinted with permission.

Treatment of quail embryos with testosterone also demasculinizes males, through aromatization of the testosterone (Balthazart et al., 2009). Both ovaries and testes of embryos produce testosterone (Schumacher et al., 1988; Ottinger et al., 2001). The embryonic testicular testosterone, while not sufficient to interfere with the normal development of copulatory potential in males, nonetheless does slightly demasculinize them. This was shown in an experiment in which males from eggs injected with the aromatase inhibitor ATD (1,4,6-androstatrien-3-,17-dione) on day 9 showed more copulatory behavior than controls as adults when exposed to short days to regress the testes and given testosterone replacement (Adkins-Regan, 1985). This slight demasculinization resulting from endogenous testosterone is due to aromatization, but again, one cannot know whether the relevant aromatase is in the gonads or the brain or both.

One of the most striking features of estrogens and quail sexual differentiation is that early embryonic exposure deletes copulatory potential, whereas adult estrogens facilitate copulation (Balthazart et al., 2009). The mechanisms responsible for this polarity reversal in the effect of estrogens on copulation are unknown. A similar phenomenon is seen in mammalian sexual differentiation, in which early estrogen exposure reduces future receptivity potential, but adult estrogen facilitates it. In rats, progress has been made identifying the cellular level mechanisms by which estradiol acts, both during sexual differentiation and in adulthood, that will eventually explain why this polarity reversal occurs (Wright et al., 2010). Some of these same mechanisms could be involved in quail sexual differentiation, and suggest hypotheses worth testing.

In adulthood males have more aromatase immunoreactive cells than females in the preoptic area and bed nucleus of the stria terminalis (Balthazart et al., 1996). The neuroestrogens produced by this aromatase facilitate copulation. Is this sex difference in adult aromatase itself hormonally organized by estrogen (embryonic aromatase)? Treatment of embryos with estradiol or an aromatase inhibitor alters the adult aromatase levels (Cornil et al., 2011). That does not seem to be the cause of the sex difference in copulation, however. All quail, regardless of sex or embryonic treatment have the same aromatase levels as normal males once they are given testosterone as adults to provide them with male-typical hormone levels, yet the females and early estrogen treated males still fail to copulate. Thus, the sex difference in adult brain aromatase is a largely activational phenomenon (upregulation by testosterone) that is not causally related to sexual differentiation of copulatory potential (Cornil et al., 2011).

ZEBRA FINCH SINGING

Male zebra finches sing frequently, and sing regardless of whether they are actively breeding at the time. Female zebra finches never sing even though they frequently emit other kinds of vocalizations (calls). The dichotomous sex difference in singing is based on an extraordinary degree of sexual dimorphism in the telencephalic neural song system, the set of interconnected nuclei dedicated to the production and perception of song (Nottebohm and Arnold, 1976). The dimorphism is apparent at all levels of neural characteristics, from the overall sizes of the nuclei to the sizes of the somas of the individual neurons and of the projections from one nucleus to another (reviewed in Balthazart and Adkins-Regan [2002] and Wade [2001]).

It has been known since the experiment of Gurney and Konishi (1980) that treatment of female zebra finch chicks (nestlings) with estradiol has a strongly masculinizing effect on the same characteristics of their song systems that are sexually dimorphic in normal birds. In their study and others carried out in several laboratories, many of these treated females were capable of singing fairly normal (male-typical) learned songs as adults (see Fig. 18–5, for example). Gurney and Konishi (1980) interpreted their results as indicating hormonal organization of the masculine song system during an early posthatching critical period, with estrogens as the organizing hormones. Treatment during the first month after hatching, especially the first week, is the most effective (Adkins-Regan et al., 1994; Balthazart and Adkins-Regan, 2002).

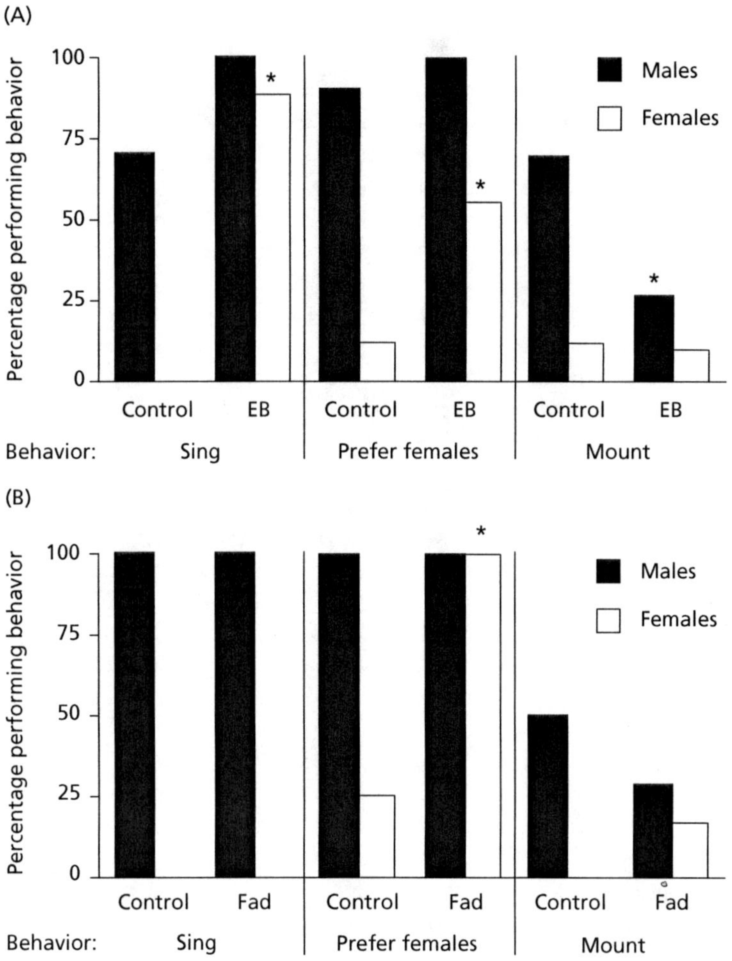

Figure 18–5. Treatment of zebra finch nestlings with estradiol benzoate (EB) masculinizes singing and sexual partner preference of females while demasculinizing the mounting behavior of males. Fadrozole (FAD) treatment during this same period does not affect the development of singing or mounting, however. Originally published in: Balthazart, J., & Adkins-Regan, E. (2002). Sexual differentiation of brain and behavior in birds. In: Pfaff, D. et al. (eds.), *Hormones, Brain and Behavior,* vol. 4. San Diego: Academic Press, pp. 223–301. © 2002 Academic Press (Elsevier). Reprinted with permission.

If this hypothesis is correct, then endogenous estradiol levels should be higher in males during the time period when exogenous estrogen treatment masculinizes females. There have been at least three studies measuring circulating estradiol levels in chicks, but two of the three have yielded negative results (Hutchison et al., 1984; Adkins-Regan et al., 1990; Schlinger and Arnold, 1992). One reason for this absence of sexual dimorphism in circulating estradiol is possibly that most of the steroid is coming from the brain, where aromatase levels are high in both sexes (Schlinger and Arnold, 1992).

Another approach to finding out if estrogens are organizing song system differentiation is to see if there are estrogen receptors in the right places at the right time. Also, if the numbers of ERs are dimorphic, that could produce organization of the sex difference through an estrogenic mechanism without requiring dimorphism in the circulating steroid level. In situ hybridization showed expression of mRNAs for ERα and ERβ in the arcopallium (where song system nucleus RA is located) and diencephalon beginning before hatching and on posthatching day 1 (when estradiol treatment will masculinize the female song system) but no sex differences were seen (Perlman and Arnold, 2003). Estrogen receptor antagonists have not worked very well to demasculinize the male song system, but that could be because they are not sufficiently good antagonists (Mathews and Arnold, 1991). Bender and Veney (2008) used a newer drug (ICI 182,780) and obtained a greater degree of demasculinization, but the effect was still quite modest relative to the large sex difference in untreated birds, and occurred in both sexes instead of being limited to males (Fig. 18–6).

Meanwhile, evidence began accumulating that gonadal steroids were not an important part of the developmental pathway for song system sexual differentiation at all. Most problematic for the idea of organization by gonadal steroids was that genetic female zebra finches with testes had song systems just like those of control females. In zebra finches, as in chickens, females with testes can be produced by injecting the eggs with fadrozole early in incubation, as the gonads are beginning to differentiate. Gong et al. (1999) and Wade et al. (1999) treated birds as embryos, nestlings, or both with fadrozole and found partial or substantial sex reversal of the gonads of females but without any masculinization of the neural song system. If sex steroids like estradiol are the signal for masculine organization, the dimorphism in the signal would have to come from somewhere else, independently of the gonads.

Because brain aromatase levels are high in zebra finches of all ages, local estrogen production in or near the song system nuclei becomes a candidate organizing signal. Support for this hypothesis came from an experiment by Holloway and Clayton (2001), in which the development of the projection from neurons in HVC (the most dorsal song system nucleus) to RA (the robust nucleus of the arcopallium) was studied in brain slices cultured in vitro. A large projection normally develops in male but not female brains. Application of estradiol or of a male brain slice to dishes containing female brain slices resulted in projections forming from HVC to RA, and application of fadrozole or the estrogen receptor antagonist tamoxifen to those containing male brain slices

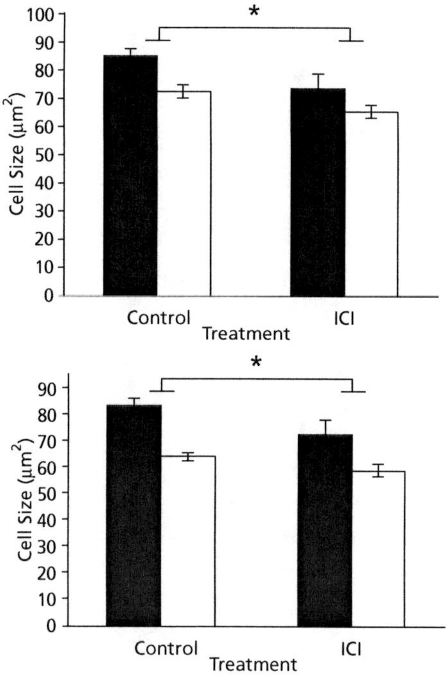

Figure 18–6. Treatment of very young zebra finches with a potent estrogen receptor antagonist (ICI 182,780) for 25 days significantly reduces neuron soma sizes in song system nuclei HVC (*top*) and RA (*bottom*) in both sexes (black bars, males; white bars, females). * $p < 0.05$ (main effect of treatment, no interaction with sex). Originally published in: Bender, A. T., & Veney, S. L. (2008). Treatment with the specific estrogen receptor antagonist ICI 182,780 demasculinizes neuron soma size in the developing zebra finch brain. *Brain Res, 1246,* 47–53. © 2008 Elsevier. Reprinted with permission.

prevented the connection from forming (Fig. 18–7). Measurements of estradiol in dishes containing untreated brain slices confirmed that more estradiol was being produced by male brain slices. These results provide strong support for the hypothesis that locally produced estradiol organizes the HVC to RA projection. Because the brain slices were collected from 25-day-old chicks, however, the experiment does not shed light on whether locally produced estrogens are responsible for the development of all the other dimorphic song system characteristics, such as overall nucleus volumes. Those have appeared before 25 days and are masculinized in females by estradiol treatment only when treatment is begun closer in time to hatching.

Nonetheless, the success of this experiment, together with the high brain estrogen production throughout life in this species, have made aromatase a continued focus of attention. Several studies have either measured aromatase activity at a finer scale both spatially (individual brain regions) and temporally (different time periods in early development), or have administered aromatase

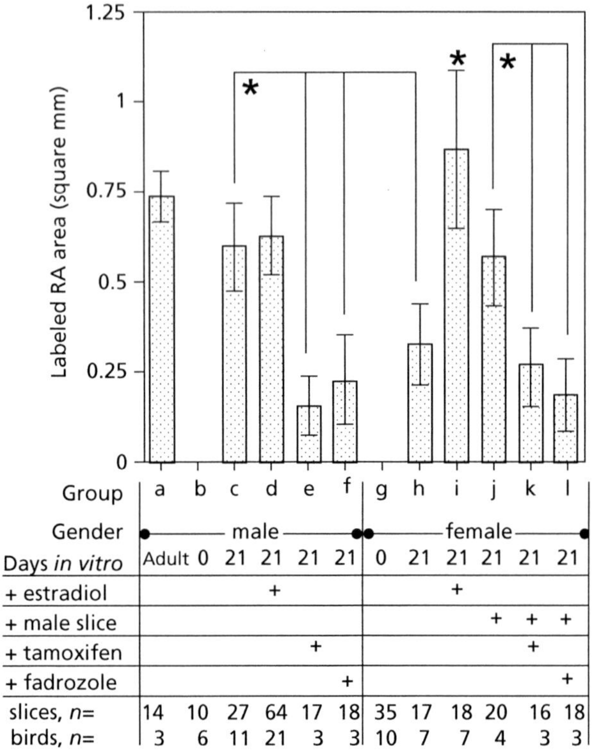

Figure 18–7. Estrogen is involved in the development of the projection from song system nucleus HVC to RA. Shown is the RA area containing fiber terminals from HVC in cultured brain slices from male or female zebra finch brains given different treatments or from uncultured brain slices from adult males (group a). * significantly different. Originally published in: Holloway, C. C., & Clayton, D. E. (2001). Estrogen synthesis in the male brain triggers development of the avian song control pathway in vitro. *Nat Neurosci, 4,* 170–175. © 2001 Nature Publishing Group. Reprinted with permission.

inhibitors during early development. In situ hybridization showed expression of mRNA for aromatase in the arcopallium and diencephalon beginning before hatching and on posthatching day 1 but with no sex differences (Perlman and Arnold, 2003). The review by London et al. (2009) concluded that although there is plenty of aromatase activity present in the diencephalon and telencephalon from posthatching day 1, it is not seen in most of the song nuclei, and levels do not seem to be different between the sexes (see also Balthazart et al., 1996).

Even with newer methods, however, it is still challenging to get enough spatial and temporal resolution in aromatase measurement to test a local estrogen production hypothesis. Thus, there is a need for experiments that give aromatase inhibitors to males to see if masculinization is prevented. More than one laboratory has done such experiments, and the results have all been resoundingly negative. Balthazart et al. (1994) gave nestlings implants of the aromatase inhibitor R76713 at 2 to 3 days posthatching. As adults the males sang less but the song system nucleus volumes were not different from those of controls. Wade and Arnold (1994) injected nestlings with fadrozole on posthatching days 1 to 30, and found no effect on song system nuclei, even though fadrozole is a very effective brain aromatase inhibitor (Wade et al., 1994). Springer and Wade (1997) injected eggs with fadrozole on day 5 of incubation. Fadrozole treated males sang more, not less, and their song system nucleus sizes were not different from those of control males. Adkins-Regan et al. (1996) injected nestlings with fadrozole for either 1 or 2 weeks after hatching and found no effect on male singing (see Fig. 18–5).

Why does exogenous estradiol work as well as it does to masculinize the female song system if fadrozole has so little effect on males? One possibility is that a very sensitive yet dimorphic estrogenic mechanism or pathway of action is involved, one requiring only a small amount of estrogen, the amount that is still produced even in fadrozole treated males. If so, then why don't females' endogenous estrogens masculinize them (Schlinger and Arnold, 1992)? Why is testosterone treatment of male nestlings so ineffective, given their high brain aromatase levels (Adkins-Regan and Ascenzi, 1990; Grisham and Arnold, 1995)?

Perhaps the sexually dimorphic component of the signal does not lie in circulating estrogen, brain aromatase, or brain ERs. Duncan & Carruth (2007) proposed that the dimorphic mechanism is an estrogen receptor coactivator. Their study identified the coactivator L7/SPA in song nuclei (among other places) and found sex differences favoring males in its mRNA and protein expression at both nestling and adult ages. Whether this mechanism plays a causal role in sexual differentiation of the song system is not yet known, and finding out will require experimental manipulation of this coactivator.

Another possibility for a dimorphic sensitive estrogenic mechanism is a membrane ER that is present at higher levels in male brains (Acharya and Veney, 2011). Or, the nongonadal signal (factors) might not be steroids at all (Schlinger, 1998). Those factors could be affected by exogenous estradiol treatment, especially by the supra-physiological brain levels those treatments probably produce. Finally, there is the intriguing possibility that the factor(s) are nonsteroidal products of genes expressed locally but only or more in one sex (such as genes on the sex chromosomes), the "direct" genetic hypothesis (Arnold et al., 2004; Chen et al., 2005).

ZEBRA FINCH MATING

Copulatory behavior is harder to study in zebra finches than in Japanese quail. Zebra finches are socially monogamous and pairs (or unpaired birds forming new pairs) often do not copulate at all during defined testing or observation periods. Nonetheless, even with lower behavior frequencies it is clear that males do more mounting than females. In addition, it seems reasonable to suppose that copulation in all bird species is evolutionarily homologous and similarly dependent on steroid sensitive neurons in the hypothalamus and preoptic area. If that is the case, then it would also be reasonable to hypothesize that similar estrogenic mechanisms would organize copulation in zebra finches and quail.

This hypothesis initially received support from a study in which zebra finch nestlings were injected with estradiol benzoate or control vehicle for the first 2 weeks post hatching (Adkins-Regan and Ascenzi, 1987). As adults they were gonadectomized and given implants of either testosterone propionate or estradiol benzoate. As expected, control males mounted more than control females. Estradiol treated females did not mount more than control females, but estradiol treated males mounted less than control males (see Fig. 18–5). This pattern of results is also seen in quail experiments. The main difference is the time period for the effect, posthatching versus embryonic, but that is consistent with the altricial versus precocial developmental mode of these two species.

Subsequent experiments have not supported a quail model for sexual differentiation of zebra finch mounting, however. Balthazart et al. (1994) gave nestlings implants of the aromatase inhibitor R76713 at 2 to 3 days post hatching. The treated females did not mount more than controls (unlike treated female quail), and male mounting was unaffected. Adkins-Regan et al. (1996) injected nestlings with fadrozole for either one or two weeks after hatching and found no effect on mounting in either sex (see Fig. 18–5). Springer and Wade (1997) injected eggs with fadrozole on day 5 of incubation. Mounting was not affected in either sex, even though the females had testes instead of ovaries, which rules out a mammalian model of sexual differentiation as well. At this time, therefore, it is quite unclear whether estrogens or aromatase are involved in the normal course of sexual differentiation of zebra finch mating.

ZEBRA FINCH SEXUAL PARTNER PREFERENCE

Sexual partner preference refers to animals' preferences for mating or pairing with the other sex versus the same sex when given a choice. Such preferences are usually sex-typical (males prefer females and females prefer males) and, therefore, highly sexually dimorphic. Most birds are socially monogamous and form pair bonds, providing opportunities for studying pairing preferences that are expressed through a repertoire of social behavior extended over time and not simply through copulation alone. Zebra finches have proven to be excellent subjects for studying this kind of sexual partner preference. They form long-lasting pair bonds that are expressed through easily observable behavior that among adults is exhibited only between paired birds, specifically clumping (sitting touching each other), allopreening (preening each other), and spending significant time in a nest box together. In wild flocks and laboratory colonies, pairs are easily identified. In the laboratory, sexual partner preference can be assessed through two-choice tests (with proximity to a male vs. female stimulus as the measure) or by observing the pair formation process among initially unpaired birds placed in an aviary together.

Estrogen and aromatase seem to be involved in the development of sexually differentiated partner preference, although exactly how is not clear and other factors also are involved in addition to steroids. In the study by Adkins-Regan and Ascenzi (1987) referred to previously in connection with mating behavior, the estradiol treated females were seen to pair with other females instead of males, as if their sexual partner preference had been masculinized (see Fig. 18–5). In a systematic investigation of this effect, females were injected with estradiol benzoate (EB) for the first 2 weeks posthatching, given testosterone implants as adults, and assessed behaviorally in both two-choice tests and aviary observations (Mansukhani et al., 1996). Half the females in each group (EB and control) lived in all-female aviaries from 40 to 100 days (from independence from the parents to testing) and half lived in mixed-sex aviaries. In two-choice tests, both EB treatment and all-female living produced preferences for other females. When allowed to pair in aviaries, however, only the females that experienced both EB and all-female living paired with other females. The results suggested that the females had been masculinized by the EB (similar to masculinization of singing) but that this masculinization also required certain social conditions during juvenile development.

As in the case of singing, a mammalian model of sexual differentiation does not turn out to fit zebra finch sexual partner preference. Adkins-Regan et al. (1996) injected nestlings with fadrozole for either 1 or 2 weeks after hatching and found that treated males did not have altered sexual partner preference as would be predicted from a mammalian hypothesis (see Fig. 18–5). They still preferred females. Instead, the one-week treatment produced females that preferred other females, but only in two-choice tests and not in aviary pairing observations. In a later study, eggs were injected on day 5 of incubation with fadrozole, producing females with testicular tissue (Adkins-Regan and Wade, 2001). As adults, these females strongly preferred to pair with other females in aviary observations (see Fig. 18–8). This is the only time this degree of sex reversal of behavior has been seen in zebra finches with a treatment that reduced estrogen production. There are several possibilities for why this effect occurred. One is that the testicular tissue produced altered (nonovarian) circulating hormone levels that were causing the result. Those hormone levels would have been altered throughout development and up to the time of testing. This makes it difficult to know whether the sex reversal in partner preference resulted from hormonal organization, activation, or both, but altered hormone levels throughout development might account for the magnitude of the effect. Another possibility is a direct effect of the fadrozole injection on brain sexual differentiation. That would require a very early critical period for the relevant brain mechanisms, however.

All told, some kind of estrogenic mechanism seems to be important for the differentiation of partner preference, either directly (on the brain) or indirectly (through altered gonadal differentiation), but exactly what is obscure. Nor is it known what the relevant brain regions are that undergo sexual differentiation for partner preference.

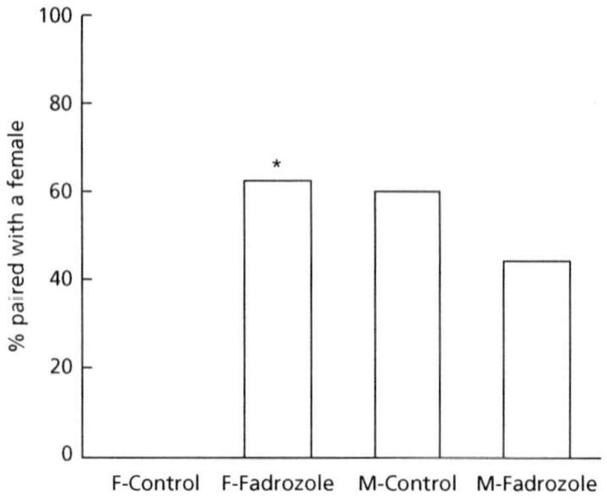

Figure 18–8. Female (F) zebra finches hatched from eggs treated with fadrozole early in incubation strongly prefer to pair with other females as adults, whereas males (M) are not significantly affected. * $p < 0.05$ compared to control females hatched from eggs injected with saline. Originally published in: Adkins-Regan, E., & Wade, J. (2001). Masculinized sexual partner preference in female zebra finches with sex-reversed gonads. *Horm Behav*, *39*, 22–28. © 2001 Elsevier. Reprinted with permission.

CONCLUSIONS

In both Japanese quail and zebra finches, early treatment with estradiol has powerful sex-reversing effects on behavior (copulatory mounting, singing, sexual partner preference) consistent with an important organizational role in sexual differentiation. In quail embryos, aromatase level is itself sexually differentiated in both gonads and brain, with more enzymatic activity or protein immunoreactivity in ovaries and female brains. Aromatase has been shown to be a key mechanism in sexual differentiation of copulatory behavior, but the role of embryonic brain aromatase (as distinct from gonadal aromatase) is not well understood. What is needed are manipulations to disable aromatase in the brains but not gonads of embryos or vice versa. Also not understood, is why embryonic estrogen reduces future copulation while adult estrogen promotes it. In zebra finches, early estrogen treatment masculinizes singing and sexual partner preference while demasculinizing copulatory mounting. Neither estradiol levels nor aromatase activity have yet been found to be dimorphic during the time when estrogen treatment is maximally effective, however. In addition, early treatment with a good aromatase inhibitor (fadrozole) does not prevent males from undergoing normal masculinization. If estrogen production by aromatase is part of the pathway for zebra finch sexual differentiation, either the mechanism(s) that turn it into a sexually dimorphic signal must lie at or after the estrogen receptor level, or the crucial estrogen production is more localized spatially and temporally than has been detected so far. There are as many questions as answers from research with these two avian species. Both have great potential for future discoveries, however—quail, because their developing brain is so sensitive to estrogen, and zebra finches, because their developing brains have so much aromatase.

REFERENCES

Acharya, K. D., & Veney, S. L. (2011). Characterization of the G-protein coupled membrane bound estrogen receptor GPR30 in the zebra finch brain reveals a sex difference in gene and protein expression. *Dev Neurobiol*, accepted manuscript online: 2011 DEC 21, DOI: 10.1002/dneu.22004

Adkins, E. K. (1975). Hormonal basis of sexual differentiation in the Japanese quail. *J Comp Physiol Psychol, 89,* 61–71.

Adkins, E. K. (1979). Effect of embryonic treatment with estradiol or testosterone on sexual differentiation of the quail brain: Critical period and dose-response relationships. *Neuroendocrinology,* 29, 178–185.

Adkins-Regan, E. (1985). Embryonic exposure to an aromatization inhibitor increases copulatory behavior of male quail. *Behav Processes, 11,* 153–158.

Adkins-Regan, E., Abdelnabi, M., Mobarak, M., & Ottinger, M. A. (1990). Sex steroid levels in developing and adult male and female zebra finches (*Poephila guttata*). *Gen Comp Endocrinol, 78,* 93–109.

Adkins-Regan, E., & Ascenzi, M. (1987). Social and sexual behaviour of male and female zebra finches treated with oestradiol during the nestling period. *Anim Behav, 35,* 1100–1112.

Adkins-Regan, E., & Ascenzi, M. (1990). Sexual differentiation of behavior in the zebra finch: effect of early gonadectomy or androgen treatment. *Horm Behav, 24,* 114–127.

Adkins-Regan, E., Mansukhani, V., Seiwert, C., & Thompson, R. (1994). Sexual differentiation of brain and behavior in the zebra finch: critical periods for effects of early estrogen treatment. *J Neurobiol, 25,* 865–877.

Adkins-Regan, E., & Wade, J. (2001). Masculinized sexual partner preference in female zebra finches with sex-reversed gonads. *Horm Behav,* 39, 22–28.

Adkins-Regan, E., Yang, S., & Mansukhani, V. (1996). Behavior of male and female zebra finches treated with an estrogen synthesis inhibitor as nestlings. *Behaviour, 133,* 847–862.

Andrews, J. E., Smith, C. A., & Sinclair, A. H. (1997). Sites of estrogen receptor and aromatase expression in the chicken embryo. *Gen Comp Endocrinol, 108,* 182–190.

Arnold, A. P., Xu, J., Grisham, W., Chen, X. Q., Kim, Y. H., & Itoh, Y. (2004). Minireview: sex chromosomes and brain sexual differentiation. *Endocrinology, 145,* 1057–1062.

Aste, N., Watanabe, Y., Harada, N., & Saito, N. (2010). Distribution and sex differences in aromatase-producing neurons in the brain of Japanese quail embryos. *J Chem Neuroanat, 39,* 272–288.

Axelsson, J., Mattsson, A., Brunstrom, B., & Halldin, K. (2007). Expression of estrogen receptor-alpha and -beta mRNA in the brain of Japanese quail embryos. *Dev Neurobiol, 67,* 1742–1750.

Balthazart, J., Absil, P., Fiasse, V., & Ball, G. F. (1994). Effects of the aromatase inhibitor R76713 on sexual-differentiation of brain and behavior in zebra finches. *Behaviour, 131,* 225–260.

Balthazart, J., Absil, P., Foidart, A., Houbart, M., Harada, N., & Ball, G. F. (1996). Distribution of aromatase-immunoreactive cells in the forebrain of zebra finches (*Taeniopygia guttata*): Implications for the neural action of steroids and nuclear definition in the avian hypothalamus. *J Neurobiol, 31,* 129–148.

Balthazart, J., & Adkins-Regan, E. (2002). Sexual differentiation of brain and behavior in birds. In: Pfaff, D. et al. (eds.), *Hormones, Brain and Behavior*, vol. 4 (pp. 223–301). San Diego: Academic Press.

Balthazart, J., Arnold, A., & Adkins-Regan, E. (2009). Sexual differentiation of brain and behavior in birds. In: Pfaff, D. et al., (eds.), *Hormones, Brain and Behavior,* 2nd ed., vol. 4 (pp. 1745–1787) San Diego: Academic Press (Elsevier).

Balthazart, J., DeClerk, A., & Foidart, A. (1992). Behavioral demasculinization of female quail is induced by estrogens—studies with new aromatase inhibitor, R76713. *Horm Behav, 26,* 179–203.

Balthazart, J., Tlemçani, O., & Harada, N. (1996). Localization of testosterone-sensitive and sexually dimorphic aromatase-immunoreactive cells in the quail preoptic area. *J Chem Neuroanat, 11,* 147–171.

Bender, A. T., & Veney, S. L. (2008). Treatment with the specific estrogen receptor antagonist ICI 182,780 demasculinizes neuron soma size in the developing zebra finch brain. *Brain Research, 1246,* 47–53.

Chen, X. Q., Agate, R. J., Itoh, Y., & Arnold, A. P. (2005). Sexually dimorphic expression of trkB, a Z-linked gene, in early posthatch zebra finch brain. *Proc Natl Acad Sci U S A, 102,* 7730–7735.

Cornil, C. A., Ball, G. F., Balthazart, J., & Charlier, T. D. (2011). Organizing effects of sex steroids on brain aromatase activity in quail. *PLoS One, 6,* Article Number: e19196. DOI: 10.1371/journal.pone.0019196

Duncan, K. A., & Carruth, L. L. (2007). The sexually dimorphic expression of L7/SPA, an estrogen receptor coactivator, in zebra finch telencephalon. *Dev Neurobiol, 67,* 1852–1866.

Elbrecht, A., & Smith, R. G. (1992). Aromatase enzyme-activity and sex determination in chickens. *Science, 255,* 467–470.

Forlano, P. M., Schlinger, B. A., & Bass, A. H. (2006). Brain aromatase: New lessons from non-mammalian model systems. *Front Neuroendocrinol, 27,* 247–274.

Gahr, M., & Balaban, E. (1996). The development of a species difference in the local distribution of brain estrogen receptive cells. *Dev Brain Res, 92,* 182–189.

Gong, A., Freking, F. W., Wingfield, J., Schlinger, B. A., & Arnold, A. P. (1999). Effects of embryonic treatment with fadrozole on phenotype of gonads, syrinx, and neural song system in zebra finches. *Gen Comp Endocrinol, 115,* 346–353.

Grisham, W., & Arnold, A. P. (1995). A direct comparison of the masculinizing effects of testosterone, androstenedione, estrogen, and progesterone on the development of the zebra finch song system. *J Neurobiol, 26,* 163–170.

Gurney, M. E., & Konishi, M. (1980). Hormone-induced sexual differentiation of brain and behavior in zebra finches. *Science, 208,* 1380–1383.

Holloway, C. C., & Clayton, D. E. (2001). Estrogen synthesis in the male brain triggers development of the avian song control pathway in vitro. *Nat Neurosci, 4,* 170–175.

Hutchison, J. B., Wingfield, J. C., & Hutchison, R. E. (1984). Sex differences in plasma concentrations of steroids during the sensitive period for brain differentiation in the zebra finch. *J Endocrinol, 103,* 363–369.

Koba, N., Mori, M., Ha, Y., Mizushima, S., Tsukada, A., Saito, N., et al. (2008). Effects of aromatase inhibitor (Fadrozole)-induced sex-reversal on gonadal differentiation and mRNA expression of p450arom, AMH and ER alpha in embryos and growth in posthatching quail. *J Poult Sci, 45,* 116–124.

London, S. E., Remage-Healey, L., & Schlinger, B. A. (2009). Neurosteroid production in the songbird brain: A re-evaluation of core principles. *Front Neuroendocrinol, 30,* 302–314.

Mansukhani, V., Adkins-Regan, E., & Yang, S. (1996). Sexual partner preference in female zebra finches: the role of early hormones and social environment. *Horm Behav, 30,* 506–513.

Mathews, G. A., & Arnold, A. P. (1991). Tamoxifen's effects on the zebra finch song system are estrogenic, not antiestrogenic. *J Neurobiol*, 22, 957–969.

Mattsson, A., & Brunström, B. (2010). Effects on differentiation of reproductive organs and sexual behaviour in Japanese quail by excessive embryonic ER alpha activation. *Reprod Fertil Dev, 22,* 416–425.

Mattsson, A., Mura, E., Brunström, B., Panzica, G., & Halldin, K. (2008). Selective activation of estrogen receptor alpha in Japanese quail embryos affects reproductive organ differentiation but not the male sexual behavior or the parvocellular vasotocin system. *Gen Comp Endocrinol, 159,* 150–157.

Mills, A. D., Crawford, L. L., Domjan, M., & Faure, J. M. (1997). The behavior of the Japanese or domestic quail (Coturnix japonica). *Neurosci Biobehav Rev, 21,* 261–281.

Nakamura, K., Shibuya, K., Saito, N., Shimada, K., Ohshima, A., Hirai, T., et al. (2008). mRNA expression of cytochrome P450 17 alpha hydroxylase, cytochrome P450 aromatase, anti-mullerian hormone, estrogen receptor alpha, and androgen receptor in developing gonads of Japanese Quail. *J Poult Sci, 45,* 298–302.

Nottebohm, F., & Arnold, A. P. (1976). Sexual dimorphism in vocal control areas of songbird brain. *Science, 194,* 211–213.

Ottinger, M. A., Pitts, S., & Abdelnabi, M. A. (2001). Steroid hormones during embryonic development in Japanese quail: plasma, gonadal, and adrenal levels. *Poult Sci, 80,* 795–799.

Panzica, G. C., Castagna, C., Viglietti-Panzica, C., Russo, C., Tlemçani, O., & Balthazart, J. (1998). Organizational effects of estrogens on brain vasotocin and sexual behavior in quail. *J Neurobiol, 37,* 684–699.

Perfito, N., Zann, R. A., Bentley, G.E., & Hau, M. (2007). Opportunism at work: habitat predictability affects reproductive readiness in free-living zebra finches. *Funct Ecol, 21,* 291–301.

Perlman, W. R., & Arnold, A. P. (2003). Expression of estrogen receptor and aromatase mRNAs in embryonic and posthatch zebra finch brain. *J Neurobiol, 55,* 204–219.

Saldanha, C. J., Schultz, J. D., London, S. E., & Schlinger, B. A. (2000). Telencephalic aromatase but not a song circuit in a sub-oscine passerine, the golden-collared manakin (Manacus vitellinus). *Brain Behav Evolut, 56,* 29–37.

Schlinger, B. A. (1998). Sexual differentiation of avian brain and behavior: current views on gonadal hormone-dependent and independent mechanisms. *Ann Rev Physiol, 60,* 407–429.

Schlinger, B. A., & Arnold, A. P. (1992). Plasma sex steroids and tissue aromatization in hatchling zebra finches—implications for the sexual-differentiation of singing behavior. *Endocrinology, 130,* 289–299.

Schlinger, B. A., & Brenowitz, E. (2002). Neural and hormonal control of birdsong. In: Pfaff, D. et al. (eds.), *Hormones, Brain and Behavior,* vol. 2. San Diego: Academic Press, pp. 799–839.

Schumacher, M., Sulon, J., & Balthazart, J. (1988). Changes in serum concentrations of steroids during embryonic and post-hatching development of male and female Japanese quail (Coturnix coturnix japonica). *J Endocrinol, 118,* 127–134.

Springer, M. L., & Wade, J. (1997). The effects of testicular tissue and prehatching inhibition of estrogen synthesis on the development of courtship and copulatory behavior in zebra finches. *Horm Behav, 32,* 46–59.

Wade, J. (2001). Zebra finch sexual differentiation: the aromatization hypothesis revisited. *Microsc Res Tech, 54,* 354–363.

Wade, J., & Arnold, A. P. (1994). Posthatching inhibition of aromatase-activity does not alter sexual-differentiation of the zebra finch song system. *Brain Res, 639,* 347–350.

Wade, J., Schlinger, B. A., Hodges, L., & Arnold, A. P. (1994). Fadrozole—a potent and specific inhibitor of aromatase in the zebra finch brain. *Gen Comp Endocrinol, 94,* 53–61.

Wade, J., Swender, D. A., & McElhinny, T. L. (1999). Sexual differentiation of the zebra finch song system parallels genetic, not gonadal, sex. *Horm Behav, 36,* 141–152.

Warren, W. C., Clayton, D. F., Ellegren, H., Arnold, A. P., Hillier, L. W. et al. (2010). The genome of a songbird. *Nature, 464,* 757–762.

Watson, J. T., Abdelnabi, M., Wersinger, S., Ottinger, M.A., & Adkins-Regan, E. (1990). Circulating estradiol and the activation of male and female copulatory behavior in Japanese quail (*Coturnix japonica*). *Gen Comp Endocrinol, 77,* 229–238.

Wright, C. L., Schwarz, J. S., Dean, S. L., & McCarthy, M. M. (2010). Cellular mechanisms of estradiol-mediated sexual differentiation of the brain. *Trends Endocrinol Metab, 21,* 553–561.

Zann, R. A. (1996). *The Zebra Finch: A Synthesis of Field and Laboratory Studies.* Oxford: Oxford University Press.

Part V

Aromatase and Brain Protection or Repair

Chapter 19

Brain Aromatase and Neuroprotection in Mammals

Iñigo Azcoitia, Maria Angeles Arevalo, and Luis Miguel Garcia-Segura

INTRODUCTION

The best characterized roles of brain aromatase and local estradiol production within the brain are those related to the control of sexual differentiation, reproduction, and behavior (Lephart, 1996; Roselli, 2007; Charlier et al., 2010). However, local estradiol production within the brain is also involved in the control of neuronal survival after brain injury. Here we examine the role of aromatase under neurodegenerative conditions in mammals. We will first examine changes in brain aromatase and activity under pathological conditions and the cell types involved in the production of estradiol within the brain under these conditions. Then we will analyze the neuroprotective activity of brain aromatase. Finally, we will consider the possible mechanisms involved in the neuroprotective actions of aromatase.

MODIFICATIONS IN THE EXPRESSION OF BRAIN AROMATASE UNDER PATHOLOGICAL CONDITIONS IN MAMMALS

The first evidence for a modification in brain aromatase expression and activity under pathological conditions was obtained using excitotoxic and mechanical lesions in the brains of rats and mice (Garcia-Segura et al., 1999). This finding was soon confirmed in zebra finches (Peterson et al., 2001; for additional references on birds, see chapter 20). Further studies have demonstrated that aromatase expression, aromatase activity and/or estradiol levels are increased in the brain of rats, mice and dogs under a variety of pathological conditions such as experimental stroke (Saleh et al., 2004, 2005; Carswell et al., 2005; Kelicen Ugur et al., 2011), injury of the olivocerebellar system (Lavaque et al., 2006a), chronic hypertension (Pietranera et al., 2011), diabetes (Burul-Bozkurt et al., 2010), and viral infection (Yarim et al., 2010). Increased estradiol levels have been also detected in dorsal root ganglia of rats after sciatic nerve injury (Schaeffer et al., 2010) and aromatase expression is increased in the sciatic nerve of diabetic rats (Burul-Bozkurt et al., 2010).

AROMATASE IS EXPRESSED BY NEURONS AND REACTIVE ASTROCYTES AFTER BRAIN INJURY IN MAMMALS

Immunohistochemical studies (Fig. 19–1) have shown that brain injury in mammals induces the expression of aromatase in reactive astrocytes (Garcia-Segura et al., 1999; Carswell et al., 2005; Yarim et al., 2010; for birds see chapter 20).

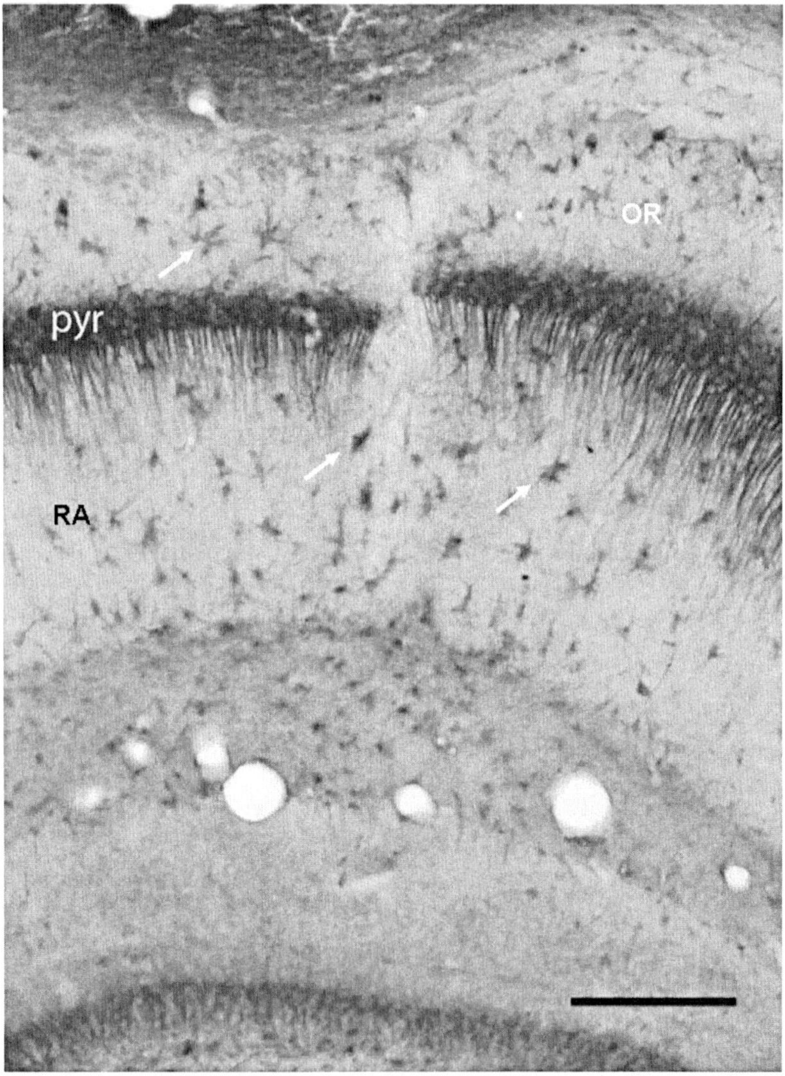

Figure 19–1. Aromatase immunoreactivity in the hippocampus of a mouse submitted to a stab wound injury. Aromatase immunoreactive pyramidal neurons are observed in the pyramidal (pyr) layer. In addition, numerous aromatase immunoreactive astrocytes (*arrows*) are observed in the stratum oriens (OR) and the stratum radiatum (RA). Scale bar, 200 μm.

Aromatase-expressing astrocytes are observed in both sexes and in all injured brain areas, including the cortex, corpus callosum, striatum, hippocampus, thalamus, hypothalamus and cerebellum (Garcia-Segura et al., 1999; Carswell et al., 2005; Yarim et al., 2010). This indicates that astrocytes from most, if not all, brain regions have the potential for expressing aromatase, and therefore presumably to produce estradiol, in response to injury. In vitro studies have confirmed that astrocytes respond to different insults by increasing aromatase expression (Azcoitia et al., 2003; Gatson et al., 2011; Lepore et al., 2011). However, in some cases brain injury is not accompanied by an increased expression of aromatase in astrocytes. For instance, aromatase immunoreactivity was not detectable in reactive astrocytes in sclerotic regions in epileptic human hippocampi (Yague et al., 2010). In addition, chronic spontaneous hypertension in rats is associated with an increased expression of aromatase in hippocampal neurons but not in hippocampal astrocytes (Pietranera et al., 2011). Therefore, astrocytes do not always express aromatase under pathological conditions and modifications in aromatase expression in the injured brain are not restricted to astrocytes.

Alterations of aromatase expression or activity in the peripheral nervous system under pathological conditions may also be restricted to neuronal compartment since estradiol levels are increased in dorsal root ganglia after sciatic nerve injury, but aromatase expression is restricted to dorsal root ganglia neurons and it is not detectable in glial cells (Schaeffer et al., 2010).

BRAIN AROMATASE IS NEUROPROTECTIVE

Several studies have analyzed the role of aromatase after brain injury in mammals by using aromatase deficient mice or aromatase inhibitors. Such studies have shown that aromatase deficient (ArKO) male mice and intact male rats chronically treated with the aromatase inhibitor fadrozole have an increased vulnerability of hippocampal neurons to excitotoxic injuries (Azcoitia et al., 2001; Garcia-Segura et al., 2003). Similar results were obtained when fadrozole was administered systemically and when it was infused in the lateral cerebral ventricle to specifically inhibit aromatase within the brain (Azcoitia et al., 2001; Garcia-Segura et al., 2003). Fadrozole also increases excitotoxic neuronal death in ovariectomized rats (Veiga et al., 2005), indicating that extragonadal aromatase is also involved in neuroprotection in females. In addition, the neuroprotective effect of testosterone and of other steroids upstream in the synthetic pathway of testosterone, such as pregnenolone and dehydroepiandrosterone (DHEA), are also impaired in the hippocampus, cerebral cortex and the olfactory bulb in animals treated with fadrozole or letrozole, another aromatase inhibitor (Veiga et al., 2003; Hoyk et al., 2004; Juhász-Vedres et al., 2006). Interestingly, the increased expression of aromatase in the injured brain is also accompanied by the upregulation in the expression of other steroidogenic molecules, such as steroidogenic acute regulatory protein (StAR) and cytochrome P450 side chain cleavage (P450scc) (Lavaque et al., 2006a). Both StAR and P450scc are involved in the first step of steroidogenesis (Sierra, 2004; Lavaque et al., 2006b). StAR facilitates the transport of cholesterol from the outer to the inner mitochondrial membrane, where P450scc is located. Then cholesterol is transformed to pregnenolone by P450scc. The upregulation of StAR and P450scc in the injured brain suggests that the nervous tissue responds to damage with an increased steroidogenesis and that the aromatization of steroid precursors to estradiol may be part of an endogenous mechanism activated in the brain tissue to cope with neurodegeneration (Azcoitia et al., 2005).

Another experimental model of neurodegeneration in which the neuroprotective actions of aromatase have been assessed is the ischemia induced by reversible middle cerebral artery occlusion (McCullough et al., 2003). Female wild-type mice that were chronically treated with fadrozole as well as female ArKO mice showed more cortical and striatal damage in this model than wild-type animals treated with vehicle. Furthermore, female ArKO mice showed more brain damage than ovariectomized wild-type animals, indicating that extragonadal synthesis of estradiol is involved in neuroprotection in this model too (McCullough et al., 2003). Aromatase is also protective for the nigrostriatal system. Aromatase deficient (ArKO) female mice are more vulnerable to nigrostriatal damage than wild-type females (Morale et al., 2008), and the intracerebral infusion of the aromatase inhibitor anastrazole in male rats has been reported to increase striatal dopamine loss after the partial neurotoxic injury of substantia nigra pars compacta (McArthur et al., 2007). The neuroprotective effect of aromatase also has been assessed in a model of cerebellar ataxia produced by the degeneration of rat inferior olivary nucleus after treatment with 3-acetylpirydine (3AP), a nicotinamide antimetabolite. Exogenous estradiol is neuroprotective for inferior olivary neurons (Sierra et al., 2003) and this hormonal action may be important for the maintenance of motor function that is under the control of the olivo-cerebellar system (Randell et al., 2001; Naessen et al., 2007). Interestingly, the inhibition of aromatase with fadrozole enhanced the injury produced by 3AP in the inferior olive of male rats, decreasing the number of neurons that survived to 3AP (Sierra et al., 2003).

The previously mentioned experimental studies clearly indicate that aromatase exerts a neuroprotective action. Furthermore, in all the experimental models studied, in mammals and birds, the neurodegenerative effect of aromatase deficiency was counterbalanced by the administration of estradiol (Azcoitia et al., 2001; McCullough et al., 2003; Sierra et al., 2003; Saldanha et al., 2005), indicating that the neuroprotective properties of aromatase lie in its ability to catalyze the formation of estradiol rather than in reducing estradiol precursor levels. By increasing local estradiol levels within the brain, aromatase may modulate local inflammation, regulate the function of the neurovascular unit, and decrease neuronal apoptosis (Azcoitia et al., 2011).

ROLE OF NEUROGENESIS IN THE NEUROPROTECTIVE ACTIONS OF AROMATASE

New neurons are generated during the entire lifespan in several regions of the vertebrate brain. These newly generated neurons migrate to their final destination and are incorporated to functional synaptic circuits (Nottebohm, 2002;

Abrous et al., 2005; Christie & Cameron, 2006; Gould, 2007). Several data indicate that adult neurogenesis is of functional significance. For instance, changes in the rate of neurogenesis in the dentate gyrus of the hippocampus in mammals are correlated with modifications in cognition and affection (Abrous et al., 2005; Dupret et al., 2007; Sahay & Hen, 2007; Trejo et al., 2008; Hernández-Rabaza et al., 2009). Neurogenesis in the subventricular zone and the olfactory bulb seems to be involved in olfactory learning and memory (Nissant & Pallotto, 2011) and to participate in paternal recognition of adult offspring (Mak & Weiss, 2010).

The role of adult neurogenesis is not simply to add new neurons to preexisting neuronal circuits but to induce a reorganization of these circuits by the substitution of cellular elements. The reorganization of the synaptic circuits may be transient, because many newly generated neurons will finally die. However, these transient synaptic circuits may be functionally relevant and there is evidence indicating that both addition and removal of new hippocampal neurons is necessary for spatial learning (Dupret et al., 2007). Neurogenesis is also regulated by endogenous factors produced under neurodegenerative conditions and may play a role in the repair of damaged neuronal circuits (Kernie and Parent, 2010; Ohira, 2011; Yoneyama et al., 2011). One of these factors is estradiol, which enhances adult neurogenesis under physiological (Tanapat et al., 2005; Galea et al., 2006) and neurodegenerative conditions (Pietranera et al., 2010; Li et al., 2011).

The brain of teleost fish maintains a high degree of neurogenesis and plasticity during adulthood (Le Page et al., 2010). Interestingly, in the adult teleost brain, aromatase is highly expressed by radial glial progenitors and participates in the regulation of neurogenesis (Forlano et al., 2001; Pellegrini et al., 2007; Diotel et al., 2010; Le Page et al., 2010). The high expression of aromatase and the high neuroplastic activity in the brain of adult teleosts may be in part be explained by the fact that the mechanisms of brain sexualization remain active during adulthood in these species (Le Page et al., 2010). In contrast, high brain aromatase expression and a high rate of neurogenesis are restricted to the developmental period in mammals, when sex brain differentiation mechanisms are active. The induction of aromatase expression in the brain under neurodegenerative conditions, discussed in previous sections of this chapter, may represent a reactivation of an early phylogenetic mechanism to regulate brain plasticity, which is mainly confined to the developmental period in mammals. This mechanism allows a continuous remodeling of the teleost brain in adulthood and may be involved in brain repair in the mammalian brain under neurodegenerative conditions. In mammals, aromatase is expressed in radial glial progenitors in the embryonic neocortex and regulates neurogenesis (Martínez-Cerdeño et al., 2006). In addition, the treatment of hippocampal cultures with the aromatase inhibitor letrozole decreases the number of proliferative cells and increases the number of apoptotic cells (Fester et al., 2006). Similar effects have been observed after the inhibition of steroidogenesis in the cultures using siRNA against StAR. Application of estradiol to the medium had no effect on proliferation and apoptosis whereas the antiproliferative and proapoptotic effects of StAR knock-down and letrozole treatment were counterbalanced by treatment of the cultures with estradiol (Fester et al., 2006). The regulation exerted by aromatase and local estradiol synthesis on neurogenesis in the developing mammalian brain may also be relevant under neurodegenerative conditions in adulthood. Indeed, recent studies have shown that aromatase deficient mice have a reduction in stroke-induced neurogenesis (Li et al., 2011). This finding is in agreement with the results obtained in experiments with zebra finch, showing that aromatase regulates neurogenesis after brain injury (Peterson et al., 2007; Walters et al., 2011; see also chapter 20).

ROLE OF SYNAPTIC PLASTICITY IN THE NEUROPROTECTIVE ACTIONS OF AROMATASE

The neuroprotective action of aromatase may also be in part related to the regulatory actions of estradiol on synaptic plasticity. The importance of synaptic homeostasis in neuroprotection has been confirmed by several authors (Hardingham, 2009; Beattie et al., 2010). Therefore, the actions of estradiol on neuritic growth and synaptic connectivity may contribute to the repair of the damaged neural tissue. Rune and collaborators have provided solid evidence indicating that local estradiol synthesis in the hippocampus by the enzyme aromatase regulates synaptic contacts in CA1 pyramidal neurons, showing that the pharmacological inhibition of aromatase activity in hippocampal slices decreases the number of dendritic spines and the number of dendritic spine synapses (Kretz et al., 2004; Rune & Frotscher, 2005; Prange-Kiel & Rune, 2006; Prange-Kiel et al., 2006). In addition, aromatase inhibition reduces axon outgrowth in hippocampal neurons (von Schassen et al., 2006). Evidence for a role of aromatase in the regulation of hippocampal synaptic plasticity in vivo has been obtained by Leranth and collaborators. They found that testosterone and dihydrotestosterone (DHT) increased the number of synapses in the dendritic spines of CA1 pyramidal cells and that the treatment of ovariectomized females with letrozole significantly reduced the effect of testosterone on the number of CA1 spine synapses (Leranth et al., 2004; MacLusky et al., 2006). As expected, the aromatase inhibitor did not affect the response of synapses to the non-aromatizable androgen DHT (Leranth et al., 2004; MacLusky et al., 2006). Further studies that assessed the effect of the adrenal androgen DHEA have confirmed the role of aromatase on hippocampal synaptic plasticity, since letrozole abolished the induction of CA1 spine synapses by DHEA in females (Hajszan et al., 2004; MacLusky et al., 2004). These findings indicate that brain aromatase modulates synaptic plasticity in brain regions related to cognition, such as the hippocampus. The enzyme may also influence synaptic plasticity in other regions of the central nervous system, since cerebellar Purkinje cells in ArKO mice show decreased dendritic growth, and decreased formation of dendritic spines and synapses (Sasahara et al., 2007).

CONCLUSIONS

The studies reviewed in this chapter indicate that brain aromatase is an important regulator of the response of neural tissue to neurodegenerative conditions.

Brain aromatase plays an important role in the protection of neural tissue after injury by increasing local estrogen levels. However, the neuroprotective role of aromatase is not limited to neurodegenerative conditions, because 1-year-old female ArKO mice show spontaneous apoptosis of pyramidal neurons in the frontal cerebral cortex (Hill et al., 2009), whereas 1-year-old male ArKO mice show spontaneous apoptosis of dopaminergic neurons in the arcuate nucleus and medial preoptic area (Hill et al., 2007). This indicates that prolonged estrogen deficiency per se causes neurodegeneration, suggesting a possible link between aromatase deficiency and human neurological and affective disorders. This is a field that remains almost completely unexplored. However, alterations in aromatase immunoreactivity have been detected in the brain of Alzheimer's disease patients (Ishunina et al., 2005) and genetic variation of the aromatase gene in humans may increase the risk of Alzheimer's disease, particularly in women (Iivonen et al., 2004; Butler et al., 2010). This latter observation, together with the different outcomes of aromatase deficiency for spontaneous apoptosis of neurons in the brain of male and female mice (Hill et al., 2007, 2009) also opens the question of possible sex differences in the neuroprotective action of aromatase. Several data suggest that the roles of brain aromatase are different in males and females. For instance, the enzyme is involved in the regulation of the number of CA1 spine synapses by testosterone and DHEA in females but not in males (Hajszan et al., 2004; Leranth et al., 2004; MacLusky et al., 2004, 2006). In addition, it has been reported that primary cultured cortical astrocytes derived from female rat pups have an enhanced aromatase activity compared with astrocytes derived from males (Liu et al., 2007). Furthermore, aromatase deprivation has opposite outcomes in males and females in murine models of Alzheimer's disease; enhancing the pathology in females and decreasing it in males (Yue et al., 2005; McAllister et al., 2010). Therefore, it is possible that sex differences in brain aromatase activity or in brain vulnerability to aromatase deprivation may be involved in the sexually dimorphic expression of different brain pathologies, such as Parkinson's and Alzheimer's diseases.

ACKNOWLEDGMENTS

Support from Ministerio de Ciencia e Innovación, Spain (grant number BFU2008-02950-C03-01/02) is acknowledged. The authors want to thank Dr. Jacques Balthazart and Dr. Gregory F. Ball for their comments, corrections, and improvements of the original manuscript.

REFERENCES

Abrous, D. N., Koehl, M., & Le Moal M. (2005). Adult neurogenesis: from precursors to network and physiology. *Physiol Rev, 85,* 523–569.

Azcoitia, I., Arevalo, M. A., De Nicola, A. F., & Garcia-Segura, L. M. (2011). Neuroprotective actions of estradiol revisited. *Trends Endocrinol Metab, 22,* 467–473.

Azcoitia, I., Sierra, A., Veiga, S., & Garcia-Segura, L. M. (2003) Aromatase expression by reactive astroglia is neuroprotective. *Ann N Y Acad Sci, 1007,* 298–305.

Azcoitia, I., Sierra, A., Veiga, S., & Garcia-Segura, L. M. (2005). Brain steroidogenesis: emerging therapeutic strategies to prevent neurodegeneration. *J Neural Transm, 112,* 171–176.

Azcoitia, I., Sierra, A., Veiga, S., Honda, S., Harada, N., & Garcia-Segura, L. M. (2001). Brain aromatase is neuroprotective. *J Neurobiol, 47,* 318–329.

Beattie, M. S., Ferguson, A. R., & Bresnahan, J. C. (2010). AMPA-receptor trafficking and injury-induced cell death. *Eur J Neurosci, 32,* 290–297.

Burul-Bozkurt, N., Pekiner, C., & Kelicen, P. (2010). Diabetes alters aromatase enzyme levels in sciatic nerve and hippocampus tissues of rats. *Cell Mol Neurobiol, 30,* 445–451.

Butler, H. T., Warden, D. R., Hogervorst, E., Ragoussis, J., Smith, A. D., & Lehmann D. J. (2010). Association of the aromatase gene with Alzheimer's disease in women. *Neurosci Lett, 468,* 202–206.

Carswell, H. V., Dominiczak, A. F., Garcia-Segura, L. M., Harada, N., Hutchison, J. B., & Macrae, I. M. (2005). Brain aromatase expression after experimental stroke: topography and time course. *J Steroid Biochem Mol Biol, 96,* 89–91.

Charlier, T. D., Cornil, C. A., Ball, G. F. & Balthazart, J. (2010). Diversity of mechanisms involved in aromatase regulation and estrogen action in the brain. *Biochimica et Biophysica Acta, 1800,* 1094–1105.

Christie, B. R., & Cameron, H. A. (2006). Neurogenesis in the adult hippocampus. *Hippocampus, 16,* 199–207.

Diotel, N., Vaillant, C., Gueguen, M. M., Mironov, S., Anglade, I., Servili, A., et al. (2010). Cxcr4 and Cxcl12 expression in radial glial cells of the brain of adult zebrafish. *J Comp Neurol,* 518, 4855–4876.

Dupret, D., Fabre, A., Döbrössy, M. D., Panatier, A., Rodríguez, J. J., Lamarque, S., et al. (2007). Spatial learning depends on both the addition and removal of new hippocampal neurons. *PLoS Biology, 5,* e214.

Fester, L., Ribeiro-Gouveia, V., Prange-Kiel, J., von Schassen, C., Bottner, M., Jarry, H., et al. (2006). Proliferation and apoptosis of hippocampal granule cells require local oestrogen synthesis. *J Neurochem, 97,* 1136–1144.

Forlano, P. M., Deitcher, D. L., Myers, D. A. & Bass, A. H. (2001). Anatomical distribution and cellular basis for high levels of aromatase activity in the brain of teleost fish: aromatase enzyme and mRNA expression identify glia as source. *J Neurosci, 21,* 8943–8955.

Galea, L. A., Spritzer, M. D., Barker, J. M., & Pawluski, J. L. (2006). Gonadal hormone modulation of hippocampal neurogenesis in the adult. *Hippocampus, 16,* 225–232.

Garcia-Segura, L.M., Veiga, S., Sierra, A., Melcangi, R.C., & Azcoitia, I. (2003). Aromatase: a neuroprotective enzyme. *Prog Neurobio, 71,* 31–41.

Garcia-Segura, L. M., Wozniak, A., Azcoitia, I., Rodriguez, J. R., Hutchison, R. E., & Hutchison, J. B. (1999). Aromatase expression by astrocytes after brain injury: implications for local estrogen formation in brain repair. *Neuroscience, 89,* 567–578.

Gatson, J. W., Simpkins, J. W., Yi, K. D., Idris, A. H., Minei, J. P., & Wigginton, J. G. (2011). Aromatase is increased in astrocytes in the presence of elevated pressure. *Endocrinology, 152,* 207–213.

Gould, E (2007). How widespread is adult neurogenesis in mammals? *Nat Rev Neurosci, 8,* 481–488.

Hajszan, T., MacLusky, N. J., & Leranth, C. (2004). Dehydroepiandrosterone increases hippocampal spine synapse density in ovariectomized female rats. *Endocrinology, 145,* 1042–1045.

Hardingham, G. E. (2009). Coupling of the NMDA receptor to neuroprotective and neurodestructive events. *Biochem Soc Trans, 37,* 1147–1160.

Hernández-Rabaza, V., Llorens-Martín, M., Velázquez-Sánchez, C., Ferragud, A., Arcusa, A., Gumus, H. G., et al. (2009). Inhibition of adult hippocampal neurogenesis disrupts contextual learning but spares spatial working memory, long-term conditional rule retention and spatial reversal. *Neuroscience, 159,* 59–68.

Hill, R. A., Chow, J., Fritzemeier, K., Simpson, E. R., & Boon, W. C. (2007). Fas/FasL-mediated apoptosis in the arcuate nucleus and medial preoptic area of male ArKO mice is ameliorated by selective estrogen receptor alpha and estrogen receptor beta agonist treatment, respectively. *Mol Cell Neurosci, 36,* 146–157.

Hill, R. A., Chua, H. K., Jones, M. E., Simpson, E. R., & Boon, W. C. (2009). Estrogen deficiency results in apoptosis in the frontal cortex of adult female aromatase knockout mice. *Mol Cell Neurosci, 41,* 1–7.

Hoyk, Z., Párducz, A., & Garcia-Segura, L. M. (2004). Dehydroepiandrosterone regulates astroglia reaction to denervation of olfactory glomeruli. *Glia, 48,* 207–216.

Iivonen, S., Corder, E., Lehtovirta, M., Helisalmi, S., Mannermaa, A., Vepsäläinen, S., et al. (2004). Polymorphisms in the CYP19 gene confer increased risk for Alzheimer disease. *Neurology, 62,* 1170–1176.

Ishunina, T. A., van Beurden, D., van der Meulen, G., Unmehopa, U. A., Hol, E. M., Huitinga, I., et al. (2005). Diminished aromatase immunoreactivity in the hypothalamus, but not in the basal forebrain nuclei in Alzheimer's disease. *Neurobiol Aging, 26,* 173–194.

Juhász-Vedres, G., Rózsa, E., Rákos, G., Dobszay, M. B., Kis, Z., Wölfling, J., et al. (2006). Dehydroepiandrosterone sulfate is neuroprotective when administered either before or after injury in a focal cortical cold lesion model. *Endocrinology, 147,* 683–686.

Kelicen Ugur, P., Lule, S., Cincioglu, M., Pekiner, C., & Gursoy-Ozdemir, Y. (2011). Megestrol acetate inhibits the expression of cytoplasmic aromatase through nuclear C/EBPβ in reperfusion injury-induced ischemic rat hippocampus. *Eur J Pharmacol, 654,* 217–225.

Kernie, S. G., & Parent, J. M. (2010). Forebrain neurogenesis after focal ischemic and traumatic brain injury. *Neurobiol Dis, 37,* 267–274.

Kretz, O., Fester, L., Wehrenberg, U., Zhou, L., Brauckmann, S., Zhao, S., et al. (2004). Hippocampal synapses depend on hippocampal estrogen synthesis. *J Neurosci, 24,* 5913–5921.

Lavaque, E., Mayen, A., Azcoitia, I., Tena-Sempere, M., & Garcia-Segura, L. M. (2006a). Sex differences, developmental changes, response to injury and cAMP regulation of the mRNA levels of steroidogenic acute regulatory protein, cytochrome p450scc, and aromatase in the olivocerebellar system. *J Neurobiol, 66,* 308–318.

Lavaque, E., Sierra, A., Azcoitia, I., & Garcia-Segura, L. M. (2006b). Steroidogenic acute regulatory protein in the brain. *Neuroscience, 138,* 741–747.

Le Page, Y., Diotel, N., Vaillant, C., Pellegrini, E., Anglade, I., Mérot, Y., et al. (2010). Aromatase, brain sexualization and plasticity: the fish paradigm. *Eur J Neurosci, 32,* 2105–2115.

Lephart, E. D. (1996). A review of brain aromatase cytochrome P450. *Brain Res Rev, 22,* 1–26.
Lepore, G., Gadau, S., Peruffo, A., Mura, A., Mura, E., Floris, A., et al. (2011). Aromatase expression in cultured fetal sheep astrocytes after nitrosative/oxidative damage. *Cell Tissue Res, 344,* 407–413.
Leranth, C., Hajszan, T., & MacLusky, N. J. (2004). Androgens increase spine synapse density in the CA1 hippocampal subfield of ovariectomized female rats. *J Neurosci, 24,* 495–499.
Li, J., Siegel, M., Yuan, M., Zeng, Z., Finnucan, L., Persky, R., Hurn, P. D., et al. (2011). Estrogen enhances neurogenesis and behavioral recovery after stroke. *J Cereb Blood Flow Metab, 31,* 413–425.
Liu, M., Hurn, P. D., Roselli, C. E., & Alkayed, N. J. (2007). Role of P450 aromatase in sex-specific astrocytic cell death. *J Cereb Blood Flow Metab, 27,* 135–141.
Mak, G. K., & Weiss, S. (2010). Paternal recognition of adult offspring mediated by newly generated CNS neurons. *Nat Neurosci, 13,* 753–758.
MacLusky, N. J., Hajszan, T., & Leranth, C. (2004). Effects of dehydroepiandrosterone and flutamide on hippocampal CA1 spine synapse density in male and female rats: implications for the role of androgens in maintenance of hippocampal structure. *Endocrinology, 145,* 4154–4161.
MacLusky, N. J., Hajszan, T., Prange-Kiel, J., & Leranth, C. (2006). Androgen modulation of hippocampal synaptic plasticity. *Neuroscience, 138,* 957–965.
Martínez-Cerdeño, V., Noctor, S. C., & Kriegstein, A. R. (2006). Estradiol stimulates progenitor cell division in the ventricular and subventricular zones of the embryonic neocortex. *Eur J Neurosci, 24,* 3475–3488.
McAllister, C., Long, J., Bowers, A., Walker, A., Cao, P., Honda, S., et al. (2010). Genetic targeting aromatase in male amyloid precursor protein transgenic mice down-regulates beta-secretase (BACE1) and prevents Alzheimer-like pathology and cognitive impairment. *J Neurosci, 30,* 7326–7334.
McArthur, S., Murray, H. E., Dhankot, A., Dexter, D. T., & Gillies, G. E. (2007). Striatal susceptibility to a dopaminergic neurotoxin is independent of sex hormone effects on cell survival and DAT expression but is exacerbated by central aromatase inhibition. *J Neurochem, 100,* 678–692.
McCullough, L. D., Blizzard, K., Simpson, E. R., Oz, O. K., & Hurn, P. D. (2003). Aromatase cytochrome P450 and extragonadal estrogen play a role in ischemic neuroprotection. *J Neurosci, 23,* 8701–8705.
Morale, M. C., L'episcopo, F., Tirolo, C., Giaquinta, G., Caniglia, S., Testa, N., et al. (2008). Loss of aromatase cytochrome P450 function as a risk factor for Parkinson's disease? *Brain Res Rev, 57,* 431–443.
Naessen, T., Lindmark, B., & Larsen, H. C. (2007). Hormone therapy and postural balance in elderly women. *Menopause, 14,* 1020–1024.
Nissant, A., & Pallotto, M. (2011). Integration and maturation of newborn neurons in the adult olfactory bulb—from synapses to function. *Eur J Neurosci, 33,* 1069–1077.
Nottebohm, F. (2002). Neuronal replacement in adult brain. *Brain Res Bull, 57,* 737–749.
Ohira, K. (2011). Injury-induced neurogenesis in the mammalian forebrain. *Cell Mol Life Sci, 68,* 1645–1656.
Pellegrini, E., Mouriec, K., Anglade, I., Menuet, A., Le Page, Y., Gueguen, M. M., et al. (2007). Identification of aromatase-positive radial glial cells as progenitor cells in the ventricular layer of the forebrain in zebrafish. *J Comp Neurol, 501,* 150–167.

Peterson, R. S., Fernando, G., Day, L., Allen, T. A., Chapleau, J. D., Menjivar, J., et al. (2007). Aromatase expression and cell proliferation following injury of the adult zebra finch hippocampus. *Dev Neurobiol, 67,* 1867–1878.

Peterson, R. S., Saldanha, C. J., & Schlinger, B. A. (2001). Rapid upregulation of aromatase mRNA and protein following neural injury in the zebra finch (Taeniopygia guttata). *J Neuroendocrinol, 13,* 317–323.

Pietranera, L., Bellini, M. J., Arévalo, M. A., Goya, R., Brocca, M. E., Garcia-Segura, L. M., et al. (2011). Increased aromatase expression in the hippocampus of spontaneously hypertensive rats: effects of estradiol administration. *Neuroscience, 174,* 151–159.

Pietranera, L., Lima, A., Roig, P., & De Nicola, A. F. (2010). Involvement of brain-derived neurotrophic factor and neurogenesis in oestradiol neuroprotection of the hippocampus of hypertensive rats. *J Neoroendocrinol, 22,* 1082–1092.

Prange-Kiel, J., Fester, L., Zhou, L., Lauke, H., Carretero, J., & Rune, G. M. (2006). Inhibition of hippocampal estrogen synthesis causes region-specific downregulation of synaptic protein expression in hippocampal neurons. *Hippocampus, 16,* 464–471.

Prange-Kiel, J., & Rune, G. M. (2006). Direct and indirect effects of estrogen on rat hippocampus. *Neuroscience, 138,* 765–772.

Randell, K. M., Honkanen, R. J., Komulainen, M. H., Tuppurainen, M. T., Kröger, H., & Saarikoski, S. (2001). Hormone replacement therapy and risk of falling in early postmenopausal women—a population-based study. *Clin Endocrinol (Oxf), 54,* 769–774.

Roselli, C. F. (2007). Brain aromatase: roles in reproduction and neuroprotection. *J Steroid Biochem Mol Biol, 106,* 143–150.

Rune, G. M., & Frotscher, M. (2005). Neurosteroid synthesis in the hippocampus: role in synaptic plasticity. *Neuroscience, 136,* 833–842.

Sahay, A., & Hen, R. (2007). Adult hippocampal neurogenesis in depression. *Nature Neuroscience, 10,* 1110–1115.

Saldanha, C. J., Rohmann, K. N., Coomaralingam, L., & Wynne, R. D. (2005). Estrogen provision by reactive glia decreases apoptosis in the zebra finch (Taeniopygia guttata). *J Neurobiol, 64,* 192–201.

Saleh, T. M., Connell, B. J., Legge, C., & Cribb, A. E. (2004). Stroke-induced changes in estrogen release and neuronal activity in the parabrachial nucleus of the male rat. *J Stroke Cerebrovasc Dis, 13,* 24–34.

Saleh, T. M., Connell, B. J., Legge, C., & Cribb, A. E. (2005). Estrogen synthesis in the central nucleus of the amygdala following middle cerebral artery occlusion: role in modulating neurotransmission. *Neuroscience, 135,* 1141–1153.

Sasahara, K., Shikimi, H., Haraguchi, S., Sakamoto, H., Honda, S., Harada, N., et al. (2007). Mode of action and functional significance of estrogen-inducing dendritic growth, spinogenesis, and synaptogenesis in the developing Purkinje cell. *J Neurosci, 27,* 7408–7417.

Schaeffer, V., Meyer, L., Patte-Mensah, C., Eckert, A., & Mensah-Nyagan, A. G. (2010). Sciatic nerve injury induces apoptosis of dorsal root ganglion satellite glial cells and selectively modifies neurosteroidogenesis in sensory neurons. *Glia, 58,* 169–180.

Sierra, A. (2004). Neurosteroids: the StAR protein in the brain. *J Neoroendocrinol, 16,* 787–793.

Sierra, A., Azcoitia, I., & Garcia-Segura, L. (2003). Endogenous estrogen formation is neuroprotective in model of cerebellar ataxia. *Endocrine, 21,* 43–51.

Tanapat, P., Hastings, N. B., & Gould, E. (2005). Ovarian steroids influence cell proliferation in the dentate gyrus of the adult female rat in a dose and time-dependent manner. *J Comp Neurol, 481,* 252–265.

Trejo, J. L., Llorens-Martín, M. V., & Torres-Alemán, I. (2008). The effects of exercise on spatial learning and anxiety-like behavior are mediated by an IGF-I-dependent mechanism related to hippocampal neurogenesis. *Mol Cell Neurosci, 37,* 402–411.

Veiga, S., Azcoitia, I., & Garcia-Segura, L. M. (2005). Extragonadal synthesis of estradiol is protective against kainic acid excitotoxic damage to the hippocampus. *Neuroreport, 16,* 1599–1603.

Veiga, S., Garcia-Segura, L. M., & Azcoitia, I. (2003). Neuroprotection by the steroids pregnenolone and dehydroepiandrosterone is mediated by the enzyme aromatase. *J Neurobiol, 56,* 398–406.

von Schassen, C., Fester, L., Prange-Kiel, J., Lohse, C., Huber, C., Böttner, M., & et al. (2006). Oestrogen synthesis in the hippocampus: role in axon outgrowth. *J Neoroendocrinol, 18,* 847–856.

Walters, B. J., Alexiades, N. G., & Saldanha, C. J. (2011). Intracerebral estrogen provision increases cytogenesis and neurogenesis in the injured zebra finch brain. *Dev Neurobiol, 71,* 170–181.

Yague, J. G., Azcoitia, I., DeFelipe, J., Garcia-Segura, L. M., & Muñoz, A. (2010). hAromatase expression in the normal and epileptic human hippocampus. *Brain Res, 1315,* 41–52.

Yarim, M., Gulbahar, M. Y., Guvenc, T., Karahan, S., Harada, N., Kabak, Y. B.,et al. (2010). Aromatase expression in the cerebellum of the dog infected with canine distemper virus. *Berliner und Munchener Tierarztliche Wochenschrift, 123,* 301–306.

Yoneyama, M., Shiba, T., Hasebe, S., & Ogita, K. (2011). Adult neurogenesis is regulated by endogenous factors produced during neurodegeneration. *J Pharmacol Sci, 115,* 425–432.

Yue, X., Lu, M., Lancaster, T., Cao, P., Honda, S., Staufenbiel, M., et al. (2005). Brain estrogen deficiency accelerates Abeta plaque formation in an Alzheimer's disease animal model. *Proc Natl Acad Sci U S A, 102,* 19198–19203.

Chapter 20

Inducible Aromatase in Astroglia: Protection and Recovery from Neural Perturbation in Birds

Kelli A. Duncan and Colin J. Saldanha

INTRODUCTION

The past several decades have witnessed a considerable widening in our understanding of how hormones affect brain structure and function. Earlier studies laid the foundation for, and the framework of, several concepts regarding how peripheral hormones, including gonadal steroids, gain access to target tissues in the brain and thereby regulate behavior. More recent work identified the brain as a source for several important steroids, thereby refining our appreciation of the complexities of steroid provision. There is perhaps no better example of this refinement than the pivotal role for central aromatization of peripherally derived androgens in the organization and activation of circuits necessary for masculine appetitive and consummatory sexual behaviors (MacLusky & Naftolin, 1981; Schumacher & Balthazart, 1983; Balthazart et al., 2009).

The role of central aromatization in the vertebrate brain has burgeoned far beyond circuits involved in copulation. Indeed, the discrete, localized expression of neural aromatase has been implicated in behaviors as diverse as the appetitive and consummatory aspects of sexual behavior, aggression, spatial memory, nociception, and motor control (Cornil, Ball, & Balthazart, 2006; Forlano, Schlinger, & Bass, 2006; Balthazart et al., 2009). Included in this expansion, is a relatively recent role for central aromatization in neuroprotection and neural repair (Roselli, 2007; Saldanha, Duncan, & Walters, 2009).

A role for estrogens in vertebrate neural plasticity is not new. However, this chapter focuses on two characteristics about the provision of estrogens and neural damage that are obvious deviations from the classical role of peripheral and neuronal steroid provision in neuroplasticity. First, the damaged homeotherm brain demonstrates *inducible* aromatase in that the aromatase gene is transcribed and translated in response to neural perturbation. This is conceptually distinct from the classical view of the regulation of central aromatase, in which changes were typically described as the up- or down-regulation of enzyme activity without clarification as to abundance or allosteric modulation of the enzymatic protein. Second, aromatase is induced in *astroglia,* not neurons. Again, this represents a departure from the classical cell type responsible for central aromatization in birds and mammals (see Forlano et al., 2006; Pellegrini et al., 2007 for counter-examples in teleosts). In the past decade, an emerging literature using birds and mammals as experimental subjects, has strongly implicated the induction of aromatase in glial cells as a novel form of central estrogen provision for the reasons described in the preceding.

The majority, if not all the experimental work on the neuroprotective actions of glial aromatase, has been conducted in birds and mammals, and this review focuses primarily on the avian model. This chapter describes the current understanding of how neural perturbation may induce glial aromatase, and further how estrogen synthesis in glia may affect neuroprotection and repair. It ends with an attempted synthesis about how cell-specific and inducible aromatase may have evolved among vertebrates as a mechanism that has captured elements of neural plasticity and co-opted these mechanisms to preserve and protect neural circuits.

INDUCED AROMATASE IN THE VERTEBRATE BRAIN

It has been known for some time now that circulating steroids, particularly estrogens, may be neuroprotective (McEwen et al., 2001; Wise, 2003; Maggi et al., 2004). Indeed, numerous studies have described a lower susceptibility to damage and a lowering of markers of cell death in the presence of high circulating estrogens in mammals (Belcredito et al., 2001; Chiueh et al., 2003; Simpkins et al., 2005; Bryant et al., 2006; Roselli, 2007; Simpkins & Dykens, 2008).

However, as highlighted in chapter 19, there is an important distinction between constitutive (peripheral and central) and inducible (glial) aromatase expression in the vertebrate central nervous system. These experiments were critical in documenting an increase in neural aromatase following brain damage and were also the first to identify the astrocyte as a cell type capable of aromatase-expression in vivo (Garcia-Segura et al., 1999). Later studies in mammals have supported these findings by showing that glial aromatase can be induced by a variety of factors that perturb the brain (Azcoitia et al., 2001; Carswell et al., 2005).

Interestingly, the expression of aromatase in glia had been documented earlier in primary cultures of the developing songbird brain (Schlinger et al., 1994). In these experiments, primary dissociated cultures from hatchling zebra finches (1–6 days old) were examined for aromatase activity every 7 days for 4 weeks in vitro. Using measures of biochemical activity and in situ hybridization with probes that targeted the aromatase transcript, these studies were able to convincingly demonstrate the aromatase expression in astrocytes. However, they were also pivotal in hinting at another, intriguing finding about glial aromatization. Although neuronal death was almost complete in about 21 days, and confluence of the glial bed as reflected in protein content reaching asymptote had occurred even earlier, aromatase activity increased with culture age (Schlinger et al., 1994). Thus, these data not only established the expression of aromatase in astrocytes in vitro, but suggested the possibility of a link between damage (cell death) and an upregulation of glial aromatase as well. However, because glial aromatase had not been observed in the avian brain in vivo, what precise physiological process was being revealed by these in vitro studies was unclear at the time.

Within the avian brain, neuronal cells expressing aromatase immunoproduct and thereby revealing their morphological characteristics are abundant and have been described in many species including, but not limited to, the zebra finch (*Taeniopygia guttata*), Japanese quail (*Coturnix japonica*), black-capped chickadee (*Peocile atricapillus*), Mallard duck (*Anas Platyrychus*), ring dove (*Streptopelia roseogrisea*), and domestic chicken (*Gallus gallus*) (Balthazart et al., 1990; Naftolin et al., 1996; Saldanha et al., 2000a; Saldanha et al., 2000b; Saldanha, Walters, & Fraley, 2010; Saldanha et al., 2011; Xie et al., 2011). However, as Garcia-Segura et al. (1999) had described in the rodent, following the induction of brain damage via a penetrating injury, aromatase positive astroglia (radial glia and astrocytes) are localized around the injury site in zebra finches (Peterson, Saldanha, & Schlinger, 2001; Peterson et al., 2004; Wynne & Saldanha, 2004; Wynne et al., 2008; Walters, Alexiades, & Saldanha, 2011). These cells have been detected as early as 6 hours following injury in the songbird brain and persist as long as 6 weeks post insult (Wynne et al., 2008).

Although it is difficult to directly compare results across laboratories, surgical protocols, and measurement techniques, we tentatively suggest that the avian brain is more responsive to brain damage in terms of induction of glial aromatase relative to the rodent brain. As mentioned earlier, astrocytic aromatase is detectable as early as 6 hours post-injury in the songbird brain. In contrast, the earliest report of glial aromatase following trauma in the rodent appears to be 24 hours following brain damage (Carswell et al., 2005). Although earlier time points were not investigated in the Carswell et al. (2005) study, glial aromatase is undetectable 30 days post-damage in rats (Carswell et al., 2005), whereas it remains readily detectable in the injured songbird brain even 40 days post damage (Wynne et al., 2008). These data suggest a protracted role for glial estrogen provision in the damaged songbird brain. It was correspondingly hypothesized that glial aromatase may have effects not only on injury-induced degeneration, but perhaps also on restorative pathways days to weeks later (Wynne et al., 2008).

Regardless of the rate and persistence of induction, however, it remains clear given the studies in songbirds and mammals, that the expression of aromatase in astroglia is a conserved physiological response to neural damage (Fig. 20-1). The remainder of this chapter focuses on the mechanisms responsible for the induction of aromatase in glia, and the downstream effects of glial estrogen provision on neuroprotection and recovery from damage.

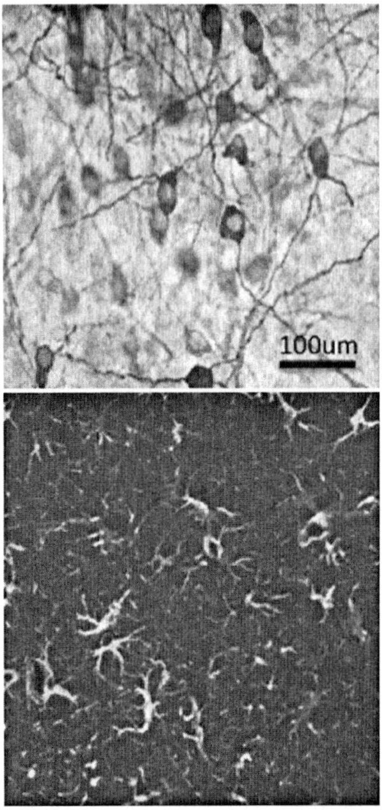

Figure 20–1. **Photomicrographs Comparing Aromatase Positive Neurons (Top) and Glial Cells (Bottom) in the Passerine Brain.** Images are modified from Saldanha, C. J., Rohmann, K. N., Coomaralingam, L., & Wynne, R. D. (2005). Estrogen provision by reactive glia decreases apoptosis in the zebra finch (Taeniopygia guttata). *J Neurobiol, 64*(2), 192–201. Saldanha, C. J., Remage-Healey, L., & Schlinger, B. A. (2011). Synaptocrine signaling: steroid synthesis and action at the synapse. *Endocr Rev, 32*(4), 532–549.

NEUROINFLAMMATION AND GLIAL AROMATASE EXPRESSION

As described, the avian and mammalian brains are dramatically responsive to brain damage in terms of the induction of aromatase in astroglia; a cell type that does not express this gene product under "normal" conditions. However, virtually nothing is known about what is responsible for the induction of aromatase in astrocytes following traumatic brain injury (TBI). In very general terms, TBI is characterized by both the physical damage and a secondary neural response that involves increased cytokine and chemokine expression (Rothwell & Strijbos, 1995; Ghirnikar, Lee, & Eng, 1998; Marciano et al., 2002; Simi et al., 2007). Among the many changes that accompany TBI, neuroinflammation caused by disruption of the blood-brain barrier may provide a plausible signal to induce aromatase transcription in reactive astrocytes (Ghirnikar et al., 1998; Allan & Rothwell, 2001; Lucas, Rothwell, & Gibson, 2006; DeLegge & Smoke, 2008). Studies in mammals of both neuronal and non-neuronal tissue demonstrate that cytokines, specifically proinflammatory cytokines, interleukin 1 beta (IL-1β), and interleukin 6 (IL-6) have the ability to upregulate aromatase expression (Reed et al., 1993; Purohit et al., 1995; Rothwell & Strijbos, 1995; Morioka et al., 2000; Allan et al., 2001; Honma et al., 2002; Purohit, Newman, & Reed, 2002; Purohit et al., 2005; Veerapaneni et al., 2009). Thus, cytokine-induced neuroinflammation from the injury is the most likely candidate for the induction of glial aromatase in the avian brain.

Using a model that creates a neuroinflammatory response without penetrating injury to the brain, we were able to ascertain the role that neuroinflammation and presumably cytokines play in mediating glial aromatase expression independent of cell death. Briefly, we unilaterally exposed the neuropil to either a neuroinflammatory agent or saline without creating any damage to the tissue (Duncan & Saldanha, 2011). Application of the inflammatory agent resulted in the upregulation of IL1β and IL-6, supporting the inference that an inflammatory response had indeed, been induced. Exposure of the brain to the neuroinflammatory agent also increased aromatase mRNA as measured by quantitative PCR, 24 hours following treatment. The induction of cytokines and glial aromatase was treatment specific, in that both were higher in the brain hemisphere exposed to the inflammatory agent. Importantly, aromatase induction was in glial cells as reflected by the appearance of numerous aromatase positive astrocytes spread throughout the hemisphere exposed to the inflammatory agent, but not in that exposed to saline (Duncan & Saldanha, 2011) (Fig. 20-2). Finally, there was no evidence of cell death in either hemisphere as ascertained by the lack of apoptotic nuclei 24 hours post treatment. These data suggest that glial aromatase is more strongly associated with immune factors and less with cell

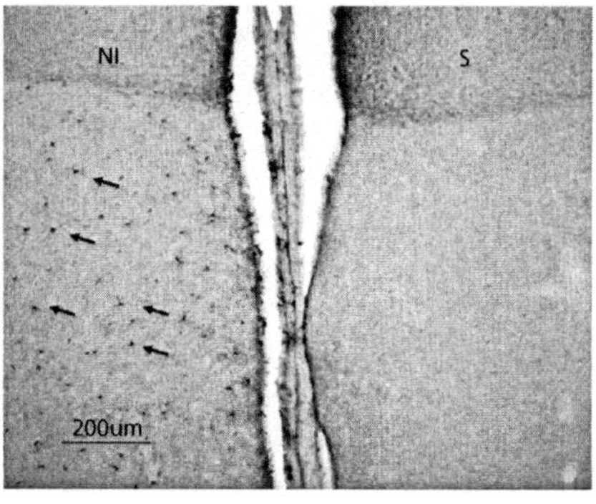

Figure 20–2. **Photomicrographs Illustrating the Induction of Aromatase Positive Glial Cells (Arrows) Following with Neuroinflammatory Treatment (NI) or Saline (S)** Neither side of the brain exhibited cell death as a result of the treatment. Image is modified from Duncan, K. A., & Saldanha, C. J. (2011). Neuroinflammation induces glial aromatase expression in the uninjured songbird brain. *J Neuroinflammation, 8*, 81.

death. It is important to acknowledge, however, that although there is reason to believe that inflammation is sufficient to induce aromatase in astroglia, we have yet to test the necessity of either inflammation or cell death in this process. However, cell death is a naturally occurring phenomenon in the passerine brain, and large numbers of cells die in the song circuit during the fall and winter nonetheless, astroglial aromatase has never been observed in the uninjured passerine brain (Alvarez-Buylla & Kirn, 1997; Smith et al., 1997).

The upregulation of glial aromatase by neuroinflammation suggests a novel interaction between the neuroimmune and neuroendocrine systems in the avian brain and supports the emerging theory that neuroinflammation and cytokine expression following injury are not always detrimental to the organism, but may be necessary for neuroprotective actions (Allan, 2000; Lenzlinger et al., 2001; Laird, Vender, & Dhandapani, 2008). Further research is needed to fully understand this complex interaction between neurotoxicity and neuroprotection, however, estrogens (the expected result of increased glial aromatase expression) and selective estrogen receptor modulators (SERM) counteract neuroinflammation by targeting expression of proinflammatory cytokines from astrocytes (Dodel et al., 1999; Dhandapani & Brann, 2002; Tapia-Gonzalez et al., 2008; Cerciat et al., 2010). These interactions suggest the tantalizing hypothesis that glial aromatization may serve as a potent and endogenous anti-inflammatory agent in the avian and mammalian brain.

AROMATASE AND NEUROPROTECTION

Effects on Secondary Degeneration

Upregulated aromatase appears to affect neurodegenerative pathways via an inhibition of apoptotic secondary degeneration (Wynne & Saldanha, 2004; Wynne et al., 2008). In the songbirds, as in the mammals (Azcoitia et al., 2001), central administration of the aromatase inhibitor fadrozole dramatically increases the volume of damage induced by a penetrating mechanical injury in the adult male zebra finch brain (Wynne & Saldanha, 2004) and replacement with E_2 prevents the exacerbation of damage by aromatase inhibition (Saldanha et al., 2005). Notably, the wave of secondary degeneration characteristic of mammalian brain damage is only detectable in the songbird brain following inhibition of injury-induced aromatase (Wynne et al., 2008). Correspondingly, TUNEL (*Terminal deoxynucleotidyl transferase* dUTP nick end labeling) in the songbird revealed that the aromatase inhibitor fadrozole increased, and concomitant E_2 administration decreased, the number of cells undergoing apoptosis 72 hours post primary insult (Wynne & Saldanha, 2004; Saldanha et al., 2005). This finding is in excellent agreement with the observed increase in apoptosis in aromatase knockout mice (Hill et al., 2009) and with molecular markers of apoptosis in which E_2 replacement is known to affect Bcl-2 expression (P. M. Wise et al., 2000).

Effects on Cytogenesis and Migration

As mentioned, damage to the vertebrate brain induces aromatase in radial glia as well; a cell type that is perhaps more involved with the regulation of cytogenesis and cell migration following neural damage (Peterson et al., 2004). Radial glia have long been studied as important scaffolds of neuronal migration during development (Rosen, Sherman, & Galaburda, 1994; Hatten & Heintz, 1995); and their potential role in restorative processes following neural injury has been intensely studied in animals as diverse as lizards (Romero-Aleman et al., 2004) and rodents (Yang et al., 1997). In the songbirds, the studies of Peterson et al. (2004) convincingly demonstrate that injury-induced mitotic cells aggregate around, possibly contact, and appear to migrate along injury-induced aromatase expressing radial glia. Thus, aromatase expression in radial glia may be key in the processes of neuronal migration towards the restoration of brain structure and function following brain damage. Indeed, systemic treatment with fadrozole does reduce the number of injury-induced mitotic cells in the subventricular zone of the songbird brain (Lee et al., 2007), suggesting that astrocytic aromatization may also contribute to cell proliferation.

Recently, we directly tested the role of encephalic estrogen provision following damage on cell proliferation in zebra finches. Our findings strongly support a role for locally produced estrogens, likely of glial origin on cytogenesis and neurogenesis in the songbird brain. More specifically, although treatment with fadrozole did not have any effect, concomitant estrogen treatment increased the number of cells double-labeled for the mitotic marker BrdU (bromodeoxyuridine) and the neuronal marker Hu in the injured songbird brain. Taken together, the data strongly suggest a pluripotent role for glial aromatization following brain injury in the songbird; one that involves a mitigating effect on secondary, apoptotic cell death and an acceleration of potentially restorative effect on neurogenesis. However, whether or not these effects have behavioral consequences is only beginning to be discovered.

Recovery of Motor and Cognitive Functions Following Injury and Induced Aromatase

To the best of our knowledge only a single study has demonstrated a behavioral role of injury-induced aromatase and estrogen provision in any vertebrate. Spence et al. (2009) created small lesions within the cerebellum of adult female zebra finches while blocking estrogen synthesis with or without estrogen replacement around the lesion site (Spence et al., 2009). Birds were tested on a spatial memory task as well as a motor task as a function of experimental group. Injury to the cerebellum resulted in the induction of glial aromatase, suggesting that increased glial aromatase is a neuroprotective mechanism throughout the brain, not just in steroid sensitive regions (Spence et al., 2009; Mirzatoni et al., 2010). Although no effects on motor function were detected, estrogen provision via injury-induced astrocytes and Bergmann glia appeared to lessen the errors made on a spatial memory task testing (Spence et al., 2009). Notably, aromatase is only detectable in the songbird cerebellum following damage (Mirzatoni et al., 2010) strongly supporting the hypothesis that the behavioral effects observed above were owing to glial aromatase and not constitutive estrogen provision (Fig. 20-3).

IS NEUROPROTECTION RELATED TO NEURODEVELOPMENT?

The songbird brain provides a unique model for the study of aromatase following brain injury and the natural mechanisms underlying neuroprotection and repair. The high expression of glial aromatase following injury coupled with the inhibition of secondary degeneration suggests that the songbird brain is naturally protected against damage (Wynne et al., 2008; Saldanha et al., 2009). We suggest that the capacity and/or ability of the songbird brain to respond to injury may be a repeated occurrence of certain profiles of gene expression typically seen during neural development (Saldanha et al., 1998, 2009). Following injury, we propose that the brain reactivates these estrogen-mediated developmental pathways as mechanisms for use during neural recovery and repair. A key finding that supports this hypothesis was the upregulation of bone morphogenetic proteins (BMPs) in the avian brain following injury (Walters & Saldanha, 2008). Bone morphogenetic proteins are well known for their roles

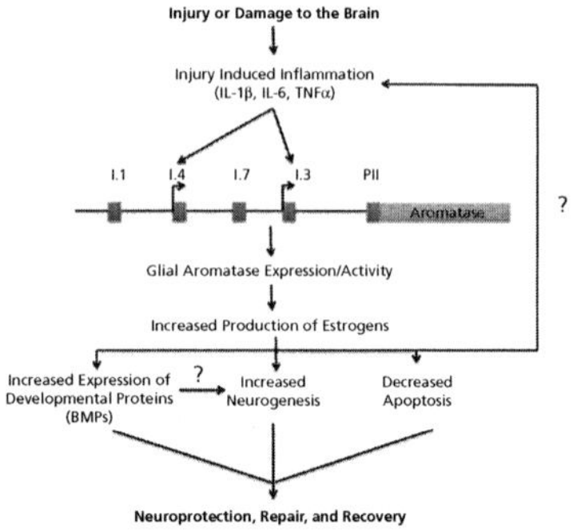

Figure 20–3. **Current Model of Injury Induced Aromatase Production on Neuroprotection in the Avian Brain** Briefly, following injury, aromatase is upregulated in glial cells around the injury site. This upregulation is hypothesized to be caused by increases in proinflammatory cytokines (IL-1β, IL-6, TNFα) that activate promoters on the aromatase gene (I.4, I.3) within glial cells. The subsequent production of estrogens promotes neuroprotection via various mechanisms including increased neurogenesis, decreased apoptosis, and increased expression of developmental proteins. Specifically, one set of developmental proteins, bone morphogenetic proteins (BMPs), like aromatase are expressed constitutively in the zebra finch brain, but are also upregulated around the injury site. The culmination of these various processes is increased neuroprotection and neural and behavioral recovery, but the precise mechanism is still unknown.

in neural development and the promotion of cellular proliferation and survival. Interestingly, when birds were given fadrozole at the time of injury, BMP2 expression was decreased (Walters & Saldanha, 2008). These data suggest that local aromatization contributes to the upregulation of BMP2 following injury and provides the link between developmental pathways and injury-induced aromatase and also expands our current model of neuroprotection in the avian brain (Walters & Saldanha, 2008). However, whether or not aromatization is observed in glial cells during development of the passerine brain remains unknown. More research is needed to fully elucidate the interactions between aromatase, developmental processes, neuroinflammation, and neuroprotection and recovery in the avian brain.

ACKNOWLEDGMENTS

The authors thank all past and present members of the laboratory for their assistance in conducting several of the experiments described here. This work was supported by NIH NS 042767.

REFERENCES

Allan, S. M. (2000). The role of pro- and antiinflammatory cytokines in neurodegeneration. *Ann N Y Acad Sci, 917,* 84–93.

Allan, S. M., Harrison, D. C., Read, S., Collins, B., Parsons, A. A., & Philpott, K. (2001). Selective increases in cytokine expression in the rat brain in response to striatal injection of alpha-amino-3-hydroxy-5-methyl-4-isoxazolepropionate and interleukin-1. *Brain Res Mol Brain Res, 93*(2), 180–189.

Allan, S. M., & Rothwell, N. J. (2001). Cytokines and acute neurodegeneration. *Nat Rev Neurosci, 2*(10), 734–744.

Alvarez-Buylla, A., & Kirn, J. R. (1997). Birth, migration, incorporation, and death of vocal control neurons in adult songbirds. *J Neurobiol, 33*(5), 585–601.

Azcoitia, I., Sierra, A., Veiga, S., Honda, S., Harada, N., & Garcia-Segura, L. M. (2001). Brain aromatase is neuroprotective. *J Neurobiol, 47,* 318–329.

Balthazart J., Foidart A., Surlemont C., Vockel A., & Harada N. (1990). Distribution of aromatase in the brain of the Japanese quail, ring dove, and zebra finch: an immunocytochemical study. *J Comp Neurol, 301*(2), 276–288.

Balthazart, J., Taziaux, M., Holloway, K., Ball, G. F., & Cornil, C. A. (2009). Behavioral effects of brain-derived estrogens in birds. *Ann N Y Acad Sci, 1163,* 31–48.

Belcredito, S., Vegeto, E., Brusadelli, A., Ghisletti, S., Mussi, P., & Ciana, P. (2001). Estrogen neuroprotection: the involvement of the Bcl-2 binding protein BNIP2. *Brain Res Brain Res Rev, 37*(1–3), 335–342.

Bryant, D. N., Sheldahl, L. C., Marriott, L. K., Shapiro, R. A., & Dorsa, D. M. (2006). Multiple pathways transmit neuroprotective effects of gonadal steroids. *Endocrine, 29*(2), 199–207.

Carswell, H. V., Dominiczak, A. F., Garcia-Segura, L. M., Harada, N., Hutchison, J. B., & Macrae, I. M. (2005). Brain aromatase expression after experimental stroke: topography and time course. *J Steroid Biochem Mol Biol, 96*(1), 89–91.

Cerciat, M., Unkila, M., Garcia-Segura, L. M., & Arevalo, M. A. (2010). Selective estrogen receptor modulators decrease the production of interleukin-6 and interferon-gamma-inducible protein-10 by astrocytes exposed to inflammatory challenge in vitro. *Glia, 58*(1), 93–102.

Chiueh, C., Lee, S., Andoh, T., & Murphy, D. (2003). Induction of antioxidative and antiapoptotic thioredoxin supports neuroprotective hypothesis of estrogen. *Endocrine, 21*(1), 27–31.

Cornil, C. A., Ball, G. F., & Balthazart, J. (2006). Functional significance of the rapid regulation of brain estrogen action: where do the estrogens come from? *Brain Res, 1126*(1), 2–26.

DeLegge, M. H., & Smoke, A. (2008). Neurodegeneration and inflammation. *Nutr Clin Pract, 23*(1), 35–41.

Dhandapani, K. M., & Brann, D. W. (2002). Protective effects of estrogen and selective estrogen receptor modulators in the brain. *Biol Reprod, 67*(5), 1379–1385.

Dodel, R. C., Du, Y., Bales, K. R., Gao, F., & Paul, S. M. (1999). Sodium salicylate and 17beta-estradiol attenuate nuclear transcription factor NF-kappaB translocation in cultured rat astroglial cultures following exposure to amyloid A beta(1–40) and lipopolysaccharides. *J Neurochem, 73*(4), 1453–1460.

Duncan, K. A., & Saldanha, C. J. (2011). Neuroinflammation induces glial aromatase expression in the uninjured songbird brain. *J Neuroinflammation, 8*, 81.

Forlano, P. M., Schlinger, B. A., & Bass, A. H. (2006). Brain aromatase: new lessons from non-mammalian model systems. *Front Neuroendocrinol, 27*(3), 247–274.

Garcia-Segura, L.M., Wozniak, A., Azcoitia, I., Rodriguez, J.R., Hutchison, R.E., Hutchison, J.B. (1999). Aromatase expression by astrocytes after brain injury: implications for local estrogen formation in brain repair.Neuroscience 89, 567–578

Ghirnikar, R. S., Lee, Y. L., & Eng, L. F. (1998). Inflammation in traumatic brain injury: role of cytokines and chemokines. *Neurochem Res, 23*(3), 329–340.

Hatten, M. E., & Heintz, N. (1995). Mechanisms of neural patterning and specification in the developing cerebellum. *Annu Rev Neurosci, 18*, 385–408.

Hill, R. A., Chua, H. K., Jones, M. E., Simpson, E. R., & Boon, W. C. (2009). Estrogen deficiency results in apoptosis in the frontal cortex of adult female aromatase knockout mice. *Mol Cell Neurosci, 41*(1), 1–7.

Honma, S., Shimodaira, K., Shimizu, Y., Tsuchiya, N., Saito, H., & Yanaihara, T. (2002). The influence of inflammatory cytokines on estrogen production and cell proliferation in human breast cancer cells. *Endocr J, 49*(3), 371–377.

Laird, M. D., Vender, J. R., & Dhandapani, K. M. (2008). Opposing roles for reactive astrocytes following traumatic brain injury. *Neurosignals, 16*(2–3), 154–164.

Lee, D. W., Fernando, G., Peterson, R. S., Allen, T. A., & Schlinger, B. A. (2007). Estrogen mediation of injury-induced cell birth in neuroproliferative regions of the adult zebra finch brain. *Dev Neurobiol, 67*(8), 1107–1117.

Lenzlinger, P. M., Morganti-Kossmann, M. C., Laurer, H. L., & McIntosh, T. K. (2001). The duality of the inflammatory response to traumatic brain injury. *Mol Neurobiol, 24*(1–3), 169–181.

Lucas, S. M., Rothwell, N. J., & Gibson, R. M. (2006). The role of inflammation in CNS injury and disease. *Br J Pharmacol, 147*(Suppl 1), S232–S240.

MacLusky, N. J., & Naftolin, F. (1981). Sexual differentiation of the central nervous system. *Science, 211*(4488), 1294–1302.

Maggi, A., Ciana, P., Belcredito, S., & Vegeto, E. (2004). Estrogens in the nervous system: mechanisms and nonreproductive functions. *Annu Rev Physiol, 66*, 291–313.

Marciano, P. G., Eberwine, J. H., Ragupathi, R., Saatman, K. E., Meaney, D. F., & McIntosh, T. K. (2002). Expression profiling following traumatic brain injury: a review. *Neurochem Res, 27*(10), 1147–1155.

McEwen, B., Akama, K., Alves, S., Brake, W. G., Bulloch, K., & Lee, S. (2001). Tracking the estrogen receptor in neurons: implications for estrogen-induced synapse formation. *Proc Natl Acad Sci U S A, 98*(13), 7093–7100.

Mirzatoni, A., Spence, R. D., Naranjo, K. C., Saldanha, C. J., & Schlinger, B. A. (2010). Injury-induced regulation of steroidogenic gene expression in the cerebellum. *J Neurotrauma, 27*(10), 1875–1882.

Morioka, M., Shimodaira, K., Kuwano, Y., Fujikawa, H., Saito, H., & Yanaihara, T. (2000). Effect of interleukin-1beta on aromatase activity and cell proliferation in human osteoblast-like cells (HOS). *Biochem Biophys Res Commun, 268*(1), 60–64.

Naftolin, F., Horvath, T. L., Jakab, R. L., Leranth, C., Harada, N., & Balthazart, J. (1996). Aromatase immunoreactivity in axon terminals of the vertebrate brain. An immunocytochemical study on quail, rat, monkey and human tissues. *Neuroendocrinology, 63*(2), 149–155.

Pellegrini, E., Mouriec, K., Anglade, I., Menuet, A., Le Page, Y., & Gueguen, M. M. (2007). Identification of aromatase-positive radial glial cells as progenitor cells in the ventricular layer of the forebrain in zebrafish. *J Comp Neurol, 501*(1), 150–167.

Peterson, R. S., Lee, D. W., Fernando, G., & Schlinger, B. A. (2004). Radial glia express aromatase in the injured zebra finch brain. *J Comp Neurol, 475*(2), 261–269.

Peterson, R. S., Saldanha, C. J., & Schlinger, B. A. (2001). Rapid upregulation of aromatase mRNA and protein following neural injury in the zebra finch (Taeniopygia guttata). *J Neuroendocrinol, 13*(4), 317–323.

Purohit, A., Ghilchik, M. W., Duncan, L., Wang, D. Y., Singh, A., & Walker, M. M. (1995). Aromatase activity and interleukin-6 production by normal and malignant breast tissues. *J Clin Endocrinol Metab, 80*(10), 3052–3058.

Purohit, A., Ghilchik, M. W., Leese, M. P., Potter, B. V., & Reed, M. J. (2005). Regulation of aromatase activity by cytokines, PGE2 and 2-methoxyoestrone-3-O-sulphamate in fibroblasts derived from normal and malignant breast tissues. *J Steroid Biochem Mol Biol, 94*(1–3), 167–172.

Purohit, A., Newman, S. P., & Reed, M. J. (2002). The role of cytokines in regulating estrogen synthesis: implications for the etiology of breast cancer. *Breast Cancer Res, 4*(2), 65–69.

Reed, M. J., Topping, L., Coldham, N. G., Purohit, A., Ghilchik, M. W., & James, V. H. (1993). Control of aromatase activity in breast cancer cells: the role of cytokines and growth factors. *J Steroid Biochem Mol Biol, 44*(4–6), 589–596.

Romero-Aleman, M. M., Monzon-Mayor, M., Yanes, C., & Lang, D. (2004). Radial glial cells, proliferating periventricular cells, and microglia might contribute to successful structural repair in the cerebral cortex of the lizard Gallotia galloti. *Exp Neurol, 188*(1), 74–85.

Roselli, C. F. (2007). Brain aromatase: roles in reproduction and neuroprotection. *J Steroid Biochem Mol Biol, 106*(1–5), 143–150.

Rosen, G. D., Sherman, G. F., & Galaburda, A. M. (1994). Radial glia in the neocortex of adult rats: effects of neonatal brain injury. *Brain Res Dev Brain Res, 82*(1–2), 127–135.

Rothwell, N. J., & Strijbos, P. J. (1995). Cytokines in neurodegeneration and repair. *Int J Dev Neurosci, 13*(3–4), 179–185.

Saldanha, C. J., Duncan, K. A., & Walters, B. J. (2009). Neuroprotective actions of brain aromatase. *Front Neuroendocrinol, 30*(2), 106–118.
Saldanha, C. J., Popper, P., Micevych, P. E., & Schlinger, B. A. (1998). The passerine hippocampus is a site of high aromatase: inter- and intraspecies comparisons. *Horm Behav, 34*(2), 85–97.
Saldanha, C. J., Remage-Healey, L., & Schlinger, B. A. (2011). Synaptocrine signaling: steroid synthesis and action at the synapse. *Endocr Rev, 32*(4), 532–549.
Saldanha, C. J., Rohmann, K. N., Coomaralingam, L., & Wynne, R. D. (2005). Estrogen provision by reactive glia decreases apoptosis in the zebra finch (Taeniopygia guttata). *J Neurobiol, 64*(2), 192–201.
Saldanha, C.J., Schultz, J.D., London, S.E., Schlinger, B.A. (2000b). Telencephalic aromatase but not a song circuit in a sub-oscine passerine, the golden collared manakin (Manacus vitellinus). *Brain Behav Evol. 56*(1), 29–37.
Saldanha, C.J., Tuerk, M.J., Kim, Y.H., Fernandes, A.O., Arnold, A.P., Schlinger, B.A. (2000a). Distribution and regulation of telencephalic aromatase expression in the zebra finch revealed with a specific antibody. *J Comp Neurol, 423*(4), 619–630.
Saldanha, C. J., Walters, B. J., & Fraley, G. S. (2010). Neurons that co-localize aromatase- and kisspeptin-like immunoreactivity may regulate the HPG axis of the Mallard drake (Anas platyrhynchos). *Gen Comp Endocrinol, 166*(3), 606–613.
Schlinger, B. A., Amur-Umarjee, S., Shen, P., Campagnoni, A. T., & Arnold, A. P. (1994). Neuronal and non-neuronal aromatase in primary cultures of developing zebra finch telencephalon. *J Neurosci, 14*(12), 7541–7552.
Schumacher, M., & Balthazart, J. (1983). The effects of testosterone and its metabolites on sexual behavior and morphology in male and female Japanese quail. *Physiol Behav, 30*(3), 335–339.
Simi, A., Tsakiri, N., Wang, P., & Rothwell, N. J. (2007). Interleukin-1 and inflammatory neurodegeneration. *Biochem Soc Trans, 35*(Pt 5), 1122–1126.
Simpkins, J. W., & Dykens, J. A. (2008). Mitochondrial mechanisms of estrogen neuroprotection. *Brain Res Rev, 57*(2), 421–430.
Simpkins, J. W., Wang, J., Wang, X., Perez, E., Prokai, L., & Dykens, J. A. (2005). Mitochondria play a central role in estrogen-induced neuroprotection. *Curr Drug Targets CNS Neurol Disord, 4*(1), 69–83.
Smith, G. T., Brenowitz, E. A., Beecher, M. D., & Wingfield, J. C. (1997). Seasonal changes in testosterone, neural attributes of song control nuclei, and song structure in wild songbirds. *J Neurosci, 17*(15), 6001–6010.
Spence, R. D., Zhen, Y., White, S., Schlinger, B. A., & Day, L. B. (2009). Recovery of motor and cognitive function after cerebellar lesions in a songbird: role of estrogens. *Eur J Neurosci, 29*(6), 1225–1234.
Tapia-Gonzalez, S., Carrero, P., Pernia, O., Garcia-Segura, L. M., & Diz-Chaves, Y. (2008). Selective oestrogen receptor (ER) modulators reduce microglia reactivity in vivo after peripheral inflammation: potential role of microglial ERs. *J Endocrinol, 198*(1), 219–230.
Veerapaneni, P., Kirma, N., Nair, H. B., Hammes, L. S., Hall, K. L., & Tekmal, R. R. (2009). Elevated aromatase expression correlates with cervical carcinoma progression. *Gynecol Oncol, 114*(3), 496–500.
Walters, B. J., Alexiades, N. G., & Saldanha, C. J. (2011). Intracerebral estrogen provision increases cytogenesis and neurogenesis in the injured zebra finch brain. *Dev Neurobiol, 71*(2), 170–181.

Walters, B. J., & Saldanha, C. J. (2008). Glial aromatization increases the expression of bone morphogenetic protein-2 in the injured zebra finch brain. *J Neurochem, 106*(1), 216–223.

Wise, P. (2003). Estradiol exerts neuroprotective actions against ischemic brain injury: insights derived from animal models. *Endocrine, 21*(1), 11–15.

Wise, P. M., Dubal, D. B., Wilson, M. E., & Rau, S. W. (2000). Estradiol is a neuroprotective factor in in vivo and in vitro models of brain injury. *J Neurocytol, 29*(5–6), 401–410.

Wynne, R. D., & Saldanha, C. J. (2004). Glial aromatization decreases neural injury in the zebra finch (Taeniopygia guttata): influence on apoptosis. *J Neuroendocrinol, 16*(8), 676–683.

Wynne, R. D., Walters, B. J., Bailey, D. J., & Saldanha, C. J. (2008). Inhibition of injury-induced glial aromatase reveals a wave of secondary degeneration in the songbird brain. *Glia, 56*(1), 97–105.

Xie, J., Kuenzel, W. J., Sharp, P. J., & Jurkevich, A. (2011). Appetitive and consummatory sexual and agonistic behaviour elicits FOS expression in aromatase and vasotocin neurones within the preoptic area and bed nucleus of the stria terminalis of male domestic chickens. *J Neuroendocrinol, 23*(3), 232–243.

Yang, H. Y., Lieska, N., Kriho, V., Wu, C. M., & Pappas, G. D. (1997). A subpopulation of reactive astrocytes at the immediate site of cerebral cortical injury. *Exp Neurol, 146*(1), 199–205.

Part VI

Rapid Changes in Aromatase Activity & Rapid Effects of Estrogens

Chapter 21

Photoperiodic Regulation of Estrogen-Dependent Aggression

Sarah A. Laredo and Brian C. Trainor

INTRODUCTION

After the discovery of the structure of deoxyribonucleic acid, endocrinologists demonstrated that steroid receptors were concentrated in the cell nucleus (Toft & Gorski, 1966; Jensen et al., 1968). The identification of estrogen receptor genes and sequences of hormone response elements allowed for an in-depth study of the molecular mechanisms responsible for hormone action (Green et al., 1986; Kuiper et al., 1996). After estrogens bind to receptors such as estrogen receptor alpha (ERα), the hormone-receptor complex enters the nucleus and binds to DNA sequences such as estrogen response elements or cyclic AMP response elements to regulate gene expression (Nilsson et al., 2001). Indeed one of the most intensely studied estrogen-dependent behaviors, lordosis, is mediated in part by ERα driving the expression of progesterone receptors in the hypothalamus (Romano et al., 1989). This upregulation of progesterone receptor protein takes at least 2 hours (Shughrue et al., 1997) and primes the brain to become sensitive to a later surge in progesterone (Blaustein et al., 1990; Pfaff et al., 1994). Although immediate early genes such as c-fos can be upregulated by steroids within minutes (Cattaneo & Maggi, 1990), changes in biologically active proteins can take an hour or longer to manifest (Zangenehpour & Chaudhuri, 2002). Most proteins that are estrogen dependent respond relatively slowly (Barnea & Gorski, 1970), sometimes not peaking until 48 hours after estrogen levels increase.

However, steroids also act in parallel with much shorter latencies and a lot of interest has recently been devoted to these more rapid effects of estrogens (see Pfaff & Levine, 2008 for review) that are usually assumed to be mediated by other cellular processes, often referred to as "nongenomic." Interest in nongenomic effects of steroids stems in part from observations showing neuronal activity changes within seconds of estradiol infusions (Kelly et al., 1976, 1977). An ongoing question since these initial observations is whether these rapid effects of estrogens are mediated by the classical estrogen receptors: ERα and ERβ. Initially this was thought to be unlikely. Standard immunohistochemistry analyses show that estrogen receptors are predominantly expressed in the cell nucleus (King & Greene, 1984), and presumably rapid effects of estrogens occur outside of the nucleus. However, several technical advances have helped to resolve this question. The integration of electron microscopy methods with estrogen receptor immunohistochemistry showed that ERα and ERβ are indeed localized to extranuclear sites such as dendritic spines, axons, and terminals (Blaustein et al., 1992; Milner et al., 2001; Towart et al., 2003; Milner et al., 2005). Cell culture studies also show that receptors encoded by the ERα and ERβ genes are present in membrane preparations (Razandi et al., 1999). These observations suggest that classical ERs are in a position to rapidly affect neuronal activity. However, these are not the only conduits for rapid effects of estrogens on the brain. The G-protein-coupled receptor 30 (GPR30) is sometimes referred to as the "nongenomic estrogen receptor." Immunohistochemical analyses show that GPR30 is present in both the membrane and cytoplasm in the forebrain, hypothalamus, and midbrain (Brailoiu et al., 2007; Sakamoto et al., 2007). An additional membrane estrogen receptor has been identified in hypothalamic neurons expressing POMC or dopamine (Qiu et al., 2008). In summary, there are several molecular pathways through which estrogens can rapidly modulate behavior.

Male sexual behavior has been shown to be rapidly regulated by estrogens. Castrated male rats showed increased sexual behaviors just 30 minutes after an estradiol injection (Cross & Roselli, 1999), whereas previous studies showed a link between estradiol and sexual behavior in males treated with implants for 1 to 3 days (Davidson, 1969; Sodersten, 1973). Rapid effects of estradiol on male sexual behavior have also been observed in quail (Cornil et al., 2006). Intriguingly, even the prototypical genomic effects of estrogens on female sexual behavior of the rat (lordosis) may be subject to modulation by nongenomic pathways (Kow & Pfaff, 2004). Membrane impermeable preparations of estradiol (which presumably acts via nongenomic mechanisms because it cannot enter the nucleus) were found to facilitate the effects of estradiol on lordosis behavior (Kow & Pfaff, 2004), an effect that is mediated in part by the activation of metabotropic glutamate receptors (Dewing et al., 2007). The effects of estrogens on gene expression in cultured neurons are also facilitated when preceded by a short pulse of membrane impermeable estradiol, suggesting that nongenomic actions of estrogens may facilitate subsequent transcriptional actions of estrogens (Vasudevan et al., 2001). Together, these data suggest that nongenomic effects of estradiol can facilitate genomic responses and that these interactions can modulate complex reproductive behaviors.

In many species, mating opportunities are linked to the outcome of competitions for limited resources. The aggressive behaviors that determine these outcomes are also steroid sensitive (Trainor & Nelson, 2012), which raises the question of whether estrogens exert rapid effects on these behaviors as well. The intensity of aggressive behavior is highly dependent on experience. Previous experience in competitive encounters (Huhman et al., 2003; Fuxjager & Marler, 2010) and reproductive status (Insel et al., 1995) are major factors regulating the intensity of aggressive behavior. In addition to these social factors, environmental cues such as population density and temperature also influence aggressive behavior (Southwick, 1955). Annual changes in photoperiod are often accurate predictors of seasonal changes in the environment, and have been found to exert important effects on estrogen-dependent aggression. Intriguingly, there is growing evidence that the effects of photoperiod on estrogen-dependent aggression are independent of changes in gonadal hormones. This chapter will discuss how photoperiod modulates estrogen-signaling systems to regulate aggressive behavior. The focus is primarily on rodents, almost all of which display a robust high aggression phenotype under winter-like short days. We review emerging evidence that the effects of photoperiod on aggression are independent of gonadal hormones, as well as evidence for the differential importance of genomic and nongenomic pathways. Finally, we consider possible brain networks that may mediate the effects of photoperiod and estrogens on aggressive behavior.

PHOTOPERIOD AND AGGRESSION

Environmental factors play an important role in driving an organism's behavioral state. Seasonal changes in photoperiod (day length) are salient cues that predict long-term changes in temperature and resource availability in nontropical locations. More specifically, in many species photoperiod drives fluctuations in physiological mechanisms that contribute to aggression, a behavior that is tightly coupled with the breeding season (Caldwell & Albers, 2004; Fiszbein et al., 2010; Sperry et al., 2010). In nontropical locations, the breeding season usually occurs in the spring or summer when food resources are usually more abundant (Bronson, 1985), and long summer-like days are sufficient to stimulate the reproductive axis (Pospichal et al., 1991). Winter is usually the nonbreeding season and is associated with regression of the reproductive axis, including testosterone production (Dark et al., 1983). Mammals enter a period of photorefractoriness in the winter months in which short days fail to inhibit the reproductive axis (Freeman & Zucker, 2001). During the spring and summer seasons (long day lengths), seasonal breeders retain stimulated reproductive systems which coincides with peak mating opportunities and rearing of offspring. As a consequence of greater daylight exposure during the breeding season, mammals no longer demonstrate photorefractoriness to short day lengths, allowing for gonadal regression upon the return of winter (Butler et al., 2010). In contrast, photorefractoriness in avian species refers to the phenomenon that occurs towards the end of summer, when long days lose their ability to stimulate the avian reproductive system (reviewed in Dawson et al., 2001).

MELATONIN AND THE HYPOTHALAMIC-PITUITARY GONADAL AXIS

Melatonin is a key regulator of an organism's behavioral and physiological response to photoperiod. It is a hormone that is released by the pineal gland with peak activation occurring at night, and basal conditions occurring during the daylight hours. Melatonin can maintain endogenous circadian rhythms, translating environmental photoperiodic cues into biological cycles. When light enters the retina, it is transmitted to the paraventricular nucleus of the hypothalamus via the suprachiasmatic nucleus, ultimately regulating the secretion of melatonin from the pineal gland (for review see Borjigin et al., 1999). Melatonin secretion peaks in the dark phase, and can thus act as a biological signal of day length to different physiological systems (Bartness et al., 1993). There is also evidence that melatonin has specific effects on estrogen signaling. Activation of the type 1 melatonin receptor (MT_1) by melatonin reduces expression of cyclic AMP (Witt-Enderby et al., 1998), and in cell culture studies, this action of MT_1 receptors reduces expression of ERα as well as estrogen-dependent gene expression (Ram et al., 1998; Yuan et al., 2002). Melatonin may also affect estrogen function independently of MT_1. Melatonin was also observed to bind directly to calmodulin (del Rio et al., 2004), which functions as a steroid receptor coactivator facilitating the binding of ERα dimers to DNA response elements. Melatonin binding to calmodulin blocks interactions with ERα, resulting in a decrease in estrogen-dependent gene expression (Rato et al., 1999). Currently it is unclear whether melatonin influences brain function or behavior through these mechanisms.

In addition to demonstrating a daily cyclicity, melatonin also varies with the onset of seasonal changes in the environment. Animals exposed to short days receive longer daily exposures to melatonin compared to animals under long days. When long day housed Syrian hamsters (*Mesocricetus auratus*) are injected with long-duration melatonin treatments (mimicking short day photoperiods), their gonads regress (Bartness & Goldman, 1988; Leitner & Bartness, 2011). This effect appears to be caused by the inhibitory effects of melatonin on the hypothalamic-pituitary gonadal (HPG) axis. Activation of the G-coupled melatonin receptors (MT_1 and MT_2) expressed on gonadotropin-releasing hormone (GnRH) neurons appears to downregulate GnRH mRNA (Roy et al., 2001), which in turn leads to less stimulation of follicle-stimulating hormone (FSH) and luteinizing hormone (LH) release (Anand et al., 2002). Both LH and FSH act directly on the ovaries and testes to synthesize estradiol and testosterone, respectively; therefore, downregulation of LH and FSH would ultimately cause diminished sex steroid release. Melatonin also may modulate estrogen levels via inhibition of aromatase activity (Cos et al., 2006) and may inhibit the production of testosterone via binding to MT_1 receptors on Leydig cells (Frungieri et al., 2005). We can therefore expect significant behavioral changes between the breeding and nonbreeding seasons because of the physiological effects of melatonin on the HPG axis.

INTERACTION BETWEEN PHOTOPERIOD AND SEX STEROIDS

A link between sex steroid hormone concentrations and aggression has been established in a wide variety of species (Cavigelli & Pereira, 2000; Book et al., 2001; Pasch et al., 2011), in many cases mediated via the aromatase-driven conversion of androgens to estrogens (Soma et al., 2000; Trainor et al., 2006). Intriguingly, the effects that steroid hormones have on aggression appear to depend on seasonal cues. Male song sparrows (*Melospiza melodia morphna*) demonstrate territorial aggression throughout the year (Wingfield, 1994b). During the summer, aggressive behaviors are testosterone dependent (Wingfield, 1994a). However, in the winter nonbreeding season aggression is observed in males with regressed gonads (Wingfield, 1993) and castration does not affect aggression (Wingfield, 1994a). It has been hypothesized that in the absence of higher levels of gonadal steroids, sex steroid precursors (such as dehydroepiandrosterone, DHEA) secreted by the adrenal gland may play an important role in driving aggressive behavior (Soma & Wingfield, 2001; Demas et al., 2004; Scotti et al., 2009). It is also a possibility that sex steroids may be synthesized de novo in the brain (Holloway & Clayton, 2001). In support of this hypothesis, engaging in aggressive behavior increases the activity of 3 beta hydroxysteroid dehydrogenase (which converts DHEA to androstenedione) in the nonbreeding season but not the breeding season (Pradhan et al., 2010). These data suggest that androgens (and potentially estrogens) are produced locally within the brain in the nonbreeding season.

A similar disconnect between gonadal steroids and aggression has been observed in rodents under winter-like short day photoperiods. Surprisingly in most species of rodents studied, short days increase male aggression levels despite inhibiting gonadal size and testosterone secretion (Jasnow et al., 2000; Demas et al., 2004; Trainor et al., 2007b). Interestingly, not all male Siberian hamsters (*Phodopus sungorus*) housed under short days have regressed gonads (Prendergast et al., 2001). These "nonresponders," however, have the same high aggression phenotype as "responders" with regressed gonads (Wen et al., 2004). Similarly, male California mice (*Peromyscus californicus*) do not have regressed gonads under short days (Nelson et al., 1995), yet are more aggressive than males housed under long days (Trainor et al., 2010; Trainor et al., 2008). Together, these data suggest that the effects of photoperiod on aggression are not dependent on changes in gonadal hormones. However, despite evidence of dissociation between gonadal hormones and aggressive behavior, estrogens synthesized within the brain may have an important role in mediating aggression under differing photoperiodic conditions.

RAPID EFFECTS OF ESTROGENS ON BEHAVIOR

Initial observations suggested that the effects of short days on aggression might be mediated by changes in ERα and ERβ expression in brain regions known to modulate aggressive behavior (Nelson & Trainor, 2007), such as the lateral septum (LS), bed nucleus of the stria terminalis (BNST), and medial amygdala (MEA). In old-field mice (*Peromysucs polionotus*) and deer mice (*Peromyscus maniculatus*) males housed in short days had more ERα-ir cells and more ERα mRNA in the LS with a corresponding decrease in ERβ-ir cells and ERβ mRNA in the posterior BNST (Trainor et al., 2007b). The number of ERα-ir cells in the LS was also positively correlated with an aggressive composite score in old-field mice. Short days also reduced the number of ERβ–ir cells in the MEA. Overall these data were consistent with knockout studies that reported that ERα deletion reduced aggression and ERβ deletion increased aggression in certain contexts. Increased numbers of ERα-positive cells in the BNST and MEA under short days was also observed in Siberian hamsters (Kramer et al., 2008). However, increased aggression under short days was observed in male California mice despite the fact that there was no effect of photoperiod on the number of ERα- or ERβ-positive cells in any brain region examined (Trainor et al., 2008). These observations in California mice were not consistent with the hypothesis that the effects of short days on aggressive behavior are mediated by changes in nuclear estrogen receptor expression.

To test whether changes in classical estrogen receptor expression mediate the effects of photoperiod on aggressive behavior, subtype specific ER agonists were administered to male old-field mice under different photoperiods (Trainor et al., 2007a). All males were castrated and most were treated with an aromatase inhibitor (fadrozole), which prevents the conversion of testosterone to estradiol. Some mice were also cotreated with estradiol, an ERα-selective agonist (propylpyrazole-triol, PPT), or an ERβ-selective agonist (diarylpropionitrile, DPN) to determine the effect of each receptor type on aggressive behavior. Fadrozole increased aggressive behaviors compared with saline treatment in long day males, whereas fadrozole decreased aggressive behaviors compared with saline in short day males. When cotreated with estradiol, the effects of fadrozole on aggressive behavior were blocked, indicating that estrogens have opposing effects on aggression under differing photoperiodic conditions. Treatment with PPT and DPN yielded equivalent behavioral results, with both ERα and ERβ agonists increasing aggression in short days and decreasing aggression in long days (Fig. 21–1).

Previous knockout studies suggested that ERα and ERβ have opposing effects on aggressive behavior (Ogawa et al., 1997, 1999). However, experiments in old-field mice suggest that the relationship between estrogen receptors and aggressive behaviors is more complex. One possibility is that behavioral phenotypes observed in knockout studies (i.e., Ogawa et al., 1997, 1999) may be explained by the lack of ERα and ERβ receptors during development, which could alter maturing neurocircuitry. Alternatively, the effects of ERα and ERβ on aggression may depend on experience. For example, the effects of ERβ on aggression are stronger in younger mice than older mice (Nomura et al., 2002). The effects of estrogen receptors on behavior may also be dependent on environmental factors. One possible mechanism for this interaction would be that the environment regulates the downstream pathways that are activated by estrogen receptors. This hypothesis was tested indirectly using microarrays to measure estrogen dependent genes in mRNA samples from the BNST of male old-field mice housed under long days or short days. The analysis of expression data focused on genes regulated by estrogen response elements. Overall, 11 of these genes were differentially expressed, of which 9 were upregulated in long days. A followup qPCR study on one of these transcripts (*XRCC1*) showed that upregulated expression under long days was estrogen dependent. Although not definitive, these data suggested that estrogen-dependent gene expression in the

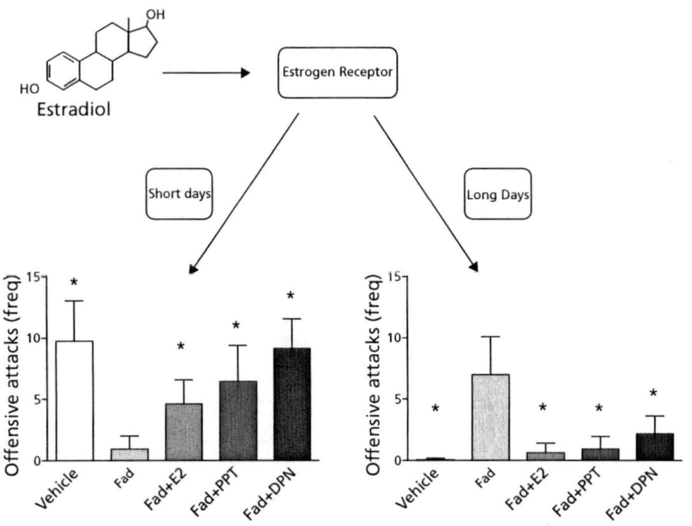

Figure 21–1. Photoperiod determines the direction of the effects of estrogens on aggressive behavior in beach mice (*Peromyscus polionotus*). Beach mice are more aggressive when exposed to short days (shown in the left graph) than when exposed to long days (shown in the right graph). Treatment with the estrogen synthesis inhibitor fadrozole (fad) decreases aggression if beach mice are tested in short days, but increases aggression if tested in long days. The effects of fad are reversed with cotreatment with estradiol (E_2). The compounds PPT (propylpyrazole-triol, an (ERα agonist) and DPN (diarylpropionitrile, an ERβ agonist) both reversed the effects of fadrozole in the same way. Both PPT and DPN treatment increased aggression under short days and decreased aggression under long days. Photoperiod apparently regulates the molecular actions of estrogens, acting rapidly on short days (presumably nongenomically, via ERα and ERβ associated with the cell membrane) and more slowly on long days (presumably genomically). Adapted from Trainor et al. (2007a) and Nelson and Trainor (2007).

BNST was weaker under short days, suggesting that short days might inhibit the transcriptional effects of estrogen receptors in the BNST. Based on these data, it was hypothesized that nongenomic pathways might be a more important class of mechanisms mediating the effects of estrogens on aggressive behavior under short days.

This hypothesis was tested by examining the rapid effects of estradiol on aggressive behavior, because nongenomic effects of hormones are known to occur rapidly (Aikey et al., 2002; Taziaux et al., 2007). First, mice were housed in either long days or short days for 8 weeks. Each mouse was then castrated and implanted with a minipump containing fadrozole. After 10 days, mice were randomly assigned to be injected with either saline or estradiol 15 minutes prior to a resident-intruder test. When injected with estradiol, short day mice displayed increased aggression, whereas no effect of estradiol was observed in long day mice (Fig. 21–2) (Trainor et al., 2007a). This same selective and rapid action of estradiol under short days

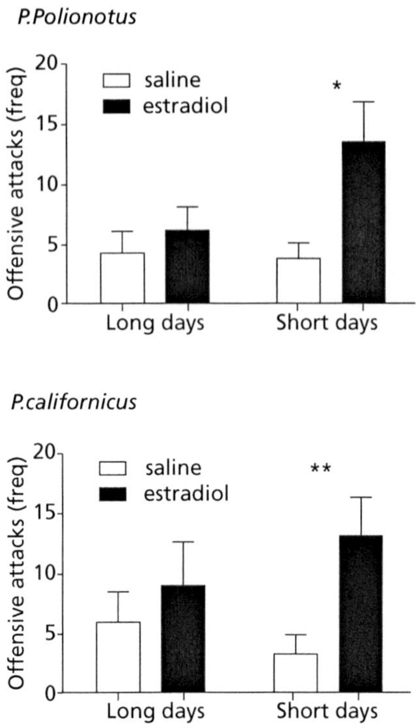

Fig. 21–2. The rapid effect of estradiol injection on aggression depends on photoperiod. Estradiol acts within 15 minutes to increase aggression when male (*A*) beach mice or (*B*) California mice are housed in short days, but not long days. $^{*}p < 0.05$, $^{**}p < 0.01$. Adapted from Trainor et al. (2007a) and Trainor et al. (2008).

was observed in California mice (see Fig. 21–2) (Trainor et al., 2008). The observations in California mice are significant because in this species short days have no effect on gonadal hormones or estrogen receptor immunoreactivity in the brain. However, despite the absence of these changes, the behavioral effects of estradiol were still modulated by short days. The California mouse data support the hypothesis that the effects of photoperiod on estrogen dependent aggression are not dependent on changes in the level of nuclear receptor expression. The rapid effects of estradiol were only observed under short days, which suggests that short days are modulating the molecular cascades that are activated by estrogen receptors.

EFFECTS OF PHOTOPERIOD AND AGGRESSION ON CORRELATES OF NEURAL ACTIVATION

Across many rodent species, males are more aggressive under short day photoperiods, yet two studies using c-fos immunohistochemistry as an indirect marker of brain activity found no effect of photoperiod on aggression-induced c-fos staining (Kramer et al., 2008; Trainor et al., 2008). Although c-fos is a widely used marker of brain activity, induction of c-fos can occur through several molecular pathways (Hoffman & Lyo, 2002). An alternative approach to assessing changes in brain activity is through measurements of phosphoproteins that can be rapidly activated or inactivated. Many of these proteins are upstream of c-fos transcription and translation and thus provide a more focused estimate of brain activity. An added benefit of focusing on phosphoproteins is that several of these proteins are known to be phosphorylated by estrogens. For example injections of 17-β estradiol induce phosphorylation of cyclic AMP response element binding protein (pCREB) in the medial preoptic nucleus and medial septum within 15 minutes of the first injection (Abraham, et al., 2004). Similarly, estradiol injections induce phosphorylation of extracellular signal-regulated kinase (ERK) (Abraham et al., 2004; Zsarnovszky et al., 2005). These rapid actions of estradiol on phosphorylation are significant because they can modulate neuronal activity and neurotransmitter release (Selcher et al., 2003).

In male California mice, engaging in resident-intruder aggression tests significantly increased the number of pERK -ir cells in the BNST and MEA under short days but not long days (Trainor et al., 2010). In addition, pERK cell counts in the BNST and MEA were positively correlated with aggressive behaviors. When these experiments were repeated using western blots to measure pERK in BNST punch samples, increased aggression-induced pERK expression was again observed in short days but not long days. These were the first data to show an effect of photoperiod on cellular activity within the brain that might be connected to the short day high aggression phenotype. These results were also intriguing in light of observations that estradiol rapidly induces phosphorylation of ERK. In contrast, observations in female California mice showed that engaging in aggression tests increased in pERK expression in the BNST and MEA regardless of photoperiod, despite increased aggression levels in short day, housed females (Silva et al., 2010). Currently it is unclear which mechanisms mediate the increased aggression phenotype in female rodents. Recent studies suggest that short day–induced aggression in females is not mediated by estrogen (Gutzler et al., 2009) or vasopressin (Gutzler et al., 2010) signaling.

HOUSING ENVIRONMENT AND ESTROGEN-DEPENDENT BEHAVIORS

Estrogen receptors are thought to be the first steroid receptor to have evolved within vertebrates (Thornton, 2001), and that this event allowed for estrogenic regulation of reproduction. However, it is becoming increasingly clear that other compounds in the environment possess the ability to alter estrogen signaling (Morito et al., 2001). One type of widely used bedding for laboratory rodents, corncob, is a significant source of phytoestrogens that have the ability to alter estrogen signaling (reviewed in Rochester & Millam, 2009). Both male and female rats housed on corncob bedding show significantly lower rates of estrogen-dependent mating behavior (Mani et al., 2005; Markaverich et al.,

2002b). Several studies reviewed above were conducted using corncob bedding (Trainor et al., 2007a, 2007b, 2008). When these studies were replicated using a cardboard bedding (Carefresh), there were some important differences. Fadrozole increased aggression if male California mice were kept on Carefresh bedding, whereas fadrozole tended to decrease aggression (consistent with previous data) when mice were kept on corncob bedding (Villalon Landeros et al., 2012). This study also provided additional insights on the potential mechanisms mediating the effects of corncob bedding on behavior. It was previously shown that the behaviorally active component of corncob bedding was tetrahydrofuran-diols (THF-diols), which do not bind to estrogen receptors (Markaverich et al., 2002a). In the California mouse study, corncob bedding decreased the number of ERα-ir cells in the BNST and VMH. These brain areas are known to mediate both aggressive and mating behaviors, suggesting that THF-diols may affect estrogen-dependent behavior not by binding to estrogen receptors, but by altering expression of receptors. Corncob bedding also suppressed pERK activity in several brain areas including the BNST, MPOA, MEA, and VMH. Many aspects of behavior and brain function are known to be mediated by ERK signaling, so these results suggest that corncob bedding influences a wide range of laboratory investigations. These results also provide insights into the relationships between ERK and aggression. The positive correlation between male California mouse aggression and pERK expression was observed in mice housed on Carefresh bedding (Trainor, et al., 2010). However, fadrozole treatment was found to increase aggression and decrease pERK expression under the same bedding conditions. Thus, although it appears that there may be an important relationship between pERK expression in the BNST and aggression, it is unlikely that phosphorylation of ERK is the mechanism through which estradiol acts rapidly to alter aggression.

Despite the lack of clarity surrounding the relationship between ERK and aggression, recent data support the notion that photoperiod alters the molecular cascades activated by estradiol (Laredo et al. unpublished). In this study, mice were tested on Carefresh bedding, and estradiol acted within 30 minutes to decrease aggression. Interestingly, this rapid effect was not blocked by cotreatment with a protein synthesis inhibitor. This indicates that in the absence of protein synthesis, short-day mice demonstrate a nongenomic behavioral effect. In long day photoperiods, rapid effects of estradiol on aggression were absent, indicating that estradiol likely regulates aggression via genomic mechanisms under long days. Thus, even though a different type of bedding was used, the previously reported findings that photoperiod modulates genomic versus nongenomic actions of estrogens were replicated.

CONCLUSIONS

Photoperiodic cues are an important factor mediating the relationship between estrogens and aggression. This has potentially important implications for researchers interested in seasonal variation in human behavior. Several reports suggest that there is seasonal variation in anger and hostility in humans, with the highest levels of negative affect recorded during winter months (Harmatz et al., 2000; Winkler et al., 2006). The human reproductive system is not dependent on variation in the light cycle and there are no prominent seasonal changes in gonadal steroid hormones concentrations, suggesting that these seasonal changes in behavior are not mediated by gonadal hormones such as estrogens. However, the revelation that the brain may be an important source of estrogens suggests the possibility that neuroestrogens could indeed have important effects on seasonal changes in behavior. For both human and animal research it will be important to develop experimental approaches that can assess rapid changes in estrogen synthesis as well as determine the rapid effects of estrogens on behavior in different environmental contexts.

ACKNOWLEDGMENTS

This research was supported by F32 MH076313 and R01 MH085069 to BCT. SAL is supported by an NSF Graduate Research Fellowship.

REFERENCES

Abraham, I. M., Todman, M. G., Korach, K. S., & Herbison, A. E. (2004). Critical in vivo roles for classical estrogen receptors in rapid estrogen actions on intracellular signaling in mouse brain. *Endocrinology, 145*(7), 3055–3061.

Aikey, J. L., Nyby, J. G., Anmuth, D. M., & James, P. J. (2002). Testosterone rapidly reduces anxiety in male house mice (*Mus musculus*). *Horm Behav, 42*(4), 448–460.

Anand, S., Losee-Olson, S., Turek, F. W., & Horton, T. H. (2002). Differential regulation of luteinizing hormone and follicle-stimulating hormone in male siberian hamsters by exposure to females and photoperiod. *Endocrinology, 143*(6), 2178–2188.

Barnea, A., & Gorski, J. (1970). Estrogen-induced protein. Time course of synthesis. *Biochem, 9,* 1899–1904.

Bartness, T. J., & Goldman, B. D. (1988). Peak duration of serum melatonin and short-day responses in adult Siberian hamsters. *Am J Physiol, 255*(5 Pt 2), 812–822.

Bartness, T. J., Powers, J. B., Hastings, M. H., Bittman, E. L., & Goldman, B. D. (1993). The timed infusion paradigm for melatonin delivery: what has it taught us about the melatonin signal, its reception, and the photoperiodic control of seasonal responses? *J Pineal Res, 15,* 161–190.

Blaustein, J. D., Lehman, M. N., Turcotte, J. C., & Greene, G. (1992). Estrogen receptors in dendrites and axon terminals in the guinea pig hypothalamus. *Endocrinology, 131*(1), 281–290.

Blaustein, J. D., Olster, D. H., Delville, Y., Nielson, K. H., Tetel, M. J., & Turcotte, J. C. (1990). Hypothalamic sex steroid hormone receptors and female sexual behavior: new insights from immunocytochemical studies. In: Balthazart, J. (ed.), *Hormones, Brain and Behaviour in Vertebrates,* vol. 2. Basel: Karger, pp. 75–90.

Book, A. S., Starzyk, K. B., & Quinsey, V. L. (2001). The relationship between testosterone and aggression: a meta-analysis. *Aggress Violent Beh, 6*(6), 579–599.

Borjigin, J., Li, X., & Snyder, S. H. (1999). The pineal gland and melatonin: molecular and pharmacologic regulation. *Annu Rev Pharmacol Toxicol, 39,* 53–65.

Brailoiu, E., Dun, S. L., Brailoiu, G. C., Mizuo, K., Sklar, L. A., Oprea, T. I., et al. (2007). Distribution and characterization of estrogen receptor G protein-coupled receptor 30 in the rat central nervous system. *J Endocrinology, 193*(2), 311–321.

Bronson, F. H. (1985). Mammalian reproduction: an ecological perspective. *Biol Reprod, 32*(1), 1–26.

Butler, M. P., Turner, K. W., Park, J. H., Schoomer, E. E., Zucker, I., & Gorman, M. R. (2010). Seasonal regulation of reproduction: altered role of melatonin under naturalistic conditions in hamsters. *Proc Biol Sci, 277*(1695), 2867–2874.

Caldwell, H. K., & Albers, H. E. (2004). Effect of photoperiod on vasopressin-induced aggression in Syrian hamsters. *Horm Behav, 46*(4), 444–449.

Cattaneo, E., & Maggi, A. (1990). C-fos induction by estrogen in specific rat brain areas. *Eur J Pharmacol: Mol Pharmacol, 188*(2–3), 153–159.

Cavigelli, S. A., & Pereira, M. E. (2000). Mating season aggression and fecal testosterone levels in male ring-tailed lemurs (*Lemur catta*). *Horm Behav, 37*(3), 246–255.

Cornil, C. A., Dalla, C., Papadopoulou-Daifoti, Z., Ballien, M., & Balthazart, J. (2006). Estradiol rapidly activates male sexual behavior and affects brain monoamine levels in the quail brain. *Behav Brain Res, 166*(1), 110–123.

Cos, S., Gonzalez, A., Guezmes, A., Mediavilla, M. D., Martinez-Campa, C., Alonso-Gonzalez, C., et al. (2006). Melatonin inhibits the growth of DMBA-induced mammary tumors by decreasing the local biosynthesis of estrogens through the modulation of aromatase activity. *Int J Cancer, 118*(2), 274–278.

Cross, E., & Roselli, C. E. (1999). 17beta-estradiol rapidly facilitates chemoinvestigation and mounting in castrated male rats. *Am J Physiol, 276*, R1346–R1350.

Dark, J., Zucker, I., & Wade, G. N. (1983). Photoperiodic regulation of body mass, food intake, and reproduction in meadow voles. *Am J Physiol, 245*(3), R334–R338.

Davidson, J. M. (1969). Effects of estrogen on the sexual behavior of male rats. *Endocrinology, 84*, 1365–1372.

Dawson, A., King, V. M., Bentley, G. E., & Ball, G. F. (2001). Photoperiodic control of seasonality in birds. *J Biol Rhythms, 16*(4), 365–380.

del Rio, B., Pedrero, J. M. G., Martinez-Campa, C., Zuazua, P., Lazo, P. S., & Ramos, S. (2004). Melatonin, an endogenous-specific inhibitor of estrogen receptor alpha via calmodulin. *J Biol Chem, 279*, 38294–38302.

Demas, G. E., Polacek, K. M., Durazzo, A., & Jasnow, A. M. (2004). Adrenal hormones mediate melatonin-induced increases in aggression in male Siberian hamsters (*Phodopus sungorus*). *Horm Behav, 46*(5), 582–591.

Dewing, P., Boulware, M. I., Sinchak, K., Christensen, A., Mermelstein, P. G., & Micevych, P. E. (2007). Membrane estrogen receptor-alpha interactions with metabotropic glutamate receptors 1a modulate sexual receptivity in rats. *J Neurosci, 27*, 9294–9300.

Fiszbein, A., Canepa, M., Vazquez, G. R., Maggese, C., & Pandolfi, M. (2010). Photoperiodic modulation of reproductive physiology and behaviour in the cichlid fish *Cichlasoma dimerus*. *Physiol Behav, 99*(4), 425–432.

Freeman, D. A., & Zucker, I. (2001). Refractoriness to melatonin occurs independently at multiple brain sites in Siberian hamsters. *Proc Natl Acad Sci U S A, 98*(11), 6447–6452.

Frungieri, M. B., Mayerhofer, A., Zitta, K., Pignataro, O. P., Calandra, R. S., & Gonzalez-Calvar, S. I. (2005). Direct effect of melatonin on Syrian hamster testes:

melatonin subtype 1a receptors, inhibition of androgen production, and interaction with the local corticotropin-releasing hormone system. *Endocrinology, 146*(3), 1541–1552.

Fuxjager, M. J., & Marler, C. A. (2010). How and why the winner effect forms: influences of contest environment and species differences. *Behav Ecol, 21*(1), 37–45.

Green, S., Walter, P., Kumar, V., Krust, A., Bornert, J. M., Argos, P., et al. (1986). Human oestrogen receptor cDNA: sequence, expression and homology to v-erb-A. *Nature, 320*(6058), 134–139.

Gutzler, S. J., Karom, M., Erwin, W. D., & Albers, H. E. (2009). Photoperiodic regulation of adrenal hormone secretion and aggression in female Syrian hamsters. *Horm Behav, 56*(4), 481–489.

Gutzler, S. J., Karom, M., Erwin, W. D., & Albers, H. E. (2010). Arginine-vasopressin and the regulation of aggression in female Syrian hamsters (*Mesocricetus auratus*). *Eur J Neurosci, 31*(9), 1655–1663.

Harmatz, M. G., Well, A. D., Overtree, C. E., Kawamura, K. Y., Rosal, M., & Ockene, I. S. (2000). Seasonal variation of depression and other moods: a longitudinal approach. *J Biol Rhythms, 15*(4), 344–350.

Hoffman, G. E., & Lyo, D. (2002). Anatomical markers of activity in neuroendocrine systems: are we all "fos-ed out"? *J Neuroendocrinol, 14*, 259–268.

Holloway, C. C., & Clayton, D. F. (2001). Estrogen synthesis in the male brain triggers development of the avian song control pathway in vitro. *Nat Neurosci, 4*(2), 170–175.

Huhman, K. L., Solomon, M. B., Janicki, M., Harmon, A. C., Lin, S. M., Israel, J. E., et al. (2003). Conditioned defeat in male and female Syrian hamsters. *Horm Behav, 44*(3), 293–299.

Insel, T. R., Preston, S., & Winslow, J. T. (1995). Mating in the monogamous male: behavioral consequences. *Physiol Behav, 57*, 615–621.

Jasnow, A. M., Huhman, K. L., Bartness, T. J., & Demas, G. E. (2000). Short-day increases in aggression are inversely related to circulating testosterone concentrations in male Siberian hamsters (*Phodopus sungorus*). *Horm Behav, 38*, 102–110.

Jensen, E. V., Suzuki, T., Kawashima, T., Stumpf, W. E., Jungblut, P. W., & DeSombre, E. R. (1968). A two-step mechanism for the interaction of estradiol with rat uterus. *Proc Natl Acad Sci U S A, 59*(2), 632–638.

Kelly, M. J., Moss, R. L., & Dudley, C. A. (1976). Differential sensitivity of preoptic-septal neurons to microelectrophoresed estrogen during the estrous cycle. *Brain Res, 114*, 152–157.

Kelly, M. J., Moss, R. L., & Dudley, C. A. (1977). The effects of microelectrophoretically applied estrogen, cortisol and acetylcholine on medial preoptic-septal unit activity throughout the estrous cycle of the female rat. *Exp Brain Res, 30*, 53–64.

King, W. J., & Greene, G. L. (1984). Monoclonal antibodies localize estrogen receptor in the nuclei of target cells. *Nature, 307*, 745–747.

Kow, L. M., & Pfaff, D. W. (2004). The membrane actions of estrogens can potentiate their lordosis behavior-facilitating genomic actions. *Proc Natl Acad Sci U S A, 101*(33), 12354–12357.

Kramer, K. M., Simmons, J. L., & Freeman, D. A. (2008). Photoperiod alters central distribution of estrogen receptor alpha in brain regions that regulate aggression. *Horm Behav, 53*, 358–365.

Kuiper, G. G. J. M., Enmark, E., Pelto-Huikko, M., Nilsson, S., & Gustaffson, J. A. (1996). Cloning of a novel estrogen receptor expressed in rat prostate and ovary. *Proc Natl Acad Sci U S A, 93*, 5925–5930.

Leitner, C., & Bartness, T. J. (2011). An intact dorsomedial hypothalamic nucleus, but not the subzona incerta or reuniens nucleus, is necessary for short-day melatonin signal-induced responses in Siberian hamsters. *Neuroendocrinology, 93*(1), 29–39.

Mani, S. K., Reyna, A. M., Alejandro, M. A., Crowley, J., & Markaverich, B. M. (2005). Disruption of male sexual behavior in rats by tetrahydrofurandiols (THF-diols). *Steroids, 70*, 750–754.

Markaverich, B. M., Alejandro, M. A., Markaverich, D., Zitzow, L., Casajuna N., Camarao, N., et al. (2002a). Identification of an endocrine disrupting agent from corn with mitogenic activity. *Biochem Biophys Res Commun, 291*, 692–700.

Markaverich, B. M., Mani, S., Alejandro, M. A., Mitchell, A., Markaverich, D., Brown, T., et al. (2002b). A novel endocrine-disrupting agent in corn with mitogenic activity in human breast and prostatic cancer cells. *Environ Health Perspect, 110*, 169–177.

Milner, T. A., Ayoola, K., Drake, C. T., Herrick, S. P., Tabori, N. E., McEwen, B. S., et al. (2005). Ultrastructural localization of estrogen receptor beta immunoreactivity in the rat hippocampal formation. *J Comp Neurol, 491*(2), 81–95.

Milner, T. A., McEwen, B. S., Hayashi, S., Li, C. J., Reagan, L. P., & Alves, S. E. (2001). Ultrastructural evidence that hippocampal alpha estrogen receptors are located at extranuclear sites. *J Comp Neurol, 429*(3), 355–371.

Morito, K., Hirose, T., Kinjo, J., Hirakawa, T., Okawa, M., Nohara, T., et al. (2001). Interaction of phytoestrogens with estrogen receptors alpha and beta. *Biol Pharm Bull, 24*(4), 351–356.

Nelson, R. J., Gubernick, D. J., & Blom, J. M. (1995). Influence of photoperiod, green food, and water availability on reproduction in male California mice (*Peromyscus californicus*). *Physiol Behav, 57*(6), 1175–1180.

Nelson, R. J., & Trainor, B. C. (2007). Neural mechanisms of aggression. *Nat Rev Neurosci, 8*(7), 536–546.

Nilsson, S., Makela, S., Treuter, E., Tujague, M., Thomsen, J. S., Andersson, G., et al. (2001). Mechanisms of estrogen action. *Physiol Rev, 81*, 1535–1565.

Nomura, M., Durbak, I., Chan, J., Gustafsson, J. A., Smithies, O., Korach, K. S., et al. (2002). Genotype/age interactions on aggressive behavior in gonadally intact estrogen receptor beta knockout (βERKO) male mice. *Horm Behav, 41*, 288–296.

Ogawa, S., Chan, J., Chester, A. E., Gustafsson, J. A., Korach, K. S., & Pfaff, D. W. (1999). Survival of reproductive behaviors in estrogen receptor beta gene-deficient (betaERKO) male and female mice. *Proc Natl Acad Sci U S A, 96*(22), 12887–12892.

Ogawa, S., Lubahn, D. B., Korach, K. S., & Pfaff, D. W. (1997). Behavioral effects of estrogen receptor gene disruption in male mice. *Proc Natl Acad Sci U S A, 94*(4), 1476–1481.

Pasch, B., George, A. S., Hamlin, H. J., Guillette, L. J., Jr., & Phelps, S. M. (2011). Androgens modulate song effort and aggression in Neotropical singing mice. *Horm Behav, 59*(1), 90–97.

Pfaff, D. W., & Levine, J. E. (2008). Reconciling molecular neuroendocrine signals and the scientists who study them. *Front Neuroendocrinol, 29*(2), 167–168.

Pfaff, D. W., Schwartz-Giblin, S., McCarthy, M. M., & Kow, L.-M. (1994). Cellular and molecular mechanisms of female reproductive behaviors. In: Knobil, E. & Neill, J. D. (eds.), *The Physiology of Reproduction*. New York: Raven, pp. 107–220.

Pospichal, M. W., Karp, J. D., & Powers, J. B. (1991). Influence of daylength on male hamster sexual behavior: masking effects of testosterone. *Physiol Behav, 49*(3), 417–422.

Pradhan, D. S., Newman, A. E., Wacker, D. W., Wingfield, J. C., Schlinger, B. A., & Soma, K. K. (2010). Aggressive interactions rapidly increase androgen synthesis in the brain during the non-breeding season. *Horm Behav, 57*(4–5), 381–389.
Prendergast, B. J., Kriegsfeld, L. J., & Nelson, R. J. (2001). Photoperiodic polyphenisms in rodents: neuroendocrine mechanisms, costs, and functions. *Quarterly Rev Biol, 76*(3), 293–325.
Qiu, J., Rønnekleiv, O. K., & Kelly, M. J. (2008). Modulation of hypothalamic neuronal activity through a novel G-protein-coupled estrogen membrane receptor. *Steroids, 73*(9–10), 985–991.
Ram, P. T., Kiefer, T., Siverman, M., Song, Y., Brown, G. M., & Hill, S. M. (1998). Estrogen receptor transactivation in MCF-7 breast cancer cells by melatonin and growth factors. *Mol Cell Endocrinol, 141*, 53–64.
Rato, A. G., Pedrero, J. G., Martinez, M. A., del Rio, B., Lazo, P. S., & Ramos, S. (1999). Melatonin blocks the activation of estrogen receptor for DNA binding. *FASEB J, 13*(8), 857–868.
Razandi, M., Pedram, A., Greene, G. L., & Levin, E. R. (1999). Cell membrans and nuclear estrogen receptors (ERs) originate from a single transcript: studies of ERα and ERβ expressed in Chinese hamster ovary cells. *Mol Endocrinol, 13*, 307–319.
Rochester, J. R., & Millam, J. R. (2009). Phytoestrogens and avian reproduction: Exploring the evolution and function of phytoestrogens and possible role of plant compounds in the breeding ecology of wild birds. *Comp Biochem Physiol A Mol Integr Physiol, 154*(3), 279–288.
Romano, G. J., Krust, A., & Pfaff, D. W. (1989). Expression and estrogen regulation of progesterone receptor mRNA in neurons of the mediobasal hypothalamus: an in situ hybridization study. *Mol Endocrinol, 3*, 1295–1300.
Roy, D., Angelini, N. L., Fujieda, H., Brown, G. M., & Belsham, D. D. (2001). Cyclical regulation of GnRH gene expression in GT1–7 GnRH-secreting neurons by melatonin. *Endocrinology, 142*(11), 4711–4720.
Sakamoto, H., Matsuda, K., Hosokawa, K., Nishi, M., Morris, J. F., Prossnitz, E. R., et al. (2007). Expression of G protein-coupled receptor-30, a G protein-coupled membrane estrogen receptor, in oxytocin neurons of the rat paraventricular and supraoptic nuclei. *Endocrinology, 148*(12), 5842–5850.
Scotti, M. A., Schmidt, K. L., Newman, A. E., Bonu, T., Soma, K. K., & Demas, G. E. (2009). Aggressive encounters differentially affect serum dehydroepiandrosterone and testosterone concentrations in male Siberian hamsters (*Phodopus sungorus*). *Horm Behav, 56*(4), 376–381.
Selcher, J. C., Weeber, E. J., Christian, J., Nekrasova, T., Landreth, G. E., & Sweatt, J. D. (2003). A role for ERK MAP kinase in physiologic temporal integration in hippocampal area CA1. *Learn Mem, 10*(1), 26–39.
Shughrue, P. J., Lane, M. V., & Merchenthaler, I. (1997). Regulation of progesterone receptor messenger ribonucleic acid in the rat medial preoptic nucleus by estrogenic and antiestrogenic compounds: an in situ hybridization study. *Endocrinology, 138*(12), 5476–5484.
Silva, A. L., Fry, W. H., Sweeney, C., & Trainor, B. C. (2010). Effects of photoperiod and experience on aggressive behavior in female California mice. *Behav Brain Res, 208*(2), 528–534.
Sodersten, P. (1973). Estrogen-activated sexual behavior in male rats. *Horm Behav, 4*, 247–256.

Soma, K. K., Sullivan, K. A., Tramontin, A. D., Saldanha, C. J., Schlinger, B. A., & Wingfield, J. C. (2000). Acute and chronic effects of an aromatase inhibitor on territorial aggression in breeding and nonbreeding male song sparrows. *J Comp Physiol A, 186*(7–8), 759–769.

Soma, K. K., & Wingfield, J. C. (2001). Dehydroepiandrosterone in songbird plasma: seasonal regulation and relationship to territorial aggression. *Gen Comp Endocrinol, 123*(2), 144–155.

Southwick, C. H. (1955). Regulatory mechanisms of house mouse populations: social behavior affecting litter survival. *Ecology, 36*(4), 627–634.

Sperry, T. S., Wacker, D. W., & Wingfield, J. C. (2010). The role of androgen receptors in regulating territorial aggression in male song sparrows. *Horm Behav, 57*(1), 86–95.

Taziaux, M., Keller, M., Bakker, J., & Balthazart, J. (2007). Sexual behavior activity tracks rapid changes in brain estrogen concentrations. *J Neurosci, 27*(24), 6563–6572.

Thornton, J. W. (2001). Evolution of vertebrate steroid receptors from an ancestral estrogen receptor by ligand exploitation and serial genome expansions. *Proc Natl Acad Sci U S A, 98*(10), 5671–5676.

Toft, D., & Gorski, J. (1966). A receptor molecule for estrogens: isolation from the rat uterus and preliminary characterization. *Proc Natl Acad Sci U S A, 55*(6), 1574–1581.

Towart, L. A., Alves, S. E., Znamensky, V., Hayashi, S., McEwen, B. S., & Milner, T. A. (2003). Subcellular relationships between cholinergic terminals and estrogen receptor-alpha in the dorsal hippocampus. *J Comp Neurol, 463*(4), 390–401.

Trainor, B. C., Crean, K. K., Fry, W. H., & Sweeney, C. (2010). Activation of extracellular signal-regulated kinases in social behavior circuits during resident-intruder aggression tests. *Neuroscience, 165*(2), 325–336.

Trainor, B. C., Finy, M. S., & Nelson, R. J. (2008). Rapid effects of estradiol on male aggression depend on photoperiod in reproductively non-responsive mice. *Horm Behav, 53*(1), 192–199.

Trainor, B. C., Kyomen, H. H., & Marler, C. A. (2006). Estrogenic encounters: how interactions between aromatase and the environment modulate aggression. *Front Neuroendocrinol, 27*, 170–179.

Trainor, B. C., Lin, S., Finy, M. S., Rowland, M. R., & Nelson, R. J. (2007a). Photoperiod reverses the effects of estrogens on male aggression via genomic and nongenomic pathways. *Proc Natl Acad Sci U S A, 104*(23), 9840–9845.

Trainor, B. C., & Nelson, R. J. (2012). Neuroendocrinology of aggressive behavior. In: Fink, G., Pfaff, D. W. & Levin, J. (eds.), *Handbook of Neuroendocrinology.* San Diego: Elsevier, pp. 509–520.

Trainor, B. C., Rowland, M. R., & Nelson, R. J. (2007b). Photoperiod affects estrogen receptor alpha, estrogen receptor beta and aggressive behavior. *Eur J Neurosci, 26*(1), 207–218.

Vasudevan, N., Kow, L. M., & Pfaff, D. W. (2001). Early membrane estrogenic effects required for full expression of slower genomic actions in a nerve cell line. *Proc Natl Acad Sci U S A, 98*, 12267–12271.

Villalon Landeros, R., Morisseau, C., Yoo, H. J., Fu, S. H., Hammock, B. D., & Trainor, B. C. (2012). Corncob bedding alters the effects of estrogens on aggressive behavior and reduces estrogen receptor alpha expression in the brain. *Endocrinology, 153*, 949–953.

Wen, J., Hotchkiss, A., Demas, G., & Nelson, R. (2004). Photoperiod affects neuronal nitric oxide synthase and aggressive behaviour in male Siberian hamsters (*Phodopus sungorus*). *J Neuroendocrinol, 16*(11), 916–921.

Wingfield, J. C. (1993). Control of testicular cycles in the song sparrow, *Melospiza melodia melodia*: interaction of photoperiod and an endogenous program? *Gen Comp Endocrinol, 92*(3), 388–401.

Wingfield, J. C. (1994a). Control of territorial aggression in a changing environment. *Psychoneuroendocrinology, 19*(5–7), 709–721.

Wingfield, J. C. (1994b). Regulation of territorial behavior in the sedentary song sparrow, *Melospiza melodia morphna. Horm Behav, 28*(1), 1–15.

Winkler, D., Pjrek, E., Konstantinidis, A., Praschak-Rieder, N., Willeit, M., Stastny, J., et al. (2006). Anger attacks in season al affective disorder. *Int J Neuropsychopharmacol, 9*(2), 215–219.

Witt-Enderby, P. A., Masana, M. I., & Dubocovich, M. L. (1998). Physiological exposure to melatonin supersensitizes the cyclic adenosine 3′,5′-monophosphate-dependent signal transduction cascade in Chinese hamster ovary cells expressing the human mt1 melatonin receptor. *Endocrinology, 139*, 3064–3071.

Yuan, L., Collins, A. R., Dai, J., Dubocovich, M. L., & Hill, S. M. (2002). MT1 melatonin receptor overexpression enhances the growth suppressive effect of melatonin in human breast cancer cells. *Mol Cell Endocrinol, 192*, 147–156.

Zangenehpour, S., & Chaudhuri, A. (2002). Differential induction and decay curves of c-fos and zif268 revealed through dual activity maps. *Mol Brain Res, 109*(1–2), 221–225.

Zsarnovszky, A., Le, H. H., Wang, H. S., & Belcher, S. M. (2005). Ontogeny of rapid estrogen-mediated extracellular signal-regulated kinase signaling in the rat cerebellar cortex: potent nongenomic agonist and endocrine disrupting activity of the xenoestrogen bisphenol A. *Endocrinology, 146*(12), 5388–5396.

Chapter 22

Cellular Mechanisms Controlling Rapid Changes in Brain Aromatase Activity

Thierry D. Charlier, Charlotte A. Cornil, Gregory F. Ball, and Jacques Balthazart

INTRODUCTION

Estrogens, such as 17β-estradiol (E2), have major effects on sexual and social behavior, aggression, and cognition. These effects usually develop somewhat slowly after hormone exposure but are long lasting. Such effects are mediated by the transcriptional induction of specific targets genes via the recruitment of estrogen receptors. However, it is now clear that estrogens also activate cellular, physiological, and behavioral effects that are too rapid to result from de novo mRNA transcription and protein synthesis (Cornil et al., 2006a; Charlier et al., 2010; Cornil & Charlier, 2010). For example, the acute elevation of estrogen concentration in vitro triggers the activation of numerous intracellular signaling pathways including the modulation of intracellular calcium concentrations (Mermelstein et al., 1996; Kenealy et al., 2011) and the phosphorylation of a variety of proteins such as the mitogen-activated protein kinase (MAPK) and cAMP response element binding protein (Moss et al., 1997; Watters et al., 1997; Wade & Dorsa, 2003; Wu et al., 2005; Roepke et al., 2011). In turn, the activation of intracellular signaling pathways can potentially lead to modulations of electrical activity (Joels, 1997; Moss et al., 1997; Rønnekleiv & Kelly, 2002) and to neuronal activation (Rønnekleiv & Kelly, 2002; Abraham et al., 2003, 2004; Boulware et al., 2005) in various brain regions. Importantly, these acute effects of estrogens described at the cellular level also influence a variety of physiological and behavioral processes in different species. Cross and Roselli first showed that in castrated male rats the acute injection of a high dose of E2 stimulates mounts and anogenital investigations within 35 minutes (Cross & Roselli, 1999). In addition, the acute administration of E2 facilitates lordosis behavior in female rats, in part via an interaction of liganded estrogen receptor alpha (ERα) with metabotropic glutamate receptors (Dewing et al., 2007; Micevych & Mermelstein, 2008). Similarly, a single injection of E2 or estrogen receptor alpha and beta agonists (PPT and DPN, respectively) facilitates the expression of most aspects of male sexual behavior within 10 to 15 minutes in quail and this effect disappears after 30 minutes (Cornil et al., 2006b; Seredynski and Cornil, unpublished data; and see chapter 23, this volume). In male plain midshipman fish, an intramuscular injection of E2 increases within 5 minutes the duration of fictive vocalizations stimulated by electrical stimuli in vitro and this effect lasts for 15 to 30 minutes (Remage-Healey & Bass, 2004). In another fish species, the goldfish, E2 significantly increased approaches toward females within 10 to 25 minutes of administration (Lord et al., 2009). Hayden-Hixson and colleagues also reported that a microinjection of E2 in the anterior hypothalamus stimulates agonistic behavior (flank marking) within 15 minutes in male hamsters (Hayden-Hixson & Ferris, 1991). Although these rapid, probably non-genomic, effects of estrogens are currently the focus of an extensive amount of research, there is much that is still unknown about the mechanisms responsible for the rapid control of estrogen release. If E2 action is in numerous instances similar to a neurotransmitter or at least to a neuromodulator (Balthazart & Ball, 2006), estrogens cannot, however, be stored in synaptic vesicles before rapid release due to their lipophilic nature. We therefore hypothesized that a rapid control of estrogen synthesis via local changes in aromatase activity underlies the fast effect of estrogens on physiology and behavior. In this chapter, we will briefly review the distribution and control of brain aromatase activity and review in more depth the rapid control of aromatase activity by phosphorylation processes.

DISTRIBUTION AND CONTROL OF BRAIN AROMATASE ACTIVITY

The enzyme aromatase catalyzes the synthesis of estrogens from androgens and is present in numerous, well-defined, brain regions. The presence of aromatase in specific brain nuclei can broaden the range of physiological and behavioral effects of testosterone thanks to a local increase in estrogens added to the effects of androgens. In the brain of reptiles, birds, and mammals, aromatase expression is restricted to specific neuronal populations mainly located in the hypothalamic/preoptic areas including the medial preoptic nucleus and in the limbic system (Roselli et al., 1985; Foidart et al., 1995a, 1995b; Wagner & Morrell, 1997; Metzdorf et al., 1999; Roselli & Resko, 2001) (see also Chapter 3, this volume). Songbirds have an additional population of aromatase expressing neurons in the caudomedial nidopallium (NCM), an area adjacent to the song control nucleus HVC (Shen et al., 1995; Saldanha et al., 2000). Aromatase is also abundantly expressed in the brain of teleost fishes but seems only present in glia (Forlano et al., 2001; Forlano et al., 2006; Diotel et al., 2010, but see Chapter 2, this volume). Interestingly, aromatase protein and activity are not only observed in the cell soma but also in presynaptic boutons in a variety of species (Naftolin et al., 1996; Saldanha et al., 2003; Peterson et al., 2005), suggesting that E2 is synthesized specifically at the synapse and acts as a neurotransmitter or neuromodulator (Balthazart & Ball, 2006) (see chapter 6, this volume). The presence of aromatase in defined brain regions means that there can be a very localized increase in estrogens in a selected brain region without an effect in the periphery and/or surrounding brain areas. The presence of this enzyme in the brain of both sexes also confirms that estrogens can have physiological effects in both males and females.

One way through which estrogen concentrations could be rapidly increased in the brain is the modulation of the concentration of androgenic substrate accessible to aromatase. This rapid modulation of testosterone has been observed in a few cases in response to specific social interactions such as territorial intrusion and sexual interactions (Saginor & Horton, 1968; Purvis & Haynes, 1974; Batty, 1978; Harding, 1981; Wingfield, 2006; Addis et al., 2010; Pradhan et al., 2010). Such rapid increases in peripheral androgen concentration could then potentially translate into rapid and transient changes in local estrogen concentrations following local conversion into estrogens by aromatase.

Rapid changes in circulating concentration of testosterone, however, might not be very common and do not probably reach a sufficient amplitude and other mechanisms such as a change of the rate of transformation of androgens into estrogens are required in order to affect brain estrogen concentrations (Cornil et al., 2006a). Changes in aromatase activity (AA) often reflect changes in aromatase concentration resulting from the slow variation of the synthesis of this protein at the transcriptional level. For example, sex steroids significantly induce the expression of the aromatase gene in the hypothalamus of all vertebrates investigated so far. Accordingly, low levels of aromatase expression are detected after gonadectomy, whereas steroid replacement markedly increases aromatase transcript, protein, and activity (Roselli & Resko, 1984; Schumacher & Balthazart, 1986; Roselli & Resko, 1989; Fusani et al., 2001). This effect appears to be mediated mainly by androgen receptors in mammals (Roselli et al., 1997; chapter 3, this volume), whereas estrogen receptors are more important in birds (Harada et al., 1993; Balthazart et al., 1994). These changes in AA generally occur after several hours or days of treatment. However, the rapid effects of E2 introduced above require a more rapid regulation of estrogen-concentration, and require mechanisms rapidly affecting the synthesis of the steroid.

Recent studies have suggested that rapid changes in AA modulate behavior. For example, the acute blockade of estrogen synthesis by systemic administration of aromatase inhibitors (Vorozole, R83842 or Androstatrienedione, ATD) significantly reduced male sexual behavior in quail and mice within 30 and 10 minutes, respectively (Cornil et al., 2005b; Taziaux et al., 2007). Similar pharmacological treatments were also shown to impair approach behavior towards females in male goldfish (Lord et al., 2009), fictive vocalizations in midshipman fish (Remage-Healey & Bass, 2004) and aggressive behavior in Beach and California mice (Trainor et al., 2007; Trainor et al., 2008). Several findings also support the involvement of local aromatization in the rapid modulation of sensory processing. Fadrozole reduced song-elicited neuronal firing rate in zebra finches and prevented the expression of hearing-driven gene expression in NCM, a brain nucleus involved in song discrimination in songbirds (Tremere et al., 2009). In addition, retrodialysis of the aromatase inhibitor fadrozole within NCM rapidly suppressed the preference for a bird's own song compared with a conspecific male song in male zebra finches (Remage-Healey et al., 2010). Another example of acute regulation of sensory processing by locally produced estrogens is the rapid modulation of nociception. Intrathecal injections of aromatase inhibitors markedly increased, in less than 5 minutes, the latency of foot withdrawal from a hot water bath. This effect was no longer observed 30 minutes after the injection and was prevented by the concurrent infusion of E2 (Evrard & Balthazart, 2004). Altogether, these experiments confirm that the rapid modulation of estrogen concentration, via the alteration of AA, has a significant impact on numerous behaviors and sensory processes.

RAPID CHANGES IN AROMATASE ACTIVITY

It is only recently that rapid changes in AA were demonstrated to occur in vivo in response to changes in the environment. Performance of sexual behavior can indeed rapidly affect AA in specific brain regions: a rapid and transient change in AA was detected in hypothalamus/preoptic area (HPOA) of quail after a 5- to 30-minute sexual encounter with a receptive female (decrease [Cornil et al., 2005a]) and after acute restraint stress (increase [Dickens et al. 2011]). These rapid changes in enzymatic activity could thus produce fast changes of local estrogen concentration in behaviorally relevant situations. Moreover in zebra finches, a brief exposure to song (30 min) resulted in an increase of AA in NCM, principally in synaptic terminals (Remage-Healey et al., 2009).

Interestingly, these rapid changes in enzymatic activity were also triggered in HPOA explants from Japanese quail by a rise in extracellular K^+ concentration (Balthazart et al., 2001b). In addition, the incubation of HPOA brain explants with thapsigargin, a sesquiterpene lactone used as a non-competitive inhibitor of endoplasmic reticulum Ca^{++} ATPase that modulates the intracellular pool of Ca^{++}, rapidly (5 min) and transiently reduced AA. These results suggested that the K^+-induced aromatase inhibition resulted from a Ca^{++} increase linked to the transient depolarization. In agreement with this hypothesis, the enzymatic inhibition was significantly reduced when the K^+-induced depolarization was produced in HPOA blocks that were incubated in a Ca^{++}-free medium (Balthazart et al., 2001b). In the zebra finch telencephalon, a K^+-induced depolarization was also shown to reduce significantly the concentration of estradiol, likely through the rapid inhibition of AA. Interestingly, this effect of the depolarization apparently required the activation of voltage-gated calcium channels at the presynaptic level, because the use of ω-conotoxin, a specific inhibitor of N-type Ca^{++} channels blocked the K^+-induced decrease in estradiol concentration (and thus presumably the inhibition of AA (Remage-Healey et al., 2011). This is consistent with the recent observation that the acute control of aromatase activity seems to prevail in the presynaptic terminals compared to microsomes (see below, [Cornil et al., 2012]).

Ca^{++} release can also be triggered by direct effects of glutamate. Interestingly, intracellular recording combined with biocytin labeling of the recorded neurons indicated that the electrical activity of aromatase positive neurons is sensitive to glutamate, a finding consistent with the idea that glutamatergic inputs on aromatase cells acutely regulate AA (Cornil et al., 2000, 2004). Accordingly, the activation of glutamate receptors in male quail HPOA explants rapidly reduced AA.

More specifically, the glutamate agonists 2-amino-3-(5-methyl-3-oxo-1,2- oxazol-4-yl) propanoic acid (AMPA), kainate and, to a lesser extent, N-methyl-D-aspartate (NMDA) rapidly (within 5 min) and reversibly downregulated AA. These effects were specific, because the addition of the glutamate receptor antagonist NBQX blocked the effect of kainate on AA (Balthazart et al., 2006) (Fig. 22–1A.). The effect of the activation of glutamate receptors on AA apparently involves the release of calcium from intracellular stores (rather than its influx into the neurons from extracellular environment) because the removal of calcium from the incubation medium at the time of kainate application had no effect on enzymatic activity inhibition (Balthazart et al., 2001b). Similarly, in vivo retrodialysis of glutamate, but not NMDA, in the zebra finch telencephalon induced a significant reduction of local E2 concentration (Remage-Healey et al., 2008). The direct links between the activation of glutamatergic receptor, calcium release, and the exact intracellular pathway(s) involved in the rapid modulation of AA, however, require further investigation. It should be noted that the inhibitory amino acid GABA did not affect the enzymatic activity in quail HPOA explants or in vivo in zebra finch telencephalon (Balthazart et al., 2001b; Remage-Healey et al., 2008) (see Fig. 22–1B).

IMPORTANCE OF PHOSPHORYLATIONS

These rapid modulations of aromatase activity are presumably not mediated by transcription-dependent changes in the concentration of the protein but rather involve rapid modification of the protein itself. Post-translational modifications are a common mechanism involved in the control of protein activity and a widely used regulatory mechanism in the brain (Nestler & Greengaard, 1984). For example, phosphorylation of NMDA receptors has been shown to increase NMDA activity, whereas dephosphorylation decreases these currents in hippocampal neurons (Chen & Roche, 2007). Similarly, tyrosine hydroxylase activity, the rate-limiting enzyme in catecholamine synthesis, is rapidly modified by phosphorylations, triggered by the presence of suitable concentrations of adenosine triphosphate (ATP) and of the divalent cation magnesium, Mg^{++} (Albert et al., 1984; Daubner et al., 1992). These phosphorylations are catalyzed by specific kinases that transfer the terminal phosphate group from ATP to specific amino acid residues (tyrosine, threonine, serine) of the protein. In all cases, kinase activity critically depends on the Mg^{++} intracellular concentration. Previous studies have implicated divalent cations in the control of AA (Ca^{++} [Onagbesan & Peddie, 1989]; Mg^{++} [Steimer & Hutchison, 1991]). We therefore wondered whether Ca^{++}, ATP, and Mg^{++}, at concentrations equivalent to those used to obtain maximal changes in tyrosine hydroxylase activity (Ames et al., 1978) could also potentially affect AA.

As predicted, AA measured in male Japanese quail HPOA homogenates was significantly inhibited within 15 minutes after exposure to high but physiological concentrations of ATP (2–8 mM), Ca^{++} (0.5–2 mM) and Mg^{++} (2–10 mM) (Balthazart et al., 2001b). Aromatase activity was not affected by a 15-minute preincubation in control conditions. No effect of ATP or Mg^{++} was observed when there was no preincubation or only a short 5-minute preincubation. These compounds also did not affect AA even after 15

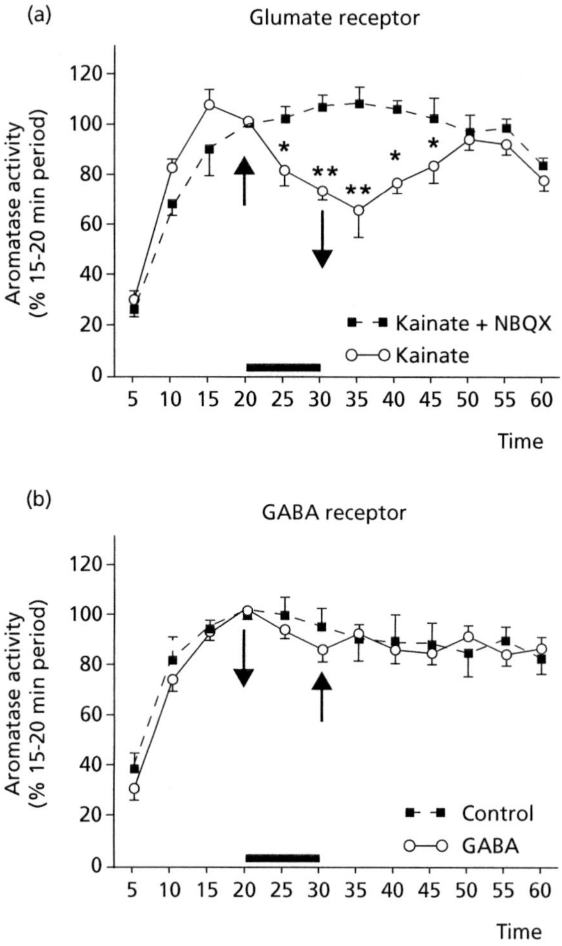

Figure 22–1. Effects of Neurotransmitters on Aromatase Activity in Quail hypothalamus/preoptic area (HPOA) Explants. A. Aromatase activity (mean ± SEM) in paired in vitro explants exposed for 10 minutes (*up arrow*) to 100 μM kainate (Glutamate receptor agonist) after a 20-minute preincubation in physiological saline containing or not the non-NMDA glutamate antagonist NBQX. Explants were again exposed to physiological saline containing or not NBQX at 30 minutes (*down arrow*) until the end of the experiment. The medium was changed every 5 minutes. B. Aromatase activity (mean ± SEM) in paired explants incubated in vitro for 20 minutes in physiological saline then exposed for 10 minutes (*up arrow*) to 100 μM GABA. Explants were returned to physiological saline at 30 minutes (*down arrow*) until the end of the experiment. All data are expressed as a percentage of the basal release, defined as the activity during the period preceding the experimental manipulation (time 15–20 min).

minutes when they were tested independently. Additional experiments demonstrated that the presence of Ca^{++} is also required to inhibit AA. The combination of ATP with Mg^{++} induced a very marked decrease of AA that was further amplified by the addition of Ca^{++}. This enzymatic inhibition was completely abolished in the presence of 1 mM ethylene glycol tetraacetic acid (EGTA), a Ca^{++} chelating agent, clearly supporting the Ca^{++}-dependence of aromatase inactivation by ATP and Mg^{++}. Furthermore, parametric studies demonstrated that Mg^{++} and Ca^{++} concentrations affect AA in the presence or absence of ATP in a dose-dependent manner. Altogether, these observations strongly suggest that phosphorylations rapidly inhibit this enzymatic activity.

THE ROLE OF KINASES AND PHOSPHATASES

The human genome is thought to contain 518 putative protein kinases divided into 90 tyrosine (Tyr) kinases and 428 serine/threonine (Ser/Thr) kinases (Manning et al., 2002; Johnson & Hunter, 2005). In addition, the reversible nature of phosphorylations suggests that there should be a large number of protein phosphatases in the human genome, although only 107 putative protein tyrosine phosphatases (Alonso et al., 2004) and around 30 protein Ser/Thr phosphatases (Shi, 2009) have been described until now.

To test whether ATP/Mg/Ca conditions induce kinase-dependent protein phosphorylations or whether these conditions lead to a nonspecific inactivation of AA, we first tested effects of the second messengers cAMP and cGMP (1mM, respectively activators of protein kinase A, PKA and protein kinase G, PKG). These kinase activators did not modify AA in the absence or presence of ATP/Mg/Ca (Balthazart et al., 2003). However, calmodulin (0.3 μM), the activator of Ca^{++}/calmodulin-dependent kinases (CAMK) significantly inhibited the AA measured in the presence or absence of ATP/Mg/Ca. We also tested the effects of numerous kinase inhibitors on the AA of preoptic-hypothalamic homogenates in the presence or absence of ATP/Mg/Ca (Balthazart et al., 2003). Several of these inhibitors substantially blocked the inhibition of AA induced by the addition of ATP/Mg/Ca while others had no effect (Fig. 22–2). Staurosporine (STAU, 10 μM), a general Ser/Thr kinase inhibitor and genistein (GEN, 50 μM), a general Tyr kinase inhibitor significantly blocked the inhibition produced by ATP/Mg/Ca while they had no effect on basal AA. The level of enzyme activity remaining in the presence of divalent ions and ATP was therefore significantly higher than in the absence of these kinase inhibitors. Similar effects (blockade of inhibition) were also observed in presence of bisindolylmaleimide (10 μM, a PKC inhibitor), ML7 (10 μM, a myosin light chain kinase inhibitor), H7 (1-(5-iso- quinolinesulfonyl)-2-methylpiperazine, 50 μM, an inhibitor of PKA and PKC), H9 (N-(2-aminoethyl)-5-isoquinoline sulfonamide, 1 μM, a specific inhibitor of PKA/PKG), H89 dihydrochloride (10 μM, A PKA inhibitor) and trifluoperazin (TRI, 100 μM, a calmodulin antagonist that should inhibit all types of CAMK). No effect was in contrast detected in the presence of KN93 (10 μM-, a calmodulin kinase II inhibitor), of PKGI (200 μM, a protein kinase G inhibitor), or chelerythrine chloride (10 μM, a specific inhibitor of PKC). Taken together, these data therefore indicate that the activity of aromatase is controlled by the phosphorylation of both Tyr and Ser/Thr residues (Balthazart et al., 2001a, 2003).

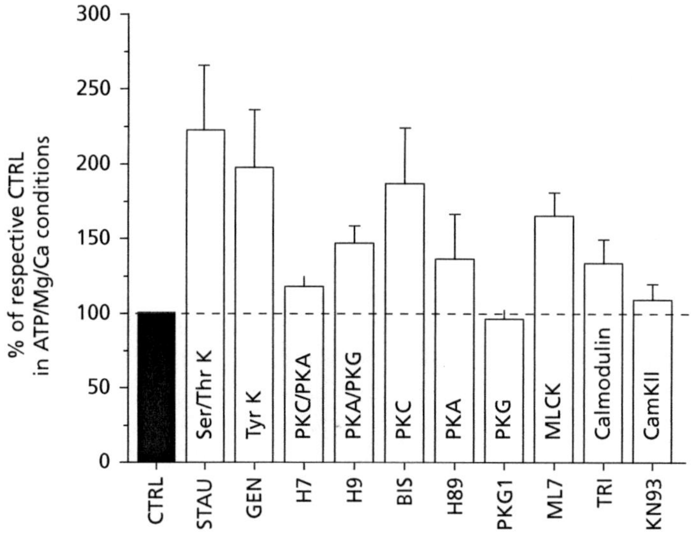

Figure 22–2. Summary of the effects of different protein kinase inhibitors on AA measured in quail hypothalamus/preoptic area (HPOA) homogenates in the presence 1 mM ATP, 1 mM Mg^{++} and 0.5 mM Ca^{++}. All data are expressed as percentage of the corresponding control values measured in the presence of ATP/Mg/Ca but with no inhibitor. See text for complete name of the compounds, their targeted group of kinases and the concentration used.

Other groups also investigated in parallel the potential influence of phosphorylations on AA in other tissues. In some cancer cell lines, E2 increases AA by a non-genomic action of ERα via crosstalk with intracellular signaling pathways. The E2-dependent increase in AA was completely inhibited after selective inhibition of the protein kinase MEK-1/2 by PD98059 or of the epidermal growth factor tyrosine kinase by PD153035 without any corresponding change the aromatase mRNA concentration (Shozu et al., 2001; Kinoshita & Chen, 2003). In addition, the blockade of MAPK or phosphoinositide 3-kinase reduced AA (Yue et al., 2003), whereas growth factors increased the enzymatic activity through the stimulation of cAMP-dependent protein kinases and PKC in vitro or via the activation of tyrosine kinase c-Src (Mendelson et al., 1986; Ryde et al., 1992; Richards et al., 2002; Catalano et al., 2009). However, these reactions were relatively slow and could not explain the very rapid changes of E2 synthesis observed in the brain.

Comparisons of AA in male and female quail HPOA homogenates suggest that the calcium-dependent phosphorylations underlying the rapid regulation of enzymatic activity are quantitatively different in the two sexes. The addition of ATP/Mg/Ca decreased AA in both sexes and this effect was blocked by the simultaneous addition of a variety of protein kinase inhibitors (Konkle & Balthazart, 2011). Interestingly however, the magnitude of some of these effects in females differed significantly from what was observed in males. These findings suggest that the control of aromatase activity by calcium may be slightly different in males and females. This observation could partially explain the sex difference in enzymatic activity observed in the quail HPOA as well as in other species. Interestingly, phosphorylating conditions do not seem to have identical affects on AA in different subcellular compartments isolated via differential ultracentrifugation from zebra finch telencephalon. The reduction of AA by phosphorylating conditions was significantly more pronounced in the P2 fraction, containing synaptosomes, as compared with the P3 fraction, containing microsomes (endoplasmic reticulum; Cornil et al., 2012).

If kinases decrease AA in a reversible manner, one could expect that the enzymatic activity increases and possibly returns to baseline levels in the presence of phosphatases. Contrary to this expectation, we found that 100 μM sodium orthovanadate (a general inhibitor of protein phosphatases) significantly decreases AA. Furthermore, specific protein phosphatases that could upregulate basal AA have not been identified to date (Balthazart et al., 2003, 2005a). In several experiments, the reversibility of AA inhibition by phosphorylations was investigated using alkaline and acid phosphatase (PPase). Unexpectedly, alkaline PPase inhibited AA in a dose-dependent manner in the presence, as well as in the absence, of ATP/Mg/Ca. In contrast, acid PPase completely blocked the inhibitory effects of ATP/Mg/Ca on AA, although at high concentrations it also moderately inhibited AA in the absence of ATP/Mg/Ca (Balthazart et al., 2005a). It should be noted, however, that the addition of acid PPase was unable to restore AA after the enzymatic activity had been inhibited by exposure to ATP/Mg/Ca. The specific enzymes involved in the phosphorylation-dephosphorylation mechanisms associated to the rapid and reversible control of AA are, therefore, still not fully defined and are currently under investigation.

We also showed that calmodulin significantly inhibits quail preoptic-hypothalamic AA both in the presence and absence of phosphorylating conditions, suggesting that calmodulin itself interacts directly with aromatase rather than through a modulation of Ca^{++}/calmodulin-dependent protein kinases (Balthazart et al., 2005b). Preincubation with 1 mM EGTA or with a monoclonal antibody raised against calmodulin immediately increased AA in the presence or absence of phosphorylating conditions. The presence of calcium-independent and calcium-dependent calmodulin binding motifs on the quail aromatase sequence, including 1–8-14a and b motifs, strongly reinforces the idea that calmodulin itself, without the activation of a calmodulin kinase could modulate AA. The specific mechanisms by which calmodulin would directly, regulate aromatase activity is currently unknown. The rapid control of brain AA thus appears to include at least two mechanisms: (1) a regulatory process that involves the Ca^{++}/calmodulin binding site, and (2) a phosphorylation by several protein kinases (PKC, PKA as well as Ca^{++}/calmodulin kinase(s)) of the aromatase molecule (see the following for additional discussion of this aspect of the results).

THE RAPID REDUCTION OF AA BY PHOSPHORYLATIONS IS A GENERAL PHENOMENON

Investigations of the rapid modulation of AA had until recently been carried out only on quail brain tissue, so the importance of phosphorylations in the rapid control AA in other tissues or species could not be evaluated. To investigate whether the rapid inhibition of AA by ATP/Mg/Ca-dependent phosphorylation processes is specific to the neuronal environment or can be observed in other aromatase-rich tissues, effects of phosphorylating conditions were quantified in ovary and ovarian follicles (F4 and F5) homogenates. These experiments demonstrated a drastic decrease in enzymatic activity within 15 minutes (Charlier et al., 2011) (Fig. 22–3A).

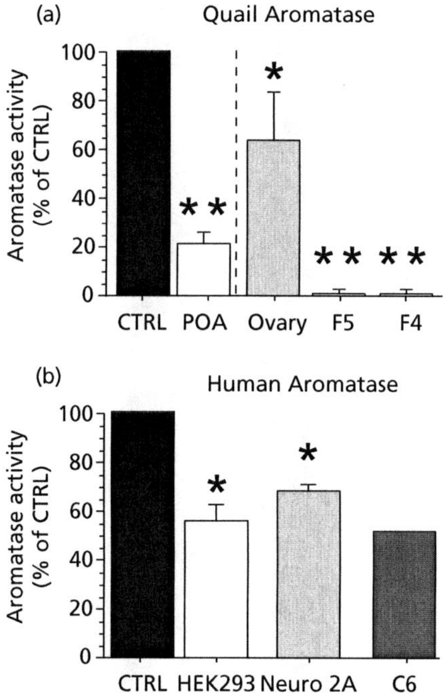

Figure 22–3. Bar graphs illustrating the inhibition of aromatase activity by phosphorylating conditions in quail hypothalamus/preotic area (HPOA) or ovarian tissue (*A*) or in human aromatase expressed by various cells lines (*B*). Aromatase activity in phosphorylating conditions is expressed as percentage of the mean activity compared with the same tissue in control conditions. A. Phosphorylating conditions were 2 mM Ca^{++}, 10 mM Mg^{++}, and 2 mM ATP for the HPOA or 2 mM Ca^{++}, 10 mM Mg^{++}, and 8 mM ATP for the ovary and follicles F 5 and F4. Results are expressed as mean ± SEM. * $p<0.05$ vs. matched tissue in CTRL conditions, ** $p < 0.01$ vs. matched tissue in CTRL condition. B. Phosphorylating conditions were 2 mM Ca^{++}, 10 mM Mg^{++}, and 2 mM ATP. Human aromatase was expressed in HEK293 (Human Embryonic Kidney), Neuro2A (Mouse neuroblastoma), C6 (Rat glioma) cell lines. Results are expressed as mean ± SEM, except for C6, where $n = 1$. * $p < 0.05$ vs. matched tissue in CTRL conditions.

We also stably expressed human aromatase in several cell lines (Fig. 22–3B), including HEK293 (Human embryonic kidney), C6 (Rat glioma), and Neuro2A (Mouse neuroblastoma). This aromatase exhibited all the biochemical and kinetic characteristics of the native human aromatase, including molecular mass on Western blot (52 kDa) and Michaelis-Menten affinity constant for androstenedione (±10 nM). Similar to what was observed in the hypothalamus/preoptic area of Japanese quail and telencephalon of zebra finch, a KCl-induced depolarization triggered a pronounced inhibition of AA expressed in HEK293 cells (Fig. 22–4A.). This effect was transient and could be fully reversed when cell cultures returned to control conditions. Importantly, we also demonstrated that the rapid inhibition of human AA in HEK293 cells does not involve aromatase degradation because the concentration of

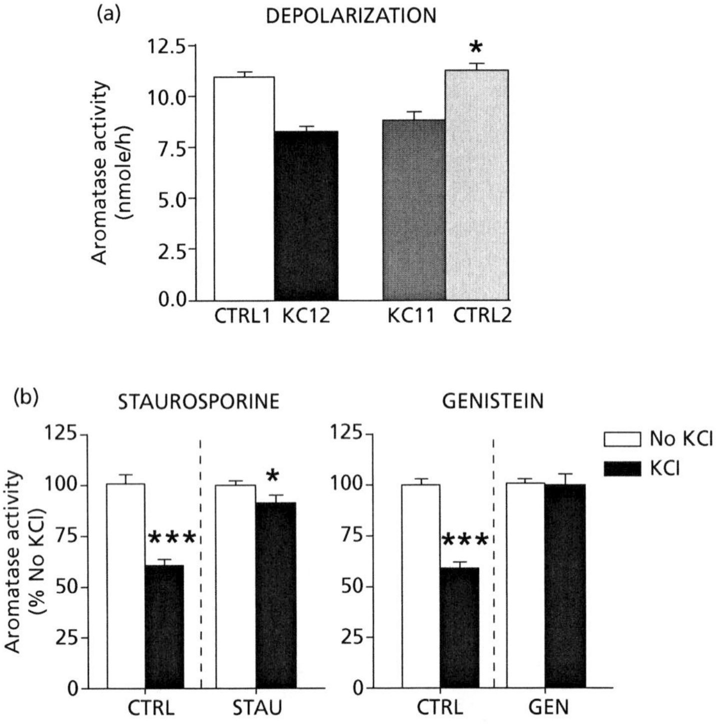

Figure 22–4. K+-Induced Depolarization Rapidly Reduces Aromatase Activity. A. Bar graph representing the effect of KCl-induced depolarization on the human aromatase activity in HEK293 intact cells (*left*). Aromatase activity completly recovers when cells are returned to normal control conditions (*right*). CTRL1 = DMEM (culture medium), KCl2 = DMEM + 100mM KCl, KCl1 = DMEM + 100mM KCl, CTRL2 = DMEM. Results are expressed as mean ± SEM, * $p < 0.05$ vs. matched conditions from treatment 1. B. The inhibition of aromatase activity by KCl, is largely blocked by the addition of 1mM staurosporin or 50 mM genistein. Results are expressed as mean ± SEM, * $p < 0.05$ vs. no KCl condition, same treatment; ***, $p < 0.001$ vs. no KCl, same treatment.

the protein was not affected: the amount of aromatase protein quantified by Western blot analysis with actin used as an internal standard was similar after depolarization and in control conditions. Interestingly, the rapid enzymatic inhibition induced by depolarization involved the activity of protein kinases. Addition of staurosporine (Ser/Thr kinase inhibitor) or genistein (Tyr kinase inhibitor) blocked the effect of KCl-induced depolarizations on AA (Charlier et al., 2011) (Fig. 22–4B).

The importance of protein phosphorylation was further confirmed by the demonstration that a 15-minute preincubation in phosphorylating conditions (ATP/Ca/Mg) significantly reduced the activity of human aromatase from cell lysates as compared with matched control samples (in HEK293, C6, and Neuro2A). These results indicate that the modulation of AA by phosphorylations is a general process, present not only in birds, but also presumably in humans and other mammals. Interestingly, we also showed that phosphorylating conditions do not affect the apparent enzyme affinity for its substrate but only change the maximum velocity of reaction. Among the 14 amino acid residues involved in the tethering of androstenedione in the binding pocket of aromatase, only one residue could potentially be affected by phosphorylation (T310, [Ghosh et al., 2009]), which is consistent with the observation that phosphorylation does not affect the affinity for the substrate. The phosphorylation of aromatase is therefore likely to concern other residues in the catalytic site of the enzyme that significantly slows down the aromatization process. Aromatization is a complex mechanism involving numerous reaction steps, including 19-hydroxylation and 10-methylation (Thompson & Siiteri, 1974). These different steps are currently not fully understood and discussing the implication of specific residues for each step would therefore be purely speculative.

AA INHIBITION IS ASSOCIATED WITH PHOSPHORYLATIONS OF THE AROMATASE PROTEIN

The experiments summarized above strongly suggest that phosphorylation processes rapidly and transiently regulate AA. However, because all assays were carried out on cell lines, brain homogenates, or in vivo and not on purified aromatase protein, they did not address the question of whether phosphorylations controlling enzymatic activity directly affect the aromatase itself or another coexisting protein that could secondarily regulate aromatase. To answer this question, aromatase from quail HPOA homogenates was purified by immunoprecipitation, and the presence of phosphorylated residues on the purified aromatase protein was investigated by Western blotting using anti- phosphoserine, -phosphothreonine, or-phosphotyrosine. In all experiments, phosphorylated Ser, Thr, or Tyr was identified by the specific antibodies at the level of the electrophoresis band corresponding to aromatase (Balthazart et al., 2003). Quantification of the blots indicated that the intensity of the phosphorylation signal was denser when homogenates were preincubated with ATP/Mg/Ca as compared with homogenates incubated in control conditions.

To further confirm that phosphorylations underlying the rapid modulation of AA targeted the aromatase protein itself, we engineered a modified human aromatase containing a c-myc tag that allows its immunoprecipitation. HEK293 cells transfected with this construct were incubated with [γ-^{32}P]-ATP in phosphorylating or non-phosphorylating (control) conditions. A ^{32}P-labeled protein was detected at the expected molecular weight for aromatase c-myc in phosphorylating conditions, whereas only a faint band was present in control conditions (Charlier et al., 2011). In parallel experiments, the immunoprecipitated protein visualized with antiphosphoserine similarly showed an immunoreactive band at the expected molecular weight after 5 minutes of incubation of the cell lysate in phosphorylating conditions (ATP/Ca/Mg; (Charlier et al., 2011)). The intensity of this band increased after a 15-minute incubation in phosphorylating conditions. That band was neither present in control conditions nor in the cells transfected with the empty vector in control or phosphorylating conditions. Altogether, these experiments demonstrate that the aromatase protein itself is rapidly phosphorylated in the presence of ATP/Mg/Ca, and strongly suggest that these phosphorylations directly cause the rapid decrease of enzymatic activity.

IDENTIFICATION OF AROMATASE RESIDUES INVOLVED IN THE RAPID CONTROL OF ACTIVITY

Pharmacological experiments on quail HPOA homogenates and HEK293-expressing human aromatase indicated that the inhibition of AA by phosphorylation is catalyzed by the activity of two Ser/Thr kinases, protein kinase A (PKA) and C (PKC) (Balthazart et al., 2003; Charlier et al., 2011). Based on this knowledge, we used bioinformatic tools to analyze the quail and human aromatase coding sequence and identified potential phosphorylation sites, highly conserved among several avian and mammalian species. The NETPHOS 2.0 program identified 15 and 19 putative phosphorylation sites for quail and human sequences respectively. They concerned 10 of the 27 serine, 5 of the 24 threonine, and 4 of the 17 tyrosine residues on the human sequence. The parallel search for consensus sequences corresponding to protein kinases that were shown to affect quail and human AA during the pharmacological experiments (i.e., PKC and PKA (Balthazart et al., 2003; Charlier et al., 2011)) pointed to six different residues, five serines (S118, S167, S247, S267, S497). and one threonine (T143) on the human sequence. Additional kinases that were not investigated during pharmacological experiments also displayed consensus sequences in the human aromatase sequence.

As a result of these analyses, we focused our attention on six different residues: S247, S267, and S497 that had high scores in both the predictive phosphorylation sites and PKA or PKC recognition consensus sequence; T462 and T493 that correspond to positions S455 and S486 in quail aromatase, two residues that were predicted to be involved in the phosphorylation of aromatase (Balthazart et al., 2003). In addition, relatively recent in vitro studies of the mouse aromatase suggest that two other amino acid residues, namely serine

S118 (Miller et al., 2008) or tyrosine Y361 (Catalano et al., 2009) can be phosphorylated and affect the stability or activity of the enzyme but after much longer latencies. Specifically, the phosphorylation of S188 by AGC-like kinases (such as PKC and PKA) seems important for the stabilization of the protein (Miller et al., 2008). Independently, the phosphorylation of Y361 triggered by the tyrosine-kinase c-Src seems involved in the E2-dependent upregulation of AA in breast cancer cell lines (Catalano et al., 2009). However, the effects of a rapid change of the phosphorylation status of S188 and Y361 residues has not been previously investigated, thus we performed experiments designed to establish whether S188 could also be involved in the rapid regulation of AA.

Using the human aromatase as template (the expression of quail aromatase in cell lines is not yet performed), six different mutants S/T to alanine (A) were produced to determine the potential importance of these amino acids in the rapid modulation of AA by phosphorylating conditions (preincubation with ATP/Ca/Mg for 15 min). All mutants still expressed AA and phosphorylating conditions markedly reduced this enzymatic activity in the six different mutants alone (S118A, S247A, S267A, T462A, T493A, S497A) or in combination (S267A-S247A, T493A-T462A-S267A, T493A-T462A-S267A-S247A, S497A-T493A-T462A-S267A-S247A) roughly to the same extent as in wild-type enzyme (Fig. 22–5). In all cases, inhibition was more pronounced after exposure to a higher concentration of ATP. Against all expectations, these single or combined mutations did not block the rapid inhibition of aromatase by phosphorylating conditions.

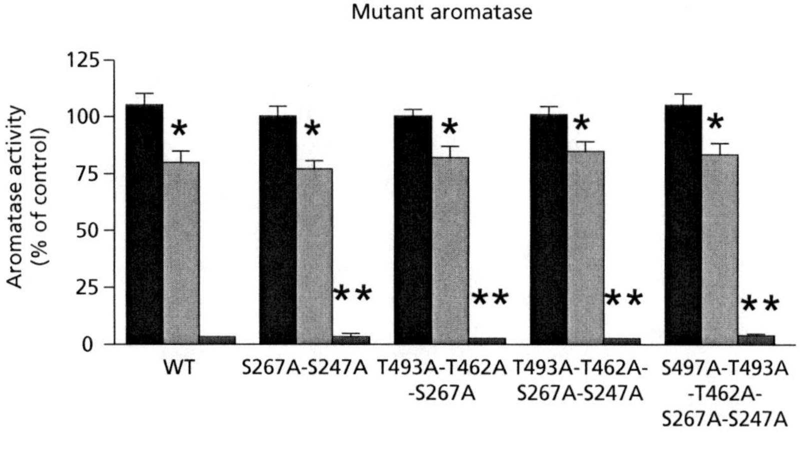

Figure 22–5. Aromatase activity in wild-type (WT) enzyme and enzyme modified by site-directed mutagenesis (multiple mutations) in control conditions (black bars) and following preincubation with 2 mM Ca^{++}, 10 mM Mg^{++}, and 2 mM (light grey bars) or 8 mM (dark grey bars) ATP. Aromatase activity in phosphorylating conditions is expressed as percentage of the mean activity in corresponding control conditions. Values are presented as mean ± SEM.

The lack of effects of these mutations on the rapid control of AA by phosphorylations could be due to numerous reasons (see Charlier et al., 2011 for detailed discussion). It is also possible that a combination of several phosphorylated residues that was not tested here is required to control AA. Multiples residues of a protein are often phosphorylated in vivo. This is, for example, the case for numerous cytochromes P450, the protein family that includes aromatase (Redlich et al., 2008). The control of AA by multiple kinases (Balthazart et al., 2003, 2005a) reinforces the idea that several amino acids must concomitantly be phosphorylated to modify AA. Although we mutated amino acids with the highest phosphorylation and kinase recognition prediction scores, other consensus sites for phosphorylations and for other types of kinases were also predicted on the quail and human aromatase sequences, suggesting that other amino acids could be involved. Alternatively, it is possible that the control of AA by phosphorylations depends on the phosphorylation of proteins interacting with aromatase in addition to the enzyme itself. Aromatase forms an electron-transfer complex with NADPH-cytochrome P450 reductase. The formation of this complex strictly conditions aromatase activity (Thompson & Siiteri, 1974; Miller, 2005; Hong et al., 2009). The phosphorylation of the reductase could be a control mechanism affecting the electron transfer and therefore AA and this idea should be further investigated.

Apart from the implication of specific residues in the rapid control of AA by phosphorylation, we also observed that two mutants, S118A and S497A, affected basal AA. More specifically, S118A aromatase had a markedly reduced enzymatic activity, while S497A mutant showed a higher AA than the wild-type and other mutants. To our knowledge, the two residues shown here to affect AA (S118 and S497) have not been directly implicated in substrate binding, reaction catalysis, or direct interaction with the NADPH reductase (Amarneh et al., 1993; Zhou et al., 1994; Kao et al., 2001; Hong et al., 2009). Miller and colleagues suggested that S118 phosphorylation by PKC could be required for stabilization of aromatase (Miller et al., 2008) but the exact function of these two residues in the control of basal AA and protein stability remains to be determined.

CONCLUSIONS

In summary, numerous biochemical and pharmacological studies confirm that AA can be rapidly modulated via post-translational modifications. The rapid modulation of AA by phosphorylating conditions is a widespread mechanism observed in several different tissues that express aromatase and is observed in a variety of species, including humans. Discussions of the relatively recent discovery that estrogens, and E2 particularly, are involved in a large number of non-genomic signaling processes have usually overlooked the fact that synthesis of these estrogens needs to be rapidly controlled for estrogens to act relatively rapidly in a manner akin to a neurotransmitter or a neuromodulator (Balthazart & Ball, 2006). Although the molecular mechanisms leading to these modifications remain poorly understood, the experiments reviewed here demonstrate that rapid changes in AA take place in the brain and these changes must result in a local rapid modulation of estrogens availability that will ultimately affect cellular events in the absence of changes in protein synthesis. The phosphorylation/dephosphorylation processes affecting AA provide a new widespread mechanism by which estrogen concentrations could be rapidly altered in the brain and other tissues.

ACKNOWLEDGMENTS

The work of the authors described in this review was supported by grants from the NIMH (R01 MH50388) to GFB and JB and from the Belgian FRFC (Nbr. 2.4537.9) and from the University of Liège (Crédits spéciaux) to JB. TDC is Research Associate at the University of Liège. CAC is F.R.S-FNRS Research Associate. We thank all members of our laboratories, past and present, who contributed to the data summarized here and are too numerous to be listed individually.

REFERENCES

Abraham, I. M., Han, S.-K., Todman, M. G., Korach, K. S., & Herbison, A. E. (2003). Estrogen receptor beta mediates rapid estrogen actions on gonadotropin-releasing hormone neurons in vivo. *J Neurosci, 23,* 5771–5777.

Abraham, I. M., Todman, M. G., Korach, K. S., & Herbison, A. E. (2004). Critical *in* vivo roles for classical estrogen receptors in rapid estrogen actions on intracellular signaling in mouse brain. *Endocrinology, 145,* 3055–3061.

Addis, E. A., Busch, D. S., Clark, A. D., & Wingfield, J. C. (2010). Seasonal and social modulation of testosterone in Costa Rican rufous-collared sparrows (Zonotrichia capensis costaricensis). *Gen Comp Endocrinol, 166,* 581–589

Albert, K. A., Helmer-Matyjek, E., Nairn, A. A., Müller, T. H., Haycock, J. W., Greene, L. A., et al. (1984). Calcium/phospholipid-dependent protein kinase (protein kinase C) phosphorylates and activates tyrosine hydroxylase. *Proc Natl Acad Sci U S A, 81,* 7713–7717.

Alonso, A., Sasin, J., Bottini, N., Friedberg, I., Osterman, A., Godzik, A., et al. (2004). Protein tyrosine phosphatases in the human genome. *Cell, 117,* 699–711.

Amarneh, B., Corbin, C. J., Peterson, J. A., Simpson, E. R., & Graham-Lorence, S. (1993). Functional domains of human aromatase cytochrome P450 characterized by linear alignment and site-directed mutagenesis. *Mol Endocrinol, 7,* 1617–1624.

Ames, M. M., Lerner, P., & Lovenberg, W. (1978). Tyrosine hydroxylase: activation by protein phosphorylation and end product inhibition. *J Biol Chem*, 253, 27–31.

Balthazart, J., Baillien, M., & Ball, G. (2001a). Phosphorylation processes mediate rapid changes of brain aromatase activity. *J Steroid Bioch Molec Biol*, 79, 261–277.

Balthazart, J., Baillien, M., & Ball, G. F. (2001b). Rapid and reversible inhibition of brain aromatase activity. *J Neuroendocrinol, 13,* 61–71.

Balthazart, J., Baillien, M., & Ball, G. F. (2005a). Interactions between kinases and phosphatases in the rapid control of brain aromatase. *J Neuroendocrinol, 17,* 553–559.

Balthazart, J., Baillien, M., & Ball, G. F. (2006). Rapid control of brain aromatase activity by glutamatergic inputs. *Endocrinology, 147,* 359–366.

Balthazart, J., Baillien, M., Charlier, T. D., & Ball, G. F. (2003). Calcium-dependent phosphorylation processes control brain aromatase in quail. *Eur J Neurosci, 17,* 1591–1606.

Balthazart, J., Baillien, M., Charlier, T. D., & Ball, G.F. (2005b). Effects of calmodulin on aromatase activity in the preoptic area. *J Neuroendocrinol, 17,* 664–671.

Balthazart, J. & Ball, G. F. (2006). Is brain estradiol a hormone or a neurotransmitter. *TINS, 29,* 241–249.

Balthazart, J., Stoop, R., Foidart, A., & Harada, N. (1994). Synergistic control by androgens and estrogens of aromatase in the quail brain. *Neuroreport, 5,* 1729–1732.

Batty, J. (1978). Acute changes in plasma testosterone levels and their relation to measures of sexual behaviour in the male house mouse (Mus musculus). *Anim Behav, 26,* 349–357.

Boulware, M. I., Weick, J. P., Becklund, B. R., Kuo, S. P., Groth, R. D., & Mermelstein, P. G. (2005). Estradiol activates group I and II metabotropic glutamate receptor signaling, leading to opposing influences on cAMP response element-binding protein. *J Neurosci, 25,* 5066–5078.

Catalano, S., Barone, I., Giordano, C., Rizza, P., Qi, H., Gu, G., et al. (2009). Rapid estradiol/ERalpha signaling enhances aromatase enzymatic activity in breast cancer cells. *Mol Endocrinol, 23,* 1634–1645.

Charlier, T. D., Cornil, C. A., Ball, G. F., & Balthazart, J. (2010). Diversity of mechanisms involved in aromatase regulation and estrogen action in the brain. *Biochim Biophys Acta, 1800,* 1094–1105.

Charlier, T. D., Harada, N., Balthazart, J., & Cornil, C. A. (2011). Human and quail aromatase activity is rapidly and reversibly inhibited by phosphorylating conditions. *Endocrinology, 152,* 4199–4210.

Chen, B. S., & Roche, K. W. (2007). Regulation of NMDA receptors by phosphorylation. *Neuropharmacology, 53,* 362–368.

Cornil, C., Foidart, A., Minet, A., & Balthazart, J. (2000). Immunocytochemical localization of ionotropic glutamate receptors subunits in the adult quail forebrain. *J Comp Neurol, 428,* 577–608.

Cornil, C. A., Ball, G. F., & Balthazart, J. (2006a). Functional significance of the rapid regulation of brain estrogen action: where do the estrogens come from? *Brain Res, 1126,* 2–26.

Cornil, C. A., & Charlier, T. D. (2010). Rapid behavioural effects of oestrogens and fast regulation of their local synthesis by brain aromatase. *J Neuroendocrinol, 22,* 664–673.

Cornil, C. A., Dalla, C., Papadopoulou-Daifoti, Z., Baillien, M., & Balthazart, J. (2006b). Estradiol rapidly activates male sexual behavior and affects brain monoamine levels in the quail brain. *Behav Brain Res, 166,* 110–123.

Cornil, C. A., Dalla, C., Papadopoulou-Daifoti, Z., Baillien, M., Dejace, C., Ball, G. F., et al. (2005a). Rapid decreases in preoptic aromatase activity and brain monoamine concentrations after engaging in male sexual behavior. *Endocrinology, 1416,* 2809–2820.

Cornil, C.A., Leung, C.H., Pletcher, E.R., Naranjo, K.C., Blauman, S.J., Saldanha, C.J. (2012) Acute and specific modulation of presynaptic aromatization in the vertebrate brain. *Endocrinology.* 153(6):2562–2567

Cornil, C. A., Seutin, V., Motte, P., & Balthazart, J. (2004). Electrophysiological and neurochemical characterization of neurons of the medial preoptic area in Japanese quail (Coturnix japonica). *Brain Res, 1029,* 224–240.

Cornil, C. A., Taziaux, M., Baillien, M., Ball, G. F., & Balthazart, J. (2005b). Rapid effects of aromatase inhibition on male reproductive behaviors in Japanese quail. *Horm Behav, 49,* 45–67.

Cross, E., & Roselli, C.E. (1999). 17b-estradiol rapidly facilitates chemoinvestigation and mounting in castrated male rats. *Am J Physiol, 276,* R1346–R1350.

Daubner, S. C., Lauriano, C., Haycock, J. W., & Fitzpatrick, P. F. (1992). Site-directed mutagenesis of serine 40 of rat tyrosine hydroxylase. Effects of dopamine and cAMP-dependent phosphorylation on enzyme activity. *J Biol Chem, 267,* 12639–12646.

Dewing, P., Boulware, M. I., Sinchak, K., Christensen, A., Mermelstein, P. G., & Micevych, P. (2007). Membrane estrogen receptor-alpha interactions with metabotropic glutamate receptor 1a modulate female sexual receptivity in rats. *J Neurosci, 27,* 9294–9300.

Dickens, M. J., Cornil, C. A., & Balthazart, J. (2011). Acute stress differentially affects aromatase activity in specific brain nuclei of adult male and female quail. *Endocrinology, 152,* 4242–4251.

Diotel, N., Le Page, Y., Mouriec, K., Tong, S. K., Pellegrini, E., Vaillant, C., et al. (2010). Aromatase in the brain of teleost fish: expression, regulation and putative functions. *Front Neuroendocrinol*, 31, 172–192.

Evrard, H. C., & Balthazart, J. (2004). Rapid regulation of pain by estrogens synthesized in spinal dorsal horn neurons. *J Neurosci, 18,* 7225–7229.

Foidart, A., Harada, N., & Balthazart, J. (1995a). Aromatase-immunoreactive cells are present in mouse brain areas that are known to express high levels of aromatase activity. *Cell Tissue Res*, 280, 561–574.

Foidart, A., Reid, J., Absil, P., Yoshimura, N., Harada, N., & Balthazart, J. (1995b). Critical re-examination of the distribution of aromatase-immunoreactive cells in the quail forebrain using antibodies raised against human placental aromatase and against the recombinant quail, mouse or human enzyme. *J Chem Neuroanat, 8,* 267–282.

Forlano, P. M., Deitcher, D. L., Myers, D. A., & Bass, A. H. (2001). Anatomical distribution and cellular basis for high levels of aromatase activity in the brain of teleost fish: aromatase enzyme and mRNA expression identify glia as a source. *J Neurosci*, 21, 8943–8955.

Forlano, P. M., Schlinger, B. A., & Bass, A. H. (2006). Brain aromatase: new lessons from non-mammalian model systems. *Front Neuroendocrinol, 27,* 247–274.

Fusani, L., Hutchison, J. B., & Gahr, M. (2001). Testosterone regulates the activity and expression of aromatase in the canary neostriatum. *J Neurobiol*, 49, 1–8.

Ghosh, D., Griswold, J., Erman, M., & Pangborn, W. (2009). Structural basis for androgen specificity and oestrogen synthesis in human aromatase. *Nature, 457,* 219–223.

Harada, N., Abe-Dohmae, S., Loeffen, R., Foidart, A., & Balthazart, J. (1993). Synergism between androgens and estrogens in the induction of aromatase and its messenger RNA in the brain. *Brain Res*, 622, 243–256.

Harding, C. F. (1981). Social modulation of circulating hormone levels in the male. *Am Zool, 21,* 223–231.

Hayden-Hixson, D. M., & Ferris, C. F. (1991). Steroid-specific regulation of agonistic responding in the anterior hypothalamus of male hamsters. *Physiol Behav, 50,* 793–799

Hong, Y., Li, H., Yuan, Y. C., & Chen, S. (2009). Sequence-function correlation of aromatase and its interaction with reductase. *J Steroid Biochem Mol Biol, 118,* 203–206,

Joels, M. (1997). Steroid hormones and excitability in the mammalian brain. *Front Neuroendocrinol, 18,* 2–48.

Johnson, S. A., & Hunter, T. (2005). Kinomics: methods for deciphering the kinome. *Nat Methods, 2,* 17–25.

Kao, Y. C., Korzekwa, K. R., Laughton, C. A., & Chen, S. (2001). Evaluation of the mechanism of aromatase cytochrome P450: a site directed mutagenesis study. *Eur J Biochem, 268,* 243–251.

Kenealy, B. P., Keen, K. L., Ronnekleiv, O. K., & Terasawa, E. (2011). STX, a novel nonsteroidal estrogenic compound, induces rapid action in primate GnRH neuronal calcium dynamics and peptide release. *Endocrinology, 152,* 3182–3191.

Kinoshita, Y., & Chen, S. (2003). Induction of aromatase (CYP19) expression in breast cancer cells through a nongenomic action of estrogen receptor alpha. *Cancer Res, 63,* 3546–3555.

Konkle, A. T., & Balthazart, J. (2011). Sex Differences in the Rapid Control of Aromatase Activity in the Quail Preoptic Area. *J Neuroendocrinol, 23,* 424–434.

Lord, L. D., Bond, J., & Thompson, R. R. (2009). Rapid steroid influences on visually guided sexual behavior in male goldfish. *Horm Behav, 56,* 519–526.

Manning, G., Whyte, D. B., Martinez, R., Hunter, T., & Sudarsanam, S. (2002). The protein kinase complement of the human genome. *Science, 298,* 1912–1934.

Mendelson, C. R., Corbin, C. J., Smith, M. E., Smith, J., & Simpson, E. R. (1986). Growth factors suppress and phorbol esters potentiate the action of dibutyryl adenosine 3′,5′-monophosphate to stimulate aromatase activity of human adipose stromal cells. *Endocrinology, 118,* 968–973.

Mermelstein, P. G., Becker, J. B., & Surmeier, D. J. (1996). Estradiol reduces calcium currents in rat neostriatal neurons via a membrane receptor. *J Neurosci, 16,* 595–604.

Metzdorf, R., Gahr, M, & Fusani, L. (1999). Distribution of aromatase, estrogen receptor, and androgen receptor mRNA in the forebrain of songbirds and nonsongbirds. *J Comp Neurol, 407,* 115–129.

Micevych, P. E., & Mermelstein, P. G. (2008). Membrane estrogen receptors acting through metabotropic glutamate receptors: an emerging mechanism of estrogen action in the brain. *Mol Neurobiol, 38,* 66–77.

Miller, T. W., Shin, I., Kagawa, N., Evans, D. B., Waterman, M.R. & Arteaga, C.L. (2008). Aromatase is phosphorylated in situ at serine-118. *J Steroid Biochem Mol Biol, 112,* 95–101.

Miller, W. L. (2005). Minireview: regulation of steroidogenesis by electron transfer. *Endocrinology, 146,* 2544–2550.

Moss, R. L., Gu, Q., & Wong, M. (1997). Estrogen: nontranscriptional signaling pathway. *Recent Prog Horm Res, 52,* 33–68; discussion 68–69.

Naftolin, F., Horvath, T. L., Jakab, R. L., Leranth, C., Harada, N., & Balthazart, J. (1996). Aromatase immunoreactivity in axon terminals of the vertebrate brain—An immunocytochemical study on quail, rat, monkey and human tissues. *Neuroendocrinol,* 63, 149–155.

Nestler, E. J., & Greengard (1984). *Protein Phosphorylation in the Central Nervous System.* : Wiley, Hoboken, NJ, USA.

Onagbesan, O. M., & Peddie, M. J. (1989). Calcium-dependent stimulation of estrogen secretion by FSH from theca cells of the domestic hen (Gallus domesticus). *Gen Comp Endocrinol, 75,* 177–186.

Peterson, R. S., Yarram, L., Schlinger, B. A., & Saldanha, C. J. (2005). Aromatase is presynaptic and sexually dimorphic in the adult zebra finch brain. *Proc Biol Sci, 272,* 2089–2096.

Pradhan, D. S., Newman, A. E., Wacker, D. W., Wingfield, J. C., Schlinger, B. A., & Soma, K. K. (2010). Aggressive interactions rapidly increase androgen synthesis in the brain during the non-breeding season. *Horm Behav,* 57, 381–389

Purvis, K., & Haynes, N. B. (1974). Short-term effects of copulation, human chorionic gonadotrophin injection and non-tactile association with a female on testosterone levels in the male rat. *J Endocrinol, 60,* 429–439.

Redlich, G., Zanger, U. M., Riedmaier, S., Bache, N., Giessing, A. B., Eisenacher, M., et al. (2008). Distinction between human cytochrome P450 (CYP) isoforms and identification of new phosphorylation sites by mass spectrometry. *J Proteome Res, 7,* 4678–4688.

Remage-Healey, L., & Bass, A. H. (2004). Rapid, hierarchical modulation of vocal patterning by steroid hormones. *J Neurosci, 24,* 5892–5900.

Remage-Healey, L., Coleman, M. J., Oyama, R. K., & Schlinger, B. A. (2010). Brain estrogens rapidly strengthen auditory encoding and guide song preference in a songbird. *Proc Natl Acad Sci U S A, 107,* 3852–3857.

Remage-Healey, L., Dong, S., Maidment, N. T., & Schlinger, B. A. (2011). Presynaptic control of rapid estrogen fluctuations in the songbird auditory forebrain. *J Neurosci, 31,* 10034–10038.

Remage-Healey, L., Maidment, N. T., & Schlinger, B. A. (2008). Forebrain steroid levels fluctuate rapidly during social interactions. *Nat Neurosci, 11,* 1327–1334.

Remage-Healey, L., Oyama, R. K., & Schlinger, B. A. (2009). Elevated aromatase activity in forebrain synaptic terminals during song. *J Neuroendocrinol, 21,* 191–199.

Richards, J. A., Petrel, T. A., & Brueggemeier, R. W. (2002). Signaling pathways regulating aromatase and cyclooxygenases in normal and malignant breast cells. *J Steroid Biochem Mol Biol, 80,* 203–212.

Roepke, T. A., Ronnekleiv, O. K., & Kelly, M. J. (2011). Physiological consequences of membrane-initiated estrogen signaling in the brain. *Front Biosci, 16,* 1560–1573.

Rønnekleiv, O. K., & Kelly, M. J. (2002). Rapid membrane effects of estrogen in the central nervous system. In: Pfaff, D.W., Arnold, A. P., Etgen, A. M., Fahrbach, S. E., & Rubin, R. T. (eds.) *Hormones, Brain and Behavior*. St Louis: Elsevier Science, pp. 361–380.

Roselli, C., & Resko, J. (2001). Cytochrome P450 aromatase (CYP19) in the non-human primate brain: distribution, regulation, and functional significance. *J. Steroid Bioch Molec Biol, 79,* 247–253.

Roselli, C. E., Abdelgadir, S. E., & Resko, J. A. (1997). Regulation of aromatase gene expression in the adult rat brain. *Brain Res Bull, 44,* 351–357.

Roselli, C. E., Horton, L. E., & Resko, J. A. (1985). Distribution and regulation of aromatase activity in the rat hypothalamus and limbic system. *Endocrinology,* 117, 2471–2477.

Roselli, C. E., & Resko, J. A. (1984). Androgens regulate brain aromatase activity in adult male through a receptor mechanism. *Endocrinology, 114,* 2183–2189.

Roselli, C. E., & Resko, J. A. (1989). Testosterone regulates aromatase activity in discrete brain areas of male rhesus macaques. *Biol Reprod, 40,* 929–934.

Ryde, C. M., Nicholls, J. E., & Dowsett, M. (1992). Steroid and growth factor modulation of aromatase activity in MCF7 and T47D breast carcinoma cell lines. *Cancer Res, 52,* 1411–1415.

Saginor, M., & Horton, R. (1968). Reflex release of gonadotropin and increased plasma testosterone concentration in male rabbits during copulation. *Endocrinology, 82,* 627–630.

Saldanha, C. J., Peterson, R. S., Yarram, L., & Schlinger, B. A. (2003). The synaptocrine hypothesis: a novel mechanism of estrogen delivery. *Horm Behav, 44,* 74.

Saldanha, C. J., Tuerk, M. J., Kim, Y. H., Fernandes, A. O., Arnold, A. P., & Schlinger, B. A. (2000). Distribution and regulation of telencephalic aromatase expression in the zebra finch revealed with a specific antibody. *J Comp Neurol, 423,* 619–630.

Schumacher, M., & Balthazart, J. (1986). Testosterone-induced brain aromatase is sexually dimorphic. *Brain Res, 370,* 285–293.

Shen, P., Schlinger, B. A., Campagnoni, A. T., & Arnold, A. P. (1995). An atlas of aromatase mRNA expression in the zebra finch brain. *J Comp Neurol, 360,* 172–184.

Shi, Y. (2009). Serine/threonine phosphatases: mechanism through structure. *Cell, 139,* 468–484.

Shozu, M., Sumitani, H., Murakami, K., Segawa, T., Yang, H. J., & Inoue, M. (2001). Regulation of aromatase activity in bone-derived cells: possible role of mitogen-activated protein kinase. *J Steroid Biochem Mol Biol, 79,* 61–65.

Steimer, T., & Hutchison, J. B. (1991). Micromethods for the in vitro study of steroid metabolism in the brain using radiolabelled tracers. In: Greenstein, B. (ed.) *Neuroendocrine Research Methods,* vol 2. Chur, Switzerland: Harwood Academic Publishers, pp. 875–919.

Taziaux, M., Keller, M., Bakker, J., & Balthazart, J. (2007). Sexual behavior activity tracks rapid changes in brain estrogen concentrations. *J Neurosci, 27,* 6563–6572.

Thompson, E. A. Jr., & Siiteri, P. K. (1974). Utilization of oxygen and reduced nicotinamide adenine dinucleotide phosphate by human placental microsomes during aromatization of androstenedione. *J Biol Chem, 249,* 5364–5372.

Trainor, B. C., Finy, M. S., & Nelson, R. J. (2008). Rapid effects of estradiol on male aggression depend on photoperiod in reproductively non-responsive mice. *Horm Behav, 53,* 192–199.

Trainor, B. C., Lin, S., Finy, M. S., Rowland, M. R., & Nelson, R. J. (2007). Photoperiod reverses the effects of estrogens on male aggression via genomic and nongenomic pathways. *Proc Natl Acad Sci U S A, 104,* 9840–9845.

Tremere, L. A., Jeong, J. K., & Pinaud, R. (2009). Estradiol shapes auditory processing in the adult brain by regulating inhibitory transmission and plasticity-associated gene expression. *J Neurosci, 29,* 5949–5963.

Wade, C. B., & Dorsa, D. M. (2003). Estrogen activation of cyclic adenosine 5′-monophosphate response element-mediated transcription requires the extracellularly regulated kinase/mitogen-activated protein kinase pathway. *Endocrinology, 144,* 832–838.

Wagner, C. K., & Morrell, J. I. (1997). Neuroanatomical distribution of aromatase mRNA in the rat brain: indications of regional regulation. *J Steroid Biochem Mol Biol, 61,* 307–314.

Watters, J. J., Campbell, J. S., Cunningham, M. J., Krebs, E. G., & Dorsa, D. M. (1997). Rapid membrane effects of steroids in neuroblastoma cells: effects of estrogen on mitogen activated protein kinase signalling cascade and c-fos immediate early gene transcription. *Endocrinology, 138,* 4030–4033.

Wingfield, J. C. (2006). Communicative behaviors, hormone-behavior interactions, and reproduction in vertebrates. In: Neill, J. D. (ed.) *Physiology of Reproduction.* New York: Academic Press, pp. 1995–2040.

Wu, T. W., Wang, J. M., Chen, S., & Brinton, R. D. (2005). 17Beta-estradiol induced Ca2+ influx via L-type calcium channels activates the Src/ERK/cyclic-AMP response element binding protein signal pathway and BCL-2 expression in rat hippocampal neurons: a potential initiation mechanism for estrogen-induced neuroprotection. *Neuroscience, 135,* 59–72.

Yue, W., Wang, J. P., Conaway, M. R., Li, Y., & Santen, R. J. (2003). Adaptive hypersensitivity following long-term estrogen deprivation: involvement of multiple signal pathways. *J Steroid Biochem Mol Biol, 86,* 265–274.

Zhou, D., Cam, L. L., Laughton, C. A., Korzekwa, K. R., & Chen, S. (1994). Mutagenesis study at a postulated hydrophobic region near the active site of aromatase cytochrome P450. *J Biol Chem, 269,* 19501–19508.

Chapter 23

Rapid Modulation of Aromatase Activity by Social and Environmental Stimuli in Quail

Charlotte A. Cornil, Molly J. Dickens, Gregory F. Ball, and Jacques Balthazart

INTRODUCTION

Since the discovery that testosterone effects on brain and behavior are often mediated through its neural conversion into an estrogen, research on the control of local estrogen synthesis by aromatase has mainly focused on changes in enzyme availability taking place over relatively long time periods following developmental or seasonal changes in circulating steroid levels. Studies conducted in a variety of species have established that the transcriptional activity of androgens and estrogens synergistically regulates the concentration of the enzymatic protein (see chapters 3 and 10, this volume). However, more recently, studies taking advantage of the high concentrations of the aromatase protein in the avian brain have revealed a novel mechanism regulating brain estrogen synthesis (see chapter 22, this volume). These in vitro studies demonstrated rapid modulations (within minutes) of preoptic/hypothalamic aromatase activity by calcium-dependent phosphorylations. Similar rapid enzymatic changes were observed following K^+-induced depolarizations or exposure to neurotransmitters such as glutamate. Such findings shifted the interest of research on aromatase towards much shorter physiological time scales and suggested that a mechanism was in place to allow changes in local estradiol concentration in a time frame compatible with the activation of fast non-genomic actions of estrogens. The present chapter provides evidence that this mechanism, discovered in vitro, is functionally significant in vivo. We will first review studies that identified rapid modulations of aromatase activity following social interactions or acute exposure to stress. Then, we will focus on the implications of such rapid controls of local estrogen availability for the control of reproduction.

AROMATASE ACTIVITY IS RAPIDLY MODULATED IN VIVO

Effect of Sexual Interactions

Multiple sets of experimental results suggest that the rapid modulation of aromatase activity measured in the hypothalamus-preoptic area (HPOA)

depends on direct action of glutamate on aromatase cells. For example, electrophysiological recordings in both rat and quail demonstrated that preoptic neurons, including aromatase neurons, are sensitive to glutamate (Karlsson et al., 1997; Cornil et al., 2002, 2004). It has also been shown in rats, based on microdialysis measurements, that extracellular glutamate concentrations gradually increase in the POA during sexual interactions and they reach a peak when the male ejaculates (Dominguez et al., 2006). Although this phenomenon has yet to be investigated in birds, one can hypothesize that such a copulation-induced rise in extracellular glutamate levels would result in a rapid down-regulation of preoptic aromatase activity similar to the decrease detected in HPOA explants (Balthazart et al., 2006).

To test this idea, an initial experiment investigated whether aromatase activity is affected immediately following the expression of sexual behavior. To this end, sexually experienced male quail were given visual access to or copulated with a receptive female for 1, 5, or 15 minutes. Brains were collected immediately after the behavioral interaction. Analysis of the enzymatic activity in the HPOA identified a time-dependent modulation of activity following visual access to or copulation with the female. Specifically, a decrease in enzymatic activity was detectable after 1 min, reached a maximum after 5 minutes, and returned to control levels after 15 minutes of interaction (Cornil et al., 2005). Although such a reduction in local estrogen synthesis might potentially influence physiological and behavioral processes, this drop in enzymatic activity was much less dramatic than that observed in vitro. This discrepancy can be explained by the fact that the HPOA contains several populations of aromatase containing cells that may contribute differently to behavioral control (Foidart et al., 1995). Therefore, the limited changes observed in enzymatic activity might reflect a dilution of effects located in discrete regions or even the summation of changes occurring in opposite directions.

Follow-up experiments have been completed that were aimed at determining which population(s) of aromatase-expressing neurons participate(s) in this rapid enzymatic change induced by sexual interactions (de Bournonville et al., 2011). The effect of various durations of sexual interaction (0, 2, 5, 10, or 15 min) was assessed on aromatase activity measured in six brain regions (medial preoptic nucleus, bed nucleus of the stria terminalis, ventromedial and tuberal hypothalamus, nucleus taeniae of the amygdala, and periaqueductal gray) microdissected by the Palkovits punch technique (Schumacher and Balthazart, 1987; Cornil et al., 2011). Based on its central role in the control of male sexual behavior (Panzica et al., 1996), it was anticipated that the medial preoptic nucleus (POM) would be primarily influenced by copulation. However, the fastest enzymatic change appeared in the tuberal hypothalamus after only 2 minutes of interaction (-25%) (Table 23–1). After 5 minutes, a marked decrease was detected in the POM. Both effects progressively developed for at least 15 minutes, at which point the maximal inhibition of activity reached 40%. In contrast, visual interaction with the female (no physical contact) resulted in a significant reduction of enzymatic activity in the medial part of the bed nucleus of the stria terminalis (BSTm) only.

Table 23–1. Summary of Acute Changes in Aromatase Activity Induced by Copulation or Stress Exposure in Six Brain Nuclei Containing Aromatase

Sex	Stimulus	POM	BSTm	TnA	VMN	Tub	PAG
Male	Copulation	↘	(↘)	-	-	↘	-
	Stress	↗	-	-	↗	(↗)	-
Female	Stress	(↗)	(↗)	-	↘	↘	-

Arrows indicate significant up- or down-regulations. Arrows within parentheses indicate statistical tendencies. Abbreviations: POM, medial preoptic nucleus; BSTm, medial bed nucleus of the stria terminalis; TnA, nucleus taeniae of the amygdala; VMN, ventromedial nucleus of the hypothalamus; Tub, tuberal hypothalamus; PAG, periaqueductal gray. de Bournonville et al. 2011; Dickens et al. 2011.

Together, these data indicate that copulating with or simply seeing a female affects male aromatase activity in different brain regions. This might thus explain why both types of interactions similarly reduced aromatase activity in the entire HPOA in previous studies. Although the functional significance of such a reduction in brain aromatase activity still remains to be determined (see below for discussion), these data demonstrate that in vivo aromatase activity is rapidly modulated in behaviorally relevant situations and in specific brain areas.

Effect of Stress

Another set of experiments investigated whether stress and the associated rise in plasma corticosterone caused by stress also acutely influence brain aromatase activity. It is generally assumed that activation of the stress response system rapidly affects reproductive processes including behavior (Wingfield and Sapolsky, 2003). In addition, although it is not known whether stress induces changes in preoptic glutamate release, corticosterone has been shown to acutely reduce glutamatergic transmission in hypothalamic neurons (Di et al., 2009). Based on the in vitro work summarized earlier (see above and chapter 22 in this volume), it was thus expected that if stress affects aromatase activity, the direction of this change would be opposite to the direction of the change induced by copulation. This hypothesis was first tested by exposing male quail to acute stress either by placing them in a small opaque bag for 15 minutes or injecting them with corticosterone 30 minutes before brain collection. Control groups consisted of birds left undisturbed in their home cage or injected with the vehicle. Brains were collected immediately after the procedure and their HPOA dissected for enzymatic assays. As hypothesized, these manipulations induced a significant rise in aromatase activity (Balthazart et al., 2009).

The regional- and sex-specificity of this enzymatic response was then examined (Dickens et al., 2011). Males and females were subjected to acute restraint stress for 0, 15, or 30 minutes. To assess reversibility, an additional group of birds was subjected to 30 minutes of stress followed by 30 minutes of recovery, during which they were left undisturbed in their home cage. All subjects were sacrificed immediately after manipulation. Aromatase activity was then assayed in the six brain regions described above. Stress induced sex- and region-specific changes in aromatase activity (see Table 23–1). Regional variation in the time-response or post-stress recovery was also observed. Interestingly, the regions affected were the same as those shown to be sensitive to sexual behavior. In males, a rapid upregulation of aromatization was detected in the POM and the mediobasal hypothalamus (MBH, consisting of the ventromedial nucleus of the hypothalamus and tuberal hypothalamus). In contrast, females exhibited a moderate increase in aromatase activity in the POM and a profound enzymatic downregulation in MBH. In addition, the two sexes presented differential time responses to the stressor, males responding quickly and reversibly, while female responses were not as fast but sustained beyond stressor cessation.

Although stress induced a significant increase in circulating corticosterone levels, individual corticosterone concentrations did not correlate with aromatase activity in any brain region considered (Dickens et al., 2011). Since circulating corticosterone cannot explain by itself the sex and regional specificity of the enzymatic responses, other candidates should be considered. The overall higher concentration of corticosterone measured in jugular compared to brachial veins suggests the existence of a brain synthesis of corticosterone that is potentially affected by stress (Dickens et al., 2011). Individual brain regions may thus be exposed to differential local concentrations of corticosterone. Multiple other factors including arginine-vasotocin, corticosterone-releasing factor, and norepinephrine may mediate these sex and anatomically specific enzymatic changes induced by acute stress and are currently under investigation.

These direct effects of stress on aromatase involve brain nuclei known for their critical role in the control of reproduction. Although the function of locally synthesized estrogens remains to be established in most of these regions, these results suggested that aromatase might mediate acute effects of stress on behavior and fertility. This hypothesis was tested in an experiment using a two-by-two design such that all individuals were either stressed or non-stressed for 15 minutes immediately before a 5-minute sexual interaction with a stressed or nonstressed partner (Dickens et al., In press). Surprisingly, behavioral endpoints were unaltered by stress although fertilization was reduced in stressed females. Stress or copulation alone induced the same effects on aromatase activity as previously described (Fig. 23–1).

However, when presented in sequence, sexual behavior reversed the stress-induced upregulation of aromatization in the male POM, suggesting that copulation may rapidly restore pre-stress enzymatic activity to actively preserve sexual behavior. Likewise, although the directionality differed from males, the apparent pattern observed in the female tuberal hypothalamus led to the same conclusion: pairing seems to "rescue" basal enzymatic activity. The pattern of the acute enzymatic response to stress and mating also shows a clear anatomical specificity. In some regions, such as the male POM or the female Tuber,

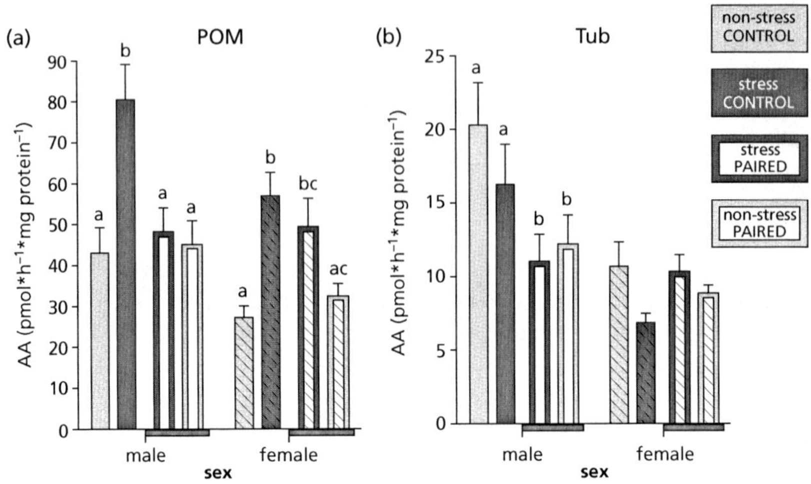

Figure 23–1. Acute effects of mate pairing and restraint stress on brain aromatase activity in (*A*) the medial preoptic nucleus (POM) and (*B*) the tuberal hypothalamus (Tub) of male and female quail (*open* and *hashed columns*, respectively). Control groups (*closed bars*) are non-stressed/unpaired or stressed/unpaired, whereas experimental groups (*open bars*) are stressed individuals paired with an unstressed partner and non-stressed individual paired with a non-stressed partner. Paired groups (as opposed to unpaired controls) are indicated by the *grey bar* under the x-axis. Groups within each sex that are significantly different from each other (determined by post hoc Bonferroni tests) are identified by different letters. Modified from (Dickens et al., In press).

aromatase seems responsive to both stimuli. However, the enzyme appears exclusively sensitive to either mating or stress in other regions: male MBH responding only to copulation while female POM and VMN responding solely to stress (Dickens et al., In press).

Comparison Between Stress and Copulation Effects

What can we infer from these observations so far? First, there appears to be layers of complexity when it comes to acute control of aromatase activity by sexual interactions or stress exposure; changes are specific to sex, brain region and stimulus type. At first, it is not surprising that copulation induced enzymatic changes in male POM and a female MBH, since they both have been recognized as key regions for the control male and female reproductive behaviors, respectively (Ball and Balthazart, 2009). Yet, nothing is known about the functional implications of local changes in estrogen synthesis in brain regions other than the POA where the role of aromatization in the activation of male sexual behavior has been extensively studied (Watson and Adkins-Regan, 1989 ab; Balthazart et al., 1990; Balthazart and Surlemont, 1990). In females, it is usually assumed that active estrogens originate from the ovary. However, the presence of substantial concentrations of active aromatase in the female brain indicates that this enzyme might also be functionally relevant in the female brain as supported by two stereotaxic studies conducted in rodents (Hsu, 1990; Veney and Rissman, 2000). The rapid effects of stress and copulation presented here further argue in favor of a functional implication of brain aromatase in females. Although the role of aromatization in the tuberal hypothalamus is not known, the rapid changes in aromatase activity identified in this region are likely related to the control of pituitary secretions. Accordingly, the drop of activity detected after copulation (de Bournonville et al., 2011) is paralleled by a reduction of circulating testosterone concentration (Cornil et al., 2009). The direct and rapid effects of stress on aromatase activity in brain nuclei critically involved in the control of reproduction suggest that aromatase might mediate acute stress effects on behavior and fertility.

Most of the stress effects were reversed within 30 min. However, the stress-induced downregulation of aromatization observed in female VMN and tuber persisted longer but were rescued following sexual interaction with a male. Conversely, copulation effects on aromatase activity seem to take slightly longer to return to baseline levels. Although it is not known yet what the functional consequences of these differential latencies are, the fast time course and the reversibility of most effects provides further evidence that the mechanism(s) responsible for these changes involve(s) post-translational modifications of the enzyme in vivo. Indeed, enzymatic changes that would result from catabolism and de novo synthesis of the enzyme would be expected to occur only after much longer latencies. In addition, the "rescue" of stress-induced changes by copulation strongly suggests that such mechanism(s) involve(s) an active process.

It must finally be noted that based on the stimulatory role played by central aromatization of testosterone on male sexual behavior and the general assumption that acute stress impairs reproduction, the direction of the enzymatic changes described in POM seem counterintuitive. Yet, these effects have been reproduced in several separate studies. Importantly, in agreement with our main hypothesis, the enzymatic changes induced by stress and copulation occur mostly in directions opposite to one another.

RAPID EFFECTS OF ESTROGENS

Estrogens have long been regarded as slow-acting messengers involved in the regulation of gene transcription through binding to specific nuclear receptors (estrogen receptors, ER) (McEwen & Alves, 1999). Besides this genomic mode of action, estrogens also activate cellular effects independent from de novo protein synthesis. Features of these non-genomic effects include a rapid time frame (sec to min), some degree of transiency, an initiation at the cell membrane and the resulting activation of a large variety of cellular signaling pathways (Maggi et al., 2004; Cornil et al., 2006a). More recently, it has become clear that these non-genomic effects have profound consequences for the control of physiological and behavioral endpoints (Cornil & Charlier, 2010). Rather surprisingly, the origin of the estrogens able to trigger these non-genomic actions is rarely addressed. The rapid changes in aromatase activity described here may serve as a mechanism to control the local concentration of estrogens in a temporal domain compatible with their non-genomic actions at the cellular level. The present section will summarize the result of behavioral studies testing the effects of rapid changes in estrogen bioavailability conducted in quail.

Male Sexual Behavior

Initial studies examined the effects of estrogenic compounds administered peripherally through intraperitoneal injections. Castrated males primed with a chronic subthreshold dose of testosterone unable to restore copulatory behavior by itself were injected with a bolus of 17beta-estradiol to mimic an increased bioavailability of estrogens. Such injections facilitated the expression of copulatory behavior within 15 minutes (Fig. 23–2A). This effect disappeared 30 minutes after the injection (Cornil et al., 2006b).

Subsequent experiments tested the effects of a rapid inhibition of estrogen synthesis by acutely injecting aromatase inhibitors into gonadally intact males or castrated males chronically treated with a high concentration of testosterone. Acute injections of Vorozole™, a non-steroidal aromatase inhibitor, produced within 30 to 45 minutes a moderate but significant inhibition of copulatory behavior (Fig. 23–2B; (Cornil et al., 2006c)). A pronounced inhibition was also observed following Vorozole™ injection on two measures of appetitive sexual behavior known to depend on testosterone aromatization: the learned social proximity response and the frequency of rhythmic cloacal sphincter movements (RCSM). The effect on RCSM was significant within 30 to 45 minutes and was no longer significant 15 minutes later (Fig. 23–2C). Androstatrienedione (ATD), a steroidal aromatase inhibitor with a different mode of action, induced a similar but weaker inhibition of RCSM frequency (Cornil et al., 2006c). The observation of similar effects of aromatase inhibition in gonadally intact males and castrates chronically treated with testosterone suggests that the behavioral effect results from a rapid change in local estrogen bioavailability rather than a rapid change in gonadal secretion. This is further confirmed by data showing rapid decreases rather than increases in testosterone following sight of or copulation with a female (Cornil et al., 2009).

A second set of studies investigated the effects of estrogenic drugs delivered through an intracerebroventricular cannula allowing for greater anatomical specificity and the use of lower concentrations and volumes. First, drugs thought to mimic an inhibition of estrogen synthesis or action were tested on gonadally intact males. Similar to previous data, Vorozole™ and, to a lesser extent ATD, produced a profound reduction of RCSM frequency within 30 minutes. Similar treatments reduced both appetitive and consummatory components of male sexual behavior. In addition, these effects of aromatase inhibition on appetitive and consummatory aspects of the behavior were prevented by co-injection of estradiol along with the aromatase inhibitor, suggesting that the behavioral inhibition does not result from toxic effects of the treatment (Seredynski and Cornil, unpublished data).

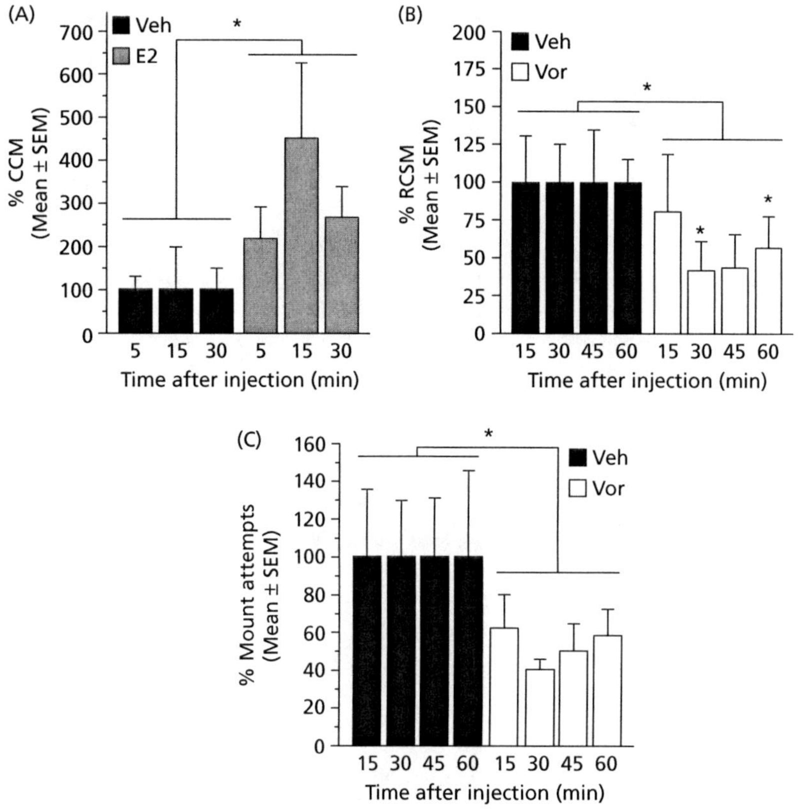

Figure 23–2. Acute Effects of Estradiol Injection or Aromatase Blockade on Male Sexual Behavior. A. Time-response curve of the acute effects of 17β-estradiol (E2, 500μg/kg) on consummatory behavior assessed by cloacal contact movement (CCM) frequency expressed as percentage of controls. B-C. Time-response curve of the acute effects of the aromatase inhibitor, vorozole™ (30mg/kg) on the frequency of rhythmic cloacal sphincter movements (RCSM; appetitive behavior) and mount attempts (consummatory behavior). All values are expressed as percentage of controls. * $p < .05$ compared with the corresponding control condition. Redrawn from (Cornil et al. 2006b) and (Cornil et al. 2006c).

There are several candidate estrogen receptors that might mediate non-genomic effects of estrogens including G-protein coupled receptors such as GPR30 (Filardo & Thomas, 2005) or Gq-mER (Roepke et al., 2009) but also classical nuclear estrogen receptors translocated to the cell membrane (Micevych & Dominguez, 2009). As a first approach to identify the receptors involved in these behavioral responses, two general antagonists of nuclear receptors, ICI 182,870 and tamoxifen, were tested. Both drugs induced a marked reduction in RCSM frequency and copulatory behavior within 15 minutes after injection (Seredynski and Cornil, unpublished data).

In order to test the effect of a rapid increase in estrogen bioavailability, castrated males were first chronically treated with testosterone associated with i.p. injections of vorozole administered twice daily for the duration of the experiment. This treatment was designed to allow expression of the full range of androgenic effects, while profoundly limiting the possibility of estrogenic effects. As predicted by previous work (Taziaux et al., 2004), this treatment nearly eliminated copulatory behavior and reduced RCSM frequency. Interestingly, an acute estradiol injection in these subjects deprived of chronic estrogen activity resulted in a doubling of the RCSM frequency within 15 minutes. Because the membrane impermeable estradiol analog, estradiol-BSA, produced the same response, this effect appears to be initiated at the membrane. In addition, DPN, a specific estrogen receptor beta (ERβ) agonist and, to a lower extent, PPT, a specific estrogen receptor alpha (ERα) agonist, had the same behavioral effect as estradiol. A specific ERα antagonist without behavioral activity when administered alone, MPP resulted in a complete blockade of the facilitating effect of estradiol on behavior when co-injected with estradiol. Together, these results suggest that both ER isoforms are likely involved in the potentiation of appetitive behavior by estradiol. Finally, estradiol had absolutely no effect on copulatory behavior sensu stricto suggesting that the transcriptional activity of estrogens is required to prime this aspect of the behavior (Vasudevan et al., 2001; Kow and Pfaff, 2004).

Nociception

Active aromatase is also present in sensory neurons located in the dorsal horn of the spinal cord in quail (Evrard et al., 2000). This spinal aromatase appears to act through genomic and non-genomic mechanisms in the modulation of pain thresholds in response to thermic stimulation. Genomic effects were demonstrated by comparing castrates chronically treated with testosterone, estradiol, or empty capsules. A 2-week hormonal treatment markedly enhanced the sensitivity to noxious stimuli as is illustrated by decreased latencies of foot withdrawal from a 54°C hot water bath compared to untreated castrates (Evrard & Balthazart, 2004a). Blockade of both estradiol and testosterone effects by tamoxifen and aromatase inhibitors demonstrated the critical role of aromatization in these effects. Similar increases in foot withdrawal latency were also observed within 1 to 5 minutes of acute aromatase inhibition induced by an intrathecal injection of an aromatase inhibitor. This acute effect was prevented by a concomitant estradiol injection thus confirming the steroid specificity of this effect. This change in pain threshold was transient and disappeared within 30 minutes postinjection. It was also anatomically defined, because intrathecal spinal but not intracerebroventricular injections mimicked the effects obtained after systemic injections (Evrard & Balthazart, 2004b). The effects of rapid changes in local estrogen production are thus not confined to male sexual behavior but also important for other behavioral/physiological traits such as responses to painful stimuli.

SUMMARY

Together these data allow a few generalizations about the acute effects of estrogens on physiological and behavioral endpoints. (1) Their latency is considerably shorter than the typical latencies observed for genomic actions suggesting that they arise from non-genomic actions, as supported by the demonstration that similar effects are reproduced by a membrane impermeable estrogen. (2) They are transient, indicating that mechanisms of inactivation exist either through catabolism or dilution of the active estrogen (Cornil et al., 2006) or receptor desensitization (Micevych & Dominguez, 2009). (3) Behavioral inhibition following aromatase blockade demonstrates that these effects depend on local aromatization. Furthermore, the persistence of these effects in castrated males chronically exposed to stable concentrations of testosterone (via Silastic implants) indicate that they do not depend on rapid changes of circulating hormones but specifically involve a local metabolism. (4) A chronic hormonal pre-treatment with low doses of androgens is required for behavioral activation indicating that these rapid effects require some genomic priming (Vasudevan et al., 2001; Kow & Pfaff, 2004). (5) Initial results suggest that at least part of the effects on sexual behavior are initiated at the membrane by the classical estrogen receptors ERα and ERβ. Further studies will verify the identity of the receptors that are implicated and tell us whether they are the sole players. Recent data obtained in the rat indicate that these two receptors and GPR30 might be involved in the spinal responses (Liu et al., 2011). (6) Effects sharing similar properties have now been described on various physiological and behavioral endpoints in different vertebrate species of both sexes. These are thus not isolated phenomena but likely participate in the regulation of numerous vital processes (Cornil & Charlier, 2010).

FUNCTIONAL SIGNIFICANCE OF THE RAPID CHANGES IN AROMATASE ACTIVITY

The enzymatic and behavioral results summarized here seem somewhat contradictory when attempting to integrate them in a simple functional model. The enzymatic data fit with predictions made based on the glutamate-induced rapid inhibition of aromatase activity in quail HPOA explants and the rise in extracellular glutamate concentration demonstrated in the POA of copulating male rats. Indeed, sexual interactions in quail, which are supposed to be accompanied by a release of glutamate in the HPOA, result in a decreased enzymatic activity in discrete preoptic/hypothalamic regions. On the other hand, enzymatic changes induced by stress and copulation occur mostly in opposite directions in agreement with the general assumption that acute stress impairs reproduction. Yet, these observations seem difficult to reconcile with the stimulatory role of estrogens on sexual behavior illustrated by the acute systemic data as well as the central injection data. Indeed, the results of behavioral experiments intuitively would lead to the prediction of a rise of local estrogen synthesis during sexual interaction to facilitate/trigger its expression. Part of the explanation may lie in the fact that, because of technical limitation, aromatase activity is measured post-mortem and therefore after, not during, the execution of the behavior. The apparent discrepancy between the positive effects of estrogens on sexual behavior and the enzymatic inhibition detected after sexual interaction could thus simply reflect the different time points at which these two effects have been measured. Injection experiments could mimic processes involved in behavior initiation (potentially a very short-lived increase in enzyme activity), whereas aromatization changes observed within minutes after copulation would reflect processes associated with the termination of copulation. It should be noted, however, that aromatase activity is inhibited in both males visually exposed to a female and males allowed to physically interact with her. It thus seems unlikely that the enzymatic inhibition reflects satiation processes unless there exists some inertia between the moment enzymatic changes are detected and when behavioral effects become apparent. Indeed, it is possible that, upon interruption of the enzymatic reaction, time is needed for the enzymatic product to either be degraded or diffuse before it stops inducing its effects. Alternatively, it is possible that a form of negative feedback impedes or decreases the activity of the enzymatic molecules immediately after they have been active (e.g., owing to liberation of the enzyme and the synthesized estrogens in the synaptic cleft, which would modify the ionic and pH environment of the enzyme).

What would then be the interpretation of stress-induced enzymatic changes? Assuming that enzymatic changes translate into similar changes in local estrogen concentrations, one would anticipate that the result of a stress-induced rise in preoptic aromatization would facilitate male sexual behavior (Dickens et al., 2011). A recent study, however, did not reveal any behavioral changes in males or in females following exposure to acute stress apparently contradicting this hypothesis (Dickens et al., In press). One could therefore speculate that the stress-induced rise in aromatase activity, presumably paralleled by a rise in local estradiol concentration, serves as a protection mechanism against the physiological consequences of stress. Alternatively, this enzymatic upregulation could simply compensate the stress-induced decrease in the enzymatic substrate (testosterone) that has been reported in plasma after stress exposure. In this case, a disconnect between local enzymatic activity and product would be predicted (higher enzymatic activity, but no change in product owing to decrease in substrate) and no behavioral change would be expected as observed in our recent experiment.

Altogether, these considerations highlight the necessity to further establish the link between aromatase activities measured post-mortem and local estrogen concentrations attained in vivo. The resolution of this issue is technically challenging. One way to get an estimation of this relationship is to compare enzymatic activity and intra-tissue concentrations of estrogens measured. This method was recently used to study the acute effects of an aggressive encounter on local estradiol production and concentration in wild male white-crowned sparrows (Charlier et al., 2011). Surprisingly, this study revealed changes occurring within 30 minutes in estradiol content in different brain regions where no changes in aromatase activity could be detected. Furthermore, there was no correlation between aromatase activity and local estradiol concentrations, suggesting that complex and poorly understood mechanisms of regulation are taking place within the 30-minute interval between behavioral events and brain collection. This approach might, however, prove useful if brains were assessed after shorter durations, especially if very pronounced changes in enzymatic activity are expected.

Although it does not provide any information on local enzymatic activity, in vivo microdialysis can also be used to monitor the local extracellular estradiol concentrations during the course of specific behavioral events. This method was recently validated to monitor estradiol synthesis in the caudomedial nidopallium of zebra finches and this approach demonstrated that glutamate decreases estradiol concentration and thus presumably synthesis in vivo (Remage-Healey et al. 2008). Future studies employing these methods should help us determine the functional significance of the rapid changes in aromatase activity for the control of behavior.

CONCLUSIONS

In summary, in vivo studies carried out in quail demonstrate that aromatase activity is rapidly modulated by behaviorally relevant conditions. Important differences between the sexes, brain regions, and effects of different stimuli were identified, suggesting that the local regulation of estrogen bioavailability is complex and potentially involved in the fine tuning of specific physiological processes. In parallel, systemic or central injections of drugs mimicking acute changes in estrogen bioavailability markedly alter behavioral outputs. Together with studies conducted in other species (Cornil & Charlier, 2010), these studies provide strong evidence that rapid changes in local testosterone metabolism profoundly impact physiological and behavioral processes. Yet, the specific relationship between behavior-induced changes in enzymatic activity and their function still remains to be firmly established. Another important challenge will be to identify the specific subcellular site(s) where these rapid enzymatic changes take place and the neurochemical nature of aromatase expressing cells that are concerned. Aromatase is expressed and active in the endoplasmic reticulum and pre-synaptic terminals (Schlinger & Callard, 1989; Naftolin et al., 1996). Non-genomic actions of estrogens identified so far are compatible with both pre- and postsynaptic modulation of synaptic transmission. But it is not yet known whether acute regulation of estrogen synthesis occurs in both cellular compartments. Recent studies conducted in songbirds suggest that presynaptic boutons might constitute a specific site for acute regulation of estrogen production (Remage-Healey et al., 2009; Cornil et al., 2012; Remage-Healey et al., 2011). Finally, it is currently unknown whether estrogen-induced changes in neurotransmission involve aromatase cells themselves or adjacent cells and what neurotransmitter(s) is/are affected. These questions are currently under investigation and will provide important insight into the functional consequences of the local modulation of estrogen synthesis.

ACKNOWLEDGMENTS

The work of the authors described in this review was supported by grants from the NIMH (R01 MH50388) to GFB and JB and from the Belgian FRFC (Nbr. 2.4537.9) and from the University of Liège (Crédits spéciaux) to JB. CAC is F.R.S-FNRS Research Associate. MJD was supported by an NSF International Research Fellowship (IRFP 0910495).We thank all members of our laboratories, past and present, who contributed to the data summarized here and are too numerous to be listed individually.

REFERENCES

Ball, G. F., & Balthazart, J. (2009). Neuroendocrine regulation of reproductive behavior in birds. In: Pfaff, D., Arnold, A. P., Etgen, A. M., Fahrbach, S. E., and Rubin, R. T. (eds.), *Hormones, Brain and Behavior*. San Diego: Academic Press, pp. 855–895.

Balthazart, J., Baillien, M., & Ball, G. F. (2006). Rapid control of brain aromatase activity by glutamatergic inputs. *Endocrinology, 147*(1), 359–366.

Balthazart, J., Cornil, C. A., Charlier, T. D., Taziaux, M., & Ball, G. F. (2009). Estradiol, a key endocrine signal in the sexual differentiation and activation of reproductive behavior in quail. *J Exp Zool A Ecol Genet Physiol, 311*(5), 323–345.

Balthazart, J., Evrard, L., & Surlemont, C. (1990). Effects of the non-steroidal inhibitor R76713 on testosterone-induced sexual behavior in the japanese quail (Coturnix coturnix japonica). *Horm Behav, 24,* 510–531.

Balthazart, J., & Surlemont, C. (1990). Androgen and estrogen action in the preoptic area and activation of copulatory behavior in quail. *Physiol Behav, 48,* 599–609.

Charlier, T. D., Po, K. W., Newman, A. E., Shah, A. H., Saldanha, C. J., & Soma, K. K. (2011). 17beta-Estradiol levels in male zebra finch brain: combining Palkovits punch and an ultrasensitive radioimmunoassay. *Gen Comp Endocrinol, 167*(1), 18–26.

Cornil, C. A., Ball, G. F., & Balthazart, J. (2006a). Functional significance of the rapid regulation of brain estrogen action: where do the estrogens come from? *Brain Res, 1126,* 2–26.

Cornil, C. A., Ball, G. F., Balthazart, J., & Charlier, T. D. (2011). Organizing effects of sex steroids on brain aromatase activity in quail. *PLoS One, 6*(4), e19196.

Cornil, C. A., Balthazart, J., Motte, P., Massotte, L., & Seutin, V. (2002). Dopamine activates noradrenergic receptors in the preoptic area. *J Neurosci, 22*(21), 9320–9330.

Cornil, C. A., & Charlier, T. D. (2010). Rapid behavioural effects of oestrogens and fast regulation of their local synthesis by brain aromatase. *J Neuroendocrinol, 22*(7), 664–673.

Cornil, C. A., Dalla, C., Papadopoulou-Daifoti, Z., Baillien, M., & Balthazart, J. (2006b). Estradiol rapidly activates male sexual behavior and affects brain monoamine levels in the quail brain. *Behav Brain Res, 66*(1), 110–123.

Cornil, C. A., Dalla, C., Papadopoulou-Daifoti, Z., Baillien, M., Dejace, C., & Balthazart, J. (2005). Sexual behavior affects preoptic aromatase activity and brain monoamines' levels. *Endocrinology, 146*(9), 3809–3820.

Cornil, C. A., Leung, C. H., Pletcher, E. R., Naranjo, K. C., & Saldanha, C. J. (2011). Synaptosomal aromatase is more rapidly inhibited by calcium dependent phosphorylations than microsomal aromatase in the zebra finch brain. *Abs Soc Neurosci, 37,* 501.03.

Cornil, C.A., Leung, C.H., Pletcher, E.R., Naranjo, K.C., Blauman, S.J., Saldanha, C.J. (2012). Acute and specific modulation of presynaptic aromatization in the vertebrate brain. *Endocrinology. 153*(6):2562–2567.

Cornil, C. A., Seutin, V., Motte, P., & Balthazart, J. (2004). Electrophysiological and neurochemical characterization of neurons of the medial preoptic area in Japanese quail (*Coturnix japonica*). *Brain Research, 1029*(2), 224–240.

Cornil, C. A., Stevenson, T. J., & Ball, G. F. (2009). Are rapid changes in gonadal testosterone release involved in the fast modulation of brain estrogen effects? *Gen Comp Endocrinol, 163*(3), 298–305.

Cornil, C. A., Taziaux, M., Baillien, M., Ball, G. F., & Balthazart, J. (2006c). Rapid effects of aromatase inhibition on male reproductive behaviors in Japanese quail. *Horm Behav, 49*(1), 45–67.

de Bournonville, C., Ball, G. F., Balthazart, J., & Cornil, C. A. (2011). Rapid changes of aromatase activity in discrete brain regions following social interactions. *Trabajos Del Institui Cajal, LXXXIII*, 147–148.

Di, S., Maxson, M. M., Franco, A., & Tasker, J. G. (2009). Glucocorticoids regulate glutamate and GABA synapse-specific retrograde transmission via divergent nongenomic signaling pathways. *J Neurosci, 29*(2), 393–401.

Dickens, M. J., Balthazart, J., & Cornil, C. A. (In press) Simple males, subtle females: differential effects of acute stress and copulation on neural aromatase and corticosterone in quail. J. Neuroendocrinol. doi: 10.1111/j.1365-2826.2012.02340.x.

Dickens, M. J., Cornil, C. A., & Balthazart, J. (2011). Acute stress differentially affects aromatase activity in specific brain nuclei of adult male and female quail. *Endocrinology, 152*(11), 4242–4251.

Dominguez, J. M., Gil, M., & Hull, E. M. (2006). Preoptic glutamate facilitates male sexual behavior. *J Neurosci, 26*(6), 1699–1703.

Evrard, H., Baillien, M., Foidart, A., Absil, P., Harada, N., & Balthazart, J. (2000). Localization and controls of aromatase in the quail spinal cord. *J Comp Neurol, 423*, 552–564.

Evrard, H. C., & Balthazart, J. (2004a). Aromatization of androgens into estrogens reduces response latency to a noxious thermal stimulus in male quail. *Horm Behav, 45*, 181–189.

Evrard, H. C., & Balthazart, J. (2004b). Rapid regulation of pain by estrogens synthesized in spinal dorsal horn neurons. *J Neurosci, 24*(33), 9225–9229.

Filardo, E. J., & Thomas, P. (2005). GPR30: a seven-transmembrane-spanning estrogen receptor that triggers EGF release. *Trends Endocrinol Metab, 16*(8), 362–367.

Foidart, A., Reid, J., Absil, P., Yoshimura, N., Harada, N., & Balthazart, J. (1995). Critical re-examination of the distribution of aromatase-immunoreactive cells in the quail forebrain using antibodies raised against human placental aromatase and against the recombinant quail, mouse or human enzyme. *J Chem Neuroanat, 8*, 267–282.

Hsu, C.-H. (1990). Blockade of lordosis by androst-1,4,6-triene-3,17-dione (ATD) and tamoxifen in female hamsters primed with testosterone propionate. *Horm Behav, 24*, 14–19.

Karlsson, U., Sundgren, A. K., Näsström, J., & Johansson, S. (1997). Glutamate-evoked currents in acutely dissociated neurons from the rat medial preoptic nucleus. *Brain Res, 759*, 270–276.

Kow, L.-M., & Pfaff, D. W. (2004). The membrane actions of estrogens can potentiate their lordosis behavior-facilitating ge,nomic actions. *PNAS, 101*(33), 12354–12357.

Liu, N. J., Chakrabarti, S., Schnell, S., Wessendorf, M., & Gintzler, A. R. (2011). Spinal synthesis of estrogen and concomitant signaling by membrane estrogen receptors regulate spinal kappa- and micro-opioid receptor heterodimerization and female-specific spinal morphine antinociception. *J Neurosci, 31*(33), 11836–11845.

Maggi, A., Ciana, P., Belcredito, S., & Vegeto, E. (2004). Estrogens in the nervous system: mechanisms and nonreproductive functions. *Ann Rev Physiol, 66,* 291–313.

McEwen, B. S., & S. Alves, E. (1999). Estrogen actions in the central nervous system. *Endocr Rev, 20,* 279–307.

Micevych, P., & Dominguez, R. (2009). Membrane estradiol signaling in the brain. *Front Neuroendocrinol, 30,* 315–327.

Naftolin, F., Horvath, T. L., Jakab, R. L., Leranth, C., Harada, N., & Balthazart, J. (1996). Aromatase immunoreactivity in axon terminals of the vertebrate brain. *Neuroendocrinol, 63,* 149–155.

Panzica, G. C., Viglietti-Panzica, C., & Balthazart, J. (1996). The sexually dimorphic medial preoptic nucleus of quail: a key brain area mediating steroid action on the male sexual behavior. *Front Neuroendocrinol, 17,* 51–125.

Remage-Healey, L., Dong, S., Maidment, N. T., & Schlinger, B. A. (2011). Presynaptic control of rapid estrogen fluctuations in the songbird auditory forebrain. *J Neurosci, 31*(27), 10034–10038.

Remage-Healey, L., Maidment, N. T., & Schlinger, B. A. (2008). Forebrain steroid levels fluctuate rapidly during social interactions. *Nat Neurosci, 11*(11), 1327–1334.

Remage-Healey, L., Oyama, R. K., & Schlinger, B. A. (2009). Elevated aromatase activity in forebrain synaptic terminals during song. *J Neuroendocrinol, 21*(3), 191–199.

Roepke, T. A., Qiu, J., Bosch, M. A., Ronnekleiv, O. K., & Kelly, M. J. (2009). Cross-talk between membrane-initiated and nuclear-initiated oestrogen signalling in the hypothalamus. *J Neuroendocrinol, 21*(4), 263–270.

Schlinger, B. A., & Callard, G. V. (1989). Localization of aromatase in synaptosomal and microsomal subfractions of quail (*Coturnix coturnix japonica)* brain. *Neuroendocrinol, 49,* 434–441.

Schumacher, M., & Balthazart, J. (1987). Neuroanatomical distribution of testosterone metabolizing enzymes in the Japanese quail. *Brain Research, 422,* 137–148.

Taziaux, M., Cornil, C. A., & Balthazart, J. (2004). Aromatase inhibition blocks the expression of sexually-motivated cloacal gland movements in male quail. *Behav Processes 67*(3), 461–469.

Vasudevan, N., Kow, L.-M., & Pfaff, D. W. (2001). Early membrane estrogenic effects required for full expression of slower genomic actions in a nerve cell line. *PNAS, 98*(21), 12267–12271.

Veney, S. L., & Rissman, E. F. (2000). Steroid implants in the medial preoptic area or ventromedial nucleus of the hypothalamus activate female sexual behaviour in the musk shrew. *J Neuroendocrinol, 12*(11), 1124–1132.

Watson, J. T., & Adkins-Regan, E. (1989a). Activation of sexual behavior by implantation of testosterone propionate and estradiol benzoate into the preoptic area of the male Japanese quail (Coturnix japonica). *Horm Behav, 23,* 251–268.

Watson, J. T., & Adkins-Regan, E. (1989b). Testosterone implanted in the preoptic area of male Japanese quail must be aromatized to activate copulation. *Horm Behav, 23,* 432–447.

Wingfield, J. C., & Sapolsky, R. M. (2003). Reproduction and resistance to stress: when and how. *J Neuroendocrinol, 15*(8), 711–724.

Chapter 24

Modulation of Central Auditory Processing by a Brain-Generated Estrogen: Rapid Effects on Neural Coding, Discrimination, and Behavior

Liisa A. Tremere and Raphael Pinaud

INTRODUCTION

Recent findings, primarily obtained in songbirds, demonstrated that brain-generated estradiol (E2) is a key functional component of central auditory circuits and markedly influence hearing-based behaviors. This chapter reviews these exciting new findings, and their implications, in four general sections: First, we will examine novel insights related to the anatomical and neurochemical organization of estrogen-related circuits, and will discuss data suggesting that estrogen-producing and -sensitive neurons are directly activated by auditory input in freely behaving animals. Second, we will review exciting new findings indicating that E2 is produced by central auditory neurons in an activity-dependent fashion, and that this brain-generated hormone instantaneously modulates central auditory processing in the awake brain by regulating local neurotransmission. Third, we will discuss the functional relevance of this neurosteroid for the computational capacity of auditory neurons, neural coding of acoustic communication signals, and how these emergent properties shape auditory behaviors. Finally, we will discuss new results suggesting that estrogen-modulation of central sensory function appears to be a general property of central auditory pathways in vertebrates, including mammals, and may transcend beyond the auditory domain into other sensory modalities.

A number of extensive reviews have appeared recently that are dedicated to, or at least discuss, general issues related to the roles played by peripheral and central estrogen in the regulation of auditory processing (Maney & Pinaud, 2011; Saldanha et al., 2011; Schlinger & Remage-Healey, 2011; Pinaud & Tremere, 2012). This chapter is, therefore, not aimed at providing a comprehensive review of these topics. Instead, we focus primarily on research conducted by our laboratory and, to a lesser extent, selected works from other groups that will facilitate discussions of these topics by providing an appropriate context.

EARLY FINDINGS SUGGESTING THAT ESTROGEN MAY INFLUENCE HEARING FUNCTION

For nearly 30 years now, correlational data obtained in humans and a variety of experimental model systems have suggested that circulating steroid hormones may influence multiple aspects of auditory function and, in certain cases, hormone insufficiency may be associated with hearing pathologies. For example, in middle aged mice, progesterone + E2 treatment designed to mimic clinically relevant hormone replacement therapy (HRT), significantly impaired auditory sensitivity relative to animals treated with E2 alone, or unmanipulated controls (Price et al., 2009). These observations agreed with results obtained in post-menopausal women where E2, when combined with progesterone as part of HRT, significantly impaired audiometric functions, hearing sensitivity, and the ability to discriminate sounds in a noisy environment relative to control or E2-monotherapy (Guimaraes et al., 2006). These findings provided direct evidence for a negative role for progesterone in hearing function and highlighted that E2 effects must be examined separately.

Significantly more precedent exists for a role of E2 in the adaptive modulation of hearing function. Curiously, in contrast to the detrimental effects of progesterone, E2 appears to enhance or improve hearing function. For instance, auditory-brainstem responses are more vigorous, exhibit shorter latencies, and are more temporally precise in women of all ages, relative to men. Moreover, event-related potentials and auditory thresholds correlate tightly with changes in plasma E2 levels across the menstrual cycle in women. Likewise, hearing function is more efficient in menopausal women subjected to hormone therapy, compared with untreated controls (Jerger & Hall, 1980; Davis & Ahroon, 1982; Walpurger et al., 2004; Hultcrantz et al., 2006). A particularly interesting example that underscores a potential role for E2 in the control of hearing function can be found in patients afflicted by Ullrich-Turner syndrome (X-monosomy). These women produce little or no ovarian estrogens and, among other pathologies, suffer from an array of hearing problems including diminished efficiency in the processing of auditory cues and early age-related hearing loss (Hultcrantz & Sylven, 1997; Hultcrantz et al., 2006; Bergamaschi et al., 2008).

Similar to, and consistent with, the findings obtained in humans, research conducted in animal models further support a role for E2 in enhancing auditory function. In mice, auditory recognition and preference behaviors are markedly modulated by E2 in mothers, but not virgins, indicating that shifts in E2 levels around parturition may influence hearing function and associated behaviors (Ehret et al., 1987; Koch & Ehret, 1989; Miranda & Liu, 2009). Ovariectomization of rodents and nonhuman primates results in abnormal auditory brainstem response profiles which, interestingly enough, are rescued by E2 replacement treatment (Coleman et al., 1994; Golub et al., 2004). Severe hearing loss that leads to premature deafness has also been reported for mice that are deficient in estrogen receptor β, suggesting that normal estrogen signaling is critical for normal hearing function and its dysfunction may be a contributing factor to presbycusis (Hultcrantz et al., 2006; Simonoska et al., 2009).

More complex auditory-based behaviors also appear to be modulated by E2 levels. For instance, ovariectomized nonreproductive midshipman fish exposed to E2 implants undergo a functional retuning of auditory processing that mirrors properties observed in animals during reproductive states, when E2 concentrations are high (Sisneros et al., 2004). Increased levels of circulating E2 also heighten behavioral responses to auditory communication signals in adult female songbirds, and increase hearing-driven responses in auditory areas as revealed by activity-dependent gene expression (Maney et al., 2006; Vyas et al., 2009; Sanford et al., 2010).

The results discussed above in both humans and animal models clearly suggest that E2 acts to enhance hearing function. Importantly, these findings have been overwhelmingly, if not exclusively, interpreted to suggest that E2 modulates auditory processing at the level of the inner and outer ear hair cells in the cochlea, a site known to express the estrogen synthetic enzyme and the classic intracellular estrogen receptors (Forlano et al., 2005; Meltser et al., 2008; Noirot et al., 2009; Maruska & Fernald, 2010). However, even though estrogens certainly impact the development and function of the peripheral organ, multiple aspects of E2's modulation of auditory processing cannot be accounted for by actions of this hormone in the cochlea. Yet, "extra cochlear" effects of E2, up until recently, had not been seriously considered. Similarly, even though it has long been known that the brain is a major site of E2 synthesis, research in hearing function has almost exclusively assumed that the gonads were the source of this hormone; the impact of E2 synthesized "extra gonadally" has virtually been ignored. In sum, even though little doubt now remains that E2 influences hearing function, both the hormone source and its site of action have been narrowly considered.

SENSORY–NEUROENDOCRINE OVERLAP IN CENTRAL AUDITORY CIRCUITS

Over the past few years, exciting data has dramatically shifted our understanding of how E2 modulates hearing function. Most of these findings have been obtained in songbirds, a remarkable experimental model that utilizes a refined auditory system to perceive and discriminate learned communication signals—songs—that are used for an array of social behaviors, including individual recognition and partner selection (Gentner, 2004; Nowicki & Searcy, 2004). One region in the songbird brain in particular, the caudomedial nidopallium (NCM), a forebrain area considered to be the analogue of the auditory association cortex in mammals (Vates et al., 1996; Mello et al., 2004; Pinaud & Terleph, 2008), has been the platform for many new studies aimed at exploring the interface between E2 and hearing function. The NCM neurons respond preferentially to same-species songs, relative to heterospecific songs and artificial stimuli, and exhibit types of experience-dependent plasticity that may underlie auditory discrimination and the formation of long-lasting song recognition memories (Mello et al., 1992; Chew et al., 1995; Mello & Clayton, 1995; Stripling et al., 2001; Pinaud & Terleph, 2008; Thompson & Gentner, 2010). Importantly, it has been known for more than 15 years that NCM neurons express high levels of aromatase (ARO) and both estrogen receptor subtypes (ERα and ERβ), suggesting that the local synthesis of neurosteroids may directly modulate auditory processing (Shen et al., 1995; Bernard et al., 1999; Metzdorf et al., 1999; Saldanha et al., 2000; Pinaud et al., 2006; Jeong et al., 2011; Maney & Pinaud, 2011 ; Pinaud & Tremere, 2012).

To push our current understanding of the neurochemical and functional organization of E2-associated circuits in NCM, we recently began to dissect in a comprehensive fashion properties of estrogen-producing and -sensitive neurons in this auditory area, in a series of studies using molecular biological approaches (Jeong et al., 2011). First, we determined that E2-associated circuits are exclusively present in NCM, but not other areas in the auditory forebrain, suggesting that neurohormonal action may selectively influence the computational properties of a higher-order auditory area. Second, we refined our understanding of estrogen-producing and -sensitive circuits by conducting stringent and unbiased quantitative analyses of these neuronal populations. These analyses revealed that nearly 68% of NCM neurons are capable of producing estrogens, as revealed by ARO expression. Moreover, whereas approximately 46% of cells in NCM express ERβ, only 7% express ERα (Jeong et al., 2011). These findings are consistent with earlier reports indicating high ARO and ERβ, but low ERα expression in NCM (Bernard et al., 1999; Metzdorf et al., 1999; Saldanha et al., 2000; Tremere et al., 2009). Hence, local signaling through the classic estrogen receptors is chiefly achieved through ERβ in NCM. Third, to gain insight into the potential overlap between estrogen-producing and -sensitive circuits, we carried out double-fluorescence in-situ hybridization (dFISH) studies and determined that nearly 56% of ARO-positive cells in NCM express ERβ (~83% of the ERβ-positive cell population express ARO) (Jeong et al., 2011). The overlap between ARO and ERα-expressing neurons is negligible, perhaps because of the low number of ERα-positive cells in NCM (Saldanha and Coomaralingam, 2005). These findings suggest that locally produced E2 may influence NCM neurons, primarily via ERβ, through autocrine and/or paracrine mechanisms. Fourth, we set out to determine the neurochemical identity of ARO and ERβ-positive cells. The populations of excitatory and inhibitory neurons in NCM are well balanced, with nearly 50% of neurons exhibiting a GABAergic phenotype, as revealed by RNA and protein studies (Pinaud et al., 2004; Pinaud & Mello, 2007; Pinaud et al., 2008). Colocalization studies using standard markers for excitatory and inhibitory neurons revealed that nearly 60% of ARO-positive cells are excitatory, with the remaining 40% associated with local GABAergic neurons. Similarly, approximately 74% of ERβ-positive neurons exhibited an excitatory phenotype, with the residual 26% being composed of inhibitory cells (Jeong et al., 2011). These findings shed the first light into the neurochemical organization of E2-associated circuits in NCM and revealed that estrogenic influences in this auditory station may exert prominent postsynaptic influence on glutamate-based signaling (but see below). Moreover, these initial efforts not only provide critical information about the functional organization of E2-associated circuits in NCM, but also help in the design and interpretation of physiological studies carried out in-vitro and in-vivo (see below).

HEARING-DEPENDENT, LOCAL E2 PRODUCTION IN THE NCM OF FREELY-BEHAVING ANIMALS

The expression of activity-dependent genes has been successfully used to identify song-responsive neurons, at single-cell resolution, in auditory forebrain areas of freely behaving songbirds (reviewed in Mello et al., 2004; Mello & Pinaud, 2006). We used the song-induced expression of the immediate early gene *zenk* (a.k.a., zif-268, egr-1, NGFI-A and krox-24) to determine if, and to what extent, playbacks of conspecific songs to behaving animals engaged estrogen-producing and -sensitive neurons in NCM. We found that nearly 96% of all estrogen-producing cells (ARO-positive) are activated by song stimuli, which reflect approximately 67% of the overall neuronal population in NCM. Our findings also revealed that 68% of the cells that are positive for ERβ, the dominant receptor subtype in NCM, are activated by song, which account for only 32% of the overall neuronal population (Jeong et al., 2011). Taken together, our results indicate that sensory experience in freely-behaving animals robustly activates E2-producing cells, but has a significantly lesser impact on the population of cells expressing estrogen receptors in NCM. The significantly smaller activity-dependent engagement of cells expressing either forms of estrogen receptor, relative to the estrogen-producing neuronal population, supports a model through which sensory input may primarily regulate local E2 production.

In support of this view, in vivo microdialysis studies carried out in freely behaving songbirds demonstrated that E2 levels nearly double in NCM, but not adjacent brain areas, as a result of socially-relevant auditory stimulation (Remage-Healey et al., 2008). Importantly, this experience-dependent increase in E2 levels occurred rapidly, in tens of minutes, which temporally allows this neurohormone to modulate central auditory processing in behaviorally-relevant timescales.

BRAIN-GENERATED E2 MODULATES AUDITORY PROCESSING IN THE AWAKE BRAIN

These earlier findings strongly suggested that E2 could locally influence the physiology of an auditory forebrain area in the adult vertebrate brain. Moreover, the convergence of the results detailed above highlighted a possible physiological role for the pronounced sensory-neuroendocrine overlap found in NCM. In return, NCM provided a unique and highly amenable preparation for investigating how locally generated E2 shapes the physiology of central auditory neurons, a question that was not systematically probed until two years ago. To address directly this issue, our group elaborated a system for use in fully awake songbirds whereby it is possible to bilaterally and simultaneously record, from multiple electrodes, hearing-driven neural activity. Importantly, these recordings can be made during bilateral, independent, and simultaneous pharmacological manipulations (Tremere et al., 2009; Tremere et al., 2010; Tremere & Pinaud, 2011). This preparation allowed our group to test if, and the extent to which, E2 influences song-evoked neural responses. To this end, we obtained neurophysiological responses to songs before and during local infusions of E2 unilaterally in NCM; vehicle was simultaneously injected contralaterally. We found that, whereas vehicle failed to change the firing pattern of NCM neurons, microinfusions of E2 markedly increased neuronal discharge rates, in a dose-dependent fashion, relative to preinfusion responses (range of ~68%–145% in our studies) (Tremere et al., 2009) (Fig. 24–1A,B). These findings indicated that E2 can directly modulate central auditory processing in the adult and awake vertebrate brain. Equally remarkable is the time course of these effects; E2 infusions enhanced spiking rates of NCM neurons virtually instantaneously, on a time scale that is congruent with rapid neuromodulators (Tremere et al., 2009; Tremere & Pinaud, 2011). Consistent with this view, we also found that local blockade of estrogen receptors with tamoxifen (TMX) or ICI 182,780 (ICI) significantly suppressed song-driven NCM neuronal responses, also in a dose-dependent way (range of 48%–71% from preinfusion levels) (Tremere et al., 2009; Tremere & Pinaud, 2011) (Fig. 24–1C). These findings suggested that interference with E2's actions in NCM neurons significantly suppressed neurophysiological responses to auditory stimuli; however, the extent to which local versus circulating hormone was acting in NCM remained unclear. We directly tackled this issue by unilaterally treating NCM with ATD or fadrozole, two selective aromatase inhibitors. This manipulation unilaterally suppressed E2 production, but allowed for normal gonadal and contralateral production of estrogens. Remarkably, we found that blockade of local E2 synthesis in NCM markedly decreased evoked neurophysiological responses, dose-dependently (range of 44%–78% from preinfusion levels) (Tremere et al., 2009; Tremere & Pinaud, 2011) (Fig. 24–1D). These findings provided

neurons. B. Local infusions of physiological E2 levels rapidly increased song-evoked firing rates of NCM neurons. E2-mediated effects on firing rates are long-lasting, and remain detectable 2 hours, but not 4 hours, following interruption of E2 infusions. C. Local blockade of estrogen receptors with ICI decreased firing rates of NCM neurons, and effect that also persists for at least 2 hours. D. Local inhibition of aromatase activity with FAD, a manipulation that suppressed unilateral production of E2 in NCM, decreased song-evoked neuronal discharge rates. The effects of FAD were detectable 2 hours, but not 4 hours, following interruption of drug infusions. These findings indicate that NCM-derived E2 facilitates song-evoked activity of NCM neurons. From Tremere & Pinaud (2011). *J Neurosci, 31*(9), 3271–3289. Reprinted with permission.

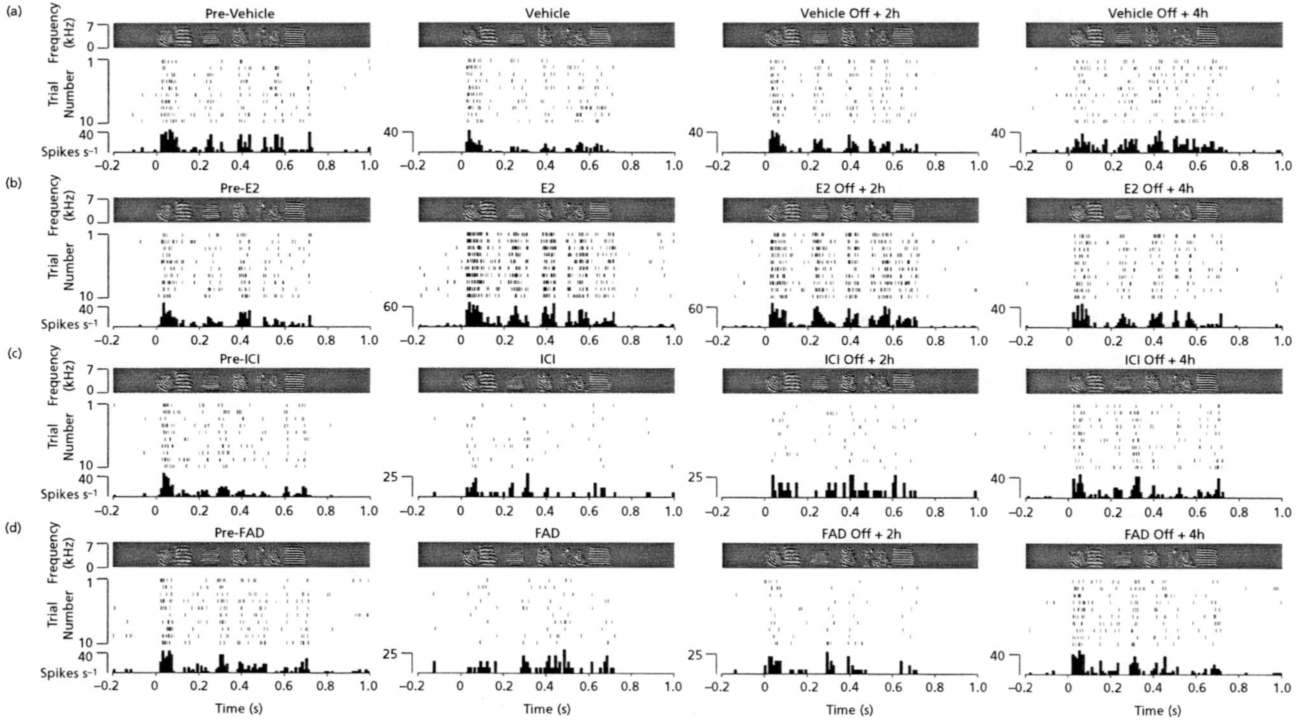

Figure 24–1. E2 Rapidly and Persistently Enhances Song-Driven Discharge Rates of NCM Neurons. A. Representative response of an NCM neuron before, during, and after (2 and 4 h) interruption of local vehicle infusion. *Top:* song spectrogram. *Middle:* Spike raster plot of the neuron during the first 10 renditions of the stimulus. *Bottom:* Peristimulus time histograms. Vehicle did not affect the short- or long-term spiking activity of NCM

the first direct evidence that brain-generated E2 acutely enhances auditory-driven responses in the awake vertebrate brain. Moreover, our results demonstrated that E2 produced by central auditory neurons robustly modulates the gain of sensory responses virtually instantaneously. Our findings that aromatase inhibition suppress song-driven neural responses in NCM were recently validated in the anesthetized songbird by an independent group (Remage-Healey et al., 2010).

We next probed whether the effects of E2 are acute or if they drive long-lasting changes in NCM neuronal responses. To this end, we recorded single-unit neurophysiological responses before and during E2 injections, as well as 2 hours and 4 hours after local infusions were interrupted (see Fig. 24–1). Remarkably, even though local E2 levels are rapidly degraded (Remage-Healey et al., 2010), we found that the E2-mediated increases in spiking activity led to sustained changes in neural responses, that lasted for several hours (Tremere & Pinaud, 2011) (see Fig. 24–1B). Similar long-lasting effects in NCM neuronal discharge rates were also observed when we blocked estrogen receptors with either TMX or ICI; that is, these agents suppressed evoked responses for several hours after interruption of their local infusion (see Fig. 24–1C). Finally, we tested whether it was the locally generated E2 that mediated long-term changes in song-driven responses in NCM neurons. Unilateral suppression of E2 production with either ATD or fadrozole led to a significant decrease in evoked single-unit activity that lasted at least 2 hours (Tremere & Pinaud, 2011) (see Fig. 24–1D). Together, these results demonstrated that brain-generated E2 may strongly redetermine operational rules of auditory networks in real-time, including global and feature-specific gain control, as well as promote long-lasting changes in neural performance.

Finally, as indicated above, NCM neurons exhibit selective responses towards conspecific songs, relative to vocalizations of heterospecifics or artificial stimuli, such as pure tones and white noise (Mello et al., 1992; Chew et al., 1996; Velho et al., 2005; Pinaud & Terleph, 2008). This selectivity is found in NCM but not lower-order auditory areas, suggesting that this property may emerge from within NCM circuitry, or from top-down mechanisms. Considering that estrogen-associated circuits are unique to NCM in the auditory forebrain, we wondered if E2 signaling shaped this process by selectively modulating responses to conspecific songs. To test this possibility we recorded responses that were evoked by an array of nonsong auditory stimuli, that retained the statistical properties of songs and its elemental units, but in which spectral and/or temporal features were gradually degraded (Tremere et al., 2009). We found that E2, blockade of estrogen receptors, and local suppression of E2's production in NCM affected the processing of each nonsong stimuli in a manner that was indifferent to the effects of these agents on responses evoked by conspecific songs (Tremere et al., 2009). These findings indicate that locally produced E2 enhances the gain of hearing-driven responses irrespective of the behavioral relevance of the stimulus; hence, although E2 is unlikely to shape local neuronal selectivity, it plays the more central function of gain-control in NCM circuitry.

ESTRADIOL CONTROLS FAST NEUROTRANSMISSION VIA A PRE-SYNAPTIC, NONGENOMIC MECHANISM

The rapidity with which NCM-derived E2 modulates the activity of auditory neurons is clearly incompatible with mechanisms that involve "classic" genomic responses (i.e., regulation of transcriptional activity through ER-dependent gene expression). Instead, the instantaneous impact of E2 on the physiology of NCM neurons led our group to hypothesize that this neurohormone may shape spiking activity via a fast, nongenomic mechanism that likely involves a direct modulation on synaptic transmission (Tremere et al., 2009). Such rapid effects of E2 on basic membrane properties, and synaptic transmission and plasticity, have been shown in multiple circuits of the vertebrate brain, including the hippocampus (reviewed in Woolley, 2007). To investigate this possibility, we carried out whole-cell patch-clamp recordings in an NCM slice preparation that our group developed previously (Pinaud et al., 2008; Tremere et al., 2009). We then tested the impact of E2, or interference with local E2 signaling, on both excitatory and inhibitory synaptic transmission. Our results demonstrated that E2 exerted a marked and selective effect on GABAergic transmission in NCM. More specifically, we found that E2 significantly decreased the frequency of miniature inhibitory postsynaptic currents (mIPSCs) on NCM neurons; no changes were observed in the amplitude of these events (Tremere et al., 2009). The effects of E2 on the inhibitory network were mediated postsynaptically on type-A GABAergic ($GABA_A$) receptors, because all recorded events were blocked by coapplication of the selective antagonist bicuculline to the recording bath. Consistent with the observation that E2 suppressed GABAergic transmission, we also found that blockade of ERs, or local inhibition of ARO activity significantly increased the frequency of mIPSCs (Tremere et al., 2009). Once again, these manipulations did not affect the amplitude of these synaptic events.

We subsequently tested if and how E2 affects excitatory synaptic transmission. To our surprise, E2, blockade of classic ERs, or local suppression of ARO activity did not affect either the frequency or the amplitude of miniature excitatory postsynaptic currents (Tremere et al., 2009). These findings suggested that E2 does not appear to affect fast glutamatergic transmission in NCM neurons.

In summary, our studies demonstrated that locally produced E2 is likely to increase spike probability and hearing-driven discharge activity of NCM neurons nongenomically, through the selective suppression of local GABAergic transmission. Moreover, the observations that E2 exclusively decreased the frequency, but not amplitude, of mIPSCs indicates that the locus of action for this neurohormone is pre-synaptic, and likely involves a modulation of GABA release probability onto NCM neurons (Tremere et al., 2009). Finally, recent, unpublished findings from our laboratory also suggest that E2 exerts robust post-synaptic effects on NCM's neuronal physiology, suggesting that E2's effects on central auditory neurons are pervasive across cellular compartments and involve multiple types of synaptic mechanisms and/or transmitter systems.

BRAIN-GENERATED E2 ENHANCES THE NEURAL DISCRIMINATION OF SENSORY SIGNALS: IMPLICATIONS FOR NEURAL CODING AND INFORMATION CAPACITY

By combining in vitro and in vivo neurophysiological recordings, we showed that brain-generated E2 suppresses GABAergic transmission presynaptically; the E2-mediated decrease in inhibitory tone putatively allows for net excitation across the network and, consequently, increases discharge rates of NCM neurons. A perhaps more critical question, however, was how does E2 affect the ability of NCM neurons to encode, decode and discriminate auditory signals? More specifically, how does this brain-generated hormone govern the ability of auditory neurons to extract and process information from auditory cues? Although our demonstration that E2 increases neuronal spiking activity in NCM provides one of the most striking demonstrations of how a brain-generated hormone affects real-time sensory processing in the adult vertebrate brain, these findings offer little insight on the functional significance of this neurohormonal modulation. For instance, from a neural coding perspective, the E2 increase in neuronal spiking may be related to increases in noise, which could consequently decrease coding efficiency. To establish in a direct and quantitative manner the functional relevance of E2's modulation of neural coding in NCM, we implemented and utilized stringent information-theoretical and decoding-based approaches to analyze neural data in conditions where E2 signaling is enhanced or suppressed (Tremere & Pinaud, 2011). Our metrics also allowed us to specifically and quantitatively assess how manipulations directed at local E2 signaling influence the ability of NCM neurons to discriminate across different auditory stimuli (i.e., the neural discrimination of auditory cues).

We found that, whereas vehicle did not alter the information capacity of NCM neurons, physiologically relevant increases in local E2 levels significantly increased the information that NCM neurons carry about stimulus structure, relative to preinfusion levels (Tremere & Pinaud, 2011) (Fig. 24–2A-B). Remarkably, this increase in the efficiency with which NCM neurons encode information was paralleled by a significant improvement in the ability of these cells to discriminate across stimuli in the awake brain (see Fig. 24–3A, D). We also discovered that the E2-mediated optimization of neural coding and discrimination was largely caused by improved efficiency in the computational capacity of NCM neurons, given that mutual information rates (between stimuli and neural responses) were increased as a result of E2 infusions (Tremere & Pinaud, 2011). Consistent with this view, blockade of local ERs with TMX or ICI significantly decreased information rates, the information that NCM neurons carry about stimulus structure and the neural discrimination of acoustic stimuli (see Figs. 24–2C; 24–3B, D). Importantly, decreased efficacy in the neural coding and discrimination of songs were replicated by local infusions of ARO inhibitors ATD and fadrozole, indicating that it is the local production of E2, as opposed to circulating E2, that shapes these computational properties in NCM (Tremere & Pinaud, 2011) (see Figs. 24–2D; 24–3C-D). Finally, infusions of estrone or 5β-dihydrotestosterone in NCM failed to affect neuronal spiking,

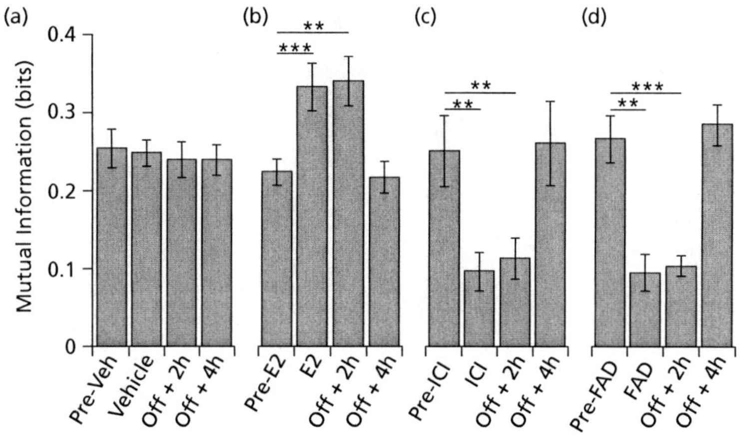

Figure 24–2. Locally-Derived E2 Increases the Information Transmitted About Songs by NCM Neurons. The mutual information between stimulus and neural responses, an index of the effectiveness of the neural encoding, is plotted as a function of the experimental session [before, during, and after (2 and 4 h) pharmacological treatment]. (*A–D*) Each graph set shows mean mutual information data obtained under different pharmacological conditions that included vehicle (*A*), E2 (*B*), estrogen receptor antagonist ICI (*C*), and aromatase inhibitor FAD (*D*). Significant and long-lasting increases in information rates are observed as a result of local E2 infusions (*B*). In contrast, these values are decreased below predrug levels after blockade of estrogen receptors (*C*) or inhibition of the local production of E2 in NCM (*D*). Similar to estrogen receptor antagonism, local inhibition of aromatase yields effects on information rates that persist for several hours after interruption of drug treatment. $^{**}p < .01$; $^{***}p < .001$. Modified from Tremere & Pinaud (2011). *J Neurosci, 31*(9), 3271–3289. Reprinted with permission.

information rates and the neural discrimination of auditory signals by single-units in the awake brain, indicating that these effects are selectively mediated by locally-derived E2 (Tremere & Pinaud, 2011).

Remarkably, and congruent with the long-lasting effects of E2 on NCM neuronal discharge rates, we also found long-term changes in the computational properties of NCM neurons that surpass the acute changes in local neurohormone levels. Specifically, even though exogenous E2 is rapidly degraded in NCM (in a scale of minutes) (Remage-Healey et al., 2010), performance enhancement in the neural coding and discrimination of acoustic stimuli lasted for at least 2 hours after interruption of local hormone infusions (Tremere & Pinaud, 2011) (see Fig. 24–2B). Similarly, pharmacological manipulations that blocked local ERs or inhibited ARO activity also led to long-lasting deficits in the efficiency of information coding and the neural discrimination of songs in the awake NCM (see Fig. 24–2C-D). Together, our findings demonstrated that a primary functional role of brain-generated E2 is to increase the information that neurons carry about stimulus structure, thereby enhancing auditory coding efficiency, to improve the neural discrimination of auditory cues.

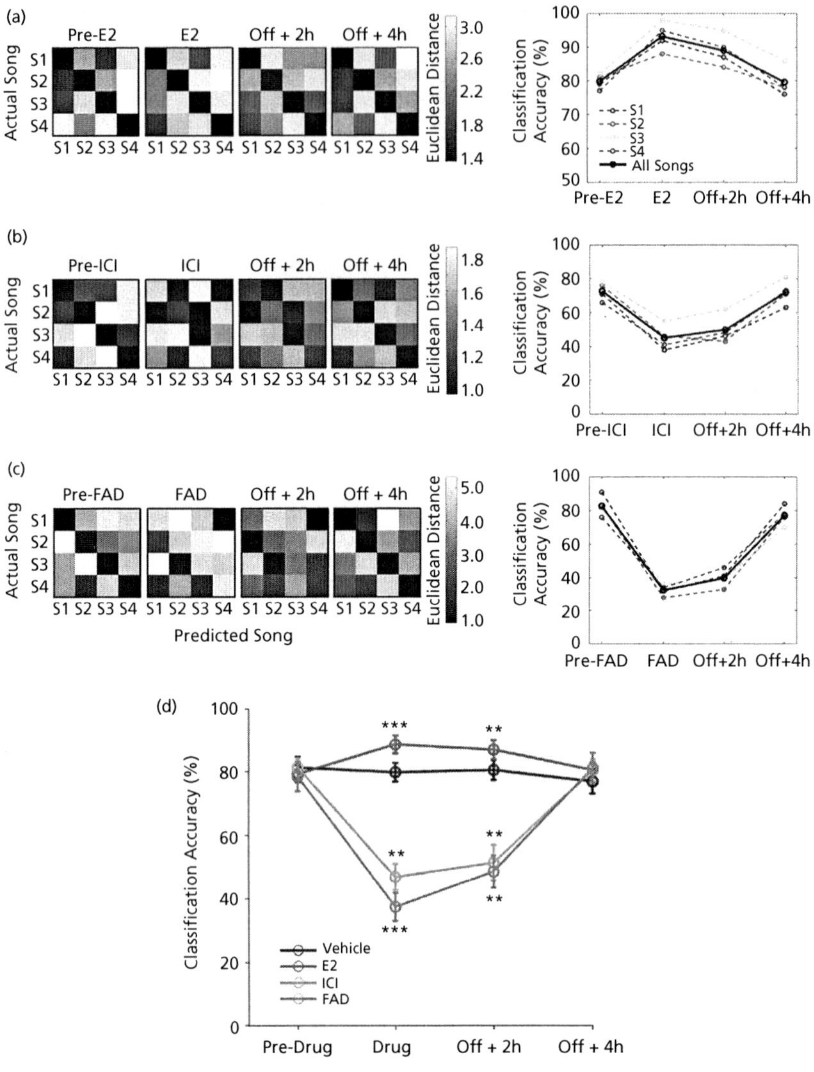

Figure 24–3. E2 Enhances the Neural Discrimination of Songs. Two-dimensional confusion matrices (*left panels*) and decoder's performance (*right panels*) for one representative NCM neuron treated with E2 (*A*), ICI (*B*), and FAD (*C*). *Top panels* show confusion matrices measured from classifications of the neural data before, during, and after (2 and 4 h) pharmacological treatment, and provide an index of the ability of NCM neurons to discriminate among the four different songs tested. Each matrix plots the Euclidean metric, coded in grayscale values, between actual (*y*-axis) and decoded (*x*-axis) songs. A perfect decoder (100% accuracy) would yield a diagonal confusion matrix, with no off-diagonal values (song misidentifications). See Tremere and Pinaud (2011) for details. Note that the pattern classifier reliably decodes song identity in the predrug condition, for all three neurons. Line graphs on the *right-most panels* show classification accuracies for each neuron; *dashed lines* indicate data obtained for each song

Although the findings above provided the first direct study probing the functional relevance of brain-generated hormone to sensory processing, the mechanistic bases underlying E2's optimization of neural coding and discrimination efficacy remains unknown. We recently initiated efforts to address this issue. That E2 coherently increases spike rates, information capacity and the neural discrimination of songs by NCM neurons, argues against an acute effect of this locally derived hormone as a noise-control mechanism. We, therefore, hypothesized that E2 may directly influence the rate and/or temporal coding of auditory stimuli. In fact, an effect of E2 on the rate code could directly account for increased mutual information rates and the enhanced ability of NCM neurons to discriminate songs. We directly addressed this issue by artificially eliminating the effects of E2 on the rate coding of songs, and re-assessing information capacity and the neural discrimination of songs in NCM (Tremere & Pinaud, 2011).

Our results demonstrated that even though the elimination of rate code effects decreased the impact of E2 on information rates and the neural discrimination of auditory signals in NCM, such effects were not, by any means, eliminated. These findings suggested that E2 primarily impacts NCM neuronal computations by impinging upon the temporal coding of auditory cues (Tremere & Pinaud, 2011). In fact, spike timing is critical for signal integration, and optimal information processing and transfer by auditory neurons; hence, E2 may directly influence these computational properties in the awake vertebrate brain. Moreover, these findings are highly consistent with our earlier results demonstrating that $GABA_A$-mediated inhibition, a target of E2 signaling

individually, and the *solid black line* shows mean data for all songs. A. Local infusions of E2 in NCM improve the neural discrimination of songs and, consequently, significantly increase classification accuracies from pre-infusion levels. These changes persist for at least 2 hours after interruption of E2 infusions. B. Local blockade of estrogen receptors with ICI leads to a significant and persistent disruption in the ability of NCM neurons to discriminate across songs, and in the performance of the decoder. These changes also persist for at least 2 hours after interruption of ICI infusions but return to predrug levels after 4 hours. C. Suppression of the local production of E2 with FAD drives a marked and lasting disruption in the structure of confusion matrices, indicating increased errors in song discrimination based on the neural responses. FAD significantly decreases the decoder's performance indicating that NCM neurons carry less information about song structure in the absence of locally-generated E2. D. Mean decoder's performance for all neurons recorded in our studies, for each pharmacological manipulation and session. Note that vehicle infusions in NCM (*blue*) do not alter classification accuracies over time. Local infusions of exogenous E2 (*red*) significantly enhance performance of the classifier, an effect that persists for at least 2 hours following termination of hormone infusion. Blockade of estrogen receptors with ICI (*green*), or through local inhibition of aromatase with FAD (*grey*), significantly decreased auditory coding. These effects were also persistent (2 h), but returned to predrug levels 4 hours following the interruption of drug infusions. $^{**}p < 0.01$; $^{***}p < 0.001$. Adapted from Tremere & Pinaud (2011). *J Neurosci, 31*(9), 3271–3289. Reprinted with permission.

in NCM, exerts dramatic effects on the temporal organization of auditory-driven neurophysiological responses in this auditory area (Pinaud et al., 2008). Current efforts in our laboratory are aimed at quantitatively establishing how E2 shapes temporal coding of songs in NCM and how these processes affect auditory discrimination at the network and behavioral levels.

BRAIN-GENERATED E2 ENHANCES NEURAL CODING TO IMPROVE BEHAVIORAL DISCRIMINATION

Our results provided direct evidence that NCM-derived E2 enhances the neural discrimination of songs by optimizing information capacity and the effectiveness of neural coding of auditory signals. To what extent, however, do these E2-mediated improvements in coding efficiency translate into enhanced auditory behavior performances in freely behaving animals? We addressed this question by bilaterally interfering with E2's actions or production in the NCM of animals subjected to a standard, previously established, auditory discrimination task (Tremere & Pinaud, 2011). This task, which explores the innate preference of zebra finches to certain behaviorally relevant communication signals, requires birds to discriminate between two vocalizations (Clayton, 1988; Riebel & Smallegange, 2003; Gobes & Bolhuis, 2007; Holveck & Riebel, 2007; Remage-Healey et al., 2010). We found that bilateral blockade of ERs in NCM with TMX or ICI largely abolished song preference, presumably by interfering with auditory discrimination, suggesting that E2 is required for higher-order auditory processing in a freely behaving vertebrate. We also found that behavior is modulated by E2 produced in NCM given that bilateral infusions of ARO inhibitors ATD and fadrozole yielded results that were identical to those found with local blockade of ERs (Tremere & Pinaud, 2011). Similar findings were independently obtained by another research group. Specifically, application of fadrozole via retrodialysis in vivo eliminated the behavioral preference of zebra finches to the birds' own song relative to other conspecific songs in a phonotaxis paradigm, similar to the one employed by our group (Remage-Healey et al., 2010). Interestingly, the authors only observed significant effects on behavior when interfering with the local hormonal production in the left, but not the right hemisphere, suggesting that E2's effects may be lateralized. Although the neural bases of these hemisphere-specific effects remain unknown, our previous data on spike rates, information rates and neural discrimination have not revealed evidence for lateralization effects on these neurophysiological parameters (Tremere et al., 2009; Tremere & Pinaud, 2011).

ESTROGEN-ASSOCIATED CIRCUITS: A GENERAL FEATURE OF AUDITORY FOREBRAIN CIRCUITS IN THE VERTEBRATE BRAIN?

Little doubt now exists that brain-generated E2 rapidly modulates auditory-driven responses in the adult songbird brain to influence auditory-based behaviors. These recent and exciting findings add to a body of evidence suggesting that experimentally induced or biological oscillations in circulating E2 levels influence various reproductive and social behaviors in birds in general, and songbirds in particular, including those that depend on hearing function (reviewed in (Balthazart et al., 2004; Balthazart & Ball, 2006; Cornil et al., 2006; Balthazart et al., 2009; Maney & Pinaud, 2011; Pinaud & Tremere, 2012). A critical question that emerges from these studies, however, is the extent to which the modulation of sensory physiology by brain-generated E2 is a conserved feature of central auditory circuits in the vertebrate brain, as opposed to a specialization of auditory forebrain networks in avian species. Even though the answer to this question does not diminish the significant contributions and novel insights offered by the songbird research described above, how widespread rapid neurohormonal modulation of sensory processing is in the vertebrate (especially the mammalian) brain remains a critical question. We recently set out to investigate this possibility by exploring whether or not estrogen-associated circuits are found in the mouse primary auditory cortex (A1) (Tremere et al., 2011). Remarkably, we found that estrogen-producing and -sensitive neurons are present in high numbers across all cortical layers of A1, with the exception of layer I (Fig. 24–4). More specifically, using unbiased stereological quantification methods, we found that ARO-positive cells are more prominently found in the supragranular layers (II/III), followed by the infragranular (V/VI) and granular

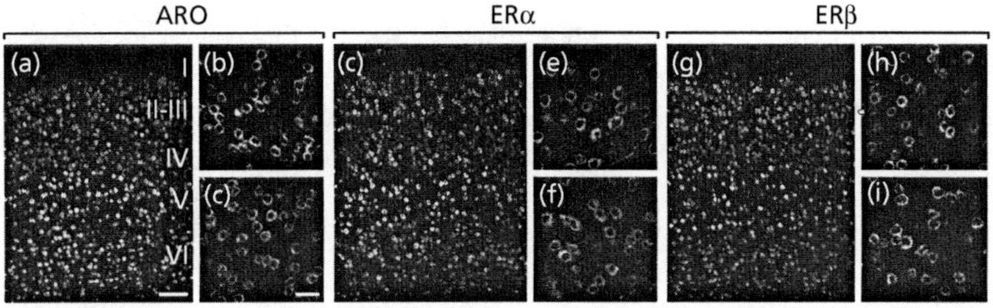

Figure 24–4. Estrogen-producing and estrogen-responsive neurons are highly prevalent and homogeneously distributed in the mouse primary auditory cortex (A1). Fluorescence in-situ hybridization for ARO (*A–C*), ERα (*D–F*) and ERβ (*G–I*) in A1. For each gene, the long panels (*A*, *D*, and *G*) depict low power photomicrographs of coronal sections obtained in A1 highlighting the distribution of cells positive for each of the probed mRNAs. Note that ARO (*A*), ERα (*D*), and ERβ (*G*) are expressed at significant levels in A1, across all cortical layers, with the exception of layer I. The set of two smaller photomicrographs for each gene (*B–C*; ARO), (*E–F*; ERα) and (*H–I*; ERβ) depict representative high-power images obtained in supragranular (*top*) and infragranular (*bottom*) layers of A1 showing detailed views of ARO-, ERα- and ERβ-positive cells. Sense strand hybridization did not yield measurable signals for each of the genes studied (not shown). Low-power and high-power images were obtained with epifluorescence and confocal microscopy, respectively. Scale bars (in μm): 100 (*A, D, G*), 25 (*B–C, E–F, H–I*). From Tremere et al. (2011). *The mouse primary visual cortex is a site of production and sensitivity to estrogens, PLoSONE*, 6(5), e20400. © 2011. Reprinted with permission.

(IV) cell layers. Overall, we discovered that nearly 67% of all neurons in A1 express ARO mRNA and, hence, can potentially produce estrogens locally. Similarly to findings obtained in songbirds, we also found that ERβ-positive cells are highly numerous in A1, accounting for approximately 57% of the overall population of neurons in this cortical area. Higher numbers of ERβ-positive neurons are found in the infragranular, supragranular and granular layers, respectively (Tremere et al., 2011). Finally, unlike the findings obtained in the songbird NCM, our results revealed that ERα is expressed in a significant population of neurons in the rodent A1 (44%), with highest expression detected in the infragranular, granular and supragranular layers, respectively.

We next assessed if these estrogen-associated circuits are activated by auditory experience in freely-behaving mice. To this end, we used dFISH directed at the activity-dependent early gene *egr-1*, a reliable marker of hearing-activated neurons (reviewed in Herdegen & Leah, 1998; Mello & Pinaud, 2006), and markers for estrogen-producing (ARO) and -sensitive cells (ERα and ERβ), in the brains of animals stimulated with pure-tones that span through the frequency range of A1 (Stiebler et al., 1997; Tremere et al., 2011). When all hearing-activated neurons are considered, whereas 83% of these cells express ARO, only 23% and 41% express ERα and ERβ mRNAs respectively. These findings suggest that auditory experience predominantly engages estrogen-producing, and to a significantly lesser extent, estrogen-sensitive neurons in the mouse A1. These observations are well aligned with the experience-dependent activation of estrogen-associated circuits in songbirds (Jeong et al., 2011), and the rapid hearing-dependent increases in E2 detected in the NCM of freely behaving songbirds (Remage-Healey et al., 2008). Finally, we conducted studies to neurochemically identify cells that participate in estrogen-associated networks in the mouse A1. We found that ARO- and ERβ-positive cells are primarily glutamatergic (~70% and 80% of labeled neurons, respectively). In contrast, the vast majority of the population of ERα-positive cells is composed of GABAergic neurons (~80%).

From our perspective, these results are exciting because they clearly demonstrate that auditory forebrain circuits of a mammalian species are replete with estrogen-producing and -sensitive neurons, and that such circuits are sensitive to sensory experience. Moreover, from an auditory neurophysiological perspective, the fact that ERα- and ERβ-positive neurons encompass neurochemically-distinct cell populations raises the possibility that, through highly selective actions on excitatory versus inhibitory networks, E2 may be well positioned to continually reshape functional receptive field properties, an effect that could predispose the network to dynamically and efficiently code for biologically informative features of sound. The primarily excitatory ERβ-positive neuronal population is likely to modulate intracortical and corticofugal excitatory connectivity, as well as thalamocortical activity. In contrast, E2 action on inhibitory neurons that are positive for ERα may directly modulate functional pruning, as well as frequency and temporal tuning (Suga et al., 1997; Chen & Jen, 2000; Wehr & Zador, 2003; Dorrn et al., 2010). Unpublished in vivo neurophysiological studies recently conducted in our laboratory reveal that, indeed, the actions of locally-generated E2 plays critical roles in the physiology and computational properties of A1 neurons through mechanisms that preserve the need of rapid modulatory action in sensory circuits (Tremere & Pinaud, unpublished observations).

CONCLUSIONS

A major recent shift has been forcing researchers to re-think roles of hormones in the brain. Discoveries that the brain produces steroid hormones, like E2, and definitive proof that these molecules can rapidly affect neuronal physiology by acting at the synaptic machinery, have transformed the pervasive dogma—that steroid hormones are slow molecules produced by the gonads to influence reproductive biology—to a modern view that brain-generated steroids (neurosteroids) are dynamic signaling molecules that can selectively and rapidly affect brain function. It is in this novel and rapidly growing area that we focus much of our research efforts. In the last 2 years, exciting new data have started to foster a new understanding of neurosteroid action in sensory circuits. Work in our laboratory has shown that brain-generated, as opposed to the gonadal, E2 directly regulates auditory-driven responses, in real-time. We further showed that this ultrafast neurohormonal influence occurs through an E2-mediated suppression of local inhibitory transmission via a presynaptic mechanism. Moreover, we recently demonstrated that E2's modulation of auditory processing increases information carried by auditory neurons about stimulus structure to enhance both the neural and the behavioral discrimination of learned acoustic signals. Our findings are consistent with, and complementary to, those of other colleagues who have demonstrated critical features of E2 production and sensitivity in hearing-associated circuits leading to altered auditory-based behavioral performance in songbirds. Recent evidence from our group also suggests that estrogen-associated circuits may be a fundamental component of central auditory networks in the vertebrate brain, as opposed to a specialization of auditory stations in the songbird; hence, alterations in hearing function known to occur concomitantly with E2 insufficiency in mammals, both in health and diseased states, may be causally related. Future research is expected to shed important light on these issues and will likely provide novel information on the mechanisms and functional roles of neurosteroids in sensory processing and perceptual learning. Finally, research at this sensory-neuroendocrine interface is expected to continue providing fundamental insights on the more global question of how neurosteroids affect neuronal and brain function.

ACKNOWLEDGMENTS

Work in the Pinaud Lab is supported by grants from the NIH (DC-010181 and DC-010033), NSF (1064684), the Schmitt Foundation, the Brain Research Foundation and a Searle Leadership Fund. Work in the Tremere Lab is supported by NIH (RR-024215). We thank Jin Kwon Jeong and Kaiping Burrows for participation in parts of the original work that forms the basis of this chapter, and Greg DeAngelis for numerous insightful discussions.

REFERENCES

Balthazart, J., & Ball, G. F. (2006). Is brain estradiol a hormone or a neurotransmitter? *Trends Neurosci, 29,* 241–249.

Balthazart, J., Baillien, M., Cornil, C. A., & Ball, G. F. (2004). Preoptic aromatase modulates male sexual behavior: slow and fast mechanisms of action. *Physiol Behav, 83,* 247–270.

Balthazart, J., Taziaux, M., Holloway, K., Ball, G. F., & Cornil, C. A. (2009). Behavioral effects of brain-derived estrogens in birds. *Ann N Y Acad Sci, 1163,* 31–48.

Bergamaschi, R., Bergonzoni, C., Mazzanti, L., Scarano, E., Mencarelli, F., Messina, F., et al. (2008). Hearing loss in Turner syndrome: results of a multicentric study. *J Endocrinol Invest, 31,* 779–783.

Bernard, D. J., Bentley, G. E., Balthazart, J., Turek, F. W., & Ball, G. F. (1999). Androgen receptor, estrogen receptor alpha, and estrogen receptor beta show distinct patterns of expression in forebrain song control nuclei of European starlings. *Endocrinology, 140,* 4633–4643.

Chen, Q. C., & Jen, P. H. (2000). Bicuculline application affects discharge patterns, rate-intensity functions, and frequency tuning characteristics of bat auditory cortical neurons. *Hear Res, 150,* 161–174.

Chew, S. J., Mello, C., Nottebohm, F., Jarvis, E., & Vicario, D. S. (1995). Decrements in auditory responses to a repeated conspecific song are long-lasting and require two periods of protein synthesis in the songbird forebrain. *Proc Natl Acad Sci U S A, 92,* 3406–3410.

Chew, S. J., Vicario, D. S., & Nottebohm, F. (1996). A large-capacity memory system that recognizes the calls and songs of individual birds. *Proc Natl Acad Sci U S A, 93,* 1950–1955.

Clayton, N. S. (1988). Song discrimination learning in zebra finches. *Anim Behav, 36,* 1016–1024.

Coleman, J. R., Campbell, D., Cooper, W. A., Welsh, M. G., & Moyer, J. (1994). Auditory brainstem responses after ovariectomy and estrogen replacement in rat. *Hear Res, 80,* 209–215.

Cornil, C. A., Ball, G. F., & Balthazart, J. (2006). Functional significance of the rapid regulation of brain estrogen action: where do the estrogens come from? *Brain Res, 1126,* 2–26.

Davis, M. J., & Ahroon, W. A. (1982). Fluctuations in susceptibility to noise-induced temporary threshold shift as influenced by the menstrual cycle. *J Aud Res, 22,* 173–187.

Dorrn, A. L., Yuan, K., Barker, A. J., Schreiner, C. E., & Froemke, R. C. (2010). Developmental sensory experience balances cortical excitation and inhibition. *Nature, 465,* 932–936.

Ehret, G., Koch, M., Haack, B., & Markl, H. (1987). Sex and parental experience determine the onset of an instinctive behavior in mice. *Naturwissenschaften, 74,* 47.

Forlano, P. M., Deitcher, D. L., & Bass, A. H. (2005). Distribution of estrogen receptor alpha mRNA in the brain and inner ear of a vocal fish with comparisons to sites of aromatase expression. *J Comp Neurol, 483,* 91–113.

Gentner, T. Q. (2004). Neural systems for individual song recognition in adult birds. *Ann NY Acad Sci U S A, 1016,* 282–302.
Gobes, S. M., & Bolhuis, J. J. (2007). Birdsong memory: a neural dissociation between song recognition and production. *Curr Biol, 17,* 789–793.
Golub, M. S., Germann, S. L., & Hogrefe, C. E. (2004). Endocrine disruption and cognitive function in adolescent female rhesus monkeys. *Neurotoxicol Teratol, 26,* 799–809.
Guimaraes, P., Frisina, S. T., Mapes, F., Tadros, S. F., Frisina, D. R., & Frisina, R. D. (2006). Progestin negatively affects hearing in aged women. *Proc Natl Acad Sci U S A, 103,* 14246–14249.
Herdegen, T, & Leah, J. D. (1998). Inducible and constitutive transcription factors in the mammalian nervous system: control of gene expression by Jun, Fos and Krox, and CREB/ATF proteins. *Brain Res Brain Res Rev, 28,* 370–490.
Holveck, M. J., & Riebel, K. (2007). Preferred songs predict preferred males: consistency and repeatability of zebra finch females across three test contexts. *Anim Behav, 74,* 297–309.
Hultcrantz, M., Simonoska, R., & Stenberg, A. E. (2006). Estrogen and hearing: a summary of recent investigations. *Acta Otolaryngol, 126,* 10–14.
Hultcrantz, M., & Sylven, L. (1997). Turner's syndrome and hearing disorders in women aged 16–34. *Hear Res, 103,* 69–74.
Jeong, J. K., Burrows, K., Tremere, L. A., & Pinaud, R. (2011). Neurochemical organization and experience-dependent activation of estrogen-associated circuits in the songbird auditory forebrain. *Eur J Neurosci, 34,* 283–291.
Jerger, J, & Hall, J. (1980). Effects of age and sex on auditory brainstem response. *Arch Otolaryngol, 106,* 387–391.
Koch, M., & Ehret, G. (1989). Estradiol and parental experience, but not prolactin are necessary for ultrasound recognition and pup-retrieving in the mouse. *Physiol Behav, 45,* 771–776.
Maney, D., & Pinaud, R. (2011). Estradiol-dependent modulation of auditory processing and selectivity in songbirds. *Front Neuroendocrinol, 32,* 287–302.
Maney, D. L., Cho, E., & Goode, C. T. (2006). Estrogen-dependent selectivity of genomic responses to birdsong. *Eur J Neurosci, 23,* 1523–1529.
Maruska, K. P., & Fernald, R. D. (2010). Steroid receptor expression in the fish inner ear varies with sex, social status, and reproductive state. *BMC Neurosci, 11,* 58.
Mello, C. V., & Clayton, D. F. (1995). Differential induction of the ZENK gene in the avian forebrain and song control circuit after metrazole-induced depolarization. *J Neurobiol, 26,* 145–161.
Mello, C. V., & Pinaud, R. (2006). Immediate early gene regulation in the auditory system. In: Pinaud, R., Tremere, L. A. (eds.), *Immediate Early Genes in Sensory Processing, Cognitive Performance and Neurological Disorders.* New York: Springer-Verlag, pp. 35–56.
Mello, C. V., Velho, T. A., & Pinaud, R. (2004). Song-induced gene expression: a window on song auditory processing and perception. *Ann N Y Acad Sci, 1016,* 263–281.
Mello, C. V., Vicario, D. S., & Clayton, D. F. (1992). Song presentation induces gene expression in the songbird forebrain. *Proc Natl Acad Sci U S A, 89,* 6818–6822.
Meltser, I., Tahera, Y., Simpson, E., Hultcrantz, M., Charitidi, K., Gustafsson, J. A., et al. (2008). Estrogen receptor beta protects against acoustic trauma in mice. *J Clin Invest, 118,* 1563–1570.

Metzdorf, R., Gahr, M., & Fusani, L. (1999). Distribution of aromatase, estrogen receptor, and androgen receptor mRNA in the forebrain of songbirds and nonsongbirds. *J Comp Neurol, 407,* 115–129.

Miranda, J. A., & Liu, R. C. (2009). Dissecting natural sensory plasticity: hormones and experience in a maternal context. *Hear Res, 252,* 21–28.

Noirot, I. C., Adler, H. J., Cornil, C. A., Harada, N., Dooling, R. J., Balthazart, J., et al. (2009). Presence of aromatase and estrogen receptor alpha in the inner ear of zebra finches. *Hear Res, 252,* 49–55.

Nowicki, S., & Searcy, W. A. (2004). Song function and the evolution of female preferences: why birds sing, why brains matter. *Ann N Y Acad Sci, 1016,* 704–723.

Pinaud, R., Fortes, A. F., Lovell, P., & Mello, C. V. (2006). Calbindin-positive neurons reveal a sexual dimorphism within the songbird analogue of the mammalian auditory cortex. *J Neurobiol, 66,* 182–195.

Pinaud, R., & Mello, C. V. (2007). GABA immunoreactivity in auditory and song control brain areas of zebra finches. *J Chem Neuroanat, 34,* 1–21.

Pinaud, R, & Terleph, T. A. (2008). A songbird forebrain area potentially involved in auditory discrimination and memory formation. *J Biosci, 33,* 145–155.

Pinaud, R., Terleph, T. A., Tremere, L. A., Phan, M. L., Dagostin, A. A., Leao, R. M., et al. (2008). Inhibitory network interactions shape the auditory processing of natural communication signals in the songbird auditory forebrain. *J Neurophysiol, 100,* 441–455.

Pinaud, R., Velho, T. A., Jeong, J. K., Tremere, L. A., Leao, R. M., von Gersdorff, H., et al. (2004). GABAergic neurons participate in the brain's response to birdsong auditory stimulation. *Eur J Neurosci, 20,* 1318–1330.

Pinaud, R, & Tremere, L. A. (2012) Control of central auditory processing by a brain-generated oestrogen.*Nature Rev Neurosci, 13,* 521–527.

Price, K., Zhu, X., Guimaraes, P. F., Vasilyeva, O. N., & Frisina, R. D. (2009). Hormone replacement therapy diminishes hearing in peri-menopausal mice. *Hear Res, 252,* 29–36.

Remage-Healey, L., Coleman, M. J., Oyama, R. K., & Schlinger, B. A. (2010). Brain estrogens rapidly strengthen auditory encoding and guide song preference in a songbird. *Proc Natl Acad Sci U S A, 107,* 3852–3857.

Remage-Healey, L., Maidment, N. T., & Schlinger, B. A. (2008). Forebrain steroid levels fluctuate rapidly during social interactions. *Nat Neurosci, 11,* 1327–1334.

Riebel, K., & Smallegange, I. M. (2003). Does zebra finch (Taeniopygia guttata) preference for the (familiar) father's song generalize to the songs of unfamiliar brothers? *J Comp Psychol, 117,* 61–66.

Saldanha, C. J., & Coomaralingam, L. (2005). Overlap and co-expression of estrogen synthetic and responsive neurons in the songbird brain—a double-label immunocytochemical study. *Gen Comp Endocrinol, 141,* 66–75.

Saldanha, C. J., Remage-Healey, L., & Schlinger, B. A. (2011). Synaptocrine signaling: steroid synthesis and action at the synapse. *Endocr Rev, 32,* 532–549.

Saldanha, C. J., Tuerk, M. J., Kim, Y. H., Fernandes, A. O., Arnold, A. P., & Schlinger B. A. (2000). Distribution and regulation of telencephalic aromatase expression in the zebra finch revealed with a specific antibody. *J Comp Neurol, 423,* 619–630.

Sanford, S. E., Lange, H. S., & Maney, D. L. (2010). Topography of estradiol-modulated genomic responses in the songbird auditory forebrain. *Dev Neurobiol, 70,* 73–86.

Schlinger, B. A., & Remage-Healey, L. (2011). Neurosteroidogenesis: insights from studies of songbirds. *J Neuroendocrinol, 24,* 16–21.

Shen, P., Schlinger, B. A., Campagnoni, A. T., & Arnold, A. P. (1995). An atlas of aromatase mRNA expression in the zebra finch brain. *J Comp Neurol, 360,* 172–184.

Simonoska, R, Stenberg, A. E., Duan, M., Yakimchuk, K., Fridberger, A., Sahlin, L., et al. (2009). Inner ear pathology and loss of hearing in estrogen receptor-beta deficient mice. *J Endocrinol, 201,* 397–406.

Sisneros, J. A., Forlano, P. M., Deitcher, D. L., & Bass, A. H. (2004). Steroid-dependent auditory plasticity leads to adaptive coupling of sender and receiver. *Science, 305,* 404–407.

Stiebler, I., Neulist, R., Fichtel, I., & Ehret, G. (1997). The auditory cortex of the house mouse: left-right differences, tonotopic organization and quantitative analysis of frequency representation. *J Comp Physiol A, 181,* 559–571.

Stripling, R., Kruse, A. A., & Clayton, D. F. (2001). Development of song responses in the zebra finch caudomedial neostriatum: role of genomic and electrophysiological activities. *J Neurobiol, 48,* 163–180.

Suga, N., Zhang, Y., & Yan, J. (1997). Sharpening of frequency tuning by inhibition in the thalamic auditory nucleus of the mustached bat. *J Neurophysiol, 77,* 2098–2114.

Thompson, J. V., & Gentner, T. Q. (2010). Song recognition learning and stimulus-specific weakening of neural responses in the avian auditory forebrain. *J Neurophysiol, 103,* 1785–1797.

Tremere, L. A., Burrows, K., Jeong, J. K., & Pinaud, R. (2011). Organization of estrogen-associated circuits in the mouse primary auditory cortex. *J Exp Neurosci, 2011,* 45–60.

Tremere, L. A., Jeong, J. K., & Pinaud, R. (2009). Estradiol shapes auditory processing in the adult brain by regulating inhibitory transmission and plasticity-associated gene expression. *J Neurosci, 29,* 5949–5963.

Tremere, L. A., & Pinaud, R. (2011). Brain-generated estradiol drives long-term optimization of auditory coding to enhance the discrimination of communication signals. *J Neurosci, 31,* 3271–3289.

Tremere, L. A., Terleph, T. A., Jeong, J. K., & Pinaud, R. (2010). Bilateral multielectrode neurophysiological recordings coupled to local pharmacology in awake songbirds. *Nat Protoc, 5,* 191–200.

Vates, G. E., Broome, B. M., Mello, C. V., & Nottebohm, F. (1996). Auditory pathways of caudal telencephalon and their relation to the song system of adult male zebra finches. *J Comp Neurol, 366,* 613–642.

Velho, T. A., Pinaud, R., Rodrigues, P. V, & Mello, C. V. (2005). Co-induction of activity-dependent genes in songbirds. *Eur J Neurosci, 22,* 1667–1678.

Vyas, A., Harding, C., Borg, L., & Bogdan, D. (2009). Acoustic characteristics, early experience, and endocrine status interact to modulate female zebra finches' behavioral responses to songs. *Horm Behav, 55,* 50–59.

Walpurger, V., Pietrowsky, R., Kirschbaum, C., & Wolf, O. T. (2004). Effects of the menstrual cycle on auditory event-related potentials. *Horm Behav, 46,* 600–606.

Wehr, M., Zador, A. M. (2003). Balanced inhibition underlies tuning and sharpens spike timing in auditory cortex. *Nature, 426,* 442–446.

Woolley, C. S. (2007). Acute effects of estrogen on neuronal physiology. *Annu Rev Pharmacol Toxicol, 47,* 657–680.

Chapter 25

Acute Synthesis and Action of Estrogens in Communication Circuits

Luke Remage-Healey

INTRODUCTION

The evolution of acoustic communication has given rise to an extraordinary diversity of acoustic signals and signal complexity throughout the animal kingdom. A hallmark feature of acoustic signals is that they are modulated both temporally and spectrally on a fast timescale. For example, the spectrotemporal features of human speech, bat echolocation, and cricket song alike are modulated on the order of milliseconds (e.g., Brown, 1976; Machens et al., 2003; Divenyi, 2009). Rapid modulation of acoustic signals provides a means for information to be continually updated on a moment-to-moment basis between sender and receiver (Bradbury & Vehrencamp, 1998). This places high demands on the temporal precision of neural circuits that support vocal production and perception. A variety of mechanisms have been uncovered within the central nervous system that ensure accurate timing of vocal output and auditory processing alike. For example, auditory neurons in the midbrain of female tree frogs are extremely sensitive to both the duration and interpulse interval of male advertisement calls (<10 millisecond sensitivity), and slight variation in the timing, duration, and interval of calls has consequences for both neural encoding and behavioral preferences (Gerhardt, 1994; Edwards et al., 2002). To achieve maximum flexibility in acoustic signal production and perception, communication networks depend critically on active "tuning" by neuromodulators. The neuropeptide arginine vasotocin (AVT) is one example of a classically defined neuromodulator identified in several vocal vertebrates, and the release of AVT in discrete neural circuits provides a mechanism for behavioral plasticity in the acoustic communication signals of frogs and teleost fishes (Boyd, 1994; Marler et al., 1999; Goodson & Bass, 2000). This section describes recent work that has shed light on a relatively newfound neuromodulatory system, brain-derived estrogen signaling, and how it has been co-opted by some vertebrates to enable rapid modulation of both the production and perception of acoustic communication signals.

This chapter explores the idea that estrogen production and action within acoustic communication circuits occurs rapidly, at the timescale of neuromodulation (seconds to minutes). This is in contrast to the "classical" timescale of transcriptional regulation (hours to weeks) for which steroids are well known. Several recently published reviews provide excellent in-depth perspectives on how estrogens can regulate acoustic communication via long-term, "classical" genomic mechanisms (Forlano et al., 2006; Gahr, 2007; Schlinger & Brenowitz, 2008; Sisneros, 2009; Wilczynski & Lynch, 2011). This chapter briefly touches on the long-term actions of estrogens on vocal and auditory processing, a literature that has been formative for the current focus. The main focus of the chapter on rapid estrogen signaling in acoustic communication networks is considered from a comparative perspective, highlighted by exciting new evidence generated from vocalizing amphibians, birds, and teleost fishes.

ESTROGENS AND ACOUSTIC COMMUNICATION

Reproduction depends heavily on steroid hormones such as estrogens to coordinate the timing of peripheral physiology (e.g., gonadal state) with the behavioral physiology of the central nervous system (CNS) (Simerly, 2002). Animals emit and process sounds primarily for use in mate attraction or resource acquisition and defense (e.g., territoriality), typically in the context of reproductive events and seasons. It is not surprising, therefore, that the expression of aromatase and/or estrogen receptors have been localized to neural pathways that govern acoustic communication (Kelley, 1981; Metzdorf et al., 1999; Gahr, 2001; Wu et al., 2003b; Sisneros, 2004; Forlano et al., 2005a; Noirot et al., 2009; Caras et al., 2010; Maruska & Fernald, 2010). Across the vertebrate lineage, estrogens have been found to play a role in shaping vocal and auditory processing by acting directly on the CNS. In human beings, for example, auditory responsiveness in women shifts during the menstrual cycle, and the most striking changes occur in the late follicular phase when circulating estradiol levels are peaking (Al-Mana et al., 2010). An extensive body of research primarily in birds and amphibians has shown that circulating estrogens can act via long-term, classically genomic mechanisms in the CNS to affect vocal production and auditory processing in support of reproduction (Yovanof & Feng, 1983; Adkins-Regan et al., 1994; Ball et al., 1999; Maney et al., 2006; Meitzen et al., 2007; Lynch & Wilczynski, 2008; Maney et al., 2008; Sanford et al., 2010; Matragrano et al., 2011; Wilczynski & Lynch, 2011; Zornik & Kelley, 2011).

In addition to long-term seasonal or developmental actions of estrogens, the timeframe for how hormones such as steroids can act within the brain to regulate behavior has been extended to exceedingly rapid events (Balthazart & Ball, 2006; Remage-Healey & Bass, 2006; Hofmann, 2010). Over the past 15 years, the timescale of steroid actions has increasingly been found to correspond to the rapid timescale of behavioral events (seconds to minutes) and even neural events (microseconds to milliseconds). It has therefore become imperative to test whether steroids such as estrogens can act as bona fide neuromodulators in the vertebrate CNS.

THREE FOUNDATIONAL CONCEPTS

A series of breakthroughs in the 1970s and 1980s provided a critical foundation for much of the work described in this chapter on acute estrogen synthesis and action in acoustic communication networks. These can be grouped into three broad foundational themes:

1. **Foremost was a series of observations that exogenous steroids could have rapid actions (i.e., within seconds to minutes) on the patterning and excitability of neurons.** Specifically, a remarkable series of studies in rodents showed that membrane excitability of neurons in the hippocampus, amygdala, pituitary, and midbrain were rapidly regulated by in vitro bath application of estrogens and other steroids (Ruf & Steiner, 1967; Pfaff & Pfaffman, 1969; Dufy et al., 1979; Teyler et al., 1980; Nabekura et al., 1986; Wong & Moss, 1992). Second, investigation of the stress-induced suppression of reproductive behavior in roughskin newts demonstrated that corticosteroids could inhibit reproductive clasping behavior via rapid membrane-specific actions on medullary neurons (Moore & Miller, 1984; Orchinik et al., 1991; Rose et al., 1998; Moore & Evans, 1999; Orchinik et al., 2002). Last, in vitro recordings in hypothalamic slices from guinea pigs showed that estrogens can have rapid actions on membrane electrophysiology via specific intracellular signaling pathways (Kelly et al., 1984; Kelly & Levin, 2001; Qiu et al., 2003; Ronnekleiv & Kelly, 2005). Thus, it became clear that estrogens and other steroids were capable of extremely fast modulation of neuronal excitability in the vertebrate CNS.
2. **In parallel with this work, clues began to emerge that changes in social behavior and communication were accompanied by fluctuating levels of circulating steroids.** Plasma androgens were observed to rapidly respond to social stimuli, such as territorial intruders or potential mates, in rodents (Batty, 1978; Leshner, 1983), teleost fishes (Heiligen & Kramer, 1972; Hannes & Franck, 1983; Cardwell & Liley, 1991; Oliveira et al., 2001), and songbirds (Harding & Follett, 1979; Harding, 1981; Wingfield, 1985), but apparently not in lizards (Thompson & Moore, 1992; Klukowski & Nelson, 1998). The widespread nature of these observations led to a formal predictive framework (the "challenge hypothesis"), which accounted for variation in plasma androgens during periods of social instability and challenge (Wingfield et al., 1990; Hirschenhauser & Oliveira, 2006). Subsequent studies have largely confirmed the basic tenets of the challenge hypothesis, albeit with some modifications (Wingfield, 2005; Goymann et al., 2007). The phenomenon that plasma androgens fluctuate in response to minute-by-minute changes in the social environment has therefore become well established among vertebrates.
3. **A third essential foundation to the work described in this chapter the observation that plasma androgens could be converted by**

the enzyme aromatase ("aromatized") within the CNS to achieve local changes in central estrogen levels. As aromatase expression in the CNS became evident in several animal species (Flores et al., 1973; Reddy et al., 1973), a formal "aromatization hypothesis" was proposed to account for long-term effects of estrogens on masculinization of the brain (Naftolin et al., 1975; Naftolin & MacLusky, 1984; also see chapters 1 and 9). A critical series of behavioral observations came in the early 1980s when Hutchison and Steimer showed that acoustic and visual sexual stimuli caused short-term changes in brain aromatase activity in doves (Steimer & Hutchison, 1981; Hutchison & Steimer, 1985). Aromatase activity within the preoptic area was elevated within days of pairing, demonstrating that sociosexual stimuli can be transduced by the CNS into altered brain aromatase activity within a relatively short timeframe. Indeed, estrogens implanted into the preoptic area of Japanese quail were shown to be essential for the expression of copulatory behavior, and aromatase inhibitors blocked copulation when administered to the preoptic area in the days leading up to testing (Watson & Adkins-Regan, 1989; Balthazart & Surlemont, 1990). Subsequently, the effects of aromatase inhibition on male copulatory behavior in Japanese quail were observed within a rapid timeframe (within 30 minutes) (Cornil et al., 2006a). In parallel, the rapid, minute-by-minute actions of injected estrogens were linked to the expression of sexual behavior in several vertebrate species. In 1999, Cross and Roselli observed that estradiol (E2) injections increased the expression of copulatory behavior in male rats within approximately 30 minutes (Cross & Roselli, 1999). This phenomenon was then extended to both male Japanese quail (Cornil et al., 2006b) and male mice (Taziaux et al., 2007). In addition, fieldwork in song sparrows showed that territoriality was suppressed within 24 hours by a single intramuscular injection of the aromatase inhibitor fadrozole (Soma et al., 2000), demonstrating that changing brain aromatase activity had important consequences for animals in their natural environments.

Together, the three critical foundations outlined above led inextricably to the hypothesis that *estrogen synthesis and action within the central nervous system drives acute, minute-by-minute changes in social communication behaviors.* Below, a sequence of studies are described that were guided by this foundational work and together provide strong support for this central hypothesis.

REGULATION OF VOCAL MOTOR NETWORKS BY ESTROGENS

Estrogens regulate the patterning and output of neural circuits involved in communication via long-term (days to weeks) mechanisms in a broad range of vertebrates, from rats (Nyby et al., 1992) to electric fishes (Schaefer & Zakon, 1996). In Mongolian gerbils, discrete estradiol implants near the hypothalamus induced an upregulation of ultrasonic vocalizations within days (Holman et al.,

1991). There are likewise reports of a persistent, long-term effect of systemic estrogen treatment (as well as contraceptive use) on vocal parameters in women (Amir et al., 2003; Firat et al., 2009). Perhaps the most complete understanding of the long-term effects of estrogens on vocal function has come from studies of songbirds. The behaviorally identified, interconnected forebrain network of vocal nuclei in songbirds is exquisitely sensitive to estrogens during development, sexual differentiation, and seasonal changes (Simpson & Vicario, 1991; Adkins-Regan et al., 1994; Grisham & Arnold, 1995; Casto & Ball, 1996; Fusani et al., 2003; Tramontin et al., 2003; Soma et al., 2004; Meitzen et al., 2007). In contrast to this wealth of data on long-term regulation, how and whether estrogens regulate vocal patterning on a more rapid, modulatory timescale (seconds to minutes) is currently unexplored in birds or mammals.

The most compelling evidence that estrogens *rapidly* regulate vocal circuits via nonclassical mechanisms comes from studies of teleosts and amphibians. One model system that has revealed a great deal about the neural and neuromuscular control of vocal motor function is the African clawed frog, *Xenopus laevis* (Potter et al., 2005; Kelley & Bass, 2010; Zornik & Kelley, 2011). On a rapid (modulatory) timescale, estradiol had relatively acute effects on synaptic strength at the laryngeal neuromuscular junction in *Xenopus* (Wu et al., 2001). The quantal content of neuromuscular events (as evoked by laryngeal nerve stimulation) was weakened by either hCG or estradiol within the first hour of treatment. This represents the only evidence to date in any vertebrate that estrogens can rapidly affect the neuromuscular junction of a vocal motor pathway.

Within the central nervous system, the rapid modulation of vocal patterning by steroid hormones (including estrogens) was uncovered in a series of experiments with a teleost fish, the plainfin midshipman, *Poricthys notatus*. Midshipmen are highly vocal during the breeding season; when male midshipmen establish territories, they vocalize in the contexts of mate attraction and during agonistic interactions (Brantley & Bass, 1994). As shown in Figure 25–1, the descending motor pathway that controls vocalizations in midshipmen is comprised of forebrain and midbrain nuclei that project to a hindbrain-spinal vocal pattern generator (VPG) comprised of three nuclei (Bass et al., 1994; Goodson & Bass, 2002). The rhythmic activity of the VPG directly patterns the duration and frequency of midshipman calls (Bass & Baker, 1990; Remage-Healey & Bass, 2004). Midshipman vocal nuclei can be electrically stimulated to generate fictive calls (patterned, rhythmic activity that mimics natural vocalizations) from the VPG using an in vivo preparation (see Fig. 25–1). Thus, the intact midshipman nervous system provides an opportunity to study the rapid actions of steroids on experimentally-elicited fictive calls from the VPG in vivo.

Taking advantage of these properties, the modulation of fictive calling was tested in vivo using intramuscular injection of steroid hormones, including androgens and estrogens. Both low and high doses of 17-beta-estradiol (E2) caused rapid (within 5 minutes) and sustained (up to 30 minutes) increases in the duration of fictive calls in adult male midshipman (Remage-Healey & Bass, 2004). Therefore, not only was the midshipman VPG rapidly sensitive

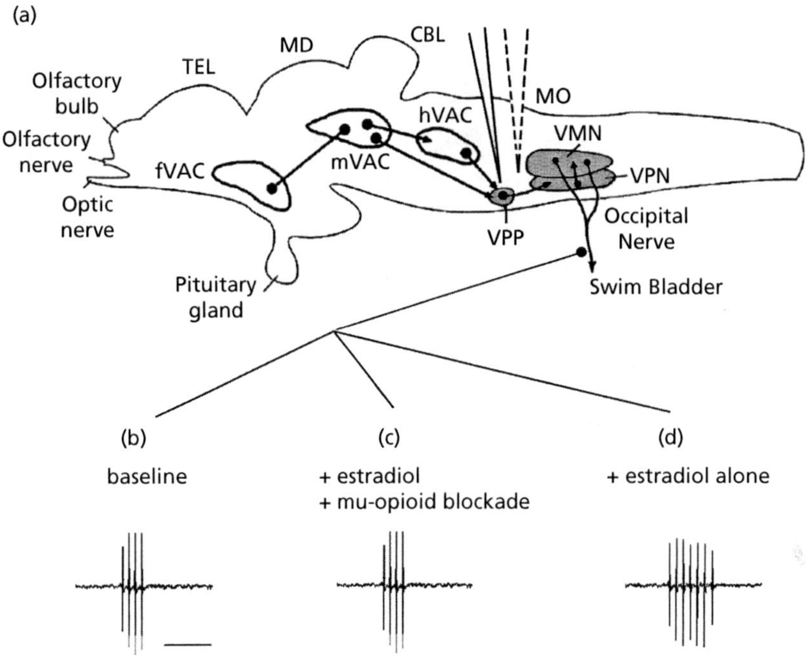

Figure 25–1. Rapid Actions of Estrogens on Vocal Patterning in Midshipmen. A. A schematic of the midshipman brain (*sagittal view*), showing rapid estrogen effects that depend on opioidergic activity. The vocal motor pathway descends from forebrain (f-s), midbrain (m-), and hindbrain (h-) vocal-acoustic centers (VACs), which together project to the hindbrain-spinal vocal pattern generator (VPG; group of three nuclei shown in gray) The VPG is composed of a vocal pre-pacemaker nucleus (VPP) that projects to a vocal pacemaker nucleus (VPN), which in turn patterns the rhythmic activity of the vocal motor nucleus (VMN). The VMN sends coordinated, rhythmic contractile discharges to the vocal muscles attached to a resonant swimbladder. These coordinated discharges can be recorded via an extracellular electrode adjacent to the occipital nerve, and patterned fictive calls (*traces below*) can be elicited by electrical microstimulation of the VPP (*solid electrode*). A drug delivery electrode (*dashed electrode*) is depicted within the fourth ventricle near the VPG. CBL, cerebellum, MD, midbrain, MO, medulla oblongata, TEL, telencephalon. B–D. Estradiol has rapid actions on fictive calling in midshipman. mu-Opioid circuits are required for rapid (within 5 minutes) estradiol-mediated increases in fictive call duration. Figure panels adapted in part from Remage-Healey, L., & Bass, A. H. (2010). Estradiol interacts with an opioidergic network to achieve rapid modulation of a vocal pattern generator. *J Comp Physiol A Neuroethol Sens Neural Behav Physiol, 196,* 137–146.

to E2, but the effect at both doses had a limited timeframe (i.e., fictive call duration returned to baseline by 45–60 minutes after the initial injection). There was no effect of the stereoisomer 17-alpha-estradiol on fictive calling, emphasizing the ligand-specificity of a putative estrogen receptor within the VPG. Together, these findings showed that the actions of estrogens on a vertebrate neural pathway that patterns vocalizations can be highly *specific, acute,* and also *transient*.

Midshipmen occur in three adult morphotypes (type I males, type II males, and females) that are examples of fixed alternative reproductive tactics (see Bass, 1996; Rhen & Crews, 2002). Type I male midshipmen build rocky shelters as nests and emit patterned, low-frequency calls to attract females to deposit eggs in their nests. Type II males steal fertilizations from type I males via female mimicry at the nest, and unlike type I males, type II males and females only appear to vocalize in agonistic contexts (Brantley & Bass, 1994). Circulating steroid concentrations differ markedly among the three adult reproductive morphs (Knapp et al., 1999; Sisneros et al., 2004), raising the possibility that the observed rapid actions of androgens and estrogens on fictive calling in type I males (outlined above) were divergent from actions in type IIs and females. Importantly, the VPG region is enshrouded in aromatase mRNA expression in all three adult morphotypes (Forlano et al., 2001, especially see Fig. 3C; Forlano & Bass, 2005), consistent with local synthesis of estrogens within the VPG.

The possibility that estrogens and androgens could differentially interact with the VPG in each morph was directly tested using the fictive calling electrophysiology preparation in type IIs and females (Remage-Healey & Bass, 2007). Just as in type Is, both low and high doses of E2 caused rapid (within 5 minutes, sustained up to 45 minutes) increases in the duration of fictive calls in type II males and females. Both type II males and females also showed a significant and rapid increase in fictive call duration in response to treatment with testosterone, unlike type I males. Interestingly, however, the rapid actions of testosterone in type II males were dependent on functional androgen receptors (i.e., sensitive to AR blockade), whereas the rapid actions of testosterone in females were dependent on enzymatic conversion to E2 (i.e., sensitive to aromatase blockade). This work established that both within- and between-sex variation in the rapid actions of steroids were linked to divergent levels of those steroids in the plasma, because males have elevated testosterone levels, whereas females have elevated estradiol levels. A primary conclusion of this work was that the morph-specific phenotype of *rapid* sensitivity to steroids is very likely organized during early development and differentiation, when the fixed midshipman reproductive morphs diverge (Remage-Healey & Bass, 2007).

The only modulatory feature common to all the three midshipman morphs was the fast upregulation of fictive calling by E2. This conserved modulatory property led to an investigation of the potential receptor-mediated mechanism(s) that could account for the rapid actions of estrogens in the midshipman brain. The VPG region exhibits significant expression of nuclear estrogen receptor alpha mRNA as revealed by in situ hybridization histochemistry, manifest as discrete clusters localized to the large vocal motor neurons within the VPG (Forlano et al., 2005b). The rapid actions of E2 on fictive calling suggested the presence of a membrane-bound receptor or "nongenomic" mechanism, yet to date no extranuclear estrogen receptors have been detected in the midshipman hindbrain–spinal region. The fictive vocal preparation was used to investigate further potential estrogen receptor mechanisms using type I male midshipmen. The selective estrogen-receptor modulator ICI 182,780 is an effective blocker of the rapid effects of estrogens on neuronal activity in mammals (Micevych et al., 2007; Grassi et al., 2009) and can act as an ER antagonist in teleost fishes (Menuet et al., 2005; Pinto et al., 2006). However, pretreatment with ICI 182,780 did not block the rapid actions of E2 in midshipman vocal patterning (Remage-Healey & Bass, 2010). Instead, the actions of E2 were dependent on recruitment of a mu-opioid network in the hindbrain–spinal VPG region. When either the broad-spectrum opioid blocker naloxone or the mu-opioid selective blocker beta-funaltrexamine were picospritzed into the VPG region the rapid actions of E2 on fictive calling were eliminated (Remage-Healey & Bass, 2010) (see also Fig. 25–1). Similar treatment with the kappa-opioid selective blocker nor-binaltorphimine had no effect on the rapid actions of E2, demonstrating the specificity of mu-opioid/E2 comodulation on within the VPG. These findings reinforced a growing body of work showing that the acute, nonclassical actions of estrogens and other steroids in the brain depend on comodulatory interactions with opioids and their receptors (Evans et al., 2000; Sinchak & Micevych, 2001; Qiu et al., 2003).

In summary, electrophysiological studies of the midshipman vocal motor system established that estrogens rapidly modulate a discrete network of neurons that directly patterns vocalizations. A key observation stemming from these studies was that the acute actions of estrogens occur within the hindbrain–spinal circuitry of both males and females. One hypothesis consistent with these findings is that local brain estrogen production could occur in a very fast timescale in support of rapid behavioral changes. These and other concurrent findings (see the following) gave rise to the idea that acoustic communication circuits could participate in their own estrogen-mediated signaling via local and acute synthesis. Subsequently, this idea was tested and refined in a series of studies from multiple labs examining brain estrogen production and action in the auditory cortex of songbirds.

RAPID AND LOCAL CHANGES IN BRAIN-DERIVED ESTROGENS

The relatively fast effects (seconds to minutes) of estrogens on neural circuits is evident not only for those circuits involved in midshipman vocal communication, but also for a number of limbic, hypothalamic, and cortical structures involved in learning, neuroendocrine feedback, and reproduction. The observation that brain circuits are rapidly responsive to estrogens has recently been combined with the notion that some of these circuits are capable of synthesizing their own ready supply of brain-derived estrogens (neuroestrogens). Fundamental insights on the rapid control of brain aromatase activity and its relationship to reproductive behavior have come from detailed and elegant studies of copulation in Japanese quail (see chapters 22 and 23). A particularly important model system for uncovering the de novo production of estrogens within the brain is a songbird, the Australian zebra finch. The zebra finch brain expresses all of the enzymes necessary for the production of estradiol from cholesterol (London et al., 2006). In some brain regions the aromatase enzyme is expressed and active at the presynaptic terminal (Peterson et al., 2005; Rohmann et al., 2007). In principle, this "synaptocrine" regulation of aromatase activity provides a mechanism for some circuits to achieve exceedingly precise temporal and spatial targeting of estrogens to discrete synapses (Remage-Healey et al., 2011a,b; reviewed in Saldanha et al., 2011; see also chapter 6).

A foundational study by Schlinger and Arnold (1992) demonstrated unequivocally that the CNS provides the major source of estrogen levels detected in the plasma of adult male zebra finches. Subsequently, the local production of estrogens within in vitro cultures of the developing zebra finch brain was implicated in the masculinization of the forebrain song motor pathway (Holloway & Clayton, 2001). Experimental determination of brain estradiol content in the forebrain of songbirds, using solid-phase extracted brain tissue homogenates coupled to immunoassays, confirmed that brain estrogens in aromatase positive regions was significantly and reliably elevated over plasma levels in male songbirds (Newman et al., 2008; Chao et al., 2011; Charlier et al., 2011). Furthermore, field studies of territorial song sparrows showed that the activity/expression of androgen-metabolizing enzymes (aromatase, 5-alpha-reductase, and 5-beta-reductase) appear to be independently controlled in adjacent brain regions (Soma et al., 2003). Together this work suggested that the songbird forebrain contains steroid "microenvironments" that can be independently controlled in a nucleus-by-nucleus manner. Songbirds therefore became an attractive model system to investigate the potential role of local and acute estrogen synthesis on neural circuit function and behavior.

When brain estradiol content and aromatase activity have been quantified in parallel using tissue derived and extracted from the same brain macroareas (tissue blocks/regions) there has generally been a lack of strict correspondence between the two measures. To some extent, these poor correlations between aromatase activity and estrogen content can be explained by methodological differences, because aromatase activity is typically expressed as a function of protein content, whereas estradiol content is expressed as a function of "wet weight" of freshly dissection micropunches or macroareas (Newman et al., 2008; Chao et al., 2011; Charlier et al., 2011). However, there may be a fundamental difference in the temporal features that are quantified by these two methods that reflects how estrogens are controlled within the CNS. The dissociation of recent past estrogen synthesis (as measured by estradiol content in brain tissue) and future capacity for estrogen synthesis (as measured by the extent of aromatase activity within brain explants in vitro) may reflect the rapid rate-of-change of estrogen synthesis in brain circuits. In other words, the integration of estradiol content as measured in brain tissue homogenates could be unrelated to estrogen fluctuations that will occur subsequently. Thus the ability to capture local and acute changes in neuroestrogens over time became necessary to fully understand how estrogens fluctuate within brain circuits.

One method to measure neurochemical fluctuations, in vivo microdialysis, was adapted specifically for the measurement of neuroestrogens in discrete forebrain regions in zebra finches (Remage-Healey et al., 2008). Microdialysis relies on surgical implantation of a small (approximately 200 μM) semipermeable probe into a brain region of interest, through which local neurochemistry can be measured and manipulated in vivo. Steroid hormones pose particular challenges to this technology because of their lipophilic chemistry, and microdialysis needed to be extensively validated before experimentation. Much of the validation of in vivo microdialysis for neuroestrogens has been outlined previously, including independent GC/MS verification of the presence of brain-generated estradiol sampled via this method (Remage-Healey et al., 2008, 2011b). Because of current detection limits of estradiol quantification assays, microdialysis is unable to integrate fluctuations that occur faster than 15- to 30-minute time intervals. This methodological limitation poses a challenge to assess the coupling of changes in neuroestrogens with abrupt changes in behavioral state (i.e., within seconds to minutes). Even so, this technology has provided the first evidence for acute fluctuations in neuroestrogen levels in awake, behaving animals. Because samples are collected from the same animal over time, direct comparisons can be made before, during, and after exposure to behavioral interactions, pharmacological agents, and environmental stimuli. The primary findings using this new technology for understanding the role of neuroestrogens in auditory processing in the songbird brain are described in below.

The first E2 microdialysis studies were conducted in male zebra finches and were focused on the caudomedial nidopallium (NCM), an auditory forebrain region considered similar structurally and functionally to the secondary auditory cortex of mammals (e.g., Bolhuis & Gahr, 2006; Wang et al., 2010). Both cell bodies and presynaptic terminals within the NCM express aromatase protein (Saldanha et al., 2000; Peterson et al., 2005; see chapter 6) and NCM is a relatively large region and therefore a good target for microdialysis (although it is important to note that NCM is not neurochemically homogeneous across its entirety; Pinaud et al., 2006; Salvante et al., 2010). Microdialysis in awake, behaving males revealed that neuroestrogens in NCM were acutely elevated when males were engaged in social interactions with females for 30 minutes, and returned to baseline after the social interaction ended (Remage-Healey et al., 2008). In two separate experiments this pattern of a female-induced elevation was observed in males with probes directed at NCM, in contrast to probes that were directed outside NCM that did not show a change during social interactions with females. Therefore, these initial experiments indicated a degree of region-specificity for the rapid changes in estradiol within an auditory circuit.

Although male zebra finches reliably sing to females during social interactions, the rapid elevation in NCM neuroestrogens was not strictly dependent on the singing behavior of each focal male undergoing microdialysis. In fact, a separate set of experiments revealed that auditory activation alone (via acoustic playbacks of male song versus control sounds) was sufficient to elicit rapid elevations in NCM neuroestrogens in males (Fig. 25–2A) (Remage-Healey et al., 2008).

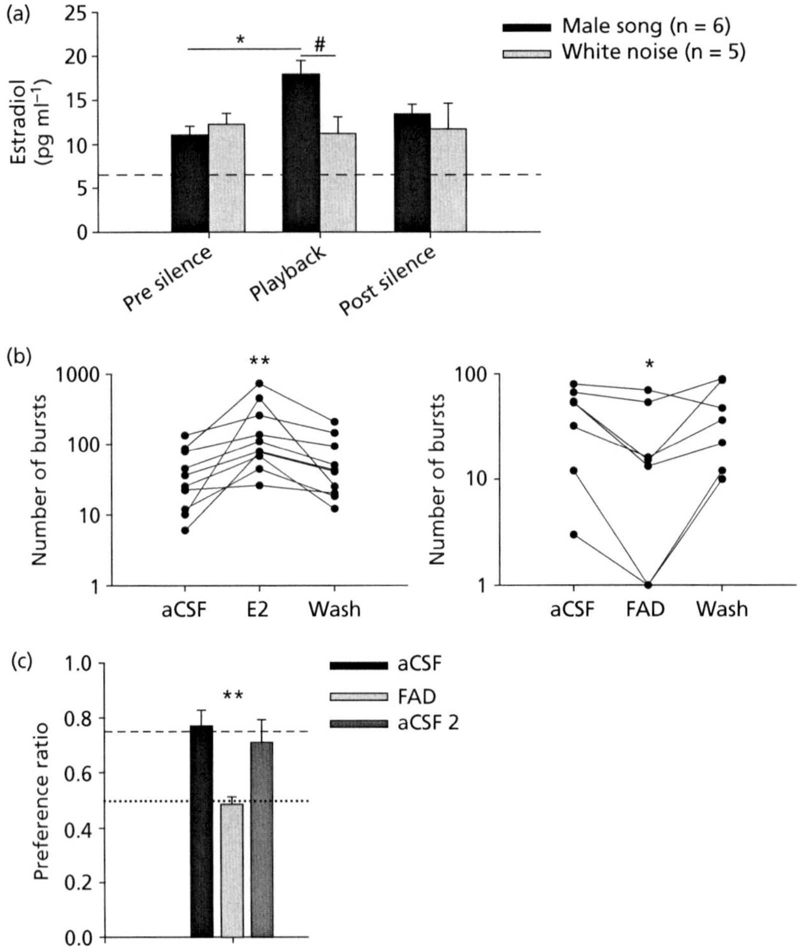

Figure 25–2. Rapid Production and Action of Estrogens in the Auditory Caudomedial Nidopallium of Zebra Finches. A. Local neuroestrogen levels are rapidly elevated in NCM during acoustic playback of male song but not control white noise sounds. $*p$ = .011 for playback versus pre-silence in the male song group; $\#p$ = .023 for male song versus white noise during playback. B. Local changes in neuroestrogens acutely modulate the auditory-evoked activity of single NCM neurons. The *left panel* shows that estradiol (E2) (30 μg/ml) increases the number of auditory-evoked single unit bursts in NCM. $**p$ = .005, n = 10 units from seven birds. By contrast, the right panel shows that the aromatase inhibitor fadrozole (FAD) (100 μM) decreases the number of auditory-evoked single unit bursts in NCM. $*p$ = .018; n = 7 units from five birds. C. Song preference ratio (time spent near bird's own song [BOS] speaker/total time spent near conspecific or BOS song speaker) is reliably expressed during aCSF retrodialysis (near species-typical level, *dashed line*), whereas the song preference ratio is reduced to chance (approximately 0.5, *dotted line*) during FAD retrodialysis (100 μM) $**p$ = .007. Figure panels adapted in part from Remage-Healey, L., Maidment, N. T., & Schlinger, B. A. (2008). Forebrain

These elevations were again specific to NCM and did not occur in nearby cortical regions, nor in the plasma. Therefore, auditory stimuli (and song in particular) were able to drive local and fast changes in estrogen levels within the songbird auditory forebrain.

Based on these findings with males, subsequent studies examined both the sex-specificity and sensory-modality specificity of E2 responses in NCM. These experiments showed that estradiol levels within the NCM of adult female zebra finches were also similarly responsive to song stimuli (Remage-Healey et al., 2012). Thus, in males and females alike, neuroestrogens were elevated during song playback within the NCM, but not outside NCM (and not in the general circulation in either sex), emphasizing neuroestrogen fluctuations specific to NCM. Also, 30 minutes of visual stimulation of a conspecific female alone (but neither a conspecific nor heterospecific male) caused a rapid elevation in NCM neuroestrogens in males (but not in females). Together this evidence supports the hypothesis that rapidly increasing estrogens in NCM could be preparatory for neurocomputational tasks that may immediately ensue for a male engaged in a social interaction involving song or other auditory signals. In summary, experiments using microdialysis in zebra finches demonstrated that neuroestrogen fluctuations are acute and highly localized within the auditory forebrain, and are responsive to natural stimuli. The neural and behavioral outcomes of these rapid changes in neuroestrogens within the auditory NCM then became a primary research focus.

RAPID AND LOCAL EFFECTS OF ESTROGENS ON AUDITORY PROCESSING AND BEHAVIOR

The rapid rise in neuroestrogens in NCM of males and females provided strong impetus to examine the consequences for auditory processing intrinsic to NCM. This question was addressed using a combination of electrophysiological and behavioral approaches. In vivo microdialysis and retrodialysis (reverse-delivery of exogenous compounds) provided a means to test questions at both levels (i.e., electrophysiological and behavioral) using the same methodological approach. The coupling of in vivo neuroestrogen microdialysis with extracellular recordings in NCM allowed the delivery of drugs (e.g., estrogens, conjugated-estrogens, and the aromatase inhibitor fadrozole) while simultaneously monitoring changes in auditory-evoked activity of NCM neurons. Second, this approach enabled a full within-subject design, in which all three phases of drug responsiveness (predrug, drug treatment, and drug washout) could be monitored and manipulated.

steroid levels fluctuate rapidly during social interactions. *Nat Neurosci, 11,* 1327–1334; Remage-Healey, L., Coleman, M. J., Oyama, R. K., & Schlinger, B. A. (2010). Brain estrogens rapidly strengthen auditory encoding and guide song preference in a songbird. *Proc Natl Acad Sci U S A, 107,* 3852–3857.

In anesthetized males, E2 retrodialysis caused rapid increases in the auditory-evoked activity of NCM neurons relative to both pre- and washout conditions (Remage-Healey et al., 2010). Using spike-sorting algorithms, recordings were analyzed for changes in activity at the level of single neurons (i.e., "units"). In response to E2 treatment, single-unit firing frequency was elevated, in part because single units had increasingly switched from tonic, isolated firing patterns to firing in bursts (three spikes or more in rapid succession) during E2 exposure as compared to aCSF pretreatment (Fig. 25–2B). In concordance with these findings for exogenous estrogens, when the aromatase inhibitor fadrozole was retrodialyzed into NCM the occurrence of single-unit bursts was significantly suppressed (see Fig. 25–2B). In both cases, the altered burst firing (upregulation in the case of E2 and downregulation in the case of FAD) returned to baseline during the posttreatment aCSF washout period. Therefore, both the firing rate and burstiness of single auditory neurons were rapidly modulated by estrogen fluctuations within the NCM. Recent investigations have formally extended these observations to females, in which estradiol similarly elevates, whereas fadrozole suppresses auditory-evoked multiunit firing rates in NCM (Remage-Healey et al., 2012).

At approximately the same time, largely identical rapid effects of estrogens on NCM auditory processing were independently generated by Pinaud, Tremere, and colleagues in adult zebra finches (Tremere et al., 2009; Tremere & Pinaud, 2011; see chapter 24). Although the methods used were slightly different (head-fixed awake versus anesthetized subjects; picospritz delivery versus retrodialysis of estrogens) E2 treatment also caused rapid increases in auditory-evoked firing rates of NCM neurons, whereas the aromatase inhibitor ATD or the estrogen receptor antagonist tamoxifen each caused rapid suppression of NCM auditory-evoked firing rates. Thus, in two independent laboratories, the same direction of effects for E2 actions were convincingly demonstrated for zebra finch NCM neurons. Further, using in vitro recordings the Pinaud group provided strong evidence that the rapid actions of estrogens on NCM neurons were mediated via suppression of inhibitory (GABAergic) currents, and that one downstream effect was to enhance the expression of MAPK-dependent immediate early genes (Tremere et al., 2009). Last, follow-up studies from this same group showed that the actions of estrogens in NCM can enhance neuronal encoding of song stimuli as well as neural discrimination among songs (Tremere & Pinaud, 2011). Together, these independent and convergent findings provided convincing evidence that estrogens produced in the NCM can exert rapid actions on the patterning and information encoded by NCM auditory neurons.

Adult zebra finches reliably exhibit behavioral preferences for familiar versus unfamiliar songs (Holveck & Riebel, 2007), and for males this preference is pronounced when the male is asked to compare his own song with that of a conspecific. The same retrodialysis techniques described in the preceding for NCM electrophysiology were also applied to awake, behaving males during song preference tests. Using the same dose as in the electrophysiology setting,

FAD retrodialyzed into the left NCM of adult male zebra finches caused an acute (within 30 minutes) and transient suppression of the behavioral expression of song preference (Fig. 25–2C) (Remage-Healey et al., 2010). When FAD was washed out with normal aCSF over 30 minutes, the song preference of individuals for a familiar song was restored to control levels (Remage-Healey et al., 2010). In parallel with the convergent electrophysiology results described in the preceding, recent experiments have provided independent confirmation of the suppressive effects of aromatase inhibition and estrogen receptor blockade on song preference in NCM, using intracranial cannula injections (Tremere & Pinaud, 2011). In summary, convergent studies from independent laboratories have shown that estrogen production in the NCM appears to be a rapid, modulatory phenomenon that can acutely boost auditory processing and behavioral preferences in adult zebra finches.

CONCLUSIONS

The central thesis of this chapter is that estrogens rapidly modify vocal patterning and auditory processing via local and acute actions in the CNS. Studies drawn from the midshipman vocal motor system and the zebra finch auditory system provide strong evidence for local and acute actions of estrogens in communication circuits. An outstanding issue for this work is identifying which receptors, ion channels and molecular mechanisms are responsible for carrying out these fast effects of estrogens. As we have learned from the study of rapid actions of estrogens in liver, brain, gonads, and muscle, there is a wide array of candidate mechanisms for estrogen signaling, including the classical estrogen receptor inserted into the membrane and newly identified G-protein coupled receptors (Kelly & Levin, 2001; Thomas et al., 2005; Vasudevan et al., 2005; Marino et al., 2006; Mhyre et al., 2006; Prossnitz et al., 2008). Expression of ERalpha has been characterized in the midshipman brain, and is localized to the hindbrain-spinal VPG region that mediates the rapid actions on vocal patterning, yet the rapid E2 actions are insensitive to classical receptor blockade using ICI-182,780 (see above). Instead, the estrogen actions in midshipman depend on cocmodulation with a mu-opioidergic network. The interaction between fast changes in estrogens and "classical" neuromodulatory systems will continue to be an exciting research avenue, and many labs are actively exploring the interrelationships between estrogens, catecholamines, and opioids in the control of brain circuit function (Riters et al., 2005; Cornil et al., 2006b; Quesada & Micevych, 2008; Schultz et al., 2009; Salvante et al., 2010; Matragrano et al., 2011).

A second outstanding issue arising from this work on the CNS is the extent to which peripheral auditory and vocal physiology is also rapidly altered by estrogens. The rapid, neurophysiological effect of estrogens on the vocal neuromuscular junction of Xenopus (Wu et al., 2001) offers a glimpse at how estrogens can modify peripheral control of vocalizations in addition to the actions on the CNS outlined above. Once again, fresh examination of the loci for long-term actions of estrogens may provide clues about where rapid modulation is likely to occur. To this end, the vertebrate vocal organs (syrinx, larynx) are almost uniformly sensitive to long-term actions of estrogens (Narbaitz et al., 1980; Holt et al., 1986; Wade et al., 2002; Wu et al., 2003a; Veney & Wade, 2005; Martin & Veney, 2008), suggesting that estrogens may yet have acute impacts on vocal patterning at the periphery in birds and mammals (the vocal musculature of teleosts appears to be insensitive to estrogens; Brantley et al., 1993). In the auditory domain, ERs are expressed throughout the ascending auditory pathway in songbirds (Walters et al., 1988; Jacobs et al., 1996; Metzdorf et al., 1999; Saldanha & Coomaralingam, 2005) and the inner ear/hair cells of the zebra finch, mouse, and midshipman contain both aromatase and estrogen receptors (Forlano et al., 2005b; Hultcrantz et al., 2006; Noirot et al., 2009; Sisneros, 2009). Thus, it will be fascinating to explore how peripheral auditory physiology is modulated by local and rapid actions of estrogens. In conclusion, as with many new frontiers in neuroscience and neuroendocrinology, comparative inquiry into a variety of model organisms has uncovered—and will hopefully continue to reveal—neuromodulatory mechanisms such as the neuroestrogen modulation of acoustic communication networks outlined in this chapter.

ACKNOWLEDGMENTS

The author thanks for Christiane Meyer, Maaya Ikeda, and Ashley Paon for comments on the manuscript, and Andrew Bass, Elizabeth Adkins-Regan, Barney Schlinger, and Art Arnold for guidance and thoughtful discussion about the topic. Research support was provided by NINDS R00NS066179.

REFERENCES

Adkins-Regan, E., Mansukhani, V., Seiwert, C., & Thompson, R. (1994). Sexual-differentiation of brain and behavior in the zebra finch—critical periods for effects of early estrogen-treatment. *J Neurobiol, 25,* 865–877.

Al-Mana, D., Ceranic, B., Djahanbakhch, O., & Luxon, L. M. (2010). Alteration in auditory function during the ovarian cycle. *Hear Res, 268,* 114–122.

Amir, O., Biron-Shental, T., Muchnik, C., & Kishon-Rabin, L. (2003). Do oral contraceptives improve vocal quality? Limited trial on low-dose formulations. *Obstet Gynecol 101,* 773–777.

Ball, G. F., Bernard, D. J., Foidart, A., Lakaye, B., & Balthazart, J. (1999). Steroid sensitive sites in the avian brain: does the distribution of the estrogen receptor alpha and beta types provide insight into their function? *Brain Behav Evol, 54,* 28–40.

Balthazart, J., & Ball, G. F. (2006). Is brain estradiol a hormone or a neurotransmitter? *Trends Neurosci, 29,* 241–249.

Balthazart, J., & Surlemont, C. (1990). Androgen and estrogen action in the preoptic area and activation of copulatory behavior in quail. *Physiol Behav, 48,* 599–609.

Bass, A. H. (1996). Shaping brain sexuality. *Am Scientist, 84,* 352–363.

Bass, A. H., & Baker, R. (1990). Sexual dimorphisms in the vocal control-system of a teleost fish—morphology of physiologically identified neurons. *J Neurobiol, 21,* 1155–1168.

Bass, A. H., Marchaterre, M. A., & Baker, R. (1994). Vocal-acoustic pathways in a teleost fish. *J Neurosci, 14,* 4025–4039.

Batty, J. (1978). Acute changes in plasma testosterone levels and their relation to measures of sexual-behavior in male house mouse (mus-musculus). *Anim Behav, 26,* 349–357.

Bolhuis, J. J., & Gahr, M. (2006). Neural mechanisms of birdsong memory. *Nat Rev Neurosci, 7,* 347–357.

Boyd, S. K. (1994). Arginine vasotocin facilitation of advertisement calling and call phonotaxis in bullfrogs. *Horm Behav, 28,* 232–240.

Bradbury, J. W., & Vehrencamp, S. L. (1998). *Principles of Animal Communication.* Sunderland, MA: Sinauer Associates.

Brantley, R. K., & Bass, A. H. (1994). Alternative male spawning tactics and acoustic-signals in the plainfin midshipman fish porichthys-notatus girard (teleostei, batrachoididae). *Ethology, 96,* 213–232.

Brantley, R. K., Marchaterre, M. A., & Bass, A. H. (1993). Androgen effects on vocal muscle structure in a teleost fish with inter-sexual and intra-sexual dimorphism. *J Morphol, 216,* 305–318.

Brown, P. (1976). Vocal communication in pallid bat, antrozous-pallidus. [Zeitschrift fur tierpsychologie.] *J Comp Ethol, 41,* 34–54.

Caras, M. L., Brenowitz, E., & Rubel, E. W. (2010). Peripheral auditory processing changes seasonally in Gambel's white-crowned sparrow. *J Comp Physiol A Neuroethol Sens Neural Behav Physiol, 196,* 581–599.

Cardwell, J. R., & Liley, N. R. (1991) Androgen control of social-status in males of a wild population of stoplight parrotfish, sparisoma-viride (scaridae). *Horm Behav, 25,* 1–18.

Casto, J. M., & Ball, G. F. (1996). Early administration of 17beta-estradiol partially masculinizes song control regions and alpha2-adrenergic receptor distribution in European starlings (Sturnus vulgaris). *Horm Behav, 30,* 387–406.

Chao, A., Schlinger, B. A., & Remage-Healey, L. (2011). Combined liquid and solid-phase extraction improves quantification of brain estrogen content. *Front Neuroanat, 5,* 57.

Charlier, T. D., Newman, A. E., Heimovics, S. A., Po, K. W., Saldanha, C. J., & Soma, K. K. (2011). Rapid effects of aggressive interactions on aromatase activity and oestradiol in discrete brain regions of wild male white-crowned sparrows. *J Neuroendocrinol, 23,* 742–753.

Cornil, C. A., Dalla, C., Papadopoulou-Daifoti, Z., Baillien, M., & Balthazart, J. (2006b). Estradiol rapidly activates male sexual behavior and affects brain monoamine levels in the quail brain. *Behav Brain Res, 166,* 110–123.

Cornil, C. A., Taziaux, M., Baillien, M., Ball, G. F., & Balthazart, J. (2006a). Rapid effects of aromatase inhibition on male reproductive behaviors in Japanese quail. *Horm Behav, 49,* 45–67.

Cross, E., & Roselli, C. E. (1999). 17 beta-estradiol rapidly facilitates chemoinvestigation and mounting in castrated male rats. *Am J Physiol-Reg I, 276,* R1346–R1350.

Divenyi, P. (2009). Perception of complete and incomplete formant transitions in vowels. *J Acoust Soc Am, 126,* 1427–1439.

Dufy, B., Vincent, J. D., Fleury, H., Pasquier, P. D., Gourdji, D., & Tixiervidal, A. (1979). Dopamine-inhibition of action potentials in a prolactin secreting cell-line is modulated by estrogen. *Nature, 282,* 855–857.

Edwards, C. J., Alder, T. B., & Rose, G. J. (2002). Auditory midbrain neurons that count. *Nat Neurosci, 5,* 934–936.

Evans, S. J., Searcy, B. T., & Moore, F. L. (2000). A subset of kappa opioid ligands bind to the membrane glucocorticoid receptor in an amphibian brain. *Endocrinology, 141,* 2294–2300.

Firat, Y., Engin-Ustun, Y., Kizilay, A., Ustun, Y., Akarcay, M., Selimoglu, E., et al. (2009). Effect of intranasal estrogen on vocal quality. *J Voice, 23,* 716–720.

Flores, F., Naftolin, F., Ryan, K. J., & White, R. J. (1973) Estrogen formation by isolated perfused rhesus-monkey brain. *Science, 180,* 1074–1075.

Forlano, P., Schlinger, B., & Bass, A. (2006). Brain aromatase: New lessons from non-mammalian model systems. *Front Neuroendocrinol, 27,* 247–274.

Forlano, P. M., & Bass, A. H. (2005) Seasonal plasticity of brain aromatase mRNA expression in glia: divergence across sex and vocal phenotypes. *J Neurobiol, 65,* 37–49.

Forlano, P. M., Deitcher, D. L., & Bass, A. H. (2005a). Distribution of estrogen receptor alpha mRNA in the brain and inner ear of a vocal fish with comparisons to sites of aromatase expression. *J Comp Neurol, 483,* 91–113.

Forlano, P. M., Deitcher, D. L., & Bass, A. R. H. (2005b). Distribution of estrogen receptor alpha mRNA in the brain and inner ear of a vocal fish with comparisons to sites of aromatase expression. *J Comp Neurol, 483,* 91–113.

Forlano, P. M., Deitcher, D. L., Myers, D. A., & Bass, A. H. (2001). Anatomical distribution and cellular basis for high levels of aromatase activity in the brain of teleost fish: aromatase enzyme and mRNA expression identify glia as source. *J Neurosci, 21,* 8943–8955.

Fusani, L., Metzdorf, R., Hutchison, J. B., & Gahr, M. (2003). Aromatase inhibition affects testosterone-induced masculinization of song and the neural song system in female canaries. *J Neurobiol, 54,* 370–379.

Gahr, M. (2001). Distribution of sex steroid hormone receptors in the avian brain: functional implications for neural sex differences and sexual behaviors. *Microsc Res Techniq, 55,* 1–11.

Gahr, M. (2007). Sexual differentiation of the vocal control system of birds. *Adv Genet, 59,* 67–105.

Gerhardt, H. C. (1994). The evolution of vocalization in frogs and toads. *Annu Rev Ecol Syst, 25,* 293–324.

Goodson, J. L., & Bass, A. H. (2000). Forebrain peptides modulate sexually polymorphic vocal circuitry. *Nature, 403,* 769–772.

Goodson, J. L., & Bass, A. H. (2002). Vocal-acoustic circuitry and descending vocal pathways in teleost fish: convergence with terrestrial vertebrates reveals conserved traits. *J Comp Neurol, 448,* 298–322.

Goymann, W., Landys, M. M., & Wingfield, J. C. (2007). Distinguishing seasonal androgen responses from male-male androgen responsiveness: revisiting the Challenge Hypothesis. *Horm Behav, 51,* 463–476.

Grassi, S., Frondaroli, A., Dieni, C., Scarduzio, M., & Pettorossi, V. E. (2009). Long-term potentiation in the rat medial vestibular nuclei depends on locally synthesized 17beta-estradiol. *J Neurosci, 29,* 10779–10783.

Grisham, W., & Arnold, A. P. (1995) A direct comparison of the masculinizing effects of testosterone, androstenedione, estrogen, and progesterone on the development of the zebra finch song system. *J Neurobiol, 26,* 163–170.

Hannes, R. P., & Franck, D. (1983). The effect of social-isolation on androgen and corticosteroid levels in a cichlid fish (haplochromis-burtoni) and in swordtails (xiphophorus-helleri). *Horm Behav, 17,* 292–301.

Harding, C. F. (1981). Social modulation of circulating hormone levels in the male. *Am Zool, 21,* 223–231.

Harding, C. F., & Follett, B. K. (1979). Hormone changes triggered by aggression in a natural population of blackbirds. *Science, 203,* 918–920.

Heiligen, W., & Kramer, U. (1972). Aggressiveness as a function of external stimulation. *J Comp Physiol Psychol, 77,* 332.

Hirschenhauser, K., & Oliveira, R. F. (2006). Social modulation of androgens in male vertebrates: meta-analyses of the challenge hypothesis. *Anim Behav, 71,* 265–277.

Hofmann, H. A. (2010). The neuroendocrine action potential. Winner of the 2008 Frank Beach Award in Behavioral Neuroendocrinology. *Horm Behav, 58,* 555–562.

Holloway, C. C., & Clayton, D. E. (2001). Estrogen synthesis in the male brain triggers development of the avian song control pathway in vitro. *Nat Neurosci, 4,* 170–175.

Holman, S. D., Hutchison, R. E., & Hutchison, J. B. (1991). Microimplants of estradiol in the sexually dimorphic area of the hypothalamus activate ultrasonic vocal behavior in male Mongolian gerbils. *Horm Behav, 25,* 531–548.

Holt, G. R., Aufdemorte, T. B., & Sheridan, P. J. (1986). Estrogen receptor in the larynx of the aged baboon (Papio cynocephalus). *Ann Otol Rhinol Laryngol, 95,* 608–617.

Holveck, M. J., & Riebel, K. (2007). Preferred songs predict preferred males: consistency and repeatability of zebra finch females across three test contexts. *Anim Behav, 74,* 297–309.

Hultcrantz, M., Simonoska, R., & Stenberg, A. E. (2006). Estrogen and hearing: a summary of recent investigations. *Acta Oto-Laryngol, 126,* 10–14.

Hutchison, J. B., & Steimer, T. (1985). Preoptic formation of 17-beta-estradiol is influenced by behavioral stimuli in the dove. *Brain Res, 360,* 366–369.

Jacobs, E. C., Arnold, A. P., & Campagnoni, A. T. (1996). Zebra finch estrogen receptor cDNA: cloning and mRNA expression. *J Steroid Biochem, 59,* 135–145.

Kelley, D. B. (1981). Locations of androgen-concentrating cells in the brain of Xenopus laevis: autoradiography with 3H-dihydrotestosterone. *J Comp Neurol, 199,* 221–231.

Kelley, D. B., & Bass, A. H. (2010). Neurobiology of vocal communication: mechanisms for sensorimotor integration and vocal patterning. *Curr Opin Neurobiol, 20,* 748–753.

Kelly, M. J., & Levin, E. R. (2001). Rapid actions of plasma membrane estrogen receptors. *Trends Endocrinol Metab, 12,* 152–156.

Kelly, M. J., Ronnekleiv, O. K., & Eskay, R. L. (1984). Identification of estrogen-responsive LHRH neurons in the guinea pig hypothalamus. *Brain Res Bull, 12,* 399–407.

Klukowski, M., & Nelson, C. E. (1998). The challenge hypothesis and seasonal changes in aggression and steroids in male northern fence lizards (Sceloporus undulatus hyacinthinus). *Horm Behav, 33,* 197–204.

Knapp, R., Wingfield, J. C., & Bass, A. H. (1999). Steroid hormones and paternal care in the plainfin midshipman fish (Porichthys notatus). *Horm Behav, 3,* 81–89.

Leshner, A. I. (1983). The hormonal responses to competition and their behavioral significance. In: Svare, B. (Ed.), *Hormones and Aggressive Behavior.* New York: Plenum Press, pp. 393–404.

London, S. E., Monks, D. A., Wade, J., & Schlinger, B. A. (2006). Widespread capacity for steroid synthesis in the avian brain and song system. *Endocrinology, 147,* 5975–5987.

Lynch, K. S., & Wilczynski, W. (2008). Reproductive hormones modify reception of species-typical communication signals in a female anuran. *Brain Behav Evol, 71,* 143–150.

Machens, C. K., Schutze, H., Franz, A., Kolesnikova, O., Stemmler, M. B., Ronacher, B., et al. (2003). Single auditory neurons rapidly discriminate conspecific communication signals. *Nat Neurosci, 6,* 341–342.

Maney, D. L., Cho, E., & Goode, C. T. (2006). Estrogen-dependent selectivity of genomic responses to birdsong. *Eur J Neurosci, 23,* 1523–1529.

Maney, D. L., Goode, C. T., Lange, H. S., Sanford, S. E., & Solomon, B. L. (2008). Estradiol modulates neural responses to song in a seasonal songbird. *J Comp Neurol, 511,* 173–186.

Marino, M., Ascenzi, P., & Acconcia, F. (2006). S-palmitoylation modulates estrogen receptor α localization and functions. *Steroids, 71,* 298–303.

Marler, C. A., Boyd, S. K., & Wilczynski, W. (1999). Forebrain arginine vasotocin correlates of alternative mating strategies in cricket frogs. *Horm Behav, 36,* 53–61.

Martin, L. C., & Veney, S. L. (2008). The specific estrogen receptor antagonist ICI 182,780 masculinizes development of the zebra finch syrinx. *Gen Comp Endocrinol, 156,* 434–439.

Maruska, K. P., & Fernald, R. D. (2010). Steroid receptor expression in the fish inner ear varies with sex, social status, and reproductive state. *BMC Neurosci, 11,* 58.

Matragrano, L. L., Sanford, S. E., Salvante, K. G., Sockman, K. W., & Maney, D. L. (2011). Estradiol-dependent catecholaminergic innervation of auditory areas in a seasonally breeding songbird. *Eur J Neurosci, 34,* 416–425.

Meitzen, J., Moore, I. T., Lent, K., Brenowitz, E. A., & Perkel, D. J. (2007). Steroid hormones act transsynaptically within the Forebrain to regulate neuronal phenotype and song stereotypy. *J Neurosci, 27,* 12045–12057.

Menuet, A., Pellegrini, E., Brion, F., Gueguen, M. M., Anglade, I., Pakdel, F., et al. (2005). Expression and estrogen-dependent regulation of the zebrafish brain aromatase gene. *J Comp Neurol, 485,* 304–320.

Metzdorf, R., Gahr, M., & Fusani, L. (1999). Distribution of aromatase, estrogen receptor, and androgen receptor mRNA in the forebrain of songbirds and nonsongbirds. *J Comp Neurol, 407,* 115–129.

Mhyre, A. J., Shapiro, R. A., & Dorsa, D. M. (2006). Estradiol reduces nonclassical transcription at cyclic adenosine 3′,5′-monophosphate response elements in glioma cells expressing estrogen receptor alpha. *Endocrinology, 147,* 1796–1804.

Micevych, P. E., Chaban, V., Ogi, J., Dewing, P., Lu, J. K., & Sinchak, K. (2007). Estradiol stimulates progesterone synthesis in hypothalamic astrocyte cultures. *Endocrinology, 148,* 782–789.

Moore, F. L., & Evans, S. J. (1999). Steroid hormones use non-genomic mechanisms to control brain functions and behaviors: a review of evidence. *Brain Behav Evol, 54,* 41–50.

Moore, F. L., & Miller, L. J. (1984). Stress-induced inhibition of sexual-behavior—corticosterone inhibits courtship behaviors of a male amphibian (taricha-granulosa). *Horm Behav, 18,* 400–410.

Nabekura, J., Oomura, Y., Minami, T., Mizuno, Y., & Fukuda, A. (1986). Mechanism of the rapid effect of 17-beta-estradiol on medial amygdala neurons. *Science, 233,* 226–228.

Naftolin, F., & MacLusky, N. J. (1984). *Aromatization Hypothesis Revisited.* New York: Raven Press, 79–92.

Naftolin, F., Ryan, K. J., Davies, I. J., Reddy, V. V., Flores, F., Petro, Z., et al. (1975). The formation of estrogens by central neuroendocrine tissues. *Recent Prog Horm Res, 31,* 295–319.

Narbaitz, R., Stumpf, W. E., & Sar, M. (1980). Estrogen target cells in the larynx: autoradiographic studies with 3H-diethylstilbestrol in fetal mice. *Horm Res, 12,* 113–117.

Newman, A. E., Chin, E. H., Schmidt, K. L., Bond, L., Wynne-Edwards, K. E., & Soma, K. K. (2008). Analysis of steroids in songbird plasma and brain by coupling solid phase extraction to radioimmunoassay. *Gen Comp Endocrinol, 155,* 503–510.
Noirot, I. C., Adler, H. J., Cornil, C. A., Harada, N., Dooling, R. J., Balthazart, J., et al. (2009). Presence of aromatase and estrogen receptor alpha in the inner ear of zebra finches. *Hear Res, 252,* 49–55.
Nyby, J., Matochik, J. A., & Barfield, R. J. (1992). Intracranial androgenic and estrogenic stimulation of male-typical behaviors in house mice (Mus domesticus). *Horm Behav, 26,* 24–45.
Oliveira, R. F., Lopes, M., Carneiro, L. A., & Canario, A. V. M. (2001). Watching fights raises fish hormone levels—Cichlid fish wrestling for dominance induce an androgen surge in male spectators. *Nature, 409,* 475–475.
Orchinik, M., Gasser, P., & Breuner, C. (2002). Rapid corticosteroid actions on behavior: Cellular mechansims and organismal consequences. In: Pfaff, D. W. (Ed.), *Hormones, Brain and Behavior*. St. Louis: Elsevier, pp. 567–600.
Orchinik, M., Murray, T. F., & Moore, F. L. (1991). A corticosteroid receptor in neuronal membranes. *Science, 252,* 1848–1851.
Peterson, R. S., Yarram, L., Schlinger, B. A., & Saldanha, C. J. (2005). Aromatase is pre-synaptic and sexually dimorphic in the adult zebra finch brain. *Proc R Soc B-Biol Sci, 272,* 2089–2096.
Pfaff, D. W., & Pfaffman, C. (1969). Olfactory and hormonal influences on basal fore-brain of male rat. *Brain Res, 15,* 137–157.
Pinaud, R., Fortes, A. F., Lovell, P., & Mello, C. V. (2006). Calbindin-positive neurons reveal a sexual dimorphism within the songbird analogue of the mammalian auditory cortex. *J Neurobiol, 66,* 182–195.
Pinto, P. I. S., Singh, P. B., Condeca, J. B., Teodosio, H. R., Power, D. M., & Canario, A. V. M. (2006). ICI 182,780 has agonistic effects and synergizes with estradiol-17 beta in fish liver, but not in testis. *Reprod Biol Endocrinol, 4,* 67–78.
Potter, K. A., Bose, T., & Yamaguchi, A. (2005). Androgen-induced vocal transformation in adult female African clawed frogs. *J Neurophysiol, 94,* 415–428.
Prossnitz, E. R., Arterburn, J. B., Smith, H. O., Oprea, T. I., Sklar, L. A., & Hathaway, H. J. (2008). Estrogen signaling through the transmembrane G protein–coupled receptor GPR30. *Annu Rev Physiol, 70,* 165–190.
Qiu, J., Bosch, M. A., Tobias, S. C., Grandy, D. K., Scanlan, T. S., Ronnekleiv, O. K., et al. (2003). Rapid signaling of estrogen in hypothalamic neurons involves a novel G-protein-coupled estrogen receptor that activates protein kinase C. *J Neurosci, 23,* 9529–9540.
Quesada, A., & Micevych, P. (2008). Estrogen and progesterone modulate [35S] GTPgammaS binding to nociceptin receptors. *Neuroendocrinology, 88,* 35–42.
Reddy, V. V., Naftolin, F., & Ryan, K. J. (1973). Aromatization in the central nervous system of rabbits: effects of castration and hormone treatment. *Endocrinology, 92,* 589–594.
Remage-Healey, L., & Bass, A. H. (2004). Rapid, hierarchical modulation of vocal patterning by steroid hormones. *J Neurosci, 24,* 5892–5900.
Remage-Healey, L., & Bass, A. (2006). A rapid neuromodulatory role for steroid hormones in the control of reproductive behavior. *Brain Res, 1126,* 27–35.
Remage-Healey, L., & Bass, A. H. (2007). Plasticity in brain sexuality is revealed by the rapid actions of steroid hormones. *J Neurosci, 27,* 1114–1122.

Remage-Healey, L., & Bass, A. H. (2010). Estradiol interacts with an opioidergic network to achieve rapid modulation of a vocal pattern generator. *J Comp Physiol A Neuroethol Sens Neural Behav Physiol, 196,* 137–146.

Remage-Healey, L., Coleman, M. J., Oyama, R. K., & Schlinger, B. A. (2010). Brain estrogens rapidly strengthen auditory encoding and guide song preference in a songbird. *Proc Natl Acad Sci U S A, 107,* 3852–3857.

Remage-Healey, L., Dong, S. M., Chao, A., & Schlinger, B. A. (2012). Sex-specific, rapid neuroestrogen fluctuations and neurophysiological actions in the songbird auditory forebrain. *J Neurophysiol, 107*(6), 1621–1631.

Remage-Healey, L., Dong, S., Maidment, N. T., & Schlinger, B. A. (2011b). Presynaptic control of rapid estrogen fluctuations in the songbird auditory forebrain. *J Neurosci, 31,*(27) 10034–10038.

Remage-Healey, L., Maidment, N. T., & Schlinger, B. A. (2008). Forebrain steroid levels fluctuate rapidly during social interactions. *Nat Neurosci, 11,* 1327–1334.

Remage-Healey, L., Saldanha, C. J., & Schlinger, B. A. (2011a). Estradiol synthesis and action at the synapse: evidence for "synaptocrine" signaling. *Front Neuroendocr Sci,* in press.

Rhen, T., & Crews, D. (2002). Variation in reproductive behaviour within a sex: neural systems and endocrine activation. *J Neuroendocrinol, 14,* 517–531.

Riters, L. V., Schroeder, M. B., Auger, C. J., Eens, M., Pinxten, R., & Ball, G. F. (2005). Evidence for opioid involvement in the regulation of song production in male European starlings (Sturnus vulgaris). *Behav Neurosci, 119,* 245–255.

Rohmann, K. N., Schlinger, B. A., & Saldanha, C. J. (2007). Subcellular compartmentalization of aromatase is sexually dimorphic in the adult zebra finch brain. *Dev Neurobiol, 67,* 1–9.

Ronnekleiv, O. K., & Kelly, M. J. (2005). Diversity of ovarian steroid signaling in the hypothalamus. *Front Neuroendocrinol, 26,* 65–84.

Rose, J. D., Marrs, G. S., & Moore, F. L. (1998). Rapid, corticosterone-induced disruption of medullary sensorimotor integration related to suppression of amplectic clasping in behaving roughskin newts (Taricha granulosa). *Horm Behav, 34,* 268–282.

Ruf, K., & Steiner, F. A. (1967). Steroid-sensitive single neurons in rat hypothalamus and midbrain—indentification by microelectrophoresis. *Science, 156,* 667–669.

Saldanha, C. J., & Coomaralingam, L. (2005). Overlap and co-expression of estrogen synthetic and responsive neurons in the songbird brain—a double-label immunocytochemical study. *Gen Comp Endocrinol, 141,*(1) 66–75.

Saldanha, C. J., Remage-Healey, L., & Schlinger, B. A. (2011). Synaptocrine signaling: steroid synthesis and action at the synapse. *Endocr Rev, 32*(4), 532–549.

Saldanha, C. J., Tuerk, M. J., Kim, Y. H., Fernandes, A. O., Arnold, A. P., & Schlinger, B. A. (2000). Distribution and regulation of telencephalic aromatase expression in the zebra finch revealed with a specific antibody. *J Comp Neurol, 423,* 619–630.

Salvante, K. G., Racke, D. M., Campbell, C. R., & Sockman, K. W. (2010). Plasticity in singing effort and its relationship with monoamine metabolism in the songbird telencephalon. *Dev Neurobiol, 70,* 41–57.

Sanford, S. E., Lange, H. S., & Maney, D. L. (2010). Topography of estradiol-modulated genomic responses in the songbird auditory forebrain. *Dev Neurobiol, 70,* 73–86.

Schaefer, J. E., & Zakon, H. H. (1996). Opposing actions of androgen and estrogen on in vitro firing frequency of neuronal oscillators in the electromotor system. *J Neurosci, 16,* 2860–2868.

Schlinger, B., & Brenowitz, E. A. (2008). Neural and hormonal control of birdsong. In: Pfaff, D. W. (Ed.), *Hormones, Brain and Behavior*. St. Louis, MO: Elsevier, pp. 897–941.

Schlinger, B. A., & Arnold, A. P. (1992). Circulating estrogens in a male songbird originate in the brain. *Proc Natl Acad Sci U S A, 89,* 7650–7653.

Schultz, K. N., von Esenwein, S. A., Hu, M., Bennett, A. L., Kennedy, R. T., & Musatov, S., et al. (2009). Viral vector-mediated overexpression of estrogen receptor-alpha in striatum enhances the estradiol-induced motor activity in female rats and estradiol-modulated GABA release. *J Neurosci, 29,* 1897–1903.

Simerly, R. B. (2002). Wired for reproduction: organization and development of sexually dimorphic circuits in the mammalian forebrain. *Annu Rev Neurosci, 25,* 507–536.

Simpson, H. B., & Vicario, D. S. (1991). Early estrogen treatment of female zebra finches masculinizes the brain pathway for learned vocalizations. *J Neurobiol, 22,* 777–793.

Sinchak, K., & Micevych, P. E. (2001). Progesterone blockade of estrogen activation of mu-opioid receptors regulates reproductive behavior. *J Neurosci, 21,* 5723–5729.

Sisneros, J. A. (2004). Steroid-dependent auditory plasticity leads to adaptive coupling of sender and receiver. *Science, 305,* 404–407.

Sisneros, J. A. (2009). Adaptive hearing in the vocal plainfin midshipman fish: getting in tune for the breeding season and implications for acoustic communication. *Integr Zool, 4,* 33–42.

Sisneros, J. A., Forlano, P. M., Knapp, R., & Bass, A. H. (2004). Seasonal variation of steroid hormone levels in an intertidal-nesting fish, the vocal plainfin midshipman. *Gen Comp Endocrinol, 136,* 101–116.

Soma, K. K., Schlinger, B. A., Wingfield, J. C., & Saldanha, C. J. (2003). Brain aromatase, 5 alpha-reductase, and 5 beta-reductase change seasonally in wild male song sparrows: relationship to aggressive and sexual behavior. *J Neurobiol, 56,* 209–221.

Soma, K. K., Sullivan, K. A., Tramontin, A. D., Saldanha, C. J., Schlinger, B. A., & Wingfield, J. C. (2000). Acute and chronic effects of an aromatase inhibitor on territorial aggression in breeding and nonbreeding male song sparrows. *J Comp Physiol A-Sens Neural Behav Physiol, 186,* 759–769.

Soma, K. K., Tramontin, A. D., Featherstone, J., Brenowitz, E. A. (2004). Estrogen contributes to seasonal plasticity of the adult avian song control system. *J Neurobiol, 58,* 413–422.

Steimer, T., & Hutchison, J. B. (1981) Androgen increases formation of behaviorally effective estrogen in dove brain. *Nature, 292,* 345–347.

Taziaux, M., Keller, M., Bakker, J., & Balthazart, J. (2007). Sexual behavior activity tracks rapid changes in brain estrogen concentrations. *J Neurosci, 27,* 6563–6572.

Teyler, T. J., Vardaris, R. M., Lewis, D., & Rawitch, A. B. (1980) Gonadal-steroids—effects on excitability of hippocampal pyramidal cells. *Science, 209,* 1017–1019.

Thomas, P., Pang, Y., Filardo, E. J., & Dong, J. (2005). Identity of an estrogen membrane receptor coupled to a G protein in human breast cancer cells. *Endocrinology, 146,* 624–632.

Thompson, C. W., & Moore, M. C. (1992). Behavioral and hormonal correlates of alternative reproductive strategies in a polygynous lizard—tests of the relative plasticity and challenge hypotheses. *Horm Behav, 26,* 568–585.

Tramontin, A. D., Wingfield, J. C., & Brenowitz, E. A. (2003). Androgens and estrogens induce seasonal-like growth of song nuclei in the adult songbird brain. *J Neurobiol, 57,* 130–140.

Tremere, L. A., Jeong, J. K., & Pinaud, R. (2009). Estradiol shapes auditory processing in the adult brain by regulating inhibitory transmission and plasticity-associated gene expression. *J Neurosci, 29,* 5949–5963.
Tremere, L. A., & Pinaud, R. (2011). Brain-generated estradiol drives long-term optimization of auditory coding to enhance the discrimination of communication signals. *J Neurosci, 31,* 3271–3289.
Vasudevan, N., Kow, L. M., & Pfaff, D. (2005). Integration of steroid hormone initiated membrane action to genomic function in the brain. *Steroids, 70,* 388–396.
Veney, S. L., & Wade, J. (2005). Post-hatching syrinx development in the zebra finch: an analysis of androgen receptor, aromatase, estrogen receptor alpha and estrogen receptor beta mRNAs. *J Comp Physiol A Neuroethol Sens Neural Behav Physiol, 191,* 97–104.
Wade, J., Buhlman, L., & Swender, D. (2002). Post-hatching hormonal modulation of a sexually dimorphic neuromuscular system controlling song in zebra finches. *Brain Res, 929,* 191–201.
Walters, M. J., McEwen, B. S., & Harding, C. F. (1988) Estrogen receptor levels in hypothalamic and vocal control nuclei in the male zebra finch. *Brain Res, 459,* 37–43.
Wang, Y., Brzozowska-Prechtl, A., & Karten, H. J. (2010). Laminar and columnar auditory cortex in avian brain. *Proc Natl Acad Sci U S A, 107,* 12676–12681.
Watson, J. T., & Adkins-Regan, E. (1989). Testosterone implanted in the preoptic area of male japanese quail must be aromatized to activate copulation. *Horm Behav, 23,* 432–447.
Wilczynski, W., & Lynch, K. S. (2011). Female sexual arousal in amphibians. *Horm Behav, 59,* 630–636.
Wingfield, J. C. (1985). Short-term changes in plasma-levels of hormones during establishment and defense of a breeding territory in male song sparrows, melospiza-melodia. *Horm Behav, 19,* 174–187.
Wingfield, J. C. (2005). A continuing saga: the role of testosterone in aggression. *Horm Behav, 48,* 253–255.
Wingfield, J. C., Hegner, R. E., Dufty, A. M., & Ball, G. F. (1990). The challenge hypothesis—theoretical implications for patterns of testosterone secretion, mating systems, and breeding strategies. *Am Nat, 136,* 829–846.
Wong, M., & Moss, R. L. (1992). Long-term and short-term electrophysiological effects of estrogen on the synaptic properties of hippocampal CA1 neurons. *J Neurosci, 12,* 3217–3225.
Wu, K. H., Tobias, M. L., & Kelley, D. B. (2001). Estrogen and laryngeal synaptic strength in Xenopus laevis: opposite effects of acute and chronic exposure. *Neuroendocrinology, 74,* 22–32.
Wu, K. H., Tobias, M. L., & Kelley, D. B. (2003a). Estrogen receptor expression in laryngeal muscle in relation to estrogen-dependent increases in synapse strength. *Neuroendocrinology, 78,* 72–80.
Wu, K. H., Tobias, M. L., Thornton, J. W., & Kelley, D. B. (2003b). Estrogen receptors in Xenopus: duplicate genes, splice variants, and tissue-specific expression. *Gen Comp Endocrinol, 133,* 38–49.
Yovanof, S., & Feng, A. S. (1983). Effects of estradiol on auditory evoked responses from the frog's auditory midbrain. *Neurosci Lett, 36,* 291–297.
Zornik, E., & Kelley, D. B. (2011). A neuroendocrine basis for the hierarchical control of frog courtship vocalizations. *Front Neuroendocrinol, 32,* 353–366.

Chapter 26

Rapid Effects of Estradiol on Behavioral and Physiological Responses Related to Reproduction in Male Goldfish

Richmond R. Thompson and Lisa A. Mangiamele

INTRODUCTION

That sex steroids can rapidly affect cell physiology and brain function is now widely accepted, but the behavioral consequences of such actions, most of which are associated with estradiol (E2), and the social contexts in which they occur have thus far only been demonstrated in a few species. Estradiol rapidly stimulates female receptivity (lordosis) in rats via actions in the hypothalamus (Dewing et al., 2007), male sexual behavior in Japanese quail, mice, and rats via actions in the preoptic area (Cross & Roselli, 1999; Cornil et al., 2006b; Taziaux et al., 2007), the production of vocalizations related to social signaling in two related teleost fish, toadfish and plainfin midshipmen, via actions in the hindbrain (Remage-Healey & Bass, 2004, 2005), and song preferences in zebra finches via actions in the telencephalon (Remage-Healey et al., 2010). The receptor mechanisms that mediate most of these behavioral effects have not yet been determined, although rapid effects of E2 on lordosis depend on the activation of membrane versions of estrogen receptor alpha (ERα), which themselves interact with metabotropic glutamate receptors (Dewing et al., 2007). Additionally, E2 may interact with opioid receptors in plainfin midshipmen to affect social vocalizations (Remage-Healey & Bass, 2010).

Recently, it has become clear that the rapid effects of E2 on behavior are associated with socially induced changes in local steroid levels in the brain that are independent of changes in circulating testosterone (T) or E2 in several avian species, including Japanese quail (Cornil et al., 2009) and zebra finches (Remage-Healey et al., 2008). In these species, local elevations in E2 appear to depend on rapid changes in aromatase activity that are induced by neurotransmitter release (Cornil et al., 2005, 2006a; Balthazart et al., 2006). Such a mechanism would make it possible to not only initiate rapid estrogenic signaling, but also to rapidly turn it off and is thus consistent with the short time course of some of estrogen's behavioral effects (reviewed in Cornil et al., 2006a).

Yet, it is also possible that steroids produced by the gonads, in particular the elevation of androgens, which often occurs in response to sexual and aggressive stimuli, could underlie rapid, nongenomic effects of E2 on behavior in some species by increasing the substrate available for aromatization. The behavioral effects of these socially induced androgen surges were originally difficult to understand within the classic steroid model, according to which steroids exclusively affect gene transcription, because the social interaction would likely be over before such a mechanism could produce enough changes in protein levels to influence the individual's behavior. Therefore, it has often been suggested that such surges may influence future social behaviors, and recent studies have indeed shown that rapid elevations in T in response to agonistic stimuli do enhance the expression of subsequent aggressive behaviors in some species (Trainor et al., 2004; Oyegbile & Marler, 2005; Oliveira et al., 2009). However, nongenomic behavioral and/or physiological effects produced by rapid elevations in circulating T, and thus local E2, may also influence an animal's behavior during social interactions. Such a mechanism would be particularly effective if stimuli that immediately precede the social interaction could drive the T surges, in which case they could prime the animal for the impending interaction and, if the elevations persist, sustain increased responsiveness for the duration of the interaction. Increased responsiveness may then dissipate as hormone levels return to normal after the interaction, although it is also possible that responsiveness could be more rapidly decreased by local mechanisms, such as aromatase inhibition, if the social context changes dramatically.

Studies in teleost fish are beginning to indicate that socially induced fluctuations in circulating sex steroids can, in fact, modulate ongoing social interactions through rapid, nongenomic mechanisms in ways that are behaviorally and ecologically important. This chapter discusses examples in several fish species in which rapid surges in circulating androgens occur in both agonistic and sexual contexts and, in some cases, in which those elevations are associated with rapid neurophysiological and behavioral effects. It then describes our recent experiments in goldfish (*Carassius auratus*), a competitively breeding fish, that begin to elucidate the mechanisms by which surges in T, via its conversion to E2, could rapidly influence the processing of social signals, the regulation of sexual behavior, and the reproductive physiology of males. Together, these studies help us understand the significance of rapid estrogen effects in activating responses that are important in ensuring reproductive success in a competitive social environment.

STEROID RESPONSES TO SOCIAL STIMULI IN FISH AND THEIR POTENTIAL ROLE IN RAPID BEHAVIORAL MODULATION

As in other vertebrates, levels of circulating androgens in teleosts are elevated in response to social challenges (Oliveira et al., 2002). For example, in *Astatotilapia burtoni,* a highly aggressive cichlid, males can rapidly and reversibly switch from a subordinate, nonterritorial behavioral phenotype to a dominant, territorial one, depending on their social environment. Circulating androgens rapidly increase in males during their ascent from subordinate to dominant social status. Within 30 minutes of perceiving an opportunity to ascend in status, both 11-ketotestosterone (11-KT), a nonaromatizable androgen found in teleosts, and the rate of territorial behaviors are significantly increased in previously subordinate males (Maruska & Fernald, 2010), suggesting that rapid androgen responses may be a mechanism that allows ascending fish to immediately begin engaging in agonistic interactions. Alternatively, socially induced androgen elevations may not directly affect overt aggression, but rather the production of social signals used during the interaction, the processing of signals produced by the other individual, or even immediate cognitive processes that will influence future interactions between the same individuals (see Oliveira et al., 2002, for a more complete discussion of how androgens may rapidly influence attentional processes associated with social learning). In fact, socially induced elevations in androgens have been causally linked to rapid changes in the production of signals used for social communication in two teleost species. Male gulf toadfish use vocalizations produced by the swimbladder for agonistic and sexual communication, and levels of 11-KT increase when they hear playbacks of those vocalizations, as do the rates at which they produce vocalizations (Remage-Healey & Bass, 2005). This finding suggested that socially induced elevations in 11-KT may underlie increased call rates in toadfish, and subsequent experiments demonstrated that elevating 11-KT (which cannot be aromatized into estradiol), but not T or E2, rapidly stimulated the production of calls and increased neuronal activity within the hindbrain neurons that directly drive calling behavior (Remage-Healey & Bass, 2006). In a related species, the plainfin midshipman, 11-KT, but not T, had a similar rapid effect on vocal motor circuitry (Remage-Healey & Bass, 2004). In contrast, T, but not 11-KT, rapidly modulates neuronal responses in the same hindbrain neurons in nonterritorial "sneaker" males, which mimic female midshipmen and steal fertilizations from courting males, as well as in females (Remage-Healey & Bass, 2007). T is the dominant circulating androgen in these phenotypes, although it is not yet known if social cues rapidly modulate its levels. Estradiol rapidly affects neuronal responses in the hindbrain in all phenotypes in this species, although the source of endogenous E2 in the hindbrain in the different phenotypes is also not yet known. Together, these studies indicate that sex steroids have rapid effects on social signaling mechanisms, and those effects likely occur, at least in some cases, in response to changes in circulating levels of sex steroids that are driven by exposure to agonistic stimuli.

In several species of fish, as in other vertebrates, sexual stimuli also influence levels of circulating androgens (e.g., Olsen & Liley, 1993; Waring et al., 1996). In male goldfish, sexual stimuli associated with spawning selectively increase T, but not 11-KT (Kobayashi et al., 1986). Therefore, we wondered if those surges might be able to produce rapid effects on behavior in this species, and took the first step toward making that determination by testing if experimentally induced elevations in T could rapidly stimulate behavioral responses toward potential mates. We did not expect such elevations to be necessary for the display of courtship and/or sexual behaviors, because a long history of research in behavioral neuroendocrinology using castration and hormone replacement has demonstrated that constant levels of circulating steroids, as long as they reach a certain threshold, can fully activate many sexual behaviors, in some cases even if the levels achieved by hormone replacement, are lower than those typically found in circulation (e.g., Grunt & Young, 1953; D'Occhio & Brooks, 1982). In fact, such studies have often raised questions about whether sexually induced elevations above such thresholds have any behavioral consequences. Furthermore, in fish 11-KT is the more potent androgen and is, in fact, completely sufficient, when delivered in constant doses over a long period of time, to support the expression of male-typical courtship behaviors in goldfish. It can even induce male-typical behaviors in female goldfish (Kobayashi & Nakanishi, 1999). Thus, we hypothesized that whereas a chronic elevation of 11-KT likely acts to sculpt the neuronal circuits that make male-typical sexual behavior possible, socially induced rapid elevations in T may act within those circuits to influence processes that increase the likelihood that a male will successfully reproduce in a competitive environment.

In goldfish, males engage in a scramble competition for access to ovulated females. Female goldfish ovulate at night, and subsequent spawning typically occurs in dim light conditions. When they spawn, many males chase, nudge, and spawn repeatedly with individual females. Although most work in goldfish has focused on the role that sex pheromones play in stimulating male reproductive responses, most notably on the ability of progesterone metabolites to prime male endocrine responses and prostaglandins to stimulate male courtship behaviors (Sorensen et al., 1989; Poling et al., 2001), visual cues also play an important role in sexual communication, as demonstrated by several lines of evidence. Males mate even when made anosmic, and they still preferentially follow ovulating versus nonovulating females (Partridge et al., 1976). Males also prefer to approach females more often than males in two-choice preference tests in which only visual cues are available (Thompson et al., 2004). Therefore, we hypothesized that the female-induced T surges that occur in males, likely in response to the preovulatory pheromone 17α,20β-dihydroxy-4-pregnen-3-one (17,20βP), which has been shown to increase levels of circulating gonadotropins (Sorensen et al., 1989), may rapidly influence responsiveness to female visual stimuli. This enhanced responsiveness would increase the likelihood a male can localize and stay close to a potential mate during the group chases that typically precede spawning, thus increasing his chances of reproductive success.

Furthermore, we predicted that any such influences of T would be mediated by its local conversion to E2, a hypothesis based on earlier work in Gloria Callard's laboratory describing the distribution of aromatase within the goldfish brain (Gelinas & Callard, 1993, 1997). In particular, those studies demonstrated that aromatase levels are high not only in regions of the brain known to regulate male sexual behaviors in vertebrates, most notably the preoptic area, but also in neuronal fibers in the retina and the optic tectum, suggesting that locally produced E2 may influence synaptic processes in these visual processing areas. For example, local elevations of E2 in the retina following surges in circulating

T could rapidly influence overall light sensitivity and/or selectively facilitate the processing of particular spectral characteristics of females, which could help males more quickly locate potential mates in dim light mating conditions. Although the specific female visual cues to which males attend are unknown, as males and females are not obviously dimorphic in the visible light spectrum, goldfish have receptor cells that are maximally sensitive to light at 360 nm and can thus see in the ultraviolet (UV) range (Bowmaker et al., 1991; Palacios et al., 1998). Therefore, it is possible that they use UV patterns for sex discrimination. In fact, we have preliminary data that suggest females may have higher UV reflectances than males. Additionally, local elevations of estradiol in the optic tectum could influence visual orientation responses toward potential mates, perhaps by enhancing responsiveness to moving stimuli or to particular female visual features.

TESTOSTERONE AND ESTRADIOL RAPIDLY INFLUENCE VISUALLY MEDIATED SEXUAL BEHAVIOR IN MALE GOLDFISH

To test whether sex steroids could rapidly modulate behavioral responses to female visual stimuli in male goldfish, we set up a testing environment in which we could measure male approach responses toward just the visual stimuli of females. This involved putting our experimental males in the center compartment of a rectangular test tank with stimulus chambers on each of the long ends. Plexiglas sheets sealed the stimulus compartments and thus prevented all olfactory communication. There is no evidence that goldfish use auditory cues for social communication, and we were unable to detect any with hydrophonic equipment. Thus, males had to rely entirely on visual cues to gain any information about the stimulus animals that were placed in the side tanks.

In our tests, we injected the experimental males intraperitoneally and then placed them in the center of the test tank. We then allowed them to habituate for 5 to 15 minutes, depending on the experiment, and used the Limelight video tracking system to record the time they spent in close proximity (within 2 inches) to both Plexiglas partitions separating the central test compartment from the side stimulus compartments for the next 15 minutes. We then put a stimulus animal in the stimulus compartment on the side where the male had spent the least amount of time. Thus, our dependent variable of interest was the time spent in close proximity to that stimulus compartment while the stimulus fish was present minus the time spent in proximity to the same compartment before a stimulus animal was added. Using this experimental paradigm, the effects of our steroid manipulations would have to occur within 1 hour, a time course for steroid effects on behavior that are typically considered too rapid to be accounted for by traditional genomic mechanisms (Balthazart & Ball, 2006).

Because there is a great deal of variability between individual fishes in this behavioral measure, we wanted to conduct within-subjects tests. However, the potential for steroid manipulations to produce long-term effects made it impossible to use a standard, within-subject counterbalanced design. Therefore, we had two groups of fishes that were each tested on 2 days, typically 48 hours apart. All fish always received a control injection on day one. Half the fish received a control injection again on the second day, whereas the experimental group received a hormone injection. Thus, it was possible to compare changes across days in individual fish in response to our steroid manipulations while controlling for any potential order effects.

In our first studies using this test paradigm, we found that intraperitoneal injections of T at a dose (3 μg per fish) that produced elevations well within the physiological range for breeding males significantly increased the amount of time that males would spend in close proximity to females 30 to 45 minutes after the injections, but did not increase general activity (Lord et al., 2009). We observed similar effects several hours after the injections, suggesting they last for as long as levels of T remain elevated (unpublished data). We then repeated the same experiment using males as stimuli, and found that T injections did not stimulate approach responses toward the male stimulus. We had previously shown that males selectively orient toward females over males when only visual cues are present, and that females treated chronically with androgens (T and 11-KT capsules) display a similar, masculine preference (Thompson et al., 2004). Thus, we already knew that males can use visual cues to detect females, and that chronic elevations of androgens masculinize responses toward those cues. We now also know that T can rapidly and selectively stimulate approach responses toward female visual stimuli. As mentioned, males and females are not visibly dimorphic, although males do develop tubercles, which are tiny bumps along the gill plates, when in full reproductive condition, and females appear swollen with eggs when ready to ovulate. It is possible that those visual cues are being used for visual sexual discrimination, and/or that there are differences in UV reflectance patterns or differences in behavior that allow males to selectively orient toward females, and that T selectively enhances the processing of those cues. Of course, it is also possible that T simply increases the motivation to approach females once those cues have been processed in the ascending visual pathways.

We next tested if T's behavioral effects depend on its conversion to E2. We injected males with a water-soluble form of estradiol at the same dose that we used for T, although the biologically active form of E2 in that dose is much smaller (3 μg water soluble E2 equals 0.14 μg of actual E2). We did observe elevations in circulating E2 after our injections, but it is impossible to relate those increases to physiological levels, as the circulating E2 would normally be irrelevant in males relative to the levels produced locally by aromatase in the brain, which is entirely unknown in this species. E2, like T, significantly increased approach responses towards the visual cues of females 30 to 45 minutes after the injections, and could even do so within 10 to 25 minutes of the injection (Fig. 26–1).

We then performed the critical experiment to see if blocking aromatase activity would block T's ability to stimulate visual approach responses. These tests were done in late winter and early spring, before the gonads had fully recrudesced, and involved a slight modification in our behavioral protocol, as

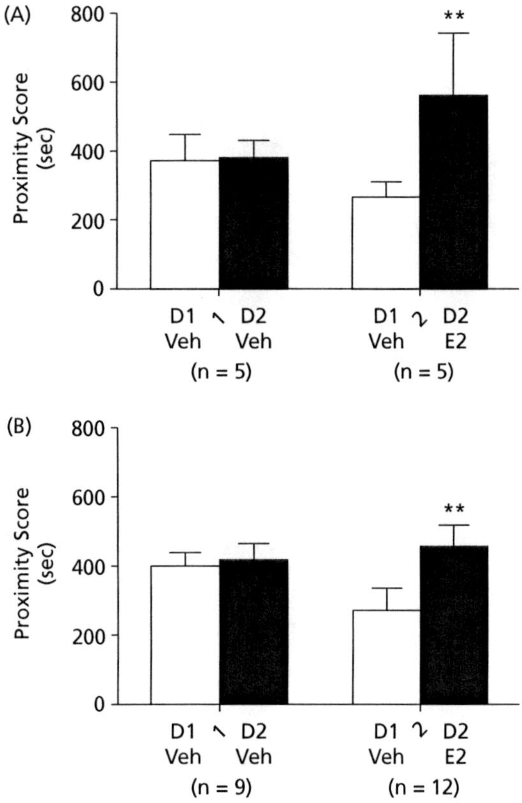

Figure 26–1. Estradiol Rapidly Stimulates Visually Guided Approach Behavior in Male Goldfish. Mean ± SEM of the time spent near a stimulus female after injections on D1, the first day of testing when all fish were injected with the vehicle, and D2, the second day of testing when half the fish again received vehicle injections and half received estradiol (E2) injections (*A*, 30–45 minutes after injections; *B*, 10–25 minutes after injections). **Within-groups difference across test days, $p < .01$, one-tailed. Reprinted with permission from Lord, L. D., et al. (2009). Rapid steroid influences on visually guided sexual behavior in male goldfish. *Horm Behav, 56*, 519–526.

fish now had to be injected twice, first with the aromatase inhibitor fadrozole (FAD) or vehicle, and then with T or vehicle. We first confirmed that T would still stimulate visual approach responses if the injections were given 15 minutes after an injection of saline (Fig. 26–2A). Next, we determined the effects of FAD on visual approach responses in fish that were not steroid injected. We chose to do our tests early in the year, before fish had come into full breeding condition, so that concentrations of circulating steroids would not be near peak levels (Pasmanik & Callard, 1988). Additionally, we kept males in all-male tanks and isolated from females before the tests, which further reduces T levels (Stacey et al., 2001b). We took these measures because we wanted to isolate the potential

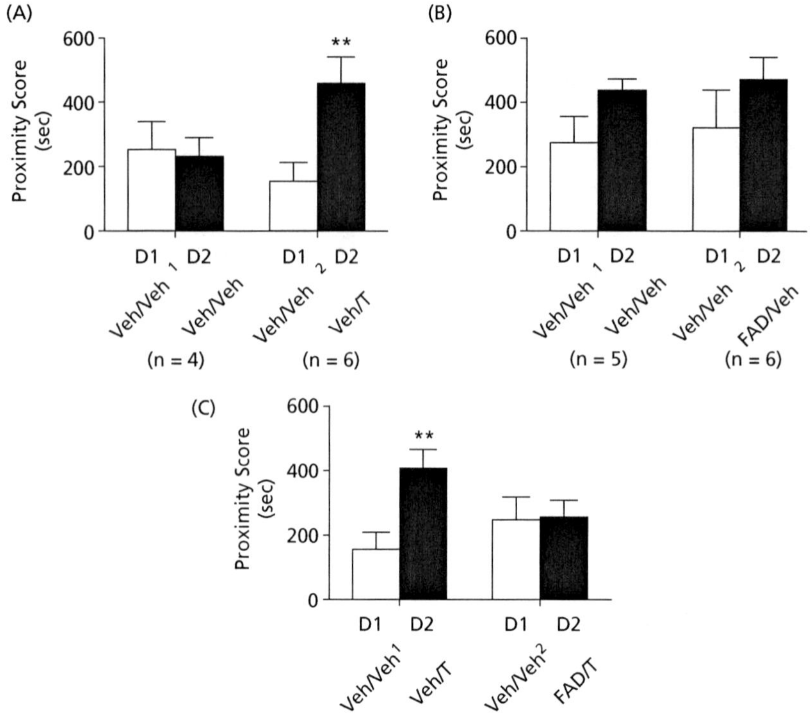

Figure 26–2. Testosterone-Induced Social Approach Behavior Is Rapidly Blocked by Fadrozole (FAD), an Aromatase Inhibitor, in Male Goldfish. Mean ± SEM of the time in proximity to a stimulus female 45 to 60 minutes after an initial injection and 30 to 45 minutes after a second injection on D1, the first day of testing, and D2, the second day of testing. In the first experiment (*A*), all fish were injected twice with vehicle on D1. Control fish again received two vehicle injections on D2, whereas experimental fish were injected with vehicle then testosterone (T). In the second experiment (*B*), all fish received two vehicle injections on D1, then control fish were again injected twice with vehicle on D2, whereas experimental fish were injected with FAD then vehicle. In the third experiment (*C*), all fish were injected twice with vehicle on D1, then positive control fish were injected with vehicle followed by T on D2, whereas experimental fish were injected with FAD then T. **Within-groups difference across test days, $p < .01$, one-tailed. Reprinted with permission from Lord, L. D., et al. (2009). Rapid steroid influences on visually guided sexual behavior in male goldfish. *Horm Behav, 56,* 519–526.

effects of FAD on the acute T elevations induced by our exogenous injections. Indeed, FAD did not decrease social approach responses in fish that were not injected with T (Fig. 26–2B). However, FAD injections given 15 minutes before T injections completely blocked T's ability to stimulate visually-guided sexual approach responses 30 to 45 minutes later (Fig. 26–2C). Together, these experiments demonstrated that E2 is sufficient to stimulate visually guided sexual approach responses and that aromatase activity is necessary for T to produce its effects, thus indicating that locally produced elevations in E2 likely mediate the rapid effects of T. However, we cannot fully resolve whether the behaviorally relevant elevations occur within the brain or in the testes, which could release a second hormonal signal in response to E2 that itself goes to the brain to induce the behavioral effect. Furthermore, we do not yet know the natural contexts in which these effects typically occur. We suspect the T elevations that occur in response to sexual stimuli may influence subsequent spawning through rapid E2 mechanisms. We will test this hypothesis by determining if the female stimuli that induce T surges, particularly the preovulatory pheromones released the night before spawning, also increase visual approach responses, and then if FAD treatments rapidly block any such behavioral effects. Finally, it will be important to determine if any behavioral effects induced by preexposure to female stimuli in intact males that exhibit T surges disappear in males preexposed to the same stimuli while peripheral hormone levels are held constant.

It should be noted that although we used full spectrum lighting for all of our experiments, the Plexiglas barriers that separated the experimental males from females block all light wavelengths below 400 nm. Thus, it is questionable if the males in our tests perceived any UV reflectance patterns from the stimulus females. However, the UV sensitive cones in goldfish have a broad tuning range and do respond to wavelengths above 400 nm (Palacios et al., 1998), and we used high light levels, so the UV system could have been activated in these tests. Thus, it remains possible that T selectively influences the transmission of social information through the UV-sensitive, ascending visual pathway. We also cannot yet determine if T might additionally (or instead) affect the motivation to approach females once they are perceived. We are currently conducting behavioral experiments to test whether T or E2 affects visual detection thresholds and if UV reflectance is important for sexual communication, as well as electrophysiological experiments to test if acute elevations in T and/or E2 affect retinal responses to light, including isolated UV wavelengths.

RAPID PHYSIOLOGICAL RESPONSES TO E2 ACCOMPANY BEHAVIORAL RESPONSES IN GOLDFISH

In the process of conducting these experiments, we also noted that expressible milt (sperm and seminal fluid) can be rapidly modulated by T. To test this possibility in a systematic fashion, we selected focal male goldfish from tanks in which they were held in small, same-sex groups. We first measured the volume of milt that could be manually expressed, then injected males intraperitoneally with either T or vehicle and returned them to their home tanks. We measured milt volume again 1 hour later. To test whether our treatments rapidly influenced the volume of milt expressed, we conducted an analysis of covariance with treatment (T or vehicle) as a between groups factor and preinjection milt volume as a covariate to control for individual differences in starting rates of milt expression. We found that, compared with controls, males injected with T had higher postinjection milt volumes, suggesting that T can rapidly influence rates of sperm production and/or release on a timescale consistent with that of its effect on social behavior (Mangiamele & Thompson, in press). Furthermore, we found that such T-induced increases in milt expression are dependent on aromatase activity, because they disappeared if fish were treated with an aromatase inhibitor before the T injections, and rapid increases in milt could also be elicited by E2 conjugated to BSA, which cannot cross cell membranes (Mangiamele & Thompson, in press). These results are intriguing because they suggest that T, via its conversion to E2, can have immediate effects on physiological as well as behavioral processes that affect reproductive output, and that those effects are dependent on a membrane estrogen receptor.

Although work in our laboratory is ongoing, the mechanism mediating this rapid increase in expressible milt is currently unknown. For example, our experiments do not allow us to conclude whether T-mediated effects on milt output are a result of steroid action at the level of the gonads, or via effects on the central nervous system. mRNAs from estrogen receptors α and β have been localized in goldfish brain and testes (Tchoudakova et al., 1999; Choi & Habibi, 2003), suggesting that both mechanisms are plausible. In addition, GPR30 (also known as GPER), a G protein–coupled estrogen receptor that localizes to the cell membrane, has been identified in the brain and gonads of some fish species (e.g., zebrafish, Atlantic croaker), suggesting that a nongenomic mechanism of steroid action mediated by this receptor is also possible at both of these sites (Pang et al., 2008; Liu et al., 2009). In goldfish, contraction of the sperm ducts can be induced by stimulating a brain pathway running from the olfactory bulbs to the preoptic area, brainstem, and rostral spinal cord (Demski & Dulka, 1984; Dulka & Demski, 1986). Although little is known about the specific innervation of the testes and sperm ducts in fish, rapid T/E2 influences on a brain–gonad pathway could explain rapid steroid effects on expressible milt. Although it is questionable if peripherally delivered E2-BSA, which did increase expressible milt in our experiments, can reach the brain, other molecules that do not readily reach the brain when injected peripherally in mammals appear to do so in fish (Santangelo & Bass, 2006), so this mechanism remains viable. Alternatively, E2's rapid effects may depend on local actions in the testes, where it could directly influence the contractile activity of the sperm ducts. The sperm ducts in teleosts contain smooth muscle cells (Weisel, 1949; Demski & Hornby, 1982) that contract in response to application of neurohypophyseal and urophyseal hormones (Berlind, 1972; Billard, 1977), although nothing is known about the influence of sex steroids on these muscles. In rats, however, T does influence the contractile properties of the seminiferous tubules (Farr & Ellis, 1980), and can rapidly (within 6 minutes) induce contractile bursts in the bulbospongiosus muscle of the penis (Sachs & Leipheimer, 1988). Thus, it is plausible that T could rapidly influence muscle excitability to facilitate the release of milt from the ducts during spawning in goldfish.

A potential adaptive value of rapid steroid effects on milt release, like rapid effects on visual responsiveness, would be to increase mating success in a competitive environment. In externally fertilizing species, an individual male's reproductive success depends not only on his access to females, but also on his sperm quantity and quality (Ball & Parker, 1996), characteristics that can change in response to the social environment. Exposure to female sexual stimuli, in particular, increases milt volume (and thus total sperm numbers), sperm motility, and paternity in competitive mating tests in goldfish (DeFraipont & Sorensen, 1993; Zheng et al., 1997). Our present results suggest that those effects likely depend, at least in part, on T surges induced by female sexual stimuli and the rapid activation of membrane estrogen receptors once the T has been aromatized. We hypothesize that this signaling mechanism is most likely initiated by exposure to the preovulatory pheromone 17,20βP, which rapidly increases gonadotropin (GtH-II) in male goldfish (Dulka et al., 1987) and thus, presumably, T, although the latter was not measured in that study (Fig. 26–3, *left side*). Interestingly, other sexual stimuli can also increase levels of expressible milt, but through a different, apparently GtH-independent mechanism. For example, exposing males to spawning females in which sexual activity has been induced by prostaglandin F2α (PGF2α) increases expressible milt, even in males without an

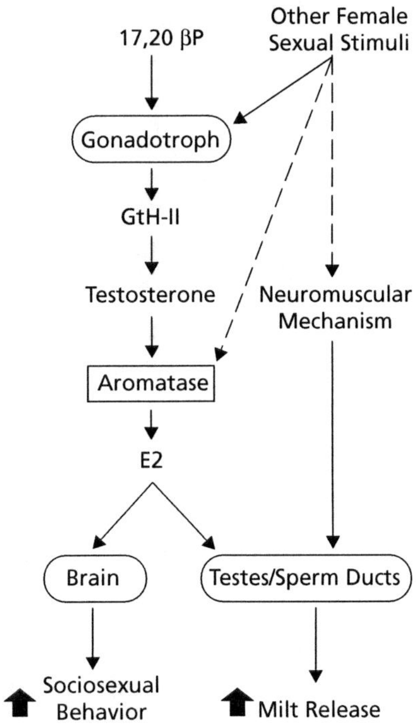

Figure 26–3. Schematic representation of potential pathways through which sexual stimuli, particularly the preovulatory pheromone 17α,20β-dihydroxy-4-pregnen-3-one (17,20 βP), may increase behavioral and physiological responses in male goldfish through rapid estrogenic signaling mechanisms. See text for a detailed description of those pathways.

intact pituitary and thus without a surge in GtH-II (Zheng & Stacey, 1996). Thus, there may be mechanisms through which social stimuli can increase expressible milt that do not depend on increases in circulating T, perhaps through neural inputs to the testes and/or sperm ducts that directly increase expressible milt (see Fig. 26–3, *right side*). Such a mechanism could still involve estrogenic signaling mechanisms if those inputs can rapidly influence aromatase kinetics and thus increase local E2 levels. Alternatively, it is possible that exposure to PGF2α-treated females increases circulating T (and as a consequence expressible milt) in male goldfish through mechanisms other than a gonadotropin-driven system. Unfortunately, T levels were not measured in those studies, so this determination cannot yet be made. Clearly, determining the mechanisms through which sex steroids mediate the effects of social stimuli on expressible milt in this species will be an interesting avenue of future research that will likely shed new light on the functional roles of rapid estrogen signaling mechanisms in sperm competition and, ultimately, reproductive success.

CONCLUSIONS

In conclusion, we have shown that E2 derived from T aromatization can rapidly modulate behavioral and physiological responses to sexual stimuli in male goldfish, effects that are likely to increase the reproductive success of males that exhibit elevations of T in spawning contexts. We are currently testing this model in studies that will determine if preexposure to the sexual stimuli that increase T levels also increase behavioral responsiveness to the visual cues of females, and if inhibiting endogenous aromatase blocks any such effects, as well as if the effects disappear in animals maintained on constant levels of circulating T. If so, it would demonstrate that socially induced steroid surges rapidly modulate processes that increase an animal's competitive abilities in complex social and ecological environments. In fact, we think such surges likely produce rapid, coordinated effects on multiple central and peripheral processes important for reproduction, perhaps enhancing sensory detection abilities, increasing motivational drive, amplifying motor output related to social signaling, courtship, and/or copulation, and simultaneously affecting peripheral effector systems that support those behaviors and ultimately increase the probability of fertilization. Although such influences are not apt to dramatically influence the probability of whether a particular behavior will occur or not as classical genomic actions do, particularly in standard laboratory tests in which individual males and females are often tested together, they are nonetheless likely critical in natural breeding contexts in which tremendous intrasexual and intersexual pressures affect an animal's ability to successfully reproduce. We therefore believe future studies on rapid steroid effects will help us better understand how animals have adapted in response to the pressures of sexual selection. Undoubtedly, such studies will also continue to elucidate novel mechanisms through which steroids affect behavior, including an ever-increasing array of receptor signaling pathways that may allow steroids to have strikingly diverse effects on behavior.

REFERENCES

Ball, M. A., & Parker, G. A. (1996). Sperm competition games: external fertilization and "adaptive" infertility. *J Theoret Biol, 180,* 141–150.

Balthazart, J., & Ball, G. F. (2006). Is brain estradiol a hormone or a neurotransmitter? *Trends Neurosci, 29,* 241–249.

Balthazart, J., et al. (2006). Rapid changes in production and behavioral action of estrogens. *Neuroscience, 138,* 783–791.

Berlind, A. (1972). Teleost caudal neurosecretory system: sperm duct contraction induced by urophyseal material. *J Endocrinol, 52,* 567–574.

Billard, R. (1977). Effect of various hormones on sperm release in the hypophysectomized goldfish. *IRCS Medical Science Library Compendium, 5,* 188.

Bowmaker, J. K., et al. (1991). Ultraviolet-sensitive cones in the goldfish. *Vis Res, 31,* 349–352.

Choi, C., & Habibi, H. (2003). Molecular cloning of estrogen receptor α and expression pattern of estrogen receptor subtypes in male and female goldfish. *Mol Cell Endocrinol, 204,* 169–177.

Cornil, C. A., et al. (2005). Rapid decreases in preoptic aromatase activity and brain monoamine concentrations after engaging in male sexual behavior. *Endocrinology, 146,* 3809–3820.

Cornil, C. A., et al. (2006a). Functional significance of the rapid regulation of brain estrogen action: where do the estrogens come from? *Brain Res, 1126,* 2–26.

Cornil, C. A., et al. (2006b). Estradiol rapidly activates male sexual behavior and affects brain monoamine levels in the quail brain. *Behav Brain Res, 166,* 110–123.

Cornil, C. A., et al. (2009). Are rapid changes in gonadal testosterone release involved in the fast modulation of brain estrogen effects? *Gen Comp Endocrinol, 163,* 298–305.

Cross, E., & Roselli, C. E. (1999). 17 beta-estradiol rapidly facilitates chemoinvestigation and mounting in castrated male rats. *Am J Physiol, Reg Int Comp Physiol, 276,* R1346–R1350.

D'Occhio, M. J., & Brooks, D. E. (1982) Threshold of plasma testosterone required for normal mating activity in male sheep. *Horm Behav, 16,* 383–394.

DeFraipont, M., & Sorensen, P. W. (1993) Exposure to the pheromone 17a,20b-dihydroxy-4-pregnen-3-one enhances the behavioral spawning success, sperm production and sperm motility of male goldfish. *Anim Behav, 46,* 245–256.

Demski, L. S., & Dulka, J. G. (1984). Functional-anatomical studies on sperm release evoked by electrical stimulation of the olfactory tract in goldfish. *Brain Res, 291,* 241–247.

Demski, L. S., & Hornby, P. J. (1982). Hormonal control of fish reproductive behavior: brain-gonadal steroid interactions. *Can J Fish Aquat Sci, 39,* 36–47.

Dewing, P., et al. (2007). Membrane estrogen receptor-alpha interactions with metabotropic glutamate receptor 1a modulate female sexual receptivity in rats. *J Neurosci, 27,* 9294–9300.

Dulka, J. G., & Demski, L. S. (1986). Sperm duct contractions mediate centrally evoked sperm release in goldfish. *J Exp Zool, 237,* 271–279.

Dulka, J. G., et al. (1987). A steroid sex pheromone synchronizes male-female spawning readiness in goldfish. *Nature, 325,* 251–253.

Farr, C. H., & Ellis, L. C. (1980). In-vitro contractility of rat seminiferous tubules in response to prostaglandins, cyclic GMP, testosterone, and 2,4′-dibromoacetophenone. *J Reprod Fertil, 58,* 37–42.

Gelinas, D., & Callard, G. V. (1993). Immunocytochemical and biochemical evidence for aromatase in neurons of the retina, optic tectum and retinotectal pathways in goldfish. *J Neuroendocrinol, 5,* 635–641.

Gelinas, D., & Callard, G. V. (1997). Immunolocalization of aromatase- and androgen receptor-positive neurons in the goldfish brain. *Gen Comp Endocrinol, 106,* 155–168.

Grunt, J. A., & Young, W. C. (1953). Consistency of sexual behavior patterns in individual male guinea pigs following castration and androgen therapy. *J Comp Physiol Psychol, 46,* 138–144.

Kobayashi, M., et al. (1986). Gonadotropin surge during spawning in male goldfish. *Gen Comp Endocrinol, 62,* 70–79.

Kobayashi, M., & Nakanishi, T. (1999). 11-ketotestosterone induces male-type sexual behavior and gonadotropin secretion in gynogenetic crucian carp, Carassius auratus langsdorfii. *Gen Comp Endocrinol, 115,* 178–187.

Liu, X., et al. (2009). Identification of a membrane estrogen receptor in zebrafish with homology to mammalian GPER and its high expression in early germ cells of the testis. *Biol Reprod, 80,* 1253–1261.

Lord, L. D., et al. (2009). Rapid steroid influences on visually guided sexual behavior in male goldfish. *Horm Behav, 56,* 519–526.

Mangiamele, L. A., & Thompson, R. R. (in press). Testosterone rapidly increases ejaculate volume and sperm density in competitively breeding goldfish through an estrogenic membrane receptor mechanism. *Horm Behav.*

Maruska, K. P., & Fernald, R. D. (2010). Behavioral and physiological plasticity: Rapid changes during social ascent in an African cichlid fish. *Horm Behav, 58,* 230–240.

Oliveira, R. F., et al. (2002). Social modulation of androgen levels in male teleost fish. *Comp Biochem Physiol, 132,* 203–215.

Oliveira, R. F., et al. (2009). Why do winners keep winning? Androgen mediation of winner but not loser effects in cichlid fish. *Proc R Soc B Biol Sci, 276,* 2249–2256.

Olsen, K. H., & Liley, N. R. (1993). The significance of olfaction and social cues in milt availability, sexual hormone status, and spawning behavior of male rainbow trout (Oncorhynchus mykiss). *Gen Comp Endocrinol, 89,* 107–118.

Oyegbile, T. O., & Marler, C. A. (2005). Winning fights elevates testosterone levels in California mice and enhances future ability to win fights. *Horm Behav, 48,* 259–267.

Palacios, A. G., et al. (1998). Spectral sensitivity of cones in the goldfish, Carassius auratus. *Vis Res, 38,* 2135–2146.

Pang, Y., et al. (2008). Estrogen signaling characteristics of Atlantic croaker G protein-coupled receptor 30 (GPR30) and evidence it is involved in maintenance of oocyte meiotic arrest. *Endocrinology, 147,* 3410–3426.

Pasmanik, M., & Callard, G. V. (1988). Changes in brain aromatase and 5 alpha-reductase activities correlate significantly with seasonal reproductive cycles in goldfish (Carassius auratus). *Endocrinology, 122,* 1349–1356.

Partridge, B. L., et al. (1976) The role of pheromones in the sexual behaviour of the goldfish. *Anim Behav, 24,* 291-299.

Poling, K. R., et al. (2001). The three steroidal components of the goldfish preovulatory pheromone signal evoke different behaviors in males. *Comp Biochem Physiol, 129,* 645–651.

Remage-Healey, L., & Bass, A. H. (2004). Rapid, hierarchical modulation of vocal patterning by steroid hormones. *J Neurosci, 24,* 5892–5900.

Remage-Healey, L., & Bass, A. H. (2005). Rapid elevations in both steroid hormones and vocal signaling during playback challenge: a field experiment in gulf toadfish. *Horm Behav, 47,* 297–305.
Remage-Healey, L., & Bass, A. H. (2006). From social behavior to neural circuitry: steroid hormones rapidly modulate advertisement calling via a vocal pattern generator. *Horm Behav, 50,* 432–441.
Remage-Healey, L., & Bass, A. H. (2007). Plasticity in brain sexuality is revealed by the rapid actions of steroid hormones. *J Neurosci, 27,* 1114–1122.
Remage-Healey, L., & Bass, A. H. (2010). Estradiol interacts with an opioidergic network to achieve rapid modulation of a vocal pattern generator. *J Comp Physiol, 196,* 137–146.
Remage-Healey, L., et al. (2008). Forebrain steroid levels fluctuate rapidly during social interactions. *Nat Neurosci, 11,* 1327–1334.
Remage-Healey, L., et al. (2010). Brain estrogens rapidly strengthen auditory encoding and guide song preference in a songbird. *Proc Natl Acad Sci U S A, 107,* 3852–3857.
Sachs, B. D., & Leipheimer, R. E. (1988). Rapid effect of testosterone on striated muscle activity in rats. *Neuroendocrinology, 48,* 453–458.
Santangelo, N. & Bass, A. H. (2006). New insights into neuropeptide modulation of aggression: field studies of arginine vasotocin in a territorial tropical damselfish. *Proc R Soc B, Biol Sci, 22,* 273(1605), 3085–3092.
Sorensen, P. W., et al. (1989). Differing behavioral and endocrinological effects of two female sex pheromones on male goldfish. *Horm Behav, 23,* 317–332.
Stacey, N., et al. (2001b). Milt production in goldfish: regulation by multiple social stimuli. *Comp Biochem Physiol Part C: Toxicol Pharmacol, 130,* 467–476.
Taziaux, M., et al. (2007). Sexual behavior activity tracks rapid changes in brain estrogen concentrations. *J Neurosci, 27,* 6563–6572.
Tchoudakova, A., et al. (1999). Molecular cloning of an estrogen receptor β subtype from the goldfish, Carassius auratus. *Gen Comp Endocrinol, 113,* 388–400.
Thompson, R. R., et al. (2004). Visual sex discrimination in goldfish: seasonal, sexual, and androgenic influences. *Horm Behav, 46,* 646–654.
Trainor, B. C., et al. (2004). Opposing hormonal mechanisms of aggression revealed through short-lived testosterone manipulations and multiple winning experiences. *Horm Behav, 45,* 115–121.
Waring, C. P., et al. (1996). Milt and endocrine responses of mature male Atlantic salmon (Salmo salar L.) parr to water-borne testosterone, 17,20 beta-dihydroxy-4-pregnen-3-one 20-sulfate, and the urines from adult female and male salmon. *Gen Comp Endocrinol, 103,* 142–149.
Weisel, G. F. (1949). The seminal vesicles and testes of Gillychthys, a marine teleost. *Copeia, 2,* 101–110.
Zheng, W., & Stacey, N. E. (1996). Two mechanisms for increasing milt volume in male goldfish, Carassius auratus. *J Exp Zool, 276,* 287–295.
Zheng, W., et al. (1997). The steroid pheromone 4-pregnen-17α,20β-diol-3-one increases fertility and paternity in goldfish. *J Exp Biol, 200,* 2833–2840.

Chapter 27

Concluding Statements and Current Challenges

Jacques Balthazart and Gregory F. Ball

Our goal in assembling these chapters was to provide a fair representation of the current state of knowledge concerning the production of brain estrogens and the role of these locally produced steroids in the control of brain physiology and in particular behavior. Immense progress has been made since the first demonstration of an active aromatase enzyme in the brain by Naftolin et al. in the early 1970s. A number of major concepts have emerged and were shown to be broadly applicable. As illustrated in this book, it was established, among other things, that:

1. Testosterone largely acts as a prohormone. This steroid is extensively metabolized in its target structures and, in particular, it is aromatized into estrogens in many tissues, including the brain.
2. In the brain, many effects of testosterone at the cellular and behavioral level are caused by locally produced estradiol (E2). This notion is now well established in many animal species from fish to mammals.
3. The specificity of androgen versus estrogen action on behavior is not linked to the nature of the steroid but rather to the sex-typical attributes of the brain substrate on which the steroids are acting.
4. E2 often synergizes with androgens to control many biological responses.
5. Aromatase is encoded by a complex gene, including 10 exons. Alternate untranslated exons I are used to control gene expression in a tissue- and species-specific manner. These processes have now been well characterized in many model systems.
6. Brain (aromatase)–derived estrogens have pleiotropic roles both in the reproductive and nonreproductive context.
7. One important newly discovered function of aromatase and brain estrogens concerns neuroprotection.
8. Brain estrogens have multiple physiological and behavioral membrane–initiated effects that occur with a time-course that is too short to be mediated by changes in transcription.
9. Aromatase is expressed and enzymatically active in presynaptic boutons. E2 acts in a paracrine (synaptocrine) manner to regulate activity of postsynaptic neurons.

These findings have transformed and continue to deeply influence the entire field of (behavioral) neuroendocrinology. However, many of these facts are still ignored outside the field and it is, for a neuroendocrinologist, quite surprising to discover that, more than 40 years after the first formulation of the aromatase hypothesis, the basic notion of testosterone action being mediated by its estrogenic metabolites has not percolated through to the public at large but also even more surprisingly in many circles of the biomedical community. We hope that this book will help to correct this problem.

At the same time, although these major conceptual changes have occurred, a number of new questions arose and have not been answered in a satisfactory manner. We list below a selected number of these significant questions that undoubtedly represent the new frontiers of neuroendocrine science.

1. The relative roles and androgens or estrogens in the control of many (physiological/behavioral) responses have been established for many species, but this dichotomy remains unclear in primates and in humans in particular.
2. Does it matter that a given response is controlled in one species by androgens binding to androgen receptors and in other species by estrogens binding to their own cognate receptors? Similarly, is it important that estrogens signal through ERα versus ERβ? In the end, behavioral control is exerted by changing the transcription of specific genes. Are different sets of genes regulated by these receptors and what are these genes? Is the control of these genes by one or another receptor in different species the result of random processes during evolution or does this represent some significant but currently unexplained constraint on hormone action?
3. Androgens and estrogens very often interact to control physiological or behavioral responses. What are the mechanistic bases of this synergy? Are these two types of hormones affecting the transcription of different or similar sets of genes? In the same or different neurons? Is one steroid affecting the action of the other? Do they act on the same cells or in separate cells (circuits) that provide afferent inputs to a common target?
4. What are the molecular mechanisms controlling the rapid changes in aromatase activity observed in vitro? Are calcium-dependent phosphorylations directly affecting the aromatase protein or do they control the activity of critical cofactors? What is (are) the specific kinase(s) implicated and how is their activity regulated? Are these phosphorylations fully reversible through the activity of the corresponding phosphatases? What is the role of calmodulin in these processes?
5. Are the rapid changes in aromatase activity observed in vivo after sexual interactions or exposure to acute stress based on the same mechanisms as those described in vitro? What are the relative roles of protein phosphorylation, changes in intracellular calcium, and the activation of glutamatergic receptors?

6. Work performed so far solely in zebra finches suggests that changes in brain aromatase activity are, as logically expected, reflected in changes in local estrogens concentrations? How widespread is this conclusion? What is the respective time-course of these two types of events? And importantly what is their functional significance?
7. What are the intracellular signaling pathways implicated in fast estrogen signaling (on behavior)?
8. What is the neural circuitry mediating the rapid nongenomic effects of estrogens on behavior? To what degree does it overlap with the circuitry mediating genomic actions?
9. Brain estrogens derived from local aromatization of testosterone were shown to exert major neuroprotective effects in several animal models. A few groups started to decipher the cellular mechanisms mediating this aspect of neuroprotection but much remains to be discovered in this domain. Quite importantly, the potential inclusion of steroid-induced neuroprotection in the treatment of human pathological processes needs to be evaluated more intensively.
10. We have identified during the last few decades a large number of physiological and behavioral responses that are controlled by estrogens locally produced by aromatase in the brain. The responses cover both long- and short-term effects of the steroids. Is this list now nearly complete, or do we have more surprises awaiting us in the near future? The recent discovery of estrogenic modulation of the processing of auditory and visual signals suggests that this might indeed be the case. In particular, it is very likely that we only have at present a very incomplete view of how widespread are the rapid nongenomic effects of brain estrogens. This topic certainly deserves additional research.

It is obvious that methods and tools available to biologists and in particular to neuroscientists are now qualitatively more sophisticated than when brain aromatase was first discovered. Many more people are also currently studying brain structure and function than in the entire history of humankind. It can thus be expected that answers to many of these questions will be become available in the not too distant future. These future results may provide an excellent excuse for producing the second volume of this book!

Index